The Western Gulf of Mexico Basin

Tectonics, Sedimentary Basins, and Petroleum Systems

Edited by
Claudio Bartolini
Richard T. Buffler
Abelardo Cantú-Chapa

AAPG Memoir 75

Published by

The American Association of Petroleum Geologists

Tulsa, Oklahoma, U.S.A.

Printed in the U.S.A.

Association Editor: Neil F. Hurley, 1997–2001; John C. Lorenz, 2001–2004
Geoscience Director: Robert C. Millspaugh
Publications Manager: Kenneth M. Wolgemuth
Special Publications Editor: Hazel Rowena Mills
Copy Editor: Sharon Mason
Cover Photo: Gulf of Mexico Basin, April 24, 2000, provided by the SeaWiFS Project, NASA/Goddard Space Flight Center, and ORBIMAGE Satellite: OrbView-2; sensor, SeaWiFS.
Cover Design: Rusty Johnson
Production: ProType Inc., Tulsa, Oklahoma
Printing: The Covington Group, Kansas City, Missouri

This and other AAPG publications are available from:

The AAPG Bookstore
P.O. Box 979
Tulsa, OK 74101-0979
Telephone: 1-918-584-2555 or 1-800-364-AAPG (U.S.A.)
Fax: 1-918-560-2652 or 1-800-898-2274 (U.S.A.)
www.aapg.org

Canadian Society of Petroleum Geologists
No. 160, 540 Fifth Avenue S.W.
Calgary, Alberta T2P 0M2
Canada
Telephone: 1-403-264-5610
Fax: 1-403-264-5898
www.cspg.org

Geological Society Publishing House
Unit 7, Brassmill Enterprise Centre
Brassmill Lane, Bath BA13JN
U.K.
Telephone: +44-1225-445046
Fax: +44-1225-442836
www.geolsoc.org.uk

Affiliated East-West Press Private Ltd.
G-1/16 Ansari Road, Darya Ganj
New Delhi 110 002
India
Telephone: +91-11-3279113
Fax: +91-11-3260538
E-mail: affiliat@nda.vsnl.in

The Western Gulf of Mexico Basin

Tectonics, Sedimentary Basins, and Petroleum Systems

Acknowledgments

The editors are grateful to the following individuals who provided constructive and critical reviews of the manuscripts and whose invaluable suggestions improved considerably the content and quality of this volume: Albert Bally, Luis Sánchez Barreda, Peter A. Bentham, Jon Blickwede, Burke Burkart, Richard Chuchla, Carlos Dengo, Tom Dignes, Samuel Eguiluz, Paul Enos, Katherine Giles, Robert Goldhammer, Gary Gray, Jan Hardenbol, Neil Hurley, Chris Johnson, G. Randy Keller, Christopher Lehmann, Nancy McMillan, Ryszard Myczynski, Harold Lang, Robert C. Laudon, Timothy F. Lawton, Jose F. Longoria, Alejandro Morelos, Bryan O'Neill, James Pindell, Gary Prost, Joshua Rosenfeld, Robert Scott, Richard Sedlock, Nimio Tristan, William Ward, Javier Meneses-Rocha, James Lee Wilson, and Zvi Sofer.

We are indebted to Jon Blickwede, Gary Prost, Joshua Rosenfeld, and Harold Lang for their detailed and valuable reviews of the content and English language of Mexican manuscripts.

We acknowledge Petróleos Mexicanos (PEMEX) and the Instituto Mexicano del Petróleo (IMP) for permitting their geoscientists to publish their research studies in this volume. We also thank the Instituto Politécnico Nacional and the Instituto de Geología of UNAM (Universidad Nacional Autonóma de México) for their contributions. We are grateful to AAPG for the opportunity to present to the geologic community the most recent advances on the geology of the western Gulf of Mexico region.

The editors and authors are grateful to Sharon Mason and Rowena Mills of the AAPG Publications Department for maintaining quality control of the book.

Dedication

We dedicate this volume to the paleontologists of Mexico, the United States, and Europe for more than 50 years of important paleontologic and biostratigraphic contributions to the geology of the Mexican Republic. This particular group of professionals has provided valuable research to both academia and industry, particularly to Petróleos Mexicanos (PEMEX), the Instituto Mexicano del Petróleo, the Instituto de Geología, and to several departments of geology at different universities in Mexico.

AAPG
wishes to thank the following
for their generous contributions

to

The Western Gulf of Mexico Basin

Tectonics, Sedimentary Basins, and Petroleum Systems

Instituto Mexicano del Petróleo (IMP)
Instituto Politécnico Nacional
The University of Texas Institute for Geophysics
Two anonymous sponsors

Contributions are applied toward the production costs of
publication, thus directly reducing the book's purchase price and
making the volume available to a larger readership.

About the Editors

Claudio Bartolini received his B.S. degree in geology from the University of Sonora, Mexico. He earned an M.S. degree in geology from the University of Arizona, and after graduation worked for four years as an exploration geologist for Gold Fields Mining Corporation in Arizona, California, Nicaragua, and Mexico. Subsequently, he enrolled at the University of Texas at El Paso, and earned a Ph.D. in geology (tectonics and sedimentary basins) in 1998. After finishing his doctoral program, he joined ARCO International Company, Latin America exploration group. Bartolini is the senior editor of the 1999 Geological Society of America Special Paper 340, *Mesozoic Sedimentary and Tectonic History of North-Central Mexico*. He has organized several symposia on the geology of Mexico, and is one of the founders of the new Latin American Association of Earth Sciences (www.alacit. org). He has focused his research on the Paleozoic and Mesozoic regional geology, plate tectonics, and sedimentation of Mexico. Bartolini is presently employed with the IHS Energy Group in Houston, Texas, where he is involved in research on oil and gas fields in Mexico, Central America, and the Caribbean.

Richard T. Buffler received his B.S. in geology from the University of Texas at Austin in 1959 and his Ph.D. in geology from the University of California, Berkeley, in 1967. After short stints with Shell Oil Company, Shell Development Co., and the University of Alaska, he joined the University of Texas Institute for Geophysics (then located in Galveston) in 1975 as a research scientist. Shortly thereafter, he joined the faculty of the Department of Geological Sciences in Austin. While in Galveston, Buffler began his career as a marine geologist/geophysicist, as well as his long-term work on the geologic history of the Gulf of Mexico Basin, which he continues today, based in Austin. His most recent Gulf projects involve an industry-sponsored basinwide synthesis of the Cenozoic history of the basin and a geophysical study of the Chicxulub K/T impact crater on Yucatán. He also recently has returned to his land roots with geologic field projects in Eritrea, Eastern Java, and northern Mexico and along the New Mexico/Arizona border.

Abelardo Cantú-Chapa is a Mexican paleontologist with experience in the paleontology and biostratigraphy of Mexico. He earned his bachelor's degree from the University of Nuevo León and the Instituto Politécnico Nacional in 1956. After two years as a micropaleontologist, working in the Burgos Basin for Petróleos Mexicanos (PEMEX), he studied sedimentology and sedimentary petrography at the Maritime Station d'Endoume of the University of Marseille and the French Petroleum Institute in Reuil Malmaison, Paris. In 1962, he earned his doctorate from the Université de France at the Sorbonne in Paris with his dissertation on the biostratigraphy of Upper Jurassic and Cretaceous ammonites of Mexico. On returning to Mexico, Cantú-Chapa was appointed director of the Macropaleontology Laboratory at the Instituto Mexicano del Petróleo. There, he was responsible for the study of Mesozoic ammonites of Mexico. From 1966 to 1987, he was a research scientist at the same institute, where he continued his studies of ammonites and of core and cuttings samples from PEMEX wells. Cantú-Chapa is currently chairman of the Graduate School of Geology and professor of paleontology and stratigraphy at the Instituto Politécnico Nacional in Mexico City. A specialist in Mesozoic ammonites, he has published numerous stratigraphic studies, utilizing logs and core and outcrop data for most regions of Mexico. He has also published a systematic study of Mexican Paleozoic cephalopods.

Table of Contents

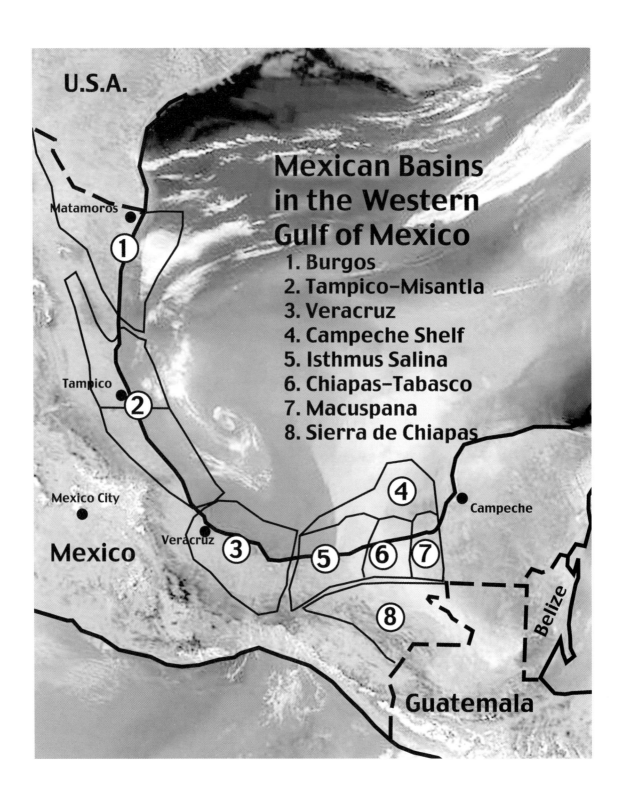

U.S.A.

Matamoros

①

**Mexican Basins
in the Western
Gulf of Mexico**
1. Burgos
2. Tampico–Misantla
3. Veracruz
4. Campeche Shelf
5. Isthmus Salina
6. Chiapas–Tabasco
7. Macuspana
8. Sierra de Chiapas

Tampico

②

Mexico City

④

Campeche

Veracruz ③ ⑤ ⑥ ⑦

Mexico

⑧

Belize

Guatemala

Introduction

The Gulf of Mexico Basin is one of the best-known and most highly explored petroleum basins in the world. The literature on the basin is enormous and continues to grow each year. The deep-water part of the basin is now one of the most prolific new provinces in the world, and it is the continued focus for oil and gas exploration. Here the industry leads the world in the application of modern drilling and production techniques. Despite all this focus on the Gulf of Mexico Basin and its successes, there is still a noticeable gap in our knowledge about a major part of the basin that lies primarily in Mexico. The literature about the geology and petroleum aspects of this part of the Gulf Basin has been sparse and slow in coming, despite some excellent work over the years by academic and petroleum workers in Mexico. This is a result partly of the proprietary nature of much of the work in company files. This attitude is now changing in Mexico, and we believe that this volume represents a milestone toward sharing some of this excellent new work with the world community. We are proud that AAPG can be part of this endeavor.

We have assembled in this memoir for the first time a set of papers that focuses entirely on the Mexican part of the Gulf of Mexico Basin. These papers provide an enormous amount of new data on all aspects of this part of the basin, ranging from structure to stratigraphy to petroleum geology, and they represent a major new contribution to the overall geology of the Gulf of Mexico Basin. Perhaps the most important contribution of this volume is that 13 of the papers are direct contributions from Mexican authors, including papers by members of PEMEX (the Mexican national oil company), the Instituto Mexicano del Petróleo (IMP), the Instituto de Geología from Universidad Nacional Autónoma de México, and the Instituto Politécnico Nacional. Also of major significance as a volume from the AAPG is the common thread that runs through most of the papers, which is the understanding of petroleum systems in Mexico and the implications for exploration and production.

The 21 chapters of this volume are grouped into five general topics: regional studies, structural geology, sedimentary basins, stratigraphy, and oil and gas fields. The papers in each topic are arranged stratigraphically, from oldest to youngest. Although only certain topics may be of specific interest to some readers, there is overlap between topics.

First are five papers of regional importance, including a new look (Cantú-Chapa, Chapter 1) at the early Mesozoic history of the Gulf as a major province of the Pacific realm, prior to its formation and opening in the latest Jurassic. Next is a geophysical look (Bartolini and Mickus, Chapter 2) at the northeastern part of the basin, which allows the delineation of crustal blocks that can be interpreted in terms of both the late Paleozoic to early Mesozoic history of the region as well as the later Laramide deformation. The next paper in this section is a massive synthesis of both the southern and western margins of the basin by Goldhammer and Johnson (Chapter 3). This paper compiles a huge amount of data from both Mexican outcrops and the literature to present working models for the Mesozoic reconstruction of these large areas, documented by sets of paleogeographic maps. This paper is destined to be the definitive synthesis of these areas for a long time to come. The fourth regional paper (Magoon et al., Chapter 4) is a summary of the southern Gulf of Mexico petroleum system. It discusses source rocks; the generation, expulsion, and migration of hydrocarbons; and the traps and reservoirs of this huge and complex system. It concludes with estimates of the enormous undiscovered reserves (23.3 billion barrels of oil and 49.3 trillion cubic feet of natural gas) and future potential global impact of this huge area. The fifth paper (Guzmán et al., Chapter 5) is an overview of the petroleum geochemistry of oils from different basins and subbasins in the Gulf of Mexico Basin. The authors present the geochemical characterization of a wide selection of the produced oils in the petroleum provinces of the Mexican Gulf Coast region. Biomarker and isotope data establish the distinction of five major genetic groups, which were derived from a wide range of ages and types of source rocks.

The second set of three papers focuses mainly on regional structural provinces and periods of deformation The first two (Carrillo et al., Chapter 6; Gray et al., Chapter 7) are concerned with the evolution of the Sierra Madre Oriental Laramide fold belts in the northwestern basins, including their burial history. Of particular significance is the documentation of their subsequent uplift and exhumation, a little-known part of the history of northeastern Mexico. The third paper (Meneses-Rocha, Chapter 8) deals with the complex tectonic history of the southern margin from the Mesozoic through the Cenozoic, which combines at different times all three types of plate motions—extensional, strike-slip, and contractional deformation. This complex deformational history, which continues today, greatly influenced the hydrocarbon systems in this part of the world.

Next we have three papers that discuss new data about the structure and stratigraphy of three major basins of eastern Mexico, all with different histories. The La Popa Basin (Lawton et al., Chapter 9) is the most fascinating, because salt domes are exposed in outcrop, and here the complex history of these structures and their influence on sedimentation can be documented. Following is a paper by Equiluz (Chapter 10) on the Sabinas Basin just to the north, an area with significant gas potential, and on which almost no new data have been published in many years. Prost and Aranda (Chapter 11) discuss the Veracruz Basin, a major hydrocarbon basin to the south. Included are some new ideas about the evolution of the basin, as well as a presentation of its petroleum systems.

Eight papers (several authors, Chapters 12–19) deal with different aspects of the Triassic through Cenozoic stratigraphy of the region. Included are work based on outcrops and, most importantly, new subsurface and well information not previously available. The papers apply modern sequence-stratigraphic principles, although there are different approaches to how the Mesozoic is subdivided, i.e., by unconformities or by regional transgressions. Several papers provide new surface and subsurface biostratigraphy and paleontology (both micro as well as macro) for the region. Of particular significance in many of these papers are not only the regional but also the global implications of some of the new data presented. The final two papers (Martínez-Castillo, Chapter 20; Williams-Rojas and Hurley, Chapter 21) are about specific oil and gas fields, with particular reference to understanding the reservoirs of these fields in southern Mexico.

Unfortunately, these 21 papers represent only part of the new works we had hoped to include. There are many more studies and papers in progress, which we hope will be published soon as complements to this volume.

Perhaps one of the most interesting revelations of this volume and one of its most important overall conclusions is that the Gulf of Mexico Basin is not strictly a passive margin, as most models have traditionally implied. It becomes extremely clear that the structural and stratigraphic history, particularly of the western margin, as outlined in this volume, has been enormously influenced by the Pacific active-plate margin realm. This influence began with the Triassic and Jurassic arc systems to the west, and it continues today with the effects of the active Polochic-Motagua plate boundary to the south and the Late Cenozoic uplift along eastern Mexico. These Pacific tectonic influences, combined with the passive-margin salt and shale tectonics and gravity systems such as the Peridido and Mexican Ridges fold belts, make the western Gulf of Mexico Basin one of the most complex and fascinating basins in the world.

This volume complements the recently published Geological Society of America Special Paper 340 on the Mesozoic geologic evolution of north-central Mexico. Most of the papers in this volume focus on the evolution of eastern Mexico and the Gulf of Mexico coastal and offshore regions, whereas the Special Paper included a variety of papers on the onshore geology of the north and central parts of the country. The one exception in this book is the paper by Goldhammer and Johnson, which overlaps with their paper in the GSA Special Paper. Their GSA paper emphasized the facies and cyclicity of the Mesozoic carbonates. It included details on facies (sedimentary structures, grain types, etc.) and also dealt with high-frequency cycles and third-order sequences. Their paper in this volume, on the other hand, is concerned strictly with second-order sequences and regional paleogeography. Although there is some overlap, we believe it is important to include this major synthesis in this volume to make it available to the AAPG readership.

Finally, it is a great pleasure to have contributed to the petroleum geology of Mexico and the Gulf of Mexico Basin. We hope this memoir will serve as a source of inspiration for future geologic research and petroleum exploration in this complex but enchanting region of North America.

Claudio Bartolini
Richard T. Buffler
Abelardo Cantú-Chapa

Regional
Studies

Cantú-Chapa, A., 2001, Mexico as the western margin of Pangea based on
biogeographic evidence from the Permian to the Lower Jurassic, *in*
C. Bartolini, R. T. Buffler, and A. Cantú-Chapa, eds., The western Gulf of
Mexico Basin: Tectonics, sedimentary basins, and petroleum systems:
AAPG Memoir 75, p. 1–27.

1

Mexico as the Western Margin of Pangea Based on Biogeographic Evidence from the Permian to the Lower Jurassic

Abelardo Cantú-Chapa
Instituto Politécnico Nacional, Mexico City, Mexico

ABSTRACT

In this paper, ammonite biostratigraphy and biogeography are used as the basis for an investigation of the origin of the Gulf of Mexico. Three key observations indicate a Pacific rather than an Atlantic origin for the Gulf of Mexico:

1) The Bajocian ammonite *Stephanoceras*, which occurs throughout the western American margin (Alaska, Canada, United States, Venezuela, Peru, Argentina, and Chile), has also been recorded in the base of the Tecocoyunca Series in Oaxaca (southern Mexico).

2) The Bathonian and Callovian transgressive cycles, which have been recorded in eastern and southeastern Mexico, have been dated on the basis of the ammonites *Wagnericeras* and *Reineckeia*, which are of East Pacific affinity and have been recorded in the subsurface of the Gulf of Mexico coastal plain. The transgressions started in Oaxaca and ended with the Gulf of Mexico opening around the Tampico and Campeche areas, as well as in locations around the margins of the Gulf (e.g., southeastern United States and Cuba) during the early Oxfordian, justifying this age for the origin of this paleogeographic province.

3) In Mexico, several groups of known cephalopods, from the Permian to the Jurassic, are related only to fauna in the Pacific province. They occur in isolated sequences located to the west of the present coastline of the Gulf of Mexico.

A paleogeographic boundary between these sedimentary sequences and red beds and metamorphic or intrusive rocks is located in eastern Mexico. It can therefore be inferred that this region constituted the western margin of Pangea from the Permian to the Middle Jurassic, and that the Gulf of Mexico did not exist during this time interval.

The Middle and Upper Jurassic marine sedimentary sequence is divided here on the basis of a series of regional transgressions that are named after their type localities, namely:

- the Bajocian Oaxaca transgression in southern Mexico
- the lower Bathonian–lower Callovian Metlaltoyuca-Huehuetla transgression in eastern Mexico

- the lower Oxfordian Mazapil transgression in central Mexico
- the lower Kimmeridgian Samalayuca transgression in northern Mexico
- the middle Callovian–upper Oxfordian Boquiapan-Balam transgression in southeastern Mexico
- the middle Callovian and upper Oxfordian Cedro-Cucurpe transgression in northwestern Mexico
- the lower Kimmeridgian Chiapas transgression in southern Mexico

INTRODUCTION:
THE ORIGIN OF THE GULF OF MEXICO

The origin and age of the Gulf of Mexico and the North Atlantic Ocean are generally considered in the context of two very different approaches, namely structural geology and ammonite paleobiogeography. The first approach attempts to understand Atlantic rifting and the formation of new oceanic crust by considering the origin of the various rock types that are present in the Gulf of Mexico and neighboring regions—including evaporites, red beds, and intrusive and metamorphic rocks. The second approach uses ammonites to date the diachronous bases of transgressive marine sequences that were deposited on top of these continental rocks. Different arguments involved in these approaches can be summarized as follows:

Tectonic Arguments

Plate Collisions

The Ouachita orogen that formed at the end of the Paleozoic resulted from the convergence of the North American and Gondwanan Plates, giving rise to the Marathon-Ouachita geosyncline (Ross, 1979; Sedlock et al. 1993; Walper, 1981). According to Ross (1979), biostratigraphic and paleogeographic fusulinid data and paleomagnetic, tectonic, and sedimentologic evidence are relevant only to the formation of the western margin of Pangea and does not shed light on the origin of the Gulf of Mexico.

The "Mojave-Sonora Megashear"

A number of authors have tentatively identified this structure, which supposedly divided north-central Mexico into two parts from Precambrian to Jurassic times (López-Infanzón, 1986; Sedlock et al., 1993, p. 69). However, the megashear's location, structure, and age are questionable when the Permian-to-Jurassic marine succession is considered. These rocks were deposited above this proposed megastructure and do not appear to have been deformed by it, although the "Mojave-Sonora Megashear" was apparently active until the Late Jurassic (López-Infanzón, 1986). It is therefore an obvious inconsistency (Figures 1a, b). For other well-documented areas and time intervals, however, it is accepted that the structure of the in-

trusive or metamorphic basement influenced the transgressive nature of subsequent phases of sedimentation; e.g., the Middle–Upper Jurassic Huasteca Series.

Likewise, examples are known in which the structural attitude of the red beds or the intrusive and metamorphic rocks controlled deposition of the overlying marine sediments; e.g., in wells in the northeast of the Poza Rica district (Cantú-Chapa, 1992, Figure 14). Other examples are discussed below.

Therefore, it appears that the existence of the "Mojave-Sonora Megashear" is questionable, although its existence is often referred to in studies of Mexican geology.

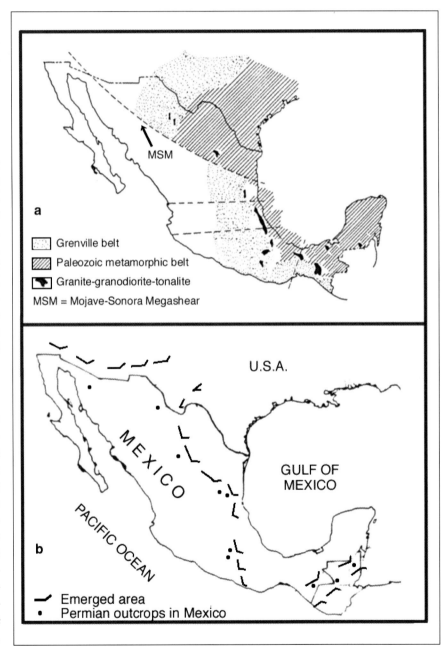

FIGURE 1. Comparison of two Permian paleogeographic models in Mexico: (a) the "Mojave-Sonora Megashear" (MSM), according to the distribution of Permian plutonic rocks (López-Infanzón, 1986); (b) the ammonoid distribution at that time (Cantú-Chapa, 1997).

Separation of North and South America from Mexico

The separation of North and South America accompanied the origin of the Gulf of Mexico. It involved both the southeastward drift of Gondwana, which was joined to North America by Mexico, and the rotation of the Yucatán Block, which was joined to present-day Texas, Louisiana, and Florida during the Upper Paleozoic–Jurassic (White, 1980; Walper, 1981).

Several plate-tectonic studies based on paleomagnetic data have sought to explain the origin of the Gulf of Mexico without specifying the timing and location of plate separation or its relationship with Middle and Upper Jurassic transgressive sedimentary cycles in Mexico (Anderson and Schmidt, 1983; Klitgord and Schouten, 1986). To explain the origin of the Gulf of Mexico, other authors have considered that rotation of the Yucatán Block occurred along the Caltam trend (Michaud, 1987). These authors tried to determine the original position, separation, and subsequent rotation of Yucatán during or before the Middle Jurassic (Anderson and Schmidt, 1983; Michaud, 1987; Sedlock et al., 1993; Walper, 1980; White, 1980).

Tectono-stratigraphic Terrains

Sedlock et al. (1993) arbitrarily divided Mexico into 17 tectono-stratigraphic terrains to which they gave pre-Hispanic names. These authors summarized more than a century of Mesozoic and Cenozoic stratigraphic studies in Mexico. Their results were, however, somewhat oversimplified. For example, Figure 2 compares these authors' model of "Coahuiltecano Terrain" in northeastern Mexico with more detailed stratigraphic models that have been published elsewhere (Cantú-Chapa, 1989a, 1989b, 1999; Imlay, 1980).

Sedlock et al. (1993) did not give any explanation for the origin of the thick sedimentary sequences; they based their arguments solely on the "tectonostratigraphic-terrains" concept. Their model of Gulf of Mexico rifting is vague in both timing and location of the Pangean rifting that occurred.

Sedimentologic and Paleogeographic Arguments

Red Beds and Evaporites

Attempts have been made to explain the origin of the Gulf of Mexico from theoretical studies of the continental sediments and salts deposited during the Late Triassic–Early Jurassic as a result of Pangean rifting (Reed, 1994; Salvador, 1991; Sedlock et al., 1993). The age and origin of these rocks have perhaps been overemphasized, because it is difficult to date them in the absence of reliable biostratigraphic data. None of the above authors considered ammonite dating of the transgressive marine series deposited above the continental rocks. The study of these marine rocks is of great importance because it defines the areas which were affected by the transgressions, and it identifies patterns of oceanic exchange, as well as ammonite migration routes. In eastern Mexico, there are many examples from the subsurface and from outcrop in which the age of evaporites and red beds has been inferred on the basis of their stratigraphic relationship with the overlying marine series. Marine sedimentary rocks have been dated as Bathonian–Lower Cretaceous from their ammonite fauna (Cantú-Chapa, 1989b, 1992).

The "Balsas Portals"

These mythic straits (Figure 3) supposedly allowed oceanic exchange to occur between the Pacific Ocean and the Gulf of Mexico from Callovian times. Some authors have located the portals in western and central Mexico without citing biostratigraphic data confirming their proposed Middle–Late Jurassic age (Imlay, 1980; Salvador, 1991; Schuchert, 1935) (Figure 3). This region of Mexico covered by volcanic rocks is called Eje Neovolcánico.

The "Hispanic Corridor"

This paleogeographic concept has been invoked to explain the transfer of marine fauna (ammonites and corals) from Europe to America during the Jurassic (Stanley and Beauvais, 1994) (Figure 4a). However, to authenticate the corridor, ammonites of different ages were correlated—Sinemurian from Mexico with Bajocian from Venezuela (Bartok et al., 1985) (Figure 4b).

End of Gulf of Mexico Opening

Michaud (1987) suggested that the opening of the Gulf of Mexico ended in the Oxfordian. This appears to have been confirmed by the occurrence of Oxfordian calcarenites and evaporites around the margins of the Gulf. However, a process of this magnitude cannot begin and end in such a short time interval as the Oxfordian. More probably, the marine transgression extended over Mexico and adjacent parts of the Gulf, although isolated emerged areas may have persisted in Poza Rica and Tampico in the east until the Berriasian (Cantú-Chapa, 1989b, 1992).

A Berriasian transgression has also been recorded elsewhere in the Gulf of Mexico. A transgressive succession resting on gneisses, conglomerates, and sandstones has been observed in some Gulf wells (Schlager and Buffler, 1984). In northeastern Mexico and Texas, a transgression covered various types of basement and continental rocks from the Kimmeridgian to the Neocomian (Cantú-Chapa, 1989a, 1989b). In southern Mexico, transgressive sedimentation persisted into the Albian (Cantú-Chapa, 1987b).

These examples of marine transgression point to the continuous opening of the Gulf of Mexico. The sedimentary cycle terminated with a regression, indicated by coal-bearing sediments, which began in northern Mexico (Chihuahua and Coahuila, Sabinas Basin) during the Late Cretaceous (Imlay, 1980). This regression persisted throughout the Cenozoic, ultimately resulting in the present Gulf of Mexico coastline, which corresponds to a marine basin in a generally regressive stage.

STRATIGRAPHIC EVIDENCE FOR THE ORIGIN OF THE GULF OF MEXICO

Drift of the South American continent and the displacement of the Yucatán Block relative to the origin of the Gulf of Mexico must be supported by biostratigraphic data. The breakup of Pangea may not have been a simple process, as has been suggested. For example, many subsurface intrusive and metamorphic structures have been found in the Gulf of Mexico coastal plain. Radiometric age dating of their emplacement gives results which vary from Permian to Early Jurassic.

In this part of Mexico, the Middle Jurassic–Lower Creta-

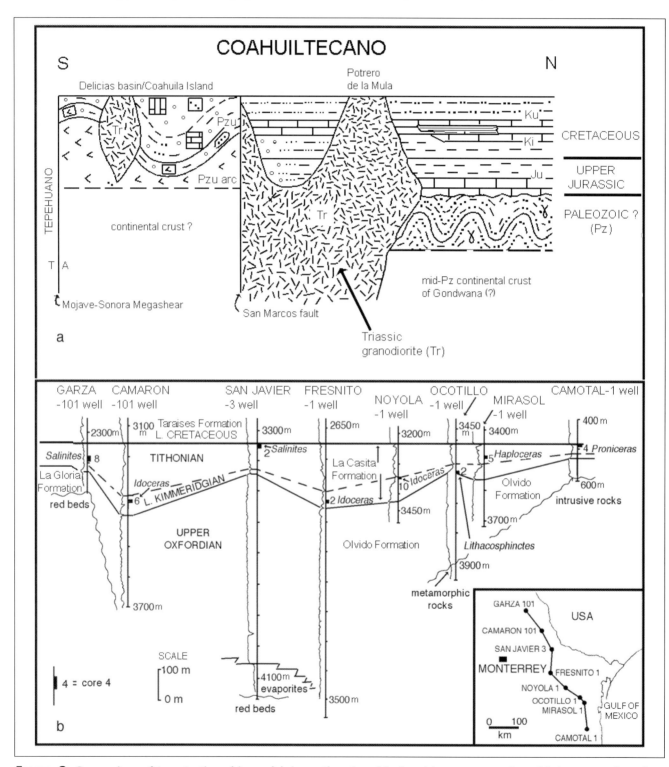

Figure 2. Comparison of two stratigraphic models in northeastern Mexico: (a) tectono-stratigraphic interpretation of Mesozoic rocks (Sedlock, et al., 1993). (b) stratigraphy interpretation based on subsurface data of the Upper Jurassic in northeastern Mexico (Cantú-Chapa, 1999).

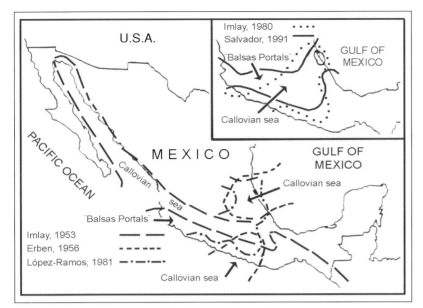

FIGURE 3. Evolution of the "Balsas Portals" and Callovian paleogeographic concepts in Mexico, according to different authors.

ceous transgression covered a variety of lithologies. Biostratigraphic dating of the base of the transgressive series may help to explain the origin and age of the Gulf of Mexico better than arguments based on the emplacement of intrusive rocks or the dating of metamorphic phases. Furthermore, the relationship between the metamorphic and intrusive remnants of eastern Mexico and those in northern South American must also be established (Sedlock et al., 1993, p. 77).

Ammonite biostratigraphy and biogeography have been used to date the bases of the sedimentary series and to establish the routes by which marine incursions occurred. This approach has also led to an understanding of the polarity of the major Middle to Upper Jurassic transgression that covered Mexico and the southern United States. A specific transgression in a particular area can be identified in terms of the ammonite groups present. Ammonites are present in strata above the evaporites and calcarenites, which were deposited from the Middle Jurassic until the origin of the Gulf of Mexico in the Oxfordian.

MEXICAN TRANSGRESSIONS AND THE PALEOBIOGEOGRAPHY OF MIDDLE AND UPPER JURASSIC AMMONITES

In this study, the principal Mexican transgressions are named after their type locations. Seven transgressive events are identified and are briefly described.

The Bajocian Oaxaca Transgression in Southern Mexico

This name is given to the Bajocian depositional event that began in the Oaxaca region and is divisible into two phases:

Bajocian **Stephanoceras** in the Eastern Pacific Province

The occurrence of *Stephanoceras* resolves paleogeographic problems associated with the opening of the Gulf of Mexico and the North Atlantic. It appeared on the eastern margins of the Pacific Ocean in the Bajocian just before invading the Oaxaca region of southern Mexico. It is present in sedimentary successions in western North America and South America, from southern Alaska to Venezuela, Chile, and Argentina, where the genus marks the advanced phases of a transgression (Bartok et al., 1985; Hillebrandt et al., 1992; Imlay, 1980). In Oaxaca, it marks the beginning of the transgressive cycle that is characterized by the base of the Tecocoyunca Series (Figure 5).

Studying the distribution of the genus *Stephanoceras* on the eastern margins of the Pacific Ocean province during the Middle Jurassic may help to explain the marine transgression into southern Mexico through this incipient oceanic route. Certain species are slightly older than those in Oaxaca.

The absence of fossils of Bajocian age from the rest of Mexico and the North Atlantic challenges the model of east-to-west migration of fauna in Mexico that has been proposed by various authors, as well as the existence of the so-called Hispanic Corridor. A Middle Jurassic transgression through the "Balsas Portals" in western Mexico (Figure 3) (López-Ramos, 1981; Sal-

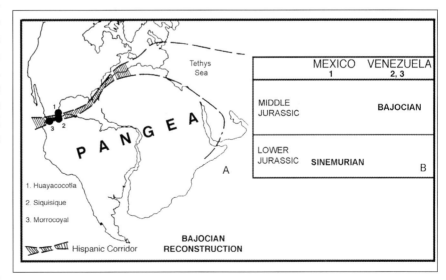

FIGURE 4. (a) The "Hispanic Corridor," a hypothetical paleogeographic concept, is improperly justified when ammonites from different ages are correlated. (b) Sinemurian of Mexico (Erben, 1956) is correlated with the Bajocian of Venezuela (Bartok et al., 1985).

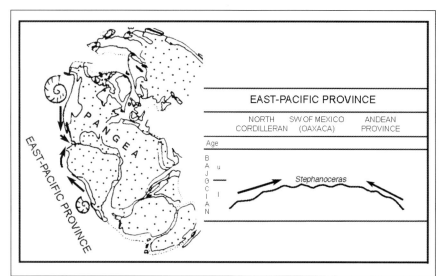

FIGURE 5. *Stephanoceras* marks the basal stage of the transgressive Bajocian Tecocoyunca Series in Oaxaca, southern Mexico. This ammonite was also present in western North and South America and migrated to the study area (Bartok et al., 1985; Hillebrandt et al., 1992; Imlay, 1980).

vador, 1991) or Central America can similarly be discarded for lack of paleontologic evidence (Schmidt-Effing, 1980).

In the Oaxaca region, the sedimentary sequence begins with the Tecocoyunca Series (Figure 6a). This consists of Upper Bajocian–Callovian sandstones, coal, siltstones, and limestones (the Taberna, Simon, Otatera, and Yucuñuti Formations). These formations are overlain by the Tlaxiaco Series, consisting of Oxfordian limestones (Chimeco Formation) and Kimmeridgian-Tithonian argillaceous limestones and shales (Sabinal Formation) (Figure 6c). A sedimentary basin formed in Oaxaca and the Gulf of Tlaxiaco (Figure 6a). From this basin, two routes of communication, still unknown, could have bifurcated, one toward the east in the lower Bathonian (Metlaltoyuca event pt.) and the other to the southeast in the Callovian (Boquiapan event pt.) (Figure 6b).

Epistrenoceras *(Bathonian)*

The biostratigraphic succession continued through the late Bathonian in the Gulf of Tlaxiaco with different ammonite genera, in which Andean forms such as *Epistrenoceras* predominate; their biogeographic affinity to the Oaxaca fauna is remarkable. There are also other genera of Mediterranean affinity; however, the South American ammonites predominate by 33% in Oaxaca (Sandoval and Westermann, 1989).

The Metlaltoyuca-Huehuetla Transgression (early Bathonian–early Callovian) in Eastern Mexico

The Metlaltoyuca-Huehuetla transgression marks the onset of a sedimentary cycle. Its age varies from Bathonian to Callovian, depending on the location. The transgression occurred in the northwest of Poza Rica, where numerous wells are located (Metlaltoyuca event pt.) (Cantú-Chapa, 1992, Figure 12) and

in the Sierra Madre Oriental (Huehuetla event pt.) (Cantú-Chapa, 1971) (Figure 7a, b). Two genera of ammonites (*Wagnericeras* and *Kepplerites*) from the Eastern Pacific province are found in Middle Jurassic sediments in eastern Mexico; they indicate migration from the Pacific to the Atlantic, in contrast to the proposed transgression from the Atlantic.

Wagnericeras *and* Kepplerites *(late Bathonian–early Callovian)*

Ammonites from wells northwest of Poza Rica indicate a very advanced connection during the late Bathonian–early Callovian between the Eastern Pacific province and an area in eastern Mexico adjacent to the future gulf. Here, the sedimentary sequence known as the Huasteca Series (Cantú-Chapa, 1969, 1992) was deposited over red beds as follows (Figure 7b) (from base to top):

*   evaporites (Huehuetepec Formation) and calcarenites (Tepexic Formation) of probable early Bathonian age
*   siltstones (Palo Blanco Formation) of late Bathonian–early Callovian age
*   middle Callovian–Oxfordian shales Santiago Formation)
*   early Kimmeridgian argillaceous limestones with shales (Taman Formation)
*   Kimmeridigan–early Tithonian calcarenites (San Andres Member)
*   late Tithonian argillaceous limestones, shales, and bentonite (Pimienta Formation) (Cantú-Chapa, 1992, Figure 12)

This transgressive sequence is usually restricted to the most recent units in areas relatively close to eastern Mexico and is discussed further in this paper.

Reineckeia *(Callovian)*

Reineckeid ammonites from the lower and middle Callovian also support the proposed marine communication from the Pacific to eastern Mexico, through Oaxaca, where they are abundant.

This group of fossils is also present in the thick calcarenites of the Tepexic Formation (*Neuqueniceras*) and the claystones of the Santiago Formation (*Reineckeia*), both in the subsurface (Poza Rica District) and in outcrop (Huehuetla) in eastern Mexico (Cantú-Chapa, 1969, 1971, 1992). Their presence suggests platform and open-marine facies, respectively. The latter indicates advanced stages of subsidence resulting from maximum transgression across the ancestral Mexican continent from the Pacific Ocean toward the future Atlantic Ocean (Figure 8).

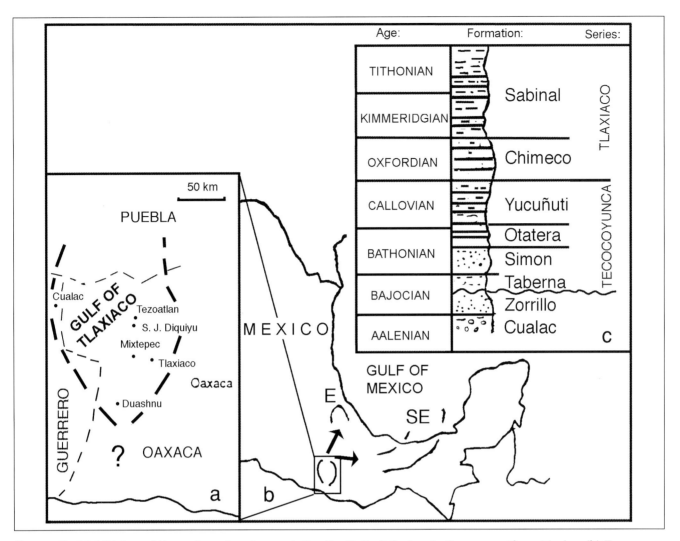

Figure 6. (a) Middle and Upper Jurassic outcrops define the Gulf of Tlaxiaco in Oaxaca, southern Mexico. (b) Two possible lower Bathonian and Callovian transgressive routes toward the east and southeast of Mexico. (c) Middle and Upper Jurassic stratigraphy in Oaxaca.

Reineckeid ammonites found in Mexico are of Eastern Pacific affinity (South America and Mexico: Oaxaca and Baja California). This group has also been found in Europe, but it has not been identified until now either in the Gulf of Mexico or in the Atlantic province.

Neuqueniceras and the bivalve *Lyogrpyphaea nebrascensis* (Meek and Hayden) characterize another part of the Middle Jurassic sedimentary cycle, which is more recent than those represented by *Wagericeras*. In Mexico, reineckeid ammonites (*Neuqueniceras*) include large specimens with bulky ornamentation, which indicates that they lived in a carbonate-rich environment. On the other hand, the bivalve forms cockles, which is an irrefutable sign of a particular deposit corresponding to restricted platform regions connected with the Pacific Ocean of that time. The presence of the early and middle Callovian reineckeids over a large area from eastern to southeastern Mexico points to the incipient communication of the Pacific Ocean with the eastern region of Mexico (Figure 8). *Rei-*

neckeia has also been found in the Pacific portion of the Baja California (Geyssant and Rangin, 1979). A bivalve characteristic of the argillaceous facies of the Santiago Formation from the early and middle Callovian is *Posidonia ornati* (Quenstedt), which has been found only in wells in northwestern Poza Rica.

Late Callovian–early Oxfordian

An ammonite group that corresponds to the upper Callovian–middle Oxfordian cardioceratids, which are characteristic of the Boreal and Alpine provinces, is absent in the thick argillaceous sequences of eastern Mexico (Santiago Formation). Manganese nodules at the top of the sequence indicate a reducing environment (Maynard et al., 1990) and would point to restricted-marine depositional conditions and the absence of communication with the Tethyan Ocean across the Paleo-Atlantic Ocean and Gulf of Mexico. However, it should be noted that the presence of the genus *Peltoceras* in Oaxaca is characteristic of the late Callovian (Imlay, 1980). Its presence is not known in the rest of the eastern regions of the country.

The Mazapil Transgression (lower Oxfordian) in Central Mexico

This transgression is recognized in the Mazapil region of central Mexico. Development of the sedimentary cycle advanced from the Tampico region and Poza Rica in eastern Mexico in two directions:

- toward the north-central area, giving rise to deposits of dolomitized limestones (La Gloria Formation, early Oxfordian) that were covered by late Oxfordian–Tithonian shales with calcareous concretions (La Casita Formation) or by phosphatized limestones (La Caja Formation) with bivalves and ammonites of Kimmeridgian to Tithonian age (Figure 7a[d])

- toward the northeast, where evaporites were deposited beginning in the late Oxfordian, either as gypsum, sandstones, and limestones (Minas Viejas Formation) or as limestones and anhydrites (Olvido Formation) over rocks of different origins that made up the preexisting continent. These were covered by highly argillaceous limestones known as the La Casita Formation of Kimmeridgian-Tithonian age (Figure 2b).

If the Yucatán Block was joined to the southern United States during the Triassic to Middle Jurassic, as some authors indicate, it must have been separated before the late Oxfordian to allow this transgression, with its corresponding fauna, to spread from the preexisting Mexican territory toward Louisiana, Campeche (southeastern Mexico), and Cuba. Another possibility is that during the late Oxfordian, various ammonite genera (*Euaspidoceras*, *Ochetoceras*, and *Discosphinctes*) may have come from the Tethyan Province through the primitive Atlantic Ocean.

These ammonites confirm a late Oxfordian age for the lithologic units located around the Gulf of Mexico. The fossils that indicate the onset of the transgressive event are recorded near the base of different sedimentary sequences:

- calcarenites in the subsurface of Louisiana (Imlay and Hermann, 1984)

- argillaceous limestones overlying sands in the subsurface of Campeche (Ek Balam Group, Angeles-A. and Cantú-Chapa, this volume)

- shales, sandstones, and limestones in outcrops in western Cuba (Judoley and Furrazola, 1968)

However, the same genera of ammonites are found only at the top of the Santiago Formation (thick shale with calcareous concretions) in eastern Mexico (subsurface of Poza Rica–Tampico and the neighboring Sierra Madre Oriental) and at the base of the La Casita Formation (shales with calcareous concretions) in the north-central part of the country (San Pedro del Gallo, Durango). In both regions, they characterize an advanced stage of sedimentation and subsidence that differs from the sedimentary regimes of the locations previously mentioned for the late Oxfordian (Figure 7a[d]).

Therefore, it is inferred that it was from eastern Mexico that one of the two marine

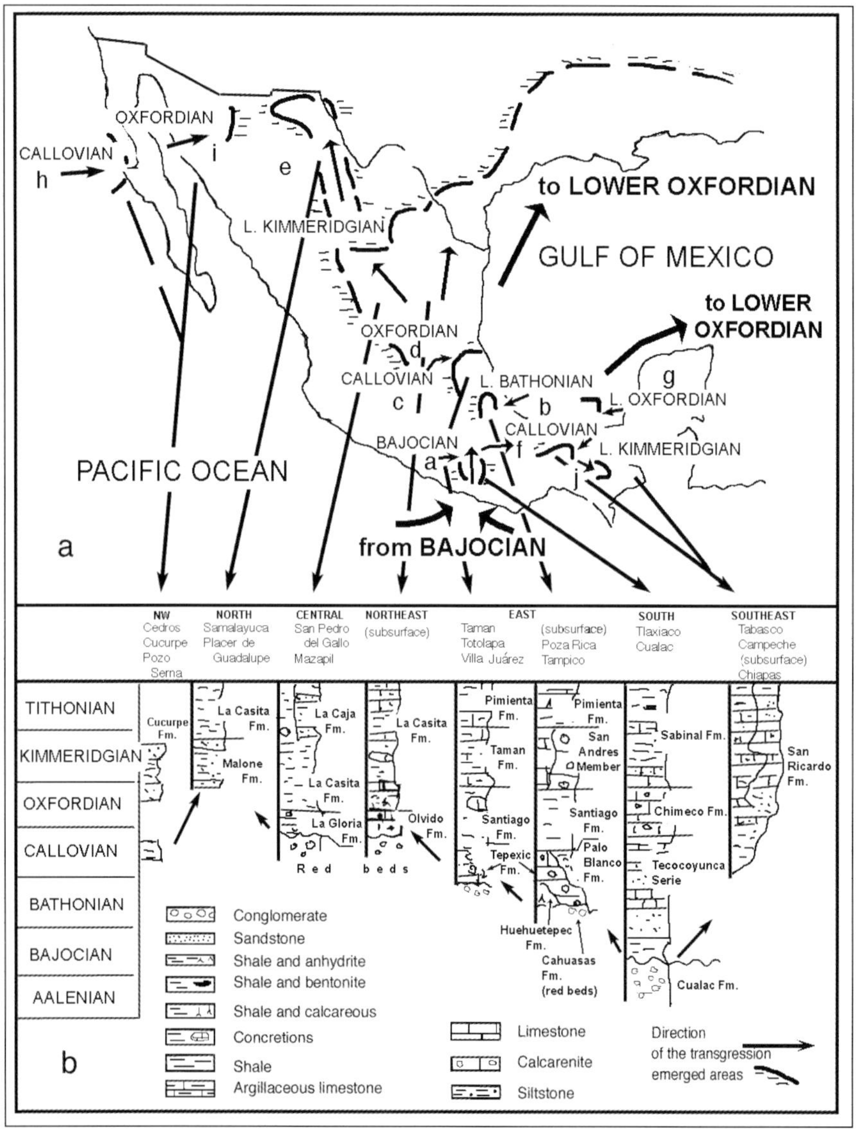

FIGURE 7. (a) Middle and Upper Jurassic transgressive events in Mexico, which originated in the ancestral Pacific Ocean during the Bajocian in the south of the country and Callovian in the northwest. Transgressive events: a, Oaxaca; b, Metlaltoyuca; c, Huehuetla; d, Mazapil; e, Samalayuca; f, Boquiapan; g, Balam; h, Cedro; i, Cucurpe; j, Chiapas. (b) Stratigraphic crosscorrelations. Direction of the transgressions toward the north and the present area of the Gulf of Mexico during the lower Kimmeridgian and the lower Oxfordian, respectively.

invasion routes, coming from Oaxaca, departed toward the region of the present Gulf of Mexico. This is discussed further below. The other route trended directly from the Oaxaca region toward southeastern Mexico and from there toward Cuba (Figure 7a[f, g]).

Ataxioceras, Idoceras, *and* Glochiceras *(lower Kimmeridgian)*

The base of this stage has been identified biostratigraphically only in the northeast and east of Mexico with *Ataxioceras* or ammonites of equivalent age (rasenids) of Tethyan affinity (Cantú-Chapa, 1992) (Figures 2b, 8, and 9). On the other hand, the genus *Idoceras*, associated with *Nebrodites* and *Aspidoceras*, is dominant in the second zone of the lower Kimmeridgian. This stage of stabilized subsidence conditions is represented by an argillaceous lithofacies with calcareous concretions (La Casita Formation). The top of the lower Kimmeridgian is characterized by *Glochiceras* throughout practically all Mexico.

Biostratigraphic data described by Imlay (1980), often ignored in later studies, include the discovery of an association between *Ameoboceras-Idoceras* and the bivalve *Aulacomyella* in Sonora, which indicates the presence of the lower Kimmeridgian rather than the upper Oxfordian in northwestern Mexico.

Mazapilites-Durangites *(Tithonian)*

The bi- or tripartite subdivision of the Tithonian in the Tethyan Province has been challenged in European countries. It is adopted here as a particular evolutionary event by which to divide this stage into the lower and upper Tithonian. The sudden occurrence and wide distribution of the genus *Mazapilites* mark the top of the lower Tithonian. This ammonite has been found in the central, eastern, and southeastern parts of the country.

Strata around the Kimmeridgian-Tithonian boundary are characterized by the ammonite *Hybonoticeras*; its distribution is limited to the center of Mexico. It is followed by *Virgatosphinctes* and the bivalve *Aulacomyella neogeae* and by *Mazapilites*. The lower Tithonian cycle represents the sedimentologic stability in central and eastern Mexico (Figure 8). *Virgatosphinctes* is a fossil with clear andine and indomalgach affinity. On the other hand, *Mazapilites* has been found outside Mexico only in Germany (Berckhemer and Holder, 1959).

The upper Tithonian is characterized by *Suarites, Kossmatia, Durangites, Substeueroceras,* and *Parodontoceras* aff. *callistoides* as follows:

- in argillaceous facies (La Casita Formation) in the north and northeast (Figure 2b)

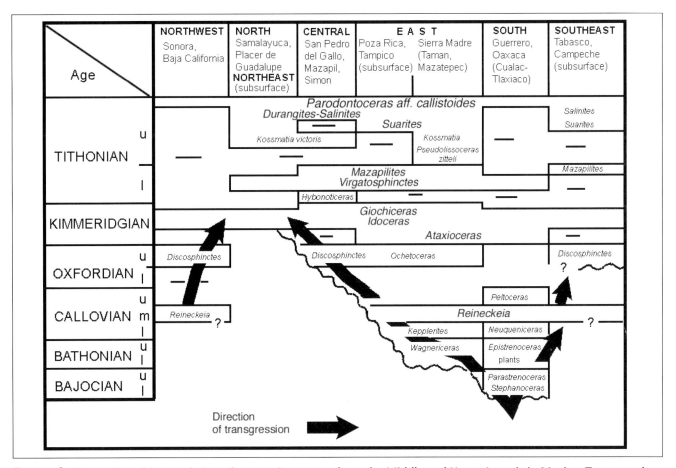

FIGURE 8. Biostratigraphic correlation of ammonite genera from the Middle and Upper Jurassic in Mexico. Transgressive cycle from the south toward the north and southeast, and most likely from the northwest toward the north-central portions of the country.

- in calcareous argillaceous facies with bentonite (Pimienta Formation) in the east (Cantú-Chapa, 1989b, 1992, Figures 8 and 12)
- in argillaceous limestones with shales (Sabinal Formation) in the south, Oaxaca (Figure 6a)
- in argillaceous limestones with bentonite (Edzna Formation) in the southeast, Campeche (Angeles-A. and Cantú-Chapa, this volume; Cantú-Chapa, 1977)

It should be noted that Tethyan ammonites predominate in the second, neo-Jurassic sedimentary series (Tlaxiaco Series) of the Kimmeridgian-Tithonian in Oaxaca (Figure 6). These fossils, also found in the east of Mexico, indicate the opening of the Gulf of Mexico and the North Atlantic. Only at the base of the Cretaceous does there exist a strong South American influence, characterized by spiticeratids and berriaselids ammonites.

The Samalayuca Transgression (Lower Kimmeridgian) in Northern Mexico

The most northerly (and youngest) of the major Jurassic transgressions in Mexico is represented in Chihuahua and in southeastern Texas by sandstones and shales containing *Idoceras* (Lower Malone Formation: lower Kimmeridgian), followed by a thick sequence of shales (La Casita Formation) with *Virgatosphinctes*, *Suarites*, and *Kossmatia* (Tithonian) (Cantú-Chapa, 1976a; Imlay, 1980) (Figures 7a[e] and 8).

The Boquiapan-Balam Transgression (Middle Callovian–Upper Oxfordian) in Southeastern Mexico

This name is derived from oil wells in the Tabasco area (Boquiapan event) in southeastern Mexico (Figure 7a[f]). There, a thick, highly carbonaceous and argillaceous sedimentary series of middle Callovian age is covered by limestones of Oxfordian to Tithonian age that are partly calcarenitic and partly argillaceous.

On the other hand, in wells in the Campeche area of the Gulf of Mexico, the sedimentary cycle began with evaporites, limestones, and sands (Balam event) (Figure 7a[g]). The ammonites *Reineckeia* (middle Callovian) and *Ochetoceras* (late Oxfordian) characterize these two transgression in southeastern Mexico, respectively (Figure 8).

The Cedro-Cucurpe Transgression (Middle Callovian and Upper Oxfordian) in Northwestern Mexico

This transgression is recorded in northwestern Mexico. In the western portion of the Baja California peninsula (Cedro event), it is of middle Callovian age (Figure 7a[h]). The sedimentary cycle continued into the Sonora region (Cucurpe event) in the late Oxfordian (Figure 7a[i]). It consists of a volcano-sedimentary sequence (lavas, sandstones, and shales). The most representative ammonites are *Reineckeia* (Callovian) and *Discosphinctes* (late Oxfordian Cucurpe Formation) (Geyssant and Rangin, 1979; Rangin, 1977).

The possibility exists of finding evidence of sedimentary sequences of marine origin dating from the late Oxfordian between Sonora (Cucurpe) and Durango (San Pedro del Gallo), because the same ammonite groups are present in both localities (Rangin, 1977; Imlay, 1980) (Figure 7a[e]).

The Chiapas Transgression (Lower Kimmeridgian) in Southern Mexico

In the central region of Chiapas, Michaud (1987) identified

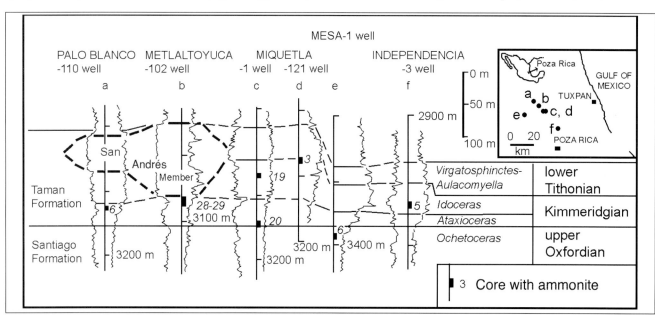

FIGURE 9. Cross section of wells in northeastern Poza Rica from gamma logs and stratigraphic interpretation showing the Oxfordian-Kimmeridgian boundary, eastern Mexico. The datum is the base of the Kimmeridgian. Arbitrary horizontal scale.

the presence of a lower Kimmeridgian sequence based on foraminifera and dascycladacean algae associated with calcareous
lithofacies that correspond to part of the San Ricardo Formation. This is the only stage in this southern region in Mexico
that has been dated definitively. As yet, there is not sufficient
biostratigraphic evidence to indicate that an older marine
incursion could have occurred there (Figure 7a[j]).

THE DIACHRONOUS BASE OF THE HUASTECA SERIES (MIDDLE AND UPPER JURASSIC) AND THE LOWER CRETACEOUS IN THE SUBSURFACE OF THE TAMPICO REGION, EASTERN MEXICO

Deposition of the Huasteca Series occurred before the opening of the Gulf of Mexico and can be divided into five main
transgressions in the Tampico and Poza Rica regions of eastern
Mexico: Bathonian, Callovian, Kimmeridgian, Tithonian, and
Berriasian-Valanginian. This age range for the
base of the sedimentary cycle was the result of
the irregular topography of the continental surface on which the series were deposited. Several
cross sections based on subsurface data illustrate the apparently isolated depositional regimes that coexisted within short distances of
one another. They illustrate the nonconformable relationship between the marine sequence
and the underlying strata, which comprise
metamorphic, intrusive, and continental rocks
(Figures 10, 11, 12, and 14). Two stratigraphic
sections are illustrated to date the nonconformable bases of the sedimentary sequence,
which are diachronous in this part of Mexico
(Figures 13 and 15).

Although the Huasteca Series in the Poza
Rica region is characterized by a progressive
northwest-to-southeast transgression, sedimentation in Tampico was more isolated in time
and space. There, marine waters spread through
narrow, fault-bound channels, and emergent
areas controlled the routes of marine communication. Some of these sections have been
mapped. Three of them were analyzed in a previous study of Poza Rica (Cantú-Chapa, 1992).
In these sections, the abrupt change of age at
the nonconformable base of the sedimentary
sequence found in neighboring wells was affected by the irregular topography. The following sections have been named according to the
most abrupt nonconformable stratigraphic elements that characterize them.

Placetas-Ixcatepec Wells

Figure 10 illustrates how the Middle and

Upper Jurassic seas invaded the southern part of the Tampico
district in different stages and over short distances through
channels that allowed the transgression to occur. In fact, the
calcarenitic base of the sedimentary sequence is of different
ages; it was affected by emergent areas that prevented its simultaneous deposition. In the Placetas-1 well and Santa Maria
Ixcatepec-3 well, the transgressive unit is of different ages:
Tithonian (San Andrés Member), deposited over intrusive
rocks in the first well, and Callovian (Tepexic Formation),
deposited over red beds in the second.

Cuachiquitla-Ixtazoquico Wells

In the south of Tampico District, the transgressive cycle
occurred in different stages of the Middle and Late Jurassic, as
follows (Figure 11):

• In the Cuachiquitla-1 well, the Huasteca Series is
complete from the lower Callovian to the Tithonian.
Here, the sedimentary cycle began with the deposition

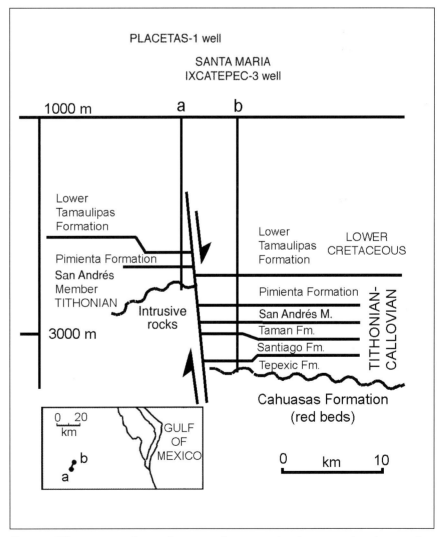

FIGURE 10. Stratigraphic and structural cross section between the Placetas-1 well and Santa María Ixcatepec-3 well in eastern Mexico, showing how the structure of the ancient continent controlled patterns of sedimentation, resulting in diachronous sequences within very short distances.

of evaporitic rocks of the Huehuetepec Formation over red beds (Cahuasas Formation).

- In the Ixtazoquico-1 well, marine sedimentation did not begin until the lower Kimmeridgian with cal carenites (San Andrés Member) also deposited above the Cahuasas Formation. At the same time, this latter unit overlies the Huayacocotla Formation of marine origin from the Sinemurian.
- In the Pilcuautla-1 well, located to the southwest of the Cuachiquitla paleochannel, the sedimentary cycle began with the Pimienta Formation of late Tithonian age, also deposited above red beds. The area where this well is located represents one of the margins of the channel.

Los Cues–Cahuayotes Wells

Figure 12 illustrates the stratigraphic-structural relationship of rocks from the Middle and Upper Jurassic in four wells in the Tampico district, as follows:

- The Los Cues-102 well was the only one which penetrated a thick Callovian-to-Tithonian marine sequence deposited in a channel above intrusive rocks, where the Tepexic, Santiago, Taman, and Pimienta Formations and the San Andrés Member were found.
- By contrast, Tamismolon-102 well identified the base of the transgression at the top of the Jurassic (Pimienta Formation).
- The Los Cues-101 well and Cahuayotes-102 well cut the base of the transgression (Lower Cretaceous). The sequence was deposit-ed on a variable substratum: red beds (Cahuasas Formation) and metamorphic and intrusive rocks, as shown in the four wells (Figure 12).

Barcodon-Topila Wells

The gamma-ray curve from three wells in Tampico District was used to correlate the stratigraphic section in Figure 13. This section emphasizes the stratigraphic relationship between the Tithonian and Berriasian transgressions and indicates the types of continental rocks over which the marine sedimentary cycle began in this part of eastern Mexico, as follows:

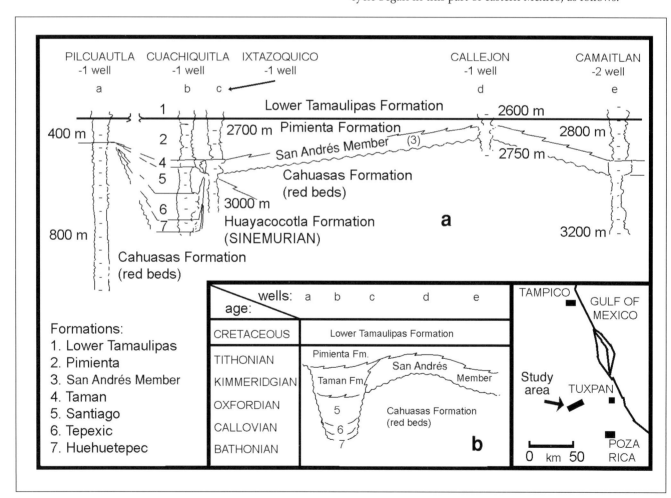

FIGURE 11. (a) Cross section and stratigraphic interpretation between Pilcuautla-1 and Camaitlan-2 wells based on radioactive logs from eastern Mexico. (b) Sedimentary rocks of Callovian-Tithonian ages were penetrated only in the Cuachiquitla-1 well (b). In contrast, the marine sedimentary cycle began at the Upper Jurassic in the other wells (a, c, d, e). Datum is the Lower Tamaulipas–Pimienta Formations.

- In the Barcodon-102 well, the calcarenites of the San Andrés Member from the Tithonian were deposited over metamorphic rocks.
- In the Topila 105-well, the marine sequence began in the Berriasian (lower Tamaulipas Formation), also on metamorphic rocks.
- In the Bejuco-6 well, the sedimentary sequence began with calcarenites (San Andrés Member) from the lower Kimmeridgian, deposited over intrusive rocks (Cantú-Chapa, 1987a, Figures 3, 7).

The stratigraphic section identifies microfossils and ammonites recovered from cores, by which the basal units of the transgressive event were dated.

Pueblo Viejo–Bocacajeta Wells

The previous section was complemented by another section perpendicular to it, also located in the Tampico region. It illustrates the stratigraphic-structural relation between the Pueblo Viejo-102 and Bocacajeta-1 wells at Middle and Upper Jurassic levels (Figure 14). A normal fault defines two areas of deposition. The basal units of the sequence are of different ages: Callovian in the Bocacajeta-1 well, and Tithonian in the Chunca-2 and Pueblo Viejo-102 wells.

In the Bocacajeta-1 well, the series lying above intrusive rocks is complete and very thick. It consists of calcarenites from the Callovian Tepexic Formation to argillaceous limestones with bentonite from the upper Tithonian Pimienta Formation. By contrast, in the Pueblo Viejo-102 and Chunca-2 wells located to the west, the Pimienta Formation (Tithonian) represents the base of the sedimentary cycle that was deposited above metamorphic rocks, which most probably formed a structural high in the Late Jurassic. Therefore, the initial phase of the marine transgression was younger in the last two wells than in the Bocacajeta-1 well. This observation is of great significance to explain the origin of the Gulf of Mexico.

Mantarraya–Los Mangles Wells

This section was compiled using offshore well data with the help of fossils and well logs (Figure 15). It shows the different ages of the late transgressive phase in the Tampico region. In this case, the unconformity between the red beds (Cahuasas Formation) and the marine sedimentary sequence also allows us to infer the direction of the transgression that took place over less than 15 km in a short period of time, as follows:

- The Mantarraya-1 well penetrated only 8 m of the Pimienta Formation (upper Tithonian), which overlies the Cahuasas

Formation. It is covered by the lower Tamaulipas Formation, in which a sample of *Dichotomites (Dichotomites) mantarraiae* Cantú-Chapa (1990), an upper Valanginian ammonite, was found very close to its base.
- By contrast, in the Los Mangles-1 well, the unconformity between the red beds and the limestones of the lower Tamaulipas Formation is slightly younger. Lack of paleontologic data from the base of the formation does not allow a conclusion to be made as to whether the Berriasian is condensed or absent. On the other hand, calpionellids and nannoconus were recovered from cores 2 and 3 of the Los Mangles-1 well. Correlating with the Mantarraya-1 well, where calpionellids were found, it appears that the age of the two successions is late Valanginian (Figure 15).

LATE JURASSIC OPENING OF THE GULF OF MEXICO THROUGH THE TAMPICO REGION

The base of the transgressive Huasteca Series was deposited during different stages of the Middle and Upper Jurassic throughout Tampico and the Sierra Madre Oriental (Tamazunchale). These two areas in eastern Mexico have paleogeographic relations with the Oxfordian transgression that initiated the Gulf of Mexico. The explanation for the opening of this basin is found in an advanced stage of the transgression, which occurred predominantly in the Tampico region bordering the gulf.

Jurassic sedimentation in the Bejuco (southern Tampico) and Poza Rica areas was discussed by Cantú-Chapa (1987a,

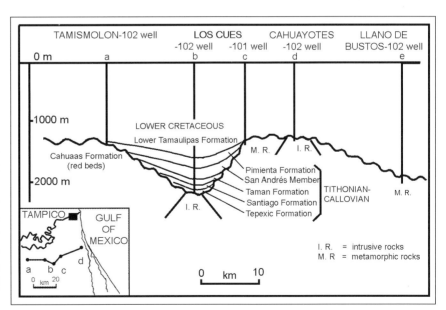

FIGURE 12. Stratigraphic-structural cross section between the Tamismolon-102 and Cahuayotes-102 wells. Sedimentary rocks of Callovian-Tithonian ages were penetrated only by the Los Cues-102 well. In contrast, the marine sedimentary cycle began at the top of the Jurassic or at the base of the Cretaceous in the other wells. Location is south of Tampico in eastern Mexico.

1992). These sections also indicate that the direction of marine incursion was consistently toward the Gulf.

Evidence for a Pre-Jurassic– Lower Cretaceous Continent

Well cuttings from the Tampico region indicate the petrography of the continental surface on which Middle Jurassic–Lower Cretaceous marine sediments were deposited. The contact shows rocks of marine (undifferentiated Upper Paleozoic and Lower Jurassic), continental (Cahuasas Formation), and intrusive and metamorphic origin (Figure 16).

Upper Paleozoic Rocks of Marine Origin

Several wells in northwestern Tampico penetrated undifferentiated Upper Paleozoic rocks, which are overlain by calcarenites (San Andrés Member) or argillaceous limestones (Pimienta Formation) of Tithonian age (Figure 16).

The Huayacocotla Formation (Sinemurian)

This formation is characterized by black shales, siltstones, and sandstones with vermiceratid ammonites of Sinemurian age obtained from several wells. It therefore represents basinal deposition. There is another sequence of shales and paralic sandstones in wells in southwestern Tampico, where benetitial ferns (Pterophyllum) of probable Pliensbachian age (or base Middle Jurassic) predominate (Flores, 1974); the sequence is informally known as the Rosario Formation. These units (the Huayacocotla and Rosario Formations) are confined to an elongate paleobay, 30 to 50 km wide and about 130 km long, which has been identified in the subsurface of Tampico. The paleobay is parallel to the present-day coastline but is without communication to the Gulf of Mexico region. It has paleogeographic relationships with outcrops of the same age located to the southwest in the Sierra Madre Oriental (Erben, 1956; Suter, 1990) (Figure 16).

Erben (1956) named the region where these rocks outcrop the "Huayacocotla paleo-bay" and suggested that there was communication with the present-day Gulf of Mexico. However, this is impossible to prove because of a lack of biostratigraphic data in the subsurface of the Gulf of Mexico margin. Calcarenites of Callovian (Tepexic) or Tithonian age (San Andrés Member) were also deposited over the Huayacocotla Formation; this nonconformable stratigraphic relationship is observed in the subsurface in Tampico and in the Sierra Madre Oriental (Alto Ixtla) (Suter, 1990).

Cahuasas Formation

These red beds have been identified in wells in Tampico and Poza Rica (Figure 16). They are overlain by, from base to top:

- evaporites assigned to the Huehuetepec Formation of Bathonian-Callovian age
- calcarenites assigned to the Callovian Tepexic Formation and Tithonian San Andrés Member
- micritic limestones (Berriasian-Valanginian) assigned to the lower Tamaulipas Formation

Based on subsurface data, the Cahuasas Formation formed islands and also constituted a large emergent area to the southeast of Tampico and east of Poza Rica, parallel to the present coastline, which was not covered until the Lower Cretaceous (Figure 16).

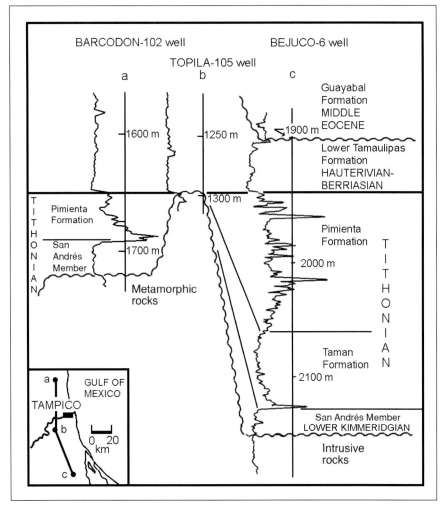

FIGURE 13. Stratigraphic cross section between the Barcodon-102 and Bejuco-6 wells based on gamma logs, ammonites, and microfossils. Datum is the Jurassic-Cretaceous boundary. Different ages exist for the base of the marine sedimentary cycle deposited over metamorphic rocks in wells near Tampico.

Metamorphic and Intrusive Rocks

These widely distributed rocks in the Tampico district were reached by the drill bit after crossing calcarenites (San Andrés Member) and argillaceous limestones (Pimienta Formation) of the Tithonian or micritic limestones (lower Tamaulipas Formation), the base of which is of Berriasian-Valanginian age (Figure 16). Metamorphic or intrusive-sedimentary rocks have also been recorded in wells in the Bejuco region (Cantú-Chapa, 1987a).

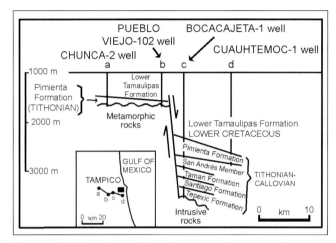

FIGURE 14. Stratigraphic and structural cross section between the Chunca-2 and Cuauhtemoc-1 wells, Tampico District, eastern Mexico. The unconformable base of the marine sedimentary series is of different ages—Callovian in the east and Tithonian in the west.

Middle-Upper Jurassic of Tamazunchale Area in the Sierra Madre Oriental

The Huasteca Series has been described from the Callovian in the Sierra Madre Oriental (Pisaflores); at this age, the sequence is already argillaceous and corresponds to the Santiago Formation in which ammonites of the *Reineckeia* genera are found. Therefore, this lithofacies represents an advanced stage of the sedimentary cycle (Cantú-Chapa, 1971). However, in the adjacent Tamazunchale area, Suter (1990) cites calcarenites from the Tepexic Formation without having determined their age from ammonites.

Therefore, we have inferred that this unit represents the base of the sedimentary cycle (Huasteca Series). Outcrop studies complement the stratigraphic data from the subsurface in eastern Mexico (Tampico and Poza Rica) regarding the age of the basal part of this series. Of particular sedimentological interest in the advanced stage of the sedimentary cycle of the Huasteca Series is the presence of:

- layers of manganese-bearing limestones with shale. These were deposited between the top of the argillaceous sequence of the Santiago Formation and the base of the argillaceous limestones of the Taman Formation in the Molango region (Maynard et al., 1990).
- In addition, the calcarenites of the Tepexic Formation contain isolated ferruginous oolites in Mesita-1 well (Poza Rica), as well as fragments of the ammonite *Neuqueniceras* (Cantú-Chapa, 1992, Figure 5).

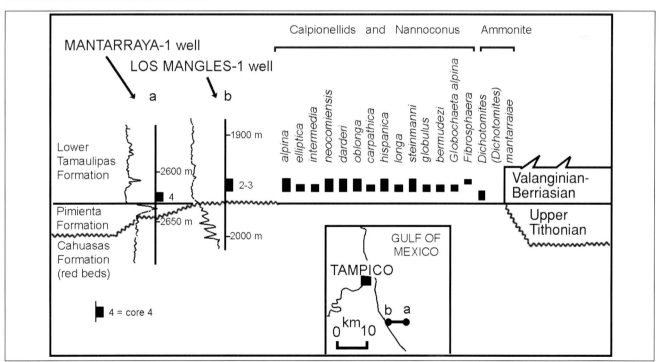

FIGURE 15. Stratigraphic cross section between the offshore Mantarraya-1 and Los Mangles-1 wells in the Tampico District, eastern Mexico. Datum is the Tithonian–Lower Cretaceous unconformity. Note the different ages of the base of the sedimentary cycle and the stratigraphic position of the microfossils and ammonites taken from the cores. Microfossils studied by F. Bonet (Petróleos Mexicanos); ammonites studied by Cantú-Chapa (1990).

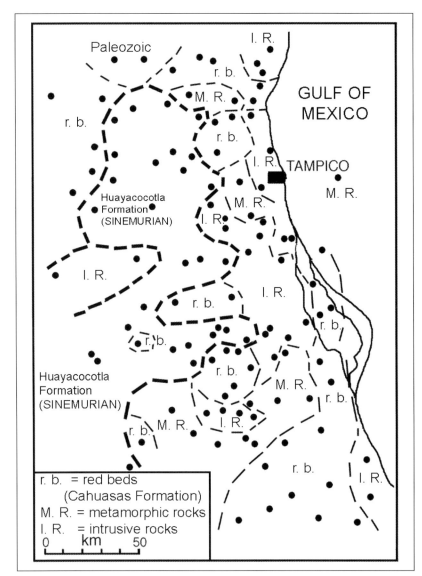

FIGURE 16. Distribution of the ancient continent in the subsurface of northwestern Poza Rica and Tampico, where the sedimentary series of the Middle–Upper Jurassic and the Lower Cretaceous were deposited. The Huayacocotla Formation (Sinemurian) forms an elongated paleogeographic element, which is parallel to the actual coastline of the Gulf of Mexico without communication with it. (This figure is based on analysis of the study wells.)

- Similarly, Cantú-Chapa (1971) reported the bivalve *Bositra buchi* (Roemer), generally associated with the *Reineckeia* ammonites, in the dark-gray to black carbonaceous with disseminated pyrite shale of the Santiago Formation, which represents the subsidence phase. Both ammonites are of early to middle Callovian age.

Trends of the Transgressive Middle and Upper Jurassic Cycle in the Subsurface of Tampico

Stratigraphic analysis of the Middle Jurassic in Mexico suggests that the transgression beginning in Oaxaca, in the south

of the country, continued northward through areas yet undefined until it became evident in the present Sierra Madre Oriental—Tamazunchale to Huauchinango (Cantú-Chapa, 1971)—where it bifurcated.

One branch advanced toward the northwest of Poza Rica, and the other covered various regions in the Tampico district; both are evident only in the subsurface. Therefore, the basal facies of the Huasteca Series may crop out in the Sierra Madre Oriental, which borders these two regions. This would explain its relation with what until now has been seen only in the subsurface of northwestern Poza Rica: the upper Bathonian.

The Huasteca Series begins with pre–upper Bathonian evaporites, deposited above the Cahuasas Formation (wells of northwestern Poza Rica). From there, the transgression moved toward the southeast of this region, covering it in successive stages during the Middle and Upper Jurassic (Cantú-Chapa, 1992). By contrast, the base of this series is nonconformable and of different ages in the Tampico region; its approximate age varies abruptly from the Callovian to the Valanginian over very small areas. The opening of the present region of the Gulf of Mexico certainly began in this region.

In fact, stratigraphic studies of Middle and Upper Jurassic sequences in the Tampico region, over approximately 1600 km^2, explain the direction of the oceanic communication between the present Mexican territories and the Gulf of Mexico. The above descriptions and a complementary isometric map covering the subsurface of that region and part of the northwestern area of Poza Rica reinforce the explanation of the opening of the Gulf of Mexico from the Tampico region (Figure 17c).

The map uses the Jurassic-Cretaceous boundary identified in Cantú-Chapa (1989b). The formational units of the Huasteca Series deposited over this old continent are suggested by short lines in each well. Their ages are abbreviated: T (Tithonian), C-T (Callovian-Tithonian), and B-T (Bathonian-Tithonian, restricted to northwestern Poza Rica where there is evidence of older sedimentation). In this last case, the Huehue-tepec Formation represents the base of the sedimentary sequence deposited over red beds (Cahuasas Formation) (Figure 17a).

The transgression advanced from this region toward the south of Poza Rica, where it covered a large region during different stages of the Upper Jurassic (Cantú-Chapa, 1992, Figure 11). Simultaneously, the sea invaded the southern part of the Tampico district through a series of closely spaced channels (Figures 10–15). The transgressive base of the Middle and Upper Jurassic sedimentary sequence is almost always calcarenitic and diachronous. Its deposition was not continuous because it was affected by emergent areas.

In the same way, marine communication routes are inferred through wells Metlaltoyuca-102 and Horcones-1001 (Figure 17a, c: 1, 2); their continuity with the other areas is apparently cut off. Another route is across wells Martinica-1, Santa Maria Ixcatepec-3, and Tanquian-1 (Figure17a, c: 3, 4, 5), which suggests an advance of the Middle Jurassic sea toward the north.

Jurassic sedimentary rocks in these wells are about 800 to 1000 m thick and appear to be of pre-Callovian age. In fact, *Hecticoceras* is present in the argillaceous sequence of the Santiago Formation and indicates an advanced stage of the sedimentary cycle in the Silozuchil-2 well. Figure 17a illustrates an area that remained emergent until the beginning of the Cretaceous, corresponding to the present coastline of the southern regions of Tampico and the whole of Poza Rica.

Furthermore, Figure17b shows the distribution of the Huayacocotla Formation from the Lower Jurassic, above which Upper Jurassic sediments were deposited unconformably. In addition, paleoislands, made up of intrusive or metamorphic rocks that remained emergent until the Early Cretaceous, are seen in the center of the study area. Some of them were covered by the calcarenites of the San Andrés Member from the Kimmeridgian-Tithonian.

The advance of the Callovian sea occurred in the center of the Tampico region across restricted areas, as seen in wells Los Cues-102, Eleja-1, and Limon-191 (Figures 12, 17a, c: 7–9, d). The nature of these communication routes is difficult to establish because of lack of data. However, these relations can be inferred from wells Acamaya-1 and Bocacajeta-1 (Figure 17a, c: 6, 10, d), which are located close to the present coastline. Both wells represent the only transgressive communication routes toward the Gulf of Mexico. However, in the case of the Acamaya-1 and Bocacajeta-1 wells bordering the marine region, the Middle and Upper Jurassic sequence is thick. Because they now have more reference elements, the last data somewhat modify the previously expressed concept in regard to the Callovian transgression in the Bejuco area (Cantú-Chapa, 1987a).

The upper Callovian–lower Oxfordian transgression advanced toward the present Gulf of Mexico in the region between the Bocacajeta-1 and Cuauhtemoc-1 wells. This first proto-Atlantic communication route, which originated in the east from the Pacific Ocean, is hereby named the "Tampico paleochannel" (Figures 14 and 17a, d).

Upper Jurassic sediments covered the northern part of Tampico District and overlie the San Andrés Member calcarenites and the argillaceous limestones containing bentonite and chert of the Tithonian Pimienta Formation. These sediments were deposited on a continent surface composed of a variety of rocks: older sedimentary rocks of Late Paleozoic and Early Jurassic age, along with continental, intrusive, and metamorphic rocks (Figures 17a, b, c).

Deposition in eastern Mexico advanced toward the northeast where, in an elongate area of more than 400 km, subsurface data show the stage of the sedimentary cycle corresponding to the upper Oxfordian (Figure 2b). The carbonates and evaporites of the Olvido Formation indicate that the transgressive platform phase that began at that age was deposited over an old continent formed by red beds in the northern part (wells Garza-101 and San Javier-3) and metamorphic (Ocotillo-1) and intrusive rocks (Camotal-1) in the south. At the same time, argillaceous-calcareous basinal deposits corresponding to the La Casita Formation were laid down over the thick sequence of the Olvido Formation. The few ammonites obtained from cores characterize only two substages: *Lithacosphinctes* and *Idoceras* (lower Kimmeridigan) and *Salinites* and *Proniceras* (upper Tithonian).

MIDDLE JURASSIC TO LOWER CRETACEOUS STRATIGRAPHY AROUND THE GULF OF MEXICO

The importance of Middle and Upper Jurassic sedimentation in the Tampico region was discussed above, based on material recovered from oil wells. Figure 16 illustrates the marine invasion of the Gulf region, which advanced from southern Mexico. Middle and Upper Jurassic lithofacies occurring around the Gulf of Mexico are shown, taking the top of the Tithonian as a reference datum.

Dating the base of the sedimentary sequences as Middle Jurassic to Lower Cretaceous in the subsurface of the Gulf of Mexico coastal plain is very significant because it shows their paleogeographic relations and changes of facies with isochronic sequences in the rest of the country. It also explains their biostratigraphic relation with adjoining parts of the Gulf (e.g., Louisiana and Cuba) and establishes chronostratigraphic hierarchies in areas of sedimentation surrounding this basin (Imlay, 1980; Judoley and Furrazola, 1968).

The stratigraphic section in Figure 18 summarizes this large area, approximately 2500 km in length. Biostratigraphic data from the subsurface of the southeastern United States have been added, based on material studied by Imlay and Hermann (1984). Ammonites from characteristic oil wells have been used in its construction, confirming the accuracy of the stratigraphic method because it allows large-scale dating and correlation of sedimentary sequences.

This section shows the different ages of the areas that allowed the passage of the Middle Jurassic transgression until a Pacific-Atlantic communication route was established during the Oxfordian. Depositional areas of different ages are outlined, and their stratigraphic and paleogeographic relationship with the opening of the Gulf of Mexico is illustrated on a regional scale. In the area bordering the Gulf of Mexico, there are no important outcrops of Middle and Upper Jurassic rocks; the only two known are from the Upper Tithonian near Cruillas in Tamaulipas and Chinameca in Veracruz (Cantú-Chapa, 1982).

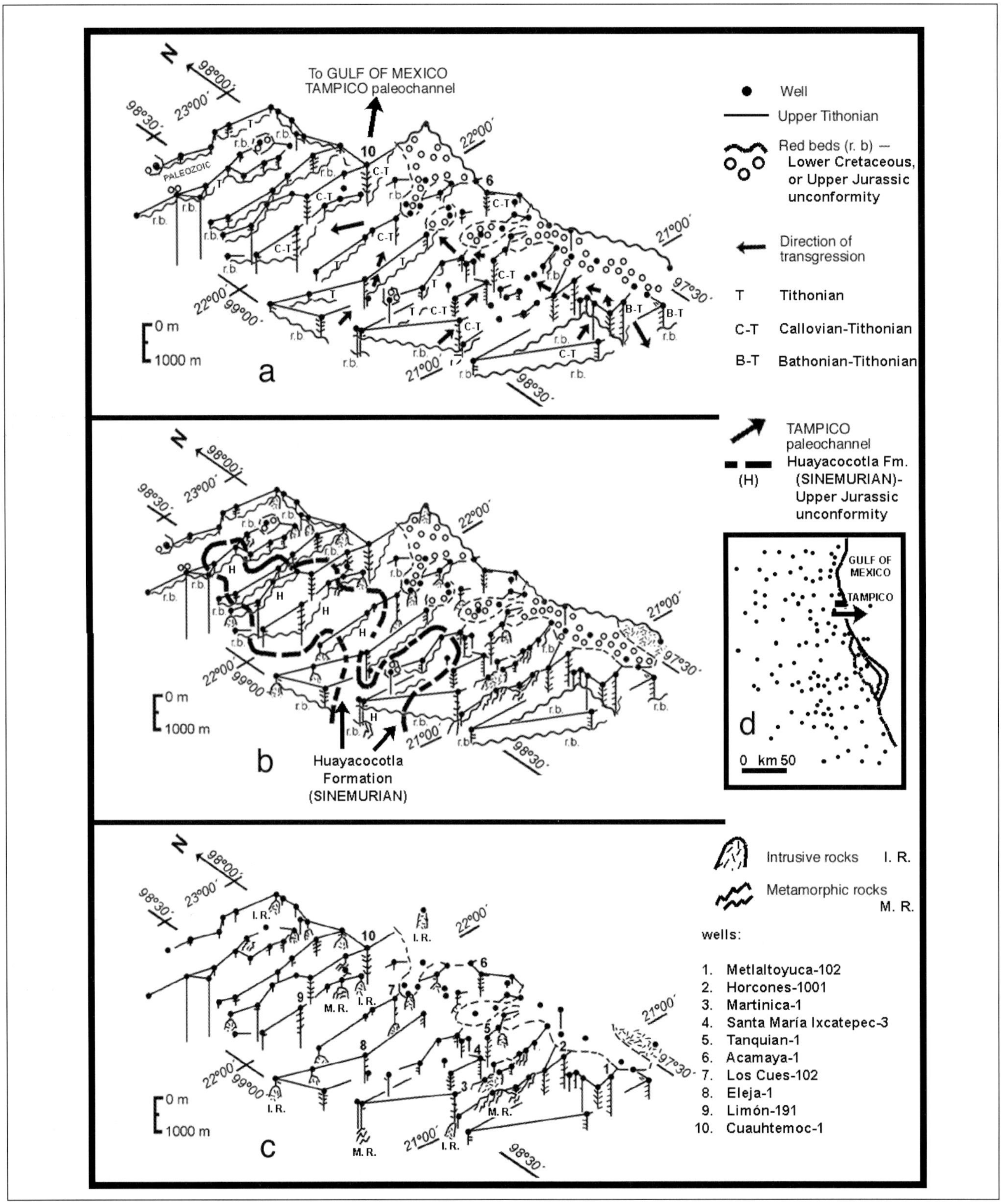

FIGURE 17. (a) Isometric map of wells in northwestern Poza Rica and Tampico regions, eastern Mexico. Reference datum: top of the Jurassic or base of the Lower Cretaceous in contact with red beds (r.b.). Only some wells show a complete sequence of the Huasteca Series (Bathonian-Tithonian) deposited over sedimentary rocks (Paleozoic, red beds, or Huayacocotla Formation), intrusive (I R), or metamorphic rocks (M R). The arrows indicate the direction of Middle Jurassic transgression toward the Gulf of Mexico. The formational units of the Huasteca Series deposited over this old continent are suggested by short lines in each well. Their ages are abbreviated: T (Tithonian), C-T (Callovian-Tithonian), and B-T (Bathonian-Tithonian). (b) Locations of the Huayacocotla Formation (Sinemurian); (c) metamorphic and intrusive rocks; and (d) studied wells.

The section depicts sedimentation beginning in the Bathonian. It defines the oldest lithostratigraphic unit of the Middle Jurassic in eastern Mexico. This constitutes the beginning of a sedimentary cycle in which final transgressive stages did not occur until the early Cretaceous. Construction of the section illustrated here was based on regional sections (Figure 2b), some of which already have been published (Cantú-Chapa, 1989b, 1992).

The Diachronous Base of the Middle Jurassic to Lower Cretaceous Sequences

The base of the sedimentary sequences present in the subsurface of the Gulf of Mexico shows various ages:

Bathonian

The oldest deposits of the Middle Jurassic in eastern Mexico, especially in northwestern Poza Rica (Palo Blanco-102 well), are lower Bathonian. Eighty meters of evaporites, shales, and calcarenites (Huehuetepec Formation) were probably deposited over red beds; these were covered by calcarenites (Tepexic Formation). Neither formation contains fossils. On the other hand, upper Bathonian–lower Callovian ammonites (*Wagnericeras* and *Kepplerites*) have been found in the siltstones (Palo Blanco Formation) that overlie the last unit. Therefore, the evaporitic and calcarenitic units should be of pre–late Bathonian age (Cantú-Chapa, 1969, 1992) (Figure 18).

The sedimentary cycle corresponding to the Huasteca Series continues in the Palo Blanco-112 well with the Santiago Formation (shales from the middle Callovian-Oxfordian), the Taman Formation (argillaceous limestones from the lower Kimmeridgian–lower Tithonian), the San Andrés Member (calcarenites), and the Pimienta Formation (argillaceous limestones with bentonite). These last two also are of Tithonian age (Figure 18).

Middle Callovian

The age of the base of the sedimentary sequence corresponding to this substage was inferred from ammonites found in the overlying unit in two wells.

In the Tampico region, the Silozuchil-2 well contains the ammonite *Hecticoceras* in core 5, characterizing the middle Callovian (Cantú-Chapa, 1963); this allows the argillaceous rocks from the Santiago Formation to be dated. In this well, the Middle Jurassic sedimentary cycle begins with calcarenites (Tepexic Formation) deposited over red beds (Cahuasas Formation). By means of stratigraphic correlation, a middle Callovian–Oxfordian age was inferred for the thick Santiago Formation. In contrast, the overlying Taman Formation is relatively thin and of early Kimmeridgian age. The Pimienta Formation, approximately 125 m thick, is assigned to the Tithonian in this region of eastern Mexico (Figure 18).

In the Tabasco region in southeastern Mexico, the Boquiapan-101 well provided a specimen of *Reineckeia* from a calcareous-argillaceous unit that indicates a middle Callovian age (Figure 18).

Because of the absence of more paleontologic material from this well, this unit (1950 m thick) was dated as middle Callovian to Kimmeridgian by stratigraphic correlation. It is covered by 250 m of argillaceous calcareous sediments with bentonite from the Tithonian. Even though this well drilled the greatest thickness of the Middle and Upper Jurassic, it did not cut the whole sequence; in fact, its base is unknown.

Oxfordian

Significant sedimentologic events took place in the lower and upper Oxfordian, according to ammonites recovered from various wells. The presence of these fossils in argillaceous or calcareous-argillaceous rocks indicates an advanced phase of the sedimentary cycle in the region bordering the Gulf of Mexico, as follows (Figure 18):

Core 3 of the Balam-101 well in southeastern Mexico provided *Ochetoceras* and *Discosphinctes* characterizing the upper Oxfordian. These ammonites come from calcareous rocks that cover sandstones, which overlie anhydrites that correspond to the base of the sedimentary sequence (Ek-Balam Formation). This sequence is overlain by calcareous-argillaceous rocks approximately 560 m thick from the Kimmeridgian (Akimpech Formation) and 150 m of argillaceous limestones with bentonite from the Tithonian (Edzna Formation) (Angeles-A. and Cantú-Chapa, this volume). The last two ages were established by stratigraphic correlation based on previously studied ammonites (Cantú-Chapa, 1977, 1982).

In wells in Louisiana, Imlay and Hermann (1984) indicated the presence of *Ochetoceras* and *Discosphinctes* that also gives a late Oxfordian age for the Smackover Formation containing these fossils. This unit was deposited over evaporitic rocks, the basal contact of which is unknown. In the two previous cases, the base of the sedimentary cycle is probably of early Oxfordian age (Figure 18).

In northeastern Mexico, several wells penetrated thick sequences of evaporitic and calcarenitic rocks almost 550 m thick. Rocks corresponding to the Olvido Formation rest above evaporites in some wells (San Javier-3, Fresnito-1), which cover red beds. None of these units deposited over red beds contains fossils. However, by their stratigraphic position, it is possible to suggest that they may be assigned a late Oxfordian age. In fact, the base of the La Casita Formation contains ammonites (*Lithacosphinctes* and *Idoceras*) from the lower Kimmeridgian (Figures 2b, 18).

In some wells in northeastern Mexico (e.g., Camotal-1), basal calcarenites were deposited over intrusive rocks. These are covered by the La Casita Formation (Kimmeridgian-Tithonian), which is less than 50 m thick. This indicates that the sedimentary cycle began in the late Oxfordian over an old continent made up of intrusive and metamorphic rocks (Ocotillo-1 well and Camotal-1 well) in this part of northeastern Mexico (Figures 2b, 18).

In another part of this section, the present author classified the ammonites from Santa Lucia-10 well in southeastern Poza Rica (Figure 18). The fossil *Discosphinctes* comes from argillaceous units more than 200 m thick assigned to the Santiago Formation and *Ataxioceras* from 220 m of argillaceous limestones assigned to the Taman Formation (lower Kimmeridgian). The Upper Jurassic sedimentary cycle ends in this well with 320 m of calcarenites (San Andrés Member) and 60 m of argillaceous limestones with bentonite (Pimienta Formation), both Tithonian.

Knowledge of the Pinar del Rio region in western Cuba, where abundant ammonites (*Discosphinctes, Dichotomosphictes, Euaspidoceras, Cubaochetoceras,* and *Ochetoceras*) of the upper Oxfordian have been described, must be added to what is already known about the sedimentation that occurred around the Gulf of Mexico. The base of the Upper Jurassic sequence in Cuba is not known (Judoley and Furrazola, 1968). These fossils appear in a sequence that is calcareous at its base, passing to argillaceous sandstones with limestones at the top, forming part of the Jagua Formation. This unit represents an advanced

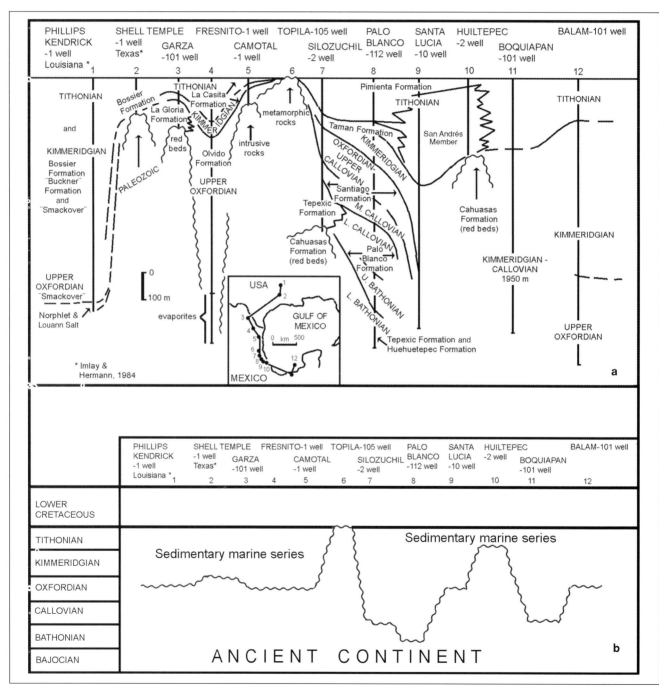

Figure 18. (a) Stratigraphic correlation of the Middle and Upper Jurassic with wells from the Gulf of Mexico coastal plain. Ages and thickness of the sedimentary series are as follows: older in the east, Bathonian (Palo Blanco-112 well), and thicker in the southeast (Boquiapan-101 well). Datum is the top of the Tithonian. (b) Stratigraphic synthesis.

event of the Upper Jurassic sedimentary cycle with clear biogeographic similarity to other regions of Mexico (San Pedro del Gallo and Taman regions, and Balam-101 well) (Imlay, 1980).

Lower Kimmeridgian

The argillaceous limestone Taman Formation was deposited over red beds in the southeast of Poza Rica. Some wells of the San Andrés field show this transgressive event of the lower Kimmeridgian because of the presence of *Idoceras*, which was found at the base of the Taman Formation (Cantú-Chapa, 1999).

Tithonian

The final Late Jurassic depositional event that took place in the Gulf of Mexico coastal plain is found in southeastern Poza Rica and is represented by Tithonian strata. The sedimentary cycle began there with thick layers of calcarenites deposited over red beds. The San Andrés Member is covered by 30 m of the Pimienta Formation in the Huiltepec-2 well area (Figure 18). Another example of Tithonian sedimentation, similar to that in southeastern Poza Rica, is present in well Barcodon-102 in the Tampico region. Certain wells in Tampico indicate a late stage of sedimentation from the Jurassic (Figures 10–15).

Lower Cretaceous

The Topila-105 well from the Tampico region is included in Figure 18 because it represents the most advanced stage of the transgressive cycle that began in the Middle Jurassic in Oaxaca. In this part of eastern Mexico, the transgression took place in the Early Cretaceous. The Topila-105 well was included in the above chronostratigraphic correlation to underline the aforementioned transgressive stage where ammonites and microfossils (calpionellids and nannoconus) have determined a Berriasian age for this event (Figure 13). The unconformity (Lower Cretaceous) separates micritic limestones of the lower Tamaulipas Formation from metamorphic rocks.

MEXICO AS THE WESTERN MARGIN OF PANGEA

Another biostratigraphic argument that proves that the opening of the Gulf of Mexico was from the Pacific to the Atlantic Ocean through the ancient Mexican territory is the presence of fossils of marine origin, from the Permian to Lower Jurassic, in this part of the country. This presence constitutes further paleogeographic evidence that allows us to delimit and place part of the Mexican territory as an oceanic province (related to the Pacific province). At the same time, this marine evidence also defines the western margin of Pangea, the subsequent rupture of which gave origin to the Gulf of Mexico and the Atlantic Ocean.

Generally, it can be seen that in conceiving the representation of this megacontinent, there is no place to establish the correct location of the present Gulf of Mexico during the time

such continental mass existed (Hsu and Bernoulli, 1978). The Tethys and a primitive Atlantic Ocean of Triassic–Lower Jurassic age have been proposed without showing paleontologic evidence of that age for an ocean (Marcoux, et al.,1982; Thierry, 1982).

The manner and time of the dismembering of Pangea constitute a controversial problem in geology. We are concerned with its analysis because it implies explaining the opening of the Gulf of Mexico. However, we think that the biostratigraphic data found in Mexico allow the inference of the marine-continental boundary (Tethys-Pangea) in this part of the American continent, especially because the area occupied an important place as the margin of Pangea in certain geologic periods.

Permian

The boundaries of the subprovince of the eastern Pacific are set in several locations of Mexico and the American continent containing ceratites (araxoceratidae and xenodiscidae) and goniatites of the Permian.

Biogeographic evidence of marine fauna of this age in Mexico and Europe does not allow the establishment of an Atlantic communication route between these two sedimentation areas, because those fossils did not extend beyond Sicily and Tunisia in the Mediterranean province, which are locations that mark the western boundary of the Tethys sea. Paleogeography of the marine Upper Permian of Mexico shows clearly that there were no links between the Pacific Ocean and the Tethys sea. A continental barrier located in the eastern part of Mexico as well as in North and Central America stresses the inexistence of the Gulf of Mexico and the Atlantic Ocean at this time (Spinosa et al., 1975) (Figure 19b).

Upper Triassic

Sequences which correspond to the upper part of this period are known only in northwest and central Mexico (Burckhardt, 1930). Westermann (1973) described the world biogeographic distribution of the bivalve *Monotis* from the marine Upper Triassic, placing them predominantly around the circum-Pacific province, that is, in the west of the American continent and in the eastern part of Asia, which was in communication with Oceania, the Mediterranean, the Middle East, and India. This fossil is found in three locations in Mexico: Baja California, Sonora, and Zacatecas (Figure 19a, c).

Westermann (1973) established the presence of a physical barrier between the eastern part of the Pacific Ocean and the Tethys, corresponding to Pangea, when he studied the distribution of the bivalves. According to Tozer (1980), the biogeographic distribution of the ceratites from the Upper Triassic is also restricted to the provinces of the Arctic, Eastern Pacific Ocean, and the Tethys sea, which coincides with the distribution of *Monotis*. Therefore, the boundary of the western edge of Pangea, which includes Mexico, is inferred by this Upper Triassic biogeographic data. The location of Zacatecas in the center of the country has more biogeographic affinity with the province of the Eastern Pacific (North and South America) than with the Tethys sea, because of the nonexistence of the Atlantic Ocean.

Lower Jurassic

It has always been a great temptation to explain the presence of the Huayacocotla Formation in eastern Mexico as a marine incursion from the Tethys during the Lower Jurassic (Sinemurian) (Erben, 1956; Schmidt-Effing, 1976; Thierry, 1982; Ziegler, 1971). However, it is common to find ammonites (vermiceratids and oxynoceratids) of this formation in the whole western region of North and South America, from Alaska to Sonora, Peru, and Chile (Hillebrandt et al., 1992; Imlay, 1980). They all correspond to the Eastern Pacific province. The same fossils are common in the Tethys province, especially in France, Italy, the northwestern part of Africa, and the Iberian peninsula (Arkell, 1956).

These paleobiogeographic elements located near the coasts of the northern African and European continents have suggested the establishment of a marine communication between eastern Mexico and regions that would correspond to the western edge of the Tethys sea through a paleogeographic province named Proto-Atlantic (Schmidt-Effing, 1976; Thierry, 1982). These ammonites were correlated without evidence of this suggested connection. A relevant fact to consider is the distribution of the Huayacocotla Formation in the subsurface of Tampico District; it is not present in the easternmost region of the Gulf of Mexico coastal plain, as seen above. In fact, this formation defines an elongated paleogeographic unit parallel to the present coastline that connects the southeast with outcrops of the Sierra Madre Oriental to the north (Figures 16 and 19d).

In spite of not having intermediate biostratigraphic data of Sinemurian age between the northwest and the east of Mexico, where vermiceratid and oxynoceratid ammonites are found, we insist on considering that these two distant marine sedimentation areas justify the definition of the Eastern Pacific province in contact with the western edge of Pangea.

Rocks of Continental Origin

Alternating with the marine-origin sequences of the Upper Permian, Upper Triassic, and Lower to Middle Jurassic are rocks of continental origin. Each of these continental events indicates the establishment of the western edge of Pangea in Mexico.

CONCLUSIONS

Incoherent and sometimes naïve geologic concepts have been used to explain the origin and age of the Gulf of Mexico. In some studies, "suspect terrains" are displaced in a manner that induces vertigo in the reader. Furthermore, the "Mojave-Sonora Megashear," which has little basis in fact, is referred to in almost every synthesis of Mexican geology. Likewise, the "Xolapa terrain" in southern Oaxaca, which is very poorly known, is always shown to have the same distribution throughout the Jurassic and Cretaceous (Meneses-R. et al., 1994).

The persistent interpretation of the distribution of the metamorphic rocks of this "Xolapa Terrain" as an immutable monolithic structure represents a problem in explaining the Pacific origin of the Gulf of Mexico. These metamorphic complexes must be studied in accordance with the subsurface model of the Tampico region presented above, whereas in nearby areas, metamorphic and intrusive rocks were covered

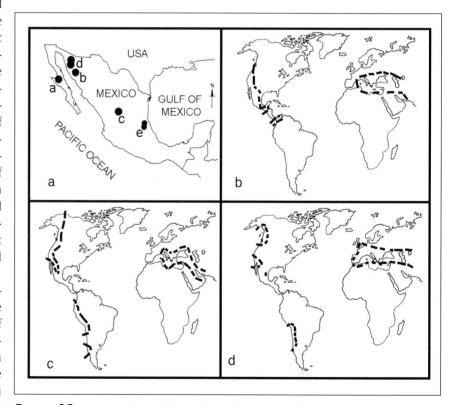

FIGURE 19. (a) Locations of the Triassic (a, b, c) and Lower Jurassic (Sinemurian, d, e) marine rocks in Mexico. Paleogeography of the Permian (b), Triassic (c), and Lower Jurassic (d) marine rocks in the American, European, and African continents, based on ammonite and bivalve distribution (after Imlay, 1980; Spinosa et al., 1975; Westermann, 1973). The absence of these fossils in some regions of eastern Mexico (d, this work), as well as in the Gulf of Mexico and the North Atlantic Ocean, suggests that the western edge of Pangea was located in this country during the Triassic to Middle Jurassic. For distribution of the Permian localities of Mexico, see Figure 1b (Cantú-Chapa, 1997).

by different phases of the Upper Jurassic and Lower Cretaceous transgression during its advance toward the Gulf of Mexico. The possibility that southern Mexico paleogeographic elements allowed the passage of the Bajocian transgression toward the future Gulf region must be considered. The presented evidence points to the consideration that this opening began through Oaxaca, and not through hypothetical passages in western Mexico.

Paleontologists always make reference to paleogeographic elements with marked fixation. They place an impossible and nonexistent communication route from the western Tethys to the eastern Pacific through a supposed "Hispanic Corridor" that would pass through Mexico, or they correlate paleogeographic provinces of different ages, e.g., the Bajocian of Venezuela with the Sinemurian of eastern Mexico (Bartok et al., 1985). All the above makes the study of the Gulf of Mexico opening illogical in many aspects.

The Bajocian-Oxfordian ammonites found at the base of the sedimentary series always show a transgressive trend from the Pacific Province (Oaxaca in southern Mexico) to the Gulf of Mexico (Louisiana and Cuba) (Figure 20).

Intermediate locations between Oaxaca and northwestern Poza Rica or in the Sierra Madre Oriental that are linked paleogeographically and which justify the upper Bajocian–middle Callovian transgression from the Pacific toward the Gulf of Mexico are still to be found. The same arguments are valid to establish communication between the subsurface of Tabasco and Campeche in southeastern Mexico and the Gulf of Tlaxiaco (Figures 21 and 22).

Biogeographic fauna-distribution maps of Mexico from the Permian to the Middle Jurassic always show a Pacific more than an Atlantic relation. Until now, there has been no biostratigraphic evidence of these ages that proves a communication route across the Gulf of Mexico coastal plain, not even in wells of the North Atlantic.

On the other hand, one cannot discard the possibility of finding, in the vast region of the North Atlantic, paleontologic material of marine origin from the Permian to the Middle Jurassic that would relate to both the Pacific origin of the Gulf of Mexico and the distribution of the western margin of the old Pangea continent. Such discoveries would substantially modify this study.

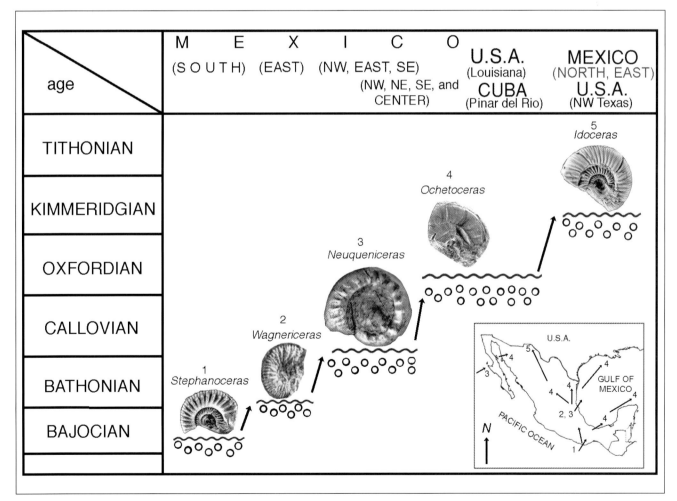

FIGURE 20. Middle and Upper Jurassic transgressive events from southern Mexico (Bajocian) to the Gulf of Mexico (United States and Cuba) (lower Oxfordian) determined by ammonites. Small map shows transgressive events explained in Figure 7a.

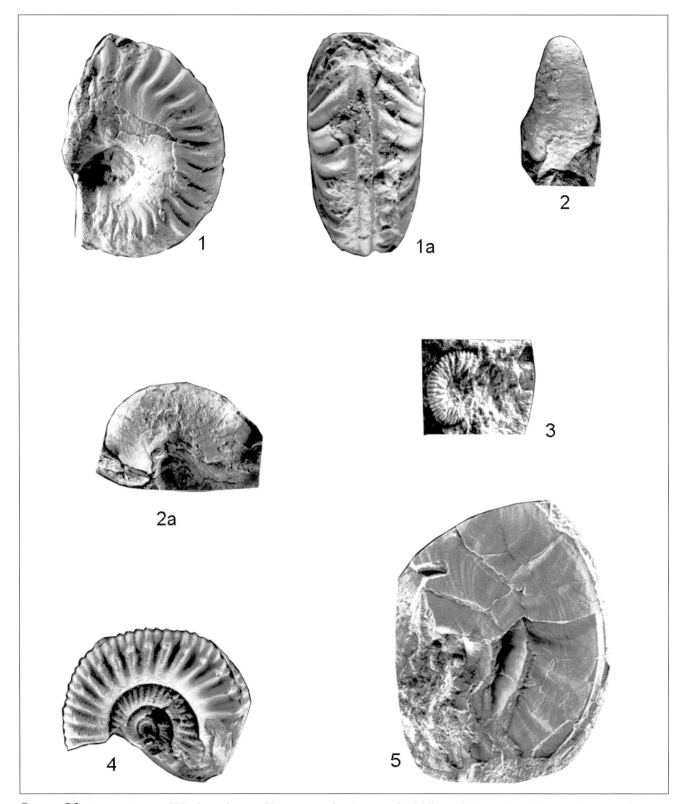

FIGURE 21. Ammonites and bivalves observed in cores and outcrops of Middle and Upper Jurassic ages, from southern (1, 1a, and 4), eastern (2, 2a, and 3), and southeastern (5) Mexico. 1, 1a. *Guerrericeras inflatum* (Westermann), lower Callovian. Coahuilote, Guerrero. 2, 2a. *Liogryphaea nebrascensis* (Meek and Hayden), middle Callovian, Poza Rica-162 well, core 5, Poza Rica District. 3. *Wagnericeras* aff. *wagneri* (Oppel), upper Bathonian, Palo Blanco-112 well, core 6, Poza Rica District (Cantú-Chapa, 1969). 4. *Stephanoceras* sp., Nuyoo, Oaxaca, middle Bajocian, Oaxaca. 5. *Ochetoceras* sp., upper Oxfordian, Balam-101 well, core 3, Campeche District, Gulf of Mexico. All specimens are 1 x.

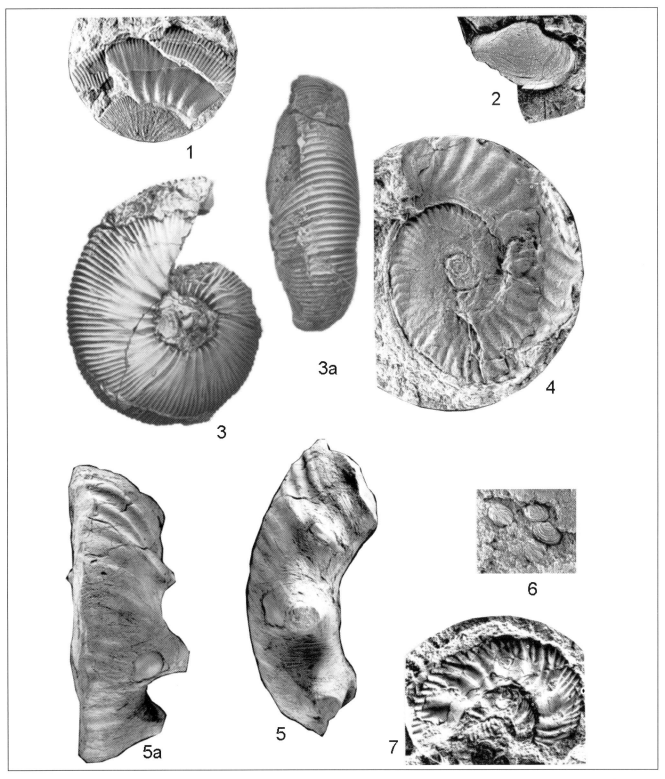

FIGURE 22. Ammonites and bivalves observed in cores and outcrops of Middle Jurassic age, from southern (5 and 5a) and eastern (1, 2, 3, 3a, 4, 6, and 7) Mexico. 1. *Ataxioceras* (*Shneidia*) sp., lower Kimmeridigian, Papantla-1 well, core 10, Poza Rica District. 2. *Posidonia ornati* (Quenstedt), middle Callovian, Talaxca-2 well, core 6, Poza Rica District. 3, 3a. *Virgatosphinctes mexicanus* (Burkhardt), lower Tithonian. Taman, San Luis Potosí. 4. *Hecticoceras* sp., middle Callovian. Silozuchil-2 well, core 5, Tampico District (Cantú-Chapa, 1963). 5, 5a. *Neuqueniceras* sp., lower Callovian, Tecocoyunca, Guerrero. 6. *Bositra buchi* (Roemer), middle Callovian, Huauchinango, Puebla (Cantú-Chapa, 1971). 7. *Reineckeia* sp., middle Callovian. Palo Blanco-112 well, core 4, Poza Rica District. Specimens 1 and 5 are x 0.5; 2, 4, 6, and 7 are x 1.5; 3 and 3a are x 1.

ACKNOWLEDGMENTS

The author would like to express his gratitude to Victor Flores for his comments and to Tonatiuh and Yaocalli Cantú for technical support. Thanks also are due to Hector Amezcua for the photographs of the specimens, to James Lee Wilson Sr. and Christopher Lehmann for providing constructive reviews of the manuscript, and to Sharon Mason for careful editing. The Consejo Nacional de Ciencia y Tecnología supports the present study (Proyecto 411300-5-2228PT).

REFERENCES CITED

Anderson, T. H., and V. A. Schmidt, 1983, The evolution of Middle America and the Gulf of Mexico–Caribbean Sea region during Mesozoic time: Geological Society of America Bulletin, v. 94, p. 941–966.

Angeles-Aquino, F., and A. Cantú-Chapa, 2001, Subsurface Upper Jurassic stratigraphy in the Campeche Shelf, Gulf of Mexico: this volume.

Arkell, W. J., 1956, Jurassic geology of the world: Edinburgh-London: Oliver and Boyd Ltd., 806 p.

Bartok, P. E., O. Renz, and G. E. G. Westermann, 1985, The Siquisique ophiolites, Northern Lara State, Venezuela: A discussion on their Middle Jurassic ammonites and tectonic implications: Geological Society of America Bulletin, v. 96, p. 1050–1055.

Berckhemer, F., and H. Holder, 1959, Ammoniten aus dem Oberen Weissen Jura Süddeutschlands: Beihefte zum Geologischen Jahrbuch, v. 35, 135 pp.

Burckhardt, C., 1930, Étudé synthétique sur le Mésozoïque méxicain: Mémoire de la Société Paléontologique Suisse, v. 49–50, 280 p.

Cantú-Chapa, A., 1963, Étudé biostratigrafique des ammonites du Centre et de l'Est du Méxique: Société géologique de France, Mémoire 99, no. 5, 103 p.

Cantú-Chapa, A., 1969, Estratigrafía del Jurásico Medio–Superior del subsuelo de Poza Rica, Veracruz (Area de Soledad-Miquetla): Revista del Instituto Mexicano del Petróleo, v. 1, no. 1, p. 3–9.

Cantú-Chapa, A., 1971, La Serie Huasteca (Jurásico Medio–Superior) del Centro-Este de México: Revista del Instituto Mexicano del Petróleo, v. 3, no. 2, p. 17–40.

Cantú-Chapa, A., 1976a, Nuevas localidades del Kimeridgiano y Titoniano en Chihuahua (Norte de México): Revista del Instituto Mexicano del Petróleo, v. 8, no. 2, p. 38–45.

Cantú-Chapa, A., 1976b, El contacto Jurásico-Cretácico, la estratigrafía del Neocomiano, el Hiato Hauteriviano Superior–Eoceno Inferior y las amonitas del pozo Bejuco 6 (Centro-Este de México): Boletín de la Sociedad Geológica Mexicana, v. 37, p. 60–83.

Cantú-Chapa, A., 1977, Las amonitas del Jurásico Superior del pozo Chac 1, Norte de Campeche (Golfo de México): Revista del Instituto Mexicano del Petróleo, v. 9, no. 2, p. 8–39.

Cantú-Chapa, A., 1982, The Jurassic-Cretaceous boundary in the subsurface of Eastern Mexico: Journal of Petroleum Geology, v. 4, no. 3, p. 311–318.

Cantú-Chapa, A., 1987a, The Bejuco Paleocanyon (Cretaceous-Paleocene) in the Tampico District, Mexico: Journal of Petroleum Geology, v. 10, no. 2, p. 207–218.

Cantú-Chapa, A., 1987b, La bioestratigrafía y la datación de discordancias fanerozoicas en México: Revista de la Sociedad Mexicana de Paleontología, v. 1, no. 1, p. 137–158.

Cantú-Chapa, A., 1989a, La Peña Formation (Aptian): A condensed limestone-shale sequence from the subsurface of NW Mexico: Journal of Petroleum Geology, v. 12, no. 1, p. 69–83.

Cantú-Chapa, A., 1989b, Precisiones sobre el límite Jurásico-Cretácico en el subsuelo del Este de México: Revista de la Sociedad Mexicana de Paleontología, v. 2, no. 1, p. 26–69.

Cantú-Chapa, A., 1990, *Dichotomites (Dichotomites) mantarraiae*, amonita del Pozo Mantarraya no. 1: Revista de la Sociedad Mexicana de Paleonotología, v. 2, no. 2, p. 43–45.

Cantú-Chapa, A., 1992, The Jurassic Huasteca Series in the subsurface of Poza Rica, Eastern Mexico: Journal of Petroleum Geology, v. 15, no. 3, p. 259–282.

Cantú-Chapa, A., 1997, Los cefalópodos del Paleozoico de México: Geociencias, Instituto Politécnico Nacional, no. 1, 127 p.

Cantú-Chapa, A., 1999, Two unconformable stratigraphic relationships between Upper Jurassic red beds and marine sequences in northeastern and eastern Mexico subsurface: Geological Society of America Special Paper 340, p. 1–5.

Erben, H. K., 1956, El Jurásico Inferior de México y sus amonitas: 20 Congreso Geológico Internacional, México, 393 p.

Flores, L. R., 1974, Datos sobre la bioestratigrafía del Jurásico Inferior y Medio del subsuelo de la región de Tampico, Tamaulipas: Revista del Instituto Mexicano del Petróleo, v. 6, no. 3, p. 6–15.

Geyssant, J. R., and C. Rangin, 1979, Découvert d'Ammonites callo-viennes dans le complexe à blocs de l'île de Cedros (Basse Californie, Méxique) et implication pour l'âge de la série ophiolitique sous-jacente: Compte-Rendus de l'Académie des Sciences de Paris, tome 289, série D, p. 521–524.

Hillebrandt, A. von, P. Smith, G. E. G. Westermann, and J. H. Callomon, 1992, Ammonite zone of the Circum-Pacific region, *in* G. E. G. Westermann, ed., The Jurassic of the Circum-Pacific: Cambridge University Press, p. 247–272.

Hsu, K. J., and D. Bernoulli, 1978, Genesis of the Tethys and the Mediterranean: Initial Reports of the Deep Sea Drilling Project: Washington, D. C., U. S. Government Printing Office, v. 42, no. 1, p. 943–949.

Imlay, R. W., 1980. Jurassic paleobiogeography of the conterminous United States in its continental setting: U.S. Geological Survey Professional Paper, v. 1062, 134 p.

Imlay, R. W., and Hermann, G., 1984, Upper Jurassic ammonites from the subsurface of Texas, Louisiana and Mississippi: Gulf Coast Section, Society for Sedimentary Geology (SEPM) Foundation, Third Annual Research Conference Proceedings, p. 149–170.

Judoley, C. M., and B. G. Furrazola, 1968, Estratigrafía y fauna del Jurásico de Cuba: Instituto Cubano de Recursos Minerales, 126 p.

Klitgord, K. D., and H. Schouten, 1986, Plate kinematics of Central Atlantic, *in* P. R. Vogt and B. Tucholke, eds., The Geology of North America, v. M, the Western North Atlantic Region: Geological Society of America, p. 351–378.

López-Infanzón, M., 1986, Petrología y radiometría de rocas ígneas y metamórficas de México: Boletín de la Asociación Mexicana de Geólogos Petroleros, v. 38, no. 2, p. 59–98.

López-Ramos, E., 1981, Paleogeografía y tectónica del Mesozoico en México: Revista Instituto de Geología, Universidad Nacional Autónoma de México, v. 5, no. 2, p. 158–177.

Marcoux, J., G. Mascle, and J.-P. Cuif, 1982, Existence de marqueurs bio-sédimentaires et structuraux téthysiens issus de la marge gondwanienne à la bordure ouest-américaine: Implications paléogeographiques et paléobiologiques: Bulletin de la Société géologique de France, v. 7, no. 24, p. 971–980.

Maynard, J. B., P. M. Okita, E. D. May, and A. Martínez-V., 1990, Paleogeographic setting of late Jurassic manganese mineralization in the Molango district, Mexico: International Association of Sedimentology Special Publication, v. 2, p. 17–30.

Meneses-R., J. J., M. E. Monroy-A., and J. C. Gómez-C., 1994, Bosquejo paleogeografico y tectonico del sur de Mexico durante el Mesozoico: Boletín de la Asociación de Geólogos Petroleros, v. 44, no. 2, p. 18–45.

Michaud, F., 1987, Stratigraphie et paléogéographie du Mésozoïque du Chiapas, Mémoire de Stratigraphie 6: Thesis, Academie de Paris, Université Pierre et Marie Curie, 301 p.

Rangin, C., 1977, Sobre la presencia del Jurásico Superior con amonitas en Sonora Septentrional: Revista Instituto de Geología, v. 1, no. 2, p. 79–156.

Reed, J. M., 1994, Probable Cretaceous to recent rifting in the Gulf of Mexico Basin, part 1: Journal of Petroleum Geology, v. 17, no. 4, p. 429–444.

Ross, C. A., 1979, Late Paleozoic collision of North and South America: Geology, v. 7, p. 41–44.

Salvador, A., 1991, Triassic-Jurassic, in A. Salvador, ed., The Gulf of Mexico Basin: Geological Society of America Journal, v. J, p. 131–180.

Sandoval, J., and G. E. G. Westermann, 1989, Bioestratigrafía y biogeografía de los ammonites del Jurásico Medio de Oaxaca y Guerrero (Sur de México): Revista de la Sociedad Mexicana de Paleontología, v. 2, no. 1, p. 18–25.

Schlager, W., and R. T. Buffler, 1984, Deep Sea Drilling Project, Leg. 77, SE Gulf of Mexico: Geological Society of America Bulletin, v. 95, p. 226–236.

Schmidt-Effing, R., 1976, El Liásico marino de México y su relación con la paleogeografía de América Central: Publicaciones Geológicas del Instituto Centroamericano de Investigación y Tecnología Industrial, Guatemala, n. 5, p. 22–23.

Schmidt-Effing, R., 1980, The Huayacocotla Aulacogen in Mexico (Lower Jurassic) and the origin of the Gulf of Mexico, in R. H. Pilger

Jr., ed., Proceedings of a Symposium at Louisiana State University, Baton Rouge, Louisiana, p. 79–86.

Schuchert, C., 1935, Historical Geology of the Antillean-Caribbean Region, or the Lands Bordering the Gulf of Mexico and the Caribbean Sea: New York, John Wiley and Sons, Inc., 881 p.

Sedlock, R. L., F. Ortega G., and R. C. Speed, 1993, Tectonostratigraphic Terranes and Tectonic Evolution of Mexico: Geological Society of America Special Paper 278, 153 p.

Spinosa, C., W. M. Furnish, and B. F. Glenister, 1975, The Xenodiscidae, Permian ceratitoid ammonoids: Journal of Paleontology, v. 40, no. 2, p. 239–283.

Stanley Jr., G. D., and L. Beauvais, 1994, Corals from an Early Jurassic coral reef in British Columbia: Refuge on an oceanic island reef: Lethaia, v. 27, p. 35–47.

Suter, M., 1990, Hoja Tamazunchale 14Q-e(5) con geología de la Hoja Tamazunchale, Estado de Hidalgo, Queretaro y San Luis Potosí: Instituto de Geología, Universidad Nacional Autónoma de México, 55 p.

Thierry, J., 1982, Téthys, Mésogée et Atlantique au Jurassique: Quelques réflexions basées sur les faunes d'Ammonites: Bulletin de la Société géologique de France, v. 7, no. 24, p. 1053-1067.

Tozer, E. T., 1980, Triassic ammonoidea: Geographic and stratigraphic distribution, in M. R. House and J. R. Senior, eds.,The Ammonoidea: London, Academic Press, p. 397–431.

Walper, J. L., 1980, Tectonic evolution of the Gulf of Mexico, in R. H. Pilger Jr., ed., Symposium on the origin of the Gulf of Mexico and the early opening of the Central North Atlantic Ocean: Louisiana State University, Baton Rouge, p. 87–98.

Walper, J. L., 1981, Geological evolution of the Gulf of Mexico—Caribbean region, in J. W. Kerr, A. J. Fergusson, and L. C. Machan, eds., Geology of the North Atlantic Borderlines: Canadian Society of Petroleum Geology, Memoir No. 7, p. 503–525.

Westermann, G. E. G., 1973, The Late Triassic bivalve *Monotis*, in A. Hallam, ed., Atlas of paleobiogeography: Amsterdam, Elsevier Scientific Publishing Company, p. 251–258.

White, G. W., 1980, Permian-Triassic continental reconstruction of the Gulf of Mexico-Caribbean area: Nature, v. 283, p. 823–826.

Ziegler, B., 1971, Biogeographie der Tethys: Jahrische Geschichte Naturkunde, Württemberg, 126, p. 229–243.

Bartolini, C., and K. Mickus, Tectonic blocks, magmatic arcs, and oceanic terrains: A preliminary interpretation based on gravity, outcrop, and subsurface data, northeast-central Mexico, *in* C. Bartolini, R. T. Buffler, and A. Cantú-Chapa, eds., The western Gulf of Mexico Basin: Tectonics, sedimentary basins, and petroleum systems: AAPG Memoir 75, p. 29–43.

Tectonic Blocks, Magmatic Arcs, and Oceanic Terrains: A Preliminary Interpretation Based on Gravity, Outcrop, and Subsurface Data, Northeast-central Mexico

Claudio Bartolini
International Geological Consultant, Houston, Texas, U.S.A.

Kevin Mickus
Department of Geosciences, Southwest Missouri State University, Springfield, Missouri, U.S.A.

ABSTRACT

Complex tectonic plate interactions at the end of the Paleozoic and early Mesozoic, particularly the undefined relationship between circum-Atlantic and circum-Pacific tectonic domains, do not permit a complete understanding of the crustal structure of north-central Mexico. Pre-Oxfordian geologic history, especially the existence of Permian–Early Triassic and Late Triassic–Middle Jurassic volcanic arcs, and general crustal structure of north-central Mexico are approached through gravity modeling and analysis of geologic and well data. Bouguer and isostatic residual gravity-anomaly maps were interpreted to illustrate anomalies caused mostly by Mesozoic and Cenozoic tectonic events, including a large-amplitude, northerly trending gradient marking the edge of Cretaceous thrusting in the Sierra Madre Oriental in Tamaulipas and Nuevo León. This gravity gradient diverges in western Nuevo León with one branch trending into southern Coahuila, which also marks the northern limit of Cretaceous thrusting. However, the other branch that trends into northern Nuevo León, may be caused partially by pre-Cretaceous intrusive and metamorphic rocks or changes in the structural style of the thrust belt north of Monterrey. Lower Bouguer and isostatic residual gravity-anomaly values in central Mexico, as compared with eastern Mexico, indicate a thicker crust formed by the addition of Mesozoic magmatic arcs and sedimentary sequences. Smaller-wavelength isostatic residual gravity-anomalies correspond to Late Permian–Early Triassic plutons or density variations in the Precambrian basement rocks in eastern Tamaulipas and Nuevo León, possible Laramide-age intrusions along the Cretaceous thrust front in Tamaulipas and Nuevo León, and Mesozoic sedimentary basins, including the Parras Basin. There is no evidence for large-scale linear anomalies that would correspond to the Late Jurassic Mojave-Sonora megashear transpressive structure across northern Mexico.

Three northeast-trending, 2.5-dimensional (2.5-D) gravity models constrained by geologic, well, and sparse regional seismic data indicate that the boundary between the Late Permian–Early Triassic and Late Triassic–Middle Jurassic volcanic arcs may lie in eastern Coahuila, south-central Nuevo León, and southwestern Tamaulipas. Subtle gravity minima along the center of each profile represent Permian granite, Triassic-Cretaceous arc rocks, or Triassic-Cretaceous oceanic rocks that extend to depths of 10–12 km. High-amplitude gravity maxima in eastern Mexico represent intermediate Permian intrusive rocks or high-grade Paleozoic metamorphic rocks that extend to 10–12 km. Short-wavelength, northwest-trending gravity anomalies may represent Laramide-age or younger intrusive bodies.

INTRODUCTION

Attempts to reconstruct the pre-Cenozoic geologic history of northern and central Mexico using surface geology alone have allowed for limited tectonic interpretation only. This is caused in great part by the Jurassic-Cretaceous strata and Tertiary volcanic rocks that cover most of this region. A Permian–Early Triassic continental magmatic arc in northeastern Mexico (López-Ramos, 1972; Damon, 1975; Damon et al., 1981; Jacobo, 1986; Torres et al., 1999), a northwest-trending Late Triassic–Jurassic continental margin volcanic arc across Mexico (Damon, 1975; Damon et al., 1981; Damon et al., 1984; López-Infanzón, 1986; Grajales et al., 1992; Jones et al., 1995; Bartolini, 1998), the Paleozoic Ouachita-Marathon trend in northern Mexico (Handschy et al., 1987; Montgomery, 1988; Moreno et al., 1992; James and Henry, 1993; Moreno et al., 1994; Mickus and Montana, 1999), and the presence of basinal successions of Late to Middle Triassic age (Burckhardt, 1930; Cantú-Chapa, 1969; Gallo et al., 1993) in central Mexico are key elements in understanding the overall architecture and composition of Mexico's upper crust. Unfortunately, disentangling the geologic evolution of north-central Mexico is a major challenge because of the complex circum-Pacific and circum-Atlantic plate interactions at the end of the Paleozoic and beginning of the Mesozoic.

Our preliminary interpretations rely on the combined analysis of regional gravity data, surface geology, and subsurface information from PEMEX boreholes. We provide a provisional assessment of the overall crustal structure of northeast-central Mexico (Figure 1). The main purposes of this

study are: (a) construction of regional gravity models to interpret the Mesozoic and Paleozoic crustal structure of north-central Mexico; (b) regional interpretation of volcanic arcs, sedimentary basins, positive crustal blocks, and their relationship in space and time; and (c) identification of possible Jurassic transpressive megastructures.

GRAVITY DATA AND PROCESSING

Approximately 34,000 gravity stations (Figure 1) obtained from the University of Texas at Dallas, University of Texas at El Paso, and Cancienne (1987) were processed into simple Bouguer gravity-anomaly values and merged into one coherent database. Terrain corrections were available only for a portion of the data, precluding calculation of complete Bouguer gravity anomaly values for the entire data set. The merged data were gridded at a 4-km posting using a minimum curvature algorithm (Briggs, 1974), and this grid was used to construct a simple Bouguer gravity-anomaly map (Figure 2). Areas with little or no gravity data were not analyzed.

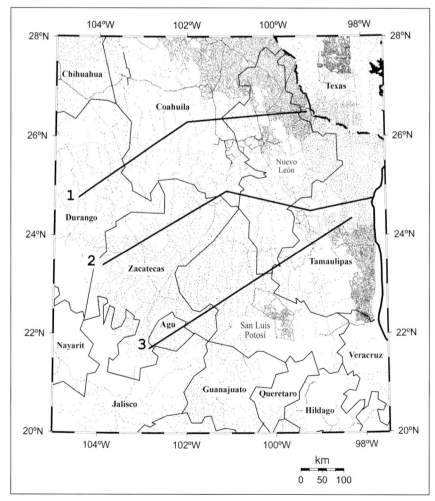

FIGURE 1. Location of gravity stations in north-central Mexico and southwest Texas. Also shown are the locations of profiles 1, 2, and 3, which were used to create regional gravity models, and the states in Mexico.

Some of the Bouguer gravity-anomaly values depicted in Figure 2 have been interpreted for regional crustal structures by Hunt (1992) and by Mickus and Montana (1999), who showed that most anomalies can be correlated with the effects of Mesozoic to Cenozoic geologic events. Because this study is concerned with crustal features, we tried to separate crustal-depth gravity anomalies from deeper-sourced gravity anomalies. There are many techniques to accomplish this, including low- and band-pass wavelength filtering, polynomial trend-surface fitting, and isostatic residual analysis. Mickus and Montana (1999) used wavelength filtering to create a series of band-pass-filtered gravity-anomaly maps to qualitatively infer the existence of Triassic-Jurassic rift basins, the extent of a Jurassic magmatic arc, and a northern extension of high-grade metamorphic rocks exposed at Huizachal-Peregrina. However, we decided to use the isostatic residual gravity technique of Simpson et al. (1986) to create our anomaly map (Figure 3).

Residual gravity-anomaly maps using techniques such as wavelength filtering or polynomial trend-surface fitting may create anomalies that are not related to actual subsurface density contrasts (Ulrych, 1968). Isostatic residual gravity-anomaly maps are based on geologically estimated parameters (e.g., density of the topography, depth of

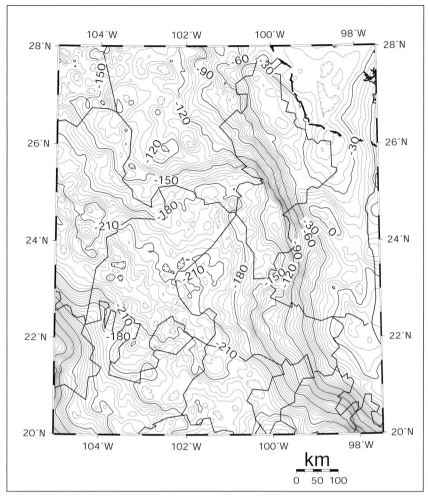

FIGURE 2. Bouguer gravity-anomaly map of the study area. Contour interval is 5 mGal. Relative gravity minima are shown by hachured contour lines.

compensation, and density between a crustal root and surrounding material), and thus are more likely to produce anomalies caused by crustal sources (Simpson et al., 1986). Regional seismic surface, wave dispersion analyses (Keller and Shurbet, 1975; Pinkerton, 1978; Gomberg et al., 1988) and regional gravity studies (Schellhorn, 1987; Mickus and Montana, 1999) were used to determine a depth of compensation (30 km), density of the topography (2.67gm/cc), and density contrast across the crustal root (0.35gm/cc). These values, along with a 4-km topography grid, were used to determine the isostatic regional gravity anomaly. This regional gravity anomaly was then subtracted from the Bouguer gravity-anomaly values (Figure 2) to create an isostatic residual gravity-anomaly map (Figure 3).

MODELING AND DISCUSSION

To quantify the interpretations of the Bouguer and isostatic residual gravity anomalies discussed below, we constructed three crustal-scale gravity models (Figures 4–6) that cross the major tectonic provinces in northern Mexico (Figure 1). The models were derived using a 2.5-D forward-modeling algorithm (Lai, 1984) in which the predicted gravity anomalies were determined by using the gravity-station elevations. Because we are using simple Bouguer gravity anomalies, the topography was included in the models. Because gravity models are nonunique, constraints (e.g., rock densities, depth to, thickness, and lateral variation of rock units) must be used to constrain geologically meaningful models. Rock densities and subsurface depths/geometries of rock units are not readily available in northeastern Mexico; however, crustal-thickness variations were available from regional seismic studies (Dorman et al., 1972; Keller and Shurbet, 1975; Pinkerton, 1978; Gomberg et al., 1988; Winker and Buffler, 1988), and average crustal and upper-mantle densities were estimated from these studies using P-wave velocities and experimental density/velocity relationships (Nafe and Drake, 1957). Density values for near-surface rocks were obtained from gravity studies by Cancienne (1987), McDonnell (1987), and Hunt (1992). Geologic maps (de Cserna, 1956; Tardy, 1980; Padilla y Sánchez, 1982; Ramírez-Ramírez, 1992; Tristán-González and Tórrez-Hernández, 1992; Bartolini, 1998; INEGI's 1:250,000-scale geologic maps), and well data (López-Ramos, 1972; Padilla y Sán-

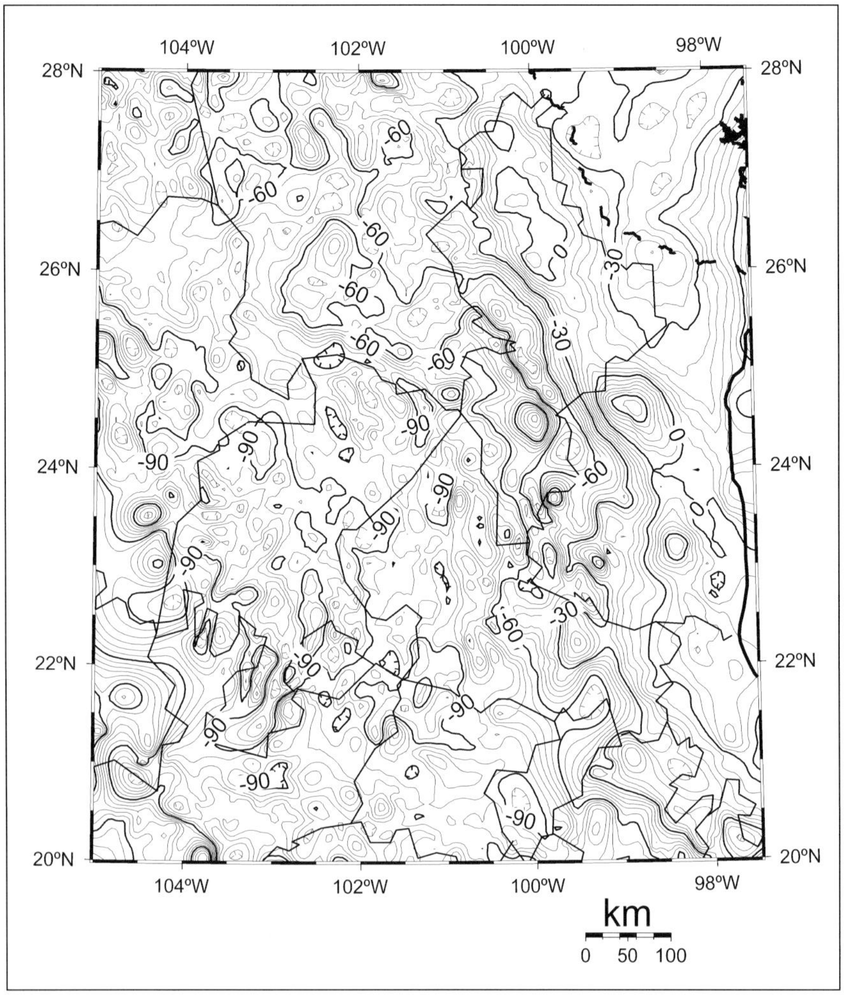

FIGURE 3. Isostatic residual gravity-anomaly map of the study area. Contour interval is 5 mGal. Relative gravity minima are shown by hachured contour lines.

chez, 1982; Jacobo, 1986; Wilson, 1990; Gonzalez-García, 1984; Grajales et al., 1992) provided constraints on the lateral locations of rock units. Upper Jurassic evaporite deposits of the Minas Viejas Formation range in thickness from 500 to 2500 m in eastern Mexico (Eguiluz, written communication, 2000). However, these strata were not separated in the modeling because of the limited subsurface data and because of their complex and heterogeneous geographic distribution.

The final models (Figures 4–6) were obtained through a trial-and-error process until the predicted gravity values matched the observed Bouguer gravity anomalies using the above constraints. The final models are not unique but represent what we feel are reasonable geologic cross sections, given the constraints that were available. These models were created mainly to aid in locating the Paleozoic and Mesozoic volcanic arcs and rocks associated with them. In addition, general crustal thickness, location of the thrust-belt front, and thickness of Jurassic and younger strata are determined. The main points in the modeling will be discussed in model 1 (Figure 4), and unique aspects of the other models are described separately.

Model 1

The Late Triassic–Middle Jurassic volcanic arc is represented by upper crustal density values of 2.67gm/cc in the Southwest, and the Permian–Early Triassic volcanic arc is represented by higher-density values of 2.71gm/cc (Figure 4). The exact location of the boundary between the two volcanic arcs is poorly determined along this profile because it may be either at 90–100 km or approximately 350 km. If we used 350 km, Permian granite would be much wider and thicker than one would expect—at least 200 km wide and 20 km thick. So based on gravity alone, we put the boundary at 90–100 km.

During the modeling process, we found it necessary to assign the upper crust containing the Permian–Early Triassic volcanic arc a higher density. What does this higher density represent? The increase in density to 2.71gm/cc under the Coahuila Block is reasonable because the basement is composed of volcanic and metamorphic rocks, and Permian igneous rocks have been encountered in wells at a distance of approximately 300 km along the profile on the model (Figure 4). In northeastern Durango state, the lower density of the upper crust within the Late Triassic–Middle Jurassic arc results partially from the fact that it is a combination of Paleozoic metasediments, Triassic-Jurassic arc rocks, and overlying sedimentary rocks. However, the overall upper-crustal density differences between the two arcs imply that there is some type of deep composition contrast between the two regions. Sparse isotopic studies within the Coahuila Block imply that the Permian granites may have an oceanic signature (Jones et al., 1984), and this may signify that the deeper upper-crustal section is more intermediate in composition than in the Late Triassic–Middle Jurassic arc region. However, based on gravity alone, this is difficult to determine, because deep seismic and well data are not available. An alternative explanation is that the higher densities are caused by transitional crust formed during the opening of the Gulf of Mexico (Winker and Buffler, 1988). Mickus and Keller (1991, 1992) showed that higher-density transitional crust was formed during the opening of the Gulf of Mexico in the south-central United States and exists throughout Texas along the buried Ouachita orogenic belt.

The 2.75-gm/cc body located at the northeastern end of the model is within a complex basement setting that includes Precambrian and Paleozoic high-grade metamorphic rocks (Ramírez-Ramírez, 1992), Permian granites, and Cretaceous

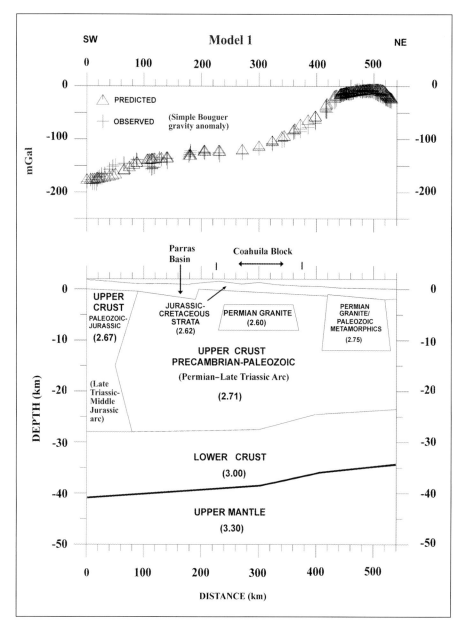

FIGURE 4. 2.5-D gravity model along profile 1 (Figure 1). This model crosses the Late Triassic–Middle Jurassic volcanic arc, the Parras Basin, the Coahuila Block, and the Permian–Early Triassic volcanic arc. The numbers in parentheses represent a body's average density in gm/cc.

show that the crust thins to 32–36 km toward the east. Given the lack of seismic studies in our study area, we cannot determine if the crustal thinning is gradual or abrupt. We tried both scenarios in the modeling and determined that abrupt thinning, coinciding with the Sierra Madre Oriental front, best fits the gravity data.

Model 2

Model 2 (Figure 5) has the same general characteristics as model 1 except that the location of the boundary between the two arcs is approximately 300 km farther east. The low-density body between 200–400 km is interpreted to have been caused by a Triassic-Jurassic arc sequence. In addition, the Jurassic-Cretaceous strata are thicker than on model 1. The Triassic-Jurassic arc sequences, which are more than 4 km thick, are overlain by 3–4 km of Upper Jurassic to Upper Cretaceous strata. These arc rocks rest on Triassic (?) strata of the Taray Formation (López-Infanzón, 1986; Bartolini, 1988); however, no pre-Triassic rocks are exposed in this region, and well data are not available. The gravity minimum at 55 km is modeled as 4-km-deep Tertiary graben containing low-density sedimentary fill. But thickness values were not available, and we could not rule out that the gravity minimum is caused by a deeper granitic intrusion.

Model 3

Model 3 (Figure 6) is similar to model 2 (Figure 5) except that the crust is thinner in the east, and the low-density body under the Jurassic-Cretaceous strata is interpreted to have been caused by the presence of Triassic-Jurassic oceanic rocks (chiefly clastic turbidites and pillow lavas). The thinner crust is expected, because crustal extension is greater in this area than in profiles in the north (Winker and Buffler, 1988). The 2.60-gm/cc block at 75–250 km is interpreted to correspond to a complex upper-crustal body consisting of Middle–Upper Triassic turbidites, Triassic-Jurassic magmatic arc rocks, Cretaceous deep-marine strata, and Upper Jurassic–Upper Cretaceous successions that formed during the active-margin tectonics associated with the Late Triassic–Middle Jurassic volcanic arc. However, we cannot rule out that this body is caused by granitic intrusions of Cretaceous and Tertiary age. This is a poorly understood region, especially at middle to lower upper-crustal levels.

sedimentary cover. No wells have penetrated this body, but both Permian granite and Precambrian metamorphic rock have been encountered in nearby wells (López-Ramos, 1972; Padilla y Sánchez, 1982; Jacobo, 1986; Wilson, 1990).

The overall crustal thickness decreases from west to east, based on seismic surface-wave studies (Keller and Shurbet, 1975; Pinkerton, 1978; Gomberg et al., 1988) and seismic refraction/reflection studies (Dorman et al., 1972; Winker and Buffler, 1988). Surface-wave analyses (Pinkerton, 1978; Gomberg et al., 1988) show that the western portion of the study area is underlain by crust with thicknesses of 42–44 km, and seismic refraction/reflection and surface-wave studies

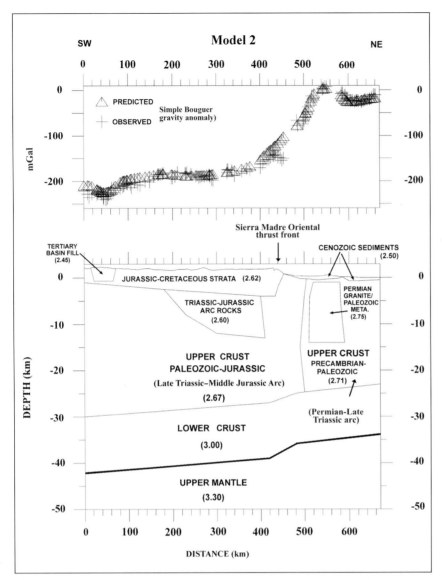

FIGURE 5. 2.5-D gravity model along profile 2 (Figure 1). This model crosses the Late Triassic–Middle Jurassic volcanic arc, the Sierra Madre Oriental front, and the Permian–Early Triassic volcanic arc. The numbers in parentheses represent a body's average density in gm/cc.

REGIONAL GEOLOGIC INTERPRETATIONS

In the west-central part of the study area, west of the fold-belt front, the crust attains a thickness of 40–45 km (Gomberg et al., 1988). This thickening is indicated on the Bouguer and isostatic residual gravity-anomaly maps (Figures 2 and 3) and in the crustal models (Figures 4–6). Gravity evidence for crustal thickening is the lower anomaly Bouguer values (~-180 to -210 mGal) and isostatic (~-90 mGal) values in the central region of the maps, as compared with higher values (~0 to -30 mGal) to the east across the Peregrina-Huizachal anticlinorium (Figures 4–6). This is caused by thinning of the crust toward the Gulf of Mexico (Winker and Buffler, 1988). Toward the north and northwest across the fold belt, Parras Basin, and the Coahuila Block, there is an increase in the Bouguer and isostatic residual

gravity values, which is probably caused by crustal thinning. The nearest seismic study (Keller and Shurbet, 1975) conducted northwest of the study area suggests crustal thinning, but the increase in gravity values could also be caused by lateral density change. The unusual thickness of the crust in this region of Mexico is the result of continuous and prolonged episodes of magmatism during the development of the Triassic-Jurassic and Cretaceous-Tertiary arcs (Damon et al., 1984; Damon et al., 1991; López-Infanzón, 1986; Grajales et al., 1992; Jones et al, 1995; Bartolini, 1998). Thickening of the crust, specifically in the Zacatecas and San Luis Potosí region (Figure 1), is complemented by the tectonic addition of Mesozoic oceanic successions along the western continental margin of Mexico at different times during the Mesozoic (Campa and Coney, 1983; Monod and Calvet, 1992; Sedlock et al., 1993; Silva-Romo et al., 1993).

The Ouachita-Marathon Belt in Mexico

Early studies of the Paleozoic continental margin of southern North America (King, 1937, 1944; Kay, 1951; Flawn et al., 1961) envisioned a prolongation of the Ouachita-Marathon orogenic belt into northern Mexico, particularly in Coahuila and Chihuahua states. The trend, width, and original configuration of the Paleozoic belt have been the subjects of continuous debate and analysis. Obviously, three of the most important issues in the light of plate-tectonics are (1) the plate-tectonic framework, (2) the timing of structural deformation, and (3) the possible connection somewhere in northwestern Mexico between the Ouachita belt and the Paleozoic orogenic belts of western North America (Poole and Madrid, 1988; Bartolini, 1988; Stewart et al., 1990; Handschy et al., 1987; Keller et al., 1989; Shurbet and Cebull, 1987; Montgomery, 1988; Moreno et al., 1992, 1994).

Debate about the Ouachita-Marathon orogenic belt is a result of the lack of Paleozoic outcrops, deep boreholes, or seismic data. Based on the association of the Ouachita interior zone with a gravity maximum that trends throughout Texas (Keller et al., 1989) and the analysis of Paleozoic outcrops in Coahuila, Handschy et al. (1987) suggested that the Ouachita belt extends from the Big Bend region in west Texas southward into Coahuila at least 100 km. Moreno et al. (1994) constructed a series of gravity models in northern Mexico and west

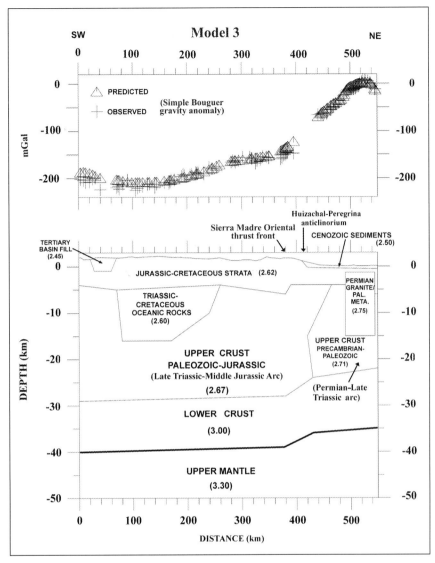

FIGURE 6. 2.5-D gravity model along profile 3 (Figure 1). This model crosses the Late Triassic–Middle Jurassic volcanic arc, the Sierra Madre Oriental front, Huizachal-Peregrina anticlinorium, and the Permian–Early Triassic volcanic arc. The numbers in parentheses represent a body's average density in gm/cc.

huahua (Torres et al., 1999) and central Coahuila (Denison et al., 1971; Jones et al., 1984; Murillo and Torres, 1987, Torres et al., 1999; Grajales et al., 1992) to Tamaulipas and Veracruz (López-Ramos, 1972; Jacobo, 1986; Wilson, 1990) along the Gulf of Mexico coast. This belt continues farther south into the states of Puebla, Oaxaca, and Chiapas and into Guatemala (Damon, 1975; Damon et al., 1981; Grajales et al., 1992; Torres et al., 1999). Woods et al. (1991) proposed that this plutonic belt extends into Belize. These plutons range in age from Permian to Early Triassic and apparently constitute the roots of a continental-margin magmatic arc (Damon et al., 1981; Grajales et al., 1992; Torres et al., 1999). The accurate mapping of the arc is difficult because it is exposed only in scattered outcrops in northern Mexico and boreholes in eastern Mexico (López-Ramos, 1972; de Cserna et al., 1970; Jacobo, 1986; Wilson, 1990; Grajales et al., 1992; Torres et al., 1999). The isotopic ages obtained from this belt, chiefly K-Ar and Rb-Sr, range from Permian to Early Triassic. Dated samples come from rocks encountered by PEMEX oil wells in the Gulf Coast region (Lopez-Ramos, 1972; Jacobo, 1986; Torres et al., 1999). However, a few dated samples come from outcrops of volcanic flows interbedded with marine fossiliferous strata in Coahuila (Bridges, 1964; Montgomery, 1988; Montgomery and Longoria, 1987; McKee et al., 1988) and Chihuahua (de Cserna et al., 1970). The Chihuahua samples may not be associated with the terrestrial Permian-Triassic arc. The prolongation of the Permian–Early Triassic arc into Chihuahua has been documented recently with the isotopic dating (K-Ar) of granite at Aldama (250 ± 20 Ma) and at Carrizalillo (267 ± 21 Ma) (Murillo and Torres, 1987; Torres et al., 1999).

The roughly north-south trend of gravity anomalies in Veracruz, Tamaulipas, and Nuevo León seems to correspond with the Permian-Triassic arc and defines a 100-km-wide structure. In Coahuila state, the width of the arc is wider. This anomalous width may be related to two events: (1) structural disruption of the arc in latest Permian time at the Ouachita-Marathon intersection—the Ouachita-Marathon belt in Coahuila apparently meets the volcanic arc at almost a right angle, and thus the rocks of the arc may have been involved in thrusting—or (2) offset of arc rocks by Jurassic transpressive activity. The traces of proposed left-lateral strike-slip faults such as the La Babia (Charleston, 1981), San Marcos (McKee et al. 1990), and Mojave-Sonora megashear (Anderson and Silver, 1979) are

Texas constrained by regional seismic data, well data, and limited Paleozoic outcrops to show that the Ouachita orogenic belt trends either south or southwestward from the Big Bend region of west Texas into Mexico. Montgomery (1988), on the other hand, proposes that the original Ouachita trend was displaced several times by Jurassic transpressive structures across northern Mexico. However, Pb isotope data (James and Henry, 1993) and possible Ouachita-Marathon facies rocks in Sonora (Poole and Madrid, 1988; Bartolini, 1988; Stewart et al., 1990) suggest that the belt may extend southwest of the Big Bend region and eventually westward into Sonora.

The Permian–Early Triassic Magmatic Arc

A north-south, northwest–southeast-trending belt (Figure 7) of granitic and granodioritic plutons extends from Chi-

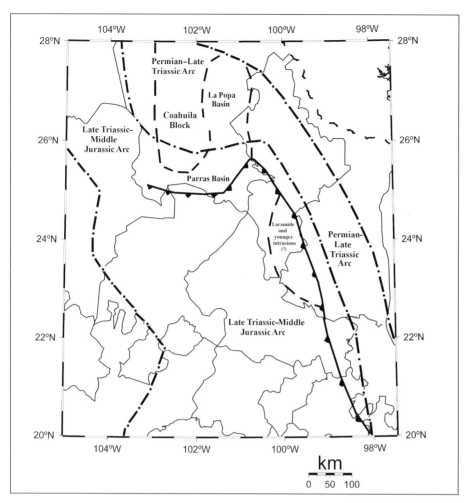

FIGURE 7. Location of major geologic features and provinces as interpreted from geologic information, drill-hole data, isostatic gravity anomalies (Figure 3), and gravity models (Figures 4–6). Bold dashed lines represent boundaries of selected geologic provinces. Bold line with triangles represents the inferred Sierra Madre Oriental front.

The lack of volcanic rocks of Permian– Early Triassic age is intriguing, and no explanation has yet been provided. One speculative explanation is that the upper volcanic part of the arc was eroded because the terrestrial arc stood along a stable southwestern margin of Pangea (Damon et al., 1981; Torres et al., 1999). Furthermore, regional crustal extension during the separation of South and North America and Jurassic extension related to the opening of the Gulf of Mexico also may have contributed to burial, attenuation, and dissection of the arc.

Bartolini et al. (1999) provisionally correlated volcanic outcrops in Nuevo León, San Luis Potosí, and Tamaulipas with the Permian–Early Triassic arc. However, their conclusions are based only on field relations, because geochemical data are limited and K-Ar isotopic ages obtained from these rocks are reset.

Geochemical studies suggest that the granitoids are calc-alkaline and are indicative of mixtures of juvenile magmas and evolved continental crust in a continental-margin volcanic arc (Torres et al., 1999). In addition, major and trace-element geochemistry of volcanic rocks whose field relations indicate a tentative pre-Mesozoic age in San Luis Potosí, Tamaulipas, and Nuevo León indicates only that volcanism was silicic (Bartolini et al., 1999).

The tectonic framework for the Permian–Early Triassic arc still is debatable, despite the existence of several proposed origins for the arc structure. Damon et al. (1981) proposed that the Permian batholith in Chiapas was associated with the closure of the proto-Atlantic and integration of Pangea during the late Paleozoic. According to Woods et al. (1991), the Permian-Triassic arc recorded plate collision and uplift during the integration of Pangea. It also has been considered an accreted volcanic arc (Grajales et al., 1992). According to Salvador (1991), the associated belt of plutons may indicate a zone of collision and subduction or may even be the magmatic core of an island arc. Subsurface and outcrop isotopic and geochemical data along the Gulf Coast of Mexico and southern Mexico indicate that the magmatic arc developed during east-dipping subduction of the paleo-Pacific oceanic lithosphere beneath a still-consolidated Pangea, and that it has no genetic relationship to circum-Atlantic tectonics (Torres et al., 1993; Torres et al., 1999).

almost parallel to the northwest trend of the arc. In addition, several other major strike-slip faults across north-central Mexico have been linked to Mesozoic oblique subduction (Longoria, 1985; Montgomery and Longoria, 1987; Montgomery, 1988).

A sharp discontinuity of the arc is evident in the isostatic residual gravity-anomaly map (Figures 3 and 7). This discontinuity occurs at the arc's northern edge, southeast of Monterrey, where the north-south trend of the arc disappears below the fold belt. Whether the arc is continuous or displaced by faults remains uncertain. The relatively small amplitude, positive gravity anomalies caused by the arc are masked by 6–10 km of Cretaceous sedimentary rocks that comprise the edge of the fold belt. On the other hand, gravity anomalies in the Coahuila Block and the Sabinas Basin are smaller in amplitude because they may be masked by thick, low-density sedimentary rocks in the Sabinas, or because the smaller size of the intrusions and/or compositional changes results in lower densities.

The Triassic-Jurassic Magmatic Arc

Reorganization of tectonic plates in the early Mesozoic led to development of a convergent margin along western North America. The subduction of oceanic lithosphere beneath the North American Plate in Late Triassic time gave rise to continental-arc magmatism from Alaska to Mexico (Coney, 1979; Damon et al., 1981).

Proposed configurations and trends of the Late Triassic–Middle Jurassic magmatic arc in northern and central Mexico (Figure 7) are based on surface exposures of volcanic, sedimentary, and intrusive rocks in seven Mexican states (Denison et al., 1969; Denison et al., 1971; Damon et al., 1981; López-Infanzón, 1986; Anderson et al., 1991; Grajales et al., 1992; Jones et al., 1990; Jones et al., 1995; Bartolini, 1997, 1998; Barboza et al., 1997). Igneous rocks also have been encountered in PEMEX boreholes (López-Ramos, 1972; Jacobo, 1986; Padilla y Sanchez; 1982; Grajales et al., 1992). The width of this magmatic arc today exceeds 150 km, although its original width and trend have probably been modified by younger tectonic events and because it is almost entirely covered by Cretaceous carbonate and Tertiary volcanic rocks of the Sierra Madre Occidental.

Transpressive Jurassic Structures in Northern Mexico

Large-scale lineaments in north-central Mexico have been interpreted as the expression of horizontal displacement along megashears or transcurrent faults active in Jurassic time (de Cserna, 1956; Murray, 1961; Sales, 1968; McBride et al., 1974; Silver and Anderson, 1974; Walper, 1977; Anderson and Silver, 1979; Charleston, 1981; Anderson and Schmidt, 1983; McKee et al., 1984; Longoria, 1985; Montgomery, 1988). The Torreón-Monterrey fracture zone was the first lineament defined in northern Mexico (de Cserna, 1956). This structure may be related to the Torreón-Saltillo fracture proposed by Murray (1961). Sales (1968) proposed the Texas lineament, which speculatively extends from the Gulf of Mexico region to eastern California. The Walper lineament apparently extends from the Gulf of Mexico to the Gulf of California (Walper, 1977). Two of the main structures in northern Mexico, the San Marcos and La Babia faults, were originally described by Charleston (1981). Silver and Anderson (1974) and Anderson and Silver (1979) proposed the Mojave-Sonora megashear, a left-lateral transform fault that extends from California to the Gulf of Mexico. Sinistral transpression within a wrench fault has been proposed to explain the present physiographic provinces of northeastern Mexico (Longoria, 1985).

With the exception of the Torreón-Monterrey and Torreón-Saltillo lineaments, which have an almost east-west orientation, all the proposed Mesozoic Jurassic lineaments, fracture zones, and megashears in north-central Mexico trend northwest-southeast. The Mojave-Sonora megashear in particular has a northwest trend from California to the Coahuila Block, where it bends to an east-west orientation, defining the southern boundary of the Coahuila Block (Anderson and Schmidt, 1983). The San Marcos fault presumably delimits the northern margin of the Coahuila Block and also divides the La Popa and Parras Basins (McBride et al., 1974; McKee et al., 1990).

The timing of motion along transform-margin-related inland transpressive structures (de Cserna, 1956; Murray, 1961; Sales, 1968; Silver and Anderson, 1974; Walper, 1977; Anderson and Silver, 1979; Charleston, 1981; Montgomery, 1988; Longoria, 1985) is a key issue in understanding the role of these structures because (1) they presumably offset the trend of the Ouachita-Marathon belt in Mexico; (2) they were supposedly active sometime during the Middle-Late Jurassic (that is, partially contemporaneous with the Jurassic arc) but mostly were active during the last stage of arc magmatism (Bartolini, 1998); and (3) they were presumably related to Late Jurassic rifting events and the opening of the Gulf of Mexico (Anderson and Schmidt, 1983; Pindell, 1985).

Perhaps the most controversial fault is the Mojave-Sonora megashear. According to Silver and Anderson (1974), a fragment of crust from California and Nevada that includes Precambrian basement, Paleozoic miogeoclinal and eugeoclinal rocks, and Triassic strata was displaced 800 to 1000 km to the southeast during Middle to Late Jurassic time. This block, known as the Caborca Block, forms most of modern northwest Sonora state.

However, the allochthonous nature of the Caborca Block has been questioned by Stewart et al. (1990), Poole (1993), and Poole et al. (1995), who point out the undisrupted nature of Paleozoic miogeoclinal and eugeoclinal belts in the western United States and Baja California (Gastil et al., 1991).

The Late Triassic–Middle Jurassic arc is apparently continuous across Mexico. Volcanic-arc-related rocks in Chihuahua, Coahuila, Durango, Zacatecas, and San Luis Potosí (Bartolini, 1988) have not lost continuity with equivalent arc rocks in Sonora (Corona, 1979; Stewart et al., 1990; Damon et al, 1984; Damon et al., 1991) and Baja California (Gastil et al., 1978).

More recently, Molina-Garza and Geissman (1999) carried out detailed paleomagnetic studies in Neoproterozoic to Cretaceous rocks on both sides of the proposed trace of the Mojave-Sonora megashear in northwestern Sonora. Their results are inconsistent with the proposed 800–1000-km displacement of the Caborca Block along the Mojave-Sonora megashear.

Gravity evidence for the existence of any of these major strike-slip faults in north-central/northeastern Mexico is weak or inconclusive (Mickus and Montana, 1999). The most definitive gravity evidence for the existence of the Mojave megashear is presented by Schellhorn (1987), who showed truncation of low-pass-filtered gravity anomalies at the proposed location of the Mojave-Sonora megashear in northwestern Mexico. Hunt (1992) used isostatic residual gravity and residual magnetic-anomaly maps to correlate linear gravity and magnetic trends with the proposed location of the Mojave-Sonora megashear. One proposed trace of the Mojave-Sonora megashear follows the northern boundary of Zacatecas state. According to Figure 3, there is a large-amplitude linear gradient in this region which extends into southwestern Coahuila. However, as shown in our computer model, this linear gradient is apparently associated with the fold belt and partially represents crustal thinning in this region. We cannot rule out that part of the anomaly is caused by the Mojave-Sonora megashear, but the lack of other linear anomalies along the proposed trace suggests that the anomaly is not caused by a structure such as the proposed Mojave-Sonora megashear.

Triassic-Cretaceous Basinal Assemblages

In Zacatecas and San Luis Potosí states, Middle–Upper Triassic (Burckhardt, 1930; Cantú-Chapa, 1969; Gallo et al., 1993) and Lower Cretaceous (Cantú-Chapa, 1974; Yta, 1992; Davila-Alcocer, 1981) deep-marine strata were deposited on slope, toe of slope, and basin-plain sites along the western passive continental margin of Pangea. Turbidity currents deposited predominantly clastic sedimentary piles, especially sandstone, shale, conglomerate, and rare limestone. Locally, pillow basalts also exist (de Cserna, 1976; McGehee ,1976; Ranson et al., 1982; Monod and Calvet, 1992; Tristán-González and Tórres-Hernández, 1992; Silva-Romo et al., 1993; Centeno-García, 1994; Barboza-Gudiño et al., 1998; Bartolini, 1998).

Northeast-eastward tectonic emplacement of the basinal strata onto the platform facies occurred sometime in the latest Triassic or earliest Jurassic (Tristán-González and Tórres-Hernández, 1992, 1994; Silva-Romo et al., 1993; Centeno-García, 1994; Barboza-Gudiño, et al., 1998; Bartolini, 1998) or during Laramide deformation (Monod and Calvert, 1992). The basinal associations are considered fragments of upper levels of oceanic crust that have been accreted onto cratonic Mexico during the Campanian-Eocene Laramide orogeny and incorporated into the Guerrero and Sierra Madre terranes (Campa and Coney, 1983; Sedlock et al., 1993).

Long-wavelength isostatic residual gravity anomalies (Figure 3) in central Mexico reflect the general crustal thickness, but no specific gravity signature can be ascribed to the region where these successions crop out (Mickus and Montana, 1999).

Laramide and/or post-Laramide intrusives (?)

A northwest-trending set of nearly concentric isostatic residual gravity maxima occurs immediately west of the Laramide Sierra Madre Oriental front (Figures 3 and 7). These relatively short wavelength anomalies are confined to the area near latitude 23°–25° N, longitude 99°–101° W. The source of these anomalies may be Laramide or younger intrusions, because granitic intrusions of Laramide age have been reported in the western Sierra Madre Oriental (de Cserna, 1989). Among these are the 78 Ma pluton that constitutes the Pico de Teyra in northern Zacatecas (López-Infanzón, 1986), the 58 ± 4 pluton at Sierra de Catorce in San Luis Potosí (Mugica and Jacobo, 1983; López-Infanzón, 1986), the 42.1 ± 1 Ma intrusive at Concepción del Oro in Zacatecas, and the 75.8 ± 1.6 Ma granodiorite near La Noria in Zacatecas (P. E. Damon, written communi-

cation, 2000). The northwesterly trend of these short-wavelength gravity anomalies is at a slight angle to the approximately north-south trend of the Permian-Triassic arc in eastern Mexico (Figure 7).

The fold-and-thrust belt

The most prominent gravity anomaly in Figure 3 is a northwest-trending gradient that generally follows the trend of the thrust front of the Sierra Madre Oriental fold-and-thrust belt. This gradient is caused by the abrupt thinning of the crust (Winker and Buffler, 1988; Mickus and Montana, 1999) and pre-Cretaceous intrusive and metamorphic rocks. This gradient masks any potential gravity anomalies resulting from smaller-scale features (e.g., the Picachos uplift). Mickus and Montana (1999) used band-pass filtering to isolate anomalies caused by these features. As the gradient trends toward the northwest, it breaks into two segments: (1) a segment that follows the trend of the Sierra Madre Oriental fold-and-thrust belt across the Parras Basin, and (2) a segment that trends to the northwest and apparently coincides with pre-Cretaceous intrusive and metamorphic rocks. The split in the gravity gradient (Figure 2) may also be caused partially by differences in structural styles, because it coincides with changes in the styles of deformation along the fold-and-thrust belt in the Saltillo-Monterrey area. The belt is conventionally divided into a western belt, where thin-skin deformation is predominant (controlled by the existence of Middle–Upper Jurassic evaporites), and an eastern belt, where basement is involved in the deformation (Tardy et al., 1975; Mitre-Salazar, 1981; Campa-Uranga, 1985). However, both structural styles have been recognized in the two regions (Padilla y Sánchez, 1982; Prost et al., 1994; Gray and Johnson, 1995).

The fold-and-thrust belt extends to central Mexico (Figure 7), where its structure and stratigraphy have been studied in detail (Campa-Uranga, 1985; Suter, 1987; Carrillo-Martínez, 1989; Tardy et al., 1975). South of 22° latitude, the trace of the Sierra Madre front cannot be defined with precision because the gravity data from that area are scarce.

CONCLUSIONS

Interpretation of Bouguer and isostatic residual gravity-anomaly maps as well as three regional gravity models, in conjunction with geologic mapping and sparse seismic and well data, suggests that although the tectonics of northeastern Mexico is influenced by Precambrian through recent events, the gravity field is interpreted to be influenced mainly by Late Paleozoic and Mesozoic tectonic events, notably events associated with formation of Late Permian–Early Triassic and Late Triassic–Middle Jurassic volcanic arcs. The regional Bouguer gravity field can be explained by crustal-thickness variations, in which large portions of northern and central Mexico consist of relatively thick crust (40–45 km) compared with northeastern Mexico, where crustal thickness varies between 32 and 36 km. The thicker crust apparently represents the magmatic additions during development of the Late Triassic–Middle Jurassic volcanic arc and/or during Laramide magmatism. Additionally, tectonic emplacement of Middle–Upper Triassic and Cretaceous basinal assemblages along the original western margin of Pangea (now in San Luis Potosí and Zacatecas states) also increased crustal thickness. Toward the west, crustal thickening is also the result of Tertiary magmatism during the evolution of the Sierra Madre Occidental.

A roughly north-south trend of small-amplitude Bouguer and isostatic residual gravity maxima in eastern Mexico may be caused by Permian–Early Triassic granitic plutons and/or density variations in the Precambrian/Paleozoic high-grade metamorphic basement complex. Northwest of these maxima, in the Coahuila Block and the Sabinas Basin, similar-sized gravity maxima may be caused by the similar Permian–Early Triassic arc intrusives. However, these anomalies are smaller in amplitude, perhaps because of the thick, overlying low-density sedimentary sequence and/or compositional changes that caused lower densities and/or smaller-sized sources.

Northwest-trending, concentric, isostatic gravity residual maxima occur immediately west of the fold-and-thrust belt in Nuevo León and Tamaulipas. These relatively short-wavelength gravity anomalies most likely record Laramide or younger intrusives whose orientation is slightly oblique to the Permian arc. Isotopic dating of numerous Laramide granitic intrusives around the area supports our interpretation.

The most prominent Bouguer gravity and isostatic residual gravity anomaly is a northwest-trending gradient that generally follows the fold-and-thrust belt front. This gradient is interpreted to be caused by abrupt thinning of the crust, pre-Cretaceous intrusive and metamorphic rocks, and thickness changes in the Jurassic-Cretaceous strata.

Gravity evidence for the existence of continental-scale, Late Jurassic transform faults is inconclusive. The only large-amplitude linear gradient coincident with the Mojave-Sonora megashear is in southwestern Coahuila. But no other linear anomalies supporting the existence of the structure were found along its proposed trace.

Three regional gravity models constrained by sparse seismic and well data, along with surface geologic maps, indicate that the Late Permian–Early Triassic volcanic arc extends from eastern Coahuila, central Nuevo León, into southwestern Tamaulipas and Veracruz, whereas the Late Triassic–Middle Jurassic arc has a general northwest trend. The exact boundaries of the arcs are difficult to define using gravity alone, especially in Coahuila, but the proposed location provides a reference for future studies. Our modeling process assumed that the Late Permian–Early Triassic arc has a higher density. Possible sources include an oceanic component in the upper crust or the higher densities caused by the formation of transitional crust during the formation of the Gulf of Mexico. In addition, low-density bodies within the trend of the Late Triassic–Middle Jurassic volcanic arc were modeled from north to south as (1)

Permian granitic intrusions, (2) Triassic-Jurassic arc rocks including intrusives (?), (3) Triassic turbidites, and (4) Laramide-age intrusions and Cretaceous deep-marine strata. The lack of deeper crustal information prevented more detailed models; however, all these units are exposed in the study area.

ACKNOWLEDGMENTS

The authors would like to express gratitude to Harold Lang and Randy Keller for their critical review of the manuscript. We also would like to thank Carlos Aiken for the use of his Mexico gravity database.

REFERENCES CITED

Anderson, T. H., and L. Silver, 1979, The role of the Mojave-Sonora megashear in the tectonic evolution of northern Sonora, Mexico, *in* T. H. Anderson and J. Roldan-Quintana, eds., Geology of northern Sonora: Geological Society of America Annual Meeting Guidebook, p. 1–22.

Anderson, T. H., J. W. McKee, and N. W. Jones, 1991, A northwest-trending Jurassic old nappe, northernmost Zacatecas, Mexico: Tectonics, v. 10, p. 383–401.

Anderson, T. H., and V. A. Schmidt, 1983, The evolution of Middle America and the Gulf of Mexico–Caribbean Sea region during Mesozoic time: Geological Society of America Bulletin, v. 94, p. 941–966.

Barboza-Gudiño, J. R., J. R. Tórrez-Hernández, and M. Tristán-González, 1997, Some pre-Oxfordian red beds and related stratigraphic units in the southern and northeastern Central Plateau, Mexico: Geological Society of America Abstracts with Programs, v. 29, p. 2.

Barboza-Gudiño, J. R., J. R. Tórrez-Hernández, and M. Tristán-González, 1998, El significado tectónico de las unidades pre-Oxfordianas del noreste de Mexico: Una visión alternativa: Resumenes, Primera Reunión Nacional de Ciencias de la Tierra, México, D. F., México, p. 159.

Bartolini, C., 1988, Regional structure and stratigraphy of Sierra El Aliso, central Sonora, Mexico: Master's thesis, University of Arizona, 189 p.

Bartolini, C., 1997, The Nazas Formation of northern and central Mexico: Mesozoic volcanic-sedimentary arc sequences: Not red beds: Geological Society of America Abstracts with Programs, v. 29, p. 3.

Bartolini, C., 1998, Stratigraphy, geochemistry, geochronology and tectonic setting of the Mesozoic Nazas Formation, north-central Mexico: Ph.D. dissertation, University of Texas at El Paso, 557 p.

Bartolini, C., H. Lang, and W. Stinnesbeck, 1999, Volcanic rock outcrops in Nuevo León, Tamaulipas and San Luis Potosí, Mexico: Remnants of the Permian–Early Triassic magmatic arc, *in* C. Bartolini, J. Wilson, and T. Lawton, eds., Mesozoic Sedimentary and Tectonic History of North-Central Mexico: Geological Society of America Special Paper 340, 388 p.

Briggs, I. C., 1974, Machine contouring using minimum curvature: Geophysics, v. 39, p. 39–48.

Bridges, L. W., 1964, Regional speculations in northern Mexico, *in* Geology of the Mina Plomosas–Placer de Guadalupe area, Chihuahua, Mexico: West Texas Geological Society Publication 64-50, p. 93–98.

Burckhardt, C., 1930, Synthetique sur le Mesozoique mexicain: Society of Paleontology Suisse Memoir, v. 49-50, p. 1–280.

Campa, M. F., and P. Coney, 1983, Tectono-stratigraphic terranes and mineral resource distributions in Mexico: Canadian Journal of Earth Sciences, v. 20, p. 1040–1051.

Campa-Uranga, M. F., 1985, The Mexican thrust belt, *in* D. G. Howell, ed., Tectonostratigraphic terranes of the Circum-Pacific region: Circum-Pacific Council for Energy and Mineral Resources, Earth Science Series, v. 1, p. 299-313.

Cancienne, G. P., 1987, Topographic corrections to gravity data in northeastern Mexico: Master's thesis, University of New Orleans, 91 p.

Cantú-Chapa, A., 1969, Una nueva localidad del Triásico Superior marino en Mexico: Revista del Instituto Mexicano del Petróleo, v. 1, p. 71–72.

Cantú-Chapa, C. M., 1974, Una nueva localidad del Cretácico Inferior en Mexico: Revista del Instituto Mexicano del Petróleo, v. 6, p. 51–54.

Carrillo-Martínez, M., 1989, Structural analysis of two juxtaposed Jurassic lithostratigraphic assemblages in the Sierra Madre Oriental fold-and-thrust belt of central Mexico: Geofísica International, v. 28, p. 1007–1028.

Centeno-García, E., 1994, Tectonic evolution of the Guerrero Terrane, western Mexico: Ph.D. dissertation, University of Arizona, Tucson, 220 p.

Charleston, S., 1981, A summary of the structural geology and tectonics of the state of Coahuila, Mexico, *in* S. B. Katz and C. I. Smith, eds., Lower Cretaceous stratigraphy and structure of northern Mexico: West Texas Geological Society Field Trip Guidebook, Publication 81-74, p. 28–36.

Coney, P. J., 1979, Mesozoic-Cenozoic Cordilleran plate tectonics: Geological Society of America Memoir 152, p. 33–50.

Corona, F. V., 1979, Preliminary reconnaissance geology of Sierra la Gloria and Cerro Basura, northwestern Sonora, Mexico, *in* T. H. Anderson and J. Roldán-Quintana, eds., Geology of northern Sonora: Geological Society of America Field Trip 27 Guidebook, p. 32–48.

Damon, P. E., 1975, Dating of Mesozoic-Cenozoic metallogenetic provinces within the Republic of Mexico (1965–1975), *in* G. P. Salas, ed., Carta y provincias metalogenéticas de la República Mexicana: Consejo de Recursos Minerales, p. 32–48.

Damon, P. E., M. Shafiqullah, and K. Clark, 1981, Age trends of igneous activity in relation to metallogenesis in the southern Cordillera: Arizona Geological Society Digest, U-14, p. 137–154.

Damon, P. E., M. Shafiqullah, and J. Roldan-Quintana, 1984, The Cordilleran Jurassic arc from Chiapas (southern Mexico) to Arizona: Geological Society of America Abstracts with Programs, v. 17, p. 140.

Damon, P. E., M. Shafiqullah, K. DeJong, and J. Roldan-Quintana, 1991, Chronology of Mesozoic magmatism in Sonora and northwest Mexico: Geological Society of America Abstracts with Programs, v. 23, p. 127.

Dávila-Alcocer, V. M., 1981, Radiolarios del Cretácico Inferior de la Formación Plateros del Distrito Minero de Fresnillo, Zacatecas: Universidad Nacional Autónoma de Mexico, Revista del Instituto de Geología, v. 5, p. 119–120.

de Cserna, Z., 1956, Tectónica de la Sierra Madre Oriental de Mexico entre Torreón y Monterrey: Contribución del Instituto Nacional para la investigación de recursos minerales de Mexico: 20° Congreso Geológico Internacional, 87 p.

de Cserna, Z., 1976, Geology of the Fresnillo area, Zacatecas, Mexico: Geological Society of America Bulletin, v. 87, p. 1191–1199.

de Cserna, Z., 1989, An outline of the geology of Mexico, *in* A. W. Bally and A. R. Palmer, eds., The geology of North America—an overview: Geological Society of America, v. A, p. 233–264.

de Cserna, Z., C. Rincón-Orta, J. Solorio-Munguía, and E. Schmmitter-Villada, 1970, Una edad radiométrica Pérmica Temprana en la región de Placer de Guadalupe, Noreste de Chihuahua: Boletín de la Sociedad Geológica Mexicana, v. 31, p. 65–73.

Denison, R. E., W. H. Burke, E. A. Hetherington, and J. B. Otto, 1971, Basement rocks framework of parts of Texas, southern New Mexico, and northern Mexico, *in* K. O. Seeward and D. A. Sundeen, eds., The geologic framework of the Chihuahua tectonic belt: West Texas Geological Society Symposium, p. 3–14.

Denison, R. E., G. S. Kenny, W. H. Burke, and E. A. Hetherington, 1969, Isotopic ages of igneous and metamorphic boulders from the Haymond Formation (Pennsylvanian), Marathon Basin, Texas, and their significance: Geological Society of America Bulletin, v. 80, p. 245–256.

Dorman, J., J. L. Worzel, R. Leyden, T. N. Crook, and M. Harziemmanuel, 1972, Crustal section from seismic refraction measurements near Victoria, Texas: Geophysics, v. 37, p. 325–336.

Flawn, P. T., A. Goldstein Jr., P. B. King, and C. E. Weaver, 1961, The Ouachita system: Bureau of Economic Geology, University of Texas at Austin, no. 6120, 401 p.

Gallo, P. I., L. Gómez, M. E. Contreras, and P. E. Cedillo, 1993, Hallazgos paleontológicos del Triásico marino en la región central de Mexico: Revista de la Sociedad Mexicana de Paleontología, v. 6, p. 1–9.

Gastil, G., G. J. Morgan, and D. Krummenacher, 1978, Mesozoic history of Peninsular California and related areas east of the Gulf of California, *in* D. G. Howell and K. A. McDougall, eds., Mesozoic paleogeography of the western United States: Pacific Section, Society for Sedimentary Geology (SEPM), Pacific Coast Paleogeography Symposium 2, p. 107–114.

Gastil, R. G., et al., 1991, The relation between the Paleozoic strata on opposite sides of the Gulf of California, *in* E. Pérez-Segura and C. Jacques-Ayala, eds., Studies of Sonoran geology: Geological Society of America Special Paper 254, p. 7–18.

Gomberg, J. S., K. F. Priestly, T. G. Masters, and J. N. Brune, 1988, The structure of the crust and upper mantle of northern Mexico: Geophysical Journal International, v. 94, p. 1–20.

Gonzalez-García, R., 1984, Petroleum exploration in the Gulf of Sabinas—a new gas province in northern Mexico, *in* J. E. Wilson, C. W. Ward, and J. M. Finneran, eds., Upper Jurassic and Lower Cretaceous carbonate platform and basin systems: Gulf Coast Section, Society for Sedimentary Geology (SEPM), Field Guidebook, p. 64–76.

Grajales, J. M., D. J. Terrell, and P. E. Damon, 1992, Evidencias de la prolongación del árco magmático Cordillerano del Triásico Tardío-Jurásico en Chihuahua, Durango y Coahuila: Boletín de la Asociación Mexicana de Geólogos Petroleros, v. 42, p. 1–18.

Gray, G. G., and C. A. Johnson, 1995, Structural and tectonic evolution of the Sierra Madre Oriental, with emphasis on the Saltillo-Monterrey corridor: AAPG Annual Convention Field Trip Guidebook no. 10, 17 p.

Handschy, J. W., G. R. Keller, and K. J. Smith, 1987, The Ouachita system in northern Mexico: Tectonics, v. 6, p. 323-330.

Hunt, R. L., 1992, A gravity and magnetic investigation of the crustal structure of south Texas/northeastern Mexico: Inference on the Mesozoic tectonics: Master's thesis, University of Texas at Dallas, 134 p.

Jacobo, A. J., 1986, El basámento del distrito de Poza Rica y su implicación en la generación de hidrocarburos: Revista del Instituto Mexicano del Petróleo, v. 18, p. 5–24.

James, E. W., and C. D. Henry, 1993, Southeastern extent of the North American craton in Texas and northern Chihuahua as revealed by Pb isotopes: Geological Society of America Bulletin, v. 105, 116–126.

Jones, N. W., J. W. McKee, T. H. Anderson, and L. T. Silver, 1990, Nazas Formation: A remnant of the Jurassic arc of western North America in north-central Mexico: Geological Society of America Abstract with Programs, v. 22, p. 327.

Jones, N. W., J. W. McKee, T. H. Anderson, and L. T. Silver, 1995, Jurassic volcanic rocks in northeastern Mexico: A possible remnant of a Cordilleran magmatic arc, in C. Jacques-Ayala, C. González-León, and J. Roldán-Quintana, eds., Studies on the Mesozoic of Sonora and adjacent areas: Geological Society of America Special Paper 301, p. 179–190.

Jones, N. W., J. W. McKee, D. B. Márquez, J. Tovar, L. E. Long, and T. S. Laudon, 1984, The Mesozoic La Mula Island, Coahuila, Mexico: Geological Society of America Bulletin, v. 94, p. 1226–1241.

Kay, M., 1951, North American Geosynclines: Geological Society of America Memoir 48, 143 p.

Keller, G. R., J. M. Kruger, K. J. Smith, and W. M. Voight, 1989, The Ouachita system: A geophysical overview, in R. Hatcher, W. Thomas, and G. Viele, eds., The Appalachian-Ouachita orogen in the United States: Geological Society of America, The geology of North America, v. F-2, p. 689–694.

Keller, G. R., and D. H. Shurbet, 1975, Crustal structure of the Texas Gulf Coast Plain: Geological Society of America Bulletin, v. 86, p. 807–810.

King, P. B., 1937, Geology of the Marathon region, Texas: U. S. Geological Survey Professional Paper 187, 148 p.

King, P. B., 1944, Geology, Part 1 of geology and paleontology of the Permian area northwest of Las Delicias, southwestern Coahuila, Mexico: Geological Society of America Special Paper 52, p. 3–33.

Lai, S. F., 1984, Generalized linear inversion of two and one-half dimensional gravity and magnetic anomalies: Ph.D. dissertation, University of Texas at Dallas, 229 p.

Longoria, J. F., 1985, Tectonic transpression in the Sierra Madre Oriental, northeastern Mexico: An alternative model: Geology, v. 13, p. 453–456.

López-Infanzón, M., 1986, Estudio petrogenético de las rocas ígneas de las Formaciones Huizachal y Nazas: Boletín de la Sociedad Geológica Mexicana, tomo 47, v. 2, p. 1–32.

López-Ramos, E., 1972, Estudio del basámento ígneo y metamórfico de las zonas norte y Poza Rica (entre Nautla, Veracruz y Jiménez, Tamaulipas): Boletín de la Asociación Mexicana de Geólogos Petroleros, v. 24, p. 265–323.

McBride, E. F., A. E. Weidie, J. A. Wolleben, and R. C. Laudon, 1974, Stratigraphy and structure of the Parras and La Popa Basins, north-eastern Mexico: Geological Society of America Bulletin, v. 84, p. 1603–1622.

McDonnell, S. L., 1987, A gravity and magnetic survey of the Parras Basin, Mexico: Master's thesis, University of New Orleans, 75 p.

McGehee, R. V., 1976, Las rocas metamórficas del Arroyo La Pimienta, Zacatecas, Zacatecas: Boletín de la Sociedad Geológica Mexicana, v. 37, p. 1–10.

McKee, J. W., N. W. Jones, and T. H. Anderson, 1988, Las Delicias Basin: A record of Late Paleozoic arc volcanism in northeastern Mexico: Geology, v. 16, p. 40–47.

McKee, J. W., N. W. Jones, and L. E. Long, 1984, History of recurrent activity along a major fault in northeastern Mexico: Geology, v. 12, p. 103–107.

McKee, J. W., N. W. Jones, and L. E. Long, 1990, Stratigraphy and provenance of strata along the San Marcos fault, central Coahuila, Mexico: Geological Society of America Bulletin, v. 102, p. 593–614.

Mickus, K. L., and G. R. Keller, 1991, Gravity constraints on Triassic rifting in the south-central USA: Geological Society of America Program with Abstracts, v. 23, p. 53

Mickus, K. L., and G. R. Keller, 1992, Lithospheric structure of the south-central United States: Geology, v. 20, p. 335–338.

Mickus, K. L., and C. Montana, 1999, Crustal structure of northeastern Mexico revealed through the analysis of gravity data, in C. Bartolini, J. Wilson, and T. Lawton, eds., Mesozoic sedimentary and tectonic history of north-central Mexico: Geological Society of America Special Paper 340, p. 357–371.

Mitre-Salazar, L. M., 1981, Las imagenes Landsat—una herramienta util en la interpretación geológico-estructural; un ejemplo en el noreste de Mexico: Universidad Nacional Autónoma de México, Revista del Instituto de Geología, v. 5, p. 37–48.

Molina-Garza, R. S., and J. W. Geissman, 1999, Paleomagnetic data from the Caborca terrane, Mexico: Implications for Cordilleran tectonics and the Mojave-Sonora megashear hypothesis: Tectonics, v. 18, p. 293–325.

Monod, O., and P. H. Calvet, 1992, Structural and stratigraphic re-interpretation of the Triassic units near Zacatecas (Zacatecas), central Mexico: Evidence of a Laramide nappe pail: Zentralblatt für Geologie und Paläontologie, Teil I, H. 6, p. 1533–1544.

Montgomery, H. A., 1988, Paleozoic paleogeography of northeastern Mexico: Ph.D. dissertation, University of Texas at Dallas, 182 p.

Montgomery, H., and J. F. Longoria, 1987, Allochthonous carbonates of the Permian volcanic arc at Las Delicias, Coahuila, Mexico: Geological Society of America Program with Abstracts, v. 19, p. 776.

Moreno, F., G. R. Keller, and K. L. Mickus, 1992, The extent of the Ouachita continental margin in northern Mexico, in F. Ortega-Gutiérrez, P. Coney, E. Centeno-García, and A. Gómez-Caballero, eds., Proceedings of the First Circum-Pacific and Circum-Atlantic Terrane Conference: Instituto de Geología, Universidad Nacional Autónoma de Mexico, p. 100–103.

Moreno, F., G. R. Keller, and K. L. Mickus, 1994, The extent of the Ouachita orogenic belt into northern Mexico: West Texas Geological Society Guidebook, n. 94-95, p. 140–148.

Mugica, R., and J. Jacobo, 1983, Estudio petrogenético de las rocas ígneas y metomórficas del Altiplano: Proyecto C-1156, Instituto Mexicano del Petróleo, Subdirección Tecnológica de Exploración (unpublished).

Murillo, M. G., and R. Torres, 1987, Mapa petrogenético y radio-métrico de la Republica Mexicana: Instituto Mexicano del Petróleo, Exploración Proyecto C-2010 (unpublished).

Murray, G. E., 1961, Geology of the Atlantic and the Gulf Coastal Province of North America: Harper and Brothers, New York, 629 p.

Nafe, J. E., and C. L. Drake, 1957, Variation with depth in shallow and deep water marine sediments of porosity, density and the velocities of compressional and shear waves: Geophysics, v. 22, p. 523–552.

Padilla y Sánchez, R. J., 1982, Geologic evolution of the Sierra Madre Oriental between Linares, Concepcion del Oro, Saltillo, and Monterrey, Mexico: Ph.D. dissertation, University of Texas at Austin, 232 p.

Pindell, J. L., 1985, Alleghenian reconstruction and subsequent evolution of the Gulf of Mexico, Bahamas, and proto-Caribbean: Tectonics, v. 4, p. 1–39.

Pinkerton, R .P., 1978, Rayleigh wave model of crustal structure of northeastern Mexico: Master's thesis, Texas Tech University, Lubbock, 58 p.

Poole, F. G., 1993, Ordovician eugeoclinal rocks on Turner Island in the Gulf of California, Sonora, Mexico: Universidad Nacional Autónoma de Mexico, Instituto de Geología, 3° Simposio de la Geología de Sonora y Areas Adyacentes, Hermosillo, Sonora, México, Abstracts, p. 103.

Poole, F. G., and R. G. Madrid, 1988, Comparison of allochthonous Paleozoic eugeoclinal rocks in the Sonoran, Marathon, and Antler orogens: Geological Society of America Abstracts with Programs, v. 20, p. 267.

Poole, F. G., J. H. Stewart, W. B. N. Berry, A. G. Harris, R. J. Repetski, R. J. Madrid, K. B. Ketner, C. Carter, and J. M. Morales-Ramirez, 1995, Ordovician ocean-basin rocks of Sonora, Mexico, in J. D. Cooper, M. L. Droser, and S. C. Finney, eds., Ordovician Odyssey, 7th International Symposium on the Ordovician System: Society for Sedimentary Geology (SEPM), v. 77, p. 267-275.

Prost, G., R. Marrett, M. Aranda, S. Eguiluz, J. Galicia, and J. Banda, 1994, Deformation history of the Sierra Madre Oriental, Mexico, and associated generation-preservation: First Joint AAPG/Asociación Mexicana de Geólogos Petroleros Research Conference Abstracts, p. 3.

Ramírez-Ramírez, C., 1992, Pre-Mesozoic geology of Huizachal-Peregrina anticlinorium, Ciudad Victoria, Tamaulipas, and adjacent parts of eastern Mexico: Ph.D. dissertation, University of Texas at Austin, 318 p.

Ranson, W. A., L. A. Fernández, W. B. Simmons, and S. Enciso de la Vega, 1982, Petrology of the metamorphic rocks of Zacatecas, Mexico: Boletín de la Sociedad Geológica Mexicana, v. 1, p. 37–59.

Sales, J. K., 1968, Crustal mechanisms of Cordilleran foreland deformation—regional and scale-model approach: AAPG Bulletin, v. 52, p. 2016–2044.

Salvador, A., 1991, Triassic-Jurassic, in A. Salvador, ed., The Gulf of Mexico Basin: Geological Society of America, The geology of North America, v. J, p. 131–180.

Schellhorn, R.W., 1987, Bouguer gravity anomalies and crustal structure of northern Mexico: Master's thesis, University of Texas at Dallas, Richardson, 167 p.

Sedlock, R.L., F. Ortega-Gutiérrez, and R. C. Speed, 1993, Tectonostratigraphic terranes and terrane evolution of Mexico: Geological Society of America Special Paper 278, 153 p.

Shurbet, D. H., and S. E. Cebull, 1987, Tectonic interpretation of the westernmost part of the Ouachita-Marathon (Hercynian) orogenic belt, west Texas–Mexico: Geology, v. 15, p. 458–461.

Silva-Romo, G., J. Arellaño-Gil, and C. C. Mendoza-Rosales, 1993, El papel de la secuencia marina Triásica en la evolución Jurásica del norte de Mexico, in F. Ortega-Gutiérrez, P. Coney, E. Centeno-García, and A. Gómez-Caballero, eds., First Circum-Pacific and Circum-Atlantic Terrane Conference: Instituto de Geología, Universidad Nacional Autónoma de México, p. 139–143.

Silver, L. T., and T. H. Anderson, 1974, Possible left-lateral early to middle Mesozoic disruption of the southwestern North American craton margin: Geological Society of America Abstracts with Programs, v. 6, p. 955

Simpson, R. W., R. C. Jachens, R. J. Blakely, and R. W. Saltus, 1986, A new isostatic residual gravity map of the conterminous United States with a discussion on the significance of isostatic residual anomalies: Journal of Geophysical Research, v. 91, p. 8348–8372.

Stewart, J. H., F. G. Poole, K. B. Ketner, R. J. Madrid, J. Roldan-Quintana, and R. Amaya-Martinez, 1990, Tectonics and stratigraphy of the Paleozoic and Triassic southern margin of North America, Sonora, Mexico, in G. Gehrels and J. Spencer, eds., Geological excursions through the Sonoran Desert region, Arizona and Sonora: Arizona Geological Survey Special Paper 7, p. 183–195.

Suter, M., 1987, Structural traverse across the Sierra Madre Oriental fold thrust belt in east-central Mexico: Geological Society of America Bulietin, v. 98, p. 249–264.

Tardy, M., 1980, Contribution a l'étude geologique de la Sierra Madre Orientale du Mexique: These de Doctorat d'Etat, L'Université Pierre et Marie Curie, Paris, 459 p.

Tardy, M., et al., 1975, Observaciones generales sobre la estructura de la Sierra Madre Oriental. La aloctonia del conjunto cadena alta-altiplano central, entre Torreón, Coahuila y San Luis Potosí, San Luis Potosí, Mexico: Universidad Nacional Autónoma de México, Instituto de Geología, v. 75, p. 1–11.

Torres, R., J. Ruiz, G. Murillo-Muñeton, and J. M. Grajales-Nishimura, 1993, The Paleozoic magmatism in Mexico: Evidences for the shift from Circum-Atlantic to Circum-Pacific tectonism, in F. Ortega-Gutiérrez, P. Coney, E. Centeno-García, and A. Gómez-Caballero, eds., First Circum-Pacific and Circum-Atlantic Terrane: Universidad Nacional Autónoma de México, Instituto de Geología, p. 154–155.

Torres, R., J. Ruiz, P. J. Patchett, and J. M. Grajales, 1999, A Permo-Triassic continental arc in eastern Mexico: Tectonic implications for reconstructions of southern North America, in C. Bartolini, J. Wilson, and T. Lawton, eds., Geological Society of America Special Paper 340, p. 191–196.

Tristán-González, M., and J. R. Tórrez-Hernández, 1992, Cartografía Geológica 1:50,000 de la Hoja Charcas, Estado de San Luis Potosí: Universidad Autónoma de San Luis Potosí, Instituto de Geología y Metalurgía, Folleto Técnico 115, 94 p.

Tristán-González, M., and J. R. Tórrez-Hernández, 1994, Geología de la Sierra de Charcas, Estado de San Luis Potosí, México: Revista Méxicana de Ciencias Geológicas, Universidad Nacional Autónoma de México, Instituto de Geología, v. 11, p. 117–138.

Ulrych, T J., 1968, Effect of wavelength filtering on the shape of the residual anomaly: Geophysics, v. 33, p. 1015–1018.

Walper, J. L., 1977, Paleozoic tectonics of the southern margin of North America: Transactions, Gulf Coast Association of Geological Societies, v. 27, p. 230–241.

Wilson, J. L., 1990, Basement structural controls on Mesozoic carbonate facies in northeastern Mexico—review, in M. E. Tucker, J. E.

Wilson, P. D. Crevello, J. R. Sarg, and J. F. Read, eds., Carbonate platforms: Facies, sequences, and evolution: International Association of Sedimentologists Special Publication 9, p. 235–255.

Winker, C. D., and R. T. Buffler, 1988, Paleogeographic evolution of early deep-water Gulf of Mexico and margins, Jurassic to Middle Cretaceous (Comanchean): AAPG Bulletin, v. 72, p. 318–346.

Woods, R. D., A. Salvador, and A. E. Miles, 1991, Pre-Triassic, *in* A. Salvador, ed., The Gulf of Mexico Basin: Geological Society of America, The geology of North America, v. J, p. 109–129.

Yta, M., 1992, Étude geodynamique metallogenetique d'un secteur de la "Faja de Plata," Mexique: La zone de Zacatecas–Francisco I. Madero-Saucito: Docteur these, Universite D'Orleans, France, 266 p.

3

Goldhammer, R. K., and C. A. Johnson, Middle Jurassic–Upper Cretaceous paleogeographic evolution and sequence-stratigraphic framework of the northwest Gulf of Mexico rim, *in* C. Bartolini, R. T. Buffler, and A. Cantú-Chapa, eds., The western Gulf of Mexico Basin: Tectonics, sedimentary basins, and petroleum systems: AAPG Memoir 75, p. 45–81.

Middle Jurassic–Upper Cretaceous Paleogeographic Evolution and Sequence-stratigraphic Framework of the Northwest Gulf of Mexico Rim

R. K. Goldhammer
Department of Geological Sciences, University of Texas at Austin, Austin, Texas, U.S.A.

C. A. Johnson
ExxonMobil Exploration Company, Houston, Texas, U.S.A.

ABSTRACT

The area of northern Mexico and Texas combines elements from two different tectono-stratigraphic provinces—(a) the Gulf of Mexico province (GOM province), located along the northwest rim of the present-day Gulf of Mexico in northeast Mexico and south Texas, and (b) the western Pacific Mexico province (WPM province), located in northwest Mexico and west Texas—and thereby enables one to compare and contrast Gulf of Mexico–driven versus Pacific-driven tectonostratigraphic processes. The area addressed in this paper (that is, the northwest rim of the Gulf of Mexico) contains elements related to both Gulf of Mexico passive-margin development (principally the stratigraphy) and to the Pacific-related convergent margin (arc) tectonism (chiefly the structure). The emphasis in the paper is on the GOM province, with particular reference made to the Sierra Madre Oriental region in the vicinity of the cities of Monterrey and Saltillo, in northeast Mexico.

In the GOM province, the Middle Jurassic through Upper Cretaceous tectonic evolution is dominated by divergent-margin development associated with the opening of the Gulf of Mexico, overprinted by nonigneous Laramide orogenic effects. The topic of salt tectonics is not addressed in this paper. The stratigraphic evolution is interpreted by us to be dominated principally by eustasy in so far as thick regional accommodation cycles can be correlated throughout the Gulf of Mexico. In this paper, we propose that the Middle Jurassic to Lower Cretaceous stratigraphy of northeast Mexico and the subsurface of the northwest GOM rim in general can be subdivided into four major second-order depositional supersequences (approxi-mately 15-myr duration), defined as large, regionally correlative, retrogradational to aggradational/progradational accommodation packages that have Gulf-wide significance. Each exhibits systematic vertical stacking patterns and associated lateral facies shifts in subordinate third-order sequences (1–3-myr duration) and component lateral/vertical facies and systems tracts. The four supersequences are: Supersequence 1—Upper Bathonian to Lower Kimmeridgian ("158.5"–"144" myr), Supersequence 2—Lower Kimmeridgian to Berriasian ("144"–"128.5" myr), Supersequence 3—Valanginian to Lower Aptian ("128.5"–"112" myr), Supersequence 4—Lower Aptian to Upper Albian ("112"–"98" myr). Throughout this paper, we place the ages of the sequences and their boundaries in quotations (e.g., "112" myr) to emphasize that the absolute ages are, of course, uncertain.

Second-order supersequence boundaries, condensed sections, transgressive surfaces, and second-order systems tracts have been identified in outcrops of the Sierra Madre Oriental, biostratigraphically dated, and correlated with the northern United States Gulf of Mexico stratigraphic section. The identification of these components is based on: (1) gross shelf-to-basin relationships of onlapping and offlapping facies; (2) stacking patterns of third-order sequences and their component high-frequency cycles; and (3) criteria for significant subaerial exposure and/or erosion of supersequence boundaries. Additional data are provided from regional seismic coverage and well-log cross sections in south Texas and extreme northeastern Mexico.

The stratigraphic evolution of the Middle Jurassic through Lower Cretaceous, northeast Mexico–Texas passive margin is interpreted to have resulted from the superimposition of four second-order relative sea-level cycles atop a first-order long-term relative sea-level rise. This first-order relative rise likely reflects a global eustatic rise driven by long-term changes in midocean ridge volume related to seafloor spreading rates associated with the opening of the Gulf of Mexico and the Atlantic. These two different orders of eustasy operated in concert with underlying thermotectonic subsidence to produce systematic changes in accommodation from the base to the top of the northeast Mexico section. Such changes, which result in long-term increase in overall accommodation, account for the overall shift from lowstand-dominated facies associations characteristic of the Middle to Upper Jurassic (red beds, evaporites, marginal-marine siliciclastics, and low-relief, shallow-marine high-energy carbonates), to highstand-dominated facies associations characteristic of the Lower Cretaceous (higher-relief, shallow-marine carbonate platforms, deep-marine shales, and pelagic carbonates). In this paper, the evolution of the northwest rim of the GOM is captured by a series of paleogeographic maps.

INTRODUCTION

This paper addresses the paleogeographic and sequence-stratigraphic evolution of the Mesozoic stratigraphy of the northwest Gulf of Mexico rim. Much of the outcrop-derived data comes from the Sierra Madre Oriental in and around the

Monterrey-Saltillo area in northeast Mexico and includes the areas south, southeast, and east of the Coahuila block (Figures 1 and 2). An in-depth treatment of the outcrop stratigraphy and facies development of the northwest GOM rim is provided by Goldhammer (1999), which discusses this area in much greater detail (with additional figures and photographs) than the summary paper presented here. This area of Mexico, with its spectacular exposures of Mesozoic stratigraphy, is significant for the following reasons:

(1) Sequence-stratigraphic models and facies models derived from outcrop studies in northeast Mexico may be used to significantly enhance our understanding of equivalent-age stratigraphy in the onshore U.S. Gulf of Mexico, for example in the East Texas Salt Basin and in south Texas (Figure 3). This is crucial in developing play concept models in exploration and production of hydrocarbons (Goldhammer, 1998a,b). Direct application of outcrop analogs to the U.S. subsurface is afforded by the fact that the pre-Santonian (pre-Laramide) paleogeographic and stratigraphic evolution of northeast Mexico and the U.S. Gulf Coast is fairly similar (e.g. Salvador, 1987, 1991a, b, c).

(2) In the same spirit, sequence-stratigraphic and facies models developed from northeast Mexico may also be used to

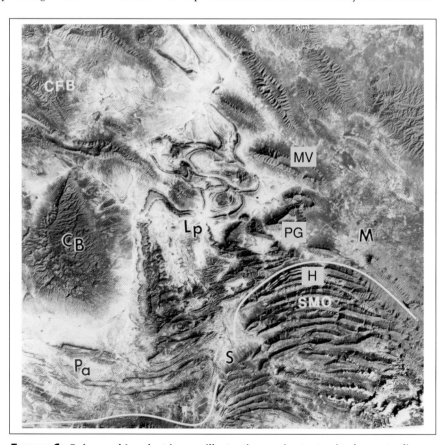

FIGURE 1. Enhanced Landsat image illustrating major tectonic elements discussed in text. M refers to the city of Monterrey, S to Saltillo, CB to the Coahuila block, SMO to the Sierra Madre Oriental fold belt, CFB to the Coahuila fold belt, Pa to the Parras basin, Lp to the La Popa basin, MV to Potrero Minas Viejas, PG to Potrero García, H to Huasteca Canyon. Scale is in km. Distance between M and S is 75 km.

enhance our understanding of other basins in Mexico (Figure 3), including the neighboring Sabinas and Burgos Basins, the Magascatzin, Tampico-Misantla Basins, and the severely deformed equivalents in the Sierra Madre Oriental (Valles–San Luis Potosí platform-basin complex). These particular basins also experienced similar patterns of stratigraphic accumulation in response to Gulf of Mexico-wide long-term accommodation cycles (Goldhammer et al., 1991) prior to the Laramide phase of deformation that affected the entire region.

(3) Of particular importance to the Mexican hydrocarbon industry, northeast Mexico also serves as a useful early-through-middle Mesozoic (pre-Cenomanian) stratigraphic and facies analog for basins to the southeast, which include most of Mexico's significant petroleum provinces. These basins include the Tampico-Veracruz region, the Reforma-Campeche trend, Macuspana Basin, and the Chiapas trough. Taking into account the pre–Upper Jurassic rifting and southerly migration of the Yucatán block, the genetic similarity between the southeast regions and northeast Mexico becomes even more apparent when one considers their original paleogeographic relationship prior to Yucatán migration (Figure 3).

(4) A complete understanding of the area is critical in linking together two somewhat disparate geologic provinces in northern Mexico (discussed below)—(a) the eastern Gulf of Mexico province (GOM province) and (b) the western Pacific Mexico province (WPM province)—and thus allows one to compare and contrast Gulf of Mexico–driven versus Pacific-driven tectonostratigraphic processes. In this context, the Monterrey-Saltillo area contains elements related to both Gulf of Mexico passive-margin development (principally the stratigraphy) and to the Pacific-related convergent margin (arc) tectonism (chiefly the structure).

The goals of this paper are to: (1) outline the tectonostratigraphic development of northeast Mexico by defining the present-day regional tectonic elements and by reviewing the tectonic evolution of the region in the context of the opening of the Gulf of Mexico; (2) propose a regional sequence-stratigraphic framework which is built on the identification and definition of four major second-order supersequences (each approximately 15 myr in duration);

(3) define the two major tectonostratigraphic provinces mentioned above and illustrate their paleogeographic evolution by introducing a series of regional paleogeographic/facies maps. These maps place the stratigraphic and tectonic development of the Monterrey-Saltillo area in a more regional context.

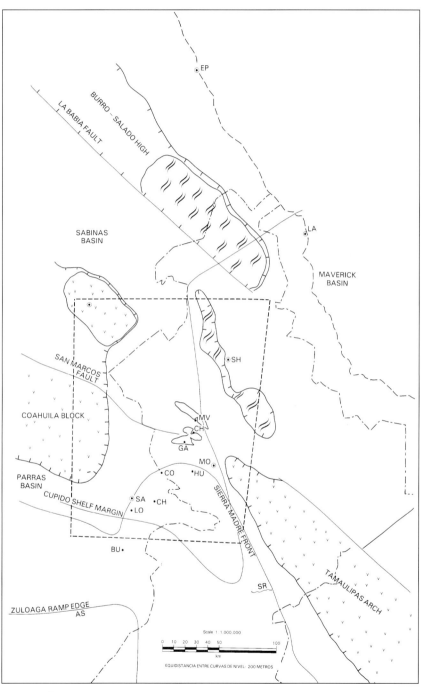

FIGURE 2. Regional tectonic elements of northeast Mexico (Tamaulipas arch, etc.; see Figure 5) and location of Landsat image shown in Figure 1 (dashed outline of rectangular polygon). Abbreviations for geographic localities are, from north to south: EP, El Paso, Texas; LA, Laredo, Texas; SH, Sabinas Hidalgo; MV, Minas Viejas; CH, Potrero Chico; GA, Potrero García; MO, Monterrey; HU, Huasteca Canyon; CO, Cortinas Canyon; SA, Saltillo; CH, Los Chorros Canyon; LO, San Lorenzo Canyon; BU, Bunuelos; SR, Santa Rosa; AS, Astillero Canyon. Scale is in km.

REGIONAL SETTING

Northeast Mexico brings together (a) the eastern Gulf of Mexico province (GOM province) and (b) the western Pacific Mexico province (WPM province). These tectonostratigraphic provinces are large subregions that have a distinctive and separate tectonic evolution, with different resulting stratigraphic packaging. They are characterized by distinctive structural belts, structural styles, and basement grain (de Cserna, 1989; Sedlock et al., 1993; Moran-Zenteno, 1994). The differing stratigraphies record a subregional response to the interaction of provincial tectonics (i.e., convergent versus divergent margins), eustatic changes in sea level, and sediment type and supply.

In northeast Mexico, the Coahuila block more or less separates the WPM to the northwest, west, and southwest from the GOM to the northeast, east, and southeast (Figure 3). In this paper, the GOM province includes the Sabinas, La Popa, Parras, Burgos, and Tampico-Misantla Basins, the "Monterrey trough" (the southward extension of the Sabinas Basin into the deformed salient of the Sierra Madre Oriental), south Texas, and the East Texas Salt Basin (Figure 3). The WPM province includes the Mesozoic-aged Chihuahua trough, the northern "Mexican geosyncline" (e.g., Imlay, 1936; López-Ramos, 1985), the Aldama peninsula, which separates the two depocenters, and portions of the Mesa Central to the south, directly west of the Coahuila block (Figure 3).

In the GOM province, the Mesozoic tectonic evolution is dominated by passive-margin development associated with the opening of the Gulf of Mexico, overprinted by nonigneous Laramide orogenic effects (Salvador, 1987, 1991a, b, c; Pindell, 1985, 1993; Ross and Scotese, 1988; Pindell et al., 1988; Pindell and Barrett, 1990; Bartok, 1993; Marton and Buffler, 1993). The stratigraphic evolution is dominated principally by eustasy (Todd and Mitchum, 1977; Vail et al., 1984; Haq et al., 1987; Scott et al., 1988; Goldhammer et al., 1991; 1998a, b; Scott, 1993;Yurewicz et al., 1993; Lehmann et al., 1998) in as far as thick regional accommodation cycles can be correlated throughout the Gulf of Mexico (Salvador, 1991a, b, c; McFarlan and Menes, 1991; Sohl et al., 1991). For example, Goldhammer et al. (1991) proposed that the Middle Jurassic to Lower Cretaceous stratigraphy of the Gulf of Mexico province can be subdivided into four major, second-order depositional supersequences (approximately 15-myr duration) that have regional Gulf-wide significance.

Second-order supersequence boundaries, condensed sections, transgressive surfaces, and second-order systems tracts have been identified in outcrops of the Sierra Madre Oriental, biostratigraphically dated, and correlated with the northern (U.S.) Gulf of Mexico stratigraphic section (Goldhammer, 1999). The identification of these components is based on: (1) gross shelf-to-basin relationships of onlapping and offlapping facies; (2) stacking patterns of third-order sequences and their component high-frequency cycles; and (3) criteria for signifi-

cant subaerial exposure and/or erosion of supersequence boundaries (Goldhammer, 1999). Additional data are provided from regional seismic coverage and well-log cross sections in south Texas and extreme northeastern Mexico.

The stratigraphic evolution of the Upper Jurassic to Lower Cretaceous Gulf of Mexico passive margin is interpreted by Goldhammer et al. (1991) to have resulted from the superimposition of four second-order relative sea-level cycles atop a first-order, long-term relative sea-level rise. This first-order relative rise likely reflects a global eustatic rise (Vail et al., 1977) driven by long-term changes in midocean ridge volume related to seafloor spreading rates associated with the opening of the Gulf and the Atlantic. These two different orders of eustasy operated in concert with underlying thermotectonic subsidence to produce systematic changes in accommodation from the base to the top of the Gulf Coast section. Such changes account for the overall shift from lowstand-dominated facies associations characteristic of the Middle to Upper Jurassic (red beds, evaporites, marginal-marine siliciclastics, and low-relief, shallow-marine, high-energy carbonates) to highstand-dominated facies associations characteristic of the Lower Cretaceous (higher-relief, shallow-marine carbonate platforms, deep-marine shales, and pelagic carbonates).

The WPM province is distinct from the GOM province in that its style of basin evolution had little to do with Gulf of Mexico tectonic evolution (i.e., rift drift of the Yucatán block) but rather patterns of stratigraphic infill that are primarily a function of tectonism related to Mesozoic Pacific tectonism and sediment supply, as opposed to the eustasy-dominated Gulf of Mexico (Córdoba, 1969; Córdoba et al., 1970, 1980; de Cserna, 1970; Seewald and Sundeen, 1971; Rangin and Córdoba, 1976; González-García, 1976; Tardy et al., 1986; Tardy, 1977; Rangin, 1978, 1979; de Cserna, 1979, 1989; Gastil, 1983; Gastil et al., 1986; Dickinson, 1981; Tóvar Rodríguez, 1981; Roldán-Quintana, 1982; Serváis et al., 1982, 1986; Brown and Handschy, 1983; Campa-Uranga and Coney, 1983; Cuévas-Pérez, 1983; Cuévas-Pérez et al., 1985; Márquez-Castañeda, 1984; Cantú-Chapa et al., 1985; Campa-Uranga, 1985; Araujo-Mendieta and Arenas-Partida, 1986; González, 1989; Dickinson et al., 1986; Brown and Dyer, 1987; Limón, 1989; Pindell and Barrett, 1990; Sedlock et al., 1993; Moran-Zenteno, 1994; Grajales-Nishimura et al., 1992). Mesozoic subduction along the Pacific margin controlled basin-specific styles of development in the WPM province where facies development was strongly affected by adjacent arc-related tectonism and sediment supply. During the Mesozoic, the Pacific margin of Mexico was the site of a long-lived plutonic-volcanic arc complex, namely the Jurassic through late Cretaceous Sinaloa arc succeeded by the latest Cretaceous Alisitos arc (Tardy, 1977; Serváis et al., 1982, 1986; Araujo-Mendieta and Arenas-Partida, 1986; Sedlock et al., 1993). The trench axis trended approximately north-northwest by south-southeast.

Based on ideas distilled from Tardy (1977), de Cserna (1979, 1989), Limón (1989), Córdoba et al. (1980), Dickinson (1981),

Serváis et al. (1982, 1986), and Araujo-Mendieta and Arenas-Partida (1986), we summarize the tectonic development of the WPM province and its relationship to the GOM province (Figure 4). Following the creation of Pangea in the latest Paleozoic, the WPM province underwent two main tectonic "cycles" of back-arc extension and back-arc closure (partial to total) driven by Pacific-related subduction. According to this model, the first phase of back-arc extension occurred in the Late Triassic to

Middle Jurassic, thus forming the Chihuahua trough and northern Mexican geosyncline west of the Coahuila block. East of the Coahuila block, counterclockwise rotation of the Yucatán out of the Gulf of Mexico resulted in the rift phase of GOM tectonic evolution. In the latest Jurassic, partial closure and inversion of the preexisting back-arc basin occurred, forming in some places thrusting (e.g., the Zacatecas-Guanajuato thrust front of de Cserna, 1989). We speculate that this event

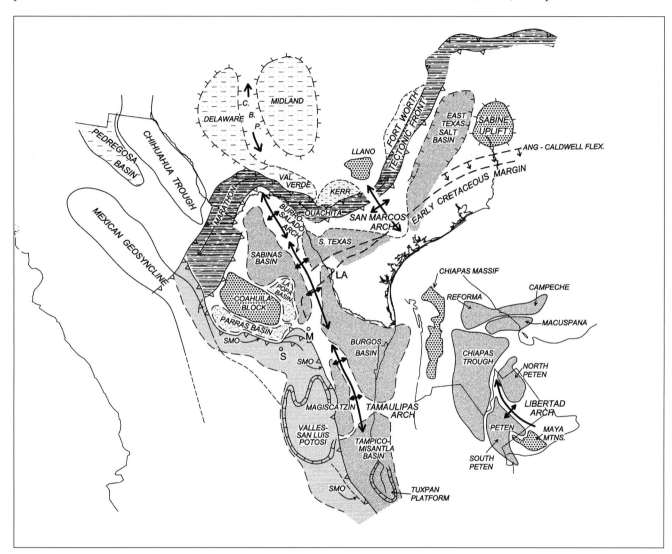

FIGURE 3. Regional base map with basin outlines for the general area discussed in this paper. Late Paleozoic foreland basins and uplifts include: Pedregosa Basin, Delaware Basin, Central Basin platform (C.B.P.), Midland Basin, Val Verde Basin, Kerr Basin, and Fort Worth Basin. These Late Paleozoic elements are bounded to the south by the Late Paleozoic Marathon-Ouachita tectonic front. The eastern Gulf of Mexico province (GOM) contains the following elements: Sabine uplift, East Texas Salt Basin, San Marcos arch, south Texas area, Sabinas Basin, Burgos Basin, Burro-Salado arch, Tamaulipas arch, Magiscatzin Basin, Tampico-Misantla Basin, Coahuila block, the eastern Sierra Madre Oriental (SMO), Valles–San Luis Potosí and Tuxpan platforms. The western Pacific Mexico province (WPM) includes the Chihuahua trough, Mexican geosyncline, and the western Sierra Madre Oriental (SMO). The Parras and La Popa Basins are Late Cretaceous to early Tertiary foreland basins genetically related to both tectonostratigraphic provinces. Yucatán elements shown include the Chiapas Massif, Reforma-Campeche trends, Macuspana Basin, Chiapas trough, north and south Peten Basins, the Maya Mountains, and the Libertad arch. The Yucatán block, which includes south-central Mexico (note outline of the present-day coast), the Yucatán Peninsula, and parts of Guatemala and Belize (note present-day country outlines), is shown in a "quasi-restored" predrift position relative to the GOM elements to emphasize their genetic similarity (refer to text). LA refers to Laredo, Texas; M to Monterrey, Mexico; and S to Saltillo, Mexico. For scale, the distance between Monterrey and Saltillo is 75 km.

induced deep-seated uplift and reactivation of the Coahuila block, an idea supported by Araujo-Mendieta and Arenas-Partida (1986). Other than Coahuila block reactivation and local triggering of coarse clastics proximal to the Coahuila block, the GOM preserves little record of this event, as it is dominated by Gulf of Mexico specific eustatic patterns.

In the Early Cretaceous, the next phase of back-arc extension occurred east of the arc and west of the Coahuila block (Figure 4), rejuvenating the Chihuahua trough and northern Mexican geosyncline, which received thick deposits of volcaniclastics (e.g., Dickinson, 1981). East of the Coahuila block, the GOM province was undergoing passive-margin decelerating subsidence. In the Late Cretaceous, the WPM underwent the Laramide phase of closure and inversion of the northern Mexican geosyncline and the Chihuahua trough, which had been the sites of early to Middle Cretaceous flysch deposition. In the latest Cretaceous, substantial uplift of the Alisitos arc coincided with regional easterly directed uplift and contractional deformation leading to the formation of the Sierra Madre Oriental fold belt (de Cserna, 1989; Sedlock et al., 1993; Moran-Zenteno, 1994).

PRESENT-DAY TECTONIC ELEMENTS OF NORTHEAST MEXICO

The stratigraphic and structural configuration of northeastern Mexico represents a complex tectonic evolution (Dickinson and Coney, 1980; López-Ramos, 1981; Padilla Sánchez, 1986; Salvador, 1987, 1991a, b, c; Pindell, 1985, 1993; Ross and Scotese, 1988; Pindell et al., 1988; Winker and Buffler, 1988; Wilson, 1990; Pindell and Barrett, 1990; Johnson et al., 1991; Bartok, 1993; Marton and Buffler, 1993; Moran-Zenteno, 1994; Gray and Johnson, 1995; Ye, 1997). It initiated in the Permian-Triassic with the Ouachita-Marathon orogenic event, followed closely by Late Triassic to Middle Jurassic rifting of Pangea, subsequent opening of the Gulf of Mexico, and passive-margin development through the Late Cretaceous. It culminated with Laramide foreland deformation through the early Tertiary, with local associated evaporite tectonism. The structural grain of northeastern Mexico consists of Triassic to Liassic basement fault blocks whose development reflects in part late Paleozoic orogenic patterns of metamorphism and igneous intrusion (Wilson, 1990). These early Mesozoic fault blocks in turn controlled Late Jurassic and Cretaceous stratigraphic patterns (Wilson et al., 1984). Additionally, these blocks strongly influenced Laramide structural patterns and foreland basin deposition (Charleston, 1981; Wilson, 1990; Johnson et al., 1991; Soegaard et al., 1997).

Tectonic Elements

Present-day tectonic elements with distinctive stratigraphic and structural characteristics are readily defined in the Monterrey-Saltillo area through analysis of Landsat images and exami-nation of regional geologic maps (Figures 1 and 2) (Humphrey, 1956; Murray, 1959; McBride et al., 1974; Charleston, 1981; Mitra-Salazar, 1981; Padilla Sánchez, 1982; Wilson, 1990; Mc-Kee et al., 1990; Goldhammer et al., 1991; Johnson et al., 1991).

Tectonic Element 1

The Coahuila block (Figures 1, 2, and 5) is characterized by a broad, southeast-plunging anticlinal dome that reflects low-intensity, Laramide age deformation of predominantly Cretaceous carbonates (Imlay, 1936; Charleston, 1981; Johnson et al., 1991; Lehmann et al., 1998). This shallow, rigid basement block is held up primarily by granite to granodiorite intrusions of Permo-Triassic age (Wilson et al., 1984). These intrusions represent the roots of an island arc system that was created south of the Ouachita-Marathon orogenic belt by closure of Gondwana and North America and stranded by subsequent rifting (Pindell and Dewey, 1982; Pindell, 1985; Wilson, 1990). Farther to the west, in the vicinity of Las Delicias, the block contains a thick (4000-m) Middle Pennsylvanian to Permian flysch and volcaniclastic succession (Wilson, 1990) that most likely represents the southern continuation of the Ouachita-Marathon orogenic trend (island arc assemblage). This trend was displaced to the southeast via left-lateral transform faulting associated with Late Triassic to Upper Jurassic extension of northeastern Mexico during the opening of the Gulf of Mexico (Anderson and Schmidt, 1983; Wilson et al., 1984; Pindell, 1985). The block is bounded on the north by the San Marcos fault (Figures 1, 2, and 5; Charleston, 1981), a left-lateral bounding structure of post-Paleozoic age, presumably active during Late Triassic to Upper Jurassic extension and rifting of northeastern Mexico. The Coahuila block was a persistent Mesozoic basement high, which strongly influenced Upper Triassic through Cretaceous facies and stratigraphy. It contains no rift-related Upper Triassic siliciclastics or Callovian evaporites (González-García, 1976; Padilla Sánchez, 1986; Wilson, 1990).

Tectonic Element 2

The Sierra Madre Oriental fold and thrust belt is of Laramide age (Late Cretaceous to Eocene). The evolution of the fold belt and structural complexities has been discussed by numerous workers (e.g., de Cserna, 1956; Humphrey, 1956; Tardy et al., 1975; Padilla Sánchez, 1982; Campa-Uranga, 1985; Quintero-Legoretta and Aranda-García, 1985; Suter, 1987; Aranda-García, 1991; Marret, 1995; Gray and Johnson, 1995; and many others). The belt is characterized by elongate anti-clines that trend east to west and arcuate, curving to the south (farther east). The anticlines have very steep, vertical limbs, and some are overturned to the north (Figures 1 and 2). Folds are arranged in a series of nappes and may be bounded by thrusts. The deformed section consists of essentially the entire Upper Triassic to Cretaceous rift to passive-margin sequence. Major topics of debate and study include: (a) the role of Callovian (?) salt as a basal detachment surface, (b) the contribution of preexisting rifted basement terrain to fold-belt geometry and structural style, (c) mechanisms of fold genesis, (d) the presence of thrusting, (e) amount of lateral shortening, and (f) overburden above the fold belt (e.g., Prost et al., 1994; Gray and Johnson; 1995; Marrett, 1995). The eastward and northward advance of thrust sheets was perhaps facilitated by the occurrence of underlying thick Middle Jurassic (Callovian) salt that was deposited in restricted rift-generated troughs (Johnson et al., 1991). One of these troughs was situated between the southeast margin of the Coahuila block and the northwestern margin of the uplifted Tamaulipas arch (Figure 3; Wilson et al., 1984).

Farther to the south in the fold-and-thrust belt, near Ciudad Victoria, displaced relics of the Ouachita-Marathon orogenic belt crop out in the form of deformed Permo-Carboniferous metasediments and lower Permian flysch (Figure 5; Anderson and Schmidt, 1983; Pindell, 1985; Wilson, 1990). To explain these outcrops and other relics of similar age in northeastern Mexico, Anderson and Schmidt (1983) postulate the existence of a major left-lateral transform fault of Upper Triassic to Middle Jurassic age—the Mojave-Sonora Megashear (MSM). This fault is hidden beneath the deformed belt but is believed to be located north of Ciudad Victoria and south of the Coahuila block (Figure 5). Fault offset on the order of 800 k is postulated to restore these displaced Ouachita-Marathon relics northwest to the orogenic front. This strike-slip fault, as well as others of similar age and displacement to the north and east (e.g., the Torreon-Monterrey fault of de Cserna, 1970; the Rio Grande lineament of Pindell, 1985), are considered to have served as major intracontinental transform faults during the Upper Triassic through Middle Jurassic attenuation of continental Mexico. These faults are essential in Pangea reconstructions of the prerift Gulf of Mexico to avoid overlap of eastern Mexico with the South American plate (Anderson and Schmidt, 1983; Pindell, 1985). The MSM is considered to be a plate boundary between the Early Mesozoic Yaquí Plate to the south (Anderson and Schmidt, 1983; Pindell, 1985) and the North American Plate to the north.

Tectonic Element 3

The Coahuila folded belt (north and northwest of the Monterrey-Saltillo corridor) is of Laramide age (Humphrey, 1956; Murray, 1959; McBride et al., 1974; Charleston, 1981; Mitra-Salazar, 1981; Padilla Sánchez, 1982) and consists of numerous isolated, northwest-southeast-oriented, elongated, and tightly compressed anticlines separated by broad synclinal valleys

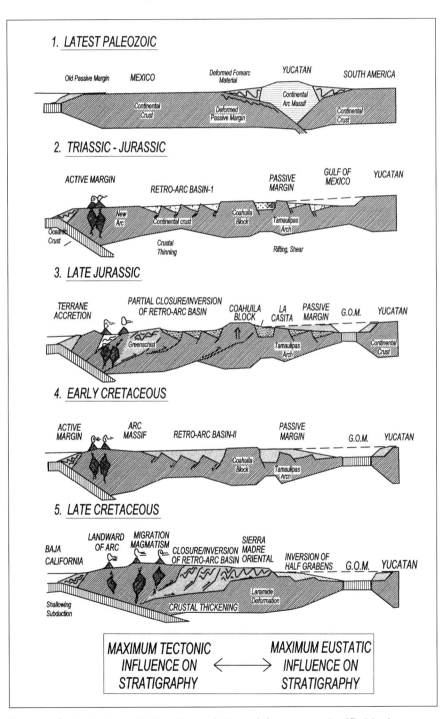

FIGURE 4. Tectonic model for the evolution of the western Pacific Mexico province (WPM) and the eastern Gulf of Mexico province (GOM) in the greater Mexico region. This schematic cross section traverses approximately west-east (from left to right) across northern Mexico. Refer to text for details.

(Figures 1 and 2). Characteristic features of this belt are doubly plunging, evaporite-cored, breached anticlines ("Potreros") that expose Callovian (?) evaporites through middle Cretaceous carbonates (Figures 1 and 2). Because intrusive effects of Middle Jurassic evaporites are observed in the anticline cores, earlier workers considered their origin partly to have resulted from evaporite diapirism (Murray, 1959; Weidie and Martinez, 1970; Laudon, 1984), as opposed to Laramide folding. The Coahuila folded belt contains important Mesozoic basement features that greatly influenced sedimentation patterns (González-García, 1976; Padilla Sánchez, 1986; Wilson, 1990). These include (Figures 1, 2, and 5): (1) a basement depression that marks the Mesozoic Sabinas Basin, located north and northeast of the Coahuila block; (2) the northwest-to-southeast-trending Burro-Salado arch to the north; and (3) the northwest-to-southeast-trending Tamaulipas arch to the east. The Sabinas Basin contains a thick (6000-m) section of Jurassic and Cretaceous strata and developed initially as a rift-related trough between two basement highs—the fault-bounded Coahuila block to the south and the Burro-Salado arch to the northeast (González-García, 1973, 1976, 1979, 1984; Padilla Sánchez, 1978, 1986; Alfonso-Zwanziger, 1978; Aranda-García and Eguiluz de Antuñano, 1983; Eguiluz de Antuñano and Aranda-García, 1983, 1984; Echanove, 1986). Within the Sabinas Basin proper, there are a few Permo-Triassic granite intrusive areas (e.g., the La Mula and Monclova uplifts; Jones et al., 1984; Wilson, 1990), which probably served as local basement highs (Figure 5). Like the Coahuila intrusives, these also reflect the remnants of a Permo-Triassic island arc.

The Burro-Salado arch (Figures 1, 2, and 5) contains basement of deformed Late Paleozoic metasediments presumed to have formed in the interior belt of the Ouachita-Marathon orogenic belt that developed south of the Late Paleozoic suture (Flawn et al., 1961; Wilson et al., 1984; Pindell, 1985). Handschy et al. (1987) suggest that the arch might be a piece of accreted Yucatán stranded during later rifting. The arch is bounded on its southwest border by a left-lateral fault, the La Babia fault (Charleston, 1981), which was most likely an active strike-slip fault during Upper Triassic to Middle Jurassic continental rifting of northestern Mexico. The Burro-Salado arch forms a slightly offset continuation of the northwest-trending Tamaulipas arch located to the southwest (Wilson, 1990). The Tamaulipas arch (San Carlos Island of Alfonso-Zwanziger, 1978) is held up by Permo-Triassic intrusive basement (remnants of the Late Paleozoic island arc) and trends parallel to the Rio Grande southeast from Monterrey to Tampico (Wilson, 1990). It is bounded on the east by a major right-lateral fault, the Tamaulipas-Chiapas fault (Figure 5; Pindell, 1985; Wilson, 1990). Pindell (1985) states that the Tamaulipas trend and this bounding fault extend all along the eastern Mexican margin from Tampico south past the Golden Lane high and offshore at Veracruz. The Arenque field at Tampico is located on this structure. Pindell and Dewey (1982) and Pindell (1985) conclude that the major bounding fault on the east was a right-lat-

eral transform fault between the Yucatán Plate and the southwestern tip of the North American Plate, which allowed migration of Yucatán away from the Texas-Louisiana margin during formation of the Gulf of Mexico (Marton and Buffler, 1993). The linearity of the trend and the fact that the Tamaulipas arch was a basement high from Upper Triassic to Upper Jurassic (Wilson et al., 1984) support a strike-slip interpretation (Pindell, 1985).

Tectonic Element 4

Late Cretaceous–Early Tertiary foreland basins include the Parras and La Popa Basins (Figures 1, 2, and 5). The Parras Basin is confined between the Coahuila block and the Sierra Madre front. The northern limit of this basin is marked by the San Marcos fault. The La Popa Basin occurs north of the Parras Basin and is flanked to the east and west by the Coahuila folded belt. The sediments in these basins consist of nearly 5000 m of Campanian to Maestrichtian shallow marine and deltaic terrigenous siliciclastics of the Difunta group (Weidie and Murray, 1967; Laudon, 1984; McBride et al., 1974; Soegaard et al., 1997; Lawton and Giles, 1997a, b). The structures in the Parras Basin are of Laramide age and are highly variable, depending on proximity to the Sierra Madre front (Weidie and Murray, 1961; Johnson et al., 1991). South of the Coahuila block, deformation is more intense, marked by highly elongate, northward-overturned, tight folds and minor thrusting, with axes parallel to the Sierra Madre front. Farther to the north and east, the intensity of deformation decreases with distance from the frontal detachment, and structures are elongated open folds. Structures in the La Popa Basin consist of broad domal uplifts, salt-involved domes and diapirs, and withdrawal synclines (Johnson et al., 1991; Lawton and Giles, 1997a,b).

MESOZOIC TECTONIC DEVELOPMENT OF NORTHEAST MEXICO

The Gulf of Mexico depicts a Mesozoic divergent margin basin formed through rifting and extension of Pangea, followed by breakup, seafloor spreading, and migration of various cooling and thermally subsiding tectonic plates (Ball and Harrison, 1969; Walper and Rowett, 1972; Dickinson and Coney, 1980; Pilger, 1981; Pindell and Dewey, 1982; Anderson and Schmidt, 1983; Buffler and Sawyer, 1985; Pindell, 1985, 1993; Salvador, 1987, 1991a; Pindell et al., 1988; Ross and Scotese, 1988; Pindell and Barrett, 1990; Sawyer et al., 1991; Marton and Buffler, 1993). The various proposed scenarios differ primarily in initial plate configurations (e.g., initial position of the Yucatán Plate), plate motions (clockwise versus counterclockwise rotation of Yucatán during breakup), the significance of major intracontinental transforms in Mexico, and the amount of attenuation of continental crust during rifting (e.g., compare Buffler and Sawyer, 1985, with Pindell, 1985). It is well beyond the purpose and scope of this paper to review or resynthesize the tectonic model for the Gulf of Mexico, but for the sake of placing northeast Mexico into a larger tectonic framework, a brief synopsis follows. Readers should refer to Pindell (1985) and Pindell and Barrett (1990, particularly Plate 12).

Late Paleozoic Reconstruction of Pangea (Figure 3)

The structural pattern of basement blocks that was to influence later Mesozoic stratigraphy in northeastern Mexico

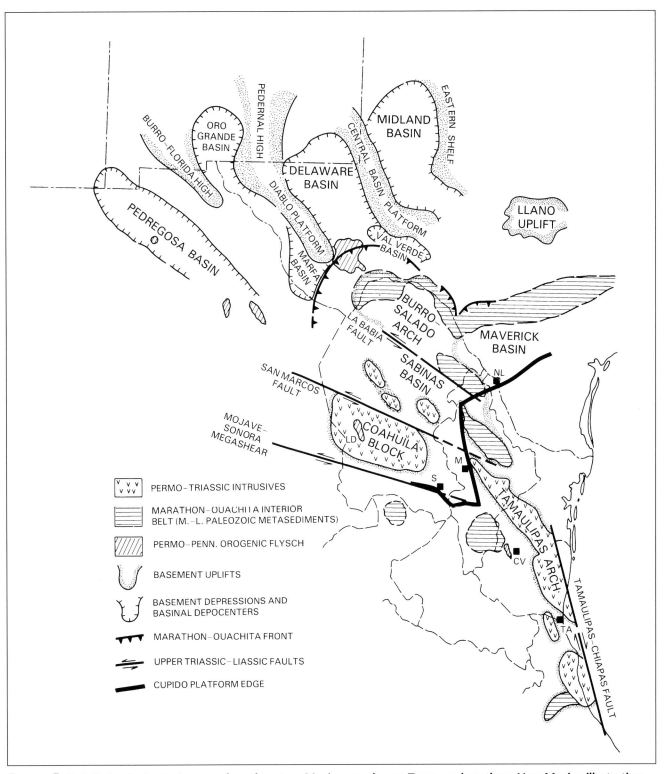

FIGURE 5. Detailed paleotectonic map of northeastern Mexico, southwest Texas, and southern New Mexico illustrating major tectonic features referred to in text. Abbreviations for Mexican cities: NL, Nuevo Laredo; TA, Tampico; CV, Ciudad Victoria; S, Saltillo; M, Monterrey. From Goldhammer et al. (1991). Note that Mexican state outlines are shown. For scale, the distance between Monterrey and Saltillo is 75 km.

largely reflects the effects of Late Mississippian to Early Permian Ouachita-Marathon suturing of the North American, South American, and Yucatán Plates (Salvador and Green, 1980; Pindell and Dewey, 1982; Wilson, 1990). Convergent orogeny was achieved with ocean closure during continent-continent collision, whereby Yucatán filled the gap between North America to the north and west and South America to the south (Pindell, 1985; Pindell and Barrett, 1990, and figures therein). Resultant accretionary complexes (e.g., Marathon-Ouachita terrains) are marked by metamorphosed and unmetamorphosed Permo-Carboniferous continental rise and slope sediments and orogenic flysch thrust up onto North American shelf sequences (Pindell and Dewey, 1982). Other complexes include displaced relicts located in northeastern Mexico (i.e., the Las Delicias and Ciudad Victoria localities mentioned above). The position of these accretionary complexes (Figures 3 and 5) and associated thrust-loaded foreland basins (e.g., the Val Verde, Marfa, and Pedregosa Basins, located north and northwest of the Monterrey-Saltillo area) suggest southeast-dipping subduction of the North American Plate (Pindell, 1985). The location of the suture zone in northeastern Mexico is believed to lie between the Marathon belt and the zone of Permo-Triassic intrusives that underlies the Coahuila block and southern Sabinas Basin (Figure 5). These granites and granodiorites are the remnant roots of an island arc system developed south of the continental suture (Pindell and Dewey, 1982; Pindell, 1985; Wilson, 1990).

Upper Triassic to Callovian Rift Stage

Rifting and initial segmentation of Pangea (Pilger, 1981; Pindell, 1985; Buffler and Sawyer, 1985) are evidenced by attenuated basement in northeastern Mexico expressed as basement highs (Coahuila block, Burro-Salado arch, Tamaulipas arch) and lows (Sabinas and Magascatzin Basins, the "Monterrey trough") (Figure 10). Basement features were often bounded by left-lateral Upper Triassic to Liassic strike-slip faults cutting across the strike of Permo-Triassic deformation and intrusion. Strike-slip faulting, in conjunction with normal faulting, generated rift grabens and half grabens that controlled the distribution of succeeding facies. Rift-related sedimentation and igneous activity accompanied this intracontinental block-faulting (Pindell and Dewey, 1982; Wilson et al., 1984; Salvador, 1987, 1991a, b, c; Wilson, 1990; Goldhammer et al., 1991). The rift sequence (discussed below) consists of a red-bed sequence whose thickness (300 m to 1000 m) is related to preservation in rift grabens (Upper Triassic to Middle Jurassic Huizachal Group), in addition to evaporites (as much as 1000 m thick), complemented by intrusive veins and dikes of rhyolitic to andesitic composition (Wilson et al., 1984; Salvador, 1987, 1991a, b, c; Wilson, 1990). The rifting stage lasted until earliest Oxfordian, where subsidence analysis (Goldhammer et al., 1991, their Figures 12 and 13) constrains the rift-to-drift transition at "158.5" myr.

Counterclockwise rotation and southerly transport of the Yucatán block initiated about this time along the dextral Tamaulipas-Chiapas transform (Pindell, 1985), and major intraplate motions occurred via left-lateral movement along the Mojave-Sonora Megashear (MSM) (Anderson and Schmidt, 1983), which trends along the southern edge of the Coahuila block (Figure 5). Bajocian to Callovian movement along the MSM (and other intracontinental zones of left-lateral offset) with displacements of several hundred kilometers allowed blocks of Cordilleran Mexico to migrate along with South America during initial breakup. This motion maintained an effective land bridge between North and South America until the Callovian, preventing any influx of Pacific seawater into rift basins. Pindell (1985) speculates that this sinistral movement in Mexico was driven by oblique subduction of the Kula/Farralon Plate beneath the Yaquí and South American Plates. This subduction zone, as evidenced by a calc-alkaline volcanic arc located to the northwest and west of northeastern Mexico (discussed below), was persistent from the Late Triassic through the Late Jurassic (Pindell and Dewey, 1982).

Upper Jurassic Drift Stage

Seafloor spreading in the Gulf of Mexico began in the earliest Oxfordian (Buffler and Sawyer, 1985; Pindell, 1985) because the Yucatán block had migrated essentially to the southern tip of the Tamaulipas arch. As motion along this fault ceased, the Tamaulipas arch subsided and was eventually onlapped by Upper Jurassic carbonates. To the west, southeastward transport of Mexican continental blocks along sinistral, intracontinental transforms ceased during the Oxfordian because the Zuloaga Formation (discussed below) masks the MSM and is not offset by the fault zone (Pindell, 1985). Gulf-wide salt deposition terminated as circulation of near-normal marine seawater was established. Pindell (1985) links this early Gulf freshening to the formation of a midoceanic ridge system, which split the Gulf-wide Callovian salt basin into two separate salt provinces (i.e., the Louann and Campeche salt provinces; Buffler and Sawyer, 1985; Pindell, 1985; Salvador, 1987, 1991a, b, c). During the drift stage, spreading accounted for continued separation between the Texas Gulf of Mexico and Yucatán and ceased during the Berriasian (Buffler and Sawyer, 1985; Pindell, 1985; Marton and Buffler, 1993), at which point separation of North and South America was concentrated in the proto-Caribbean, where seafloor spreading continued (Pindell and Barrett, 1990).

Deposition of Oxfordian strata above the breakup unconformity at "150.5" myr was not controlled by primary rifting, but rather by differential subsidence between adjacent basement blocks in a thermally subsiding margin. For example, depositional patterns of Oxfordian and Kimmeridgian near-shore clastic facies (La Gloria Formation, discussed below) and offshore shoal-water carbonate facies (Zuloaga and San Andrés Formations) were a direct function of proximity to basement highs and topographic islands. The Tamaulipas arch in particular subsided into an irregular mosaic of islands surrounded by nearshore siliciclastics and high-energy carbonate grainstones that grade offshore into lower-energy, argillaceous micritic carbonates that mark basement lows (Wilson, 1990). Differential subsidence was also influenced in part by migration of underlying evaporite, analagous to the U.S. Gulf Coast (Pindell, 1985). This is confirmed in northeastern and eastern Mexico by consideration of regional Callovian to Kimmeridgian stratigraphic development and facies patterns, characterized by rapid lateral changes in facies complemented by abrupt variations in thickness (Salvador, 1987, 1991a, b, c; Wilson, 1990). Tithonian to Berriasian sedimentation primarily reflected a reduced influence of preexisting basement highs, with the exception of the Coahuila block in northeastern Mexico. There, extensive clastic deposits (nonmarine to nearshore-marine La Casita Formation, discussed below) derived from the Coahuila block filled in remaining depositional lows, whereas offshore areas distal from the Coahuila block accumulated shales and deeper-water carbonates (La Caja and Taraises Formations). At this time, the Tamaulipas arch and Burro-Salado high had subsided, were no longer yielding exposed islands, and were only indirectly influencing facies distribution.

Cretaceous Cooling Stage

Horizontal plate motions associated with the opening of the Gulf of Mexico were completed by the Berriasian. At that time, the northeastern Mexican passive margin underwent continued decelerating tectonic subsidence with crustal cooling. Throughout most of that period, extensive carbonate platforms with cumulative shelfal thicknesses on the order of 2000 m developed around the entire Gulf of Mexico. This trend was punctuated by minor pulses of clastic sedimentation in the Sabinas Basin in the Aptian (Patula Arkose; La Peña shale) and in the Maverick Basin in the Albian (McKnight facies) related to second-order, relative sea-level effects, as will be discussed below. In the larger scheme, however, terrigenous siliciclastics were restricted to the arc systems bounding the Pacific realms (Pindell, 1985). The Coahuila block remained a basement high but was no longer exposed and supplying siliciclastics. Instead, it controlled the distribution and progradational patterns of carbonate facies (Wilson et al., 1984; Lehmann, 1997).

PALEOGEOGRAPHY AND STRATIGRAPHIC FRAMEWORK

In this section, we review the stratigraphic and paleogeographic evolution of the northeast Mexico region by presenting a series of paleogeographic maps coupled with an integrated chronostratigraphic framework for the Mesozoic of both northeast Mexico and subsurface equivalents in the Texas Gulf Coast. The chronostratigraphic chart (Figure 6) is modified from the version published by Goldhammer et al. (1991). In addition to the 44 sources originally cited by Goldhammer et al. (1991) to construct the chart, the following additional references were incorporated in modifying it to its present state

(Márquez, 1979; Scott et al., 1988; Salvador, 1991a, b, c; Salvador and Muñeton, 1991; McFarlan and Menes, 1991; Sohl et al., 1991; Wilson and Ward, 1993; Basáñez-Loyola et al., 1993; Cantú-Chapa, 1993; Scott and Warzeski, 1993; Yurewicz et al., 1993; Scott, 1993; Marton and Buffler, 1994; Moran-Zenteno, 1994; Lehmann, 1997; Lehmann et al., 1998; Michalzik and Schumann, 1994; Kerans et al., 1995; Lehmann et al., 2000). The chronostratigraphic chart (Figure 6) depicts formational stratigraphy, major facies relations, and the second-order sequence-stratigraphic framework used in this study. As summarized above, the Middle Jurassic to Lower Cretaceous stratigraphy of the GOM province can be subdivided into four major second-order depositional supersequences (approximately 15-myr duration) that have regional Gulf-wide significance. The chronostratigraphy is tied to the Haq et al. (1987) time scale, and throughout the paper, we place the ages of the sequences and their boundaries in quotations (e.g., "112" myr) to emphasize that the absolute ages are, of course, uncertain.

Because of the structural deformation in the Monterrey-Saltillo area, original large-scale stratigraphic geometries are lacking. However, in south Texas and in the East Texas Salt Basin, ample 2-D seismic coverage allows one to visualize essentially undisturbed stratigraphic architecture of age-equivalent strata to reconstruct original stratigraphic geometries and relations in northeast Mexico (e.g., Todd and Mitchum, 1977; Bebout and Loucks, 1977; Goldhammer, 1998a, b). Using these age-equivalent subsurface analogs from the Texas Gulf Coast, one can schematically portray the predeformed, general sequence architecture and facies relationships for northeast Mexico (Figure 7). This schematic is based on regional seismic data from south Texas and east Texas, and biostratigraphically constrained, regional well-log cross sections from south Texas and extreme northeast Mexico (refer to Figures 31 to 37 of Goldhammer et al., 1991; see also regional cross sections in McFarlan and Menes, 1991).

An excellent example of such data is displayed in Figure 8, which shows a relatively undeformed regional seismic dip line from a 2-D survey in south Texas. This northwest-southeast-trending line is less than 15 miles from the Texas-Mexico border and parallels the Rio Grande. Figure 9 is a line drawing of the seismic with the stratigraphic interpretation, constrained by unpublished biostratigraphic data and well-log control. The implications of this seismic section are discussed in full later, but it is introduced here to support the reconstructed stratigraphy proposed for northeast Mexico in Figure 7.

With the framework provided by Figures 6 and 7, we now investigate in some detail the Mesozoic stratigraphy and facies development of northeast Mexico, principally by a series of paleogeographic maps. These maps are based primarily on a distillation of many of the literature sources cited earlier in the text and those cited below. In some instances, unpublished subsurface data were incorporated, particularly for some of the Texas Gulf Coast portions of the maps, in which unpublished gravity and magnetic data, seismic data, and well-log information have been incorporated. A great number of authors have published Mesozoic paleogeographic reconstructions for portions of Mexico and the U.S. Gulf Coast in the past 50 years, from Imlay (1936) to Salvador (1991a) and Moran-Zenteno

(1994), and each author has emphasized different aspects of the reconstruction. The primary goal of our maps is to place the tectonic and sequence-stratigraphic evolution of northeast Mexico in a more regional framework.

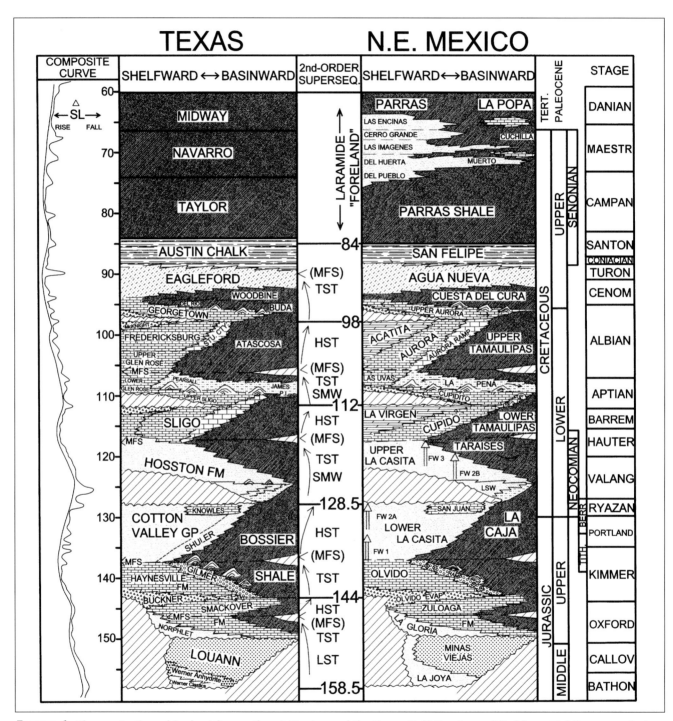

FIGURE 6. Chronostratigraphic chart for northeast Mexico and the Texas Gulf Coast, modified from Goldhammer et al. (1991). Composite eustatic sea-level curve is from Haq et al. (1987). Second-order supersequences are defined in text. Abbreviations are as follows: LST refers to lowstand systems tract, TST to transgressive systems tract, MFS to maximum flooding surface, HST to highstand systems tract, LSW to lowstand wedge, SMW to shelf margin wedge. Second-order sequence boundary ages (e.g., "112") are approximations, and ages are in millions of years before present. Lithologies for formations are discussed in text. Pinnacle reef and isolated carbonate buildups are depicted as small domes with internal concentric lines (e.g., top of Gilmer lime in the TST of the "128.5" myr supersequence). FW1, FW2a, etc., refer to parts of the La Casita Formation referenced by Fortunato and Ward (1982). Refer to text for details.

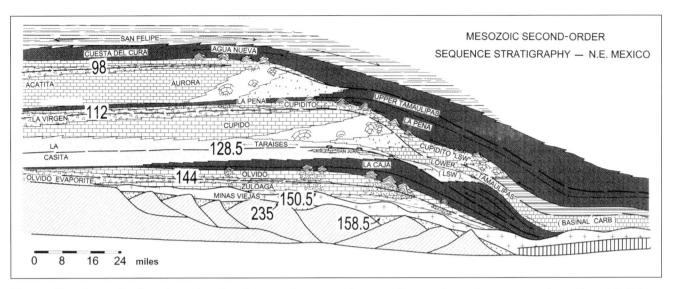

FIGURE 7. Schematic, dip-oriented regional restored cross section trending north-northwest to south-southeast (left to right) in the southern Sabinas Basin to south Texas area. Horizontal scale is approximate and no vertical scale is implied, although vertical thicknesses are relatively correct. Formational stratigraphy and lithologies are described in text. Major second-order supersequence boundaries are shown (with approximate ages in millions of years). Major second-order supersequences are defined as large-scale basin-fill cycles marked by regionally correlative facies patterns of retrogradation and progradation. Refer to text for details.

Late Triassic to Middle Jurassic (pre-Callovian)

In the Late Triassic to Middle Jurassic, the WPM and GOM provinces were characterized by a complex pattern of basement highs and lows (Figure 10). Mexican basement trends are based on published sources, whereas the Texas basement grain is simplified from a more detailed, unpublished gravity and magnetic interpretation provided to the authors. Basement highs occurred in the form of (1) fault-bounded uplifted blocks composed of Permo-Triassic basement (e.g., the Coahuila block); (2) positive anticlinoria cored by Precambrian basement (e.g., Huizachal-Peregrina anticline; Woods et al., 1991); (3) regional north-northwest- to south-southeast-trending anticlinal arches, cored by complex metamorphosed Paleozoic rocks (e.g., the Burro-Salado to Picachos trend); (4) regional north-northwest- to south-southeast-trending anticlinal arches cored by Permo-Triassic intrusives (e.g., the Tamaulipas arch; López-Ramos, 1972); and (5) broad, widespread positive areas such as the Llano uplift. Basement lows occurred as (1) fault-bounded rift grabens with synrift fill (e.g., red-bed facies in the Huayacocotla anticlinorium; Salvador, 1991b); (2) regional north-northwest- to south-southeast-trending lows or "troughs" (e.g., the "Monterrey trough"); or (3) irregularly shaped depressions bounded by smaller basement highs (e.g., the East Texas Salt Basin). As discussed above, the Chihuahua trough and incipient (?) Mexican geosyncline are interpreted to be back-arc lows or extensional basins driven by Pacific arc effects. The basement lows northeast, east, and southeast of the Coahuila block are interpreted to owe their genesis to Gulf of Mexico rifting.

During this period, red beds and associated volcanics accumulated in fault-bounded graben systems around the entire Gulf (Stone, 1975; Todd and Mitchum, 1977; Salvador, 1987, 1991a,b). In northeastern Mexico, these deposits comprise the Huizachal Group, to which Mixon et al. (1959) assigned a Late Triassic age. Mixon et al. (1959) subdivided the Huizachal Group into a lower formation, the La Boca, and an upper formation, the La Joya (Figure 6). The La Boca (late Carnian to mid-Pliensbachian) correlates to the Eagle Mills (Stone, 1975; Todd and Mitchum, 1977; Salvador, 1987, 1991a, b) and consists of nonmarine red beds, arkoses with volcanic flows, and igneous dikes and sills of rhyolite to andesite or diabase/basalt composition (Figures 11a to d; Corpstein, 1974; Padilla Sánchez, 1982). The red beds portray nonmarine alluvial fan, fluvial, and lacustrine depositional settings (Corpstein, 1974; Padilla Sánchez, 1982; Salvador, 1987, 1991a, b; Michalzik, 1988). These deposits unconformably overlie Late Paleozoic metasedimentary or Permo-Triassic granite basement. Thicknesses are on the order of 300 to 2000 m, but preservation is restricted to rift basins (Stone, 1975; Wilson, 1990).

The La Joya (latest Bathonian to Callovian, Figure 6; Bracken, 1984; Wilson, 1990) unconformably overlies the La Boca, with an angular relationship preserved far south of Saltillo in Peregrina Canyon, west of Ciudad Victoria (Corpstein, 1974). The La Joya (55 to 120 m thick) consists of lacustrine and coastal plain, nonmarine to marginal-marine siliciclastics with subordinate freshwater limestones (Corpstein, 1974; Padilla Sánchez, 1982; Michalzik, 1988). The red beds include shales, siltstones, and coarser sandstones and conglomerates (volcanic and feldspathic litharenites; Bracken, 1984) that record continued infilling of areally restricted rift basins. The La Joya laps out onto basement highs and is in part coeval with the Callovian evaporites. As such, the La Joya equates with the thin red-bed section (restricted to updip positions; Stone, 1975) at the base of the Werner Anhydrite (Figure 6).

Callovian to Early Oxfordian

In the Callovian to early Oxfordian, the WPM province was bounded to the west by the San Andrés-Sinaloa magmatic arc complex (Figure 11; Tardy, 1977; Serváis et al., 1982, 1986; Araujo-Mendieta and Arenas-Partida, 1986; Sedlock et al., 1993). West of the Aldama peninsula and Coahuila block, the Mexican geosyncline accumulated shallow-marine clastics and volcaniclastics derived from the arc, whereas the Chihuahua trough was a restricted marine basin, perhaps in part evaporitic (Córdoba, 1969; Córdoba et al., 1970, 1980; de Cserna, 1970, 1979, 1989; Rangin and Córdoba, 1976; González-García, 1976; González, 1989; Tardy, 1977; Rangin, 1978; 1979; Gastil, 1983; Gastil et al., 1986; Dickinson, 1981; Tóvar Rodríguez, 1981; Roldán-Quintana, 1982; Serváis et al., 1982, 1986; Brown and Handschy, 1983; Cuévas-Pérez, 1983; Cuévas-Pérez et al., 1985; Márquez-Castañeda, 1984; Cantú-Chapa et al., 1985; Campa-Uranga, 1985; Araujo-Mendieta and Arenas-Partida, 1986; Moran-Zenteno, 1994).

In the GOM province, widespread evaporite deposition occurred from the East Texas Salt Basin through south Texas and, as transgression continued, into the more restricted portions of the Sabinas Basin and "Monterrey trough" (Figures 6 and 11; González-García, 1976; Madrid, 1976; Zwanziger, 1979; Padilla Sánchez, 1986; Salvador, 1987, 1991a, b; Moran-Zenteno, 1994). In the Monterrey-Saltillo area, the Minas Viejas evaporite outcrops as deformed masses of gypsum (Weidie and Martinez, 1970; Laudon, 1984) unconformably overlying the Huizachal red beds and/or Paleozoic basement (Figure 6). The Minas Viejas is a marginal-marine deposit that marks the initial marine incursion into restricted, landlocked rift basins. The exact predeformed thickness (estimates of about 1000 m),

lateral distribution, and depositional setting are unknown. The gypsum lithofacies probably repesents the more landward fringe of these evaporite basins, with thick halite accumulating in basin centers.

The age and northern Gulf Coast equivalent of the Minas Viejas are somewhat problematic. Many workers (Humphrey, 1956; Weidie and Wolleben, 1969; Oivanki, 1974; Stone, 1975; Meyer and Ward, 1984; Wilson et al., 1984; Finneran, 1986; Winker and Buffler, 1988) consider the Minas Viejas to be equivalent to the Werner-Louann of the northern Gulf (Figure 6). Although a Callovian age for the Werner-Louann cannot be demonstrated unequivocally, the majority of evidence indicates a Callovian–earliest Oxfordian age (Imlay, 1980; Salvador and Green, 1980; Scott, 1984; Salvador, 1987, 1991a, b). Thus the Minas Viejas would also be Callovian. Other authors (Longoria, 1984), however, either do not recognize an outcropping Louann equivalent in northeastern Mexico or use the term *Minas Viejas* to refer to Oxfordian to Kimmeridgian evaporites that are presumed to be updip, restricted equivalents to the Zuloaga Formation. In this paper, the Minas Viejas is assigned a Callovian age (Louann equivalent) and the younger sequence of evaporites (Oxfordian to Kimmeridgian) is designated the Olvido evaporite (Buckner equivalent; Figure 6).

Middle Oxfordian to Kimmeridgian

During this time frame (Figure 12), in the WPM province, the tectonically active Sinaloa Terrain was uplifted such that contractional deformation occurred to the east, thus closing the "ancestral" Mexican geosyncline west of the Aldama peninsula and south and west of the Coahuila block (Figure 4; Tardy, 1977; de Cserna, 1979, 1989; Córdoba et al., 1980; Dickinson, 1981; Serváis et al., 1982, 1986; Araujo-Mendieta and Arenas-Partida, 1986). Earlier deposits of the Mexican geosyncline were deformed along a north-south zone termed the

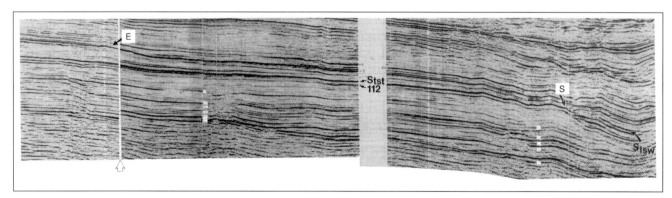

Figure 8. Regional 2-D seismic line from south Texas showing relatively undeformed dip-oriented stratigraphy. This northwest-to-southeast-trending line is less than 15 miles (about 24 km) from the Texas-Mexico border and parallels the Río Grande. In detail, the line consists of two closely spaced parallel lines spliced together. The splice is located just to the right of the "E" on the line, accounting for the lack of continuity in some reflectors in the lower half of the line across the splice. Vertical scale is in two-way traveltime in 0.10-second intervals. Total vertical scale is 2.7 seconds. Because of the proprietary nature of this industry line, two-way time and horizontal shot-point information are not shown. "E" points to the location of the Albian Edwards terminal shelf margin, "S" to Barremian to Lower Aptian Sligo margin, "Stst" to top Sligo second-order TST of the "98"-myr supersequence, "112" to 112-myr supersequence boundary, "Slsw" to Sligo second-order lowstand wedge. See line drawing in Figure 9 and refer to text.

Zacatecas-Guanajuato thrust front by de Cserna (1979). Within the Chihuahua trough, restricted shallow-marine carbonates (low-energy, lagoonal) and fine-grained clastics in the trough center were flanked by coarser-grained shallow-marine clastics (Córdoba, 1969; Córdoba et al., 1970, 1980; de Cserna, 1979; González, 1976; Tardy, 1977; Araujo-Mendieta and Arenas-Partida, 1986; Moran-Zenteno, 1994).

In the GOM province (Figure 12), shallow-marine carbonate ramps with ramp-crest high-energy grainstone facies were flanked landward by nearshore marginal-marine clastics and graded downdip into off-ramp deeper-marine shales (González-García, 1976; Zwanziger, 1979; Padilla Sánchez, 1986; Salvador, 1987, 1991a, b). In east Texas and south Texas, the classic Buckner to Smackover ramp system rimmed a regionally exposed landmass to the north (e.g., Budd and Loucks, 1981; Presley, 1984; Goldhammer, 1998a, b). In northeast Mexico and adjacent to the south, analogous carbonate ramp systems developed, nucleating on preexisting, exposed landmasses.

In northeast Mexico, the La Gloria Formation (Figure 6) is early Oxfordian in age (Imlay, 1936) and essentially represents the updip, transgressive clastic interval of the Zuloaga Formation (Figure 6; Imlay, 1936; Stone, 1975; Oivanki, 1974). It is the first prominent marine sandstones, overstepping the breakup unconformity at "150.5" myr. It equates in part to the Norphlet Formation of the northern Gulf (Wilson et al., 1984; Salvador, 1987, 1991a, b) and forms a basinward-thinning wedge of fine to coarse feldspathic quartz sandstones that laps onto exposed basement highs (Coahuila block, Tamaulipas arch; Oivanki, 1974). Thicknesses are typically < 50 to 100 m, but they exceed 600 to 700 m proximal to basement blocks (e.g., southwest flank of the Coahuila block; Oivanki, 1974). The basal La Gloria unconformably overlies the Minas Viejas, Huizachal red beds, or Paleozoic basement. It is overstepped conformably by Zuloaga carbonates (Figure 6). Depositional environments range from marginal marine (playa) to nearshore, shallow marine, where detrital lithologies are intercalated with normal-marine carbonates (Oivanki, 1974). Downdip, the La Gloria grades into Zuloaga ramp carbonates.

In northeast Mexico, the Zuloaga Formation (Figure 6) marks the establishment of open-marine conditions (González-García, 1976; Zwanziger, 1979; Padilla Sánchez, 1986), with the transition from the rift-to-drift stage of passive-margin development. The Zuloaga is Oxfordian (Imlay, 1943) and is correlated with the Smackover Formation of the northern Gulf Coast (Figure 6). Grain types and depositional facies are very similar to the Smackover in south Texas (Stone, 1975; Budd and Loucks, 1981; Johnson, 1991). It unconformably overlies Huizachal red beds or Minas Viejas evaporites. This transgressive carbonate formed an extensive low-angle, gently dipping ramp that nucleated proximal to exposed basement highs (Oivanki, 1974; Johnson, 1991). Regional facies patterns and isopachs (Oivanki, 1974; Sandstrom, 1982; Meyer and Ward, 1984; Finneran, 1986; Johnson, 1991) suggest that the Coahuila block was a prominent topographic high, and that the

Tamaulipas arch was a mosaic of islands forming an archipelago trending north-northwest to south-southeast (Figure 14; Oivanki, 1974; Todd, 1972; Stone, 1975). Proximal siliciclastics (La Gloria) rimmed the exposed islands, passing offshore into sabkha, tidal-flat, and restricted lagoon environments of the inner ramp, in turn rimmed by extensive high-energy carbonate sand shoals at the ramp edge (Oivanki, 1974; Finneran, 1986; Johnson, 1991). Deeper, low-energy subtidal environments characterize the outer ramp (Meyer and Ward, 1984).

Antecedent topography resulted in marked lateral variations in thickness and depositional facies (Johnson, 1991). Depositional textures vary across the ramp profile from muddy, peloidal mudstones to wackestones updip (peritidal to restricted lagoon) to ooid-pellet packstones and grainstones at the ramp edge (shoals) to mudstones and wackestones downdip (outer ramp). In addition, evaporites (calcium sulfates) were an important component, as evidenced by numerous crystallotopic molds, solution-collapse breccias, and intercalated layers of calcite-replaced anhydrite and/or gypsum (Oivanki, 1974; Michalzic, 1988; Johnson, 1991). Thicknesses vary from 150 to 500 m updip to greater than 450 m downdip (Oivanki, 1974).

The Olvido Formation consists of a lower, early Kimmeridgian portion of evaporites (anhydrite, gypsum) and red shales and an upper, Kimmeridgian unit of predominantly carbonate, but with varying admixtures of siliciclastics, depending on proximity to exposed paleohighs and clastic source areas (Figures 6 and 7; Carrillo-Bravo, 1963; Todd, 1972; Padilla Sánchez, 1982; Salvador, 1987, 1991a, b; Wilson, 1990). Following Goldhammer et al. (1991), the lower Olvido is termed the Olvido evaporite, and the upper Olvido is termed the Olvido lime (Figure 6). Stone (1975) indicates that the Olvido evaporite correlates with the Buckner anhydrite of the northern Gulf, for which an early Kimmeridgian age has been determined (Stone, 1975; Todd and Mitchum, 1977; Salvador, 1987, 1991a, b). The Olvido evaporite is apparently conformable atop the Zuloaga and is overstepped conformably by transgressive Kimmeridgian Olvido lime.

In the Monterrey-Saltillo area, the Olvido evaporite is typically on the order of 20 to 50 m thick, and the Olvido lime varies from 100 to 200 m thick; but in the northern Sabinas Basin, the evaporitic interval reaches a thickness of 100 to 300 m, with the overlying Kimmeridgian carbonate approximating 100 to 200 m in thickness (Stone, 1975; González-García, 1976; Padilla Sánchez, 1986). This variation in thickness reflects differences in subsidence rates between the two areas. The Sabinas Basin was an original basement low that subsided more in relation to the shelfal regime flanking the southern side of the Coahuila block. The lower evaporitic Olvido presumably records deposition in a very restricted, marginal-marine setting, implying a brief but significant phase of regression in the overall Oxfordian to Kimmeridgian transgressive trend (González-García, 1976; Padilla Sánchez, 1986).

In the vicinity of Monterrey and Saltillo, the Olvido lime is

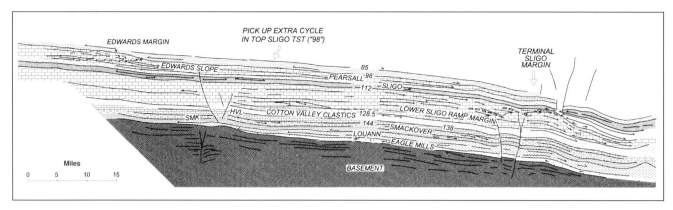

FIGURE 9. Line drawing of the interpretation of the seismic line shown in Figure 8. Stratigraphic interpretation is constrained by unpublished biostratigraphic data and industry well-log control. "SMK" refers to Smackover Formation, "HVL" to Haynesville; "144" and other numbers are in millions of years.

similar to the underlying Zuloaga in terms of grain types, depositional textures, carbonate facies, and cycle development, and is interpreted to depict a carbonate ramp regime. The Olvido lime correlates to the Kimmeridgian Haynesville Formation of south Texas, to which it is lithologically very similar (Stone, 1975), as well as other Kimmeridgian carbonates in the Gulf (e.g., the Gilmer Limestone of east Texas; Ahr, 1981; Steffensen, 1982; Goldhammer, 1998a). For example, in the northern and western reaches of the Sabinas Basin, clean carbonate packstones to grainstones full of ooids, hardened pellets, and oncolites predominate (Stone, 1975; Figures 31 to 35 in Goldhammer et al., 1991). To the south, in the Tampico-Misantla area, the Kimmeridgian section consists of a carbonate ramp system that nucleated on exposed basement highs along the Tamaulipas arch (Cantú-Chapa, 1992; Todd, 1972; González, 1977). Here, tidal flat facies with cryptalgal laminites grade offshore into thick (100-m) oolitic grainstones that mark the ramp edge (Todd, 1972). The underlying Oxfordian section consists largely of nearshore siliciclastics and carbonates, unlike the typical Zuloaga to the north. These examples serve to illustrate the role that exposed basement highs played in controlling Oxfordian to Kimmeridgian thicknesses and facies evolution.

Tithonian to Portlandian

In the Tithonian to Portlandian, the Sinaloa arc was active in the WPM province and continued to influence eastward-propagating contractional deformation with thrusting at the leading edge (Figure 13; Tardy, 1977; de Cserna, 1979, 1989; Córdoba et al., 1980; Dickinson, 1981; Serváis et al., 1982, 1986; Araujo-Mendieta and Arenas-Partida, 1986). Late Jurassic inversion and total closure of the Mexican geosyncline were achieved west and southwest of the Coahuila block, whereas to the north the Mexican geosyncline, as well as the Chihuahua trough, was the site of shallow-marine clastic deposition, largely volcaniclastic (Córdoba, 1969; Córdoba et al., 1970, 1980; de Cserna, 1979; González, 1976; Tardy, 1977; Araujo-Mendieta and Arenas-Partida, 1986; Moran-Zenteno, 1994).

In the GOM, regional facies relationships are driven primar-

ily by a major second-order Gulf-wide transgression such that antecedent carbonate ramp systems are drowned and inundated with fine-grained marine clastics, such as the Bossier Shale in Texas (Figures 6 and 7; Salvador, 1987, 1991a, b). In the East Texas Salt Basin, gas-bearing pinnacle reefs and back-stepped ooid shoal complexes developed atop the retreating ramp systems (Goldhammer, 1998a, b). Much of northeast Mexico was likewise inundated with fine-grained marine shales and siltstones of the La Caja and Pimienta Shale as deeper-marine facies lapped onto preexisting basement highs (Figure 13; González-García, 1976; Padilla Sánchez, 1986; Echanove, 1986). Despite this significant marine transgression, not all land areas were covered. In particular, the Coahuila block may have been uplifted tectonically at this point (Figure 4; Araujo-Mendieta and Arenas-Partida, 1986; Limón, 1989), as indicated by a significant influx of coarse, proximal clastics into areas proximal to the block, particularly in the Sabinas Basin (González-García, 1976; Zwanziger, 1979; Padilla Sánchez, 1986; Aranda-García and Eguiluz de Antuñano, 1983; Eguiluz de Antuñano and Aranda-García, 1983, 1984; Echanove, 1986).

In northeast Mexico, the La Caja Formation is Kimmeridgian to mid-Berriasian (Figure 6) and consists of rhythmically bedded, thin calcareous shales, siltstones, and fine sandstones, with thin limestones toward the base (Fortunato, 1982; Salvador, 1987, 1991a). Phosphatic beds and large concretions of phosphatic micrite with ammonites are conspicuous components. Thickness is variable, ranging from 25 to 150 m (Padilla Sánchez, 1982). The La Caja is equivalent to the Bossier Shale of the northern Gulf and records offshore somewhat starved basinal depositional conditions. In south Texas and elsewhere around the northern Gulf, the Tithonian Bossier onlaps the underlying Kimmeridgian Haynesville, as shown by seismic data and well-log pattern correlation (Stone, 1975; Figures 31 to 35 in Goldhammer et al., 1991). The La Caja overlaps the Burro-Salado arch, and in the Tampico-Misantla area, it overlaps the Tamaulipas arch (Cantú-Chapa, 1992; Todd, 1972; Stone, 1975; González, 1977). The Kimmeridgian portion of the La Caja is the deeper, off-ramp basinal equivalent to the Olvido lime, and hence the formation boundary with the Olvido lime updip is a time-transgressive facies change. The Tithonian to mid-Berriasian part is the deeper offshore equivalent to the updip La Casita. Although the lower portion of La Casita is Tithonian to Portlandian in age, it is discussed fully below, in the context of the Neocomian paleogeography.

Neocomian

During earliest Cretaceous (Figure 14), in the WPM prov-

ince, arc-related tectonism induced back-arc extension in the Mexican geosyncline and rejuvenated subsidence in the Chihuahua trough (Figure 4; Córdoba, 1969; Córdoba et al., 1970, 1980; DeFord and Haenggi, 1971; Seewald and Sundeen, 1971; González-García, 1976; González, 1989; Cantú-Chapa, 1976; Tardy, 1977; de Cserna, 1970, 1979, 1989; Dickinson; 1981; Tóvar Rodríguez, 1981; Roldán-Quintana, 1982; Serváis et al., 1982, 1986; Cantú-Chapa et al., 1985; Araujo-Mendieta and Arenas-Partida, 1986; Brown and Dyer, 1987; Sedlock et al., 1993; Moran-Zenteno, 1994). Proximal marginal-marine to

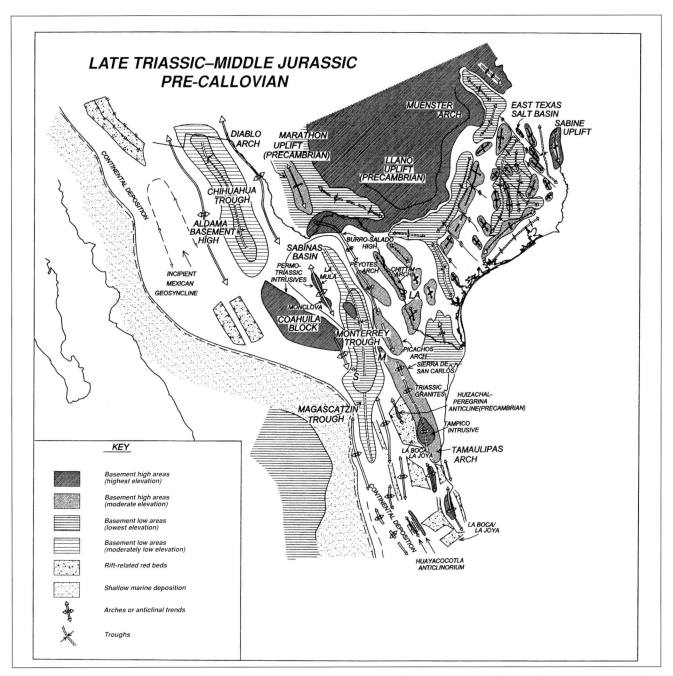

FIGURE 10. Late Triassic to Middle Jurassic (pre-Callovian) paleogeography. Texas and Mexico outlines are shown. "M" refers to Monterrey, Mexico; "S" to Saltillo, Mexico; and "LA" to Laredo, Texas. Note that the key for this reconstruction is different from the key used in remaining paleogeographic maps. Tampico intrusive is a post–Lower Cretaceous feature. Refer to text. For scale, the distance between Monterrey and Saltillo is 75 km.

KEY TO MAPS

a

▨	Land (nondeposition)
▨	Arc terrane
▨	Nonmarine to marginal-marine coarse clastics
▨	Restricted marine evaporite (sabkha) and minor supratidal carbonate
▨	Open-marine, shallow, nearshore medium- to fine-grained clastics
▨	Peritidal carbonate
▨	Muddy, lagoonal carbonate
▨	High-energy carbonate grainstones (shoal)
▨	Open-marine limestone
▨	Outer-ramp lime mudstone and shale
▨	Deep-marine, off-ramp lime mudstone and shale
▨	Moderately deep marine calcareous shale, siltstone (phosphatic lime mudstone)
▨	Coarse nonmarine to marginal-marine volcaniclastics

b

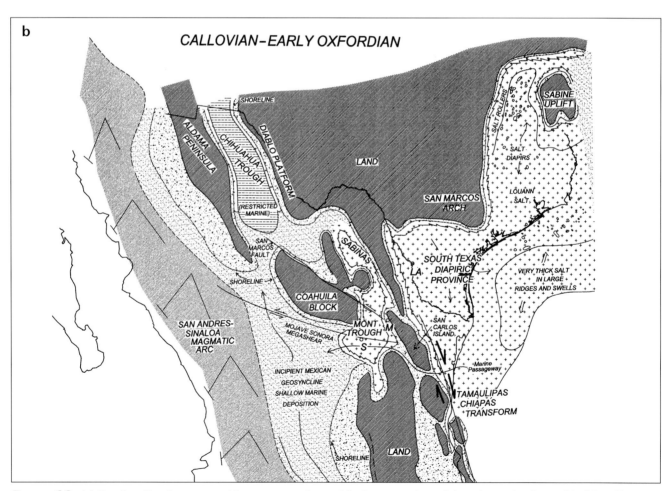

CALLOVIAN–EARLY OXFORDIAN

FIGURE 11. (a) Key for all paleogeographic reconstructions with the exception of that shown in Figure 10. (b) Callovian to early Oxfordian paleogeography. Refer to text.

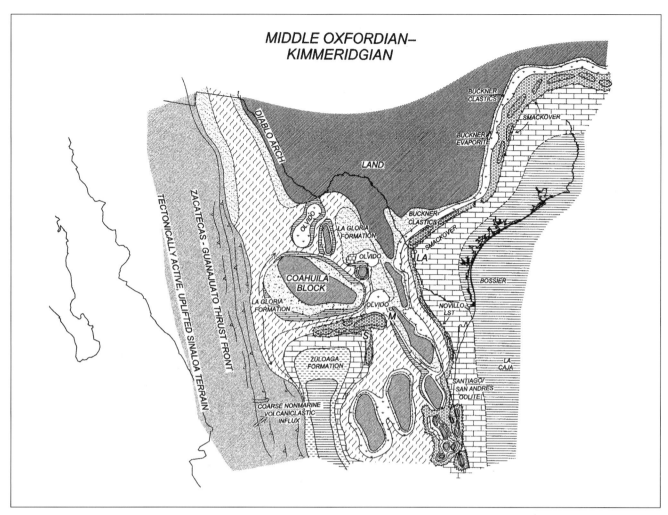

FIGURE 12. Middle Oxfordian to Kimmeridgian paleogeography. Refer to text. Refer to Figure 11a for key. Southern-most extension of the Zuloaga is partly speculative and may actually be largely coarse siliciclastics of the Las Trancas Formation (turbiditic sands).

nonmarine volcaniclastics (Mezcalero Formation) draped the upper reaches of the Mexican geosyncline flanking the exposed Aldama landmass (Figure 14), whereas outer-ramp, deeper-water shales and lime mudstones accumulated in the central part of the Mexican geosyncline. In the Chihuahua trough, coarse proximal clastics prograded southwest, flanking the Diablo arch in Texas, where they interfinger with slightly more open marine clastics in the northwestern reaches of the Sabinas Basin (González-García, 1976; Zwanziger, 1979; Márquez, 1979; Padilla Sánchez, 1986).

In the GOM province, coarse, marginal-marine to shallow-marine clastic facies (Hosston Formation) with local carbonate accumulation (Knowles Formation; McFarlan and Menes, 1991) rimmed a large exposed landmass in Texas (Figure 14), changing facies downdip into offshore shales and siltstones. This pattern of terrigenous influx was common all along the periphery of the Gulf of Mexico at that time (Stone, 1975; Todd and Mitchum, 1977; Salvador, 1987 1991a, b; see Figures 36 and 37 in Goldhammer et al., 1991). In the Sabinas Basin, coarse mar-

ginal-marine to shallow-marine clastics (Barril Viejo Formation and La Casita) accumulated north of the Coahuila block, passing downdip into finer-grained, more open marine facies of the Menchaca and Taraises Formations in the Burgos Basin (González-García,1976; Padilla Sánchez, 1986; Aranda-García and Eguiluz de Antuñano, 1983; Eguiluz de Antuñano and Aranda-García, 1983, 1984; Echanove, 1986). South, in the Tampico-Misantla area, low-relief carbonate platforms were established atop old basement-high areas, forming a complex of aerially restricted platforms that together comprise the Guaxcama carbonate complex (González, 1976, 1977; Enos, 1983; McFarlan and Menes, 1991). These low-relief banks attained thicknesses of as much as a few hundred meters during that time (McFarlan and Menes, 1991) and display a zoned lateral facies profile from restricted peritidal carbonate (dolomites) to muddy, low-energy, lagoonal carbonate to normal marine-bank facies at their periphery. West and east of the Guaxcama complex, deeper-marine, outer-ramp lime mudstone and shale facies rimmed the complex. To the east, Lower

Tamaulipas shales and deep-marine carbonates rimmed the entire Gulf.

Locally, in northeast Mexico, the La Casita Formation (of late Kimmeridgian to Hauterivian age (Figure 6) represents a period of major clastic influx (Stone, 1975; Fortunato, 1982; Fortunato and Ward, 1982; Smith, 1987; Salvador, 1987, 1991a, b; Michalzik and Schumann, 1994). The age and thickness (650 to 800 m) of the La Casita vary geographically, in part a function of proximity to the exposed Coahuila block from which most of the detrital material was derived (Fortunato, 1982). The La Casita correlates with the Cotton Valley group (Tithonian to Berriasian; nearshore Schuler and offshore Bossier Formations) as well as the overlying Hauterivian Hosston Formation (Stone, 1975; McFarlan and Stone, 1977; Todd and Mitchum, 1977; Salvador, 1987, 1991a). Of major significance in portions of the Gulf is the apparent absence of any Valanginian strata in a shelfal position separating the Cotton Valley

from the Hosston (Figure 6; Stone, 1975; Todd and Mitchum, 1977; McFarlan and Stone, 1977; Goldhammer et al., 1991). This unconformity at the top of the Cotton Valley is marked by extensive subaerial erosion in shelfal positions and lowstand clastic wedges downdip of the Cotton Valley shelf margin (McFarlan and Stone, 1977; see Figures 35 and 36 in Goldhammer et al., 1991). Typically, in a basinward position, the upper part of the Schuler Formation is characterized by an interval of limestones that thickens seaward (approximately 20 to 30 m thick; Stone, 1975), the Cotton Valley Lime or Knowles Limestone (Mann and Thomas, 1964) that rests essentially subjacent to the top Berriasian unconformity (Stone, 1975; Todd and Mitchum, 1977).

In the Monterrey-Saltillo area, the La Casita has been subdivided into three regionally pervasive stratigraphic units that appear to correlate to the northern Gulf section, both lithologically and biostratigraphically (Fortunato, 1982; Fortunato and

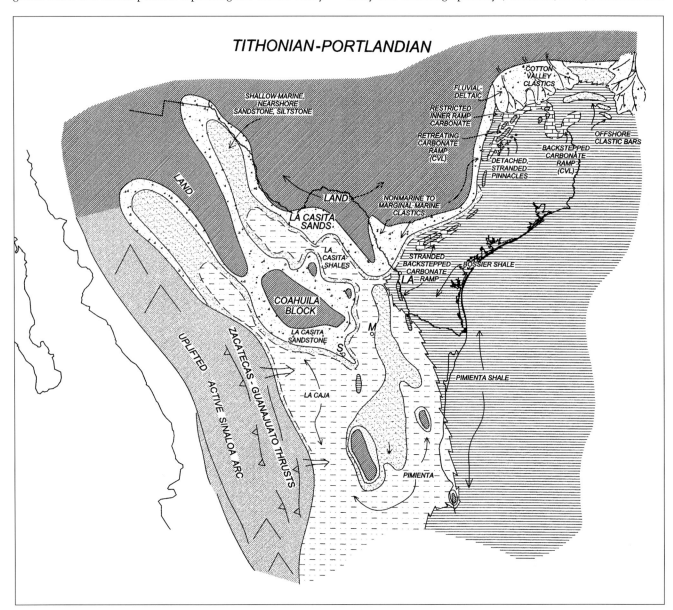

FIGURE 13. Tithonian to Portlandian paleogeography. Refer to text. Symbols are as in Figure 11a.

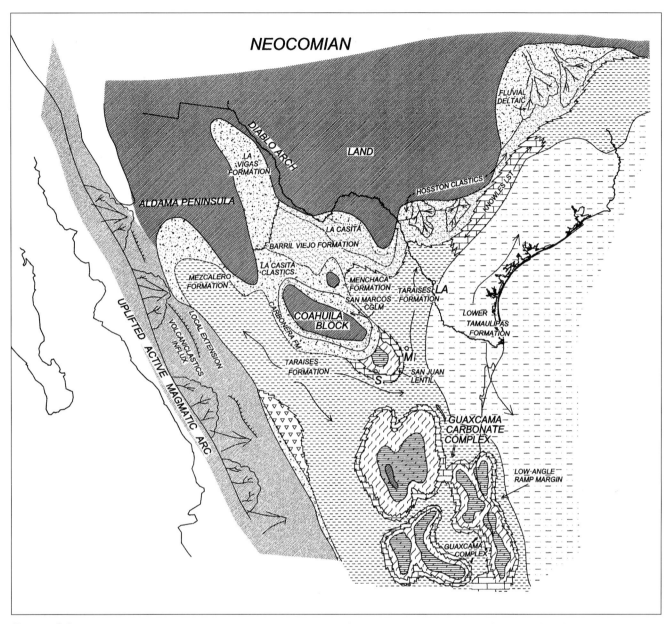

FIGURE 14. Neocomian paleogeography. Refer to text. Symbols are as in Figure 11a. Much of the detail in the Guaxcama carbonate platform is speculative, because so little is known of it for this time interval.

Ward, 1982). The three units are interpreted to depict main stages in the progradation and retreat of an extensive fan-delta complex. Unit 1 (137 to 148 m thick; "FW 1" in Figure 6) of Fortunato and Ward (1982) consists of black carbonaceous mudstone and siltstone with some burrow mottling and scattered layers rich in oysterlike pelecypods. This unit, interpreted as a prodelta deposit on a deep subtidal shelf, equates to the lower Cotton Valley and is probably Tithonian in age, although there are no biostratigraphic data to support this contention. Unit 2 (as much as 350 m thick; "FW 2A" and "FW 2B" in Figure 6) is predominantly fine- to coarse-grained sandstone (texturally and mineralogically immature arkose and lithic arkose) with many conglomerate intervals in the lower and upper parts (Fortunato and Ward, 1982). Burrow mottling,

trough crossbedding, and channelling are common features. Both coarsening-upward cycles (meter scale; mudstones and siltstones to coarse sandstone or conglomerate), and fining-upward cycles (meter scale; pebble conglomerate to medium sandstone) occur in the lower and upper parts.

In Canyon del Chorro, approximately 120 m from the top of unit 2, calcareous beds reveal a Kimmeridgian to Portlandian foram-ostracode assemblage (Fortunato, 1982; 230 m from the base of the Canyon del Chorro measured section of Fortunato, 1982). Approximately 40 m above this, ammonites of Hauterivian to Valanginian age have been found by J. L. Wilson (personal communication, 1988) (270 m from the base of the Canyon del Chorro measured section of Fortunato, 1982). This suggests that the lower 230 m of unit 2 at this locality is Upper Jurassic and equivalent to the Cotton Valley, and the remaining 120 m or so is Lower Cretaceous and equivalent in part to the Hosston, although additional biostratigraphic data are needed to ensure these correlations. For the sake of clarity, the Upper Jurassic portion of unit 2 is termed "unit 2A" on Figure 6, the Lower Cretaceous interval is "unit 2B" on Figure 6. Units 2A and 2B represent lower alluvial-fan to shallow-marine deposition during the maximum seaward advance of the fan-delta complex (Fortunato and Ward, 1982). Fining-upward cycles demarcate fluvial deposition on distal subaerial alluvial fans. Coarsening-upward cycles depict progradation of delta lobes in coastal and submarine-shelf environments. The overall nature of these coarse siliciclastics suggests rapid depostion of immature sediment in a fluvially dominated domain without substantial marine reworking.

Unit 3 (115 to 310 m thick; "FW 3" in Figure 6) consists of siltstones or trough crossbedded or burrowed sandstones, intercalated with thin carbonates. This upper unit of the La Casita at the Canyon del Chorro section (Figure 6) is probably Hauterivian (based on the above biostratigraphic control) and equates to the upper Hosston of the U.S. Gulf Coast. The transition from units 2 to 3 is marked by the first occurrence of micritic dolomite. Overall, the unit becomes finer grained, thinner bedded, and more calcareous upward where it grades conformably into the Taraises Formation (Fortunato and Ward, 1982). A few dark marine shales are also prominent toward the top. This unit marks the waning of coarse siliciclastics and the landward retreat of the fan-delta complex, accompanying a Gulf-wide marine transgression. With diminished siliciclastic supply, carbonate production was initiated, yielding this mixed clastic-carbonate unit.

The Taraises Formation is the Lower Cretaceous (mid-Berriasian through Hauterivian) deeper-water, offshore facies equivalent to the middle and upper units of the La Casita (Figure 6; Smith, 1981; Blauser, 1981) and correlating with the basinal Hosston of the Gulf (McFarlan and Stone, 1977; McFarlan and Menes, 1991). It conformably overlies the offshore La Caja, where the distinction between the two units is based on biostratigraphic control (Blauser, 1981). The Taraises, which thickens to the south and east (135 to 500 m thick) away from the main La Casita depocenter, consists of rhythmic-bedded, black, cherty, pelagic lime mudstones and intercalated shales (Blauser, 1981). Fossils include coccoliths, nannoconids, calpionellids, and radiolaria. These pelagic lime mudstones accumulated in a deep offshore to basinal setting (400-m to 500-m water depth) primarily associated with the La Casita retrogradation in the Hauterivian. The Taraises laps out updip where only a few meters of the formation separate La Casita from overlying Cupido (e.g., at Canyon del Chorro; Fortunato, 1982).

Locally, a thin (20-m-thick) limestone unit, the San Juan lentil, occurs in the upper portion of the La Caja (Figure 6). At Potrero Minas Viejas and Cortinas Canyon, this unit forms a persistent ridge in the La Caja shales and fine-grained siltstones. This normal open-marine limestone deposit loses its definition shelfward where the La Caja changes facies to the La Casita. The lower part of this limestone lentil contains a Berriasian fauna (J. L. Wilson and R. L. Scott, personal communication, 1989) which would equate the San Juan with the Knowles of the northern Gulf.

Barremian to Lower Aptian

In the mid- to Lower Cretaceous, the previously distinct and separate Mexican geosyncline and Chihuahua trough lose their individual outlines and merge into one major depocenter northwest of the Coahuila block as the subsiding Aldama peninsula was overstepped by regional sea-level highstand conditions (Figure 15; Córdoba, 1969; Córdoba et al., 1970, 1980; DeFord and Haenggi, 1971; Seewald and Sundeen, 1971; González-García, 1976; González, 1989; Cantú-Chapa, 1976; Tardy, 1977; de Cserna, 1970, 1979, 1989; Dickinson; 1981; Tóvar Rodríguez, 1981; Roldán-Quintana, 1982; Serváis et al., 1982, 1986; Cantú-Chapa et al., 1985; Araujo-Mendieta and Arenas-Partida, 1986; Brown and Dyer, 1987; Sedlock et al., 1993; Moran-Zenteno, 1994). At that time, this northwest-to-southeast-trending depocenter was bounded on the west by the Sinaloa magmatic arc. Proximal to the active arc system, coarse, nonmarine volcaniclastics prograded eastward into what was previously the Mexican geosyncline. East of this volcaniclastic apron and west of the Coahuila block, marine facies accumulated as moderately deep marine carbonates, and calcareous shales formed a weakly developed, low-angle carbonate ramp system. This north-northwest- to south-southeast-striking carbonate ramp rimmed an elongate deeper-marine trough in which fine-grained, clastic-rich siltsones and lime mudsones accumulated (Córdoba, 1969; Córdoba et al., 1970; 1980; de Cserna, 1979; González, 1976; Tardy, 1977; Araujo-Mendieta and Arenas-Partida, 1986; Moran-Zenteno, 1994). In northernmost Chihuahua, coarse nonmarine to marginal-marine clastics infilled shallower, updip portions of the Chihuahua trough and graded southeast into shallow-marine carbonate facies (Cuchilla Formation; Córdoba, 1969) which extended into the Sabinas Basin (La Virgen, Padilla Formations; González-García, 1973, 1976, 1979, 1984; Padilla Sánchez, 1978, 1986; Alfonso-Zwanziger, 1978; Márquez, 1979; Aranda-García and Eguiluz de Antuñano, 1983; Eguiluz de Antuñano and Aranda-García, 1983, 1984; Echanove, 1986). Complicated facies patterns and consequent formational nomenclature reflect the interaction of arc-driven clastic influx and Gulf-driven marine inundation and carbonate deposition.

To the north, northeast, and east of the Coahuila block, proximal coarse clastics (Patula Arkose, La Mula) record continued stripping of the Coahuila basement high. These fringing clastics change facies downdip, principally to the north, south, and east, into restricted marine evaporite and restricted peritidal carbonate of the La Virgen (Figure 6; González-Gar-

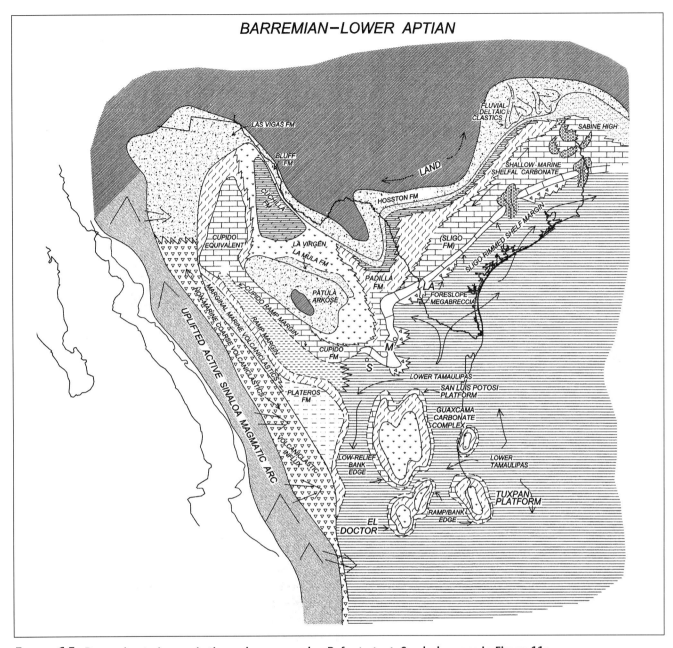

FIGURE 15. Barremian to Lower Aptian paleogeography. Refer to text. Symbols are as in Figure 11a.

cía, 1976; Zwanziger, 1979; Padilla Sánchez, 1986; Lehmann, 1997; Lehmann et al., 1998; Lehmann et al., 2000) that form the vast platform interior of the Cupido-Sligo carbonate platform. To the northeast in the GOM province, this enormous carbonate platform system maintains a low-relief, reef-rimmed shelf margin (Figures 8 and 9) with myriad platform-interior shallow-marine facies (high-energy grainstone shoals, open-shelf skeletal lime sands, muddy lagoonal carbonates, and restricted tidal-flat facies; e.g., Bebout and Loucks, 1977; Winker and Buffler, 1988; McFarlan and Menes, 1991). Seaward of the Sligo reef margin, foreslope aprons composed of platform-derived debris pass downdip into basinal lime mudstones and shales of the Lower Tamaulipas Formation (Figures

6 and 7). Rimming the exposed landmass of central Texas, coarse nonmarine to shallow-marine clastics (Hosston Formation) interfingered with Sligo carbonates. Locally, high-energy grainstone complexes developed over older regional basement highs (e.g., Sabine high).

To the south in the Tampico-Misantla area and in the southern Sierra Madre Oriental, major carbonate platforms persisted, notably the Tuxpan, El Doctor, and San Luis Potosí (Viniegra-Osario and Castillo-Tejero, 1970; Enos, 1974, 1977, 1983; González, 1976, 1977; González, 1977; Carrasco-V, 1977; Viniegra-Osario, 1981; Winker and Buffler, 1988; McFarlan and Menes, 1991; Wilson and Ward, 1993). According to the above authors, these platforms had moderately elevated rims that lacked the significant relief that was to develop in the Albian. Not as much has been written about the facies development, diagenesis, and petroleum significance of these Lower Cretaceous platforms compared with their younger mid-Cretaceous counterparts.

In northeast Mexico, the landscape of the Sierra Madre Oriental is dominated by the dramatic carbonate strata of the Cupido Formation (700 to 1200 m thick), which is Hauterivian to early Aptian in age (Figure 6; McFarlan and Stone, 1977; Conklin and Moore, 1977; Wilson and Pialli, 1977; Goldhammer et al., 1991; Lehmann, 1997; Lehmann et al., 1998; Lehmann et al., 2000). The Cupido is made up of a prograded, low-angle (low-relief), carbonate bank that steps up and over the underlying Taraises and the basinal equivalent to the Cupido, the Lower Tamaulipas (Figure 7; McFarlan and Stone, 1977; Conklin and Moore, 1977; Wilson and Selvius, 1984). The Cupido conformably overlies the La Casita or Taraises in updip positions (Smith, 1981; Wilson and Selvius, 1984). In downdip positions, the contact between the Cupido foreslope/basinal facies and the subjacent Lower Tamaulipas is a time-transgressive facies boundary. The top of the Cupido is marked by onlapping deeper-water shales and argillaceous carbonates of the La Peña, which conformably drape the Cupido bank (Figures 6–9; Tinker, 1982).

Shelf-to-basin depositional relief was on the order of a few hundred feet (less than 100 m) at the end of Cupido deposition (Tinker, 1982; Selvius, 1982). Behind the Cupido margin (which consisted of a biostromal facies, see below), a paleotopographic, depositional high persisted in the form of high-energy carbonate sand shoals, as evidenced by isopach maps of the draping La Peña Shale (Tinker, 1982). The trend of the Cupido margin, essentially paralleling the outline of the Coahuila block, and the east-to-west direction of progradation clearly underscore the influence of this positive feature in controlling Cupido facies development (Wilson, 1981; Wilson and Selvius, 1984). In contrast, the Cupido trend crosscuts the Burro-Salado arch to the northwest, indicating that this structure no longer affected depositional patterns (Wilson, 1981). The Burro-Salado arch was reactivated and uplifted by the Laramide orogeny.

A generalized platform-to-basin facies model developed by previous workers (Ekdale et al., 1976; Conklin and Moore, 1977; Loucks, 1977; Wilson and Pialli, 1977; Wilson, 1981; Selvius, 1982; Wilson and Selvius, 1984) indicates that the Cupido consists of essentially six lithofacies (units A-F of Conklin and Moore, 1977; lithofacies 1–6 of Selvius, 1982) that comprise one large-scale, shallowing-upward package. These are: (1) thin-bedded, argillaceous, pelagic lime mudstones (basinal environment—unit A, as much as 400 m thick in more basinward sections, e.g., Potrero Minas Viejas); (2) thick-bedded, intraclastic-bioclastic wackestones to packstones (ramp-slope and fore-reef environments—unit B); (3) massive rudist and coral-dominated packstones, grainstones, and boundstones with stromatoporoids and marine cements (biostromal shelf

margin—unit C, as much as 250 m thick, e.g., at Potrero Chico); (4) crossbedded skeletal to peloidal packstones to grainstones with ooids and oncolites (back-margin sand shoals—unit D); (5) thin-bedded, mudstones-packstones with cryptalgal laminites and evidence for evaporites (cyclic, peritidal platform interior—unit E, as much as 500 m at Los Chorros Arteaga); (6) medium-bedded, black, peloid-foram wackestones to packstones with pelecypods and green algae (back-margin subtidal lagoon—unit F). In practical terms (i.e., in an interpretative sense), units 1–5 comprise the large-scale prograding and shallowing succession, and unit 6 depicts a relative deepening and initiation of retrogradation back over the Cupido bank (Figures 6, 7, and 8). This transgressive unit is informally termed the *Cupidito* (Figures 6 and 7; Wilson and Pialli, 1977; Wilson, 1981). Its thickness varies laterally across the bank top, from approximately 100 m near the Cupido margin (e.g., Potrero García and Potrero Chico) to only a few meters updip in platform interior positions (e.g., Huasteca Canyon, Los Chorros Arteaga).

The Lower Tamaulipas (late Hauterivian to early Aptian, Figures 6 and 7; Ross, 1981) is the downdip, basinal equivalent of the Cupido bank and outcrops primarily to the south and east of the Monterrey-Saltillo area (Smith, 1981; Wilson and Selvius, 1984). Its lower contact with the Taraises is conformable, as is its upper contact with the La Peña (Ross, 1981). It consists of about 600 m of dark gray to black, thin- to medium-bedded, cherty lime mudstones to wackestones that display burrow mottling (Ross, 1981). It contains a pelagic fauna of radiolaria, nannoconids, coccoliths, and rare mollusks and echinoids (Ross, 1981). This pelagic deposit is interpreted to have accumulated in a dysaerobic, quiet basinal setting at depths of 50 to 150 m (Wilson, 1969; Byers, 1977; Ross, 1981).

Mid- to Upper Aptian

During this time interval, in the WPM province, the major facies trends outlined above for the Barremian to lower Aptian are fairly similar to those preserved in the mid- to upper Aptian (Figure 16; DeFord and Haenggi, 1970; Seewald and Sundeen, 1971; González-García, 1976; González, 1989; Cantú-Chapa, 1976; Tardy, 1977; de Cserna, 1970, 1979, 1989; Dickinson; 1981; Roldán-Quintana, 1982; Cantú-Chapa et al., 1985; Araujo-Mendieta and Arenas-Partida, 1986; Brown and Dyer, 1987; Sedlock et al., 1993; Moran-Zenteno, 1994). Proximal to the Sinaloa arc, volcaniclastics flanked the western edge of the WPM depocenter, changing facies to the south into deeper-marine lime mudstones and shales of the Plateros Formation, and to the north into shallow-water carbonates and offshore shales (Parral Formation; Córdoba, 1969; Córdoba et al., 1980; Tóvar Rodríguez, 1981; Cuévas-Pérez, 1983; Cuévas-Pérez et al., 1985; Limón, 1989; Serváis et al., 1986). To the northwest in the expanded Chihuahua embayment, a fairly extensive carbonate platform with slope breccias and blocks (e.g., Mural Limestone, Scott and Warzeski, 1993) existed downdip from shallow-marine clastics (Figure 16). This shallow-marine carbonate ramp was marked by shoal development at its outer edge. To the southeast into the Sabinas Basin, this carbonate system graded into outer-ramp lime mudstones and shales (Cuchilla, La Peña; Figure 6; González-García, 1976; Padilla Sánchez, 1986; Moran-Zenteno, 1994).

Outside the WPM province, the entire GOM province from east Texas to Tampico experienced a major second-order marine transgression (e.g., Bebout and Loucks, 1977; Winker and Buffler, 1988; McFarlan and Menes, 1991), which effectively backstepped and/or drowned the older Sligo-Cupido-Guaxcama carbonate system, inundating it with deeper-water shales and fine-grained terrigenous siliciclastics derived from distal highlands to the north and west. In Texas, this transgressive event is well documented by the Pearsall Group (Figures 6–9), which includes some patches of carbonate buildup (James and Pettit limestones; Bebout and Loucks, 1977) that managed to keep up somewhat with this major sea-level rise.

Likewise, southeast of the Coahuila block, scattered carbonate buildups occur in the basal La Peña at the top of the Cupidito. The Coahuila block was nearly inundated, and shallow, submerged portions of the block became sites of restricted carbonate deposition (Las Uvas facies; Lehmann et al., 1998). To the South in the Tampico-Misantla area, the areal extent of the Tuxpan, El Doctor, and San Luis Potosí platforms is diminished, because they retrograded somewhat (McFarlan and Menes, 1991).

In the Monterrey-Saltillo area, the La Peña Formation is late Aptian (Figure 6; Smith, 1981; Tinker, 1982) and correlates to the northern Gulf Aptian tripartite of the Pine Island Shale–James Limestone–Bexar Shale, which together constitute the Pearsall Group (McFarlan and Stone, 1977; Tinker, 1982). The La Peña drapes the underlying Cupido in apparent conformity, preserving underlying topographic relief at the end of Cupido time (Tinker, 1982). This is also observed on high-resolution 3-D seismic data from south Texas. Bio-

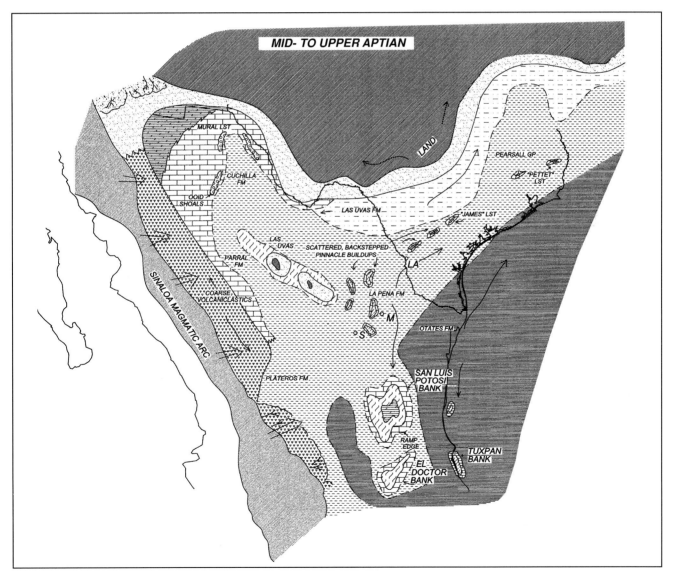

FIGURE 16. Mid- to Upper Aptian paleogeography. Refer to text. Symbols are as in Figure 11a. Note that the orientation of the El Doctor Bank is speculative.

stratigraphic control of Tinker (1982) indicates progressive onlap of La Peña over the Cupidito unit of the Cupido. The base of the La Peña rises biostratigraphically from lower to upper Aptian in an east-to-west direction across the platform (Lehmann et al., 1998). In addition, this biostratigraphic control indicates that the top of the La Peña is not an isochron but rather a time-transgressive formation boundary with overlying Albian carbonates. The formation varies in thickness from a few meters to 200 m, depending on antecedent Cupido/Lower Tamaulipas depositional relief (Tinker, 1982). For example, the La Peña is thinnest (a few meters) above the Cupido paleotopographic high (carbonate sand shoals) located behind the shelf margin and thickens behind this above the Cupido back-margin lagoon facies (as much as 150 m), perhaps enhanced by differential compaction of subjacent lithologies. Out in front of the Cupido margin, the La Peña thickens drastically into the basin (200 m; compare with Figures 7 and 8). Off to the south and east, the deep basinal equivalent to the La Peña, the Otates Formation, consists of thin-bedded, dark, argillaceous, cherty limestones and black shales (Tinker, 1982).

Albian

In the Albian, the WPM province was the site of an extensive shallow, open-marine carbonate shelf (e.g., Finlay Limestone; DeFord and Haenggi, 1971), which extended northeast essentially uninterrupted through the East Texas Salt Basin and beyond (Figure 17; Córdoba, 1969; Tóvar Rodríguez, 1981; Enos, 1983; McFarlan and Menes, 1991; Cantú-Chapa et el, 1985; Buffler and Winker, 1988). To the west, the Alisitos arc persisted (Dickinson, 1981; Serváis et al., 1986; Araujo-Mendieta and Arenas-Partida, 1986; Limón, 1989), and coarse volcaniclastics offlapped this high belt eastward, infilling an Albian back-arc basin (Dickinson, 1981). An area of shallow-marine nearshore clastic (Mezcalera Formation; Cantú-Chapa et al., 1985) and, to the south, turbidites occupied a narrow belt east of the volcaniclastic belt and west of the Coahuila peninsula (Figure 17). Southeast from Chihuahua, the basin boundary with the northwestern reaches of the Sabinas Basin is indistinguishable where shallow-marine carbonates of the Aurora Formation (Figure 6) onlapped and rimmed the Coahuila block (Cantú-Chapa et al., 1985; Tóvar Rodríguez, 1981; González-García, 1976; Zwanziger, 1979; Padilla Sánchez, 1986; Wilson and Ward, 1993; Lehmann et al., 1998). The southeastern margin of the Sabinas formed an embayment (Ocampo embayment; Cantú-Chapa et al., 1985) separating Aurora carbonates of the Coahuila peninsula from the Glen Rose–Fredericksburg–Stuart City complex of south-central Texas (e.g., Wilson, 1975; Bebout and Loucks, 1977; Enos, 1983; Buffler and Winker, 1988; McFarlan and Menes, 1991).

To the northeast in Texas, this enormous carbonate system passed updip into nearshore clastics (Washita group) and overstepped most of the previously exposed landmass of Texas, with the exception of small portions of the Llano and Mara-

thon uplifts. The Stuart City–Edwards shelf margin in Texas formed a low-relief, reef-rimmed margin (e.g., Wilson, 1975) or ramp margin (Kerans et al., 1995; Zahm, 1997) located updip from the underlying Sligo margin in south Texas (Figures 8 and 9; Wooten and Dunaway, 1977; Winker and Buffler, 1988). South of the Coahuila block, the Tampico-Misantla area was the site of the classic Albian high-relief, reef-rimmed platforms that have served as excellent petroleum reservoirs (Tuxpan, El Doctor, Valles–San Luis Potosí; Viniegra-Osario and Castillo-Tejero, 1970; Enos, 1974, 1977, 1983; González, 1976, 1977; González, 1977; Carrasco-V, 1977; Viniegra-Osario, 1981; Winker and Buffler, 1988; McFarlan and Menes, 1991; Wilson and Ward, 1993). Downdip of the Albian-rimmed shelf and isolated platforms, deep-marine shales and lime mudstones of the Upper Tamaulipas accumulated (Figure 6).

The Aurora Formation demarcates the second major phase of Cretaceous carbonate platform development in northeastern Mexico (Wilson, 1975; Smith, 1981; Lehmann, 1997; Lehmann et al., 1998; Lehmann et al., 2000). The Aurora and its basinal facies equivalent, the Upper Tamaulipas, are Albian in age (Figures 6 and 7), and in contrast to the high-relief, steep-sided reef margins of the Valles Platform to the south, Aurora carbonates were deposited on a gently dipping, low-angle carbonate ramp which rimmed the Coahuila block in the southern Sabinas Basin (Smith, 1981; Lehmann, 1997; Lehmann et al., 1998). A prominent rudist bank facies marks the ramp edge, behind which 500 to 700 m of normal-marine carbonates accumulated (Vinet, 1975; Lehmann, 1997; Lehmann et al., 1998; Lehmann et al., 2000). The Aurora rests conformably on the La Peña and is probably disconformable beneath the Cuesta del Cura.

Also prominent in northeast Mexico exposures, the upper Tamaulipas (Albian; 100 to 200 m thick), which is the basinal equivalent of the Aurora, correlates with the Atascosa Formation in south Texas (Figure 6; McFarlan and Stone, 1977; Smith, 1981; Ross, 1981). It outcrops south and east of the Aurora margin and is conformable atop the La Peña and beneath the Cuesta del Cura (Smith, 1981). There it consists of thin-thick-bedded, cherty, dark pelagic mudstone to wackestone (Ross, 1981), interpreted as deep-water anaerobic to dysaerobic off-ramp deposits. Toward the top, burrow-mottled fabrics are evidence for a slight shallowing. The top of the formation is marked by a prominent conglomerate bed a few meters thick containing cemented clasts of shallow-water Aurora lithologies derived from updip (Ross, 1981).

Cenomanian

In the earliest Upper Cretaceous, both the WPM and GOM provinces were inundated by a major superimposed first-order and second-order eustatic flooding event that connected the Gulf of Mexico with the Western Interior seaway of the Rocky Mountains (e.g., McFarlan and Menes, 1991). In the WPM province, the Alisitos arc was active, and arc magmatism had migrated eastward as compared with the Albian, a hint of things to follow in the latest Cretaceous (Tardy, 1977; de Cserna, 1979, 1989; Limón, 1989; Córdoba et al., 1980; Dickinson, 1981; Serváis et al., 1982, 1986; Araujo-Mendieta and Arenas-Partida, 1986; Sedlock et al., 1993; Moran-Zenteno, 1994). East of the arc, a back-arc basin persisted and was the site of a variety of volcaniclastic-related facies (Figure 18), which passed eastward into deeper-marine lime mudstones and shales of the Indidura and Cuesta del Cura. To the north, the last vestiges of the underlying Albian platform persisted in the northern extremes of the Chihuahua trough and to the south as calcareous turbidites (DeFord and Haenggi, 1971; Córdoba, 1969; Tóvar Rodríguez, 1981; Enos, 1983; Cantú-Chapa et el, 1985). This narrow band of shallow-marine carbonates linked up with thin, latest Albian to early Cenomanian carbonates of the backstepped Aurora platform (Figure 18). Locally in northeast Mexico and south Texas, a few pinnacle buildups of latest Albian to earliest Cenomanian age attempted to keep up with the overall transgression in the region. These are small and isolated, yet prominent on 2-D regional seismic

data in south Texas. To the southeast, the Albian platforms were backstepped and drowned as well.

In the Monterrey-Saltillo area, outcropping basinal facies are assigned to the Cuesta del Cura, which is latest Albian to Cenomanian (Figures 6 and 7; Enos, 1974; Smith, 1981; Ice, 1981) and consists of deep-water pelagic carbonates and shales that accumulated in front of elevated shallow-water, reef-rimmed, middle Cretaceous platforms (Stuart City, Aurora, Valles platforms). It equates to the Georgetown–Del Rio–Buda of south Texas (Smith, 1981). In exposures of northeast Mexico, it is composed of approximately 60 m of dark, thin- to medium-bedded, laminated lime wackestones to packstones intercalated with thin shales (rhythmites) and rare lithoclastic conglomerates (platform-derived megabreccias). Grain types include peloids, calcispheres, radiolaria, planktonic forams, echinoids, and sponge spicules (Ice, 1981). The upper formational contact is conformable (Ice, 1981).

The Indidura Formation (Cenomanian to Santonian) contains two members—the Agua Nueva and the San Felipe (Fig-

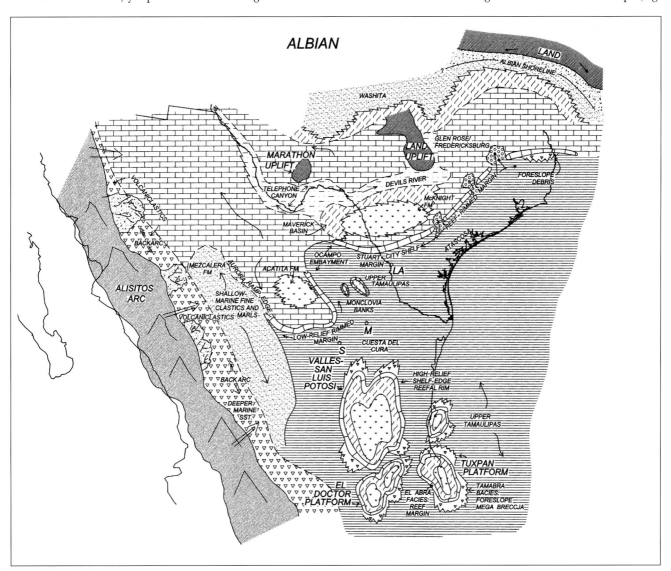

FIGURE 17. Albian paleogeography. Refer to text. Symbols are as in Figure 11a.

ures 6 and 7), equivalent to the Eagleford Shale and Austin Chalk (Smith, 1981; Winker and Buffler, 1988; Ice, 1981). This package of deep-water pelagic deposits averages 300 to 400 m in thickness and consists of pelagic lime mudstones to wackestones (radiolaria, globigerinids) and intercalated thin shales (rhythmites) that essentially blanket the entire southwest Gulf area, signifying the peak of the middle Cretaceous highstand.

Upper Cretaceous (Maestrichtian)

From the end of the Cenomanian and throughout the Maestrichtian, paleogeographic and facies relations in both the WPM and GOM provinces are drastically changed as the result of the diachronous Laramide phase of deformation (Figures 4, 6, and 19). It is well beyond the scope of this paper to treat the Laramide phase in any detail (see, for example, de Cserna, 1989; Sedlock et al., 1993; Moran-Zenteno, 1994). However, for the sake of completeness, it is summarized briefly, because it is responsible for the present-day structural relations in the Monterrey-Saltillo area. In the latest Cretaceous, the WPM province experienced Laramide uplift and easterly directed contractional deformation because the Alisitos arc migrated inboard (e.g., Dickinson, 1981). During that period, the WPM province was uplifted and eroded, and despite local foreland basin accumulations, much of the record is an extended hiatus. At that time, the Sierra Madre Oriental fold belt developed and migrated west to east (e.g., Suter, 1987). In northeast Mexico, Maestrichtian foreland basins developed in front of the advancing Sierra Madre (e.g., Sabinas/Olmos area; La Popa and Parras Basins; González-García, 1976; Padilla Sánchez, 1986; Soegaard et al., 1997).

SECOND-ORDER SUPERSEQUENCE DEVELOPMENT

In northeast Mexico, the Triassic to middle Cretaceous section can be divided into two tectonic phases of passive-margin development: (1) middle Carnian to lower Oxfordian rift stage

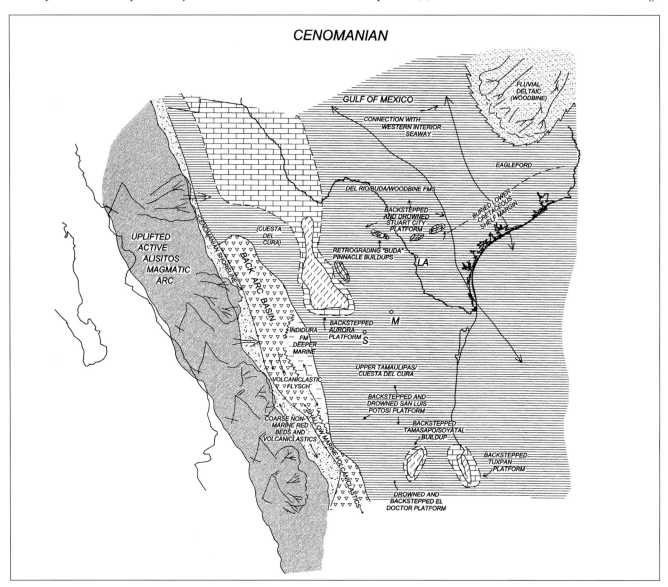

FIGURE 18. Cenomanian paleogeography. Refer to text. Symbols are as in Figure 11a.

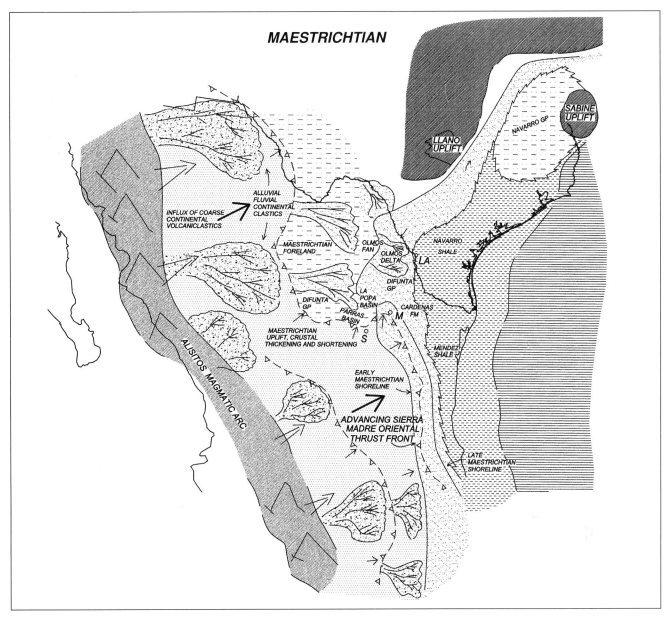

Figure 19. Maestrichtian paleogeography. Refer to text. Symbols are as in Figure 11a.

composed of 2400 m of rift-related continental to marginal-marine red beds and evaporites; and (2) lower Oxfordian to Santonian thermal cooling and drift stage consisting of 3700 m of shallow-marine to deep-marine carbonates with subordinate clastics (Goldhammer at al., 1991). During the drift stage of passive-margin development, the stratigraphic evolution is interpreted to be dominated principally by eustasy (Todd and Mitchum, 1977; Vail et al., 1984; Haq et al., 1987; Scott et al., 1988; Goldhammer et al., 1991; Scott, 1993; Yurewicz et al., 1993; Lehmann et al., 1998) for two reasons: (1) Thick regional accommodation cycles of second-order supersequence development can be correlated throughout the Gulf of Mexico (Salvador, 1991a, b, c; McFarlan and Menes, 1991; Sohl et al., 1991), and (2) the rates of background tectonic subsidence are

much too slow to account for the accommodation space required to accumulate the thick supersequences.

With regard to the packaging of the drift-related stratigraphy, the Middle and Upper Jurassic through the Lower Cretaceous stratigraphy has been subdivided into four major, second-order (i.e., 10–100-myr duration) depositional supersequences that have regional Gulf-wide significance (Figures 6–9; Goldhammer et al., 1991). Major second-order sequence boundaries, condensed sections, transgressive surfaces, and second-order systems tracts were identified in outcrop (summarized in Goldhammer, 1999), were dated using available biostratigraphy, and were correlated with the northern Gulf of Mexico stratigraphic section. The identification of these components is based on (1) gross shelf-to-basin relationships of onlapping and offlapping facies; (2) stacking patterns of third-order (1–10-myr duration) sequences and their component high-frequency (fourth- and fifth-order) cycles; and (3) in the case of sequence boundaries, criteria for significant subaerial exposure and/or erosion. Additional stratigraphic data came from unpublished regional seismic coverage in east Texas and south Texas (e.g., Figures 8 and 9) and regional well-log cross sections (Figures 31–37 in Goldhammer et al., 1991; Goldhammer, 1998a, b) in east and south Texas and extreme northeastern Mexico. In the discussion below, second-order sequence boundaries are in quotations because of the uncertainties of the absolute ages.

Upper Bathonian to Lower Kimmeridgian ("158.5"–"144" myr)

This supersequence is dominated by second-order lowstand deposition (La Joya clastics and Minas Viejas evaporites; Figures 11 and 13a) with the second-order transgressive surface at "150.5" myr (Figures 6 and 7). The La Gloria Formation and the lower portion of the Zuloaga Formation are largely transgressive, with the upper portion of the Zuloaga Formation and the Olvido evaporite unit demarcating the second-order highstand. The supersequence boundary at "144" myr falls in the Olvido evaporite unit (Figures 6 and 7). The Upper Triassic to Lower Jurassic La Boca and Middle Jurassic La Joya are nonmarine red beds with some volcanics and are areally restricted to rift basins, resting unconformably on Late Paleozoic basement. The Minas Viejas (gypsum, halite) is mostly deformed with a nonuniform distribution in the area representing marginal-marine deposition prior to the second-order marine transgression. The La Gloria consists of shallow, nearshore-marine to marginal-marine (playa) siliciclastics located peripherally about margins of exposed Paleozoic landmasses. The upper part of the La Gloria is the equivalent to the carbonate Zuloaga, which consists of shallow-water, low-angle ramp deposits that grade downdip to outer ramp mudstones, changing facies to basinal shales and thin limestone beds of the La Caja Formation. The Olvido Formation consists in the lower portion of evaporitic deposits and associated marginal-marine clastics. The upper portion deepens upward overall into carbonates.

Lower Kimmeridgian to Berriasian ("144"–"128.5" myr)

Within this supersequence, the Olvido evaporite, Olvido limestone, and basinal equivalent La Caja comprise the second-order transgressive systems tract with the second-order maximum flooding surface (MFS) placed at the top of the Olvido lime (138 myr; Figures 6 and 7). The highstand part of the supersequence contains the lower two-thirds of the siliciclastic La Casita Formation, the upper portion of the La Caja, and the lower portion of the Taraises. The La Casita Formation, in this supersequence, consists of a basal unit of carbonaceous siltstone and mudstone (interpreted as deeper-water "prodelta"

deposits), capped by an upper unit of coarse sandstones and pebble conglomerates (interpreted as an overall regressive, coarse-clastic, fan-delta system). The supersequence boundary occurs in this upper regressive interval, as reflected by the maximum seaward advance of the fan-delta system. The lower portion of the Taraises is the deeper-water, offshore (basinal) equivalent to the La Casita, consisting of argillaceous, rhythmic-bedded limestones and shales. The San Juan lentil, a persistent carbonate unit in the uppermost La Caja to lower part of the Taraises, most likely depicts a third-order, relative sea-level cycle in the overall second-order highstand.

Upper Valanginian to Lower Aptian ("128.5"–"112" myr)

In this supersequence, the upper part of the La Casita consists of fine-grained clastics and thin carbonates, which are interpreted as transgressive marine deposits recording the waning input of clastics and retreat of the fan-delta complex (Figure 6). This and the onlapping basinal equivalent of the upper portion of the Taraises depict the second-order transgressive systems tract with the second-order MFS ("118" myr) placed at the base of the overlying Cupido/Lower Tamaulipas carbonate package, which records the second-order highstand systems tract (Figures 6 and 7). The Cupido comprises a shallow-water, low-angle ramp package that overall progrades basinward and shoals from basinal/forereef strata through peritidal cyclical laminites. The offshore basinal equivalent to the Cupido is the Lower Tamaulipas, a thick succession of pelagic lime mudstones. There are numerous well-developed, higher-frequency sequences in the Cupido/Lower Tamaulipas, and the supersequence is capped by a prominent evaporite-solution collapse breccia ("112" myr) which separates the Cupido from the transgressive Cupidito.

Lower Aptian to Upper Albian ("112"–"98" myr)

The Cupidito and La Peña mark the second-order transgressive systems tract, and the second-order maximum flooding surface is placed near the La Peña/Aurora limestone contact ("107" myr). The Cupidito is a deepening-upward carbonate unit (shallow-water, low-angle ramp) that grades from peritidal cyclic carbonates at the base through shoal-water subtidal carbonates toward the top. The upper contact with the La Peña is gradational, and the La Peña consists of deep-water, transgressive, fine-grained siliciclastic shales, mudrocks, and thin-bedded argillaceous limestones. The Aurora/Upper Tamaulipas carbonate package comprises the second-order highstand systems tract, with the Aurora representing shallow open-shelf to ramp carbonate deposits that are overall aggradational to retrogradational. The pelagic lime mudstones of the Upper Tamaulipas make up the equivalent basinal strata. The upper supersequence boundary ("98" myr) is recognized from the basinward progradation of the McKnight evaporite in the Maverick basin, a nearshore- to marginal-marine unit that records the second-order relative fall in sea level at the top of the Aurora/Upper Tamaulipas. The overlying pelagic carbonates consisting of the Cuesta del Cura, Agua Nueva, and San Felipe Formations are all deep-marine, basinal rhythmites. They reflect the overall, prolonged relative rise in sea level that was characteristic of the Early to middle Cretaceous. These deposits onlap, updip, and drape much of the underlying Mesozoic section. The Middle Cretaceous Sequence Boundary (MCSB) recognized by Buffler (1991) on seismic data in offshore areas of the Gulf of Mexico is expressed as a drowning surface (Schlager, 1989) on top of the retrogradational early Cenomanian shallow-water platform carbonates.

SUMMARY

In summary, we have synthesized subsurface and outcrop data, both published and unpublished, in an attempt to provide a sequence-stratigraphic framework for the Mesozoic paleogeographic and tectonic evolution of northeast Mexico. A series of paleogeographic maps combined with an integrated chronostratigraphic framework for the Mesozoic of both northeast Mexico and subsurface equivalents in the Texas Gulf Coast is presented to place this critical area in a regional context and to provide a framework for understanding hydrocarbon systems in the onshore Gulf of Mexico. With regard to basin evolution, the Monterrey-Saltillo area contains elements related to both Gulf of Mexico passive-margin development (principally the stratigraphy) and to the Pacific-related convergent margin (arc) tectonism (chiefly the structure), and therefore allows one to compare and contrast Gulf of Mexico–driven versus Pacific-driven tectonostratigraphic processes. The differing stratigraphies record a subregional response to the interaction of provincial tectonics (i.e., convergent versus divergent margins), eustatic changes in sea level, and sediment type and supply.

In the GOM province, the tectonic evolution is dominated by passive-margin development associated with the opening of the Gulf of Mexico, overprinted by nonigneous Laramide orogenic effects. The stratigraphic evolution is dominated principally by eustasy, in as far as thick regional accommodation cycles can be correlated throughout the Gulf of Mexico. In this paper, it is proposed that the Middle Jurassic to Lower Cretaceous stratigraphy of northeast Mexico and the GOM in general can be subdivided into four major, second-order depositional supersequences (approximately 15-myr duration) that have regional Gulf-wide significance.

The WPM province is distinct from the GOM province in that its style of basin evolution had little to do with Gulf of Mexico tectonic evolution (i.e., rift-drift of the Yucatán block); rather, patterns of stratigraphic infill are primarily a function of tectonism related to Mesozoic Pacific tectonism and sediment supply, as opposed to the eustasy-dominated Gulf of Mexico. Mesozoic subduction along the Pacific margin controlled basin-specific styles of development in the WPM province where facies development was affected strongly by adjacent arc-related tectonism and sediment supply.

ACKNOWLEDGMENTS

We gratefully acknowledge the critical and thorough reviews of the manuscript provided by Paul Enos, James Lee Wilson, Claudio Bartolini, Jory Pacht, Neil Hurley, and Dick Buffler. Much of this paper is based on work done by the senior author while employed at Exxon Production Research Co. RKG would like to thank Robert G. Todd for creating and supporting the original field program in Mexico. All the illustrations were drafted by Sylvie Tsang of Sonat Exploration (Houston, Texas). RKG wishes to thank Ken Pfau for supporting additional fieldwork during 1997–1998 and for his encouragement during the writing of the manuscript. In addition, we gratefully acknowledge critical input from several geoscientists during various field trips over a period of several years, in particular Dave McLean, Kate Giles, Gerhart Westra, Art Ross, and Felipe Pontigo.

REFERENCES CITED

Ahr, W. M., 1981, The Gilmer Limestone: Oolite tidal bars on the Sabine Uplift: Gulf Coast Association of Geological Societies Transactions, v. 31, p. 1–6.

Alfonso-Zwanziger, J., 1978, Geología regional del sistema sedimentario Cupido: Boletín de la Asociación Mexicana de Geólogos Petroleros, v. 30, p. 1–56.

Anderson, T. H., and V. A. Schmidt, 1983, The evolution of middle America and the Gulf of Mexico–Caribbean Sea region during Mesozoic time: Geological Society of America Bulletin, v. 94, p. 941–966.

Aranda-García, M., 1991, El segmento San Felipe del Cinturón cabalgado, Sierra Madre Oriental, estado de Durango, Mexico: Boletín de la Asociación Mexicana de Geólogos Petroleros, v. 41, p. 18–36.

Aranda-García, M., and S. Eguiluz de Antuñano, 1983, Posibilidades Económico-Petroleras en Rocas Clásticas del Neocomiano en la Magen Sur de la Paleoisla de Coahuila: Congreso Nacional de la Asociación de Ingenieros Petroleros de México, p. 5–13.

Araujo-Mendieta, J., and R. Arenas-Partida, 1986, Estudio tectónico-sedimentario en el Mar Mexicano, estados de Chihuahua y Durango: Sociedad Geológica Mexicana Boletín, v. 47, p. 43–88.

Ball, M. M., and C. G. A.. Harrison, 1969, Origin of the Gulf and Caribbean and implications regarding ocean ridge extension, migration and shear: Gulf Coast Association of Geological Societies Transactions, v. 19, p. 287–294.

Bartok, P., 1993, Prebreakup geology of the Gulf of Mexico–Caribbean: Its relation to Triassic and Jurassic rift systems of the region: Tectonics, v. 12, p. 441–459.

Basáñez-Loyola, M. A., R. Fernandez-Turner, and C. Rosales-Dominguez, 1993, Cretaceous platforms of Valles–San Luis Potosí, northeastern central Mexico, in J. A. T. Simo, R. W. Scott, and J.-P., Masse, eds., Cretaceous carbonate platforms: AAPG Memoir 56, p. 51–60.

Bebout, D. G., and R. G. Loucks, eds., 1977, Cretaceous carbonates of Texas and Mexico: University of Texas, Bureau of Economic Geology Report of Investigations 89, 332 p.

Blauser, W. H., 1981, The stratigraphy of the Taraises Formation, Lower Cretaceous, Mexico, *in* C. I. Smith, ed., Lower Cretaceous stratigraphy and structure, northern Mexico: West Texas Geological Society Publication 81-74, p. 37–42.

Bracken, B., 1984, Environments of deposition and early diagenesis, La Joya Formation, Huizachal Group red beds, northeastern Mexico, *in* W. P. S. Ventress et al., eds., The Jurassic of the Gulf rim: Gulf Coast Section, Society for Sedimentary Geology (SEPM), Proceedings of the Third Annual Research Conference, p. 19–26.

Brown, M. L., and R. Dyer, 1987, Mesozoic geology of northwestern Chihuahua, Mexico, *in* W. R. Dickinson, ed., Mesozoic rocks of southern Arizona and adjacent areas: Arizona Geological Society Digest, v. 18, p. 381–393.

Brown, M. L., and J. W. Handschy, 1983, Tectonic framework of Chihuahua, Mexico, *in* Geology and petroleum potential of Chihuahua, Mexico: West Texas Geological Society Field Trip Guidebook, Publication 84-80, p. 161–173.

Budd, D. A., and R. G. Loucks, 1981, Smackover and lower Buckner formations, south Texas: Depositional systems on a Jurassic carbonate ramp: University of Texas, Bureau of Economic Geology Report of Investigations 112, 38 p.

Buffler, R. T., 1991, Seismic stratigraphy of the deep Gulf of Mexico basin and adjacent margins, *in* A. Salvador, ed., The Gulf of Mexico Basin: Geological Society of America, The Geology of North America, v. J, p. 353–387.

Buffler, R. T., and D. S. Sawyer, 1985, Distribution of crust and early history, Gulf of Mexico Basin: Gulf Coast Association of Geological Societies Transactions, v. 35, p. 333–344.

Byers, C. W., 1977, Biofacies patterns in euxinic basins: A general model, *in* H. E. Cook and P. Enos, eds., Deep-water carbonate environments: Society for Sedimentary Geology (SEPM) Special Publication no. 25, p. 5–18.

Campa-Uranga, M. F., 1985, The Mexican thrust belt, *in* D. G. Howell, ed., Tectonostratigraphic terranes of the circum-Pacific region: Circum-Pacific Council for Energy and Mineral Resources Earth Science Series, n. 1, p. 199–213.

Campa-Uranga, M. F., and P. Coney, 1983, Tectonostratigraphic terranes and mineral resource distributions in Mexico: Canadian Journal of Earth Sciences, v. 20, p. 1040–1051.

Cantú-Chapa, A., 1976, Nuevas localidades del Kimeridgiano y Titoniano en Chihuahua (Norte de Mexico): Revista del Instituto Mexicano del Petróleo, v. 8(2) p. 38–49.

Cantú-Chapa, A., 1992, The Jurassic Huasteca series in the sub-surface of Poza Rica, eastern Mexico: Journal of Petroleum Geology, v. 15, p. 259–282.

Cantú-Chapa, C. M., 1993, Sedimentation and tectonic subsidence during the Albian-Cenomanian in the Chihuahua Basin, Mexico, *in* J. A. T. Simo, R. W. Scott, and J.-P. Masse, eds., Cretaceous Carbonate Platforms: AAPG Memoir 56, p. 61–70.

Cantú-Chapa, C. M., R. Sandoval-Silva, and R. Arenas-Partida, 1985, Evolución sedimentaria del Cretático Inferior en el norte de México: Revista del Instituto Mexicano del Petróleo, v. 17, p. 14–37.

Carrasco-V, B., 1977, Albian sedimentation of submarine authochthonous and allochthonous carbonates, east edge of the Valles–San Luis Potosí Platform, Mexico, *in* H. E. Cook and P. E. Enos, eds., Deep-water carbonate environments: Society for Sedimentary Geology (SEPM) Special Publication 25, p. 263–272.

Carrillo-Bravo, J., 1963, Geology of the Huizachal-Peregrina anticlinorium northwest of Ciudad Victoria, Tamaulipas, *in* F. Bonet et al., eds., Geology of Peregrina Canyon and Sierra de El Abra, Mexico: Corpus Christi Geological Society Annual Field Trip, p. 11–23.

Charleston, S., 1981, A summary of the structural geology and tectonics of the state of Coahuila, Mexico, *in* C. I. Smith, ed., Lower Cretaceous stratigraphy and structure, northern Mexico: West Texas Geological Society Publication 81-74, p. 28–36.

Conklin, J., and and C. M. Moore, 1977, Environmental analysis of the Lower Cretaceous Cupido Formation, northeast Mexico, *in* D. G. Bebout and R. G. Loucks, eds., Cretaceous carbonates of Texas and Mexico: University of Texas, Bureau of Economic Geology Report of Investigations 89, p. 302–323.

Córdoba, D.A., 1969, Mesozoic stratigraphy of northeastern Chihuahua, Mexico in the border region: New Mexico Geological Society 20th Field Conference Guidebook of the Border Region, p. 91–96.

Córdoba, D. A., R. Rodríguez-Torres, and J. Guerrero-García, 1970, Mesozoic stratigraphy of the northern portion of the Chihuahua Trough, *in* K. Seewald and D. Sundeen, eds., The geologic framework of the Chihuahua tectonic belt: West Texas Geological Society, p. 83–98.

Córdoba, D. A., M. Tardy, J. C. Carfantan, M. F. Campa, and C. Rangin, 1980, Le Méxique mésogeen et le passage au système cordillérain de type Californie, *in* J. Aubouin, J. Debelmas, and M. Latreille, eds., Géologie des chaînes alpines issues de la Téthys: Mémoire du Bureau de Recherches Géologiques et Minierès, no. 115, p. 18–29.

Corpstein, P., 1974, The La Joya and La Boca Formations, *in* J. Conklin et al., eds., Geology of Huizachal-Peregrina anticlinorium: Pan American Geological Society Field Trip Guidebook, p. 81–90.

Cuévas-Pérez, E., 1983, Evolución geológica mesozoica del estado de Zacatecas, Mexico: Zentralblatt Geologische Paläontologische Teil I (3/4), p. 190–201.

Cuévas-Pérez, E., M. Serváis, W. Vortisch, and O. Monad, 1985, Una interpretación tectónica de Sinaloa a San Luis Potosí, México: Boletín de la Asociación Mexicana de Geólogos Petroleros, v. 37, p. 31–57.

de Cserna, Z., 1956, Tectónica de la Sierra Madre Oriental de México, entre Torreon y Monterrey: 20.º Congreso Internacional, 87 p.

de Cserna, Z., 1970, Mesozoic sedimentation, magmatic activity, and deformation in northern Mexico, *in* K. Seewald and D. Sundeen, eds., The geologic framework of the Chihuahua tectonic belt: West Texas Geological Society, p. 99–117.

de Cserna, Z., 1979, Cuadro tectónico de la sedimentación y magmatismo en algunas regiones de México durante el Mesozoico: Programas y resumenes del V Simposio sobre la Evolución Tectónica de México: Revista del Instituto de Geología, Universidad Nacional Autónoma de México, p. 11–14.

de Cserna, Z., 1989, An outline of the geology of Mexico, *in* A. W. Bally and A. R. Palmer, eds., The Geology of North America—An overview, v. A: Geological Society of America, p. 233–264.

DeFord, R. K., and W. T. Haenggi, 1970, Stratigraphic nomenclature of Cretaceous rocks in northeastern Chihuahua, *in* K. Seewald and D. Sundeen, eds., The geologic framework of the Chihuahua tectonic belt: West Texas Geological Society, p. 175–196.

Dickinson, W. R., 1981, Plate tectonic evolution of the Southern Cordillera, *in* W. R. Dickinson and W. D. Payne, eds., Relations of tectonics to ore deposits in the southern Cordillera: Arizona Geological Society Digest, v. 14, p. 113–135.

Dickinson, W. R., and P. J. Coney, 1980, Plate-tectonic constraints on the origin of the Gulf of Mexico, in R. H. Pilger, ed, The origin of the Gulf of Mexico and the early opening of the central north Atlantic: Louisiana State University, Baton Rouge, p. 27–36.

Dickinson, W. R., M. A. Klute, and P. N. Swift, 1986, The Bisbee Basin and its bearing on late Mesozoic paleogeographic and paleotectonic relations between the Cordilleran and Caribbean regions, in P. L. Abbott, ed., Cretaceous stratigraphy of western North America: Pacific Section, Society for Sedimentary Geology (SEPM) Book 46, p. 51–62.

Echanove, Oscar, 1986, Geología Petrolera de la Cuenca de Burgos: Coordinación Regional de Exploración, Zona Noreste: Petroleos Mexicanos, v. 38, p. 3–69.

Ekdale, A., S. R. Ekdale, and J. L. Wilson, 1976, Numerical analysis of carbonate microfacies in the Cupido Limestone (Neocomian-Aptian), Coahuila, Mexico: Journal of Sedimentary Petrology, v. 46, p. 362–368.

Enos, Paul, 1974, Reefs, platforms, and basins of middle Cretaceous in northeast Mexico: AAPG Bulletin, v. 58, p. 800–809.

Enos, Paul, 1977, Tamabra Limestone of the Poza Rica trend, Cretaceous, Mexico, in H. E. Cook and P. Enos, eds., Deep-water carbonate environments: Society for Sedimentary Geology (SEPM) Special Publication no. 25, p. 273–314.

Enos, Paul, 1983, Late Mesozoic paleogeography of Mexico, in M. W. Reynolds and E. D. Dolly, eds., Mesozoic paleogeography of west-central United States: Rocky Mountain Section, Society for Sedimentary Geology (SEPM), p. 133–157.

Eguiluz de Antuñano, S., and M. Aranda-García, 1983, Posibilidades económico-petroleras en rocas clásticas del Neocomiano en la margen sur de la Paleoisla de Coahuila: 21.º Congreso Nacional de la Asociación Ingenierios Petroleos Mexicanos, p. 5–13.

Eguiluz de Antuñano, S., and M. Aranda-García 1984, Economic oil possibilities in clastic rocks of the Neocomian along the southern margin of the Coahuila Island, in J. L. Wilson, W. C. Ward, and J. Finneran, eds., A field guide to Upper Jurassic and Lower Cretaceous carbonate platform and basin systems, Monterrey-Saltillo area, northeast Mexico: Gulf Coast Section, Society for Sedimentary Geology (SEPM), p. 43–51.

Finneran, J., 1986, Carbonate petrography and depositional environments of the Upper Jurassic Zuloaga Formation, Sierra de Enfrente, Coahuila, Mexico: Master's thesis, Stephen F. Austin State University, Nacogdoches, Texas, 225 p.

Flawn, P. T., A. Goldstein Jr., P. B. King, and C. E. Weaver, 1961, The Ouachita System: University of Texas, Bureau of Economic Geology Publication 6120, 401 p.

Fortunato, K. S., 1982, Depositional framework of the La Casita Formation (Upper Jurassic–lowermost Cretaceous) near Saltillo, Coahuila, Mexico: Master's thesis, University of New Orleans, 198 p.

Fortunato, K. S., and W. C. Ward, 1982, Upper Jurassic–Lower Cretaceous fan-delta complex, La Casita Formation of the Saltillo area, Coahuila, Mexico: Gulf Coast Association of Geological Societies Transactions, v. 32, p. 473–482.

Gastil, R. G., 1983, Mesozoic and Cenozoic granitic rocks of southern California and western Mexico: Geological Society of America Memoir 159, p. 265–275.

Gastil, R. G., R. H. Miller, and M. F. Campa-Uranga, 1986, The Cretaceous paleogeography of peninsular California and adjacent Mexico, in P. L. Abbott, ed., Cretaceous stratigraphy of western North America: Pacific Section, Society for Sedimentary Geology (SEPM), p. 41–50.

Goldhammer, R. K., 1998a, Second-order accommodation cycles and points of "stratigraphic turnaround": Implications for high-resolution sequence stratigraphy and facies architecture of the Cotton Valley Lime/Haynesville of the East Texas Salt Basin (abs): AAPG 1998 Annual Convention, Salt Lake City, Utah.

Goldhammer, R. K., 1998b, Second-order accommodation cycles and points of "stratigraphic turnaround": Implications for carbonate buildup reservoirs in Mesozoic carbonate systems of the East Texas Salt Basin and south Texas, in W. D. Demis and M. K. Nelis, eds., West Texas Geological Society Publication 98-105, p. 11–27.

Goldhammer, R. K., 1999, Mesozoic sequence stratigraphy and paleogeographic evolution of northeast Mexico, in C. Bartolini, J. L. Wilson, and T. F. Lawton, eds., Mesozoic sedimentary and tectonic history of north-central Mexico: Geological Society of America Special Paper 340, p. 1–58.

Goldhammer R. K., P. J. Lehmann, R. G. Todd, J. L. Wilson, W. C. Ward, and C. R. Johnson, 1991, Sequence stratigraphy and cyclostratigraphy of the Mesozoic of the Sierra Madre Oriental, northeast Mexico, a field guidebook: Gulf Coast Section, Society for Sedimentary Geology (SEPM), 85 p.

González, J. A., 1976, Resultados obtenidos en la exploración de la Platform de Córdoba y principales campos productores: Boletín de la Asociación Mexicana de Geólogos Petroleros, v. 37, no. 2, p. 23–56.

González, J. A., 1977, Estudio geológico económico del Distrito Papaloapan. Memoria 15: Congréso Nacional de la Asociación de Ingenieros Petroloros de Mexico-Tampico, Tamaulipas., 67 p.

González, M. L., 1989, Evaluación geológico-geoquímica de la provincia de Chihuahua: Boletín de la Asociación de Geólogos Petroleros, v. 38, n. 2, p. 3–58 (dated as July-December, 1986).

González-García, R., 1973, Modelo sedimentario del Albiano-Cenomaniano en la porción sureste de la plataforma de Coahuila (Prospecto Parra, Edo. de Coahuila): Boletín de la Asociación Mexicana de Geólogos Petroleros, v. 25, no. 7–9, p. 309–339.

González-García, R., 1976, Bosquejo Geológico de la Zona Noreste: Boletín de la Asociación Mexicana de Geólogos Petroleros, v. 28, no. 1–2, p. 1–50.

González-García, R., 1979, Exploracion petrolera en el "Golfo de Sabinas" nueva provincia productora de gas: Ingenieria Petrolera, p. 28–36.

González-García, R., 1984, Petroleum exploration in the "Gulf of Sabinas"—A new gas province in northern Mexico, in J. L. Wilson, W. C. Ward, and J. Finneran, eds., Upper Jurassic and Lower Cretaceous carbonate platform and basin systems, Monterrey-Saltillo area, northeast Mexico: Gulf Coast Section, Society for Sedimentary Geology (SEPM) Field Guide, p. 64–76.

Grajales-Nishimura, J. M., D. J. Terrell, and P. E. Damon, 1992, Evidencias de la prolongación del arco magmático cordillerano del Triásico Tardío-Jurásico en Chihuahua, Durango y Coahuila: Boletín de la Asociación Mexicana de Geólogos Petroleros, v. 42, p. 1–18.

Gray, G. G., and C. A. Johnson, 1995, Structural and tectonic evolution of the Sierra Madre Oriental, with emphasis on the Saltillo-Monterrey Corridor: A field guidebook, AAPG Annual Convention, Houston, Texas, p. 1–17.

Handschy, J. W., G. R. Keller, and K. J. Smith, 1987, The Ouachita system in northern Mexico: Tectonics, v. 6, p.112–126.

Haq, B. U., J. Hardenbol, and P. R. Vail, 1987, Chronology of fluctuating sea levels since the Triassic: Science, v. 235, p. 1156–1166.

Humphrey, W. E., 1956, Tectonic framework of northeast Mexico: Gulf Coast Association of Geological Societies Transactions, v. 6, p. 25–35.

Ice, R. G., 1981, The Cuesta del Cura Formation in North-Central Mexico, *in* C. I. Smith, ed., Lower Cretaceous stratigraphy and structure, northern Mexico: West Texas Geological Society Publication 81-74, p. 58–74.

Imlay, R. W., 1936, Evolution of the Coahuila Peninsula, Mexico, part IV, geology of the western part of the Sierra de Parras: Geological Society of America Bulletin, v. 47, p. 1091–1152.

Imlay, R. W., 1943, Jurassic formations of the Gulf regions: AAPG Bulletin, v. 27, p. 1407–1533.

Imlay, R. W., 1980, Jurassic paleobiography of the conterminous United States in its continental setting: U.S. Geological Survey Professional Paper 1062, 134 p.

Johnson, C. A., G. G. Gray, and R. K. Goldhammer, 1991, Structure and tectonics of the Sierra Madre Oriental fold-thrust belt near Monterrey, northeastern Mexico (abs.): AAPG Bulletin, v. 75, p. 603.

Johnson, C. R., 1991, Depositional cycles of the Zuloaga (Oxfordian-Kimmeridgian?) Formation, Sierra de Bunuelos, Coahuila, Mexico: Master's thesis, University of New Orleans, 242 p.

Jones, N. W., J. W. McKee, D. B. Marquez, J. Tovar, L. E. Long, and T. S. Laudon, 1984, The Mesozoic La Mula island, Coahuila, Mexico: Geological Society of America Bulletin, v. 95, p. 1226–1241.

Kerans, C., W. M. Fitchen, L. C. Zahm, and K. Kemper, 1995, High-frequency sequence framework of Cretaceous (Albian) carbonate ramp reservoir analog outcrops, Pecos River Canyon, northwestern Gulf of Mexico Basin: Field guidebook for the Carbonate Reservoir Characterization Research Laboratory, the Bureau of Economic Geology, University of Texas at Austin, 67 p.

Laudon, R. C., 1984, Evaporite diapirs in the La Popa Basin, Nuevo Leon, Mexico: Geological Society of America Bulletin, v. 95, p. 1219–1225.

Lawton, T. F., and K. A. Giles, 1997a, El Papalote diapir, La Popa Basin, *in* Structure, stratigraphy and paleontology of Late Cretaceous–early Tertiary Parras–La Popa foreland basin near Monterrey, northeast Mexico: AAPG Field Trip no. 10, 1997 AAPG Annual Convention, Dallas, Texas, p. 55–74.

Lawton, T. F., and K. A. Giles, 1997b, Influence of intermittent salt diapirism and Madrean thrusting on Late Cretaceous–Paleogene sedimentation patterns, La Popa Basin, Mexico (abs.) AAPG Annual Convention, Dallas, Texas.

Lehmann, C. T., 1997, Sequence stratigraphy and platform evolution of Lower Cretaceous Barremian-Albian carbonates of northeast Mexico: Ph.D. dissertation, University of California at Riverside, 261 p.

Lehmann, C., D. A. Osleger, and I. P. Montanez, 1998, Controls on cyclostratigraphy of Lower Cretaceous carbonates and evaporites, Cupido and Coahuila platforms, northeastern Mexico: Journal of Sedimentary Research, v. 68, p. 1109–1130.

Lehmann, C., D. A. Osleger, and I. P. Montanez, 2000, Sequence stratigraphy of Lower Cretaceous (Barremian-Albian) carbonate platforms of northeastern Mexico: Regional and global correlations: Journal of Sedimentary Research, v. 70, p. 373–391.

Limón G., M., 1989, Evaluación geológico-geoquímica de la provincia de Chihuahua: Boletín de la Asociación Mexicana de Geólogos Petroleros, v. 38, p. 3–58.

Longoria, J. F., 1984, Stratigraphic studies in the Jurassic of northeastern Mexico: Evidence for the origin of the Sabinas Basin, *in* W. P. S.

Ventress et al., eds., The Jurassic of the Gulf rim: Gulf Coast Section, Society for Sedimentary Geology (SEPM), Proceedings of the Third Annual Research Conference, p. 171–193.

López-Ramos, E., 1972, Estudio del basámento ígneo y metamórfico de las zonas Norte y Poza Rica (entre Nautla, Veracruz y Jimenez, Tamaulipas): Boletín de la Asociación Mexicana de Geólogos Petroleros, v. 24, p. 266–323.

López-Ramos, E., 1981, Paleogeografía y tectónica del Mesozoico en México: Universidad Nacional Autónoma de México, Revista del Instituto de Geología Revista, v. 5, no. 2, p.158–177.

López-Ramos, E., 1985, Geología de México (3a Edición), Tomo 2: México City, Librerias Conacyt, 454 p.

Loucks, R. G., 1977, Porosity development and distribution in shoal-water carbonate complexes—subsurface Pearsall Formation (Lower Cretaceous), south Texas, *in* D. G. Bebout and R. G. Loucks, eds., Cretaceous carbonates of Texas and Mexico: University of Texas, Bureau of Economic Geology Report of Investigations no. 89, p. 97–126.

Madrid, A., 1976, Consideraciones geologico-economicas del Jurasico Superior, NE de Mexico: Tercero Simposium de Geología de Subsuelo, Zona Noreste, Petroleos Mexicanos, p. 193–216.

Mann, C. J., and W. A. Thomas, 1964, Cotton Valley Group (Jurassic) nomenclature, Louisiana and Arkansas: Gulf Coast Association of Geological Societies Transactions, v. 14, p. 143–152.

Márquez, D. B., 1979, Evaluación petrolera de sedimentos carbonatados del Cretácico en el Golfo de Sabinas, NE de México: Ingenieria Petrolera, v. 19, no. 8, p. 28–37.

Márquez-Castañeda, B., 1984, Estudio geológico del área de Santa Barbara, Chihuahua: Universidad Nacional Autónoma de México, report of the faculty of engineering (cited in Moran-Zenteno, 1994).

Marrett, R., 1995, Structure, kinematics, and development of the Sierra Madre Oriental salient, Mexico (abs.): Geological Society of America Abstract with Programs, A73.

Marton, G., and R. T. Buffler, 1993, The southeastern Gulf of Mexico in the framework of the opening of the Gulf of Mexico Basin, *in* J. L. Pindell and B. F. Perkins, eds., Mesozoic and Early Cenozoic development of the Gulf of Mexico and Caribbean region, a context for hydrocarbon exploration: Gulf Coast Section, Society for Sedimentary Geology (SEPM), 13th Annual Research Conference Proceedings, p. 51–67.

Marton, G., and R. T. Buffler, 1994, Jurassic reconstruction of the Gulf of Mexico Basin: International Geology Review, v. 36, p. 545–586.

McBride, E. F., A. E. Weidie, J. A. Wolleben, and R. C. Laudon, 1974, Stratigraphy and structure of the Parras and La Popa Basins, northeastern Mexico: Geological Society of America Bulletin, v. 84, p. 1603–1622.

McFarlan, E., and L. S. Menes, 1991, Lower Cretaceous, *in* A. Salvador, ed., The Geology of North America, v. J : Geological Society of America, p. 181–204.

McFarlan, E. Jr., and S. W. Stone, 1977, Petroleum exploration potential of Lower Cretaceous sediments, U. S. Gulf Coast: Unpublished propietary Exxon Production Research Company report, 56 p.

McKee, J. W., N. W. Jones, and L. E. Long, 1990, Stratigraphy and provenance of strata along the San Marcos fault, central Coahuila, Mexico: Geological Society of America Bulletin, v. 102, p. 593–614.

Meyer, M. G., and W. C. Ward, 1984, Outer-ramp limestones of the Zuloaga Formation, Astillero Canyon, Zacatecas, Mexico, *in* W. P. S. Ventress et al., eds., The Jurassic of the Gulf rim: Gulf Coast Section, Society for Sedimentary Geology (SEPM), Proceedings of the Third Annual Research Conference, p. 275–282.

Michalzik, D., 1988, Trias bis tiefste Unter-Kreide der Nordostlichen Sierra Madre Oriental, Mexiko—Fazielle Entwicklung eines passiven Kontinental randes: Ph.D. dissertation, Technischen Hochschule Darmstadt, DBR, 247 p.

Michalzik, D., and D. Schumann, 1994, Lithofacies relationships and paleoecology of a Late Jurassic–Early Cretaceous fan delta to shelf depositional system in the Sierra Madre Oriental of northeast Mexico: Sedimentology, v. 41, p. 463–477.

Mitra-Salazar, L. M., 1981, Las imagenes Landsat—una herramienta útil en la interpretación geológico-estructural; en ejemplo en el noreste de México: Revista del Instituto de Geología, Universidad Nacional Autónoma de México, v. 5, p. 37–46.

Mixon, R. B., G. E. Murray, and G. T. Diaz, 1959, Age and correlation of Huizachal Group (Mesozoic), state of Tamaulipas, Mexico: AAPG Bulletin, v. 43, p. 757–771.

Moran-Zenteno, D., 1994, The geology of the Mexican republic: AAPG Studies in Geology no. 39, 160 p.

Murray, G. E., 1959, Introduction and regional geologic summary of field trip area, southeastern Coahuila and western Nuevo Leon: South Texas Geological Society Field Trip Guidebook, p. A1–A4.

Oivanki, S. M., 1974, Paleodepositional environments in the Upper Jurassic Zuloaga Formation (Smackover), northeast Mexico: Gulf Coast Association of Geological Societies Transactions, v. 24, p. 258–278.

Padilla Sánchez, R. J., 1978, Geología y Estratigrafía (Cretático Superior) del límite suroeste del estado de Nuevo León: Revista del Instituto de Geología, Universidad Nacional Autónoma de México, v. 2, no. 1, p. 37–44.

Padilla Sánchez, R. J., 1982, Geologic evolution of the Sierra Madre Oriental between Linares, Concepcion del Oro, Saltillo, and Monterrey, Mexico: Ph.D. dissertation, University of Texas at Austin, 217 p.

Padilla Sánchez, R. J., 1986, Post-Paleozoic tectonics of northeast Mexico and its role in the evolution of the Gulf of Mexico: Geofisica Internacional, v. 25, no. 1, p. 157–206.

Pilger, R. H. Jr., 1981, The opening of the Gulf of Mexico: Implications for the tectonic evolution of the northern Gulf Coast: Gulf Coast Association of Geological Societies Transactions, v. 31, p. 377–381.

Pindell, J. L., 1985, Alleghenian reconstructions and subsequent evolution of the Gulf of Mexico, Bahamas, and proto-Caribbean: Tectonics, v. 4, p. 1–39.

Pindell, J. L., 1993, Regional synopsis of Gulf of Mexico and Caribbean evolution, in J. L. Pindell and B. F. Perkins, eds., Mesozoic and Early Cenozoic development of the Gulf of Mexico and Caribbean region, a context for hydrocarbon exploration: Gulf Coast Section, Society for Sedimentary Geology (SEPM) Foundation, 13th Annual Research Conference Proceedings, p. 251–274.

Pindell, J. L., and S. F. Barrett, 1990, Geological evolution of the Caribbean region: A plate-tectonic perspective, in J. E. Case and G. Dengo, eds., Caribbean Region: Geological Society of America, The geology of North America, v. H., p. 405–432.

Pindell, J. L., and J. F. Dewey, 1982, Permo-Triassic reconstruction of western Pangaea and the evolution of the Gulf of Mexico/Caribbean region: Tectonics, v. 1, p. 179–211.

Pindell, J. L., S. C. Cande, W. C. Pitman III, D. B. Rowley, J. F. Dewey, J. Labrecque, and W. Haxby, 1988, A plate kinematic framework for models of Caribbean evolution: Tectonophysics, v. 155, p. 121–138.

Presley, M. W., ed., 1984, The Jurassic of east Texas: East Texas Geological Society, 303 p.

Prost, G., R. Marrett, M. Aranda-García, S. Eguiluz-Antuñano, J. Gali-

cia, and J. Banda, 1994, Deformation history of the Sierra Madre Oriental, Mexico, and associated hydrocarbon generation-preservation: First joint AAPG/Society of Mexican Petroleum Geologists Research Conference, Abstracts, v. 3.

Quintero-Legoretta, O., and M. Aranda-García, 1985, Relaciones estructurales entre el Anticlinorio de Parras y el Anticlinorio (Sierra Madre Oriental) en la región de Agua Nueva, Coahuila: Revista del Instituto de Geología, Universidad Nacional Autónoma de México, v. 6, p. 21–36.

Rangin, C., 1978, Consideraciones sobre la evolucion geológicas de la parte septentrional del estado de Sonora. Liberto Guía del Primer Simposio sobre la Geología Y Potential Minero del Estado de Sonora, Hermosillo, Sonora: Revista del Instituto de Geología, Universidad Nacional Autónoma de México, v. 6, p. 35–56.

Rangin, C., 1979, Evidence for superimposed subduction and collision processes during Jurassic-Cretaceous time along Baja California continental borderland, in Field guides and papers of Baja California: Geological Society of America Annual Meeting in San Diego, California, p. 37–52.

Rangin, C., and D. A. Córdoba, 1976, Extensión de la cuenca Cretácica Chihuahuense en Sonora sepentrional y sus deformaciones: Memoria del Tercer Congreso Latinoamericano de Geología, México, 14 p.

Roldán-Quintana, J., 1982, Evolución tectónica del estado de Sonora: Revista del Instituto de Geología, Universidad Nacional Autónoma de México, v. 5, no. 2, p. 178–185.

Ross, M. A., 1981, Stratigraphy of the Tamaulipas Limestone, Lower Cretaceous, Mexico, in C. I. Smith, ed., Lower Cretaceous stratigraphy and structure, northern Mexico: West Texas Geological Society Publication 81-74, p. 43–54.

Ross, M. I., and C. R. Scotese, 1988, A hierarchical tectonic model of the Gulf of Mexico and Carribean region: Tectonophysics, v. 155, p. 139–168.

Salvador, A., 1987, Late Triassic–Jurassic paleogeography and origin of Gulf of Mexico Basin: AAPG Bulletin, v. 71, p 419–451.

Salvador, A., ed., 1991a, The Gulf of Mexico Basin: Geological Society of America, The geology of North America, v. J, 568 p.

Salvador, A., 1991b, Triassic-Jurassic, in A. Salvador, ed., The Gulf of Mexico Basin: Geological Society of America, The geology of North America, v. J, p. 131–180.

Salvador, A., 1991c, Origin and development of the Gulf of Mexico basin, in A. Salvador, ed., The Gulf of Mexico Basin: Geological Society of America, The geology of North America, v. J, p. 389–444.

Salvador, A., and J. M. Q. Muñeton, 1991, Stratigraphic correlation chart, Gulf of Mexico basin, in A. Salvador, ed., 1991, The Gulf of Mexico Basin: Geological Society of America, The geology of North America, v. J, 568 p.

Salvador, A., and A. R. Green, 1980, Opening of the Caribbean Tethys, in J. Aubouin et al., coordinators, Geologie de Chaines Alpines Issues de la Tethys—Geology of the Alpine chains born of the Tethys: Bureau Recherche Geologique Minerieres, Memoir no. 115, p. 224–229.

Sandstrom, M., 1982, Stratigraphy and environments of deposition of the Zuloaga Group, Victoria, Tamaulipas, Mexico: Gulf Coast Section, Society for Sedimentary Geology (SEPM), Third Annual Research Conference, Programs and Abstracts, p. 94–97.

Sawyer, D. S., R. T. Buffler, and R. H. Pilger Jr,. 1991, The crust under the Gulf of Mexico basins, in A. Salvador, ed., The Gulf of Mexico Basin: Geological Society of America, The geology of North America, v. J, p. 53–72.

Schlager, W., 1989, Drowning unconformities on carbonate platforms, in P. Crevello, J. F. Sarg, J. F. Read, and J. L. Wilson, eds., Controls on carbonate platform to basin development: Society for Sedimentary Geology (SEPM) Special Publication no. 44, p. 15–25.

Scott, R. W., 1984, Mesozoic biota and depositional systems of the Gulf of Mexico–Caribbean region, in G. E. G. Westerman, ed., Jurassic-Cretaceous biochronology and paleogeography of North America: Geological Association of Canada Special Paper 27, p. 49–64.

Scott, R. W., 1993, Cretaceous carbonate platform, U.S. Gulf Coast, in J. A. T. Simo, R. W. Scott, and J.-P. Masse, eds., Cretaceous Carbonate Platforms: AAPG Memoir 56, p. 97–110.

Scott, R. W., S. H. Frost, and B. L. Shaffer, 1988, Early Cretaceous sea-level curves, Gulf Coast and southeastern Arabia, in S. Wilgus et al., eds., Sea level changes—an integrated approach: Society for Sedimentary Geology (SEPM) Special Publication no. 42, p. 275–284.

Scott, R. W., and E. R. Warzeski, 1993, An Aptian-Albian shelf ramp, Arizona and Sonora, in J. A. T. Simo, R. W. Scott, and J.-P. Masse, eds., Cretaceous Carbonate Platforms: AAPG Memoir 56, p. 71–80.

Sedlock, R. L., Fernando Ortega-Gutierrez, and R. C. Speed, 1993, Tectonostratigraphic terranes and tectonic evolution of Mexico: Geological Society of America Special Paper 278, 153 p.

Seewald, K., and D. Sundeen eds., 1971, The geologic framework of the Chihuahua tectonic belt: West Texas Geological Society, 165 p.

Selvius, D. B., 1982, Lithostratigraphy and algal-foraminiferal biostratigraphy of the Cupido Formation, Lower Cretaceous, in Bustamante Canyon and Potrero García, northeast Mexico: Master's thesis, University of Michigan, Ann Arbor, 68 p.

Serváis, M., E. Cuévas-Pérez, and O. Monod, 1986, Une section de Sinaloa a San Luis Potosí: Nouvelle approche de l'evolution du Mexique nord-occidental: Bulletin of the Geological Society of France, v. 8, no. 2, p. 1033–1047.

Serváis, M., R. Rojo, and D. Colorado-Lievano, 1982, Estudio de las rocas básicas y ultrabásicas de Sinaloa y Guanajuato: Postulación de un paleo-golfo de Baja-California y de una digitación tethysian en México central, México: Geomimet Revista no. 115 (3ª epoca), p. 53–71.

Smith, C. R., 1987, Provenance and depositional environments of the La Casita Formation, Sierra Madre Oriental southwest of Monterrey, northeastern Mexico: Master's thesis, University of New Orleans, 143 p.

Smith, I. C., 1981, Review of the geologic setting, stratigraphy, and facies distribution of the Lower Cretaceous in northern Mexico, in C. I. Smith, ed., Lower Cretaceous stratigraphy and structure, northern Mexico: West Texas Geological Society Publication 81-74, p. 1–27.

Soegaard, K., K. Giles, F. Vega, T. Lawton, 1997, Structure, stratigraphy and paleontology of Late Cretaceous–early Tertiary Parras–La Popa Foreland Basin near Monterrey, Mexico: AAPG Field Trip Guidebook, Field Trip no. 10, AAPG Annual Meeting, Dallas, Texas, 133 p.

Sohl, N. F., R. E. Martinez, P. Salmeron-Urena, and F. Soto-Jaramillo, 1991, Upper Cretaceous, in A. Salvador, ed., The Gulf of Mexico Basin: Geological Society of America, The geology of North America, v. J, p. 205–244.

Steffensen, C. K., 1982, Diagenetic history and the evolution of porosity in the Cotton Valley Limestone, Teague Townsite field, Freestone, Texas: Master's thesis, Texas A & M University, College Station, 134 p.

Stone, S. W., 1975, The Jurassic of the Gulf Coast: Unpublished Exxon Production Research Company report, 57 p.

Suter, M., 1987, Structural traverse across the Sierra Madre Oriental fold and thrust belt in east-central Mexico: Geological Society of America Bulletin, v. 98, p. 249–264.

Tardy, M., 1977, Essai sur la constitution de l'évolution paléogéographique et structurale de la partie septentrionale du Mexique au cours du Mésozoique et du Cénozoique: Bulletin Geological Society of France, v. 19, p. 1297–1308.

Tardy, M., J.-C. Carfantan, and C. Rangin, 1986, Essai de synthèse sur la structure du Méxique: Bulletin Geological Society of France, v. 6, p. 1025–1031.

Tardy, M., J. F. Longoria, J. Martínez-Réyes, L. M. Mitra-S, A. M. Patiño, R. Padilla y S., and R. C. Ramírez, 1975. Observaciones generales sobre la estructura de la Sierra Madre Oriental: La aloctonía del conjunto cadena Alta-Altiplano Central, entre Torreón, Coahuila y San Luis Potosí, México: Revista del Instituto de Geología, Universidad Nacional Autónoma de México, v. 1, p. 1–11.

Tinker, S. W., 1982, Lithostratigraphy and biostratigraphy of the Aptian La Peña Formation, northeast Mexico and South Texas (Part 1), and the depositional setting of the Aptian Pearsall–La Peña formations, Texas subsurface and northeast Mexico: Why is there not another Fairway Field? (Part 2): Master's thesis, University of Michigan, Ann Arbor, 80 p.

Todd, R. G., 1972, Depositional environment of Arenque Field, Tampico, Mexico: Unpublished Exxon Production Research Company report, 21 p.

Todd, R. G., and R. M. Mitchum Jr., 1977, Seismic stratigraphy and global changes of sea level, part 8: Identification of Upper Triassic, Jurassic, and Lower Cretaceous seismic sequences in Gulf of Mexico and offshore west Africa, in C. E. Payton, ed., Seismic stratigraphy—applications to hydrocarbon exploration: AAPG Memoir 26, p. 145–163.

Tóvar Rodríguez, J. C., 1981, Provincias con posibilidades petroleras en el distrito Chihuahua: Boletín de la Asociación Mexicana de Geólogos Petroleros, v. 38, p. 25–52.

Vail, P., R. J. Hardenbol, and R. G. Todd, 1984, Jurassic unconformities, chronostratigraphy, and sea-level changes from seismic stratigraphy and biostratigraphy, in J. S. Schlee, ed., Interregional unconformities and hydrocarbon accumulation: AAPG Memoir 36, p. 129–144.

Vail, P., R. M. Mitchum Jr., and S. Thompson III, 1977, Seismic stratigraphy and global changes of sea level, in C. E. Payton, ed., Seismic stratigraphy: Applications to hydrocarbon exploration: AAPG Memoir 26, p. 83–97.

Vinet, M. J., 1975, Geology of the Sierra Balauartes and Sierra de Pajaros Azules: Ph.D. dissertation, Louisiana State University, Baton Rouge, 124 p.

Viniegra Osario, F., 1981, Great carbonate bank of Yucatán, southern Mexico: Journal of Petroleum Geology, v. 3, p. 247–278.

Viniegra Osario, F., and C. Castillo-Tejero, 1970, Golden Lane Fields, Veracruz, Mexico: AAPG Memoir 14, p. 309–325.

Walper, J. L., and C. L. Rowett, 1972, Plate tectonics and the origin of the Caribbean Sea and the Gulf of Mexico: Gulf Coast Association of Geological Societies Transactions, v. 22, p. 105–116.

Weidie, A. E., and J. D. Martinez, 1970, Evidence for evaporite diapirism in northeastern Mexico: AAPG Bulletin, v. 54, p. 655–661.

Weidie, A. E., and G. E. Murray, 1961, Tectonics of Parras Basin, states of Coahuila and Nuevo Leon, Mexico: Gulf Coast Association of Geological Societies Transactions, v. 2, p. 47–56.

Weidie, A. E., and G. E. Murray, 1967, Geology of the Parras Basin and adjacent areas of northeastern Mexico: AAPG Bulletin, v. 51, p. 678–695.

Weidie, A. E., and J. A. Wolleben, 1969, Upper Jurassic stratigraphic relations near Monterrey, Nuevo Leon, Mexico: AAPG Bulletin, v. 53, p. 2418–2420.

Wilson, J. L., 1969, Microfacies and sedimentary structures in "deeper water" lime mudstones, *in* G. M. Friedman, ed., Depositional environments in carbonate rocks: Society for Sedimentary Geology (SEPM) Special Publication no. 14, p. 4–17.

Wilson, J. L., 1975, Carbonate facies in geologic history: Springer-Verlag, New York, 471 p.

Wilson, J. L., 1981, Lower Cretaceous stratigraphy in the Monterrey-Saltillo area, *in* C. I. Smith, ed., Lower Cretaceous stratigraphy and structure, northern Mexico: West Texas Geological Society Publication 81-74, p. 78–84.

Wilson, J. L., 1990, Basement structural controls on Mesozoic carbonate facies in northeastern Mexico: A review, *in* M. Tucker et al., eds., Carbonate platforms, facies, sequences and evolution: International Association of Sedimentologists Special Publication no. 9, p. 235–255.

Wilson, J. L., and G. Pialli, 1977, A Lower Cretaceous shelf margin in northern Mexico, *in* D. G. Bebout and R. G. Loucks, eds., Cretaceous carbonates of Texas and Mexico: University of Texas Bureau of Economic Geology Report of Investigations no. 89, p. 286–294.

Wilson, J. L., and D. B. Selvius, 1984, Early Cretaceous in the Monterrey-Saltillo area of northern Mexico, *in* J. L. Wilson, W. C. Ward, and J. Finneran, eds., A field guide to Upper Jurassic and Lower Cretaceous carbonate platform and basin systems, Monterrey-Saltillo area, northeast Mexico: Gulf Coast Section, Society for Sedimentary Geology (SEPM), 76 p.

Wilson, J. L., and W. C. Ward, 1993, Early Cretaceous carbonate platforms of northeastern and east-central Mexico, *in* J. A. T. Simo, R. W. Scott, J.-P. Masse, eds., Cretaceous Carbonate platforms: AAPG Memoir 56, p. 35–50.

Wilson, J. L., W. C. Ward, and J. Finneran, eds., 1984, A field guide to Upper Jurassic and Lower Cretaceous carbonate platform and basin systems, Monterrey-Saltillo area, northeast Mexico: Gulf Coast Section, Society for Sedimentary Geology (SEPM), 76 p.

Winker, C. D., and R. T. Buffler, 1988, Paleogeographic evolution of early deep-water Gulf of Mexico and margins, Jurassic to Middle Cretaceous (Comanchean): AAPG Bulletin, v. 72, p. 318–346.

Woods, R. D., A. Salvador, and A. E. Miles, 1991, Pre-Triassic, *in* A. Salvador, ed., The Gulf of Mexico Basin: Geological Society of America, The geology of North America, v. J, p. 109–130.

Wooten, J. W., and W. R. Dunaway, 1977, Lower Cretaceous carbonates of central south Texas: A shelf-marine study, *in* D. G. Bebout and R. G. Loucks, eds., Cretaceous carbonates of Texas and Mexico: University of Texas Bureau of Economic Geology Report of Investigations n. 89, p. 71–78.

Ye, H., 1997, The arcuate Sierra Madre Oriental orogenic belt, NE Mexico: Tectonic infilling of a recess along the southwestern North American continental margin, *in* K. Soegaard, K. Giles, F. Vega, and T. Lawton, eds., Structure, stratigraphy and paleontology of Late Cretaceous–early Tertiary Parras–La Popa Foreland Basin near Monterrey, Mexico: AAPG Field Trip Guidebook, Field Trip no. 10, AAPG Annual Meeting, Dallas, Texas, p. 82–115.

Yurewicz, D. A., T. B. Marler, K. A. Meyerholtz, and F. X. Siroky, 1993, Early Cretaceous carbonate platforms, north rim of the Gulf of Mexico, Mississippi and Louisiana, *in* J. A. T. Simo, R. W. Scott, and J.-P. Masse, eds., Cretaceous Carbonate platforms: AAPG Memoir 56, p. 35–50.

Zahm, L. C., 1997, Depositional model and sequence stratigraphic framework for upper Albian/lower Cenomanian carbonate ramp, western Comanche shelf, Texas: Master's thesis, University of Texas at Austin, 134 p.

Zwanzinger, J. A., 1979, Provincias Mesozoicas Productoras en el Norteste de Mexico: Ingenieria Petrolera, v. 19, no. 3, p. 35–40.

Magoon, L. B., T. L. Hudson, and H. E. Cook, 2001, Pimienta-Tamabra(!)—A giant supercharged petroleum system in the southern Gulf of Mexico, onshore and offshore Mexico, *in* C. Bartolini, R. T. Buffler, and A. Cantú-Chapa, eds., The western Gulf of Mexico Basin: Tectonics, sedimentary basins, and petroleum systems: AAPG Memoir 75, p. 83–125.

4

Pimienta-Tamabra(!)—A Giant Supercharged Petroleum System in the Southern Gulf of Mexico, Onshore and Offshore Mexico

Leslie B. Magoon
U.S. Geological Survey, Menlo Park, California, U.S.A.

Travis L. Hudson
Applied Geology, Sequim, Washington, U.S.A.

Harry E. Cook
U.S. Geological Survey, Menlo Park, California, U.S.A.

ABSTRACT

Pimienta-Tamabra(!) is a giant supercharged petroleum system in the southern Gulf of Mexico with cumulative production and total reserves of 66.3 billion barrels of oil and 103.7 tcf of natural gas, or 83.6 billion barrels of oil equivalent (BOE). The effectiveness of this system results largely from the widespread distribution of good to excellent thermally mature, Upper Jurassic source rock underlying numerous stratigraphic and structural traps that contain excellent carbonate reservoirs. Expulsion of oil and gas as a supercritical fluid from Upper Jurassic source rock occurred when the thickness of overburden rock exceeded 5 km. This burial event started in the Eocene, culminated in the Miocene, and continues to a lesser extent today. The expelled hydrocarbons started migrating laterally and then upward as a gas-saturated 35–40°API oil with less than 1 wt.% sulfur and a gas-to-oil ratio (GOR) of 500–1000 ft^3/BO. The generation-accumulation efficiency is about 6%.

Cretaceous carbonate rocks, especially carbonate slope, base-of-slope, and basinal carbonate debris flow facies such as the Tamabra Limestone of the Tuxpan area, contain considerable oil. A similar but younger reservoir facies contains most of the oil developed in the latest Cretaceous in the Villahermosa and Bay of Campeche areas in response to regional shaking that accompanied impact of the Chicxulub meteor on the Yucatán Peninsula. In the Tuxpan and Salina areas, siliciclastic sandstones are important reservoir rocks. Trapping mecha-

nisms include Upper Jurassic and Cretaceous stratigraphic traps in carbonate reef facies, especially on the western side, and numerous structural traps on the eastern side associated with Neogene movement of pre–Upper Jurassic salt. Here, lateral and vertical oil and gas migration through fractured limestone is important. On the western side, lateral migration is most important. The Pimienta-Tamabra(!) total petroleum system (TPS) is subdivided into seven assessment units which were evaluated for undiscovered petroleum. A total of 23.3 billion barrels of oil with natural gas liquid (NGL) and 49.3 tcf of natural gas are estimated to be undiscovered.

INTRODUCTION

Pimienta-Tamabra(!), the most important petroleum system in the southern Gulf of Mexico and adjacent onshore Mexico, has a cumulative production and remaining reserves of 66.3 billion barrels of known oil and 103.7 tcf of natural gas (PEMEX, 1999a). Offshore, this petroleum system includes the large area from the Bay of Campeche on the eastern side to the south onshore of the Villahermosa and Salina areas and westward beyond the coastline near Veracruz and Tampico on the western side (Figures 1 and 2; Table 1). This petroleum system continues north and east into the northern Gulf of Mexico, but that part of the system is omitted in this study. This giant $(20–100 \times 10^9$ BOE, Klemme, 1994) and highly effective petroleum system owes much of its success to the large traps that overlie the widespread, high-quality, thermally mature, Upper Jurassic source rock. Cretaceous carbonate rocks, especially platform slope facies such as that of the Tamabra Limestone of the Tampico and Tuxpan areas, are important reservoir rocks in these overlying traps. Both structural and stratigraphic traps are present. Movement of pre–Upper Jurassic salt during the Neogene fractured overlying reservoir and source rocks and created many of the structural traps that now contain upwardly migrating oil and gas. The source rock became thermally mature in some areas as early as the Eocene and over large areas in the Miocene, and generation continues today in a few places. The evaluation of the Pimienta-Tamabra(!) for undiscovered oil and gas resources indicates that there is considerable petroleum left to find.

Pimienta-Tamabra(!) was included in the World Petroleum Assessment 2000 effort to evaluate the world for undiscovered oil and gas, exclusive of the United States (USGS, 2000). This evaluation used the total petroleum system (TPS) to determine the area over which petroleum could migrate from a pod of active source rock and the assessment unit (AU) to evaluate the size and numbers of undiscovered accumulations (Magoon and Schmoker, 2000). The TPS map and the outline of the seven AUs and evaluation results are discussed at the end of this paper. For additional information about the world assessment, the reader is referred to the final report of the World Petroleum Assessment 2000 (USGS, 2000). The USGS province outlines in Mexico are omitted here, but are available on a CD-ROM (Klett et al., 1997).

The number and size of fields in the Petroconsultants (1996) database used for World Petroleum Assessment 2000 (USGS, 2000) are confidential. Therefore, the authors compiled the oil-field data into Tables 2 and 3 from the literature (Viniegra O. and Castillo-Tejero, 1970; Enos, 1977; Acevedo, 1980; Peterson, 1983; Busch, 1992; Santiago and Baro, 1992; and PEMEX, 1999a, b). The oil-field maps also were constructed from published information (Figures 2, 3, and 4;

FIGURE 1. Map of the Gulf of Mexico showing the location of the geographic extent of Pimienta-Tamabra(!), selected geographic areas, and the Chicxulub impact crater.

Viniegra O. and Castillo-Tejero, 1970; Enos, 1977; Acevedo, 1980; Peterson, 1983; Santiago and Baro, 1992; and PEMEX, 1999a). For these tables and figures, the numbered fields are those fields reported in PEMEX (1999a, b), and the lettered fields are the abandoned fields in the Tuxpan area omitted from PEMEX (1999a). In Table 2, the "total reserves" column includes the proved, probable, and possible oil and gas volumes in accordance with SPE (Society of Petroleum Engineers) and WPC (World Petroleum Congresses) definitions (PEMEX, 1999a). Table 4 is a summary of hydrocarbons in the Pimienta-Tamabra(!).

The ability to characterize, name, and map Pimienta-Tamabra(!) is possible because of the information that has become available in the literature in the last 25 years. Oil- and gas-field summaries and regional stratigraphic and structural syntheses, combined with many recent reports on the geochemistry of oil and source-rock samples in Mexico, have provided the technical foundation for this paper. With this information, it is possible to characterize the volume and chemistry of the oil accumulated in Pimienta-Tamabra(!); the source rock from which it was generated; the timing of its generation, migration, and accumulation; and the size of the generating cell. The compilation and synthesis of information on the most important aspects of hydrocarbon charge, trap, reservoir, and seal complete this regional analysis of Pimienta-Tamabra(!).

REGIONAL SETTING

Pimienta-Tamabra(!) covers more than 427,220 km² and includes production from the Tampico, Tuxpan, Veracruz, Salina, Villahermosa, and Bay of Campeche areas (Figure 2,

TABLE 1. Geographic names that correspond to the names used in this study.

This paper	Same area	Reference
Tampico area	Ebano-Panuco Fields	Enos, 1977
	Tamaulipas Field	Enos, 1977
	Tampico-Mislanta Basin (5301)	Klett et al., 1997
	Northern Region	PEMEX, 1999a
Tuxpan area	Old Golden Lane	Viniegra O. and Castillo-Tejero, 1970
	New Golden Lane	Viniegra O. and Castillo-Tejero, 1970
	Marine Golden Lane	Viniegra O. and Castillo-Tejero, 1970
	Golden Lane	Viniegra O. and Castillo-Tejero, 1970
	Poza Rica (Tamabra) Fields	Viniegra O. and Castillo-Tejero, 1970
	Golden Lane Atoll	Enos, 1977
	Poza Rica Trend	Enos, 1977
	Tuxpan Platform	McFarlan and Menes, 1991
	Tampico-Mislanta Basin	McFarlan and Menes, 1991
	Tampico-Mislanta	Nehring, 1991
	Chicontepec field and basin	Busch, 1992
	Tampico-Mislanta	Guzmán-Vega and Mello, 1996
	Tampico-Mislanta	Mello and Guzmán-Vega, 1996
	Tampico-Mislanta	Mello et al., 1996
	Tampico-Mislanta Basin (5301)	Klett et al., 1997
	Northern Region	PEMEX, 1999a
Veracruz area	Vera Cruz Basin	Enos, 1977
	Veracruz Basin	Peterson, 1983
	Cordoba Platform	McFarlan and Menes, 1991
	Veracruz Basin	McFarlan and Menes, 1991
	Veracruz	Nehring, 1991
	Vera Cruz	Guzmán-Vega and Mello, 1996
	Vera Cruz	Mello and Guzmán-Vega, 1996
	Veracruz Basin (5302)	Klett et al., 1997
	Northern Region	PEMEX, 1999a
Salina area	Salina del Istmo	Acevedo, 1980
	Saline basin	Acevedo, 1980
	Isthmus Saline Basin	Peterson, 1983
	Salina	Peterson, 1983
	Isthmus Saline	Nehring, 1991
	Salina Basin	Guzmán-Vega and Mello, 1996
	Salina Basin	Mello and Guzmán-Vega, 1996
	Salina Basin	Mello et al., 1996
	Salina-Comalcalco Basin (5304)	Klett et al., 1997
	Southern Region	PEMEX, 1999a
Villahermosa area	Chiapas-Tabasco Mesozoic Province	Acevedo, 1980
	Comalcalco Basin	Peterson, 1983
	Reforma Shelf	Peterson, 1983
	Villahermosa Horst	Peterson, 1983
	Reforma	Nehring, 1991
	Chiapas Tabasco area	Santiago and Baro, 1992
	Chiapas Tabasco	Guzmán-Vega and Mello, 1996
	Chiapas Tabasco	Mello and Guzmán-Vega, 1996
	Chiapas Tabasco	Mello et al., 1996
	Villahermosa Uplift (5305)	Klett et al., 1997
	Southern Region	PEMEX, 1999a
Bay of Campeche	Campeche (Marine) Platform	Acevedo, 1980
	Campeche Shelf	Peterson, 1983
	Bay of Campeche	McFarlan and Menes, 1991
	Campeche	Nehring, 1991
	Bay of Campeche	Salvador, 1991a
	Sonda de Campeche area	Santiago and Baro, 1992
	Campeche Sound	Guzmán-Vega and Mello, 1996
	Campeche Sound	Mello and Guzmán-Vega, 1996
	Campeche Sound	Mello et al., 1996
	Villahermosa Uplift (5305)	Klett et al., 1997
	Northeast Offshore Region	PEMEX, 1999a
	Southwest Offshore Region	PEMEX, 1999a

Table 2). It includes all southern Gulf of Mexico offshore areas south of an arbitrary line between the coastline near Tampico on the western side and the western portion of the present Yucatán shelf on the eastern side. Equivalent stratigraphic elements of this system are expected to continue to the north and east and have important implications for the petroleum potential of the northern Gulf of Mexico. However, because available oil and source-rock data important to defining Pimienta-Tamabra(!) are from onshore and offshore Mexico, this study is focused on the southern Gulf of Mexico (Figure 1). Pimienta-Tamabra(!) is a giant but relatively simple petroleum system. It owes its great extent and continuity to a stable tectonic setting and depositional sequence that began in the Jurassic with deposition of the source rock and continues to this day with the deposition of overburden rock.

Tectonic Setting

A stable, Late Jurassic tectonic setting developed after an extended period of extensional tectonics going back to the Late Triassic (Figure 5). The Late Triassic through Jurassic tectonic evolution of the region involved three general phases: (1) Late Triassic to Middle Jurassic continental rifting, (2) Middle Jurassic to early Late Jurassic opening of the Gulf of Mexico basin, and (3) Late Jurassic to present-day regional subsidence (Salvador, 1991b, c). The Late Triassic to Middle Jurassic rifting developed several large grabens with sedimentary sections varying from Upper Triassic nonmarine red beds to Middle Jurassic salt deposits. The Middle Jurassic salt deposits are very widespread in the northern and southern Gulf of Mexico. The distribution of Middle Jurassic (mainly Callovian) salt suggests widespread continuity of depositional settings in a shallow interior sea with only minor connections westward to the Pacific Ocean (Salvador, 1991b, p. 149). The presence of thicker salt deposits over large parts of both the southern and northern Gulf of Mexico indicates some continued subsidence in this region prior to opening of the Gulf of Mexico basin (Salvador, 1991b, Figure 11, p. 148). In the late Middle Jurassic or early Late Jurassic, when the Yucatán platform was translated southward to near its present position, seafloor spreading is inferred to have developed oceanic crust in what is now the central Gulf of Mexico basin. The resulting configuration of the Gulf of Mexico basin has persisted essentially to this day. Extensive salt deposition ceased in the Late Jurassic upon opening of the Gulf of Mexico basin, perhaps because of greater influx and deepening of marine waters.

Depositional Setting

The stable tectonic setting that developed in the Late Jurassic led to a stable depositional regime that continues to the present. The Late Jurassic interior sea was bordered by continental shelves, ramps, and platforms that experienced a general marine transgression as the central basin gradually subsided. Even though the Upper Jurassic contains a significant amount of shallow-water oolitic limestone deposits, this was the time of major source-rock deposition in the region; Oxfordian, Kimmeridgian, and Tithonian source-rock intervals are known from east-central and southeastern Mexico and are potentially widespread throughout the southern Gulf of Mexico (Figure 6). In general, the Upper

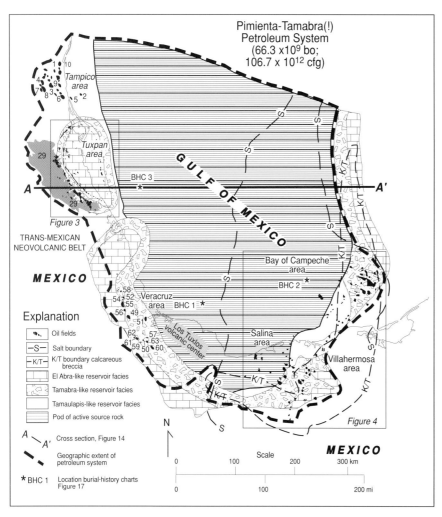

FIGURE 2. Map of Pimienta-Tamabra(!), a giant supercharged petroleum system, showing distribution of oil fields, cross-section location (Figure 14), burial-history chart locations (Figure 16), reservoir rock facies, salt boundary, K/T calcareous breccia boundary, and active source-rock pod (estimated from overburden rock thickness of 5 km or more). The Tuxpan area and the Salina, Villa-hermosa, and Bay of Campeche areas are shown in more detail in Figures 3 and 4, respectively. The northern boundary of Pimienta-Tamabra(!) is arbitrary; the petroleum system continues to the north and east.

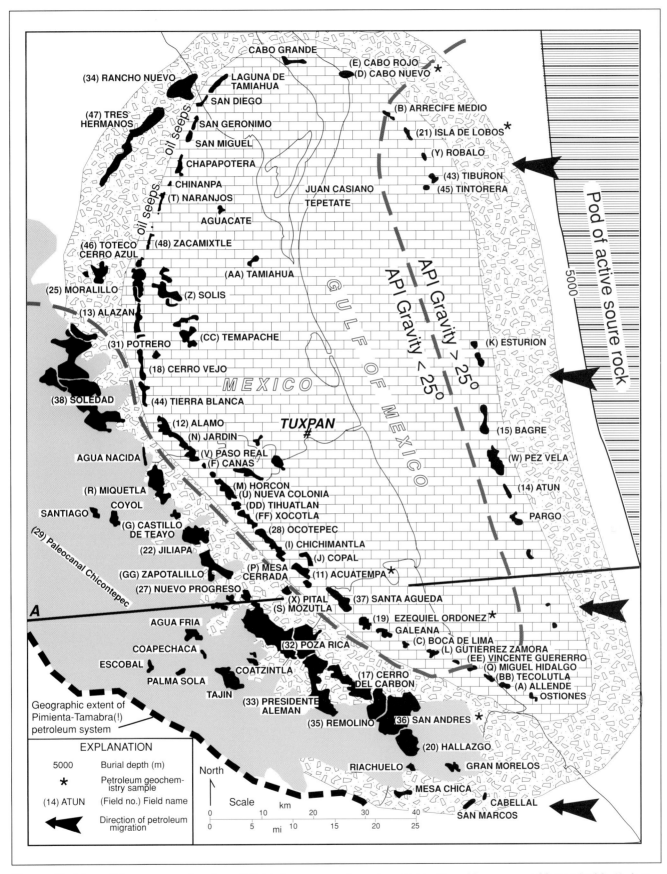

FIGURE 3. Map of Tuxpan area showing oil fields by name and number (Table 2) and by name and letter (Table 3), location of oil seeps, 25° API line, reservoir facies, cross-section location, geographic extent of petroleum system, active source-rock pod, and direction of petroleum migration.

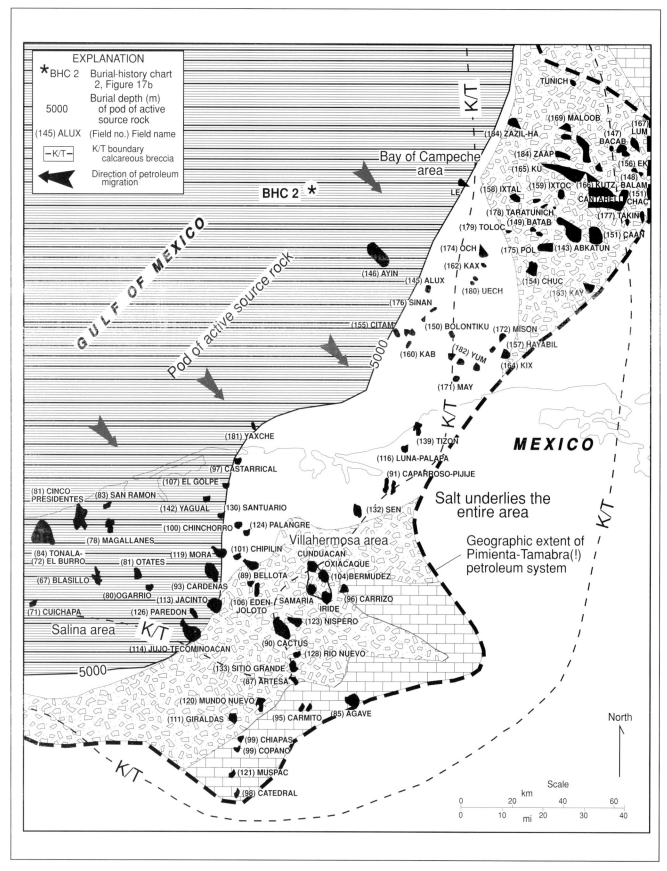

FIGURE 4. Map of the Salina, Villahermosa, and Bay of Campeche areas showing oil fields by name and number (Table 2), reservoir facies, K/T calcareous breccia boundary, salt presence, burial-history chart location (Figure 16b), geographic extent of petroleum system, active source-rock pod, and direction of petroleum migration.

Jurassic marine sediments are calcareous shale that include thin-bedded, dark gray to black limestone, argillaceous limestone, and dark gray shale deposited in various shelf, ramp, and basin settings (Salvador, 1991b). These depositional settings continued into the Early Cretaceous, but by the mid-Cretaceous, the important carbonate buildups of the Tuxpan and the Yucatán platforms were well developed (McFarlan and Menes, 1991; Wilson and Ward, 1993).

Although depositional settings conducive to good to excellent source-rock development continued into the Cretaceous (Patton et al., 1984; Thompson et al., 1990), this time is better known for the reservoir rocks that developed. The principal reservoirs of the southern Gulf of Mexico are Cretaceous carbonate rocks that developed in various platform margin, ramp, and basinal settings (Figure 7). The stability of the Yucatán platform is noteworthy, because platform margin and slope deposits were developed here throughout a large part of the Cretaceous and into the Paleocene (Galloway et al., 1991; McFarlan and Menes, 1991; Sohl et al., 1991; Santiago and Baro, 1992).

Therefore, the 100 million years of tectonic and depositional stability between the Oxfordian and the Paleocene led to development of the source, reservoir, and seal rocks of the Pimienta-Tamabra(!). Cenozoic siliciclastic deposition, which continues today, produced the overburden rock necessary to activate the petroleum system.

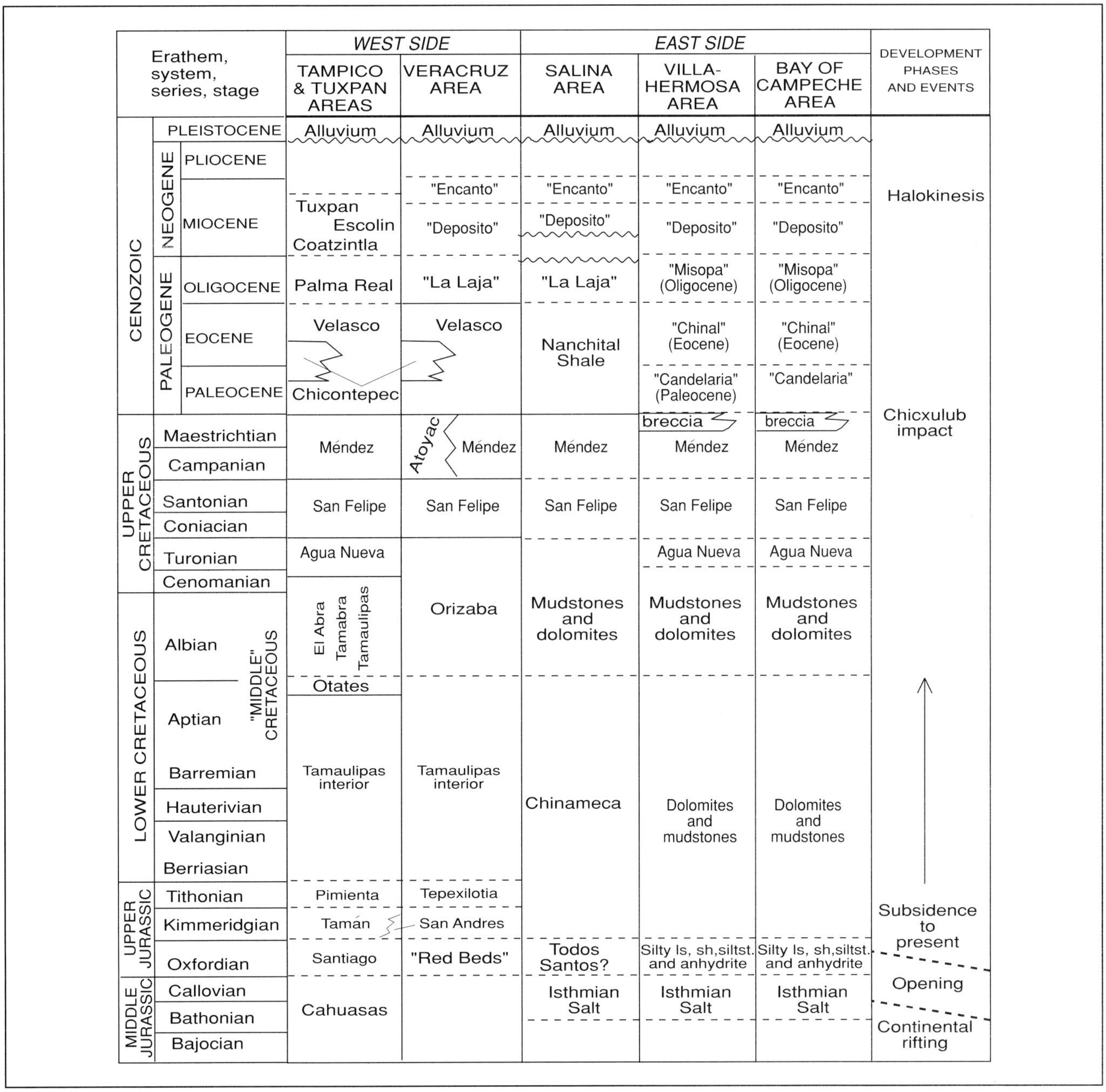

Erathem, system, series, stage		WEST SIDE		EAST SIDE			DEVELOPMENT PHASES AND EVENTS
		TAMPICO & TUXPAN AREAS	VERACRUZ AREA	SALINA AREA	VILLA-HERMOSA AREA	BAY OF CAMPECHE AREA	
CENOZOIC — NEOGENE	PLEISTOCENE	Alluvium	Alluvium	Alluvium	Alluvium	Alluvium	Halokinesis
	PLIOCENE		"Encanto"	"Encanto"	"Encanto"	"Encanto"	
	MIOCENE	Tuxpan Escolin Coatzintla	"Deposito"	"Deposito"	"Deposito"	"Deposito"	
CENOZOIC — PALEOGENE	OLIGOCENE	Palma Real	"La Laja"	"La Laja"	"Misopa" (Oligocene)	"Misopa" (Oligocene)	
	EOCENE	Velasco	Velasco	Nanchital Shale	"Chinal" (Eocene)	"Chinal" (Eocene)	
	PALEOCENE	Chicontepec			"Candelaria" (Paleocene)	"Candelaria"	Chicxulub impact
UPPER CRETACEOUS	Maestrichtian	Méndez	Atoyac / Méndez	Méndez	breccia / Méndez	breccia / Méndez	
	Campanian						
	Santonian	San Felipe	San Felipe	San Felipe	San Felipe	San Felipe	
	Coniacian						
	Turonian	Agua Nueva			Agua Nueva	Agua Nueva	
	Cenomanian	El Abra / Tamabra / Tamaulipas	Orizaba	Mudstones and dolomites	Mudstones and dolomites	Mudstones and dolomites	
LOWER CRETACEOUS ("MIDDLE" CRETACEOUS)	Albian						
	Aptian	Otates					
	Barremian	Tamaulipas interior	Tamaulipas interior	Chinameca	Dolomites and mudstones	Dolomites and mudstones	
	Hauterivian						
	Valanginian						
	Berriasian						
UPPER JURASSIC	Tithonian	Pimienta	Tepexilotia				Subsidence to present
	Kimmeridgian	Tamán / San Andres	San Andres				
	Oxfordian	Santiago	"Red Beds"	Todos Santos?	Silty ls, sh,siltst. and anhydrite	Silty ls, sh,siltst. and anhydrite	Opening
MIDDLE JURASSIC	Callovian	Cahuasas		Isthmian Salt	Isthmian Salt	Isthmian Salt	
	Bathonian						Continental rifting
	Bajocian						

FIGURE 5. Stratigraphic chart (modified from Salvador, 1991a, Plate 5) and development phases of the southern Gulf of Mexico.

TABLE 2. Oil-field information for the Pimienta-Tamabra(!). Information from PEMEX (1999a).

Field number	Field name	Crude oil (106 bbl)	Natural gas (109 cf)	GOR (ft³/bbl)
			In-place petroleum	
Tampico area				
1	Altamira	103.2	159.3	1544
2	Arenque	1190.3	1363.1	1145
3	Cacalilao	3216.5	9746	3030
4	Ebano Chapacao	1891.1	2778.8	1469
5	Jurel	421.4	709.7	1684
6	La Laja	12.1	7.2	595
7	Limon	144.2	222.7	1544
8	Pánuco	3649.4	10,613.8	2908
9	Salinas-Barco-Caracol	170.3	697.5	4096
10	Tamaulipas-Constituciones	2605.9	1351.9	519
NA	Sum of small fields	1020.7	1773.6	1738
Tuxpan area				
11	Acuatempa	101.7	70.7	695
12	Alamo San Isidro	173.5	81	467
13	Alazán	66.3	39.5	596
14	Atún	309.4	578.2	1869
15	Bagre	215.0	189.4	881
16	Carpa	174.8	54	309
17	Cerro del Carbon	49.6	31.6	637
18	Cerro Viejo	89.2	47.6	534
19	Ezequiel Ordóñez	179.1	107.3	599
20	Hallazgo	257.2	164.9	641
21	Isla de Lobos	57.7	15.5	269
22	Jiliapa	232.9	146.9	631
23	Juan Felipe	48.4	27.2	562
24	Marsopa	82.3	43	522
25	Moralillo	64.0	41.7	652
26	Muro	73.9	16.7	226
27	Nuevo Progreso	46.2	24.4	528
28	Ocotepec	81.6	45.7	560
29	Paleocanal Chicontepec	139,163.5	49,231.7	354
30	Placetas	31.5	11.3	359
31	Potrero Llano Horcones	401.1	213	531
32	Poza Rica	4809.7	5502.2	1144
33	Presidente Alemán	23.6	36.4	1542
34	Rancho Nuevo	40.6	14.7	362
35	Remolino	158.2	127.5	806
36	San Andrés	1422.0	968.4	681
37	Santa Agueda	386.5	245.4	635
38	Soledad	36.6	33.5	915
39	Sur Chinampa Norte Amatlán	718.8	422.2	587
40	Sur de Amatlán	451.5	304.2	674
41	Tejada	13.4	7.3	545
42	Tepetate Norte Chinampa	560.6	283.3	505

Total reserves		Cumulative production		Known petroleum		Recovery factor	
ude oil *)6 bbl)*	*Natural gas* *(109 cf)*	*Crude oil* *(106 bbl)*	*Natural gas* *(109 cf)*	*Crude oil* *(106 bbl)*	*Natural gas* *(109 cf)*	*Crude oil* *(%)*	*Natural gas* *(%)*
Tampico area							
2.5	0.1	9.5	17.3	12	17.4	11.6	10.9
123.2	181.5	104.5	236.4	227.7	417.9	11.6	30.7
7.8	1	330.9	972.3	338.7	973.3	10.5	10
10.9	0	187.4	277.7	198.3	277.7	10.5	10
21.3	33.3	0	0	21.3	33.3	5.1	4.7
1.4	0.8	0.1	0.1	1.5	0.9	12.4	12.5
3.6	0.1	14.1	25	17.7	25.1	12.3	11.3
9	8.4	362.8	1056.7	371.8	1065.1	10.2	10
1.1	2.2	17	68.2	18.1	70.4	10.6	10.1
139.2	63.5	214.9	211.4	354.1	274.9	13.6	20.3
2.2	15	121.6	222.3	123.8	237.3	12.1	13.4
Tuxpan area							
2.5	1.2	27.4	37	29.9	38.2	29.4	54
0.7	0.6	71.3	32.2	72	2.8	1.5	40.5
1.2	2	18.7	9.8	19.9	11.8	30	29.9
4.7	1.6	39.7	195.8	44.4	197.4	14.4	34.1
23.8	37.3	52	64.4	75.8	101.7	35.3	53.7
61.2	18.9	0	0	61.2	18.9	35	35
0.6	1.3	6.8	8.9	7.4	10.2	14.9	32.3
1.2	1.3	25.6	13	26.8	14.3	30	30
7.6	3.2	63.4	41.1	71	44.3	39.6	41.3
0.7	1.4	88.4	84.1	89.1	85.5	34.6	51.8
1.3	0.5	22.5	5.8	23.8	6.3	41.2	40.6
7.6	7.3	33.3	41.8	40.9	49.1	17.6	33.4
2.3	1.5	12.2	6.6	14.5	8.1	30	29.8
3.7	1	17.1	5.8	20.8	6.8	25.3	15.8
1.2	1	18	11.5	19.2	12.5	30	30
1	0.2	21.6	7.2	22.6	7.4	30.6	44.3
1.8	3.8	8.6	10.8	10.4	14.6	22.5	59.8
4.8	2.1	18.9	7.8	23.7	9.9	29	21.7
2,324.6	36,061.6	104.4	182	12,429	36,243.6	8.9	73.6
3.1	1.7	0.1	0	3.2	1.7	10.2	15
2.5	1.1	118.2	63.5	120.7	64.6	30.1	30.3
715.2	1864.6	1354.3	1853.6	2069.5	3718.2	43	67.6
0.5	0.4	1.5	2.7	2	3.1	8.5	8.5
3.1	0.8	10.8	3.7	13.9	4.5	34.2	30.6
2.4	3.2	26.7	36.7	29.1	39.9	18.4	31.3
104.2	90.3	378.9	319.2	483.1	409.5	34	42.3
7.9	4.3	115	119.9	122.9	124.2	31.8	50.6
0.6	0.5	10.6	11.1	11.2	11.6	30.6	34.6
4.5	2	211.1	123.8	215.6	125.8	30	29.8
6	2.7	129.9	69.7	135.9	72.4	30.1	23.8
0.6	0.5	3.2	3.4	3.8	3.9	28.4	53.4
2.2	1.4	166	83.6	168.2	85	30	30

TABLE 2. Oil-field information for the Pimienta-Tamabra(!). Information from PEMEX (1999a) (con

| Field number | Field name | In-place petroleum | | |
		Crude oil (106 bbl)	Natural gas (109 cf)	GOR (ft³/bbl)
	Tuxpan area (cont'd.)			
43	Tiburón	23.1	8.1	351
44	Tierra Blanca Chapopote Núñez	352.5	197.1	559
45	Tintorera	82.6	18.6	225
46	Toteco Cerro Azul	1256.9	705.7	561
47	Tres Hermanos	423.2	747.2	1766
48	Zacamixtle	61.4	32.8	534
NA	Sum of small fields	497.8	334.3	672
	Veracruz area			
49	Angostura	117.3	19.8	169
50*	Coapa	0.0	7.9	NA
51	Cocuite	0.0	311.7	NA
52	Cópite	27.3	440.8	16,147
53*	Macuile	0.0	6.3	NA
54	Manuel Rodriguez Aguilar	15.5	2.6	168
55	Mata Pionche	250.2	671.7	2685
56	Mecayucan	53.3	414.2	7771
57	Mirador	0.0	204.3	NA
58	Miralejos	20.1	113.9	5667
59	Nopaltepec	0.0	50	NA
60	Novillero	0.0	189.1	NA
61	Rincón Pacheco	0.0	100.5	NA
62	San Pablo	0.0	95.7	NA
63	Veinte	0.0	72.6	NA
NA	Sum of small fields	7.8	1.3	167
	Salina area			
64*	Agata	32.0	53.3	1666
65*	Arroyo Prieto	247.7	214.7	867
66*	Bacal	230.2	291.9	1268
67	Blasillo	299.1	341.7	1142
68*	Cerro Nanchital	62.9	7.4	118
69	Cinco Presidentes	971.6	935.2	963
70	Concepción	43.7	71.8	1643
71	Cuichapa-Poniente	452.6	709.9	1568
72	El Burro	108.9	21.4	197
73*	El Plan	391.4	75.9	194
74*	Ixhuatlán Oriente	66.3	64.5	973
75*	La Central	6.5	19.1	2938
76*	Lacamango	107.2	147.8	1379
77*	Los Soldados	133.2	306.7	2303
78	Magallanes-Tucán-Pajonal	1277.8	1301.8	1019
79*	Moloacán	221.7	121.9	550
80	Ogarrio	938.6	1080.8	1152
81	Otates	213.8	485.9	2273
82*	Rodador	104.2	169.6	1628

Total reserves		Cumulative production		Known petroleum		Recovery factor	
Crude oil (106 bbl)	Natural gas (109 cf)	Crude oil (106 bbl)	Natural gas (109 cf)	Crude oil (106 bbl)	Natural gas (109 cf)	Crude oil (%)	Natural gas (%)
Tuxpan area (cont'd.)							
1.9	2.1	8.6	2.1	10.5	4.2	45.5	51.9
3.1	1.8	102.6	57.4	105.7	59.2	30	30
24.7	4.6	0.1	0	24.8	4.6	30	24.7
9.4	7.9	369.1	205.2	378.5	213.1	30.1	30.2
24.1	136.2	125.7	239	149.8	375.2	35.4	50.2
0.7	0.4	17.8	9.4	18.5	9.8	30.1	29.9
6.1	24	105	98.7	111.1	122.7	22.3	36.7
Veracruz area							
5.1	1.1	27.3	4.1	32.4	5.2	27.6	26.3
0	6.6	0	0.4	0	7	NA	88.6
0	217.1	0	49.5	0	266.6	NA	85.5
1.9	136.9	3.3	271.6	5.2	408.5	19	92.7
0	5.5	0	0	0	5.5	NA	87.3
1.4	0.2	0.2	0	1.6	0.2	10.3	7.7
1.1	214.9	33.6	113.4	34.7	328.3	13.9	48.9
1	178.4	4	89.9	5	268.3	9.4	64.8
0	180.8	0	2.6	0	183.4	NA	89.8
0	29.1	3	68.9	3	98	14.9	86
0	21.5	0	17.4	0	38.9	NA	77.8
0	24.8	0	67.3	0	92.1	NA	48.7
0	17.7	0	71.8	0	89.5	NA	89.1
0	40.4	0	43.9	0	84.3	NA	88.1
0	48.1	0	16.7	0	64.8	NA	89.3
0.2	0.1	1	0.4	.2	0.5	15.4	8.5
Salina area							
0.3	1.7	14.1	22.1	14.4	23.8	45	44.7
32.4	52	1.2	1.1	33.6	53.1	13.6	24.7
13.2	12.4	102.7	140.6	115.9	153	50.3	2.4
26.3	55.5	48.5	88.2	74.8	143.7	25	42.1
22.2	3.2	7	1.8	29.2	5	6.4	67.6
40.8	41.3	284.6	401.3	325.4	442.6	33.5	47.3
2	3.2	6.8	10.4	8.8	13.6	20.1	18.9
3.4	25.7	155.5	305.1	158.9	330.8	35.1	46.6
1.6	0.3	35.7	7.8	37.3	8.1	34.3	37.9
1	0.6	161.6	31.5	162.6	32.1	41.5	42.3
0.7	0.7	14.5	14.1	15.2	14.8	22.9	22.9
1.6	8.2	0.2	9	1.8	17.2	27.7	90.1
6.9	8.5	19.8	37.8	26.7	46.3	24.9	31.3
5	12	30.6	75.7	35.6	87.7	26.7	28.6
264.7	260.8	168.2	188.3	432.9	449.1	33.9	34.5
2.9	1.8	32.4	17.8	35.3	19.6	15.9	16.1
114.3	130	169.2	264.4	283.5	394.4	30.2	36.5
2.8	2.6	31.6	71.5	34.4	74.1	16.1	15.3
11.3	5.6	17.4	27.5	28.7	33.1	27.5	19.5

TABLE 2. Oil-field information for the Pimienta-Tamabra(!). Information from PEMEX (1999a) (cont'd.

Field number	Field name	In-place petroleum		
		Crude oil (106 bbl)	Natural gas (109 cf)	GOR (ft³/bbl)
	Salina area (cont'd.)			
83	San Ramón	457.0	205.3	449
84	Tonalá–El Burro	167.1	76.1	455
NA	Sum of small fields	535.3	466.1	871
	Villahermosa area			
85	Agave	318.5	2927.1	9190
86*	Arroyo Zanapa	58.4	162.9	2789
87	Artesa	199.7	264.9	1326
88*	Ayapa	27.6	13.3	482
89	Bellota	628.0	1026.1	1634
90	Cactus	2068.6	4193.9	2027
91	Caparroso-Pijije-Escuintle	716.5	1937.6	2704
92*	Caracol	31.3	34.4	1099
93	Cárdenas	1576.4	2931.8	1860
94*	Cardo	6.4	17.9	2797
95	Carmito	154.8	1916.5	12,380
96	Carrizo	145.0	93.4	644
97	Castarrical	181.2	96	530
98	Catedral	154.0	873.4	5671
99	Chiapas-Copanó	236.4	1613.2	6824
100	Chinchorro	421.1	411.4	977
101	Chipilín	80.0	122	1525
102*	Chirimoyo	18.9	188.9	9995
103*	Comoapa	171.8	215.3	1253
104	Complejo Bermudez	11,015.5	14,290.8	1297
105*	Crisol	45.1	41.9	929
106	Edén-Jolote	514.6	886.2	1722
107	El Golpe	607.6	402.2	662
108*	Escarbado	126.9	317	2498
109*	Fénix	103.8	346	3333
110*	Gaucho	236.9	256.6	1083
111	Giraldas	464.7	2744.2	5905
112*	Iris	27.0	203.7	7544
113	Jacinto	187.3	650.5	3473
114	Jujo-Tecominoacán	4897.0	6003.5	1226
115*	Juspi	29.3	96.2	3283
116	Luna-Palapa	259.5	1432.7	5521
117*	Manea	4.8	12.9	2688
118*	Mayacaste	38.5	15.1	392
119	Mora	480.0	741.5	1545
120	Mundo Nuevo	73.8	490.8	6650
121	Múspac	363.1	3120.9	8595
122*	Nicapa	25.1	40.6	1618
123	Nispero	661.8	1043.2	1576
124	Palangre	310.6	357.5	1151

Total reserves		Cumulative production		Known petroleum		Recovery factor	
Crude oil (106 bbl)	*Natural gas* (109 cf)	*Crude oil* (106 bbl)	*Natural gas* (109 cf)	*Crude oil* (106 bbl)	*Natural gas* (109 cf)	*Crude oil* (%)	*Natural gas* (%)
	**Salina area (cont'd.)**
63.6	29.8	75.8	32.8	139.4	62.6	30.5	30.5
1	0.8	72.2	34.3	73.2	35.1	43.8	46.1
0.5	1.5	146.6	127.1	147.1	128.6	27.5	27.6
	**Villahermosa area**
33.1	480.9	123.6	1136.6	156.7	1617.5	49.2	55.3
11	36.8	14.5	34.4	25.5	71.2	43.7	43.7
40.6	9.7	35.6	51.9	76.2	61.6	38.2	23.3
0.9	0.2	6.4	3.4	7.3	3.6	26.4	27.1
67.7	450.1	142.6	265.8	210.3	715.9	33.5	69.8
107.6	412.2	296.7	557.7	404.3	969.9	19.5	23.1
234.3	629.5	68.6	197.2	302.9	826.7	42.3	42.7
7	3.1	4	5.5	11	8.6	35.1	25
246.9	545.5	373.1	737.1	620	1282.6	39.3	43.7
2.3	6.1	0.2	0.6	2.5	6.7	39.1	37.4
32.9	1216.2	15.3	176	48.2	1392.2	31.1	72.6
0.9	0.7	14.5	9.3	15.4	10	10.6	0.7
3.7	3.5	41.9	41.5	45.6	45	25.2	46.9
19.1	539.5	8.6	148.3	27.7	687.8	18	78.7
9.9	149.7	96.3	946.8	106.2	1096.5	44.9	68
97.7	182.6	26.5	22.2	124.2	204.8	29.5	49.8
25.6	38.8	14.3	21.7	39.9	60.5	49.9	49.6
3.8	54.4	2.2	41.2	6	95.6	31.7	50.6
12.1	36.5	32.9	72.5	45	109	26.2	50.6
2110.5	3846.1	2225	2894	4335.5	6740.1	39.4	47.2
1.8	5.3	0	0	1.8	5.3	4	12.6
79.9	198.6	125.5	212.2	205.4	410.8	39.9	46.4
13.1	8.1	88.4	98.9	101.5	107	16.7	26.6
32.9	101.6	7.7	18.9	40.6	120.5	32	38
18.5	58	54.1	204.9	72.6	262.9	69.9	76
92.5	104.5	0	0	92.5	104.5	39	40.7
22.5	813.3	164.8	1520.6	187.3	2333.9	40.3	85
2.1	24.9	17.5	132.2	19.6	157.1	72.6	77.1
55.8	280.1	46.7	160.3	102.5	440.4	54.7	67.7
1001.1	2390.4	810.8	1049.6	1811.9	3440	37	57.3
4.6	40.7	3.2	28.2	7.8	68.9	26.6	71.6
72.9	418.6	84.5	456.9	157.4	875.5	60.7	61.1
1.8	4.8	0	0.1	1.8	4.9	37.5	38
8.5	3.3	0	0	8.5	3.3	22.1	21.9
93.4	381.9	88	137.2	181.4	519.1	37.8	70
0.6	6.1	31.2	279	31.8	285.1	43.1	58.1
48.4	1330.7	78.4	1037.6	126.8	2368.3	34.9	75.9
7.2	14.4	0.3	0.5	7.5	14.9	29.9	36.7
56.5	88.5	134	210.3	190.5	298.8	28.8	28.6
71.8	172	6.1	6.9	77.9	178.9	25.1	50

TABLE 2. Oil-field information for the Pimienta-Tamabra(!). Information from PEMEX (1999a) (cont'd.

Field number	Field name	In-place petroleum		
		Crude oil (106 bbl)	Natural gas (109 cf)	GOR (ft³/bbl)
Villahermosa area (cont'd.)				
125*	Paraíso	28.4	14.2	500
126	Paredón	600.0	1542.6	2571
127*	Puerto Ceiba	218.4	139	636
128	Rio Nuevo	445.2	908.5	2041
129*	Sabancuy	5.2	2.8	538
130	Santuario	240.9	139.9	581
131*	Secadero	12.9	11	853
132	Sen	992.2	3284	3310
133	Sitio Grande	1152.6	2271.6	1971
134*	Sunuapa	88.7	179.7	2026
135*	Tapijulapa	4.5	27.9	6200
136*	Tepeyil	24.6	59.3	2411
137*	Tierra Colorada	39.9	36.5	915
138*	Tintal	71.7	27.3	381
139	Tizón	17.1	109	6374
140*	Topén	40.4	72.4	1792
141*	Tupilco	176.7	123.6	699
142	Yagual	335.2	270.8	808
NA	Sum of small fields	164.8	252.1	1497
Bay of Campeche area				
143	Abkatún	5514.0	4025.2	730
144*	Akal (Cantarell)	32,086.6	15,490.3	483
145	Alux	97.1	32.2	332
146	Ayín	911.6	214.3	235
147	Bacab	325.5	96.1	295
148	Balam	1356.5	287.6	212
149	Batab	184.7	170	920
150	Bolontikú	277.9	577	2076
151	Caan	1512.8	2701	1785
152	Chac (Cantarell)	529.6	255.7	483
153*	Ché	53.0	595.6	11,238
154	Chuc	1961.0	2113.5	1078
155	Citam	509.3	299.4	588
156	Ek	977.0	164.3	168
157	Hayabil	3.2	25.5	7969
158	Ixtal	514.9	607.5	1180
159	Ixtoc	143.4	163.3	1139
160	Kab	52.1	108.5	2083
161*	Kanaab	118.7	99.7	840
162	Kax	117.4	264	2249
163	Kay	2.4	13.4	5583
164	Kix	196.1	599.4	3057
165	Ku	4096.1	2092.7	511
166	Kutz (Cantarell)	695.5	335.8	483

Total reserves		Cumulative production		Known petroleum		Recovery factor	
Crude oil (106 bbl)	*Natural gas (109 cf)*	*Crude oil (106 bbl)*	*Natural gas (109 cf)*	*Crude oil (106 bbl)*	*Natural gas (109 cf)*	*Crude oil (%)*	*Natural gas (%)*
Villahermosa area (cont'd.)							
6.5	3.3	0.1	0.1	6.6	3.4	23.2	23.9
61.3	710.4	167	450.7	228.3	1161.1	38.1	75.3
55.5	32.6	10.7	5.8	66.2	38.4	30.3	27.6
31.4	66.7	79.1	160.2	110.5	226.9	24.8	25
0.4	0.6	0.5	0.3	0.9	0.9	17.3	32.1
17.1	8.5	46.5	36.4	63.6	44.9	26.4	32.1
1.4	2.4	1.2	1.8	2.6	4.2	20.2	38.2
273.3	931.2	117	316.7	390.3	1247.9	39.3	38
90.1	173.7	351.1	531.4	441.2	705.1	38.3	31
0.4	0.3	24.8	63.4	25.2	63.7	28.4	35.4
1.5	18.1	0.3	4.1	1.8	22.2	40	79.6
1.2	4.6	1.9	6.1	3.1	10.7	12.6	18
1	1.4	0	0	1	1.4	2.5	3.8
1.2	0.7	6	2.9	7.2	3.6	10	13.2
7.3	48.6	2.8	17.2	10.1	65.8	59.1	60.4
1.3	4.5	5.5	11.4	6.8	15.9	16.8	22
2.2	2.4	54.7	38.4	56.9	40.8	32.2	33
60.7	57.3	11.8	13.6	72.5	70.9	.6	26.2
0.2	0	42.0	40.7	42.2	40.7	25.6	16.1
Bay of Campeche area							
412.8	377	1967.8	1601.4	2380.6	1978.4	43.2	49.2
2,179.9	5674.2	6345.3	2690.6	18,525.2	8364.8	57.7	54
32.1	10.6	0	0	32.1	10.6	33.1	32.9
334.6	78.6	0	0	334.6	78.6	36.7	6.7
39.9	10.8	25.2	10.3	65.1	21.1	20	22
150.6	31	52.5	13.3	203.1	44.3	15	15.4
28.5	17.1	29.4	39.4	57.9	56.5	31.3	33.2
202.8	490.4	0	0	202.8	490.4	73	85
412.8	994.9	391.4	686.2	804.2	1681.1	53.2	62.2
123.2	65.4	115.1	49.7	238.3	115.1	45	45
12.2	416.9	0	0	12.2	416.9	23	70
356.7	500.1	452.9	489.4	809.6	989.5	41.3	46.8
136.1	72.9	0	0	136.1	72.9	26.7	24.3
104	15.8	24.8	5.9	128.8	21.7	13.2	13.2
1.1	17.9	0	0	1.1	17.9	34.4	70.2
200.8	236.6	0	0	200.8	236.6	39	38.9
34	44.3	36.3	35.7	70.3	80	49	49
26	73.8	0	0	26	73.8	49.9	68
18.8	51.6	3.6	2.3	22.4	53.9	18.9	54.1
25.5	59.2	14.4	30.6	39.9	89.8	34	34
1.1	6	0	0	1.1	6	45.8	44.8
66.5	287.3	0	0	66.5	287.3	33.9	47.9
1308.8	801	1149.9	612	2458.7	1413	60	67.5
295.6	189.9	0	0	295.6	189.9	42.5	56.6

TABLE 2. Oil-field information for the Pimienta-Tamabra(!). Information from PEMEX (1999a) (cont'd

Field number	Field name	In-place petroleum		
		Crude oil (106 bbl)	Natural gas (109 cf)	GOR (ft³/bbl)
Bay of Campeche area (cont'd.)				
167	Lum	649.4	87.3	134
168*	Makech	137.3	46.1	336
169	Maloob	4178.7	1458.3	349
170*	Manik	205.0	128	624
171	May	185.9	1419.3	7635
172	Misón	68.7	209.9	3055
173*	Nohoch (Cantarell)	2011.2	929.2	462
174	Och	111.4	246.5	21,623
175	Pol	2253.0	2445.4	1085
176	Sinán	587.3	711.7	1212
177	Takín	35	2.6	74
178	Taratunich	802.4	839.8	1047
179	Toloc	57.5	17.4	303
180	Uech	176.2	410.9	2332
181	Yaxché	141.2	55.5	393
182	Yum	59.7	196	3283
183	Zaap	4810.5	1679.8	349
184	Zazil-Ha	219.8	88	400

*Field locations not shown in any figures.

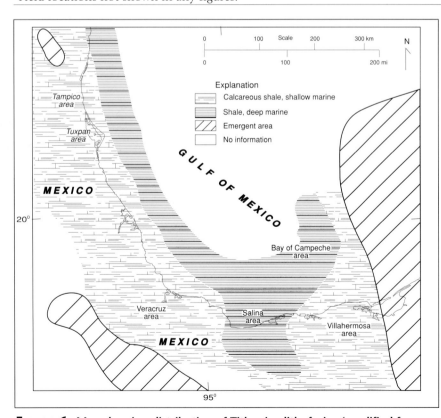

FIGURE 6. Map showing distribution of Tithonian lithofacies (modified from Salvador, 1991b, Figure 20).

Exploration History

Tar paint on pre-Hispanic artifacts indicates that Native Americans used asphalt from oil seeps in the region. Since at least the mid-19th century, written records indicate that surface seeps were known to occur onshore in the Tampico, Tuxpan, and Veracruz areas; most recently, some have been noted in the Bay of Campeche (Viniegra O. and Castillo-Tejero, 1970; Santiago and Baro, 1992; Hernandez, 1994). A surface seep was sampled as early as 1863, but the first significant oil discovery was in 1904 when the La Pez No. 1 flowed 1,500 barrels of oil per day (BOPD) from the Ebano-Panuco field in the Tampico area (Field no. 4, Tables 2 and 3). Exploration drilling in the early 1900s targeted surface seeps, and commercial discoveries were made in the Veracruz area and in the "old" Golden Lane in the Tuxpan area (Viniegra O. and Castillo-Tejero, 1970; Acevedo, 1980). Exceptional productivity marked the initial Golden Lane discoveries; several wells blew out and

Total reserves		Cumulative production		Known petroleum		Recovery factor	
Crude oil (106 bbl)	Natural gas (109 cf)	Crude oil (106 bbl)	Natural gas (109 cf)	Crude oil (106 bbl)	Natural gas (109 cf)	Crude oil (%)	Natural gas (%)
Bay of Campeche area (cont'd.)							
159.1	21.4	0	0	159.1	21.4	24.5	24.5
27.5	9.2	0	0	27.5	9.2	20	20
1637.2	567.1	128.8	50.7	1766	617.8	42.3	42.4
49.4	31.3	0	0	49.4	31.3	24.1	24.5
95.1	709.6	0	0	95.1	709.6	51.2	50
24.1	94.5	0	0	24.1	94.5	35.1	45
290.3	157.3	474	214.5	764.3	371.8	38	40
15.7	38.5	31.1	66	46.8	104.5	42	42.4
282	291.3	788.8	723	1070.8	1014.3	47.5	41.5
319	491	0	0	319	491	54.3	69
15.9	1.2	0	0	15.9	1.2	45.4	46.2
145.9	172.1	120	130	265.9	302.1	33.1	36
23	7	0	0	23	7	40	40.2
47.5	119.7	34.4	81.8	81.9	201.5	46.5	49
56.5	22.2	0	0	56.5	22.2	40	40
20	88.9	1.3	4.9	21.3	93.8	35.7	47.9
1515.6	707.2	24.9	17.4	1540.5	724.6	32	43.1
63.5	25.4	0	0	63.5	25.4	28.9	28.9

were uncontrolled for as much as a few months. Initial flow rates were estimated to be as high as 260,000 BOPD from the Cerro Azul No. 4 which had produced 87 million barrels by 1969 (Viniegra O. and Castillo-Tejero, 1970). By 1919, salt-water invasion had affected many of these highly prolific fields. Continuing exploration using gravity methods (1920) led to oil discoveries around the salt domes in the Saline Basin and at a basement high at Poza Rica (1930, Field no. 32, Tables 2 and 3), which produces oil from the Tamabra Limestone (Salas, 1949; Viniegra O. and Castillo-Tejero, 1970; Carrasco, 1977; Enos, 1977).

The oil industry in Mexico was nationalized in 1938. At that time, Petroleos Méxicanos (PEMEX), Mexico's national oil company, was formed. PEMEX applied reflection seismic surveys (1948) in the "new," or southern, Golden Lane along the southwestern side of the Cretaceous Tuxpan platform. By 1957, an offshore seismic survey had been conducted in the Tuxpan area. This survey led to the drilling of the Isla de Lobos No. 1-B well in 1963, which discovered oil in the "Marine" Golden Lane (Field no. 21, Tables 2 and 3). Further exploration in the Villahermosa area discovered the onshore Cactus and Sitio Grande fields in Cretaceous carbonates in 1972. By 1972, 225,900 BOPD were being produced from Miocene sandstone reservoirs in the Salina area (Santiago and Baro, 1992). Marine seismic surveys extended exploration offshore, and the Chac

No. 1 well discovered oil in the Bay of Campeche in 1974. Since then, exploration of the Bay of Campeche area has discovered 10 giant fields that contain many billion barrels of oil and gas, making it Mexico's premier producing area (Santiago and Baro, 1992).

PETROLEUM

Oil-oil Correlation

Many studies have characterized the oil chemistry from the Tampico, Tuxpan, Salina, Villahermosa, and Bay of Campeche areas (e.g., Bertrand et al., 1994; Bertrand et al., 1996; Medrano et al., 1996a; Mello and Guzmán-Vega, 1996; Mello et al., 1996; Román-Ramos et al., 1996; and Serano-Bello et al., 1996). A recent compilation and synthesis of data for oil samples from the Salina, Villahermosa, and Bay of Campeche areas by Guzmán-Vega and Mello (1999) are especially helpful in defining and characterizing four oil families in the southern Gulf of Mexico and Mexico. They found that one oil family (family 2) from a Tithonian source rock makes up the great majority of the petroleum in these areas. A second, closely related oil family (family 1) from Oxfordian source rock contributes thus far a minor amount of petroleum found. Families 3 and 4 are from Early Cretaceous and Tertiary source rocks, respectively, and are therefore excluded from the Pimienta-Tamabra(!). Family 2 oil is compositionally similar to oil samples from the Tampico

Table 3. Oil-field information for the Pimienta-Tamabra(!). Information from Viniegra and Castillo-Tejero (1970), Enos (1977), Acevedo (1980), Santiago and Baro (1992), and PEMEX (1999a, b).

Field number	Field name	Discovery year	Prod. depth (m)	API gravity (o)	Known oil (10⁶ bbl)	Known gas (10⁹ cfg)	GOR (ft³/bbl)	Column (m)	H2S (%)

$API\ gravity\ (o)$, $Known\ oil\ (10^6\ bbl)$, $Known\ gas\ (10^9\ cfg)$, $GOR\ (ft^3/bbl)$

Field number	Field name	Discovery year	Prod. depth (m)	API gravity (o)	Known oil (10⁶ bbl)	Known gas (10⁹ cfg)	GOR (ft³/bbl)	Column (m)	H2S (%)
Tampico area									
2	Arenque	1966	3700	34	227.7	417.9	1145	80	nd
4	Ebano Chapacao	1901	200	12	198.3	277.7	1469	nd	nd
9	Salinas-Barco-Caracol	1959	1250	17	18.1	70.4	4096	nd	nd
10	Tamaulipas-Constituciones	1956	1300	18	354.1	274.9	519	nd	nd
Tuxpan area									
11	Acuatempa	1955	1171	23	29.9	38.2	695	nd	nd
12	Alamo San Isidro	1913	603	20	72	32.8	467	nd	nd
13	Alazan	1949	582	23	19.9	11.8	569	nd	nd
A	Allende	1955	2365	28	nd	nd	nd	nd	nd
B	Arrecife Medio	1963	2045	24	nd	nd	nd	nd	nd
14	Atun	1967	2833	38	44.4	197.4	1869	nd	nd
15	Bagre	1966	2693	39	75.8	101.7	881	nd	nd
C	Boca de Lima	1956	1681	15	nd	nd	nd	nd	nd
D	Cabo Nuevo	1966	1655	20	nd	nd	nd	nd	nd
E	Cabo Rojo	1953	1198	13	nd	nd	nd	nd	nd
F	Canas	1956	828	21	nd	nd	nd	nd	nd
G	Castillo de Teayo	1963	2217	32	0.3	nd	nd	nd	nd
17	Cerro del Carbon	1959	2834	28	7.4	10.2	637	nd	nd
I	Chichimantla	1955	1071	22	3.9	nd	nd	nd	nd
J	Copal	1957	1313	18	nd	nd	nd	nd	nd
K	Esturion	1965	2597	34	nd	nd	nd	nd	nd
19	Ezequiul Ordóñez	1952	1400	21	71	44.3	599	nd	nd
L	Gutierrez Zamora	1956	1927	20	0.8	nd	nd	nd	nd
M	Horcon	1952	849	23	3.5	nd	nd	nd	nd
21	Isla de Lobos	1963	2087	40	23.8	6.3	269	nd	nd
N	Jardin	1913	648	20	nd	nd	nd	nd	nd
22	Jiliapa	1958	2240	34	40.9	49.1	631	nd	nd
P	Mesa Cerrada	1956	1084	20	12.5	nd	nd	nd	nd
Q	Miguel Hidalgo	1954	2333	31	9.6	nd	nd	nd	nd
R	Miquetla	1959	1970	35	62.9	nd	nd	nd	nd
25	Moralillo	1948	1385	21	19.2	12.5	652	nd	nd
S	Mozutla	1953	1189	22	10.9	nd	nd	nd	nd
T	Naranjos	1913	600	20	1202.5	nd	nd	nd	nd
U	Nueva Colonia	1956	859	22	nd	nd	nd	nd	nd
27	Nuevo Progresso	1955	2188	31	10.4	14.6	528	nd	nd
28	Ocotepec	1953	973	23	23.7	9.9	560	nd	nd
29	Paleocanal de Chicontepec	1935	1850	30	12,429	36,243.6	354	nd	nd
V	Paso Real	1913	628	27	48	nd	nd	nd	nd
W	Pez Vela	1968	2745	13	nd	nd	nd	nd	nd
X	Pital	nd	1300	23	0.9	nd	nd	nd	nd
32	Poza Rica	1930	2000	35	2069.5	3718.2	1144	175	nd
Y	Robalo	1964	2183	34	nd	nd	nd	nd	nd
36	San Andrés	1956	2906	29	483.1	409.5	681	nd	nd
37	Santa Agueda	1953	1361	16	122.9	124.2	635	nd	nd

		Reservoir rock properties					Trap type
Age	Lithology	Facies	Res. temp. (°C)	Porosity (%)	Permeability (md)	Area (km²)	
Tampico area							
Cret./Jur.	Carbonate	Tamaulipas	125	18	10	34	Anticline
Cretaceous	Carbonate	Tamaulipas	nd	nd	nd	nd	Anticline
Cretaceous	Carbonate	Tamaulipas	nd	nd	nd	nd	Anticline
Cretaceous	Carbonate	Tamaulipas	nd	nd	nd	nd	Anticline
Tuxpan area							
Cretaceous	Carbonate	El Abra	nd	nd	nd	nd	Reef
Cretaceous	Carbonate	El Abra	nd	nd	nd	nd	Reef
Cretaceous	Carbonate	El Abra	nd	nd	nd	nd	Reef
Cretaceous	Carbonate	El Abra	nd	nd	nd	nd	Reef
Cretaceous	Carbonate	El Abra	nd	nd	nd	nd	Reef
Cretaceous	Carbonate	El Abra	nd	nd	nd	nd	Reef
Cretaceous	Carbonate	El Abra	nd	nd	nd	nd	Reef
Cretaceous	Carbonate	El Abra	nd	nd	nd	nd	Reef
Cretaceous	Carbonate	El Abra	nd	nd	nd	nd	Reef
Cretaceous	Carbonate	El Abra	nd	nd	nd	nd	Reef
Cretaceous	Carbonate	Tamabra	nd	nd	nd	nd	Reef talus
Cretaceous	Carbonate	Tamabra	nd	nd	nd	nd	Reef talus
Cretaceous	Carbonate	El Abra	nd	nd	nd	nd	Reef
Cretaceous	Carbonate	El Abra	nd	nd	nd	nd	Reef
Cretaceous	Carbonate	El Abra	nd	nd	nd	nd	Reef
Cretaceous	Carbonate	El Abra	nd	nd	nd	nd	Reef
Cretaceous	Carbonate	El Abra	nd	nd	nd	nd	Reef
Cretaceous	Carbonate	El Abra	nd	nd	nd	nd	Reef
Cretaceous	Carbonate	Tamabra	nd	nd	nd	nd	Reef talus
Cretaceous	Carbonate	El Abra	nd	nd	nd	nd	Reef
Cretaceous	Carbonate	El Abra	nd	nd	nd	nd	Reef
Cretaceous	Carbonate	Tamabra	nd	nd	nd	nd	Reef talus
Cretaceous	Carbonate	Tamabra	nd	nd	nd	nd	Reef talus
Cretaceous	Carbonate	El Abra	nd	nd	nd	nd	Reef
Cretaceous	Carbonate	El Abra	nd	nd	nd	nd	Reef
Cretaceous	Carbonate	El Abra	nd	nd	nd	nd	Reef
Cretaceous	Carbonate	Tamabra	nd	nd	nd	nd	Reef talus
Cretaceous	Carbonate	El Abra	nd	nd	nd	nd	Reef
Eoc./Pal.	Sandstone	Channel	70	nd	nd	3731	Stratigraphic
Cretaceous	Carbonate	El Abra	nd	nd	nd	nd	Reef
Cretaceous	Carbonate	El Abra	nd	nd	nd	nd	Reef
Cretaceous	Carbonate	El Abra	nd	nd	nd	nd	Reef
Cretaceous	Carbonate	Tamabra	nd	8	700	120	Reef talus
Cretaceous	Carbonate	El Abra	nd	nd	nd	nd	Reef
Jurassic	Oolitic	nd	nd	nd	nd	nd	nd
Cretaceous	Carbonate	El Abra	nd	nd	nd	nd	Reef

TABLE 3. Oil-field information for the Pimienta-Tamabra(!). Information from Viniegra and Castillo-Tejero (1970), Enos (1977), Acevedo (1980), Santiago and Baro (1992), and PEMEX(1999a, b) (cont'd.).

Field number	Field name	Discovery year	Prod. depth (m)	API gravity (o)	Known oil (10⁶ bbl)	Known gas (10⁹ cfg)	GOR (ft³/bbl)	Column (m)	
									Fluid properties
Tuxpan area (cont'd.)									
38	Soledad	1961	1935	33	11.2	11.6	915	nd	
Z	Solis	1952	650	22	9.5	nd	nd	nd	
AA	Tamiahua	1951	1045	20	nd	nd	nd	nd	
BB	Tecolutla	1956	2303	23	0.8	nd	nd	nd	
CC	Temapache	1955	752	21	nd	nd	nd	nd	
43	Tiburón	1965	2228	30	10.5	4.2	351	nd	
DD	Tihuatlan	1957	848	22	0.3	nd	nd	nd	
45	Tintorera	1968	2227	20	24.8	4.6	225	nd	
46	Toteco Cerro Azul	1916	350	20	378.5	213.1	561	nd	
47	Tres Hermano	1959	1911	21	149.8	375.2	1766	nd	
EE	Vicente Guerrero	1955	2007	25	4.9	nd	nd	nd	
FF	Xocotla	1953	901	21	1.1	nd	nd	nd	
GG	Zapotalillo	1961	2282	27	0.1	nd	nd	nd	
Salina area									
69	Cinco Presidentes	1946	2092	34	325.4	442.6	963	nd	
73	El Plan	1931	665	26	162.6	32.1	194	nd	
78	Magallanes-Tucan-Pajonal	1957	1167	32	432.9	449.1	1019	nd	
80	Ogarrio	1957	1744	35	283.5	394.4	1152	nd	
84	Tonalá–El Burro	1928	800	32	73.2	35.1	455	nd	
Villahermosa area									
85	Agave	1976	3800	41	156.7	1617.5	9190	925	
89	Belota	1982	5009	39	210.3	715.9	1634	990	
90	Cactus	1972	3740	36	404.3	969.9	2027	nd	2.
93	Cárdenas	1980	5115	38	620	1282.6	1860	845	
100	Chincorro	1991	5100	37	124.2	204.8	977	277	
104	Complejo Bermudez	1973	4100	29	4335.5	6740.1	1297	nd	3
111	Giraldas	1977	4424	40	187.3	2333.9	5905	586	
112	Iris	1979	4454	41	19.6	157.1	7544	576	
114	Jujo-Tecominoacan	1980	5081	38	1811.9	3440	1226	1344	
116	Luna-Palapa	1985	5141	41	157.4	875.5	5521	386	
119	Mora	1981	5300	38	181.4	519.1	1545	76	
121	Múspac	1982	2525	51	126.8	2368.3	8595	188	
126	Paredón	1977	4830	38	228.3	1161.1	2571	1270	
132	Sen	1984	4759	37	390.3	1247.9	3310	280	
133	Sitio Grande	1972	4130	35	441.2	705.1	1971	nd	3
142	Yagual	1988	4900	35	72.5	70.9	808	96	
Bay of Campeche area									
143	Abkatún	1979	3100	30	2380.6	1978.4	730	670	
144	Akal (Cantarell)	1976	2492	24	18525.2	8364.8	483	nd	
147	Bacab	1977	3210	19	65.1	21.1	295	185	
148	Balam	1992	4360	28	203.1	44.3	212	nd	
149	Batab	1984	3610	32	57.9	56.5	920	300	

| Age | Lithology | Facies | Reservoir rock properties | | | | Trap type |
			Res. temp. (°C)	Porosity (%)	Permeability (md)	Area (km²)	
Tuxpan area (cont'd.)							
Cretaceous	Carbonate	Tamabra	nd	nd	nd	nd	Reef talus
Cretaceous	Carbonate	El Abra	nd	nd	nd	nd	Reef
Cretaceous	Carbonate	El Abra	nd	nd	nd	nd	Reef
Cretaceous	Carbonate	El Abra	nd	nd	nd	nd	Reef
Cretaceous	Carbonate	El Abra	nd	nd	nd	nd	Reef
Cretaceous	Carbonate	El Abra	nd	nd	nd	nd	Reef
Cretaceous	Carbonate	El Abra	nd	nd	nd	nd	Reef
Cretaceous	Carbonate	El Abra	nd	nd	nd	nd	Reef
Cretaceous	Carbonate	El Abra	nd	nd	nd	nd	Reef
Cretaceous	Carbonate	Tamabra	nd	nd	nd	nd	Reef talus
Cretaceous	Carbonate	El Abra	nd	nd	nd	nd	Reef
Cretaceous	Carbonate	El Abra	nd	nd	nd	nd	Reef
Cretaceous	Carbonate	Tamabra	nd	nd	nd	nd	Reef talus
Salina area							
Miocene	Sandstone		nd	20	176	nd	Salt dome
Miocene	Sandstone		nd	23	180	nd	Salt dome
Miocene	Sandstone		nd	19	180	nd	Salt dome
Miocene	Sandstone		nd	19	177	nd	Salt dome
Miocene	Sandstone		nd	23	175	nd	Salt dome
Villahermosa area							
Cretaceous	Dolomite		137	5	nd	48.1	Anticline
Cret./Jur.	Dolomite		148	3	nd	32.1	Anticline
Cretaceous	Carbonate		nd	7	6228	nd	Dome
Cret./Jur.	Dolomite		159	3	nd	41.1	Anticline
Jurassic	Dolomite		137	9	1007	nd	Faulted anticline
Cretaceous	Carbonate		nd	8	7800	nd	Faulted anticline
Cretaceous	Limestone		134	7	nd	36.1	Anticline
Cretaceous	Limestone		133	5	nd	8.4	Anticline
Jurassic	Dolomite		154	3	nd	60.3	Anticline
Jurassic	Dolomite		168	6	nd	15	Anticline
Jurassic	Dolomite		139	3	27	45	Faulted anticline
Cretaceous	Limestone		109	12	nd	17.8	Anticline
Jurassic	Dolomite		115	4	nd	24	Anticline
Cretaceous	Dolomite		155	7	nd	25	Anticline
Cretaceous	Carbonate		nd	8	3080	nd	Anticline
Miocene/Cret.	Carbonate		133	8	0.3	nd	Faulted anticline
Bay of Campeche area							
Pal./Cret.	Dolomite		120	10	nd	90	Anticline
Jurassic	nd	Breccia	nd	nd	nd	nd	nd
Pal./Cret.	Dolomite		120	12	nd	12	Anticline
Pal./Cret./Jur.	Sandstone		115	21	400	43	Anticline
Pal./Cret./Jur.	Dolomite		132	7	nd	15	Anticline

TABLE 3. Oil-field information for the Pimienta-Tamabra(!). Information from Viniegra and Castillo-Tejero (1970), Enos (1977), Acevedo (1980), Santiago and Baro (1992), and PEMEX (1999a, b) (cont'd.).

Field number	Field name	Discovery year	Prod. depth (m)	API gravity (o)	Known oil (10^6 bbl)	Known gas (10^9 cfg)	GOR (ft^3/bbl)	Column (m)	H2S (%)
151	Caan	1984	3300	36	804.2	1681.1	1785	500	nd
152	Chac (Cantarell)	1976	3545	20	238.3	115.1	483	nd	nd
154	Chuc	1982	3500	30	809.6	989.5	1078	420	nd
156	Ek	1980	3075	20	128.8	21.7	168	336	nd
159	Ixtoc	1979	3600	28	70.3	80	1139	40	nd
165	Ku	1979	2550	20	2458.7	1413	511	800	nd
166	Kutz (Cantarell)	1976	3100	22	295.6	189.9	483	nd	nd
169	Maloob	1979	2890	20	1766	617.8	349	400	nd
173	Nohoch (Cantarell)	1976	2350	24	764.3	371.8	462	nd	nd
175	Pol	1980	3730	31	1070.8	1014.3	1085	400	nd
180	Uech	1986	4800	38	81.9	201.5	2332	400	nd
183	Zaap	1975	2941	20	1540.5	724.6	349	nd	nd

and Tuxpan areas (Table 5; Román-Ramos et al., 1996; Geo-Mark, 1999). Analytical data for six oil samples from the Geo-Mark (1999) database that represent the Tampico and Tuxpan areas are included in Table 5.

As reported by Guzmán-Vega and Mello (1999) from carbon isotopic and biomarker data, the great majority of the oil samples in the Salina, Villahermosa, and Bay of Campeche areas

are in oil family 2 (Table 5). Biomarker characteristics, such as the dominance of low molecular weight n-alkanes, a pristane/phytane ratio less than or equal to one, a Ts/Tm ratio less than one, abundant extended hopanes, high C_{35}/C_{34} hopane ratios, abundant C_{29} steranes relative to C_{27} steranes, presence of 17 a(H)-29,30 bisnorhopane, and low to intermediate abundance of tricyclic terpanes relative to pentacyclic terpanes,

TABLE 4. Summary of field information shown in Tables 2 and 3.

	In–place petroleum			Remaining reserves	
Producing area	Crude oil (10^6 bbl)	Natural gas (10^9 cf)	GOR (ft^3/bbl)	Crude oil (10^6 bbl)	Natural gas (10^9 cf)
West side					
Tampico	14,425	29,424	2040	322	306
Tuxpan	–	–	–	–	–
Numbered fields	153,218	61,170	399	13,375	38,298
Lettered fields	–	–	–	–	–
Veracruz	492	2702	5498	11	1123
Total	168,135	93,296	555	13,708	39,727
East side					
Salina	7069	7169	1014	619	658
Villahermosa	32,327	62,006	1918	5368	17,155
Bay of Campeche	68,927	42,304	614	21,292	14,078
Total	108,322	111,479	1029	27,278	31,892
Grand total	**276,457**	**204,775**	**741**	**40,986**	**71,619**

	Reservoir rock properties						Trap type
Age	Lithology	Facies	Res. temp. (°C)	Porosity (%)	Permeability (md)	Area (km²)	
Pal./Cret.	Dolomite		147	9	nd	38	Anticline
Paleocene	Carbonate	Breccia	nd	nd	nd	nd	nd
Pal./Cret.	Dolomite		127	9	nd	23	Anticline
Pal./Cret.	Dolomite		112	9	nd	11	Anticline
Pal./Cret.	Dolomite	Breccia	142	6	2000	nd	Anticline
Eoc./Pal./Cret.	Dolomite		90	10	nd	50	Anticline
Eoc./Pal./Cret.	Dolomite	Breccia	nd	10	4000	nd	Faulted anticline
Pal./Cret.	Dolomite		89	9	nd	30	Anticline
Paleocene	Carbonate	Breccia	nd	nd	nd	nd	nd
Pal./Cret.	Dolomite		129	7	nd	44	Anticline
Jurassic	Dolomite		140	15	nd	12	Anticline
Eoc./Pal./Cret./Jur.	Dolomite	Breccia	nd	nd	nd	nd	Faulted anticline

indicate that these oil families were derived from a marine carbonate source rock (Guzmán-Vega and Mello, 1999). Carbon isotopic results separate these families; family 2 ranges from –26.4 to –28.1‰, and oil family 1 ranges from –25.4 to –25.7‰. The depositional setting of the source rock for both of these oil families is inferred to be shallow to neritic, distal carbonate ramps with local shale that was deposited in an anoxic condition, a setting that was well developed in the Upper Jurassic of the southern Gulf of Mexico. In conclusion, oil families 1 and 2 are included in the Pimienta-Tamabra(!), whereas oil families 3 and 4 are excluded (Guzmán-Vega and Mello, 1999).

Based on bulk oil properties, such as API gravity and sulfur, vanadium, and nickel concentrations, the oil samples in Table

Cumulative production		Known petroleum		Recovery factor	
Crude oil (10⁶ bbl)	Natural gas (10⁹ cf)	Crude oil (10⁶ bbl)	Natural gas (10⁹ cf)	Crude oil (%)	Natural gas (%)
West side					
1363	3087	1685	3393	11.7	11.5
–	–	–	–	–	–
3905	4068	17,280	42,367	11.3	69.3
1373	–	1373	–	–	–
72	818	83	1941	16.9	71.8
6713	7974	20,422	47,701	12.1	51.1
East side					
1596	1910	1138	2568	31.3	35.8
6207	14,619	11,575	31,774	35.8	51.2
12,212	7555	33,184	21,633	48.6	51.1
20,015	24,085	45,897	55,976	42.4	50.2
26,728	**32,058**	**66,319**	**103,677**	**24.0**	**50.6**

TABLE 5. Petroleum-geochemical information on oil samples in the Pimienta-Tamabra(!). Information from Guzmán-Vega and Mello (1999) and from GeoMark (1999).

GP	Oil family	Area	Spl ID	Field	Reservoir age	Depth (m)	API gravity (o)	S (%)	V (ppm)	Ni (ppm)	Ts/Tm (Ratio)	Pr/Ph (Ratio)	$\delta^{13}Cw$ (0/00)	$\delta^{13}Cs$ (0/00)	$\delta^{13}Ca$ (0/00)
C	nd	Tuxpan	MX087	Cacalilao	Middle Cretaceous	500	12.8	4.4	550	72	0.4	nd	nd	-27.64	-27.60
C	2b	Villahermosa	CHI-2	nd	Upper Pliocene	2028	12.9	4.2	429	70	0.3	0.6	-27.7	nd	nd
C	2c	Salina	SAL-7	nd	Upper Miocene	2591	12.9	1.2	30	21	1.2	1.1	-27.5	nd	nd
C	2a	Salina	SA-1	nd	Upper Miocene	1539	15.2	3.6	110	31	0.5	0.7	-27.9	nd	nd
C	2c	Salina	SAL-8	nd	Upper Miocene	620	15.5	2.1	38	19	1.0	nd	-27.2	nd	nd
C	2b	Campeche	CA-4	nd	Middle Eocene	3855	16.2	2.4	296	59	0.5	0.8	-27.5	nd	nd
C	2a	Salina	SA-2	nd	Upper Miocene	1540	16.3	6.0	nd	nd	0.5	0.7	-28.1	nd	nd
C	nd	Tampico	MX052	Barcodon	Middle Cretaceous	1257	17.9	5.1	446	49	0.3	0.7	nd	-27.25	-27.41
C	nd	Tuxpan	MX055	Cabo Nuevo	Middle Cretaceous	1681	20.0	3.6	542	53	0.4	1.1	nd	-26.65	-27.15
C	2c	Campeche	CA-6	nd	Lower Cretaceous	5820	20.7	3.6	nd	5	0.9	1.0	-27.3	nd	nd
C	2b	Campeche	CA-5	nd	Lower Cretaceous	4245	22.3	1.8	215	63	0.7	0.8	-27.5	nd	nd
C	nd	Tuxpan	MX089	Acuate mpa	Middle Cretaceous	1171	22.5	3.1	472	61	0.3	0.9	nd	-28.29	-27.81
B	2a	Salina	SA-3	nd	Upper Miocene	385	23.9	2.2	49	23	0.7	0.8	-27.3	nd	nd
B	2b	Campeche	CA-3	nd	Upper Cretaceous	4317	24.7	1.5	75	19	0.5	0.9	-27.4	nd	nd
B	2a	Salina	SA-4	nd	Upper Miocene	1805	25.0	1.7	20	7	1.0	1.0	-27.4	nd	nd
B	1	Campeche	CA-1	nd	Oxfordian	4400	25.8	2.2	18	13	0.4	0.6	-25.4	nd	nd
B	1	Campeche	Ca-2	nd	Oxfordian	4540	26.0	2.6	15	12	0.5	0.6	-25.7	nd	nd
B	2c	Salina	SAL-9	nd	Upper Miocene	3383	27.0	2.1	nd	nd	1.0	1.0	-27.8	nd	nd
B	2c	Salina	SAL-10	nd	Upper Miocene	2685	27.2	1.3	33	21	1.4	1.0	-27.3	nd	nd
B	nd	Tuxpan	MX091	San Andrés	Middle Cretaceous	3174	28.0	1.8	103	23	0.5	0.9	nd	-28.53	-27.78
A	2a	Salina	SA-5	nd	Upper Cretaceous	550	33.6	1.3	16	4	0.7	1.1	-26.8	nd	nd
A	2a	Salina	SA-6	nd	Middle Cretaceous	490	33.8	1.3	16	4	0.7	0.9	-26.8	nd	nd
A	2b	Villahermosa	CHI-1	nd	Middle Cretaceous	4035	35.5	0.6	5	1	0.9	1.2	-26.4	nd	nd
A	2c	Salina	SAL-11	nd	Upper Miocene	2997	36.4	0.9	7	4	1.2	0.9	-27.1	nd	nd
A	2c	Salina	SAL-12	nd	Lower Pliocene	1250	37.2	0.8	11	4	1.3	1.1	-27.4	nd	nd
A	nd	Tuxpan	MX053	Isla de Lobos	Middle Cretaceous	2096	40.0	0.7	18	11	0.7	1.6	nd	-26.17	-26.43

5 can be divided into three groups (Figure 8; Table 6). Generally, high-gravity oil has low trace-metal and sulfur contents, and low-gravity oil has higher concentrations of trace metal and sulfur. These variations are usually attributed to one or more of the following: (1) kerogen type, (2) thermal maturity of the source rock at expulsion, (3) cracking of oil in the reservoir rock, and (4) biodegradation. Similar bulk-oil-property variation is present in the greater Ekofisk area in the North Sea (di Primio et al., 1998, their Figure 11). Di Primio et al. (1998) attributed this pattern to phase separation of oil and gas during migration to shallower depths after expulsion from thermally mature source rock. This phase separation may explain the doubling of sulfur content between groups while the nickel content triples and vanadium content is three and seven times greater (Table 6). The pristane-to-phytane ratio is marginally changed from 1.1 to 0.8.

This evidence suggests that phase separation, rather than thermal maturity or biodegradation, is the cause of the decrease in API gravity. This evidence further suggests that gas and oil are a single-phase or a supercritical fluid when expelled from Upper Jurassic source rock. At standard temperature and

TABLE 6. Mean values for oil groups A-C of Table 5.

GP	API gravity (o)	S (%)	V (ppm) V	Ni (ppm)	Ts/Tm ratio	Pr/Ph ratio
C	17	3.4	313	46	0.58	0.8
B	26	1.9	45	17	0.75	0.8
A	36	0.9	12	5	0.92	1.1

TABLE 7. Primary recovery factors for carbonate reservoirs in large oil fields. Data from Roehl and Choquette (1985).

Field name	Recovery factor (%)
Bibi Hakimeh	14.5
Killdeer	14.5
West Cat Canyon	15.6
Blalock Lake East	20.0
Glenburn	20.0
Pennel	20.8
Greater Ekofisk	22.2
Fukubezawa	23.6
Mount Everette and SW Reeding	25.0
Gachsaran	25.0
Reeves	28.1
Seminole SE	28.6
Cabin Creek	33.5
Mount Vernon	34.9
Poza Rica	42.3
Fateh	43.2
Rainbow "A" Pool	50.0
Sunniland	50.0
NW Lisbon	56.1
Fairway	58.2
Median:	25%
Average:	31%

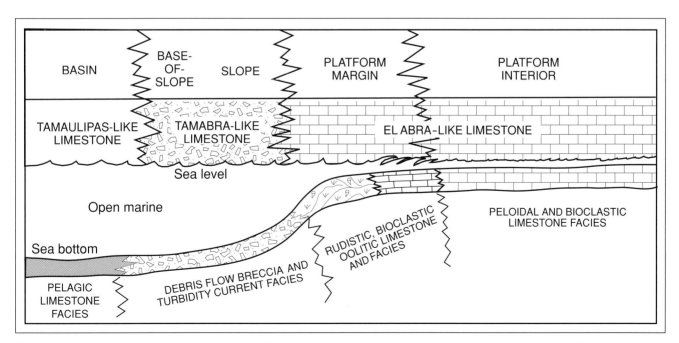

FIGURE 7. Schematic cross section showing mid-Cretaceous carbonate reservoir facies in the southern Gulf of Mexico. Distribution of these three facies is shown in Figures 2, 3, 4, and 18.

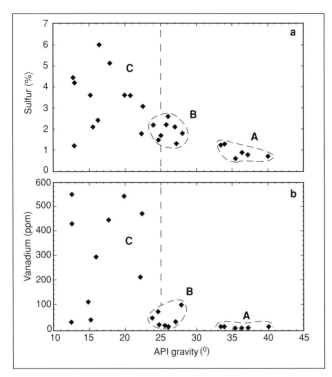

FIGURE 8. Graphs showing (a) sulfur, in wt.%, and (b) vanadium, in ppm by API gravity. A, B, and C refer to the groups of geochemical analysis in Tables 5 and 6. Dashed line is the 25° API line.

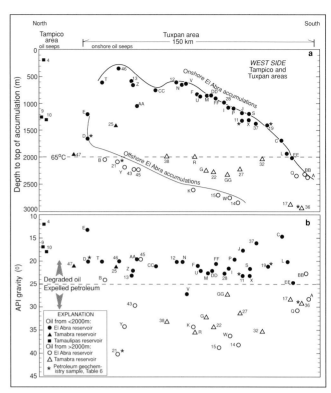

FIGURE 9. North-to-south profile from the Tampico to Tuxpan areas showing (a) depth to the top of oil accumulations in the El Abra, Tamabra, and Tamaulipas reservoir facies, and (b) API gravity of the oil in each accumulation. Numbers and letters refer to fields in Tables 2 and 3, and asterisk (*) indicates oil samples analyzed in Table 5. Depth to 65°C is from Figure 15.

pressure, this oil averages 36° API with a sulfur content of 0.9 wt.%, vanadium content of 12 ppm, nickel content of 5 ppm, and a pristane-to-phytane ratio of 1.

Petroleum Properties

API gravity of oil and depth of production from fields along north-south profiles are shown for the western and eastern sides of the Pimienta-Tamabra(!) petroleum system (Figures 9 and 10; Table 3). In the Tampico and Tuxpan areas on the western side, the API gravity of the oil ranges from 12° to 40° and the depth of production is as deep as 3000 m (Figure 9; Table 3). The El Abra, Tamabra, and Tamaulipas reservoir rocks are depicted as circles, triangles, and squares, respectively. Oil production at less than 2000 m is represented by solid symbols, whereas production at greater than 2000 m is shown by outlined symbols. The onshore and offshore El Abra reef-trend accumulations are deepest to the southeast, toward the offshore. From 300 to 2000 m, the API oil gravity is 20° to 25° with only a few exceptions, and at depths greater than 2000 m, the gravity is as high as 40°.

In the Bay of Campeche, Villahermosa, and Salina areas on the eastern side, the API gravity ranges from 19° to 41°, and the depth of oil production ranges from 800 to 5200 m (Figure 10; Table 3). Only the Miocene reservoir rock, shown as a solid square, produces from depths less than 2000 m, whereas the Cretaceous and older fields, shown as open circles and triangles, produce at depths greater than 2000 m. In contrast to the

western side, the shallow production from the Miocene reservoir rocks is oil with API gravity from 26° to 35°, suggesting that it only recently migrated to these shallow depths. In addition, the oil in the Bay of Campeche area is producing oil with API gravity from 19° to 24° from depths greater than 2000 m, suggesting that burial of the producing interval occurred subsequent to emplacement of oil.

Histograms of known oil volume in 5° API increments ("bins") and in 500 gas-to-oil ratio (GOR in ft³/BO) units for those accumulations on the western and eastern sides of the petroleum system were determined from data in Table 3 and are shown in Figure 11. The known oil and gas volumes and GORs from numbered fields in Table 3 are reproduced from Table 2. The GOR in Table 2 is derived from in-place petroleum. On the western side, 25 fields include information on API gravity, GOR, and known oil volume. Here, the largest volume of oil—more than 12 billion barrels from the Chicontepec field (Field no. 29, Table 3) — occurs in the 30° to 35° API bin and in the 0 to 500 GOR bin (Figure 11a and b). On the eastern side, 38 fields contain the necessary information, and about 25 billion barrels of oil are in the 20° to 25° bin and, again, in the 0 to 500 GOR bin (Figure 11c and d). Here, the Cantarell complex represents more than 19 billion barrels of

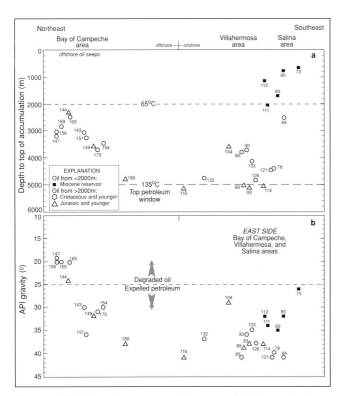

FIGURE 10. Northeast-to-southwest profile from the Bay of Campeche to the Salina areas showing (a) depth to the top of oil accumulations in the Miocene, Cretaceous and younger, and Jurassic and younger reservoir rocks, and (b) API gravity of the oil in each accumulation. Numbers refer to fields in Table 2. Depths to 65°C and 135°C are from Figure 15.

oil (Field nos. 144, 152, 166, and 173, Table 3). Even though a single large accumulation on each side represents almost all the oil in two different bins, the GOR for both sides is mostly below 2,000 ft³/BO.

The depth of production and API gravity of oil are compared to GOR on the eastern and western sides of the petroleum system to determine if phase separation could account for some of the GORs above 2,000 ft³/BO. As single-phase petroleum or a supercritical fluid migrates to a lower temperature and pressure, some of the dissolved gas comes out of solution, which could then take a different migration path from the remaining oil with a lower GOR. The result is a small volume of higher-gravity oil with a higher GOR at a shallower depth than the deepest oil accumulations. On the western side, there are 25 fields, and on the eastern side there are 39 fields, with information available to make this comparison (Figure 12a-d). The western side shows a gradual decrease in both the API gravity and GOR where the GOR is below 2,000 ft³/BO, which is an expected trend. However, on the eastern side, a similar trend is generally present for GORs below 2,000 ft³/BO, although for higher GORs the API oil gravity increases, even though the depth of occurrence decreases. Apparently, phase separation has occurred on the eastern side of the Pimienta-Tamabra(!).

Petroleum Occurrence

On the western and eastern sides, the petroleum volume, GOR, and recovery factors are compared by area (Table 4). The volume of in-place petroleum is 276.5 billion barrels of crude oil and 204.8 tcf of natural gas, which calculates as a GOR of 741. The volume of in-place oil on the western side is greater than on the eastern side, but the recovery factor for the west is 12.1 percent, quite different from the eastern side, which is 42.2 percent, or almost four time greater. This range of oil-recovery factors is typical of large fields with carbonate reservoir rocks (Table 7). The known volume of petroleum includes remaining reserves and cumulative production. Volumetrically, most known oil, that is, more than 25 billion barrels, occurs in the depth interval of 2 to 3 km (Figure 13). Half again as much known oil occurs in depth intervals of 1 to 2 km on the western side. On the eastern side, the volume of oil gradually decreases with depth to the range of 5 to 6 km (Figure 13).

Accumulation depth and oil-column height are also compared. On the western side, oil occurs in three mid-Cretaceous reef facies on the Tuxpan platform in the Tampico and Tuxpan areas. Here, the Tamalaupis, Tamabra, and El Abra facies (Figure 3) contain 20.3 billion barrels of oil (Table 8; Viniegra O. and Castillo-Tejero, 1970; Enos, 1977; and PEMEX, 1999a). The depth of the accumulations range from 200 m to 3000 m, with the shallowest accumulations onshore in the Tamalaupis Formation and the onshore El Abra. As the Tuxpan platform dips southeast, the offshore El Abra accumulations are deepest, ranging from 1200 to 2800 m. The Poza Rica field (Field no. 32, Figure 3) is a stratigraphic trap that produces from the Tamabra facies, covers 120 km², and is about 175 m thick (Enos, 1977, 1985).

The Salina, Villahermosa, and Bay of Campeche areas are underlain by pre–Upper Jurassic salt that created structural traps and fracture permeability upon movement (Figures 2 and 4; Peterson, 1983; Salvador, 1991b). In the Salina area, the producing interval ranges from 600 to 2000 m in depth and is from Miocene siliciclastic sandstone (Figure 4; Tables 2 and 3). These fields produce from faulted salt domes that formed in the late Miocene or later (Neogene).

The structural traps in the Villahermosa and Bay of Campeche areas are from 8 to 90 km² in area and average 30 km² (Table 3). The depth range of the oil accumulations is from 2500 m to 5200 m, with reservoir temperatures ranging from 89° to 168°C and averaging 132°C (Table 3). Thick oil-bearing columns vary from 185 m to 1345 m, with an average of 625 m. Reservoir porosity ranges from 3% to 15%, with an average of 7.4% (Tables 1 and 2; Santiago and Baro, 1992).

UPPER JURASSIC SOURCE ROCK

Pimienta Formation

The importance of the Upper Jurassic source rock in the southern Gulf of Mexico has been known for some time

TABLE 8. Known volume of oil by reservoir facies. Data from Tables 2 and 3.

Producing area	Reservoir facies				Known oil volume	Percent
	Carbonate			Sandstone		
	Tamaulipas-like (x10⁶ BO)	Tamabra-like (x10⁶ BO)	El Abra–like (x10⁶ BO)	All reservoirs (x10⁶ BO)	(x10⁶ BO)	(%)
Tampico	1685	0	0	0	1685	2.5
Tuxpan	–	–	–	–	–	–
Numbered fields	0	4061	790	12,429	17,280	26.1
Lettered fields	0	63	1309	0	1373	2.1
Veracruz	0	0	0	83	83	0.1
Salina	0	0	0	1138	1138	1.7
Villahermosa	4330	6779	466	0	11,575	17.5
Bay of Campeche	1097	32,087	0	0	33,184	50.0
Total:	**7112**	**42,990**	**2565**	**13,567**	**66,318**	**100.0**
Percent:	10.7%	64.8%	3.9%	20%	100%	

(Viniegra O., 1981; Holguin Quiñones, 1985; Nehring, 1991). Although other source-rock intervals have been identified in the Cretaceous and Tertiary, many recent studies continue to emphasize the importance of the Upper Jurassic (Holguin et al., 1994; Bertrand et al., 1996; Becceril et al., 1996; Lucach et al., 1996; Medrano et al., 1996b; Mello et al., 1996; Mello and Guzmán-Vega, 1996; and Guzmán-Vega and Mello, 1994, 1996, 1999). Carbon isotopic and biomarker data from samples in oil families 1 and 2 and rock extracts from the Oxfordian and Tithonian source-rock intervals show that the known oil in the Tampico, Tuxpan, Salina, Villahermosa, and Bay of Campeche areas are derived from an Upper Jurassic

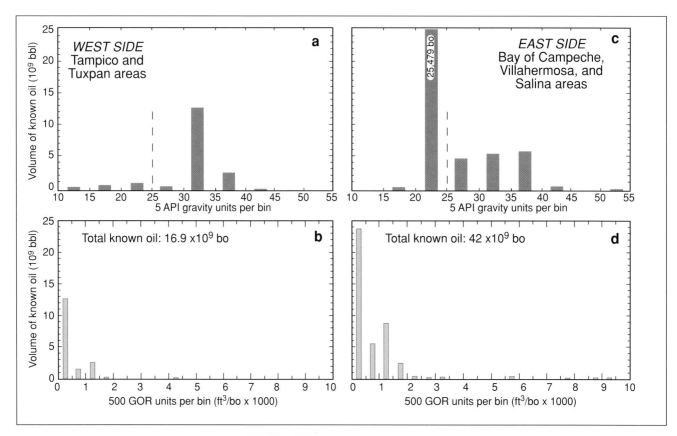

FIGURE 11. Histograms showing cumulative volume of oil in 5° API gravity bins and 500 GOR bins for (a and b) the western side of Pimienta-Tamabra(!), where most of the oil is more than 25° API, and (c and d) the eastern side of Pimienta-Tamabra(!), where most of the oil is less than 25° API. Data are from Table 3.

source rock that includes the Oxfordian and Tithonian intervals.

Many formation names have been applied to Upper Jurassic rock units in the southern Gulf of Mexico (Figure 5). The Oxfordian and Tithonian are time-stratigraphic units that contain the richest source-rock intervals in the Upper Jurassic. Since most of the oil originated from rocks of Tithonian age, we focused on this time interval for the source-rock name. Because a formation name is lacking in the east, and little oil occurs in the Veracruz area, we selected the Pimienta Formation in the Tampico and Tuxpan areas as the source-rock name for this petroleum system, even though the entire Upper Jurassic section contains potential source rock.

Oil–Source Rock Correlation

Samples from the dominant oil family in the Salina, Villahermosa, and Bay of Campeche areas have been correlated with extracts from an Upper Jurassic (Oxfordian and Tithonian) source rock. Biomarker characterization of oils and extracts from these source-rock intervals are shown in Guzmán-Vega and Mello (1999, their Figures 8, 13, and 15). One of these oil samples from the Salina area was shown by Guzmán-Vega and Mello (1999) to have biomarker compositions very similar to those extracted from Tithonian organic marl from the Tuxpan area, more than 500 km to the northwest of the Salina area. This correlation is one indication of the widespread distribution of similar depositional settings for the Upper Jurassic source rock. In general, the Late Jurassic depositional setting was favorable for development of a high-quality source rock throughout this large area. The excellent correlation of oil samples with Upper Jurassic source-rock extract is reason to make the level of certainty of this petroleum system known and included in the name by using the symbol (!) (Magoon and Dow, 1994).

Source-rock Potential and Quality

Types I and II kerogen dominate in the Upper Jurassic source-rock interval of southeastern Mexico. Source-rock sample data reported by Medrano et al. (1996b) indicate that:

1) The Oxfordian source-rock interval has good to excellent source-rock potential (richness) with total-organic-carbon content (TOC) that ranges from 0.5 to 5 wt.% (average 1.7 wt.%) and S_2 values from Rock Eval that range from 2 to 19 mg hydrocarbons (HC)/g rock (average 8 mg HC/g rock).

2) The Kimmeridgian source-rock interval has fair to good source-rock potential with TOC ranging from 0.5 to 2 wt.% (average about 0.8 wt.%) and S_2 values ranging from 2 to 6 mg HC/g rock (average 3 mg HC/g rock).

3) The Tithonian source-rock interval has very good to excellent source-rock potential, with TOC ranging from 0.5 to 16 wt.% (average 3 wt.%) and S_2 ranging from 2 to 85 mg HC/g rock (average about 14 mg HC/g rock).

Reported sample data for the Upper Jurassic source rock from the Tampico and Tuxpan areas (Bertrand et al., 1996) indicate fair source-rock potential with TOC values that range from 0.1 to 2.3 wt.% and S_2 values that range from 0.5 to 7.8

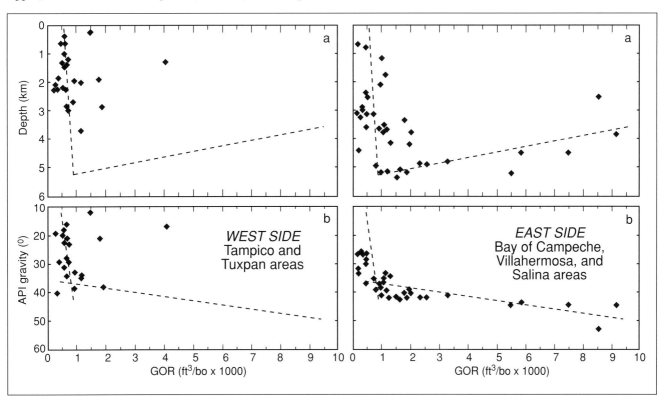

FIGURE 12. Graphs showing the relationship of (a) depth and (b) API oil gravity to GOR in ft³/BO. Data are from Table 3.

mg HC/g rock. The reported sample data from the Tampico, Tuxpan, Villahermosa, and Bay of Campeche areas are generally consistent with data summaries provided by other studies in these regions (Holguin et al., 1994; Raedeke et al. 1994; Bertrand et al., 1996). As summarized by Holguin et al. (1994), the Upper Jurassic source rock commonly has S_2 values that range from 5 to 36 mg HC/g rock and a hydrogen index (HI) between 500 and 650 mg HC/g TOC; the source-potential index (SPI) for the source-rock interval of Tithonian age in southeastern Mexico is 15 metric tons (t) HC/m². For comparison, the Upper Jurassic source rocks in central Arabia (Hanifa Formation) and the North Sea (Mandel Formation) have SPIs of 14 and 15 t HC/m², respectively (Demaison and Huizinga, 1994, their Table 4.1).

Character, Distribution, and Thickness

The character of the Upper Jurassic source rock is primarily known from well penetrations in the Tampico and Tuxpan areas (Holguin et al., 1994; Bertrand et al., 1996; Garcia, 1996; Pottorf et al., 1996; Román-Ramos et al., 1996) and the Salina, Villahermosa, and Bay of Campeche areas (Holguin et al., 1994; Medrano et al., 1996b; Mello and Guzmán-Vega, 1996; Mello et al., 1996; Guzmán-Vega and Mello, 1999). In general, the Upper Jurassic source rock includes the Oxfordian, Kimmeridgian, and Tithonian intervals that are primarily marine calcareous and shale-rich rocks, including thin-bedded, dark gray to black limestone, argillaceous limestone, calcareous shale, and dark shale deposited in various shelf, ramp, and basin settings (Salvador, 1991b). The general distribution of lithofacies for the Tithonian interval is summarized in Figure 6. The Upper Jurassic was deposited in various shelf and ramp settings that regionally merge, to the east and north, with basinal settings in the Gulf of Mexico. It is important to note that the Late Jurassic lacked siliciclastic deposition in this region.

In the Tampico and Tuxpan areas, similar lithofacies and source-rock characteristics are developed in Oxfordian, Kimmeridgian, and Tithonian intervals; oil and source-rock stud-

ies determined that the source-rock intervals were indistinguishable in the Upper Jurassic. However, in the Salina, Villahermosa, and Bay of Campeche areas, there appears to be more variation in lithofacies, both laterally and vertically (Raedeke et al., 1994). There, oil and source-rock studies locally distinguish specific source intervals, such as the Oxfordian in the northeastern Bay of Campeche (Gonzales and Cruz, 1994), but most emphasis is placed on the Tithonian section's generative potential and geochemical and spatial ties to nearby oil. The Tithonian interval in the Salina and Villahermosa areas, mostly shale-rich rocks deposited in deeper-marine environments, has excellent source-rock characteristics (Table 9).

The Upper Jurassic stratigraphy of the southern Gulf of Mexico basin is marked by a lack of siliciclastic sedimentation, but it has a characteristic open-marine shelf, ramp, and basin where shale-rich limestone and calcareous shale were deposited. The lithofacies distribution through the Late Jurassic reflects the general marine trangression and regional subsi-

TABLE 9. Thickness range of Upper Jurassic (Tithonian) source rock in Bay of Campeche area. Information from PEMEX (1999b).

Field number	Field name	Thickness (m)
143-175-154	Abkatun-Pol-Chuc	80–200
148	Balam	122
151	Caan	200–400
156	Ek	109
159	Ixtoc	100–180
169-183-165	Maloob-Zaap-Ku	80–200
171	May	130–310
174	Och, Uech, Kax	200
176	Sinan	110–190

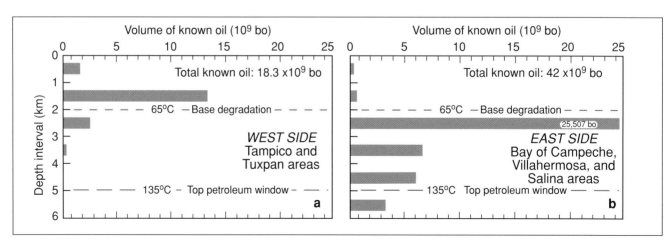

FIGURE 13. Histograms showing volume of known oil for each km depth interval in the petroleum system. (a) Western-side production is from the depth interval of 1 to 2 km, and (b) on the eastern side, most oil is being produced between 2 and 3 km. Data are from Table 3.

dence that characterized that time (Salvador, 1991b). The expansion of open-marine conditions and the wider distribution of shale-rich basinal facies that accompanied this transgression correspond to the development of the highest-quality source rock in the younger part of the Upper Jurassic (Tithonian). The general shift to a more shale-rich, basinal facies and better source-rock quality from the Oxfordian to the Tithonian (Figure 6) is also reflected in the Tithonian section. Raedeke et al. (1994) emphasize the importance of lateral lithofacies variation in the Tithonian of southeastern Mexico to source-rock potential. One variation includes a shift from a low-potential source rock deposited locally on a Tithonian submarine high to a high-potential source rock being deposited in a nearby depositional low. A second variation is a regional shift from low to high potential from the periphery to the center of the depositional basin. Last, the source-rock potential appears to follow the regional shift from a predominantly calcareous to a more shale-rich Tithonian lithofacies (Figure 6).

The lateral variation in source-rock potential that accompanies Tithonian lithofacies variation suggests that a higher-potential source rock could have developed throughout the Upper Jurassic in more basinal settings; more basinal Oxfordian and Kimmerigian lithofacies could have better source-rock quality than their shelf and ramp counterparts. Therefore, as presently known, the entire Upper Jurassic interval of the southern Gulf of Mexico contains sections with at least fair source-rock potential, but lithofacies that developed basinward of the shelf and ramp settings could have higher source-rock potential.

Thickness data summarized and mapped by Salvador (1991b) show that the Oxfordian, Kimmeridgian, and Tithonian sections in the southern Gulf of Mexico can be several hundred meters thick each and, combined, more than 1 km in thickness. Seismic studies basinward from coastal Mexico have identified an Upper Jurassic to Lower Cretaceous section that averages 2 to 3 km in thickness over deep parts of the Gulf of Mexico basin (Challenger Unit of Buffler, 1991). Upper Jurassic

sedimentary rock sections are also interpreted to be present in the Sigsbee Salt Basin along the northern flank of the Yucatán platform and in the southeastern Gulf of Mexico (Buffler, 1991). The seismic character and refraction velocities of some of these intervals suggest that they contain fine-grained carbonate sediments. Significant siliciclastic sedimentation did not occur in the southern Gulf of Mexico during the Late Jurassic, and basinal shale and shale-rich carbonate can be present throughout that region. The Upper Jurassic is considered the source-rock interval for deep-water plays in the northern Gulf of Mexico (Weimer et al., 1999; Fiduk et al., 1999). In the vicinity of nine fields in the Bay of Campeche area, the Tithonian source-rock interval ranges in thickness from 80 to 400 m (Table 9; PEMEX, 1999b).

In summary, the Upper Jurassic source rock is well developed in the southern Gulf of Mexico, with a composite thickness that ranges from several hundred meters to more than 1 km (Figure 14). More basinal depositional settings appear to have favored development of higher source-rock potential in the Tithonian, and similar settings may have resulted in higher-quality Oxfordian and Kimmeridgian source-rock intervals than these presently known. Oxfordian, Kimmerigian, and Tithonian intervals may merge basinward into vertically more continuous lithofacies typical of a high-potential source rock. The southern Gulf of Mexico basin could therefore have from several tens of meters to several hundreds of meters of fair to excellent Upper Jurassic source rock.

THERMAL MATURATION FROM OVERBURDEN ROCKS

The tectonic and depositional stability that characterized the southern Gulf of Mexico during the late Mesozoic was terminated by compressional deformation and related foredeep sedimentation along its western (Sierra Madre Oriental) and southwestern (Sierra de Chiapas) margins. This compressional deformation may have started as early as the Late Cretaceous

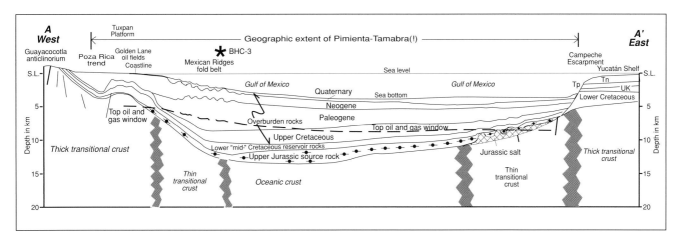

FIGURE 14. Regional cross section, southern Gulf of Mexico, showing geographic extent of Pimienta-Tamabra(!), location of burial-history chart 3 (Figure 16c), and essential elements of the petroleum system. This section combines segments from sections A-A' and B-B' of Salvador (1991d). See Figure 2 for location.

in the Sierra Madre Oriental, but it is primarily early Tertiary in age (Pottorf et al., 1996). In the Sierra de Chiapas, significant foredeep sedimentation continued into the Miocene (Schoell-kopf et al., 1994). About 4 to 5 km of Paleogene siliciclastic sediments buried Cretaceous rocks in the Sierra Madre Oriental and Sierra de Chiapas foredeeps (Pottorf et al., 1996; Peterson, 1983). This sandstone, siltstone, and shale prograded eastward and began filling the southern Gulf of Mexico. In general, siliciclastic sedimentation continued into the Neogene, when 4 to 6 km of additional overburden marine conglomerate, sandstone, siltstone, and shale were deposited in the Veracruz, Salina, and Bay of Campeche areas (Peterson, 1983). Regional cross sections (Figure 14; Peterson, 1983; Salvador, 1991d) illustrate that Cretaceous and Tertiary overburden rock is as thick as 12 km in Pimienta-Tamabra(!).

Based on present-day subsurface temperatures and thermal-maturity information from Rock Eval (T_{max}), oil generation begins in Pimienta-Tamabra(!) when the source rock is buried to 5 km (Figure 15). The subsurface oil-field temperatures from the Villahermosa and Bay of Campeche areas are used to determine the present-day geothermal gradient for the southern Gulf of Mexico (Santiago and Baro, 1992, their Table 3). Using 20°C as the surface temperature, a visual best-fit line was drawn that defines a 23°C/km (1.28°F/100 ft) geothermal gradient. Four Rock Eval T_{max} data points from Guzmán-Vega and Mello (1999) are shown on the left of Figure 15, with the deepest value being 422°C, which is less than the 435°C threshold for thermal maturity. In addition, the deepest oil accumulations occur from 5 to 6 km (Figure 13). At 5 km, the subsurface temperature is presently 135°C; for this study, the depth and temperature are considered reasonable for oil and gas generation (top petroleum window) and for initial expulsion from the source rock. Within 3 to 4 km of additional burial, the source rock is assumed to be depleted of all petroleum potential.

Three burial-history charts (Figure 16a, b, c) illustrate overburden development on the Upper Jurassic source rock and enable estimates for the time of oil and gas generation at 5-km burial depth in the Pimienta-Tamabra(!). The three charts represent burial history at: (1) a setting somewhat proximal to the Tertiary fold-and-thrust belt (Figure 16a, Veracruz area); (2) an intermediate shelf setting (Figure 16b, Bay of Campeche area); and (3) a basinal Gulf of Mexico setting (Figure 16c, offshore the Tuxpan area).

Post-Jurassic overburden rock proximal to the Tertiary fold-and-thrust belt in the Veracruz area includes the relatively thin 1 km-thick carbonate shelf deposit characteristic of the Cretaceous, followed by 3.7 km of Paleogene and 4.3 km of Neogene siliciclastic sediments. The Paleogene and some Neogene sedimentation was sufficient to place the Upper Jurassic source rock at 5 km by the early Miocene. Neogene sedimentation led to at least 4 more km of total overburden thickness and complete thermal maturation of the Upper Jurassic source rock; this source rock probably became depleted of oil-generating potential by the late Pliocene.

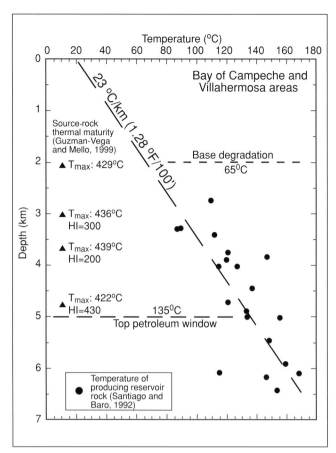

FIGURE 15. Graph showing subsurface reservoir temperatures on the eastern side to determine geothermal gradient, base of degradation, and top of petroleum window. Data are from Table 3. Shown also are Rock Eval T_{max} temperatures for the Upper Jurassic source rock (Guzmán-Vega and Mello, 1999).

In an intermediate shelf setting (Figure 16b), Cretaceous calcareous and Paleogene siliciclastic sedimentary rocks are each about 1.6 km thick. However, Neogene sedimentation was rapid, and by the early Miocene, 5 km of overburden thickness had developed on the Upper Jurassic source rock. The source rock is presently at 8.7 km of burial, where it is almost depleted of petroleum potential.

In a basinal Gulf of Mexico setting (Figure 16c), post-Jurassic overburden thickness developed more continuously up to the present. Basinal Cretaceous sediments reached 3.3 km in thickness, Paleogene sediments added another 4.3 km, and the Neogene sediments are 2.1 km thick. Five km of overburden thickness was attained in the Middle Eocene. By the early Oligocene, the Upper Jurassic was depleted of petroleum potential.

The loading caused by Paleogene siliciclastic deposition led to some Jurassic salt movement and development of related seafloor topography. This topography locally influenced Neogene sedimentation and resulted in rapid changes in sediment thickness. Such thickness variations likely have led to local

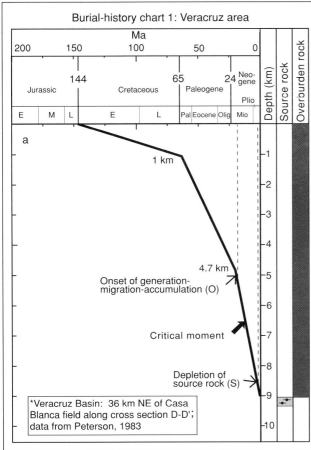

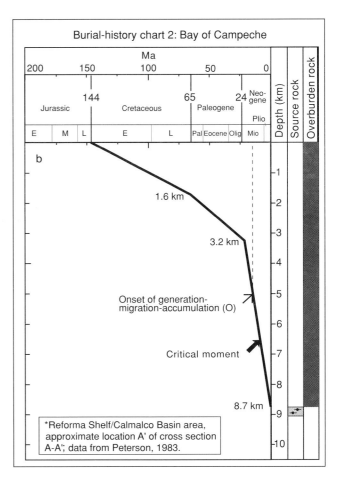

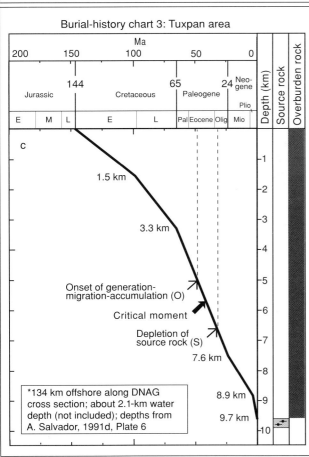

FIGURE 16. Burial-history charts in the southern Gulf of Mexico for three locations: (a) Veracruz area, (b) Bay of Campeche area, and (c) Tuxpan area. Interval thicknesses are from cross sections of Salvador (1991d) and Peterson (1983). Locations are shown in Figure 2.

variations in the time and extent of thermal maturity of the Upper Jurassic source rock (Schoellkopf et al., 1994).

In summary, Cenozoic siliciclastic sedimentation created overburden thickness sufficient to thermally mature the Upper Jurassic source rock. These sediments were derived from fold-and-thrust belts and regional uplifts to the west and southwest of the southern Gulf of Mexico. Assuming that oil generation was initiated when 5 km of total overburden thickness had developed, the onset of generation-migration-accumulation could have started in the Eocene in some foredeep and off-shore basinal settings (Figure 16c). However, large parts of the southern Gulf of Mexico developed this overburden thickness later in the Neogene, so that petroleum generation and some expulsion is ongoing today (Figures 16a, b).

PETROLEUM CHARGE

Petroleum charge can be determined in several ways, two of which have been used in this study. Demaison and Huizinga (1994) developed the source-potential index (SPI) to quickly determine the generative capacity of a pod of active source rock. The SPI is defined as "the maximum quantity of hydrocarbons (in metric tons) that can be generated within a column of source rock under 1 m^2 of surface area." The maximum volume generated is the product of the SPI for the Upper Jurassic source rock (15 t/m^2) and the area of the active pod (333 x 10^{9}m^2) or 5 x 10^{12} t of petroleum. If all the petroleum generated is 40° API, then 38,000 billion barrels of oil has been generated (7.62 BO/t of 40° API oil). Assuming 10% migration and trapping efficiency (approximately 4000 billion barrels of oil) and 25% recovery from each trap, about 1000 billion barrels of recoverable oil is the estimated petroleum charge for Pimienta-Tamabra(!), or 20 times more oil (and gas) than has been discovered.

Schmoker (1994) used a different approach to calculate the amount of oil generated in a pod of active source rock (Table 10). Although there can be local variation of source-rock quality, a compilation of data for 462 samples indicates an average TOC content of about 3% for Tithonian source rocks (Medrano et al., 1996b). Bertrand et al. (1996) reports sample data from 9 m of source-rock thickness that average more than 2 wt.% TOC. In the Bay of Campeche area, the source-rock thickness ranges from 80 to 400 m (Table 9). Based on this information, a minimum source-rock thickness of 100 m (scenario 1 in Table 10), an active source-rock pod area of 333 x 10^{9}m^2, an average TOC of 2 wt.%, a density of 2.4 gm/cm^3, an original HI of 500, and a current HI of 150 indicate an estimated generative capacity for Pimienta-Tamabra(!) of 4266 billion barrels of oil (scenario 1 in Table 10). Assuming a 10% migration and trapping efficiency and 25% recovery from each trap, 107 billion barrels of oil is estimated as the petroleum charge of the Pimienta-Tamabra(!).

Klemme (1994) classified petroleum systems on the volume of recoverable petroleum in BOE, where a barrel of oil is equivalent to 6000 ft^3 of gas. The volume of known oil is 66.3 x 10^6 barrels of oil and 103.7 x 10^9 ft^3 of gas, which is equivalent to 17.3 x 10^6 barrels of oil. So the BOE for the Pimienta-Tamabra(!) is 83.6 x 10^6. Klemme (1994) designates petroleum systems with 20–100 x 10^6 BOE as giant. Hence, the Pimienta-Tamabra(!) is a giant petroleum system.

The generation-accumulation efficiency (GAE; Magoon and Valin, 1994) for the Pimienta-Tamabra(!) is 6.5%; thus, for every 100 barrels of oil generated, about 6.5 barrels make it to the trap as in-place oil. The volume of oil generated is the denominator, or as mentioned above, 4266 x 10^9 barrels, and the numerator is the volume of in-place oil, or 276.5 x 10^9 barrels (Table 4). According to Magoon and Valin (1994), a GAE of 6.5% is representative of a moderately efficient petroleum system, i.e., between 1% and 10%.

Demaison and Huizinga (1994) classified petroleum systems using the charge factor, migration drainage style, and entrapment style. Based on their scheme, the Pimienta-Tamabra(!) is a supercharged, vertically drained, high-impedance petroleum system (Demaison and Huizinga, 1994, their Figure 4.14).

RESERVOIR ROCKS

Many stratigraphic horizons in the southern Gulf of Mexico have reservoir-quality rocks. Of all oil occurrences reported by Petroconsultants (1996), 10% are in Upper Jurassic reservoir

TABLE 10. Hydrocarbon charge calculation for Upper Jurassic source rock (HI, hydrogen index; TOC, total organic carbon; HC, hydrocarbon).

Given:

Source-rock richness, TOC:	2.0 wt. %
Source-rock quality:	
Original, HI$_o$:	500 mg HC/g TOC
Present-day, HI$_p$:	150 mg HC/g TOC
Source-rock density, ρ:	2.4 g/cm^3
Pod of active source-rock area, A:	333 x 10^9 m^2

Calculations:

			Pod of active source rock volume			Barrels of oil generated			Recoverable oil
Scenario	Thickness (m)	Area (m^2)	Volume (m^3)	Volume (cm^3)	M (g TOC)	R (mg HC/g TOC)	HCG (kg HC)	Oil generated (10^9 bbl)	(10^9 bbl)
1	100	A	33.3 x 10^{12}	33.3 x 10^{18}	1.6 x 10^{18}	350	560 x 10^{12}	4266	107
2	200	A	66.6 x 10^{12}	66.60 x 10^{18}	3.2 x 10^{18}	350	1119 x 10^{12}	8531	213
3	500	A	166.5 x 10^{12}	166.5 x 10^{18}	8 x 10^{18}	350	2797 x 10^{12}	21,328	533

Equations:

$$M \text{ (g TOC)} = [\text{TOC (wt.\%)}/100] \times \rho \text{ (g/cm}^3\text{)} \times V \text{ (cm}^3\text{)}$$

$$R \text{ (mg HC/g TOC)} = HI_o \text{ (mg HC/g TOC)} - HI_p \text{ (mg HC/g TOC)}$$

$$HCG \text{ (kg HC)} = R \text{ (mg HC/g TOC} \times M \text{ (g TOC)} \times 10^{-6} \text{ (kg/mg)}$$

$$\text{Oil (10}^{12} \text{ bbl)} = HCG \text{ (kg HC)} \times 131.15 \text{ (kg/bbl of 40 API° oil)}$$

rocks, 38% are in the mid-Cretaceous, and 28% are in the Miocene (Table 11). However, carbonate reservoirs, especially the debris-flow facies (Tamabra Limestone facies) on the flanks of platforms, are the most important (Figures 3, 4, and 7). The largest fields by oil volume (64.8%) are in carbonate debris-flow breccias and related sediment gravity-flow rocks derived from the carbonate platform margins and transported seaward into deeper-water slope and basinal settings (Table 8). Carbonate debris-flow breccias were first recognized as a unique but potentially important type of reservoir facies by Cook et al. (1972) in the Permian of Texas and later by Cook (1983a, b) and Cook and Mullins (1983). An excellent example of these types of reservoirs is the Poza Rica field.

The Poza Rica reservoir rocks are comprised of allochthonous carbonate breccia, grainstone, wackestone, and packstone facies of the mid-Cretaceous Tamabra Limestone (Viniegra O. and Castillo-Tejero, 1970; Enos, 1977, 1985). This field, on the southwestern flank of the Tuxpan platform, contained more than 2 billion barrels of recoverable oil (Enos, 1977, 1985). There is considerable range in reservoir properties at Poza Rica; porosity varies from 1% to 25% and averages 8%, and permeability varies from 0.1 to 700 md (Enos, 1977, 1985). Carbonate platform margin facies such as the Tamabra Limestone are widely distributed around the southern Gulf of Mexico (Figures 2, 3, and 4) and are especially well developed along the Yucatán platform margin, where similar depositional settings persisted into the Paleocene (Galloway et al., 1991; Santiago and Baro, 1992; Grajales-Nishimura et al., 2000). There, this facies is a very important reservoir in the huge Cantarell field, which contains at least 19.8 billion barrels of recoverable oil (Table 2).

Allochthonous carbonate debris-flow breccias and carbonate turbidity-flow deposits, which contain 64.8% of the oil in the Pimienta-Tamabra(!), can be formed in at least two ways (Cook et al., 1972). First, eustatic sea-level fluctuations of tens of meters periodically expose the reef edge to erosion and diagenetic alteration. In this situation, highstanding reef edges are more easily broken off and cascade down the reef escarpment onto the slope and to the base of slope. Second, earthquakes and related tsunamis can shake and brecciate reef-edge or platform debris and enable it to be displaced onto the slope and base of slope. In Pimienta-Tamabra(!), regional shaking from the Chicxulub meteorite impact dislodged and relocated reef material onto the slope and base of slope (Grajales-Nishimura et al., 2000).

The basinal carbonates that developed outboard of the platform margin sediments are also important reservoirs in the region. For example, more than 1 billion barrels of oil has been produced from fractured wackestone of the mid-Cretaceous Upper Tamaulipas Limestone in the Ebano-Panuco field in the Tampico area. This carbonate facies also forms important reservoirs in several fields of the Villahermosa area that are distal to the carbonate platforms that bordered the southern Gulf of Mexico in the Cretaceous (Santiago and Baro, 1992).

TABLE 11. Percent of reservoir rocks by age occurrence that contain oil by assessment unit. Information from Petroconsultants (1996).

Reservoir rock	1 (%)	2 (%)	3 (%)	4 (%)	5 (%)	6 (%)	Total (%)
Pliocene	1	0	0	0	0	0	<1
Miocene	16	7	16	0	1	74	28
Oligocene	3	0	16	0	1	0	2
Eocene	5	2	4	21	2	0	4
Paleocene	2	31	0	0	0	0	5
Upper Cretaceous	1	7	48	0	24	3	9
Middle Cretaceous	71	35	12	54	42	13	38
Lower Cretaceous	1	4	4	2	13	1	4
Upper Jurassic	0	14	0	23	17	9	10

Reef buildups at platform edges are also important reservoirs. The best examples are the wackestones, grainstones, and packstones, commonly with rudist fragments, of the mid-Cretaceous El Abra Limestone that are the major reservoirs in the Golden Lane fields around the periphery of the Tuxpan platform (Figures 2 and 3). These rocks formed in various reef and shelf-edge sand shoals, as well as semirestricted tidal-flat depositional settings (Enos, 1977). Tertiary erosion and karst development led to vuggy and cavernous porosity in the upper parts of these reefs (Viniegra O. and Castillo-Tejero, 1970).

In general, the carbonate reservoirs of the southern Gulf of Mexico are variably dolomitized and fractured. Secondary vuggy porosity is common, although average matrix porosities are low, between 3% and 10% (Santiago and Baro, 1992). Permeability varies widely from a few millidarcys to 1000 md or more (Nehring, 1991; Santiago and Baro, 1992). Maximum production figures for some fields in the Bay of Campeche indicate that individual wells produce from 2700 to more than 13,000 barrels per day (Santiago and Baro, 1992).

Cenozoic siliciclastic sedimentation developed some important sandstone reservoirs. Late Paleocene and Eocene turbidites of the Chicontepec Formation contain a large amount of unrecoverable oil in the Chicontepec field in the Tuxpan area (Tables 2, 3, and 8; Busch, 1992). Another important Tertiary sandstone reservoir rock is the Miocene Encanto Formation in the Salina area (Acevedo, 1980). Encanto Formation reservoirs contain more than a billion barrels of recoverable oil (Table 8). Thickness varies considerably, because seafloor topography related to movements of Jurassic salt influenced sediment distribution. Encanto Formation thicknesses of 1000 m or more can be present on flanks of salt structures in the Salina area. The reservoir properties of Encanto Formation sandstones are good; porosities average 19% to 25%, and permeabilities aver-

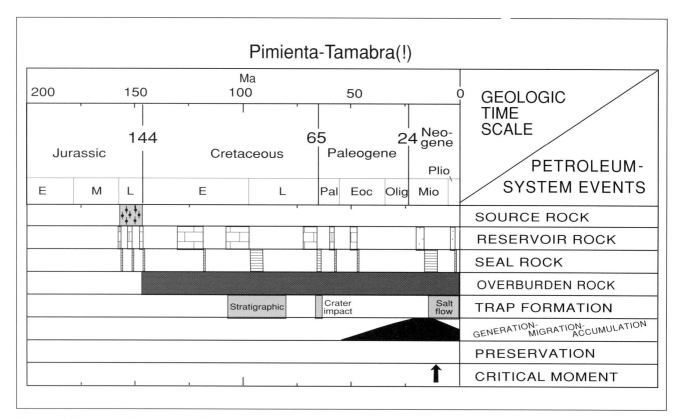

FIGURE 17. Events chart summarizing the essential elements, processes, and critical moment for the Pimienta-Tamabra(!). Names and ages of stratigraphic units are shown in Figure 5.

age about 180 md (Table 3; Acevedo, 1980). Although these Tertiary sandstone reservoirs contain significant oil in some areas, Tertiary sediments are more commonly overburden rocks.

More oil is known to occur in the Tamabra-like reservoir facies than any other facies in this petroleum system (Table 8). In decreasing order, the volumes of known oil and percentages are as follows: Tamabra-like (64.8%), siliciclastic sandstone (20%), Tamaulipas-like (10.7%), and El Abra (3.9%). For this reason, Tamabra is used as the reservoir rock name in this petroleum system (Magoon and Dow, 1994).

SEAL ROCKS

Marine calcareous or shale-rich facies capable of forming an effective seal were deposited at many times in the southern Gulf of Mexico (Figure 17). Other facies, such as evaporites (Gonzalez and Cruz, 1994), can be locally important, but shale, calcareous shale, and pelagic limestone are the most important seals. Facies changes are important seals in Cretaceous carbonate sediments, and Tertiary shales are important in structural traps of the region.

Basinal pelagic wackestone and other fine-grained pelagic limestone are important lateral and vertical seals in Cretaceous sediments. Laterally, basinward changes from platform-margin and platform-slope facies (Tamabara Limestone facies) to basinal pelagic limestone are important sealing elements at the

Poza Rica field (Enos, 1977). This field also demonstrates that sealing can develop between the platform-slope facies and various updip platform facies. These lateral carbonate facies changes are the key elements of stratigraphic trapping mechanisms in the region. The development of various fine-grained pelagic limestone, calcareous shale, and calcareous mudstone was widespread in the region during the Late Cretaceous (Agua Nueva, San Felipe, and Mendez Formations; Peterson, 1983), and an important bentonitic seal rock developed as a result of the Chicxulub impact crater (Grajales-Nishimura et al., 2000).

Late Paleocene and Eocene siliciclastic turbidites are oil bearing and laterally sealed by facies changes to pelitic rocks in the Chicontepec Formation in the Tuxpan area (Busch, 1992). Other younger Tertiary turbidites could be present in Pimienta-Tamabra(!) with similar sealing configurations. However, the principal Miocene sandstone reservoir of the Salina area is sealed by overlying, thick Neogene shale section (Acevedo, 1980; Peterson, 1983). Oligocene marine sediments fill caverns and provide vertical seals for some Golden Lane fields along the northwest margin of the Tuxpan carbonate platform (Viniegra O. and Castillo-Tejero, 1970).

In summary, facies changes have produced effective lateral seals in mid-Cretaceous carbonate rocks and in some Tertiary siliciclastic rocks. The two most important vertical seals are the widespread pelagic limestone and related rocks of the Upper Cretaceous and the widespread, thick shaly rocks of the Tertiary.

TRAP FORMATION

Both structural and stratigraphic traps are important in Pimienta-Tamabra(!) (Figure 17). Most previous work has emphasized the importance of structural traps, and examples are provided by Acevedo (1980) and by Santiago and Baro (1992). These traps are primarily developed by middle to late Tertiary movements of Jurassic salt. In many cases, these traps are complicated, highly faulted structures. The structural traps in the Bay of Campeche and Villahermosa areas are commonly 10 to 50 km^2 in area and have thick, oil-bearing columns that vary from about 200 m to more than 1300 m (Table 3; Santiago and Baro, 1992). Structural or combination structural-stratigraphic traps can be expected throughout a large part of the southern Gulf of Mexico, where salt diapirs, pillows, or other structures are present (Ewing and Lopez, 1991). To date, exploration drilling in this structural province has been primarily in water depths of 200 m or less.

The most important stratigraphic traps are those related to facies changes in carbonate sediments (Figure 17). An example is the Poza Rica field in the mid-Cretaceous Tamabra Limestone (Viniegra O. and Castillo-Tejero, 1970; Enos, 1977). Although some Tertiary tilting of this reservoir has occurred, it is the effective lateral seals caused by facies changes, both basinward and updip toward the Tuxpan platform, that are essential for trapping. Other important stratigraphic traps are present on the Tuxpan platform. The offshore Golden Lane fields, where Upper Cretaceous pelagic limestone drapes over and seals the mid-Cretaceous reef facies, are examples of unconformity traps. Another unconformity trap seems to be present where Tertiary sediments overlap and provide vertical seals in some "old" Golden Lane fields along the northwestern margin of the Tuxpan platform. Both types of stratigraphic traps could be present along other platform margins in the southern Gulf of Mexico, as shown in the events chart (Figure 17). Movements of Jurassic salt could destroy the integrity of any of these older stratigraphic traps; therefore, they are most likely to be viable where significant amounts of Jurassic salt are absent.

TIMING

The size and character of Pimienta-Tamabra(!) in the southern Gulf of Mexico are controlled by the depositional sequence of the essential elements of this petroleum system (Figure 17). A key ingredient in the size of Pimienta-Tamabra(!) was the development of a large but semirestricted epeiric sea at the end of the Jurassic that covered the southern Gulf of Mexico. This sea lacked siliciclastic deposition in its southern areas and maintained its general open-marine but semirestricted character from the Late Jurassic through the Cretaceous. The tectonic and depositional stability that has characterized this area to the present day led to the orderly development, in proper sequence, of the essential elements of Pimienta-Tamabra(!), the Upper Jurassic source rock, Cretaceous reservoir and seal rocks, and Cretaceous through Cenozoic overburden rocks (Figure 17).

The Upper Jurassic source, reservoir, and seal rocks were the first essential elements to be deposited (Figure 17). The Late Jurassic semirestricted but open-marine depositional environments, without siliciclastic deposition, enabled the organic-rich shale, calcareous shale, and marl so well represented by the Tithonian in the southern Gulf of Mexico to be deposited. Fortunately, the general transgression that took place in the Tithonian led to the present-day distribution of higher-quality source rock in areas accessible to drilling (onshore and shallow-water offshore areas). As a result, southeastern Mexico provides an observational window into the controls on Tithonian source-rock distribution, richness, and quality that is critical to understanding Pimienta-Tamabra(!). More basinal depositional settings favored development of a very high potential source rock. Similar depositional settings are inferred over large parts of the southern Gulf of Mexico in the Oxfordian and Kimmeridgian. The basinward merging of Oxfordian, Kimmeridgian, and Tithonian source-rock lithofacies suggests that the thickness of high-potential source rock in the Pimienta-Tamabra(!) is substantial. The deposition of Upper Jurassic carbonate rock, which in some areas is interbedded with the source rock, is an important reservoir rock in certain traps and acts as a carrier bed for lateral migration of petroleum to eventually access overlying Cretaceous reservoir rocks. Just south of the Poza Rica field, oil is produced from the Upper Jurassic San Andrés Formation (Enos, 1977). Upper Jurassic source-rock intervals act as seals for reservoir rocks of the same age and, together, are carrier beds (Figure 17).

The Cretaceous carbonate reservoir rocks in the Tuxpan and Tampico areas on the western side and the Villahermosa and Bay of Campeche areas on the eastern side contain 80% of the oil in the Pimienta-Tamabra(!) (Table 8). Cretaceous seals are mostly finer-grained facies of the reservoir facies, except for the impermeable, dolomitized, bentonitic bed deposited as a result of the Chicxulub meteor impact (Grajales-Nishimura et al., 2000). The substantial thickness of reservoir and seal rocks is also part of the overburden rock.

Siliciclastic sandstone reservoir rocks of Paleogene age in the Tuxpan area contain more than 18% by volume of the known oil, and the Miocene reservoirs in the Salina area contain less than 2% of known oil (Table 8). Mostly, the very thick Cenozoic section is overburden rock that is many kilometers thick (Figure 16a, b, c).

Traps formed at three times (Figure 17). Stratigraphic traps formed at the time of deposition of Cretaceous carbonate rock, as in the Tampico and Tuxpan areas. Structural traps formed when previously deposited reservoir and seal rocks were folded in three ways: (1) tectontic deformation, as in the Veracruz area; (2) by impact of a meteorite, as in the Bay of Campeche and Villahermosa areas during the Late Cretaceous to early Paleocene; and (3) by salt movement under the eastern side during the Neogene.

The time of generation-migration-accumulation of hydrocarbons is dependent on the deposition of the overburden rock

(Figure 17). Hydrocarbons began moving toward traps as early as the Eocene, but most of the petroleum accumulated during the Miocene. Most of the Upper Jurassic source rock is depleted today, but oil and gas are being expelled in small areas, which is why a preservation time is lacking. Because most of the oil and gas was generated and trapped in the Miocene, that is the critical moment (Figure 17).

The timing of generation-migration-accumulation and trap development are favorable; that is, the traps formed before the charge arrived (Figure 17). Significant Cretaceous platform-margin stratigraphic traps existed around the periphery of Pimienta-Tamabra(!) throughout the time of hydrocarbon charge. The control on Upper Jurassic source-rock thermal maturation, Tertiary siliciclastic deposition, also triggered movement of Jurassic salt and development of the important structural traps of the region. As development of the hydrocarbon charge is ongoing, effective trap filling would seem to be fostered in the system. The presence of oil-filled reservoirs from very shallow levels to depths of more than 5 km is a further indication that the timing of migration and trap filling has been optimal.

The movement of Jurassic salt has led to complex faulting that facilitated vertical migration; it is common for many reservoirs to be vertically stacked in the large structural traps of the region. However, lateral migration is also important. The very important oil fields around the Tuxpan platform (Golden Lane fields) are updip from the thermally mature Upper Jurassic source rock in an area that lacks many faults. The important stratigraphic traps there are thought to have been charged by lateral migration from generating areas to the east (Garcia, 1996). Other areas undisturbed by salt movement, such as the northern margin of the Yucatán platform, could also be charged by lateral migration.

In general, exploration of Pimienta-Tamabra(!) has focused on structural traps, both onshore and offshore in 200 m water depth. The regionwide distribution of both source and reservoir rocks (Figure 2), the presence of long-lived stratigraphic traps around the southern Gulf of Mexico reef margin, and the very large area of unexplored salt structures in offshore deeper waters (Ewing and Lopez, 1991) suggest that there are many exploration opportunities for the future. In light of the tremendous generative capacity of Upper Jurassic active source rock, it seems likely that significant new petroleum reserves await discovery.

PETROLEUM ASSESSMENT AND POTENTIAL

The U. S. Geological Survey World Assessment Team's work was carried out as a separate activity after the Pimienta-Tamabra(!) TPS was investigated. The senior author and members of the World Energy Assessment Team determined the assessment unit (AU) boundaries on the basis of geologic setting, petroleum occurrence, operational difficulty, and petroleum potential. Several months later, the same people carried out the assessment. The methodology and conclusions of this assessment are included in a CD-ROM series (USGS, 2000).

The Pimienta-Tamabra(!) TPS was evaluated for undiscovered oil and gas resources by the World Energy Assessment Team (USGS, 2000). Based on reservoir rocks, trap types, and exploration intensity, seven AUs were identified (Figure 18). Only the mean values of undiscovered oil and gas resources for each AU are listed in Table 12.

The El Abra–like reef and back-reef limestone (AU1) has a low undiscovered potential of 317 x 10^6 BO that includes NGL and 507 x 10^9 cfg (Table 12), because most oil fields are presumed already discovered. The Golden Lane and interior platform fields have a known volume of 2099 x 10^6 BO (Table 8).

The Tamabra-like debris-flow breccia limestone overlying evaporites (AU2) includes the K/T breccia and debris flows. This AU has the highest undiscovered potential of 11,072 x 10^6 BO that includes NGL and 20,449 x 10^9 cfg (Table 12), because this is a very large area where the reservoir rocks and traps are well developed and fully charged with petroleum. The Villahermosa and Bay of Campeche area contains about 67.5% of the known oil (Table 8).

The Tamabra-like debris-flow breccia limestone and overlying strata (AU3) lack a thick, mobile underlying salt layer and the K/T breccia, but the area is tectonically disrupted in the Veracruz area (Figure 18). This AU was assessed even though little information is available. This AU is a large, underexplored trend. The mean volume of undiscovered oil is 1106 x 10^6 BO that includes NGL and 6147 x 10^9 cfg (Table 12). Presently, the known volume of oil is 83 x 10^6 BO, or about 0.1% of the known oil for this TPS.

The Tamabra-like debris-flow breccia limestone of the Golden Lane (AU4) includes the Poza Rica trend that produces from the Tamabra Formation, San Andrés Formation, and some overlying Paleogene reservoir rocks with a known volume of 16.5 x 10^9 BO (Table 8). The highest potential in this area is the offshore side, where deeper stratigraphic traps may be charged; it is estimated to contain 624 x 10^6 BO that includes NGL and 1113 x 10^9 cfg (Table 12).

The Tamaulipas-like basinal limestone and Tertiary strata without underlying evaporites (AU5) include the Tamaulapis Formation, which produces where it has been fractured by tectonic activity. This AU has low potential because the analogue traps in the Tampico area had a poor production history, with a total production to date of 1685 x 10^6 BO (Table 8). Mean undiscovered volumes are 413 x 10^6 BO and 804 x 10^9 cfg (Table 12).

The Tamaulipas-like basinal limestone and Tertiary strata overlying evaporites (AU6) include the deep-water siliciclastic Miocene sandstone reservoirs and the underlying carbonate reservoir rocks as their exploration objective. The high potential results from the high quality of these sandstone reservoirs and the lack of apparent exploration in the underlying carbonate rocks. Known oil volumes are 1138 x 10^6 BO that includes

TABLE 12. Mean values of undiscovered oil and gas resources by assessment unit for the Pimienta-Tamabra(!) total petroleum system. Information from USGS (2000) (bbl, barrels; N/A, not applicable; NGL, natural-gas liquid).

AU no.	Assessment unit name	AU No.	Liquid			Gas
			Oil (10⁶ bbl)	NGL (10⁶ bbl)	Total (10⁶ bbl)	Natural gas (10⁹ ft³)
1	El Abra–like reef and back-reef limestone	53050101	289	28	317	507
2	Tamabra-like debris-flow breccia limestone overlying evaporites	53050102	9857	1215	11,072	20,449
3	Tamabra-like debris-flow breccia limestone and overlying strata	53050103	807	299	1106	6147
4	Tamabra-like debris-flow breccia limestone of the Golden Lane	53050104	557	67	624	1113
5	Tamaulipas-like basinal limestone and Tertiary strata without underlying evaporites	53050105	365	48	413	804
6	Tamaulipas-like basinal limestone and Tertiary strata overlying evaporites	53050106	7824	991	8815	18,516
7	Tamabra-like debris-flow breccia limestone north of Campeche	53050107	871	104	975	1735
		TOTAL:	20,570	2752	23,322	49,271

TABLE 13. Pimienta-Tamabra(!) TPS size from known volumes (Magoon and Schmoker, 2000) and mean values of undiscovered oil and gas resources (USGS, 2000) (B, barrels; BOE, barrel of oil equivalent to 6000 ft³ gas; N/A, not applicable; NGL, natural-gas liquid).

Item	Liquid			Gas	BOE (10⁶ B)
	Oil (10⁶ B)	NGL (10⁶ B)	Total (10⁶ B)	Natural gas (10⁹ ft³)	
Known volumes (Magoon and Schmoker, 2000)	44,412	95	44,507	50,822	52,977
Mean undiscovered resources from Table 12 (USGS, 2000)	20,570	2752	23,322	49,271	31,534
TOTAL PETROLEUM SYSTEM	**64,982**	**2847**	**67,829**	**100,093**	**84,511**

NGL (Table 8), and mean undiscovered volumes are 8815 x 10⁶ BO and 18,516 x 10⁹ cfg (Table 12).

The Tamabra-like debris-flow breccia limestone north of Campeche (AU7) is adjacent to the Campeche Escarpment in the K/T facies. Presently, this AU lacks known petroleum accumulations. The appropriate reservoir facies and proximity to the petroleum charge suggest good potential. Undiscovered mean volumes are 975 x 10⁶ BO that includes NGL and 1735 x 10⁹ cfg (Table 12).

The size and number of oil fields used in this assessment, which includes data through 1995, came from Petroconsultants (1996). Adding the total volume of oil from these fields gives a total of 44.5 x 10⁹ BO that includes NGL and 50.8 x 10¹² cfg. The mean undiscovered potential of the Pimienta-Tamabra(!) is 23.3 x 10⁹ BO that includes NGL and 49.3 x 10¹² cfg. For this assessment, the size of the TPS, using data through 1995, is 84.5 x 10⁹ BOE (Table 13). The greatest undiscovered potential is on the eastern side and includes AU2, AU6, and AU7 (Table 12).

After this assessment was completed, two volumes by the staff of PEMEX's Exploration and Development Department became available (PEMEX, 1999a, b) that contained cumulative production and total reserves for all of Mexico. Table 2 includes information on fields in the Pimienta-Tamabra(!) that

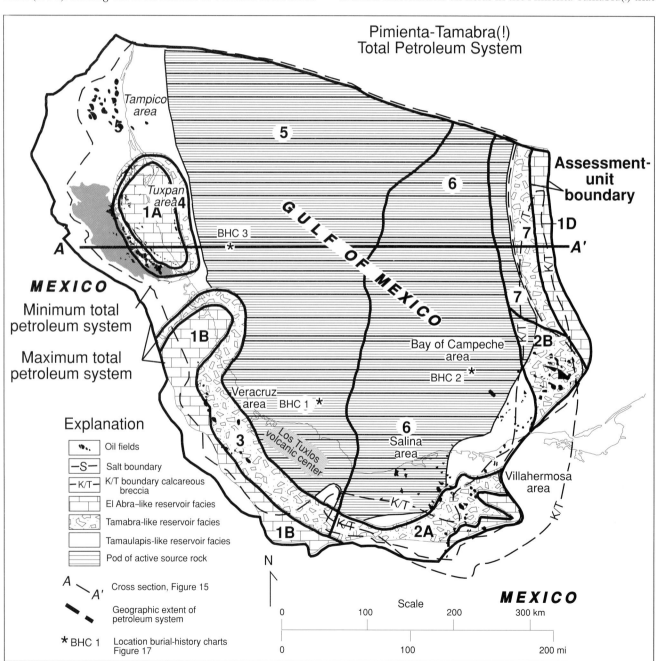

FIGURE 18. Map showing the Pimienta-Tamabra(!) TPS with the assessment units (AU) that were evaluated for undiscovered oil and gas. Evaluation results are shown in Tables 12 and 13.

have a known petroleum potential of 83.6 x 10^9 BOE (Table 4). The similarity of these two numbers — 84.5 x 10^9 and 83.6 x 10^9 — does, however, suggest that this TPS has more undiscovered new field resources than were assessed, so additional new field discoveries are likely.

SUMMARY

Pimienta-Tamabra(!) is a giant, supercharged petroleum system in the southern Gulf of Mexico. Virtually all of Mexico's oil fields in the Tampico-Misantla, Salina, Chiapas-Tabasco, and Bay of Compeche areas produce from this petroleum system. The tremendous generative capacity of the Pimienta-Tamabra(!) is a result of the very large area of thick, high-quality, deeply buried Upper Jurassic source rock. The richness of this high-quality source rock is at least 2 wt.% TOC with a hydrogen index (HI) of at least 500 mg HC/g TOC. The source-potential index (SPI) of the source rock is 15 t HC/m^2, which makes this a supercharged petroleum system. This petroleum system has a generation-accumulation efficiency (GAE) of 6.5% and is drained laterally with high impedance.

The petroleum charge started migrating as a supercritical fluid during the Eocene into stratigraphic and structural traps; it peaked in the Miocene and, to a lesser extent, is ongoing today. The oil and gas began their migration from the Upper Jurassic source rock at a depth of 5 km or more as a 35–40°API oil with a sulfur content of less than 1% and a GOR of 500–1000 ft^3/BO. At shallower depths, the oil and gas phases separated, causing the API gravity to decrease, the sulfur content to increase, and the gas to migrate to other accumulations because of its mobility. Most of the oil accumulated at a depth of 1 to 3 km, with some accumulations more than 5 km in depth.

ACKNOWLEDGMENTS

Timothy R. Klett, Feliks Persits, and Zenon C. Valin were helpful in compiling and organizing the large petroleum-occurrence data files used in this study. The compilations and syntheses provided by the Geological Society of America's Decade of North American Geology volume on the Gulf of Mexico (Salvador, 1991a) were an exceptional resource. Special appreciation is due the many geoscience professionals in Mexico who have diligently pursued the understanding of petroleum systems of the area. It is their many recent and timely contributions, insufficiently acknowledged here, that made this compilation and synthesis possible. The authors thank Petroconsultants, Inc., and GeoMark Research, Inc., for permission to include some information from their databases. The authors especially thank Colin Stabler, M. E. Henry, and A. C. Huffman for their insightful reviews and helpful suggestions that improved this manuscript. The authors accept full responsibility for the synthesis and interpretation of published information used in this petroleum-system study.

REFERENCES CITED

Acevedo, J. S., 1980, Giant fields of the southern zone—Mexico, *in* M. T. Halbouty, ed., Giant oil and gas fields of the decade 1968–1978: AAPG Memoir 30, p. 339–385.

Becerril, F. G., A. A. Sosa Patrón, M. L. Clara Valdez, N. Holguín Quiñones, F. R. Guzmán, and A. A. Rodríguez, 1996, El subsistema generador Tithoniano del sureste de México, *in* E. Gómez Luna and A. Martinez Cortés, eds., Memorias del V Congreso Latino-Americano de Geoquímica Orgánica [Proceedings of the 5th Latin American Congress on Organic Geochemistry]: Cancún, México, p. 59–60.

Bertrand, C. R., N. Cañipa, T. Carrillo, M. Espinosa, F. García, S. Lucach, and E. Olivo, 1994, Geochemical methods applied to the characterization and correlation of oils and source rocks of Mexico, *in* N. Schneidermann, O. Cruz, and R. Sanchez, eds., Abstracts, First joint AAPG/Asociación Mexicana de Geólogos Petroleros Hedberg Research Conference, Geological aspects of petroleum systems: Mexico City, Mexico, 5 p.

Bertrand, C., N. Cañipa, L. Castro, and C. León, 1996, Caracterización geoquímica de aceite y rocas de diferentes sistemas generadores de la región norte de México, *in* E. Gómez Luna and A. Martinez Cortés, eds., Memorias del V Congreso Latino-Americano de Geoquímica Orgánica [Proceedings of the 5th Latin American Congress on Organic Geochemistry]: Cancún, México, p. 39–41.

Buffler, R. T., 1991, Seismic stratigraphy of the deep Gulf of Mexico basin and adjacent margins, *in* A. Salvador, ed., The Gulf of Mexico basin: Geological Society of America, The geology of North America, v. J, p. 353–387.

Busch, D. A., 1992, Chicontepec field—Mexico, Tampico-Misantla basin, *in* N. H. Foster and E. A. Beaumont, comp., Stratigraphic traps III: AAPG, Treatise of petroleum geology, atlas of oil and gas fields, p. 113–128.

Carrasco, V. B., 1977, Albian sedimentation of submarine autochthonous and allochthonous carbonates, east edge of the Valles–San Luis Potosí platform, Mexico, *in* H. E. Cook and Paul Enos, eds., Deep-water carbonate environments: Society for Sedimentary Geology (SEPM) Special Publication 25, p. 263–272.

Cook, H. E., 1983a, Sedimentology of some allochthonous deep-water carbonate reservoirs, Lower Permian, west Texas: Carbonate debris sheets, aprons, or submarine fans? (abs.): AAPG Bulletin, v. 67, p. 442.

Cook, H. E., 1983b, Ancient carbonate platform margins, slopes, and basins, *in* H. E. Cook, A. C. Hine, and H. T. Mullins, eds., Platform margin and deep water carbonates: Society for Sedimentary Geology (SEPM), Short Course No. 12, p. 5–189.

Cook, H. E., P. N. McDaniel, E. W. Mountjoy, and L. C. Pray, 1972, Allochthonous carbonate debris flows at Devonian bank ("reef") margins, Alberta, Canada: Canadian Society of Petroleum Geology Bulletin, v. 20, no. 3, p. 439–497.

Cook, H. E., and H. T. Mullins, 1983, Basin margin environment, *in* P. A. Scholle, D. G. Bebout, and C. H. Mooreeds, eds., Carbonate depositional environments: AAPG Memoir 33, p. 539–617.

Demaison, G., and B. J. Huizinga, 1994, Genetic classification of petroleum systems using three factors: Charge, migration, and entrapment, *in* L. B. Magoon and W. G. Dow, eds., The petroleum system—From source to trap: AAPG Memoir 60, p. 3–24.

di Primio, R., V. Dieckmann, and N. Mills, 1998, PVT and phase behaviour analysis in petroleum exploration: Organic Geochemistry, v. 29, no. 103, p. 207–222.

Enos, P., 1977, Tamara limestone of the Poza Rica trend, Cretaceous, Mexico, in H. E. Cook and Paul Enos, eds., Deep-water carbonate environments: Society for Sedimentary Geology (SEPM) Special Publication 25, p. 273–314.

Enos, P., 1985, Cretaceous debris reservoirs, Poza Rica field, Veracruz, Mexico, in P. O. Roehl and P. W Choquette, eds., Carbonate petroleum reservoirs: Springer Verlag, Inc., New York, p. 459–469.

Ewing, T. E., and R. F. Lopez, 1991, Principal structural features, Gulf of Mexico basin, in A. Salvador, ed., The Gulf of Mexico basin: Geological Society of America, The geology of North America, v. J, Plate 2, scale 1:2,500,000.

Fiduk, J. C., and 10 others, 1999, The Perdido fold belt, northwestern deep Gulf of Mexico, Part 2: Seismic stratigraphy and petroleum systems: AAPG Bulletin, v. 83, p. 578–611.

Galloway, W. E., D. G. Bebout, W. L. Fisher, J. B. Dunlap Jr., R. Cabrera-Castro, J. E. Lugo-Rivera, and T. M. Scott, 1991, Cenozoic, in A. Salvador, ed., The Gulf of Mexico basin: Geological Society of America, The geology of North America, v. J, Plate 2, scale 1:2,500,000.

García, J. A. M., 1996, Oils and source rocks of the southern portion of the Tampico-Misantla basin, Mexico: Evidence for the great petroleum potential of the western Gulf of Mexico, in E. Gómez Luna and A. Martinez Cortés, eds., Memorias del V Congreso Latino-Americano de Geoquímica Orgánica [Proceedings of the 5th Latin American Congress on Organic Geochemistry]: Cancún, México, p. 94.

GeoMark, 1999, Oils 99, Oil information library system database, Houston, Texas, GeoMark Research, Inc. (database available from GeoMark Research, Inc., 9748 Whithorn Drive, Houston, Texas 77095).

Gonzalez, R., and P. Cruz, 1994, A new petroleum system in offshore Campeche, Mexico, in N. Schneidermann, O. Cruz, and R. Sanchez, eds., Abstracts, First joint AAPG/Asociación Mexicana de Geólogos Petroleros Hedberg Research Conference, Geological aspects of petroleum systems: Mexico City, Mexico, 2 p.

Grajales-Nishimura, J. M., E. Cedillo-Pardo, C. Rosales-Domínguez, D. J. Morán-Zenteno, W. Alvarez, P. Claeys, J. Ruíz-Morales, J. García-Hernández, P. Padilla-Avila, and A. Sánchez-Ríos, 2000, Chicxulub impact: The origin of reservoir and seal facies in the southeastern Mexico fields: Geology, v. 28, p. 307–310.

Guzmán-Vega, M., and M. R. Mello, 1994, Genetic assessment of hydrocarbons in southeastern Mexico, in N. Schneidermann, O. Cruz, and R. Sanchez, eds., Abstracts, First joint AAPG/Asociación Mexicana de Geólogos Petroleros Hedberg Research Conference, Geological aspects of petroleum systems: Mexico City, Mexico, 2 p.

Guzmán-Vega, M. A., and M. R. Mello, 1996, Macuspana basin, southeast Mexico: An example of a Tertiary marine deltaic petroleum habitat: Revista Latino-Americana de Geoquimíca Orgánica, v. 2., p. 15–28.

Guzmán-Vega, M. A., and M. R. Mello, 1999, Origin of oil in the Sureste basin, Mexico: AAPG Bulletin v. 83, p. 1068–1095.

Hernandez, R., 1994, Cretaceous carbonate breccia reservoirs of the Campeche area, Mexico, in N. Schneidermann, O. Cruz, and R. Sanchez, eds., Abstracts, First joint AAPG/Asociación Mexicana de Geólogos Petroleros Hedberg Research Conference, Geological aspects of petroleum systems: Mexico City, Mexico, 4 p.

Holguín Quiñones, N., 1985, Evaluacion geoquimica del sureste de Mexico: Asociación Mexicana de Geólogos Petroleros Boletín, v. 37, no. 1, p. 3–48.

Holguín, N., G. Demaison, E. Serrano, F. Galindo, A. Romero, A. Sosa, and G. Martinez, 1994, The HC generative subsystems of the productive Mexican basins, in N. Schneidermann, O. Cruz, and R. Sanchez, eds., Abstracts, First joint AAPG/Asociación Mexicana de Geólogos Petroleros Hedberg Research Conference, Geological aspects of petroleum systems: Mexico City, Mexico, 4 p.

Klemme, H. D., 1994, Petroleum systems of the world involving Upper Jurassic source rocks, in L. B. Magoon and W. G. Dow, eds., The petroleum system—From source to trap: AAPG Memoir 60, p. 51–72.

Klett, T. R., T. S. Ahlbrandt, J. W. Schmoker, and G. L. Dolton, 1997, Ranking of the world's oil and gas provinces by known petroleum volumes: U.S. Geological Survey Open File Report 97-463, 1 CD-ROM.

Lucach, S. O., M. A. Guzmán-Vega, L. M. M. Medrano, and L. Clara-Valdez, 1996, La ventana de generación del subsistema generador Tithoniano en la Sonda de Campeche, México, in E. Gómez Luna and A. Martinez Cortés, eds., Memorias del V Congreso Latino-Americano de Geoquímica Orgánica [Proceedings of the 5th Latin American Congress on Organic Geochemistry]: Cancún, México, p. 203–204.

Magoon, L. B., and W. G. Dow, 1994, The Petroleum System, in L. B. Magoon and W. G. Dow, eds., The petroleum system—From source to trap: AAPG Memoir 60, p. 3–24.

Magoon, L. B., and J. W. Schmoker, 2000, The total petroleum system—The natural fluid network that constrains the assessment unit, in U.S. Geological Survey World Energy Assessment Team, eds., World Petroleum Assessment 2000: U.S. Geological Survey DDS-60, 2 CD-ROMs.

Magoon, L. B., and Z. C. Valin, 1994, Overview of petroleum system case studies, in L. B. Magoon and W. G. Dow, eds., The petroleum system—From source to trap: AAPG Memoir 60, p. 329–338.

McFarlan, E. Jr., and L. S. Menes, 1991, Lower Creatceous, in A. Salvador, ed., The Gulf of Mexico basin: Geological Society of America, The geology of North America, v. J, p. 181–204.

Medrano, M. L., M. A. Guzmán-Vega, and L. Clara-Valdez, 1996a, Los aceites del Oxfordiano en la Sonda de Campeche, México, in E. Gómez Luna and A. Martinez Cortés, eds., Memorias del V Congreso Latino-Americano de Geoquímica Orgánica [Proceedings of the 5th Latin American Congress on Organic Geochemistry]: Cancún, México, p. 91–93.

Medrano, M. L., I. M. A. Romero, and R. Maldonado, 1996b, Los subsistemas generadores de la Sonda de Campeche, in E. Gómez Luna and A. Martinez Cortés, eds., Memorias del V Congreso Latino-Americano de Geoquímica Orgánica [Proceedings of the 5th Latin American Congress on Organic Geochemistry]: Cancún, México, p. 85–90.

Mello, M. R., and M. A. Guzmán-Vega, 1996, Biomarker characterization of oils in the Salina basin, México: Revista Latino-Americano de Geoquímica Orgánica, v. 2, p. 29–37.

Mello, M. R., et al., 1996, Geochemical characterization of oils from the Gulf of Mexico and southeastern Mexican basins: A petroleum system approach to predict deep water probes, in E. Gómez Luna and A. Martinez Cortés, eds., Memorias del V Congreso Latino-Americano de Geoquímica Orgánica [Proceedings of the 5th Latin American Congress on Organic Geochemistry]: Cancún, México, p. 106–109.

Nehring, R., 1991, Oil and gas resources, in A. Salvador, ed., The Gulf of Mexico basin: Geological Society of America, The geology of North America, v. J, p. 445–494.

Patton, J. W., P. W. Choquette, G. K. Guennel, A. J. Kaltenback, and A. Moore, 1984, Organic geochemistry and sedimentology of Lower to mid-Cretaceous deep-sea carbonates, Sites 535 and 540, Leg 77, in R. T. Buffler et. al., eds., Initial reports of the Deep-Sea Drilling Project: U. S. Government Printing Office, v. 77, p. 417–443.

PEMEX Exploration and Production, 1999a, Hydrocarbon reserves of Mexico, v. I, 193 p.

PEMEX Exploration and Production, 1999b, Major oil and gas fields of Mexico, v. II, 375 p.

Peterson, J. A., 1983, Petroleum geology and resources of southeastern Mexico, northern Guatemala, and Belize: U. S. Geological Survey Circular 760, 44 p.

Petroconsultants, 1996, Petroleum exploration and production database: Houston, Texas, Petroconsultants, Inc. (database available from Petroconsultants, Inc., P.O. Box 740619, Houston, Texas 77274-0619 USA).

Pottorf, R. J., G. G. Gray, M. G. Kozar, W. M. Fitchen, M. Richardson, R. J. Chuchla, and D. A. Yurewicz, 1996, Hydrocarbon generation and migration in the Tampico segment of the Sierra Madre Oriental fold-thrust belt: Evidence from an exhumed oil field in the Sierra de El Abra, in E. Gómez Luna and A. Martinez Cortés, eds., Memorias del V Congreso Latino-Americano de Geoquímica Orgánica [Proceedings of the 5th Latin American Congress on Organic Geochemistry]: Cancún, México, p. 100–101.

Raedeke, L. D., N. Holguín Quiñones, R. C. Haack, and J. E. Dahl, 1994, Petroleum systems of the Campeche area, in N. Schneidermann, O. Cruz, and R. Sanchez, eds., Abstracts, First joint AAPG/Asociación Mexicana de Geólogos Petroleros Hedberg Research Conference, Geological aspects of petroleum systems: Mexico City, Mexico, 3 p.

Roehl, P. O., and P. W. Choquette, eds., 1985, Carbonate petroleum reservoirs: Springer-Verlag, New York, 622 p.

Román-Ramos, J. R., E. Mena-Sánchez, and L. Bernal-Vegas, 1996, Evaluación geoquímica de los recursos petrolíferos de la cuenca Tampico-Misantla, México, in E. Gómez Luna and A. Martinez Cortés, eds., Memorias del V Congreso Latino-Americano de Geoquímica Orgánica [Proceedings of the 5th Latin American Congress on Organic Geochemistry]: Cancún, México, p. 112–114.

Salas, G. P., 1949, Geology and development of the Poza Rica oil field, Veracruz, Mexico: AAPG Bulletin, v. 33, no. 8, p. 1385–1409.

Salvador, A., ed., 1991a, The Gulf of Mexico basin: Geological Society of America, The geology of North America, v. J, 568 p.

Salvador, A., 1991b, Triassic-Jurassic, in A. Salvador, ed., The Gulf of Mexico basin: Geological Society of America, The geology of North America, v. J, p. 131–180.

Salvador, A., 1991c, Structure at the base, and subcrop below, Mesozoic marine section, Gulf of Mexico basin, in A. Salvador, ed., The Gulf of Mexico basin: Geological Society of America, The geology of North America, v. J, Plate 3, scale 1: 2,500,000.

Salvador, A., 1991d, Cross sections of the Gulf of Mexico basin, in A. Salvador, ed., The Gulf of Mexico basin: Geological Society of America, The geology of North America, v. J, Plate 6.

Santiago, J., and A. Baro, 1992, Mexico's giant fields, 1978–1988 decade, in M. T. Halbouty, ed., Giant oil and gas fields of the decade 1978–1988: AAPG Memoir 54, p. 73–99.

Schmoker, J. W., 1994, Volumetric calculation of hydrocarbons generated, in L. B. Magoon and W. G. Dow, eds., The petroleum system — From source to trap: AAPG Memoir 60, p. 323–328.

Schoellkopf, N. B., C. A. Atallah, S. R. Friend, M. Campos-Madrigal, L. D. Raedeke, and R. C. Haack, 1994, Timing of generation and migration in Campeche basin, in N. Schneidermann, O. Cruz, and R. Sanchez, eds., Abstracts, First joint AAPG/Asociación Mexicana de Geólogos Petroleros Hedberg Research Conference, Geological aspects of petroleum systems: Mexico City, Mexico, 4 p.

Serrano-Bello, E., J. R. Román-Ramos, N. Holguín Quiñones, E. Vásquez-Covarrubias, A. Galindo-Hernández, and D. Grass, 1996, Subsistemas generadores de la cuenca de Veracruz, México, in E. Gómez Luna, and A. Martinez Cortés, eds., Memorias del V Congreso Latino-Americano de Geoquímica Orgánica [Proceedings of the 5th Latin American Congress on Organic Geochemistry]: Cancún, México, p. 127–129.

Sohl, N. F., R. E. Martinez, P. Salmeron-Urena, and F. Soto-Jaramillo, 1991, Upper Cretaceous, in A. Salvador, ed., The Gulf of Mexico basin: Geological Society of America, The geology of North America, v. J, p. 205–244.

Thompson, K. F. M., M. C. Kennicutt II, and J. M. Brooks, 1990, Classification of offshore Gulf of Mexico oils and gas condensates: AAPG Bulletin, v. 74, p. 187–198.

U.S. Geological Survey World Energy Assessment Team, 2000, World Petroleum Assessment 2000: U.S. Geological Survey DDS-60, 2 CD-ROMs.

Viniegra O., F., 1981, Great carbonate bank of Yucatán, southern Mexico: Journal of Petroleum Geology, v. 3, p. 247–278.

Viniegra O., F., and C. Castillo-Tejero, 1970, Golden Lane fields, Veracruz, Mexico, in T. Halbouty, ed., Geology of the giant petroleum fields: AAPG Memoir 14, p. 309–325.

Weimer, P., M. G. Rowan, B. C. McBride, and R. M. Kligfield, 1999, Evaluating the petroleum systems of the northern deep Gulf of Mexico through integrated basin analysis: An overview: AAPG Bulletin, v. 82, no. 5B, p. 865–877.

Wilson, J. L., and W. C. Ward, 1993, Early Cretaceous carbonate platforms of northeastern and east-central Mexico, in J. A. T. Simo, R. W. Scott, and Jean-Pierre Masse, eds., Cretaceous carbonate platforms: AAPG Memoir 56, p. 35–49.

Guzmán-Vega, M. A., L. Castro Ortíz, J. R. Román-Ramos, L. Medrano-Morales, L. C. Valdéz, E. Vázquez-Covarrubias, and G. Ziga-Rodríguez, 2001, Classification and origin of petroleum in the Mexican Gulf Coast Basin: An overview, in C. Bartolini, R. T. Buffler, and A. Cantú-Chapa, eds., The western Gulf of Mexico Basin: Tectonics, sedimentary basins, and petroleum systems: AAPG Memoir 75, p. 127–142.

5

Classification and Origin of Petroleum in the Mexican Gulf Coast Basin: An Overview

Mario A. Guzmán-Vega and Lilia Castro Ortíz
Instituto Mexicano del Petróleo, Mexico City, Mexico

Juan R. Román-Ramos, Luis Medrano-Morales, Lourdes Clara Valdéz, Emilio Vázquez-Covarrubias, and Genaro Ziga-Rodríguez
Petróleos Mexicanos, Mexico City, Mexico

ABSTRACT

A geochemical and isotopic characterization of a wide selection of the produced oils in the petroleum subprovinces from the Mexican Gulf Coast Basin has revealed five major genetic groups. Their distribution and chemical features appear to reflect multiple sources, facies variations, maturation, and postfilling alteration processes. Each group is correlated with a specific generative source, namely (1) Oxfordian marine marl-dominated, (2) Oxfordian marine carbonate-dominated, (3) Tithonian marine marl-dominated, (4) Cretaceous marine carbonate-evaporitic, and (5) Tertiary marine deltaic siliciclastics.

Biomarker and isotope differences observed in the Tithonian oils can be interpreted in terms of facies variations. The Tithonian generative subsystem has produced more than 80% of all oil reserves from the Mexican Gulf Coast Basin. Oil reserves have accumulated both onshore and offshore and throughout the stratigraphic column from Kimmeridgian to Pleistocene in marine-siliciclastic and carbonate reservoirs, suggesting that vertical pathways are an important secondary migration mechanism.

INTRODUCTION

The greater Gulf Coast Basin is one of the largest petroleum regions in the world. At the end of the 1980s, it had an estimated ultimate known recovery production of 112.7 billion barrels of crude oil, 22.5 billion barrels of natural-gas liquids, and 523.8 trillion ft^3 of natural gas, for a total of 222.5 billion barrels of oil equivalent (BOE) (Nehring, 1991). The petroleum subprovinces in the Mexican southern portion are highly oil prone and are estimated to contain more than 80% of the total petroleum reserves of that region (Brooks, 1990). The southern

portion of the Mexican Gulf Coast Basin has been divided into several subprovinces for evaluation of hydrocarbon potential. Listed in a counter-clockwise direction from north to south, they are the Burgos, Tampico-Misantla, Veracruz, Isthmus Salina, Chiapas-Tabasco, Sierra de Chiapas, Macuspana and Campeche Shelf subprovinces (Figure 1). With the exception of the Burgos and Macuspana, which are mainly gas prone, all the subprovinces are oil prone.

Previous geologic and geochemical work has shown that several potential source rocks occur in the Mexican petroleum subprovinces of the Gulf Coast Basin. The Upper Jurassic

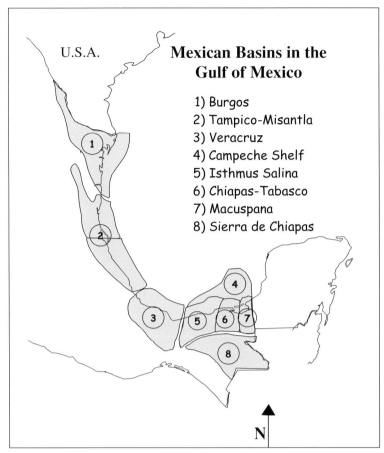

FIGURE 1. Location map.

Mexico related by age and source-rock depositional environment to the Oxfordian, Tithonian, Early Cretaceous, and Tertiary.

We present geochemical data for a number of oil and organic extract samples in an effort to classify the oil types present in the Mexican petroleum subprovinces of the Gulf Coast Basin and to assess the depositional environment of the source rocks that gave rise to them. This paper summarizes and extends previous geochemical research in the Mexican oil subprovinces (Guzmán-Vega and Mello, 1999).

METHODS

Geochemical analysis was done on oil samples from 85 oil fields, two oil seeps, and 30 potential source rocks. For the purpose of presentation, data come from only the most representative oils and source rock extracts from each subprovince (Tables 1, 2, and 3). Oil samples representing reservoirs ranging in age from Oxfordian to Pleistocene (Table 1) were first analyzed for API gravity, carbon isotopes, and Ni, V, and S content. Bitumen from representative source rocks ranging in age from Oxfordian to middle Miocene were extracted by Soxhlet. Carbon isotopes of whole oils and bitumens are reported relative to the PDB. The oils and organic extracts were fractionated via solid-liquid chromatography and the resulting saturate fractions were analyzed by gas chromatography (HP 5890) with flame-ionization detection (GC/FID). Gas chromatography–mass spectrometry (GC/MS) analyses were performed with a Hewlett-Packard 5890 gas chromatograph interfaced to a Hewlett-Packard 5790A MSD.

RESULTS

Based on their characteristics, five major source-related oil groups in the Mexican Gulf Coast Basin can be identified. Each group is correlated with a specific generative source: (1) Oxfordian marine marl-dominated, (2) Oxfordian marine carbonate-dominated, (3) Tithonian marine marl-dominated, (4) Cretaceous marine carbonate-evaporitic, and (5) Tertiary marine deltaic siliciclastic (Figure 2).

Oxfordian Marine Carbonate-dominated Source

Oils from the Oxfordian marine carbonate source are found in eolian sandstone reservoirs of Oxfordian age and in Kimmeridgian oolitic carbonate bank reservoirs in the Campeche Shelf subprovince (Figure 3). Their bulk features include API gravities of approximately 26°, sulfur contents greater than 1%, V/Ni ratios of approximately 1.5, and whole-oil ^{13}C values of approximately –25‰ (Table 1). Diagnostic features include

source rocks are considered to be the most important generative subsystem in the Mexican Gulf Coast Basin. They have long been thought to be the main source of the giant oil and gas fields of the Tampico-Misantla, Reforma, and Campeche petroleum subprovinces. Viniegra (1981) discussed geologic evidence that indicates that the Kimmeridgian-Tithonian section is the most likely oil source for much of the oil in the Tampico-Misantla and Chiapas-Tabasco subprovinces. Santiago (1979) also suggested that the Upper Jurassic section contains the main source rocks for the petroleum reserves found in southeastern Mexico. Holguin (1987) and González and Holguin (1991) summarized the extensive geochemical work carried out in Mexico, and concluded that Tithonian black calcareous shales and shaly limestones are the main source rock of oil and gas in southern Mexico.

The ability of organic geochemistry to use the molecular chemistry of oil samples for assessing the depositional environments of these source rocks that gave rise to them has allowed us to conclude that Tithonian organic-rich shales are not the only source of oil and gas in the Mexican petroleum subprovinces. Dahl et al. (1993), using biomarker analyses, postulated the existence of three oil groups: Upper Jurassic–Berriasian sourced oils, Lower Cretaceous sourced oils, and Tertiary sourced oils. Guzmán-Vega and Mello (1999) documented the existence of four major oil families in southeastern

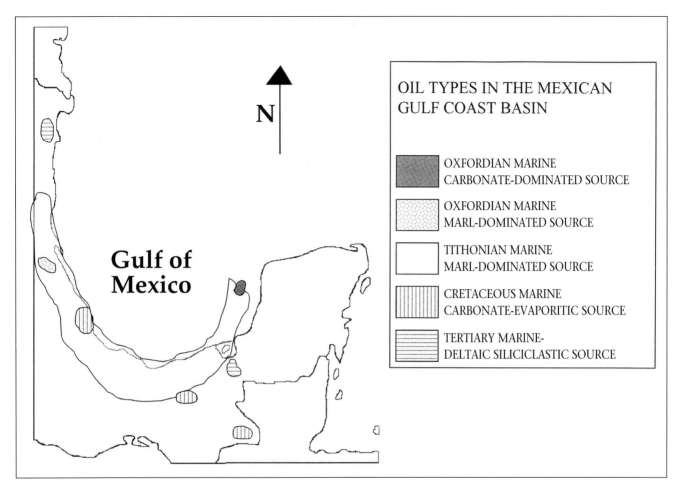

FIGURE 2. Oil-group distribution in the Mexican Gulf Coast Basin.

a predominance of low-molecular-weight n-alkanes, pristane/phytane <1, Ts/Tm <<1, abundant extended hopanes, very high C_{35}/C_{34} hopane ratios, low abundance of tricyclic terpanes relative to pentacyclics, presence of $17\alpha(H)$-29,30-bisnorhopane, high abundance of C_{29} steranes relative to their C_{27} counterparts, low relative abundance of diasteranes, and presence of C_{30} steranes (Table 2). Biomarker features from these group are similar to those described in some Smackover oils in the northern Gulf of Mexico (Wenger et al., 1990; Kennicutt et al., 1992; Comet et al., 1993). Biomarker distributions in this group reflect high bacterial input (i.e., high relative abundance of extended hopanes with high C_{35}/C_{34} ratios) and low clay input (i.e., low relative abundance of diasteranes). A marine carbonate source rock deposited under anoxic conditions, possibly hypersaline, has been interpreted as the source of this oil (Guzmán-Vega and Mello, 1999).

An oil/source-rock correlation using m/z 191 (terpanes) and m/z 217 (steranes) mass chromatograms is shown in Figure 4. The biological markers show a good match between an oil representative of this group and an Oxfordian organic-rich marl extract taken from a well (confidential) in the Campeche Shelf. Sedimentological data for the Oxfordian in this area suggest the deposition of marls in a marine environment and are con-

sistent with geochemical results (Cruz and Gonzalez, 1994). The potential source rock selected for oil/source-rock correlation corresponds to an Oxfordian laminated, shaly-marine limestone interpreted as having been deposited in a restricted environment near the shoreline. Total-organic-carbon (TOC) (approximately 2%) and hydrogen-index (430 mg HC/g TOC) values characterize a type II kerogen with good source potential.

In summary, the geochemical characteristics of this oil group are diagnostic of marine-carbonate, anoxic conditions of the source-rock depositional environment (ten Haven et al., 1985; Connan and Dessort, 1987; Mello et al., 1988, 1993, 1995; Peters and Moldowan, 1993).

Oxfordian Marine Marl-dominated Source

Oxfordian marine marl-dominated sourced oils have been discovered in the Kimmeridgian reservoirs related to the San Andrés Member of the Taman Formation in a few fields both south and north of the Tuxpan Platform in the Tampico-Misantla Basin (Figures 1 and 3). These oils have low sulfur and carbon-isotope values for whole oils of approximately –28‰ (Table 1). Diagnostic biomarker features include pristane/phytane ratios ≤1, abundant extended hopanes, low to me-

TABLE 1. Identification of and bulk data for oils from the Mexican Gulf Coast Basin.

Oil group	Subprovince	Depth (m)
Oxfordian marine carbonate-dominated source	Campeche Shelf	4400–4428
Oxfordian marine carbonate-dominated source	Campeche Shelf	4540–4565
Oxfordian marine marl-dominated source	Tampico-Misantla	2275–2345
Oxfordian marine marl-dominated source	Tampico-Misantla	1958–2005
Tithonian marine marl-dominated source (Group A)	Tampico-Misantla	1330–1340
Tithonian marine marl-dominated source (Group A)	Veracruz	2544–2553
Tithonian marine marl-dominated source (Group A)	Isthmus Salina	385–390
Tithonian marine marl-dominated source (Group A)	Chiapas-Tabasco	4035–4047
Tithonian marine marl-dominated source (Group A)	Campeche Shelf	4317–4400
Tithonian marine marl-dominated source (Group B)	Isthmus Salina	620–627
Tithonian marine marl-dominated source (Group B)	Campeche Shelf	5820–5860
Cretaceous marine carbonate-evaporitic source	Veracruz	2921–2931
Cretaceous marine carbonate-evaporitic source	Chiapas-Tabasco	2787–2805
Cretaceous marine carbonate-evaporitic source	Sierra de Chiapas	3389–3375
Tertiary marine-deltaic siliciclastic source	Burgos	2134–2142
Tertiary marine-deltaic siliciclastic source	Burgos	1299–1306
Tertiary marine-deltaic siliciclastic source	Macuspana	1585–1590
Tertiary marine-deltaic siliciclastic source	Macuspana	1496–1499

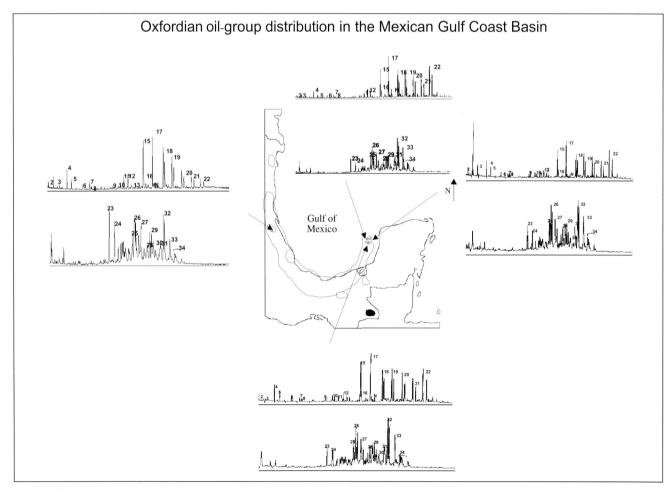

FIGURE 3. Oxfordian oil-group distribution in the Mexican Gulf Coast Basin.

Reservoir age	Lithology	δ13C (0/00)	API gravity	S (%)	V (ppm)	Ni (ppm)	V/Ni
Oxfordian	Limestones	-25.4	25.8	2.2	18	13	1.4
Oxfordian	Limestones	-25.7	26.0	2.6	15	12	1.3
Kimmeridgian	Limestones	-28.6	24.7	2.8	30	21	1.4
Kimmeridgian	Limestones	-28.4	23.5	2.6	11	4	2.75
Lower Cretaceous	Limestones	-27.57	13.3	1.8	114	33	3.5
Middle Miocene	Sandstones	-27.8	16.3	6.0	–	–	–
Upper Miocene	Sandstones	-27.3	23.9	2.2	40	23	2.1
Middle Cretaceous	Limestones	-26.4	35.5	0.6	>5	1	–
Upper Cretaceous	Limestones	-27.4	24.7	1.5	75	19	3.9
Upper Pliocene	Sandstones	-27.2	15.5	2.1	38	19	2
Lower Cretaceous	Limestones	-27.3	20.7	3.6	–	5	
Upper Cretaceous	Limestones	-23.9	31.5	1.4	>5	>1	–
Lower Cretaceous	Limestones	-24.3	22.1	2.8	48	11	4.4
Lower Cretaceous	Limestones	-22.9	18.4	2.9	21	38	0.6
Upper Miocene	Sandstones	-25.7	46.2	0.2	>5	3	–
Upper Miocene	Sandstones	-27.14	32.5	0.1	>5	>1	–
Pliocene	Sandstones	-22.5	45.4	0.1	>5	6	–
Pliocene	Sandstones	-23.7	45.1	0.1	>5	4	–

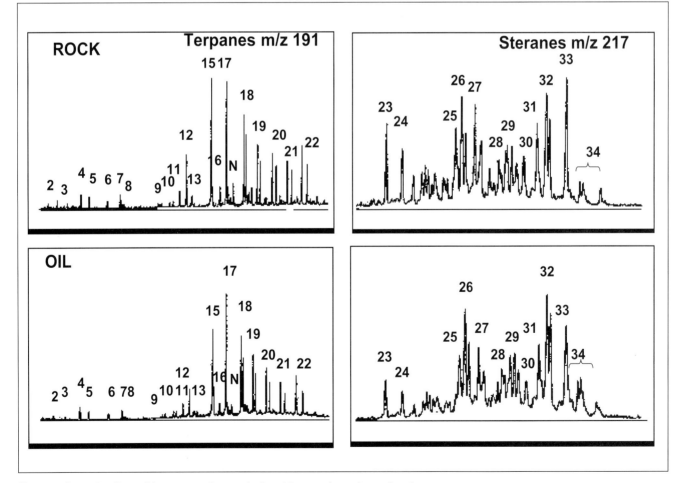

FIGURE 4. Oxfordian oil/source-rock correlation (Campeche subprovince).

TABLE 2. Source-related biomarker parameters for oils from the Mexican Gulf Coast Basin.

Oil group	Subprovince E	C_{29}/C_{30} (a)	C_{35}/C_{34} (b)	% C_{31} (c)	% C_{32} (c)
Oxfordian marine carbonate-dominated source	Campeche Shelf	0.8	1.4	30.2	23.7
Oxfordian marine carbonate-dominated source	Campeche Shelf	0.8	1.5	26.1	22.8
Oxfordian marine marl-dominated source	Tampico-Misantla	0.8	0.9	46.7	28.3
Oxfordian marine marl-dominated source	Tampico-Misantla	0.8	1.0	46.7	28.2
Tithonian marine marl-dominated source (Group A)	Tampico-Misantla	1.1	1.2	42.9	24.3
Tithonian marine marl-dominated source (Group A)	Veracruz	1.2	0.8	39.1	25.0
Tithonian marine marl-dominated source (Group A)	Isthmus Salina	0.9	1.2	43.3	25.5
Tithonian marine marl-dominated source (Group A)	Chiapas-Tabasco	1.2	1.1	39.0	25.1
Tithonian marine marl-dominated source (Group A)	Campeche Shelf	1.1	0.9	40.6	25.6
Tithonian marine marl-dominated source (Group B)	Isthmus Salina	0.8	1.0	39.3	25.4
Tithonian marine marl-dominated source (Group B)	Campeche Shelf	0.7	1.1	35.8	24.5
Cretaceous marine carbonate-evaporitic source	Veracruz	1.8	0.9	37.8	23.8
Cretaceous marine carbonate-evaporitic source	Chiapas-Tabasco	1.8	1.1	37.8	22.8
Cretaceous marine carbonate-evaporitic source	Sierra de Chiapas	1.4	1.0	29.0	26.2
Tertiary marine-deltaic siliciclastic source	Burgos	0.5	0.6	41.9	27.4
Tertiary marine-deltaic siliciclastic source	Burgos	0.6	0.6	43.5	27.8
Tertiary marine-deltaic siliciclastic source	Macuspana	0.6	0.3	49.0	29.4
Tertiary marine-deltaic siliciclastic source	Macuspana	0.6	0.4	50.8	29.5

(a) C_{29} = peak area of 17α, 21β (H)-30-norhopane in m/z 191 chromatogram. C_{30} = peak area of 17α, 21β (H)-hopane in m/z 191 chromatogram.

(b) C_{35} = peak area of 17α, 21β (H)-30-pentakishomohopane 22S + 22R in m/z 191 chromatogram. C_{34} = peak area of 17α, 21β (H)-30-tetrakishomohopane 22S+22R in m/z 191 chromatogram.

(c) %C_{31} = [C_{31}/(C_{31} to C_{35} homohopanes)] x 100; %C_{32} = [C_{32}/(C_{31} to C_{35} homohopanes)] x 100; %C_{33} = [C_{33}/(C_{31} to C_{35} homohopanes)] x 100; %C_{34} = [C_{34}/(C_{31} to C_{35} homohopanes)] x 100; %C_{35} = [C_{35}/(C_{31} to C_{35} homohopanes)] x 100; Epimers 22S + 22R measured for each carbon number in m/z 191 chromatogram.

(d) Oleanane index = peak area of 18α (H)-oleanane in m/z 191 chromatogram over peak area of 17α, 21β (H)-hopane in m/z 191 chromatogram.

(e) 24/4= peak area of tetracyclic terpane (C_{24}) in m/z 191 chromatogram. 26/3= peak area of tricyclic terpane (C_{26}) in m/z 191 chromatogram.

dium abundance of tricyclic terpanes relative to pentacyclics, C_{29}/C_{30} and C_{35}/C_{34} extended hopane ratios ≤1, Ts/Tm ratios <1, presence of C_{30} steranes, and high relative abundance of diasteranes compared with regular steranes (Table 2). The presence of C_{30} steranes indicates a marine depositional environment (Moldowan et al., 1985). High diasteranes/steranes ratios in the oils from this group suggest an important clay input or oxidizing conditions in the source-rock sedimentary environment (Peters and Moldowan, 1993).

Figure 5 shows a good correlation, using biological marker distributions, between an oil from this group and an organic extract from an Oxfordian organic-rich marl from the Santiago Formation recovered in a well (confidential) in the Tampico-Misantla subprovince. This sample is characterized by moderate organic-matter content of poor quality, as reflected by a TOC of approximately 2%, a low hydrogen index of 183 mg HC/g TOC, and a high content of terrestrial organic matter. These features suggest that this sample is poor source rock for liquid hydrocarbons. The Santiago Formation is thought to have been deposited under relatively deep-water marine conditions with no influx of coarse sediments in a clay-rich sedimentary environment (Salvador, 1991).

In summary, the biomarker features of the oils from this group provide a strong link to Oxfordian marl source rocks from the Santiago Formation.

Tithonian Marine Marl-dominated Source

This study confirms previous recognition of Tithonian source rocks as the most important generative subsystem in the Mexican Gulf Coast Basin. The oils belonging to this group

% C_{33} (c)	% C_{34} (c)	% C_{35} (c)	Oleanane index (d)	24/4/26/3 (e)	$C^*_{30}/C_{29}Ts$	Ts/Tm (g)	Dias index (h)	% C_{27} (i)	% C_{28} (i)	% C_{29} (i)	Steranes/hopanes (j)	Pr/Ph (k)
17.0	12.5	17.5	0.0	1.4	0.1	0.4	0.4	32.7	22.0	45.4	5.7	0.6
18.7	13.3	20.2	0.0	1.6	0.0	0.5	0.4	29.0	21.9	49.1	4.4	0.6
18.0	9.6	8.0	0.0	1.4	0.4	1.1	1.7	38.0	26.6	45.6	5.6	1.1
17.8	9.6	8.4	0.0	1.9	0.4	1.5	1.7	37.7	24.5	48.1	5.3	1.1
14.3	8.9	10.5	0.0	2.9	0.2	0.3	0.3	32.8	27.1	40.6	4.9	0.6
16.1	11.5	9.2	0.0	3.1	0.2	0.9	0.9	27.8	32.2	40.5	6.2	1.2
16.3	7.8	9.6	0.0	2.1	0.3	1.0	0.6	31.4	25.6	43.6	6.6	1.0
15.3	10.6	11.3	0.0	2.3	0.2	0.6	0.4	30.5	25.8	44.1	4.7	0.7
16.0	9.8	8.9	0.0	1.6	0.2	0.5	0.1	26.5	34.5	39.4	3.1	0.9
17.7	9.3	9.3	0.0	1.6	0.4	1.2	1.5	32.2	25.7	42.6	5.8	1.1
17.4	11.4	12.1	0.0	1.5	0.3	0.9	0.9	29.2	26.2	45.1	5.4	1.0
14.9	12.9	11.7	0.0	8.9	0.1	0.3	0.4	30.4	31.3	38.9	8.5	0.5
14.2	12.4	13.8	0.0	11.0	0.2	0.2	0.4	28.8	32.9	38.8	9.4	0.8
11.5	18.1	17.3	0.0	22.8	0.2	0.5	0.1	37.0	19.5	44.0	12.4	0.4
16.6	9.3	5.8	0.7	0.5	0.3	1.1	0.4	23.7	36.8	40.0	9.3	2.1
16.2	8.2	5.1	0.6	0.6	0.3	0.9	0.5	26.5	37.7	36.3	10.2	2.1
13.6	6.5	2.3	0.6	1.1	0.4	1.3	0.9	31.4	32.5	36.8	11.1	3.2
12.9	5.2	2.3	0.8	0.9	0.6	1.1	0.7	28.5	35.7	36.3	8.8	3.4

(f) C^*_{30} = peak area of 17α (H)-diahopane. 29Ts = peak area of 18a (H)-30 -norneohopane in m/z 191 chromatogram.

(g) Ts = peak area of 18α (H)-22,29,30-trisnorneohopane in m/z 191 chromatogram. Tm = peak area of 17α (H)-22,29,30-trisnorhopane in m/z 191 chromatogram.

(h) Dias index = sum of peak areas of C_{27} 20R and 20S $13\beta,17\alpha$ (H)-diasteranes in m/z 217 chromatogram, over sum of peak areas of C_{29} 20S and 20R 5α (H),$14\beta(H)$,$17\beta(H)$,24-ethylcholestane in m/z 217 chromatogram.

(i) % C_{27} = [$C_{27}/(C_{27}$ to C_{29})] x 100; % C_{28} = [$C_{28}/(C_{27}$ to C_{29})] x 100; % C_{29} = [$C_{29}/(C_{27}$ to C_{29})] x 100. Peak area of $5\alpha(H)$,$14\beta(H)$,$17\beta(H)$-cholestane 22S + 22R in m/z 218 chromatogram. Peak area of $5\alpha(H)$,$14\beta(H)$,$17\beta(H)$-methylcholestane 22S + 22R in m/z 218 chromatogram. Peak area of $5\alpha(H)$,$14\beta(H)$,$17\beta(H)$-ethylcholestane 22S+22R in m/z 218 chromatogram.

(j) Intensity of regular steranes in m/z 217 chromatogram over intesity of homohopanes in m/z 191 chromatogram.

(k) Pr = peak area of pristane. Ph = peak area of phytane. Peak areas of Pr and Ph are measured in whole-oil gas chromatogram.

are found both onshore and offshore in marine siliciclastic and carbonate reservoirs, ranging in age from Kimmeridgian to Pliocene, in the Tampico-Misantla, Veracruz, Isthmus Salina, Chiapas-Tabasco, and Campeche Shelf subprovinces (Figure 6). These oils represent more than 80% of all oil reserves in the Mexican Gulf Coast Basin.

Bulk geochemical features of these oils include saturate hydrocarbon contents ranging from 21% to 63%, low to high API gravities ranging from 15° to 47°, sulfur content from 0.1% to 6%, whole-oil $\delta^{13}C$ values of –25.6‰ to –27.8‰ with most values approximately –27‰, and V/Ni ratios ranging from 1% to 5% (Table 1). The broad variation in these geochemical features may be explained by slight source-rock organic facies variations and by a wide range of thermal evolution (from normal oil to condensate).

The suite of Tithonian oils analyzed in this study can be subdivided into two groups, designated A and B (Tables 1 and 2). Group A is the most widely spread in the Mexican petroleum subprovinces. It is present in wells in the Tampico-Misantla, Veracruz, Isthmus Salina, Chiapas-Tabasco, and Campeche Shelf subprovinces. These oils are associated with marine clastic reservoirs ranging in age from Late Cretaceous to the late Miocene (Table 1). Group B is restricted to some wells from the Campeche Shelf and the western part of the Salina subprovinces. These oils are associated with upper Miocene marine siliciclastic and Lower Cretaceous carbonate reservoirs (Table 1). Data from this study suggest that the primary difference between group A and group B oils can be ascribed to the intensity or persistence of anoxia and clay input in the depositional environment. Both groups present C_{30} steranes and a high abundance of C_{29} steranes relative to C_{27} (Table 2).

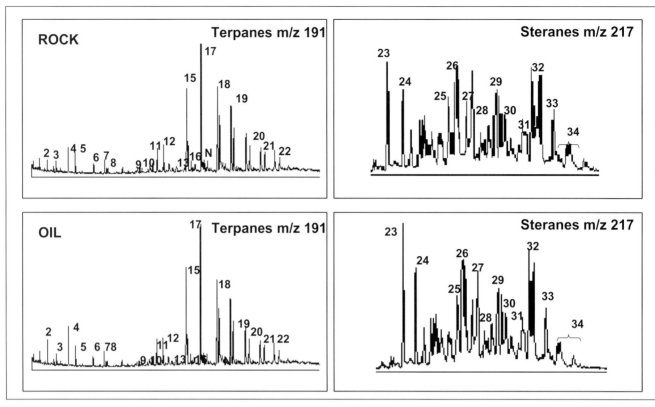

FIGURE 5. Oxfordian oil/source-rock correlation (Tampico-Misantla subprovince).

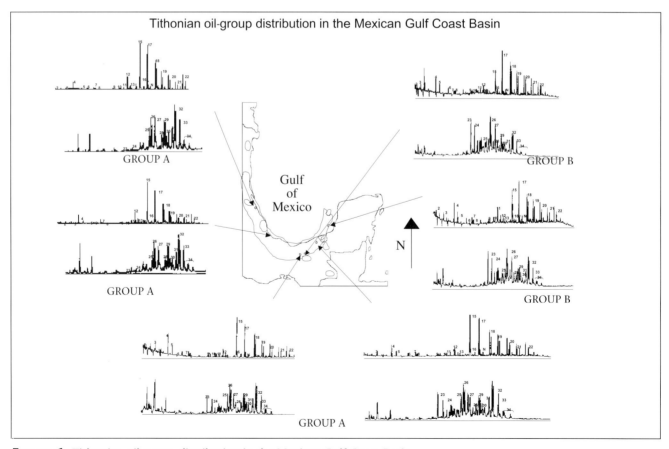

FIGURE 6. Tithonian oil-group distribution in the Mexican Gulf Coast Basin.

Group A oils have pristane/phytane ratios >1, abundant extended hopanes, C_{29}/C_{30} $\alpha\beta$-hopane ratios <1, Ts/Tm ratios <1, low to medium abundance of tricyclic terpanes relative to pentacyclics, high C_{35}/C_{34} extended hopane ratios, presence of hexahydrobenzohopanes, $17\alpha(H)$-29,30-bisnorhopane, and very low abundance of diasteranes (Table 2). The relatively high abundance of extended hopanes, the elevated C_{35} hopanes, and the very low abundance of diasteranes are commonly associated with carbonate environments possibly under hypersaline (?) conditions (Connan et al., 1986; Mello et al., 1988). The presence of hexahydrobenzohopanes and $17\alpha(H)$-29,30-bisnorhopane also is consistent with carbonate depositional environments (Connan and Dessort, 1987; Peters and Moldowan, 1993). High C_{29}/C_{30} hopane ratios have been related to oils derived from organic-rich carbonate and some evaporitic source rocks (Zumberge, 1984; Connan et al, 1986).

Group B oils present low sulfur and low V/Ni ratios when compared with oils from group A (Table 1). Although group B oils share common molecular features with oils from group A, they have characteristic molecular features. The most important of these features are pristane/phytane ratios, C_{29}/C_{30} hopanes, and C_{35}/C_{34} extended hopane ratios <1, Ts/Tm <1, high relative abundance of diasteranes compared with regular steranes, and in particular, high to very high relative abundance of a series of C_{28}-C_{34} $17\alpha(H)$-diahopanes (Table 2). Moldowan et al. (1991) suggested that a high abundance of $17\alpha(H)$-diahopane (C^*_{30}) is related to bacterial input in a clay-rich sedimentary environment under oxic or suboxic conditions. They proposed the C^*_{30}/C_{29} Ts ratio as an indicator of clay content and oxygen depletion in the paleoenvironment of deposition of the source rock.

In comparison with group A oils, group B oils have characteristics consistent with less persistent anoxia and more clay supply during deposition.

Biomarker features similar to those described in group B oils have been reported by Kennicutt et al. (1992) in the Flexure Trend oils. These authors suggested an Upper Jurassic/ Lower Cretaceous source for these oils.

The distribution of C_{26} steranes provided a key parameter and the most striking differentiation between Oxfordian and Tithonian oils in the Tampico-Misantla subprovince. The Tithonian oils can be distinguished easily by a strong predominance of 21-norcholestanes and the Oxfordian oils by a predominance of 27-norcholestanes (Guzmán-Vega et al., 1997; Figure 7). Mixing of Oxfordian and Tithonian oils has been detected at the geographic boundaries of these genetic groups.

Figure 8 illustrates a good correlation between the representative oil from group A and an organic-rich Tithonian marl from a well in the Tampico-Misantla Basin. This marl contains radiolarians and calpionellids that suggest deposition in a bathyal environment. This source rock has a TOC content of approximately 2% and a hydrogen index value of approximately 300 mg HC/g TOC. The organic matter is predominantly algal and presents a thermal evolution stage around

TABLE 3. Biomarker peak identification.

Triterpanes (m/z 191)

1	C_{19} tricyclic terpane
2	C_{20} tricyclic terpane
3	C_{21} tricyclic terpane
4	C_{23} tricyclic terpane
5	C_{24} tricyclic terpane
6	C_{25} tricyclic terpane
7	C_{24} (Des-E) tetracyclic terpane
8	C_{26} tricyclic terpane (S/R)
9	C_{28} tricyclic terpane (S/R)
10	C_{29} tricyclic terpane (S/R)
11	C_{27} $18\alpha(H)$-22,29,30-trisnorneohopane (C_{27} – Ts)
12	C_{27} $17\alpha(H)$-22,29,30-trisnorhopane (Tm)
13	C_{28} $17\alpha(H)$-21β(H)-29,30-bisnorhopane
14	C^*_{30} $17\alpha(H)$-diahopane
15	C_{29} $17\alpha(H)$-21β(H)-norhopane
16	C_{29} $18\alpha(H)$-30-norneohopane (C_{29}Ts)

O Oleanane

17	C_{30} $17\alpha(H)$-21β(H)-hopane
18	C_{31} $17\alpha(H)$-21β(H)-homohopane (22S + 22R)
19	C_{32} $17\alpha(H)$-21β(H)-bishomohopane (22S + 22R)
20	C_{33} $17\alpha(H)$-21β(H)-trishomohopane (22S + 22R)
21	C_{34} $17\alpha(H)$-21β(H)-tetrakishomohopane (22S + 22R)
22	C_{35} $17\alpha(H)$-21β(H)-pentakishomohopane (22S + 22R)

Steranes (m/z 217)

23	13β(H),17α(H)-diacholestane, 20S (C_{27} – diasterane)
24	13β(H),17α(H)-diacholestane, 20R (C_{27} – diasterane)
25	5α(H),14α(H),17α(H),20S (C_{27} – cholestane)
26	5α(H),14β(H),17β(H),20R + 20S (C_{27} – cholestane)
27	5α(H),14α(H),17α(H),20R (C_{27} – cholestane)
28	5α(H),14α(H),17α(H),20S (C_{28} – methylcholestane)
29	5α(H),14β(H),17β(H),20R + 20S (C_{28} – methylcholestane)
30	5α(H),14α(H),17α(H),20R (C_{28} – methylcholestane)
31	5α(H),14α(H),17α(H),20S (C_{28} – ethylcholestane)
32	5α(H),14β(H),17β(H),20R + 20S (C_{29} – ethylcholestane)
33	5α(H),14α(H),17α(H),20R (C_{29} – ethylcholestane)
34	C_{30} – steranes

peak oil generation (Ro approximately 0.8%). Among the potential source rocks studied here, none could be correlated with group B oils.

In summary, Tithonian oils are related to a carbonate source rock deposited under anoxic/suboxic conditions (Guzmán-Vega et al., 1991, 1992). Biomarker and isotope differences observed in the Tithonian oils can be interpreted in terms of variations in salinity, clay content, and oxygen depletion of the depositional environment of Tithonian source rocks (Guzmán-Vega et. al., 1995).

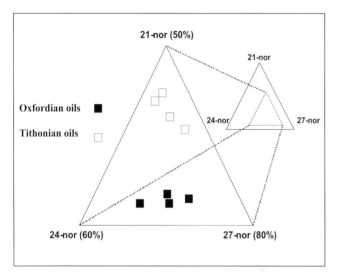

FIGURE 7. C_{26} steranes Oxfordian versus Tithonian (Tampico-Misantla subprovince).

Cretaceous Marine Carbonate-evaporitic Source

Several oil shows and fields have been discovered in thick carbonate sections associated with Lower Cretaceous platforms in the Veracruz, Chiapas-Tabasco, and Sierra de Chiapas subprovinces (Figure 9; Table 1). These platforms consist mainly of sabkha-type sequences linked to evaporitic and restricted shelf deposits. These oils have API gravities that range from 18° to 36°, sulfur contents that vary from 0.7% to 3%, V/Ni ratios that range from 0.6 to 4.4, and carbon-isotope values for the whole oils that vary from –24.4‰ to –22.9‰ (Table 1).

The oils from this group present specific biomarker features previously described in oils related to carbonate-evaporite source rocks deposited under hypersaline, highly restricted conditions. They have pristane/phytane ratios ranging between 0.3 and 0.8, occurrence of high relative abundances of C_{25} and C_{30} long-chain isoprenoids, hopane/sterane ratios <0.2, abnormally high terpane-to-sterane ratios (>15), abundant extended hopanes with high C_{34}/C_{33} ratios (>1), very high relative abundances of C_{24}–C_{27} tetracyclic Des-E terpanes, low to very low abundances of tricyclic terpanes, the presence of 17α(H)-29,30-bisnorhopane, and low abundance of diasteranes (Table 2). The abnormally high terpane/sterane ratios of these oils suggest that the organic matter is dominated by a prevailing bacterial supply (Connan et al., 1986).

The isotopic and molecular features observed in these oils are similar to those reported from carbonate-evaporitic sequences from Guatemala (Connan et al., 1986, 1995) and south Florida (Palacas et al., 1984).

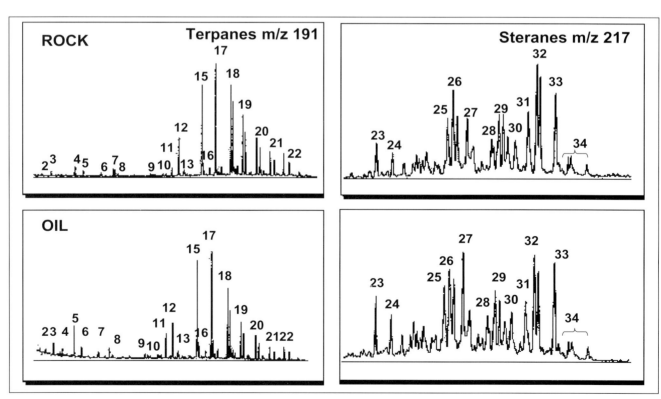

FIGURE 8. Tithonian oil/source-rock correlation.

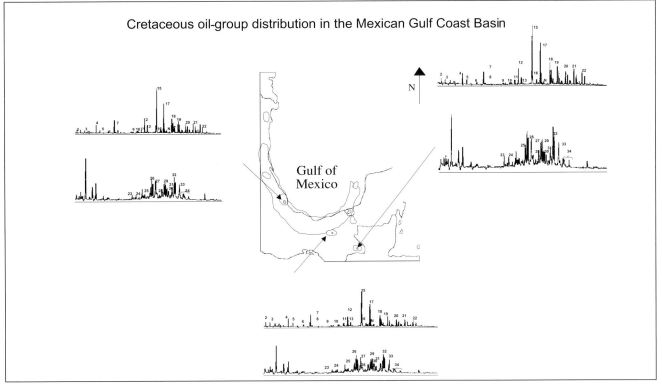

FIGURE 9. Cretaceous oil-group distribution in the Mexican Gulf Coast Basin.

Figure 10 shows a good correlation, using biological marker distributions, between an oil from this group and the organic extract obtained from a Lower Cretaceous organic-rich limestone intercalated in massive anhydrite recovered from a core in a well (confidential) from the Sierra de Chiapas area. The sample is characterized by laminated microdolomites, algal mats, and mosaic anhydrite. These features are similar to those reported in the Lower Cretaceous of Guatemala and are interpreted as typical of coastal sabkha environments (Connan et al., 1986). Total-organic-carbon and hydrogen-index values are approximately 1.5% and 450 mg HC/g TOC, respectively. The organic matter is composed mainly of type II kerogen with a predominance of bacterial and algal material.

Detailed geochemical analyses on Lower Cretaceous cores from wells in the Sierra de Chiapas have shown that each lithofacies displays a typical biomarker fingerprint. In particular, most biomarker parameters are affected by facies changes. These particular characteristics, alongside the carbonate-evaporitic Lower Cretaceous sediments, render the search for the real source beds difficult. The hydrocarbon potential of evaporitic sequences commonly is limited because of the erratic distribution of source rocks, which were deposited in thin-bedded, laterally discontinuous strata.

Tertiary Marine-deltaic Siliciclastic Source

A suite of high-API gravity fluids with heavy, stable, carbon isotopic composition compared to the other Mexican oils can be recognized as a distinct oil group in the gas-prone Burgos and Macuspana subprovinces (Figure 11). These oils are found

in Paleocene to Pleistocene deltaic and delta front sandstones (Table 1).

Condensates from the Burgos and Macuspana subprovinces typically have medium to high API gravities ranging from 28° to 45°, only traces of Ni and V, very low sulfur content, and whole-oil carbon isotope values ranging from –3.7‰ to –21.9‰ (Table 1). The most important molecular features of these oils include high pristane/phytane ratios, low abundance of extended hopanes, high relative abundance of 18α(H)-oleanane, the presence of Des-A and Des-E C_{24} tetracyclic terpanes, a dominance of C_{29} steranes, a high relative abundance of diasteranes, and the presence of C_{30} steranes, albeit in low abundance (Table 2). Guzmán-Vega and Mello (1999) interpreted these features in the Macuspana Basin as being related to source rocks deposited in an anaerobic, marine deltaic siliciclastic environment. The presence of high amounts of 18α(H)-oleanane and Des-A and Des-E C_{24} tetracyclic terpanes has been used to support the hypothesis of a significant Tertiary contribution of higher plant material to petroleum source in many Tertiary oils worldwide (Ekweozor et al., 1979; Grantham et al., 1983; Brooks, 1986; Philp and Gilbert, 1986; Abdullah et al., 1988; Mello et al.,1988; Moldowan et al., 1994; Murray et al., 1994).

Two subgroups are recognizable based on their stable carbon isotopic compositions (Table 1). Heavy $\delta^{13}C$ values are noted in the Macuspana oils ($\delta^{13}C$ of approximately –22‰). In contrast, $\delta^{13}C$ values become lighter in the Burgos oils ($\delta^{13}C$ of approximately –26‰). Both subgroups are associated with thermogenic gases generated by primary cracking probably related to a kerogen type III and expulsed at the end of the oil window (VR of approximately 0.8–0.9%; Prinzhofer et al., 2000; Carrillo-Hernández, 1999). In the Macuspana and Burgos subprovinces, the chemical and isotopic compositions of the gases are homogeneous, suggesting a single source and a very narrow range of maturity in each subprovince (Prinzhofer et al., 2000; Carrillo-Hernández, 1999). The $\delta^{13}C$ values of propane and butane from these gases (approximately –23‰ and –22‰ in the Macuspana subgroup and approximately –27‰ and –26‰ in the Burgos subgroup), associated with

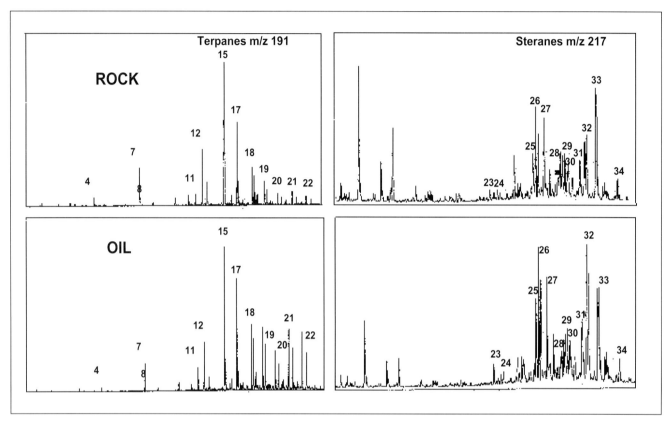

FIGURE 10. Cretaceous oil/source-rock correlation.

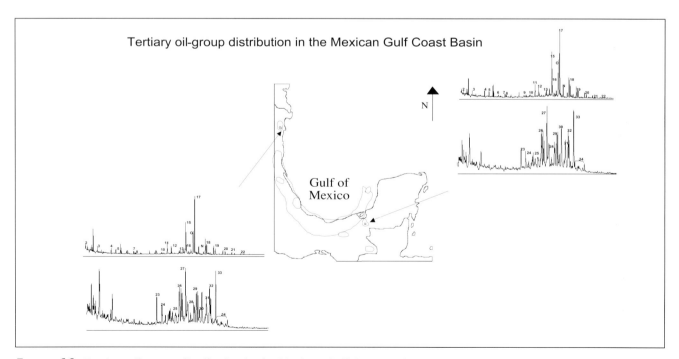

FIGURE 11. Tertiary oil-group distribution in the Mexican Gulf Coast Basin.

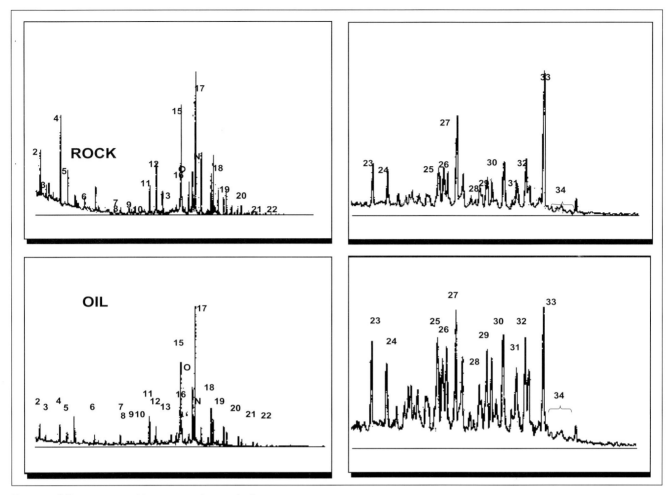

FIGURE 12. Tertiary oil/source-rock correlation.

their generation having occurred in the oil window (i.e., without secondary cracking), suggests an isotopic signature of approximately –22‰ for their related kerogen in the Macuspana subprovince and approximately –27‰ in the Burgos Basin. Similar carbon-isotopic values between the putative source kerogen of gases and condensates in each subprovince suggest a genetic association in both the Macuspana and Burgos subprovinces.

Paleogene oils described in the literature (Kennicutt et al., 1992; Sassen and Chinn, 1990) present similar chemical and isotopic features to those from the Burgos subprovince, and those from the Macuspana subprovince share several chemical and isotopic features with those described in the offshore Texas condensates group (Kennicutt et al.,1992).

The associated source rocks generating the gases and condensates in the Macuspana and Burgos subprovinces have not yet been fully identified. Through pyrolysis and optical methods, middle Eocene, lower Oligocene, and middle Miocene argillaceous sequences have shown high organic carbon and potential hydrocarbon contents, and they have been considered potential source rocks (González and Holguin, 1991).

Figure 12 illustrates an attempt at an oil/source-rock corre-

lation using an oil from this group and an organic extract from a middle Miocene organic-rich shale recovered from a well (confidential) in the Macuspana subprovince. TOC values are low (approximately 0.5%) and consist mainly of type II–III kerogen (hydrogen indices are as high as 154 mg HC/g TOC). Although the biomarker features of both Tertiary organic extracts and oils have a good correlation, there are significant differences when their respective carbon isotope compositions are compared (Table 2). Carbon-isotopic values of the oils range from –24‰ to –22‰. In contrast, the Tertiary organic extracts analyzed in this study have values of approximately –27‰. The only extracts and oils known in southern Mexico with compatible $\delta^{13}C$ to these oils are in the Cretaceous formations (Connan et al., 1995; Guzmán-Vega and Mello, 1999). However, the presence of large amounts of oleananes in these condensates makes it difficult to correlate these accumulated fluids.

An alternative hypothesis to explain the presence of oleanane in these oils is that this compound is a contaminant introduced into the oil during the pooling in Tertiary reservoirs. Reservoir overprinting could significantly alter the biomarker fingerprint of accumulated oils (Peters and Moldowan,

1993). An unusual marginally mature sterane pattern observed in these condensates, expressed as abundant 20R steranes with 27 to 29 carbon, is consistent with this pooling-contamination scenario.

CONCLUSIONS

Despite molecular variations resulting from maturation and organic facies changes, five major, discrete groups of oils can be recognized in the petroleum subprovinces of the Mexican Gulf Coast Basin. Each group is correlated with the following specific generative source: (1) Oxfordian marine marl-dominated, (2) Oxfordian marine carbonate-dominated, (3) Tithonian marine marl-dominated, (4) Cretaceous marine carbonate-evaporitic, and (5) Tertiary marine deltaic siliciclastics. Diagnostic criteria can be established to differentiate oils derived from a wide range of ages and types of source rocks. Oil groups related to Jurassic and Cretaceous source rocks are typical of carbonate-sourced oils. However, facies changes led to variations in salinity, clay content, and oxygen depletion levels between and within oil groups.

The Tithonian generative subsystem has produced more than 80% of all oil reserves from the Mexican Gulf Coast Basin. These oils are accumulated both onshore and offshore and throughout the stratigraphic column from Kimmeridgian to Pleistocene in marine siliciclastic and carbonate reservoirs, suggesting vertical pathways as the secondary migration mechanism.

The aerial extent of effective source rocks is the main control on the distribution of the oil groups identified in the Mexican Gulf Coast Basin. Cenozoic sedimentation accelerated burial of source rocks to considerable depth and provided secondary control on the distribution of oil groups.

Evidence for Mesozoic source presence in the deep-water Gulf of Mexico has been derived from the geochemistry of discovered petroleum and seabed seepage, as well as from regional extrapolation of data from the Gulf of Mexico onshore margins. At an early stage of exploration maturity, these units have already yielded approximately 5 billion BOE reserves (Duppenbecker et al., 1998). It has been suggested that the molecular features of these fluids are similar to some Jurassic oils from the northern Gulf of Mexico (Kennicutt et al., 1992). The specific source-rock interval in the deep-water Gulf of Mexico Basin remains speculative, but the existence of a world-class source-rock sequence is indicated by widespread seabed hydrocarbon seepage and by several deep-water to ultradeep-water giant to supergiant accumulations. Cole et al. (1999) have proposed that the primary source-rock sequence for the deep-water Gulf of Mexico Basin is most likely centered on the Tithonian and includes the Neocomian. The occurrence of persistent seepage suggests an active Tithonian/Neogene/Quaternary petroleum system. These ideas are consistent with the volumetric importance of the Tithonian generative subsystem in the Mexican Gulf Coast Basin.

ACKNOWLEDGMENTS

The authors are indebted to PEMEX and IMP for support and permission to publish this work.

REFERENCES CITED

Abdullah, W. H., D. Murchinson, J. M. Jones, N. Telnaes, and J. Gjelberg, 1988, Lower Carboniferous coal depositional environments on Spitsbergen, Svalbard, in L. Mattavelli and L. Novelli, eds., Advances in organic geochemistry 1987: Pergamon Press, Inc., New York, p. 953–965.

Brooks, J., 1990, Classic petroleum provinces, in J. Brooks, ed., Classic petroleum provinces: Geological Society Special Publication no. 50, p. 1–8.

Brooks, P. W., 1986, Unusual biological marker geochemistry of oils and possible source rocks, offshore Beaufort-Mackenzie Delta, Canada, in D. Leythaeuser and J. Rullkötter, eds., Advances in organic geochemistry 1985: Pergamon Press, Inc., Oxford, U.K., p. 401–406.

Carrillo-Hernández, T., 1999, Étude geochimique des gaz et condensats du Bassin de Burgos (Mexique): Rapport de stage de Diplome des Études Approfondies, Institut Francais du Petrole, Rueil Malmiason, France, 61 p.

Cole, G. A., et al., 1999, The geochemical and basin modeling aspects of the Jurassic to Lower Cretaceous sourced petroleum system, deepwater to ultra-deepwater Gulf of Mexico, offshore Louisiana: Proceedings Third Joint AAPG/Asociación Mexicana de Geólogos Petroleros (AMGP) Hedberg Research Conference, Veracruz, Mexico.

Comet, P. A., L. K. Rafalaska, and J. M. Brooks, 1993, Sterane and triterpane patterns as diagnostic tools in the mapping of oils, condensates and source rocks of the Gulf of Mexico region: Organic Geochemistry, v. 20, no. 8, p. 1265–1296.

Connan, J., and D. Dessort, 1987, Novel family of hexacyclic hopanoid alkanes (C_{32}-C_{35}) occurring in sediments and oils from anoxic paleoenvironments: Organic Geochemistry, v. 11, p. 103–113.

Connan, J., J. Bouroullec, T. D. Dessort, and P. Albrecht, 1986, The microbial input in carbonate-anhydrite facies of a sabkha palaeoenvironment from Guatemala; a molecular approach, in D. Leythaeuser and J. Rullkötter, eds., Advances in organic geochemistry 1985: Pergamon Press, Inc., Oxford, U. K., p. 29–50.

Connan, J., and D. Dessort, 1987, Novel family of hexacyclic hopanoid alkanes (C_{32}–C_{35}) ocurring in sediments and oils from anoxic paleoenviroments: Organic Geochemistry, v. 11, p. 103–113.

Connan, J., D. Dessort, Y. Poirier, and M. A. Guzmán-Vega, 1995, Unusual bacterially derived biomarker spectra in crude oils and source rocks from southern Chiapas–Guatemala area, in J. O. Grimalt and C. Dorronso, eds., Organic geochemistry: Developments and applications to energy, climate, environment and human history: Selected papers from 17th International Meeting on Organic Geochemistry, San Sebastián–Donostia, Spain, p. 198–201.

Crúz, P. H., and R. González, 1994, A new petroleum system in Mexico: Proceedings of the First Joint AAPG/Asociación Mexicana de

Geólogos Petroleros (AMPG) Hedberg Research Conference, Mexico City, Mexico, p. 13–15.

Dahl, J. E., R. C. Kaack, B. J. Huizinga, N. Holguín Quiñones, L. D. Raedeke, and N. Schneidermann, 1993, Sistemas petroleros del Sureste de México, in El sistema petrolífero y su evaluación: I Simposio de Geología de Subsuelo Asociación Mexicana de Geólogos Petroleros, Ciudad del Carmen, Mexico, p. 10–13.

Duppenbecker, S. J., A. S. Pepper, J. Wendebourg, J. M. Gaulier, and F. Schneider, 1998, Modeling petroleum system dynamics in the deep-water Gulf of Mexico: Asociacion Brasileira de Geologos Petroleros/AAPG International Conference and Exhibition, November 8–11, Rio de Janeiro, Brazil.

Ekweozor, C. M., J. I. Okogun, D. E. U. Ekong, and J. R. Maxwell, 1979, Preliminary organic geochemical studies of samples from the Niger Delta (Nigeria) II: Analyses of shale for triterpenoid derivatives: Chemical Geology, v. 27, p. 29–37.

González, R., and N. Holguin, 1991, Geology of the source rocks of Mexico, in L. Montadert and A. Ortiz, eds., Advances in exploration technology: Proceedings of the Thirteenth World Petroleum Congress, Buenos Aires, Argentina, p. 95–104.

Grantham, P. J., J. Posthuma, and A. Baak, 1983, Triterpanes in a number of Far Eastern crude oils, in M. Bjorøy et al., eds., Advances in organic geochemistry 1982: John Wiley and Sons, Chichester, U. K., p. 675–683.

Guzmán-Vega, M. A., J. Espitalié, and J. Delfaud, 1991, Source rocks characterization of Upper Jurassic of Tampico-Tuxpan Basin, east of Mexico, in L. Montadert and A. Ortiz, conveners, Advances in exploration technology, source-rock geology: Proceedings of the Thirteenth World Petroleum Congress, Buenos Aires, Argentina, p. 137–138.

Guzmán-Vega, M. A., F. J. Martínez Concha, C. V. De Araujo, and M. R. Mello, 1992, Geochemical evaluation of oils and source rocks from Tampico-Tuxpan basin, eastern Mexico, in M. R. Mello and L. A. F. Trindade, eds., Extended abstracts from the Third Latin American Congress on Organic Geochemistry: Asociación Latinoamericana de Geoquimica Organica, Manaus, Brazil, p. 51–52.

Guzmán-Vega, M. A., M. R. Mello, C. León, and N. Holguin, 1995, Tithonian oils from the Sureste Basin, Mexico: The facies variability of their source rocks, in J. O. Grimalt and C. Dorronsoro, eds., Organic geochemistry: Developments and applications to energy, climate, environment and human history: Selected papers from 17th International Meeting on Organic Geochemistry, San Sebastián–Donostia, Spain, p. 229–231.

Guzmán-Vega M. A., J. M. Moldowan, and F. Fago, 1997, Upper Jurassic organic-rich depositional environments in the Tampico-Misantla Basin, eastern Mexico: 18th International Meeting on Organic Geochemistry, Maastrich, Netherlands, p. 157–158.

Guzmán-Vega, M. A., and M. R. Mello, 1999, Origin of oil in the Sureste Basin, Mexico: AAPG Bulletin, v. 83, no. 6, p. 1068–1095.

Holguín Q., N., 1987, Evaluación geoquímica del Sureste de México: Boletín de la Asociación Mexicana de Geólogos Petroleros, v. 37, p. 3–48.

Kennicutt, I. M. C., T. J. McDonald, P. A. Comet, G. J. Denoux, and J. M. Brooks, 1992, The origins of petroleum in the northern Gulf of Mexico: Geochimica et Cosmochimica Acta, v. 56, p. 1259–1280.

Mello, M. R., P. C. Gaglianone, S. C. Brassell, and J. R. Maxwell, 1988, Geochemical and biological marker assessment of depositional environment using Brazilian offshore oils: Marine and Petroleum Geology, v. 5, p. 205–223.

Mello, M. R., E. A. Koutsoukos, E. V. Santos Neto, and A. Silva Telles Jr., 1993, Geochemical and micropaleontological characterization of lacustrine and marine hypersaline environments from Brazilian basins, in B. J. Katz and L. Pratt, eds., Source rocks in a sequence stratigraphic framework: AAPG Memoir 68, p. 17–34.

Mello, M. R., N. Telnaes, and J. R. Maxwell, 1995, The hydrocarbon source potential in the Brazilian marginal basins: A geochemical and paleoenvironmental assessment, in A. Y. Huc, ed., Paleogeography, paleoclimate, and source rocks: AAPG Studies in Geology 40, p. 233–272.

Moldowan, J. M., W. K. Seifert, and E. J. Gallegos, 1985, Relationship between petroleum composition and depositional environment of petroleum source rocks: AAPG Bulletin, v. 69. p. 1255–1268.

Moldowan, J. M., F. J. Fago, R. M. K. Carlson, D. C. Young, G. V. Duyne, J. Clardy, M. Schoell, C. T. Pillinger, and D. S. Watt, 1991, Rearranged hopanes in sediments and petroleum: Geochimica et Cosmochimica Acta, 55, p. 1065–1081.

Moldowan, J. M., J. E. Dahl, B. J. Huizinga, D. W. Taylor, L. J. Hickey, and T. M. Peakman, 1994, The molecular fossil record of oleanane and its relation to angiosperms: Science, v. 265, p. 768–771.

Murray, A. P., R. E. Summons, C. J. Boreham, and L. M. Dowling, 1994, Biomarker and n-alkane isotope profiles for Tertiary oils: Relationship to source rock depositional setting, in N. Telnaes, G. van Grass, and K. Oygard, eds., Advances in organic geochemistry 1993: Organic Geochemistry, v. 22, p. 521–542.

Nehring, J., 1991, Oil and gas resources, in A. Salvador, ed., The Gulf of Mexico Basin: Geological Society of America, Decade of North American Geology, v. J, p. 445–494.

Palacas, J. G., D. E. Anders, and J. D. King, 1984, South Florida Basin—a prime example of carbonate source rocks of petroleum, in J. G. Palacas, ed., Petroleum geochemistry and source rock potential of carbonate rocks: AAPG Studies in Geology 18, p. 71–96.

Peters, K. E., and J. M. Moldowan, 1993, The biomarker guide: Interpreting molecular fossils in petroleum and ancient sediments: Engelwood Cliffs, New Jersey, Prentice-Hall, Inc., 363 p.

Philp, R. P., and T. D. Gilbert, 1986, Biomarker distributions in Australian oils predominantly derived from terrigenous source material, in D. Leythaeuser and J. Rullkötter, eds., Advances in organic geochemistry 1985: Pergamon Press, Inc., Oxford, U.K., p. 73–84.

Prinzhofer, A., M. A. Guzmán Vega, A. Battani, and M. Escudero, 2000, Gas Geochemistry of the Macuspana Basin (Mexico): Thermogenic accumulations in sediments impregnated by bacterial gas: Marine and Petroleum Geology, v. 17, p. 1029–1040.

Salvador, A., 1991, Origin and development of the Gulf of Mexico Basin, in A. Salvador, ed., The Gulf of Mexico Basin: Geological Society of America, Decade of North American Geology, v. J, p. 389–444.

Santiago, A. J., 1979, Provincias y areas petroleras del Sureste de México: Boletín de la Asociación de Geólogos Petroleros, v. 31, nos. 1 and 2, p. 1–28,

Sassen, R., and E. W. Chinn, 1990, Implications of Lower Tertiary source rocks in South Lousiana to the origin of crude oil, offshore Louisiana, in D. Schumacher and B. F. Perkins, eds., Gulf Coast oils and gases: Proceedings of the Ninth Annual Research Conference, Gulf Coast Section, Society for Sedimentary Geology (SEPM) Foundation, p. 175–179.

ten Haven, H. L., J. W. De Leeuw, and P. A. Schenck, 1985, Organic geochemical studies of a Messinian evaporitic basin, northern Apennines (Italy), I; Hydrocarbon biological markers for a hypersaline environment: Geochimica et Cosmochimica Acta, v. 49, p. 2181–2191.

Viniegra, F., 1981, Great carbonate bank of Yucatán, southern Mexico: Journal of Petroleum Geology, v. 3, p. 247–278.

Wenger, L. M., R. Sassen, and D. Schumacher, 1990, Molecular characterization of Smackover, Wilcox, and Tuscaloosa-reservoired oils in the eastern Gulf Coast: Characteristics, origin, distribution, and exploration and production significance: Proceedings of the Ninth Annual Research Conference, Gulf Coast Section, Society for Sedimentary Geology (SEPM) Foundation, p. 37–57.

Zumberge, J. E., 1984, Source rocks of the La Luna Formation (Upper Cretaceous) in the Middle Magdalena Valley, Colombia, *in* J. G. Palacas, ed., Petroleum geochemistry and source rock potential of carbonate rocks: AAPG Studies in Geology 18, p. 127–134.

Structural Geology

Carrillo M., M., J. J. Valencia I., and M. E. Vázquez, Geology of the
southwestern Sierra Madre Oriental fold-and-thrust belt, east-central
Mexico: A review, in C. Bartolini, R. T. Buffler, and A. Cantú-Chapa, eds.,
The western Gulf of Mexico Basin: Tectonics, sedimentary basins, and
petroleum systems: AAPG Memoir 75, p. 145–158.

6

Geology of the Southwestern Sierra Madre Oriental Fold-and-thrust Belt, East-central Mexico: A Review

Miguel Carrillo M.
Instituto de Geología, Universidad Nacional Autónoma de México, Mexico City, Mexico

Juan José Valencia I.
Instituto Mexicano del Petróleo, Mexico City, Mexico

Mario E. Vázquez
Instituto Mexicano del Petróleo, Mexico City, Mexico

ABSTRACT

This study presents a regional structural analysis and a stratigraphic study in Hidalgo and Querétaro states, east-central Mexico, where strata of the Sierra Madre Oriental foreland fold-and-thrust belt are structurally juxtaposed against Middle (?) and Upper Jurassic back-arc sequences.

Rocks of the San Juan de la Rosa Formation were apparently affected by major compressive deformation during the Early Cretaceous (Nevadian?) and Laramide orogenies. Toward the east, the style of deformation is that of a typical foreland fold-and-thrust belt, whose major thrust faults are located at the Zimapán Basin margins flanked by the El Doctor and Valles–San Luis Potosí carbonate platforms. Below the El Doctor thrust, at least seven thrust faults are exposed that crosscut the middle Cretaceous rocks and the enveloping Upper Jurassic–Lower Cretaceous and Upper Cretaceous marly shaly beds in the form of a duplex.

The style of deformation between the two carbonate platforms is probably thin-skinned. Speculative, semibalanced cross sections indicate an average shortening between the Zimapán Basin margins of 39.4% of its initial width. Depth to the sole thrust is estimated to be about 2000 m below the present sea level.

In the western part of the study area, the stratigraphy consists of Middle (?) to Upper Jurassic San Juan de la Rosa Formation (volcanic-sedimentary strata), unconformably overlain by Lower Cretaceous La Peña Azul Formation. Middle Cretaceous carbonates of the Tamaulipas Formation and Upper Cretaceous marly shaly rocks of the Soyatal-Méndez Formations are the youngest rocks deformed by the Laramide dynamics in this area. Across the eastern part of the study area, the stratigraphy is more representative of the Sierra Madre Oriental fold-and-thrust belt and is dominated by the middle Cretaceous El Doctor and Valles–San Luis Potosí carbonate platforms, which are mapped as the El Doctor Formation (synonymous with the El Abra Formation). The equivalent basinal unit is the Tamaulipas Formation. The El Doctor and Tamaulipas Formations overlie the marly shaly rocks of the Upper Jurassic–Lower Cretaceous Las Trancas Formation and lie beneath the Upper Cretaceous Soyatal-Méndez Formations.

INTRODUCTION

This paper presents a synopsis of the geology of the Laramide Sierra Madre Oriental fold-and-thrust belt at its southwestern outcrops in the border area of Hidalgo and Querétaro states (latitude ~21° N; Figure 1). In this region, the Laramide trend axis is nearly perpendicular to the Moctezuma, Extorax, and Santa Maria Rivers, with a topographic relief of as much as 2 km between the Moctezuma River, situated at 1200 m above sea level, and the nearby El Espolón range, which reaches 3260 m above sea level. The well-exposed geology reveals the structural geometry in this region, including Upper Jurassic back-arc volcanic-sedimentary rock sequences (Carrillo and Suter, 1982; Barrera and Guzmán, 1984; Chauve et al., 1985) that are strikingly different from the largely carbonate rocks of the Sierra Madre Oriental foreland fold-and-thrust belt.

GEOLOGIC SETTING

The Sierra Madre Oriental fold-and-thrust belt is defined as the continuously exposed eastern part of the western North American Cordillera, which extends from Alaska to southeastern Mexico (Tardy, 1980; Suter, 1980, 1984, 1987; de Cserna, 1989). The study area consists mainly of the middle Cretaceous El Doctor Formation platform carbonates and coeval basinal rocks of the Tamaulipas Formation sandwiched by the Upper Jurassic–Lower Cretaceous and Upper Cretaceous marly shaly rocks of the Las Trancas and Soyatal-Méndez Formations (Wilson et al., 1955; Segerstrom, 1961, 1962; Enos, 1974; Martínez, 1979; Ward, 1979; Enos and Moore, 1983). In the westernmost area, Middle Jurassic (?) to Upper Jurassic volcanic-sedimentary back-arc rocks (Carrillo and Suter, 1982; Barrera and Guzmán, 1984; Carrillo et al., 1986) of the San Juan de la Rosa Formation show evidence of a compressional event during the Early Cretaceous (Nevadian orogeny?) (Chauve et al., 1985; Carrillo, 1989a). Similar rocks and timing of deformation were reported to the west in Guanajuato state (Monod et al., 1990; Ortíz et al., 1992; Lapierre et al., 1992).

Laramide dynamics caused the most prominent structural features, as described in this paper. Other deformational events include Late Jurassic to Barremian and middle Cretaceous synsedimentary faulting. Mid-Cretaceous faulting is suggested by widespread lithologic heterogeneity of the eastern part of the San Juan de la Rosa Formation. There, 1-m-thick breccia blocks of alkaline pyroclastic rocks, polymictic conglomerates, and sandstone reflect local erosion of a highstanding source area. Tectonic reactivation of source areas from Late Jurassic to Barremian along a well-defined zone was probably related to normal faulting. Synsedimentary faulting also could have been active during the middle Cretaceous, because there are numerous intercalations of conglomerates and siliciclastic rocks with brecciated limestone that form fans near the El Doctor carbonate platform (Carrillo, 1989a). These lithofacies are similar to those near the Faja de Oro carbonate platform, which show characteristics of subaerial exposure (Guzmán, 1967) and may be related to active faulting during middle Cretaceous sedimentation, as in the modern analog in the Gulf of Elat (Aqaba) (Epstein and Friedman, 1983).

In the westernmost outcrops of the study area, the San Juan de la Rosa Formation is overlain with angular unconformity by the La Peña Azul Formation. These rocks show variable stratal disruption, ranging from well-bedded outcrops to layers displaying pinch-and-swell structures, and boudinage in the most intensely deformed rocks. The latter consist of isolated lenses of varying size derived from dismembered graywackes, chert, and silicified rocks surrounded by a mudstone matrix, suggesting a compressive episode during the Early Cretaceous (Chauve et al., 1985; Carrillo, 1989a). Both deformational episodes were reported in Guanajuato state (Monod et al., 1990). A tectonic event in the study area deformed middle Miocene rocks into low-amplitude folds with axial trends parallel to Laramide orientations (Suter, 1982a; Suter et al., 1997). Sparse Neogene normal faulting was also documented (Suter, 1982b; Carrillo and Suter, 1982; Palacios-Nieto, 1982; Arvizu and Alcántara, 1989; Carrillo, 1997).

STRATIGRAPHY

In the study area, two distinct lithostratigraphic domains record the geologic evolution during the Mesozoic (Figures 1 and 2). The western domain consists of a volcanic-sedimentary sequence at least 800 m thick, composed mainly of schist and phyllitic shale, chert, rhyolite and rhyodacite tuff, and undifferentiated lavas. K/Ar isotopic ages on rhyolite (Chauve et al., 1985, p. 336) give 77 ± 4 Ma and 84.9 ± 4 Ma. Also, graywackes and silicified fine-grained rocks with radiolaria *Pseudodictyomitra* sp., *Acanthocircus carinatus* Foreman, and *Hagiastridae* or *Patulabracciidae* and *Mirifusus* sp. indicate a Dogger- Malm age (Chauve et al., 1985, p. 336). These rocks are also exposed on the western flank of the El Chilar anticline. In the eastern area of the El Chilar anticline and below the El Doctor thrust, outcrops consist of polymictic conglomerate, breccia, arkose, lithic and feldspathic graywacke sandstone, slaty shale, and tuff containing Perisphintidé ammonites of Kimmeridgian and Tithonian age (Martínez, 1979; Chauve et al., 1985). These strata are assigned to the Middle (?) to Upper Jurassic San Juan de la Rosa Formation (Chauve et al., 1985). The 77 ± 4 Ma and 84.9 ± 4 Ma isotopic ages are interpreted to have been caused by metamorphism related to the earliest Laramide activity. The phyllitic volcanic-sedimentary rocks that crop out in the western zone of the core of the El Chilar anticline exhibit alignment of sericite and phrenite minerals and show anastomosing cleavage at both microscopic and megascopic scales (Chauve et

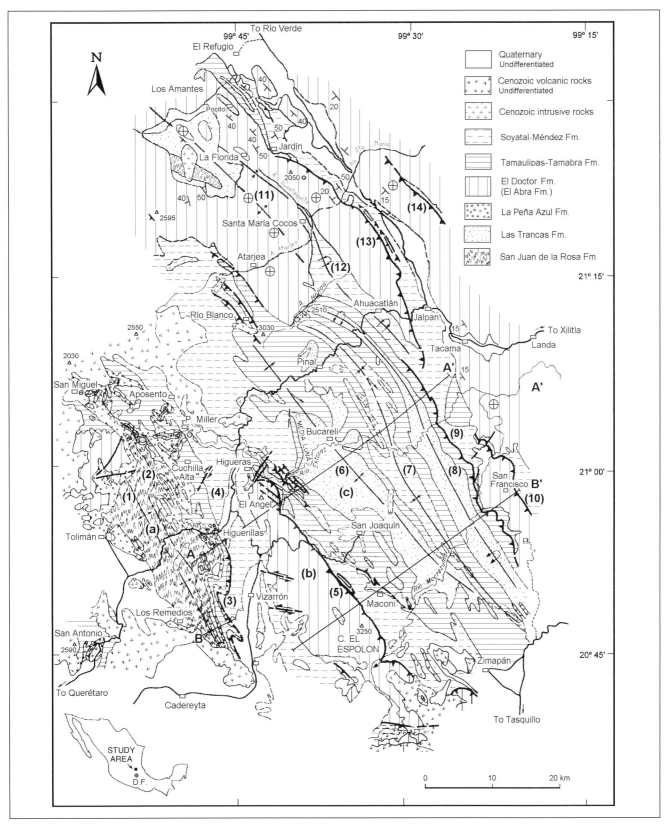

FIGURE 1. Location and geologic map of the study area. Numbers in brackets indicate: (1) El Chilar anticline, (2) El Frontón lineament, (3) Higuerillas thrust, (4) San Lorenzo syncline, (5) El Doctor thrust, (6) El Piñón anticlinorium, (7) El Aguacate syncline, (8) Bonanza fold nappe, (9) El Volantín thrust, (10) Jiliapan thrust, (11) La Yesca anticline, (12) Ahuacatlán syncline, (13) Puente de Dios thrust, (14) San Juan Buenaventura thrust. The locations of the structural sections are also shown (after Segerstrom, 1961, 1962; Carrillo and Suter, 1982; Suter, 1982a, 1982b, 1987; Lazcano-Sahagún, 1986; Carrillo, 1989b).

al., 1985). In addition, megascopic deformational features suggest two shortening events—Nevadian (?) and Laramide. Conversely, the eastern area of the core of the El Chilar anticline does not show penetrative foliation but rather contains a well-defined breccia zone several kilometers long and a few meters wide, which exposes conglomerate and breccia of Late Jurassic to Barremian age interpreted to have been deposited during normal faulting (Carrillo, 1989a). One alternative paleogeographic interpretation for these deposits is that this basin was located east of an Andean-type continental magmatic arc (Bonneau, 1972; Carfantan 1986; Suter et al., 1997). These rocks are overlain by the La Peña Azul Formation of Early Cretaceous age, and they consist of a basal conglomerate transitionally overlain by as much as 400 m of alternating conglomerate, sandstone, tuff, shale, marl, gypsum lenses, and limestone (Chauve et al., 1985; González and Carrillo, 1986).

The eastern domain is stratigraphically more representative of the Sierra Madre Oriental fold-and-thrust belt. The Upper Jurassic to Lower Cretaceous Las Trancas Formation (Segerstrom, 1961, 1962; Kiyokawa, 1981; Carrillo and Suter, 1982; González and Carrillo, 1986; Suter, 1984, 1987; Bernabé, 1994; Suter et al., 1997) consists of 1000 m of exposed alternating marl and shale. Subordinate outcrops of breccia, conglomerate, sandstone, and pyroclastics reflect facies change toward the east, where these coarser-grained rocks tend to disappear in the core of the El Piñón anticlinorium. East of Peña Blanca, these strata contain the fossil *Cassiduloidea*, and in the core of the Bonanza fold, ammonites—probably of the genus *Paradontoceras*, *Protancyloceras*, and *Parallelodon*—have been identified (Segerstrom, 1961). In Puerto de Las Trancas, rocks of this formation were deposited during the late Kimmeridgian to Portlandian (Segerstrom, 1961).

In the Zimapán Basin, the top of the Las Trancas Formation contains numerous ammonites similar to those found in the La Peña Azul Formation; they are possibly Barremian in age (González-Arreola and Carrillo-Martínez, 1986). In Barranca de Martínez (Piñón anticlinorium core), González-Arreola and Carrillo-Martínez (1986) reported: *Protancyloceras* Spath, 1924, *Protancyloceras hondense* (Imlay), family *Bochianitidae* Spath, 1922; *Karsteniceras* Royo and Gómez, 1945, *Karsteniceras beyrichii* (Karsten), family *Ancyloceratidae* Gill, 1871; *Anahamulina* Hyatt, 1900, *Anahamulina* cf. *loneli* (Uhlig), 1883; and *Silecitis* sp. family *Silecitidae* Hyatt, 1900. The later specimens range in age from late Tithonian to late Hauterivian–Barremian and middle Barremian. These ages are concurrent with similar ages reported by Kiyokawa et al. (1981) in the core of the Bonanza fold nappe. The Las Trancas Formation is coeval with the Pimienta Formation and at least the upper part of the Tamán Formation, which is located east and below the Valles–San Luis Potosí platform (Cantú-Chapa, 1971, 1998; Suter, 1984, 1990). In both domains, the La Peña Azul and the Las Trancas Formations are overlain in slight angular discordance by Aptian to Cenomanian carbonate rocks. These carbonates represent two main facies: marine platform carbonates

of the El Doctor Formation (Wilson et al., 1955; Enos, 1974; Wilson, 1975; Ward, 1979; Wilson and Ward, 1993) and basinal rocks assigned to the Tamaulipas Formation (Carrillo and Suter, 1982).

The El Doctor Formation, which is equivalent to the El Abra Formation (Carrillo, 1971; Basañez-Loyola et al., 1993), is 1500 m thick and is the dominant unit on the middle Cretaceous El Doctor and Valles–San Luis Potosí carbonate platforms (Carrasco, 1970). These shallow-water deposits consist of rudist knolls and massive, bioclastic limestone grain-rudstones (Enos, 1974; Wilson, 1975; Enos and Moore, 1983; Ward, 1979). The platform inner part is mainly well-bedded limestone. North of the Santa Maria River, these deposits contain gypsum. South of this river, the Valles–San Luis Potosí Platform overlies marly shaly rocks that were grouped into the Las Trancas Formation and volcanic-sedimentary rocks similar to those in the San Juan de La Rosa and La Peña Azul Formations (Carrillo, 1989b). The El Doctor Formation contains *Caprinulaidea gracilis* and *Mexicaprina* of Albian-Cenomanian age (Ward, 1979). We collected *Nerinea luttikei Blamchenhorn* and *Nerinea* sp. south of the El Doctor bank; in the southern part of Sierra La Peña Azul, we collected *Caprinuloidea* sp., *Kimbleia* sp., and *Toucasia* sp. of Aptian (?) to Albian age.

The term *Tamaulipas Formation* is used to designate both basinal and foreslope middle Cretaceous carbonate deposits (Carrillo and Suter, 1982; Suter, 1987). The basinal facies consist of 400 m of well-bedded Aptian to Cenomanian (?) pelagic lime mudstone-wackestone with chert lenses and bands. The foreslope facies consist of variable thicknesses of lime, silt, and bioclastic wackestone-packstone lithoclastics and dolomites. These deposits are wedge shaped, thinning toward the toe of the slope, where the dominant lithology is fine-grained limestone with chert lenses and bands. Slope facies change abruptly in thickness, from basically nonexistent (northwest of Peña Miller) to > 800 m thick at the Barranca de Tolimán. This facies corresponds to submarine fan deposits (Eckbert Seibert, personal communication, 1997). Fossils we collected from the Tamaulipas Formation limestones were identified as *Planocrapina* sp., *Nummoloculina heimi*, *Fausella* sp., *F. washintesis*, *Thaumatoprella* sp., and *Colombiceras* sp. of Aptian age (Gloria Alencaster, personal communication, 1981).

The late Turonian to Campanian Soyatal-Méndez Formations (Wilson et al., 1955; Bodenlos, 1956a, b; Segerstrom, 1961; Kiyokawa, 1981) are the youngest stratigraphic units affected by the Laramide orogeny in this area. The exposed thickness of the Soyatal-Méndez Formations is about 1000 m, but below the El Doctor thrust in the El Angel range it is thin, probably because of deposition at the proximity of a reef structure. The tops of both the El Doctor and Valles–San Luis Potosí platforms show paleokarst filled with debris from the Soyatal-Méndez Formations (Eckbert Seibertz, personal communication, 1997). In the San Francisco area, these paleokarsts were filled with phosphorite deposits (Quintus-Bosz, 1982).

The Soyatal-Méndez Formations are considered to be of

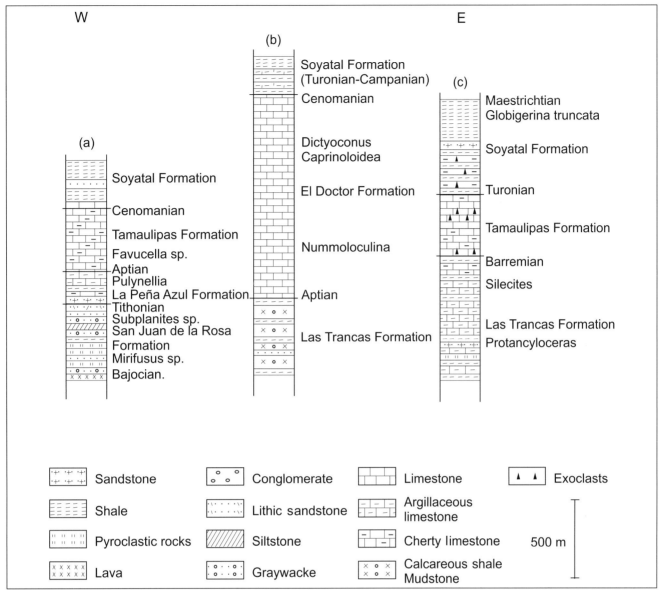

FIGURE 2. Stratigraphic units exposed in northeastern Querétaro State. (a) West of El Doctor bank, (b) El Doctor bank, (c) between the El Doctor bank and the Valles–San Luis Potosí carbonate platform (Suter, 1987). Locations are marked in Figure 1.

Turonian age (Wilson et al., 1955). Ward (1979) reported *Hippurites mexicana* sp., and Segerstrom (1961) found the ammonites *Nowakites* (near Zimapán) and *Texanites;* the latter species is typical of the Austin Chalk in Texas. The age of these ammonites ranges from Coniacian to Campanian (Simmons and Mapes, 1956, p. 13). In the Soyatal-Méndez Formations below the El Doctor thrust east of Moctezuma River, Simmons and Mapes (1955) reported the following Campanian microfossils: *Globotruncana stuarti* Lapparent (Campanian to Maestrichtian), *Globotruncana navanensis* (Upper Cretaceous), *Globotruncana fornicata* (?) D'Orbigny (Campanian), and *Globigerina truncata* D'Orbigny (Santonian to Paleocene).

Tertiary igneous rocks are the youngest outcrops exposed in the south and west of the study area. Typical intrusive rocks

are stocks, sills, dikes, porphyritic to aphanitic diorite, monzonite, and quartz-monzonite. K/Ar dating of dikes at La Negra mine yielded a 38.7 ± 0.8 Ma age (Gaytán-Rueda, 1975). Ammonites collected in the Bonanza fold nappe (Segerstrom, 1962; Suter,1987) that bracket the age of Laramide dynamics include *Nowakites* sp. and *Inoceramus* (*Pathyceramus*) ex. gr. *cyclaides* Wegner (Campanian). In addition, moderate deformed andesitic lavas of 38.1 ± 1.9 Ma and quartz-diorite intrusions ranging from 40.5 ± 2 to 50.9 ± 2.5 Ma are the youngest intrusions contemporaneous with the last Laramide activity (Kiyokawa, 1981). Along the eastern edge of the Valles–San Luis Potosí Platform, other intrusions that may form part of a continental magmatic arc (Clark et al., 1979, 1982) yielded ages ranging from 62.2 to 68 Ma (Suter, 1984). Thus, they are contemporaneous with the Laramide deformation.

GEOMETRY

No subsurface information is available for the study area. However, more than 2 km of topographic relief and continuous outcrops should be sufficient to understand the structural geometry of this portion of the belt (Figures 1 and 3). In addition, our interpretations are aided by a reflection seismic profile of the Sierra Madre Oriental fold-and-thrust belt near Concá, east of the study area. The style of deformation is interpreted as thin-skinned, with the sole thrust about 3 km below sea level (Ernesto López-Ramos, personal communication, 1993). This interpretation is in agreement with previous interpretations (Suter, 1987; Suter et al., 1997).

The rigid El Doctor and Valles–San Luis Potosí carbonate platforms, greater than 1500 m thick, had a large influence on the style of deformation, which consists of thrust nappes juxtaposed against the Zimapán Basin. Mechanical models show stress concentration and instabilities that are explained by lithology and thickness changes across the platform margins (Suter, 1984; Contreras, 1991, 1993; Contreras and Suter, 1990). Another observation that could imply mechanical and chemical consideration is the nearly systemic presence of dolomite (elephant skin) very close to the plane-thrust faults (Michael Matthes and Antonio Vélez-Bolio, oral communication, 1997). Except for drag folds, the platforms are gently folded, and imbricate compressional faults are associated with the El Doctor and San Juan Buenaventura thrusts (Lazcano-Sahagún, 1986; Carrillo, 1989b, 1990), which are the northwest continuation of the Agua Fria thrust in the Valles–San Luis Potosí Platform (Suter, 1987, 1990). In the Zimapán Basin, folds range in amplitude from kilometer to centimeter scale and are associated with minor thrusting (Suter, 1987).

Structural elements from southwest to northeast, which is the approximate transport direction, include the El Chilar anticline, the San Lorenzo syncline, the El Doctor thrust, the El Piñón anticlinorium, the El Aguacate syncline, the Bonanza fold nappe, the El Fraile syncline, the Jiliapan thrust, and the Puente de Dios thrust (Figures 1 and 3). The El Chilar anticline (Segerstrom, 1961) is a broad, open fold cored by the San Juan de la Rosa Formation and with flanks defined by the Tamaulipas Formation. Its western limb dips from 0° to 20° toward the west, and its crest is subhorizontal, as evidenced by the contact between the La Peña Azul and Tamaulipas Formations. South of the town of Higuerillas (Figure 1), the eastern side of this structure was formed by the north-northwest-trending Higuerillas thrust (Carrillo and Suter, 1982; Suter, 1987), which emplaced the La Peña Azul Formation above the Soyatal-Méndez Formations (Figures 1 and 3). There, horizontal separation along the fault perpendicular to strike is approximately 1100 m (Suter, 1987). North of Higuerillas, the thrust is not noticeable in the field. Along the Sierra Cuchilla Alta, a gentle undulation on the order of tens of meters occurs at the base of the Tamaulipas Formation. These second-order folds show an en-echelon style, suggesting a minor strike-slip fault in the direction of tectonic transport. Southwest of Peña Miller, in the valley of the Extorax River, two minor reverse faults are exposed close to the contact between the Tamaulipas and Soyatal-Méndez Formations. This suggests that the El Chilar anticline was caused by hanging-wall deformation above the Higuerillas thrust in its southern segment, whereas toward the north, the Higuerillas thrust displacement is transferred to folding. The depth of the sole thrust in this area was estimated by extending the better-constrained El Doctor thrust nappe westward. The ramp dip was interpreted to be the same as that of the northern thrust faults near Peña Miller. In Figure 3, we estimate a horizontal displacement of about 3.6 km, or 64%, for the base of the El Doctor Formation (A-A´). The El Doctor thrust nappe (Carrillo and Suter, 1982; Suter, 1987; Suter et al., 1997) (Figures 1 and 3) juxtaposes the carbonate El Doctor bank against the Zimapán Basin strata. The trace of the leading-edge thrust nappe can be followed continuously for more than 40 km to the southeast, where it is buried below the Cenozoic continental deposits of the Transmexican Volcanic Belt. The El Doctor thrust strikes roughly northwest-southeast and curves progressively toward the west in the vicinity of the El Angel Range. In the Moctezuma River canyon, the thrust fault brings the El Doctor Formation over the Soyatal-Méndez Formations with an angle of 24°, a minimum horizontal displacement of 2800 m perpendicular to strike, and a linear shortening of 5550 m (Suter, 1987). At El Angel Mountain, the El Doctor Formation is juxtaposed against the Soyatal-Méndez and Tamaulipas Formations (Figure 4). There, the thrust shows an average dip of 11° to the west, a minimum horizontal displacement of 1250 m perpendicular to the strike (Figure 1), a hanging-wall anticline, and two thrust-associated folds.

Our interpretation of the buried structure down to the structural basement (Figure 3) concurs with previous interpretations (Suter, 1987). Because the top of the original stratigraphic level of the El Doctor Formation is 800 m above present sea level, we assume a depositional relief of 600 m between the carbonate bank and the Zimapán Basin. In the method employed here, we established two pin lines—one between the axial planes of the San Lorenzo and Maconí synclines and one between the San Lorenzo and Ahuacatlán synclines. Assuming plane strain and thin-skinned deformation, we estimated the depth to décollement (Z) from $Z = A/s$, where A is the structural relief area and s is the linear shortening (Dahlstrom, 1969). By referencing the top of the El Doctor Formation, extrapolating the exposed dip of the thrust, and increasing their dip close to the contact between the El Doctor and Las Trancas Formations, we calculated a linear shortening of 1.6 km, or 20%, between the axis of the San Lorenzo syncline and the lower flank of the El Doctor thrust (Maconí syncline) and an areal shortening of 6.9 (10^6m^2) (Figure 3). These two parameters allow a reasonable estimation of the depth of the basal décollement measured from the level of reference, if we assume plane strain and thin-skinned structural deformation. The operation results in a depth to detachment of 4.3 km, which corresponds to a sole

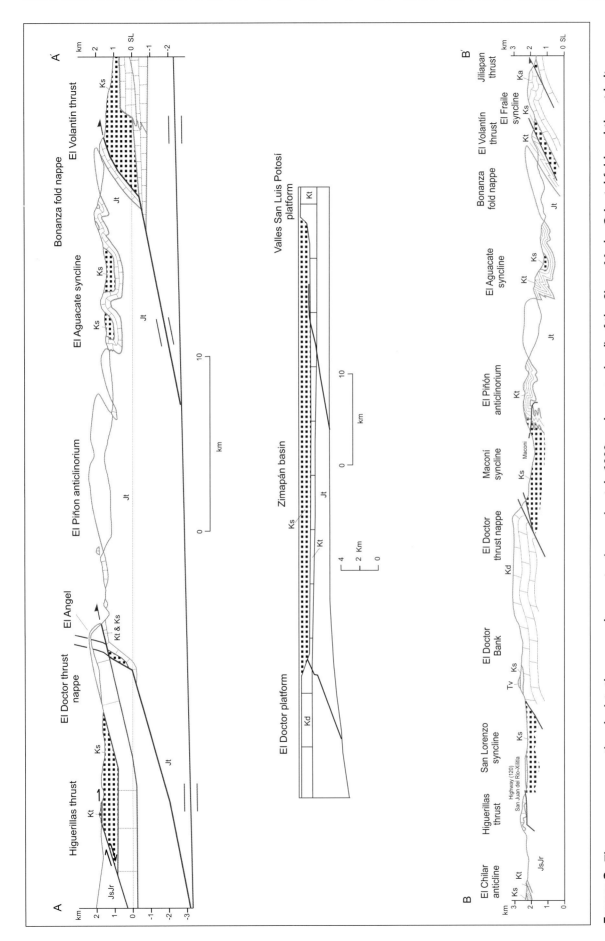

Figure 3. The upper cross section depicts the exposed geometry (approximately 1000 m above sea level) of the Sierra Madre Oriental fold-and-thrust belt between Higuerillas and Jalpan, northeastern Querétaro State. The A-A´ cross section was interpreted using the method which estimates depths of the décollement of the top and base of the El Doctor and Tamaulipas Formations by dividing the structural relief area by the shortening of these contacts (Dalsthrom, 1969). The counterpart palinspastic cross section has been semirestored using the combined equal-area and key-bed method (Mitra and Namson, 1989). The lower cross section represents the exposed geometry across the southern Zimapán Basin (from Carrillo and Suter, 1982). Explanations are in the text.

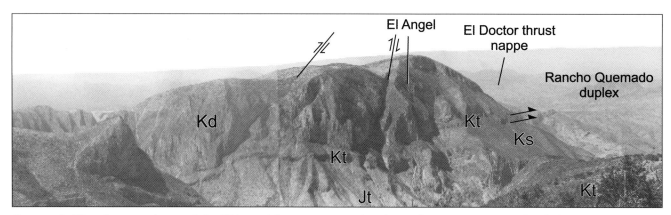

FIGURE 4. View from southeast of the El Doctor thrust nappe at the El Angel range. Ks = Soyatal-Méndez Formations, Kd = El Doctor Formation, Kt = Tamaulipas Formation, Jt = Las Trancas Formation. The thrust dips about 11° and juxtaposes carbonate platform-edge facies (Ward, 1979) against both Tamaulipas and Soyatal-Méndez Formations. Two faults located in the hanging wall offset the El Doctor Formation but do not cut the Tamaulipas (foreslope facies of variable thickness) and Soyatal-Méndez Formations.

fault depth of 3.5 km below present sea level. If we use the top of the Las Trancas Formation as a reference horizon and assume a 1000-m thickness for the El Doctor Formation, the same linear shortening of 1.6 km and an areal shortening of 5.6 (10^6m^2) give Z = 3.5 km below the reference level. This is 1.0 (10^6m^2) short of the area needed to balance the section. The 8° dip of the lower segment of the thrust increases the area deficit. Another interpretation of the buried structure of the El Doctor thrust nappe near the Moctezuma River is a flat-ramp-flat model (Suter et al., 1997). This model is based mainly on one outcrop of the fault plane located 1375 m above sea level at the Zimapán hydroelectric-dam site (Palacios, 1982) and implies a flat at least 4.5 km long directly southwest of the exposed ramp (Suter et al., 1997). The El Piñón anticlinorium (Segerstrom, 1962) is bounded by the Maconí and El Aguacate synclines. The Maconí syncline is broader and deeper in its southern portion and becomes narrower and rises northwestward as the El Doctor thrust approaches the southwest limb of the El Piñón anticlinorium (Figures 1 and 3). At this location, below the El Angel Range, several faulted blocks converge to produce a tectonic imbrication (Figures 5, 6, 7, 8, and 9). The thrusts often follow bedding of the Las Trancas Formation. At the scale of several meters, they also ramp into higher stratigraphic levels in the Tamaulipas Formation (Figures 9 and 10). Upsection, the angle of the thrusts decreases near the base of the Soyatal-Méndez Formations, which is mostly eroded. In the La Media Luna range, there is a structural discordance (Figures 3 and 11) (Segerstrom, 1961) in the form of a flat, which we interpret as a remnant of the roof thrust of a duplex here referred to as the Rancho Quemado duplex (Figure 7). Northwest of the Rancho Quemado duplex, the El Piñón anticlinorium is bounded on the west by the San Lorenzo syncline (Figure 1). In addition to the Rancho Quemado duplex and minor reverse faulting, the basinal formations—particularly the Tamaulipas Formation—exhibit second-order folding that is overturned to subhorizontal in the southwestern limb (Figure 1), whereas the axial

planes of these folds at the crest of the El Piñón anticlinorium are subvertical. Surprisingly, the contact between the Tamaulipas and Las Trancas Formations does not reflect this intense second-order folding. One interpretation is that the more competent rocks of the Tamaulipas Formation are detached and slipped over the incompetent rocks of the Las Trancas Formation (Carrillo and Suter, 1982).

The geometry along the backlimb of the El Piñón anticlinorium can be divided into two segments. The southeastern limb dips 60° toward the southwest and contains second-order subhorizontal and hinterland verging folds (Segerstrom, 1961; Carrillo and Suter, 1982). The northwestern limb is the Rancho Quemado duplex. The forelimb of the El Piñón anticlinorium is composed of a syncline, the axial plane of which dips 27° toward the southwest. This fold is linked to the northeast with a second-order 3.5-km-wide anticline with more than 1 km of structural relief (Figure 3). These intricate, complex folds and minor reverse faults are characteristic of all the basinal rocks, whereas the thick, massive rocks in both carbonate platforms have been deformed by external rotation to form broad folds, except for drag folds at the leading edge, as in the case of the El Doctor thrust nappe (Suter, 1987).

The width of the El Piñón anticlinorium between the Maconí and El Aguacate synclines is 16.5 km. The structural relief between the crest of the anticlinorium and the Maconí syncline is greater than 1.8 km. The exposed linear shortening of the El Piñón anticlinorium is 9.45 km. We have not yet modeled the Rancho Quemado duplex segment, but it could accommodate approximately 3 km of shortening.

To the southwest, the El Aguacate syncline is linked with the El Piñón anticlinorium, and to the northeast, it merges with the Bonanza fold nappe (Suter, 1987; Carrillo, 1989b). This structural unit contains a pair of folds >100 m wide. The southwestern limb is upright, and the folds in the core show a subvertical axial surface, whereas the northeastern limb has a dip of 60° to 70°, slightly overturned toward the southwest.

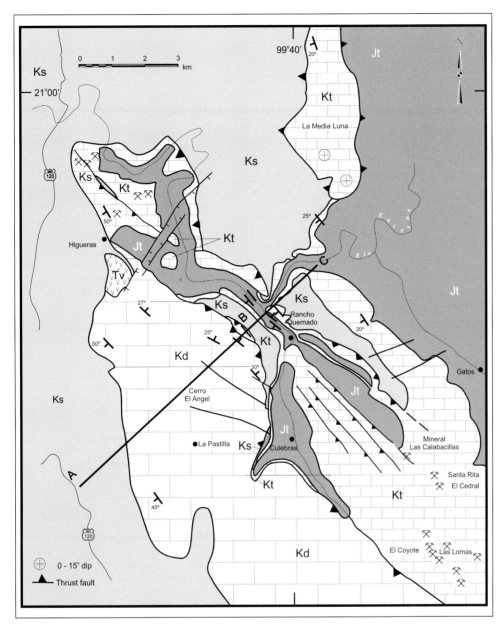

FIGURE 5. Geologic map of the El Angel–La Media Luna ranges. Positions of the structural cross sections are indicated. La Media Luna range is the structural discordance reported by Segerstrom (1961).

The second-order folds are more overturned in this northeastern portion. The El Aguacate syncline appears elevated in comparison with the Maconí and Ahuacatlán synclines, probably because of more resistant units.

The Bonanza fold nappe, approximately 2.3 km wide, extends from Arroyo Hondo, northwest of the map area, and more than 80 km to the southeast (Segerstrom, 1961, 1962; Carrillo and Suter, 1982; Suter, 1982a,1987; Carrillo, 1989b). The folding becomes overturned and dies out abruptly in a salient of the Valles–San Luis Potosí carbonate platform. The southwestern limb dips 60° to 70° toward the southwest, and the northeastern limb is cut by the El Volantín thrust (Suter, 1987). The structure is strongly overturned (Figure 12). The thrust crosses the Extorax River valley (Figure 3) but is no longer

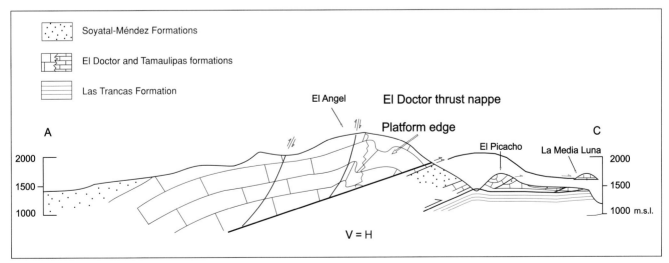

FIGURE 6. Structural geometry across the northern area of the El Doctor carbonate bank. The platform edge overrides a thin outcrop of the foreslope Tamaulipas Formation, which is thrust over the Soyatal-Méndez Formations.

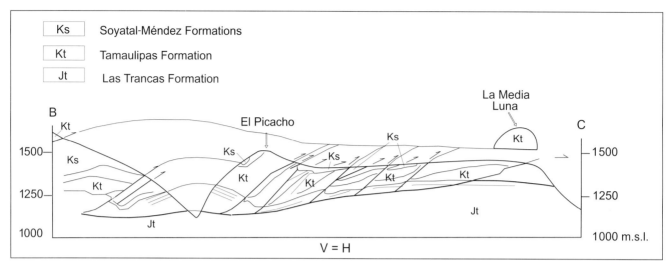

Kt   Tamaulipas Formation

Jt   Las Trancas Formation

FIGURE 7. Exposed structural geometry at the footwall of the El Doctor thrust. We speculated that the tongue of the Tamaulipas Formation below the El Doctor could be correlated with the structural discordance at the La Media Luna range reported by Segerstrom (1961), suggesting the roof thrust of a duplex. Faults located near the base of the northeastern slope of El Picacho imply some faulting that predates Soyatal-Méndez deposition.

visible in the next valley to the north; instead, there are two reverse faults of several meters' combined displacement, and the fold is no longer overturned. Apparently the El Volantín thrust transfers its displacement to the neighboring Puente de Dios thrust (Figure 1).

The El Volantín thrust fault places rocks of the Tamaulipas Formation over the Soyatal-Méndez Formations (Suter, 1982b) (Figures 1 and 3). The fault dips 17° toward the southwest but increases to 33° downdip (Figures 3 and 12). The thrust sheet shows numerous second-order folds. A rough estimate of the amount of shortening across the Zimapán Basin between the Maconí and Ahuacatlán synclines was obtained by measuring the distance along the top of the Las Trancas Formation (Figure 3) (47.0 km) and comparing it with the horizontal distance (28.5 km). This indicates that the zone has been shortened 18.5 km, or approximately 39% of its initial width. The structural relief area below the top of the Las Trancas Formation is 53.4 (10^6m^2). The semibalanced section places the depth of the sole thrust at 2900 m, or an average of 2100 m below the present sea level, and yields a dip of less than 1° for the sole fault toward the hinterland.

Restoration of our cross section was interpreted by the combined equal-area and key-bed method to find the geometry of the El Doctor thrust when the El Volantín thrust was restored. From

$$L_1 = 4A/(t_1+t_2) - Lo$$
(Mitra and Namson, 1989)

where Lo is the average bed length of the top of the Las Trancas Formation unchanged during deformation, A is the structural relief area, t_1 and t_2 are the thicknesses of the Las Trancas Formation at both cross section ends, and L_1 is the distance between these thrusts at the interception of the detachment.

FIGURE 8. View toward the east-southeast of El Picacho. Two faults offset the top of the Tamaulipas Formation, but no offset at their base is observed. This Kt/Jt contact dips toward the northeast, representing a forelimb of a folded floor thrust. Ks = Soyatal-Méndez Formations, Kt = Tamaulipas Formation, Jt = Las Trancas Formation.

$$L_1 = 4(53.4) \text{ km}^2/ \ 2.7 \text{ km}^2 - 47 \text{ km} = 32 \text{ km}.$$

We measured about 56 km^2 of the area of the assumed initial semitrapezoid, which is approximately 5% different from the 53.4 km^2 structural relief area.

BRACKETING THE TIMING OF LARAMIDE DEFORMATION

Multiple lines of evidence suggest that Laramide dynamics occurred between the early Maestrichtian and middle Eocene.

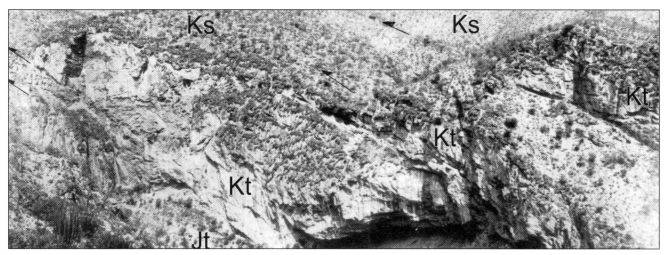

FIGURE 9. Slightly oblique view of three imbricated faults northeast of El Picacho. Note a fold-related fault on the left of the photograph. Ks = Soyatal-Méndez Formations, Kt = Tamaulipas Formation, Jt = Las Trancas Formation.

In the southern Zimapán Basin, the youngest pre-Laramide strata are those of the Soyatal-Méndez Formations, which contains *Nowakites* sp. *Inoceramus* of late Campanian age, whereas andesite lavas not affected by the Laramide deformation are dated at 38 ± 1 Ma (Kiyokawa, 1981). In this area, middle Miocene rocks are deformed by low-amplitude folds with axial trends parallel to Laramide orientations (Suter et al., 1997). In the front (east) of the Sierra Madre Oriental fold-and-thrust belt, deformation reportedly occurred between the late Maestrichtian to Paleocene (Suter, 1984); in the Tampico Misantla foredeep, Eocene beds were also affected by Laramide deformation (Suter et al., 1997), as well as the historic basement (Ochoa-Camarillo, 1997).

CONCLUSIONS

The Sierra Madre Oriental fold-and-thrust belt in Hidalgo and Querétaro states consists mainly of middle Cretaceous carbonate platforms and coeval slope and basinal facies of the El Doctor (synonymous with El Abra) and Tamaulipas Formations. These units are sandwiched by the Upper Jurassic–Lower Cretaceous Las Trancas and Upper Cretaceous Soyatal-Méndez marly shaly formations. In the western part of the area, the Middle (?) to Upper Jurassic San Juan de la Rosa Formation is volcanic-sedimentary and, with the Lower Cretaceous La Peña Azul Formation, suggests a highstanding source of varied clastic deposits along a well-defined segment that is interpreted to be a result of synsedimentary faulting during Late Jurassic to Barremian and even during the middle Cretaceous. West of the study area, an angular unconformity indicates a compressive episode during the Early Cretaceous (Chauve et al., 1985; Carrillo, 1989a).

Balanced cross sections show most of the southern Sierra Madre Oriental to be wedge shaped (Suter, 1987). The style of deformation was controlled by the strong carbonate platforms acting as buttresses to thrusting in the Zimapán Basin. Semi-balanced cross-section estimates of the distance between the two Zimapán Basin margins are approximately 47 km before deformation, which amounts to 18.55 km of shortening, or 39.4% of the basin's initial width, and an average depth to the sole thrust of 2840 m below present sea level. To the southeast, a balanced cross section estimates that the distance between the two platforms originally must have measured approximately 48 km, which was reduced to 21 km, or 44% (Suter, 1987). The depth of this décollement ranges between ~2500 and 3000 m below the Zimapán Basin (Suter, 1987). These estimates suggest a 2.4° dip of the sole thrust (Suter, 1987), less than 1° toward the southwest.

FIGURE 10. View toward the southeast of a compressional fault in the Rancho Quemado duplex. Kt = Tamaulipas Formation, Jt = Las Trancas Formation.

FIGURE 11. View toward the north-northeast, showing in the backround the structural discordance of the La Media Luna range (Segerstrom, 1961). Ks = Soyatal-Méndez Formation, Kt = Tamaulipas Formation, Jt = Las Trancas Formation.

ACKNOWLEDGMENTS

This research was supported predominantly by Project FIES 96-01-1, a joint project of the Instituto de Geología, Universidad Nacional Autónoma de Mexico (UNAM) and the Instituto Mexicano del Petróleo, in Mexico. We are grateful to Gary Gray and Gary Prost for their generous and helpful improvements to this paper. We are also grateful to the editors of this book. The author thanks Max Suter, Thierry Addate, Carlos González, Ulrike Kering, Michael Matthes, Juan C. Medina, Sergio Rodríguez, Eckbert Seibert, and Antonio Vélez for recent observations in the field, and Ernesto López-Ramos for information he provided.

Thanks to Ricardo Hernández, Jorge Mengelle, and René Téllez for field and laboratory cooperation in the FIES Project.

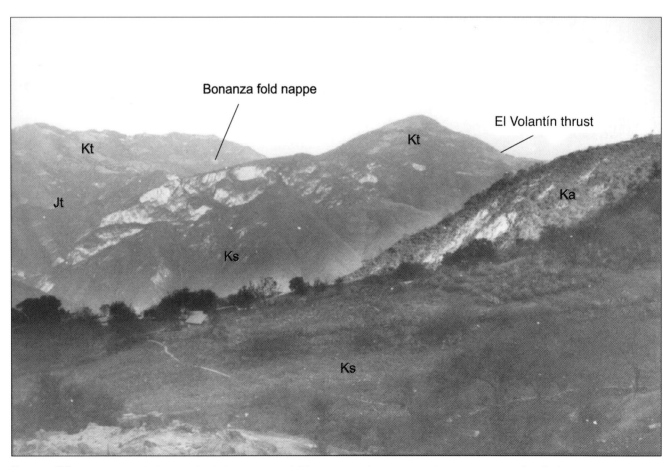

FIGURE 12. View toward the north of the Bonanza fold nappe at the Extorax River. Ks = Soyatal-Méndez Formations, Kt = Tamaulipas Formation, Jt = Las Trancas Formation, Ka = El Abra Formation.

REFERENCES CITED

Arvizu, L. G., and D. J. M. Alcántara, 1989, Factibilidad Geológica del Proyecto Hidro-Eléctrico Zimapán Hidalgo: Comisión Federal de Electricidad, Subdirección de Construcción, 102 p.

Barrera, D., and V. M. Guzmán, 1984, Petrología e implicaciones genéticas de las areniscas de la Formación Las Trancas, Jurásico Superior–Cretácico Inferior, Estados de Querétaro e Hidalgo: Professional thesis, Instituto Politécnico Nacional, Escuela Superior de Ingeniería y Arquitectura, 64 p.

Basañez-Loyola, M. A., R. Fernández-Turner, and C. Rosales-Domínguez, 1993, Cretaceous platform of Valles–San Luis Potosí, northeastern central Mexico: AAPG Memoir 56, p. 51–59.

Bernabé, M. M. G., 1994, Microfacies de la Formación Las Trancas en el anticlinal de Bonanza: Universidad Autónoma de Hidalgo, Revista de Investigaciones, v. 1, p. 50–55.

Bodenlos, A. J., 1956a, Itinerario Zimapán, Hidalgo-Tamazunchale, San Luis Potosí: 20th International Geological Congress, Excursions A-14 and C-6, guidebook, p. 179–215.

Bodenlos, A. J., 1956b, Notas sobre la geología de la Sierra Madre en la sección Zimapán-Tamazunchale: 20th International Geological Congress, Excursions A-14 and C-6, guidebook, pages 293–309.

Bonneau, M., 1972. Données nouvelles sur les séries crétacées de la côte Pacifique du Méxique: Bulletin de la Société Géologique de France, v. 7, no. 14, p. 55–65.

Cantú-Chapa, A., 1971, La serie Huasteca (Jurásico Medio–Superior) del centro este de Mexico: Revista del Instituto Mexicano del Petróleo, v. 3, p. 17–40.

Cantú-Chapa, A., 1998, Las transgresiones Jurásicas en Mexico: Revista Mexicana de Ciencias Geológicas, Universidad Nacional Autónoma de México, v. 15, no. 1, p. 25–37.

Carfantan, J. C., 1986, Du système Cordillerain Nord-Americain au domaine Caraïbe: Doctorat ès Sciences thesis, University of Savoie, France, 557 p.

Carrasco, B., 1970, La Formación El Abra (Formación El Doctor) en la Plataforma de Valles–San Luis Potosí: Revista del Instituto Mexicano del Petróleo, v. 2, p. 97–99.

Carrillo, B. J., 1975, La plataforma de Valles–San Luis: Boletín de la Asociación Mexicana de Geólogos Petroleros, v. 23, p. 1-102.

Carrillo M., M., 1989a, Structural analysis of two juxtaposed Jurassic lithostratigraphic assemblages in the Sierra Madre Oriental fold and thrust belt of central Mexico: Geofísica Internacional, v. 28, p. 1007–1028.

Carrillo M., M., 1989b, Estratigrafía y tectónica de la parte centrooriental del Estado de Querétaro: Universidad Nacional Autónoma de México, Instituto de Geología, Revista, v. 8, p. 188–193.

Carrillo M., M., 1990, Geometría estructural de la Sierra Madre Oriental entre Peñamiller y Jalpan, Estado de Querétaro: Universidad Nacional Autónoma de México, Instituto de Geología, Revista, v. 9, p. 62–70.

Carrillo M., M., 1997, Hoja Zimapán 14 Q (e) 7, Estados de Querétaro e Hidalgo: Universidad Nacional Autónoma de México, Instituto de Geología, Carta geológica de Mexico, serie 1:100,000.

Carrillo M., M., and M. Suter, 1982, Tectónica de los alrededores de Zimapán, Hidalgo y Querétaro, *in* M. Alcayde and Z. de Cserna, eds., Libro-guía de la excursión geológica a la región de Zimapán y áreas circundantes, Estados de Hidalgo y Querétaro: Sociedad Geológica Mexicana, 6.° Convención Geológica Nacional, p. 1–20.

Carrillo M., M., G. G. Velázquez, and L. Cepeda-Dávila, 1986, Contribución al estudio petrográfico y químico de areniscas del Jurásico Superior, Estados de Querétaro e Hidalgo: Universidad Nacional Autónoma de México, Instituto de Geología, Revista, v. 6, p. 269–271.

Chauve, et al., 1985, Les rapports structuraux entre les domaines cordillérain et mésogéen dans la partie centrale du Méxique: Comptes-Rendus à l'Académie, Paris, France, s. II, v. 301, p. 335–340.

Clark, K., P. Damon, S. Shutter, and M. Shafiqullah, 1979, Magmatismo en el norte de México en relación a los yacimientos metaliferos: Asociación de Ingenieros Mineros, Metalurgistas y Geólogos, Convención Nacional 13, Acapulco, Memoria, p. 8–57.

Clark, F. K. P., C. T. Foster, and P. E. Damon, 1982. Cenozoic mineral deposits and subduction-related magmatic arcs in Mexico: Geological Society of America Bulletin, v. 93, p. 533–544.

Contreras, P. J., 1991, Kinematic modeling of cross-sectional deformation sequences by computer simulation: Coding and implementation of the algorithm: Computers & Geosciences, v. 17, p. 1197–1217.

Contreras, P. J., 1993, Distribución de esfuerzos y estabilidad mecánica en las márgenes de plataformas carbonatadas: Master's thesis, Universidad Nacional Autónoma de México, 56 p.

Contreras, P. J., and M. Suter, 1990, Kinematic modeling of cross-sectional deformation sequences by computer simulation: Journal of Geophysical Research, v. 95, p. 21,913–21,929.

Dahlstrom, C. D., 1969, Balanced cross sections: Canadian Journal of Earth Sciences, v. 6, p. 743–757.

de Cserna, Z., 1989, An outline of the geology of Mexico, *in* A. W. Bally and A. R. Palmer, eds., The Geology of North America–An overview: Geological Society of America, Decade of North American Geology (DNAG), v. A, p. 233–264.

Enos, P., 1974, Reefs, platforms and basins of middle Cretaceous in Northeast Mexico: AAPG Bulletin, v. 58, p. 800–809.

Enos, P., and C. H. Moore, 1983, Fore-reef slope environment, *in* P. A. Scholle, D. G. Bebout, and C. H. Moore, eds., Carbonate depositional enviroments: AAPG Memoir 33, p. 507–537.

Epstein, A. S., and M. G. Friedman, 1983, Depositional and diagenetic relationships between Gulf of Elat (Aqaba) and Mesozoic of United States East Coast Offshore: AAPG Bulletin, v. 67, p. 953–962.

Gaytán-Rueda, J. E., 1975, Exploration and development at the La Negra mine, Maconí, Querétaro, Mexico: Master's thesis, University of Arizona, Tucson, 92 p.

González A., C., and M. Carrillo M., 1986, Presencia de amonitas heteromorfos del Jurásico Superior (Titoniano superior) y del Cretácico Inferior (Hauteriviano-Barremiano) del área de San Joaquín–Vizarrón, Estado de Querétaro: Universidad Nacional Autónoma de México, Instituto de Geología, Revista, v. 6, p. 171–177.

Guzmán, E. J., 1967, Reef type stratigraphic traps in México, *in* Origin of oil, geology and geophysics: Proceedings of the Seventh World Petroleum Congress, v. 2, p. 461–470; Autónoma de México, Instituto de Geología, Revista, v. 6, p. 171–177.

Kiyokawa, M., ed., 1981, Geological survey of the Pachuca-Zimapán area, central Mexico: Metal Mining Agency of Japan and Consejo de Recursos Minerales, Estados Unidos Mexicanos, 194 p.

Lapierre, H., L. Ortiz, W. Abouctiami, O. Monod, C. Coulon; J. A. Zimmerman, 1992, Crustal section of an intra oceanic island arc: the Late Jurassic–Cretaceous Guanajuato magmatic sequence, central Mexico: Amsterdam: Earth and Planetary Sciences Letters, v. 108 (1–3), p. 61–77.

Lazcano-Sahagún, C., 1986, Las cavernas de la Sierra Gorda: Universidad Autónoma de Querétaro, Colección Encuentro, n. 2, 177 p.

Martínez, H. S., 1979, Contribución al estudio geológico de una porción del sector Vizarrón-Tolimán, Estado de Querétaro: Professional thesis, Universidad Nacional Autónoma de México, Facultad de Ingeniería, 61 p.

Mitra, S., and J. S. Namson, 1989, Equal area balancing: American Journal of Science, v. 289, p. 563–599.

Monod, O., H. Lapierre, M. Chiodi, J. Martínez, P. Calvet, E. Ortíz, and J. L. Zimmermann, 1990, Reconstitution d´un arc insulaire intraoceanique au Mexique central: la sequence volcano-plutonique de Guanajuato (Crétacé Inferieur) Paris: Comptes-Rendus a L´Academie des Sciences, Paris S. II, v. 310 (1) p. 45–51.

Ochoa-Camarillo, H., 1997, Geología del anticlinorio de Huayacocotla en la región de Molango, Hidalgo, México: Instituto de Investigaciones en Ciencias de la Tierra Universidad Autónoma de Hidalgo, e Instituto de Geología, 2.° Convención sobre la Evolución Geológica de México y Recursos Asociados, Pachuca, Hidalgo, Libro guía de las Excursiones geológicas, Excursión 2, p. 45–63.

Ortiz, L., M. Chiodi, H. Lapierre, O. Monod, Ph. Caliet, 1992, El arco intraoceánico alóctono (Cretácico Inferior) de Guanajuato-características petrográficas, geoquímica, estructurales e isotópicas del complejo filononiano y de las lavas basálticas asociadas: Implicaciones geodinámicas México: Revista del Instituto de Geología, Universidad Nacional Autónoma de Mexico, v.9 (2), p. 126–145.

Palacios-Nieto, M., 1982, Geología y geotecnica del proyecto hidroeléctrico Zimapán, Estados de Hidalgo y Querétaro, in M. Alcayde and Z. de Cserna, eds., Libro-guía de la excursión geológica a la región de Zimapán y áreas circundantes, Estados de Hidalgo y Querétaro: Sociedad Geológica Mexicana, 6° Convención Geológica Nacional, p. 33–41.

Quintus-Bosz, R. L., 1982, Estudio geológico de la zona fosforítica de San Francisco, del municipio de Pacula, Estado de Hidalgo: Sociedad Geológica Mexcana, 6° Convención Geológica Nacional, p. 21–27.

Segerstrom, K., 1961, Geology of the Bernal-Jalpan area, Estado de Querétaro, México: U.S. Geological Survey Bulletin 1104-B, p. 19–86.

Segerstrom, K., 1962, Geology of south-central Hidalgo and northeastern Mexico: U.S. Geological Survey Bulletin 1104-C, p. 87–162.

Simons, F. S., and E. Mapes, 1956, Geology and ore deposits of the Zimapán mining district, state of Hidalgo, Mexico: U.S. Geological Survey Professional Paper 284, 128 p.

Suter, M., 1980, Tectonics of the external part of the Sierra Madre Oriental thrust-and-fold belt between Xilitla and the Moctezuma River (Hidalgo and San Luis Potosí states): Universidad Nacional Autónoma de México, Instituto de Geología, Revista, v. 4, p. 19–31.

Suter, M., 1982a, Itinerario, primer día, Zimapán-Jiliapan, in M. Alcayde and Z. de Cserna, eds., Libro-guía de la excursión geológica a la región de Zimapán y áreas circundantes, Estados de Hidalgo y Querétaro: Sociedad Geológica Mexicana, 6° Convención Geológica Nacional, p. 43–51.

Suter, M., 1982b, Itinerario, segundo día; tarde, Puerto de la Estancia-Jaguey Colorado-Zimapán, in M. Alcayde and Z. de Cserna, eds., Libro-guía de la excursión geológica a la región de Zimapán y áreas circundantes, Estados de Hidalgo y Querétaro: Sociedad Geológica Mexicana, 6° Convención Geológica Nacional, p. 56–64.

Suter, M., 1984, Cordilleran deformation along the eastern edge of the Valles–San Luis Potosí carbonate platform, Sierra Madre Oriental fold-thrust belt, east-central Mexico: Geological Society of America Bulletin, v. 95, p. 1387–1397.

Suter, M., 1987, Structural traverse across the Sierra Madre Oriental fold-thrust belt in east-central Mexico: Geological Society of America Bulletin, v. 98, p. 249–264.

Suter, M., 1990, Hoja Tamazunchale 14Q-e(5), Estados de Hidalgo, Querétaro y San Luis Potosí, con Geología de la Hoja 14Q-e(5) Tamazunchale: Universidad Nacional Autónoma de Mexico, Instituto de Geología, Carta geológica de México, serie de 1:100,000, 56 p.

Suter, M., P. J. Contreras, and C. H. Ochoa, 1997, Structure of the Sierra Madre Oriental fold-thrust belt in east-central Mexico: Instituto de Investigaciones en Ciencias de la Tierra, Universidad Autónoma de Hidalgo, e Instituto de Geología, Secondo Convención sobre la Evolución Geológica de México y Recursos Asociados, Pachuca, Hidalgo, Libro guía de las Excursiones geológicas, Excursión 2, p 45-63.

Tardy, M., 1980, Contribution a L´étude Géologique de la Sierra Madre Orientale du Mexique: Thesis de Doctoeur d´Etat, Université Pierre et Marie Curie, Paris, France, 459 p.

Ward, J. A., 1979, Stratigraphy, depositional environments and diagenesis of the El Doctor platform, Querétaro, Mexico: Ph.D. dissertation, New York State University, Binghamton, 183 p.

Wilson, B. W., M. J. P. Hernández, and T. E. Meave, 1955, Un banco calizo del Cretácico en la parte oriental del Estado de Querétaro, México: Boletín de la Sociedad de Geológica Mexicana, V 18, p. 1–10.

Wilson, J. L., 1975, Carbonate facies in geologic history: Berlin, Springer-Verlag, 471 p.

Wilson, J. L., and W. C. Ward, 1993, Early Cretaceous carbonate platforms of northeastern and east-central Mexico: AAPG Memoir 56, p. 35–49.

Gray, G. G., R. J. Pottorf, D. A. Yurewicz, K. I. Mahon, D. R. Pevear, and R. J.
Chuchla, 2001, Thermal and chronological record of syn- to post-Laramide
burial and exhumation, Sierra Madre Oriental, Mexico, *in* C. Bartolini, R. T.
Buffler, and A. Cantú-Chapa, eds., The western Gulf of Mexico Basin: Tectonics,
sedimentary basins, and petroleum systems: AAPG Memoir 75, p. 159–181.

7

Thermal and Chronological Record of Syn- to Post-Laramide Burial and Exhumation, Sierra Madre Oriental, Mexico

Gary G. Gray
*ExxonMobil Upstream Research Company,
Houston, Texas, U.S.A.*

Robert J. Pottorf
*ExxonMobil Upstream Research Company,
Houston, Texas, U.S.A.*

Donald A. Yurewicz
*ExxonMobil Exploration Company,
Houston, Texas, U.S.A.*

Keith I. Mahon
*ExxonMobil Upstream Research Company,
Houston, Texas, U.S.A.*

David R. Pevear
*ExxonMobil Upstream Research Company,
Houston, Texas, U.S.A.*

Richard J. Chuchla
*ExxonMobil Exploration Company,
Houston, Texas, U.S.A.*

ABSTRACT

Thermal and thermochronological data from the Sierra Madre Oriental indicate that a foreland basin formed on the leading edge of the fold belt during and after the late stages of the Laramide orogeny. This basin appears to have been at least 4 km thick by 50 Ma and to have reached maximum temperatures of 165° to 170°C under 5 to 7 km of burial. In areas such as the Parras and La Popa Basins, these burial amounts are in addition to the presently preserved Upper Cretaceous–Lower Tertiary section.

Estimates of maximum burial depths are derived from aqueous fluid-inclusion homogenization temperatures. The distribution of temperatures across more than 10 km of preserved stratigraphy indicates that this burial generally postdated folding. K/Ar dating of bentonites interbedded with Upper Cretaceous pelagic carbonates suggests that portions of this region had as much as 4 km of overburden by the end of the Laramide orogeny, at 50 to 45 Ma. The axis of this basin was coincident with the leading edge of the Sierra Madre Oriental, with the thickest portion centered over the Monterrey salient and the thinnest portion to the south. This basin covered the region now occupied by the Parras–La Popa, Magiscatzin, and Tampico-Misantla Basins, and may have extended northward into the Sabinas Basin. Although data are limited, this now-eroded basin appears to have had a classic foreland-basin asymmetry, thick adjacent to the fold belt on the west and thinning to the east. The magnitude and distribution of the burial event represented by this basin suggest that the frontal portion of the fold belt is prospective only for gas, whereas the interior portions may still be prospective for oil.

This basin began to invert soon after reaching maximum thickness, at 40 to 35 Ma. Estimates of the cooling rates during uplift range from 5° to 10°C/my. The entire region cooled below the apatite fission-track annealing temperature of approximately 115°C by 22 Ma. The region further cooled below the apatite U-Th/He closure temperature of approximately 75°C by 15 Ma. The timing of the attendant uplift is coincident with a pronounced shift in depositional axis to the east of the Tamaulipas and Tuxpan arches.

TABLE 1. Sample location data. Localities are plotted in Figure 3.

Sample	Locality	Latitude	Longitude	Formation	Age
1	El Papalote	26°, 03' 59"	100°, 43' 33"	Potrerillos	Paleocene
2	Delgado	25°, 59' 13"	100°, 48' 05"	Viento Fm.	Early Eocene
3	PLP-11A	25°, 58' 27"	100°, 48' 40"	Potrerillos	Maestrichtian
4	PLP-3	25°, 57' 58"	101°, 15' 15"	Cerro del Pueblo	Campanian
5	PLP-11X	25°, 56' 22"	100°, 48' 55"	Muerto	Maestrichtian
6	PLP-6	25°, 55' 43"	101°, 09' 44"	Las Encinas	Paleocene
7	PLP-5	25°, 55' 36"	101°, 11' 36"	Las Imagines	Maestrichtian
8	Fraustro	25°, 54' 26"	101°, 06' 16"	Rancho Nuevo	Paleocene
9	Potrero García	25°, 50' 46"	100°, 31' 47"	Cupido	Haut.-Barrem.
10	El Pantamo	25°, 41' 01"	101°, 17' 35"	Cerro del Pueblo	Campanian
11	Microhondas	25°, 40' 00"	100°, 44' 35"	Cañon del Tule	Maestrichtian
12	Huasteca Canyon	25°, 38' 41"	100°, 27' 48"	Cupido	Haut.-Barrem.
13	Cortinas Canyon	25°, 38' 14"	100°, 42' 40"	Cupido	Haut.-Barrem.
14	PLP-7	25°, 33' 35"	101°, 02' 45"	Cañon del Tule	Maestrichtian
15	Lunch Stop Anticline	25°, 33' 31"	100°, 23' 29"	Cupido	Haut.-Barrem.
16	El Pajonal	25°, 29' 47"	100°, 26' 12"	Cupido	Haut.-Barrem.
17	Las Tinajas	25°, 29' 34"	100°, 21' 28"	Cupido	Haut.-Barrem.
18	Santa Cruz	25°, 28' 33"	100°, 26' 04"	Cupido	Haut.-Barrem.
19	Los Chorros	25°, 24' 44"	100°, 48' 03"	Carbonera	Berriasian
20	Canyon Prieta	25°, 19' 08"	101°, 36' 50"	Cerro del Pueblo	Campanian
21	Canyon Prieta	25°, 17' 27"	101°, 37' 04"	Parras	Santonian
22	Buñuelos	25°, 02' 50"	101°, 03' 53"	Zuloaga	Oxfordian
23	Joya Verde	23°, 35' 14"	99°, 13' 11"	L. Tamaulipas	Lower Cret.
24	Valencia-Islas 1	23°, 15' 57"	99°, 09' 03"	El Abra?	Aptian-Cenom.
25	Valencia-Islas 2	23°, 13' 40"	99°, 09' 03"	El Abra?	Aptian-Cenom.
26	Rancho la Boquilla	22°, 57' 23"	99°, 36' 26"	El Abra	Aptian-Cenom.
27	Tantobal	22°, 48' 46"	98°, 57' 08"	U. Tamaulipas	Albian-Aptian
28	El Porvenir	22°, 37' 54"	99°, 33' 52"	El Abra	Aptian-Cenom.
29	Sierra Chuchara	22°, 35' 48"	99°, 02' 12"	El Abra	Aptian-Cenom.
30	Sierra Chuchara	22°, 35' 34"	99°, 02' 46"	San Felipe	Maestrichtian
31	Sierra Tamalave	22°, 33' 16"	99°, 09' 29"	El Abra	Aptian-Cenom.
32	Fortines	22°, 33' 08"	99°, 08' 10"	U. Tamaulipas	Albian-Aptian
33	Sierra la Colmena	22°, 31' 32"	99°, 16' 14"	El Abra	Aptian-Cenom.
34	Chupaderos	22°, 29' 47"	99°, 22' 45"	El Abra	Aptian-Cenom.
35	Peña el Llano	22°, 28' 46"	99°, 34' 29"	El Abra	Aptian-Cenom.
36	Puerta Santa Gertrudis	22°, 27' 37"	99°, 41' 15"	San Felipe	Maestrichtian
37	Sierra el Toro	22°, 26' 39"	99°, 17' 04"	El Abra	Aptian-Cenom.
38	Sierra Algodón	22°, 25' 53"	99°, 35' 10"	El Abra	Aptian-Cenom.
39	SierraTamalave	22°, 15' 26"	98°, 49' 43"	U. Tamaulipas	Albian-Aptian
40	Cementos Anahuac	22°, 02' 45"	98°, 52' 45"	El Abra	Aptian-Cenom.
41	San Felipe	21°, 58' 39"	98°, 57' 05"	San Felipe	Maestrichtian
42	Cementos Mexicanos	21°, 58' 24"	98°, 54' 05"	El Abra	Aptian-Cenom.
43	Taninul Quarry	21°, 57' 49"	98°, 53' 21"	El Abra	Aptian-Cenom.
44	Valencia-Islas 3	21°, 57' 18"	100°, 04' 56"	El Abra?	Aptian-Cenom.
45	Axitla	21°, 25' 17"	98°, 53' 06"	Chicontopec	Danian
46	Tamazunchale	21°, 15' 27"	98°, 49' 44"	El Abra	Aptian-Cenom.
47	Pisaflores	21°, 11' 26"	98°' 59' 55"	U. Tamaulipas	Albian-Aptian
48	La Pachuga	21°, 11' 03"	99°, 01' 37"	Taman	Kimmeridgian
49	La Cuentrada	21°, 03.764"	99°, 6.647"	San Felipe	Maestrichtian

TABLE 1. Sample location data. Localities are plotted in Figure 3 (cont'd.).

Sample	Locality	Latitude	Longitude	Formation	Age
50	San Francisco	20°, 57' 19"	99°, 21' 14"	El Abra	Aptian-Cenom.
51	El Fraile	20°, 56' 54"	99°, 17' 08"	L. Tamaulipas	Lower Cret.
52	El Doctor	20°, 50' 16"	99°, 34' 22"	El Abra	Aptian-Cenom.
53	Las Trancas	20°, 47' 52"	99°, 16' 00"	Las Trancas	Upper Jurassic
54	Cerro del Muí	20°, 45' 00"	99°, 24' 38"	U. Tamaulipas	Albian-Aptian
55	Soledad-124	20°, 42' 16"	97°, 40' 47"	Tamabra	Lower Cret.
56	Ordoñez-5	20°, 33' 33"	97°, 15' 39"	El Abra	Aptian-Cenom.
57	P. Aleman-94	20°, 29' 19"	97°, 18' 50"	Tamabra	Lower Cret.
58	San Andrés-143	20°, 24' 16"	97°, 10' 44"	Tamabra	Lower Cret.
59	ZFB-10	18°, 49' 53"	96°, 43' 21"	Guzmántla	Sant.-Camp.
60	ZFB-1	18°, 47' 21"	97°, 10' 22"	Tecamaluacan	Sant.-Camp.
61	ZFB-8b	18°, 46' 04"	96°, 59' 57"	Tecamaluacan	Sant.-Camp.
62	ZFB-4	18°, 41' 11"	97°, 03' 24"	Tecamaluacan	Sant.-Camp.
63	ZFB-14	18°, 12' 04"	96°, 29' 27"	Chicontopec	Paleocene

INTRODUCTION

Hydrocarbon exploration in fold-thrust belts is a challenging task. This is because the prospectivity of an area (assuming constant source character) is a complex interplay among the timing of deformation, local deposition, erosion, and the variation in heat flow to the base of the strata. Exploration is obviously most difficult in frontier areas where there are few data. An example of one such frontier area is the Sierra Madre Oriental in eastern Mexico (Figure 1). There are few published articles on the prospectivity of the area, and there is disagreement about whether the area has potential for liquid hydrocarbons (compare Valencia-Islas, 1993, and Morelos-García, 1996). To address these controversies, we studied the outcropping section for evidence of thermal history and trap timing. Regional data on the distribution of source rocks are presented in Yurewicz et al., personal communication, (2000).

This study spans a large portion of the Sierra Madre Oriental fold belt (Figure 1). The northernmost area investigated is near the town of San Jose de La Popa, in the La Popa Basin. The southernmost area is near the southern terminus of the fold belt, in the Sierra de Zongolica. The Coahuila block in the Parras Basin, Ciudad del Maiz, and Zimapán define the western edge. The eastern border is the Golden Lane and Poza Rica trends on the Tuxpan platform. This study concentrates on the fold belt but also includes portions of Parras, La Popa, Magiscatzin, and Tampico-Misantla Basins. The distribution of sample localities is shown in Figure 2, and a list of all sample locations is presented in Table 1.

REGIONAL SETTING

The Sierra Madre Oriental is located along the eastern edge of Mexico and is the southern continuation of the Cordilleran deformation belt in northern North America (Dickinson and Snyder, 1977; Coney, 1978; Campa-Uranga, 1985; de Cserna, 1989) (Figure 3). Like the Cordilleran system to the north, the Mexican orogen can be separated into a western, thin-skinned belt and an eastern, basement-involved belt. There are, however, many basement-involved structures in the thin-skinned portion of the Sierra Madre Oriental. In part, these basement-involved structures have a prior history (de Cserna, 1956; Guzmán and de Cserna, 1963; Gray et al., 1997).

Large-amplitude buckle folds with very long strike lengths dominate the structural style of the Sierra Madre Oriental (de Cserna, 1956; Padilla y Sánchez, 1982; Gray et al., 1997). A cross section through the Monterrey salient highlights the predominance of folding over faulting (Figure 4). Thrust faulting is common only around the frontal edge of this mountain chain. For this reason, the Sierra Madre Oriental is more correctly referred to as a fold belt than a fold-thrust belt. This structural style results in tectonic thickening of the overburden in the deformed zone while preventing hot, more deeply buried rocks from overriding cooler rocks. Therefore, structural culminations in the deformed zone may never have been deeply buried.

Structural styles vary somewhat along strike, with thrust faults becoming more prevalent to the south where the stratigraphy is characterized by alternating carbonate platforms and basins (Figure 5). Thrust faults are well developed in the region between Zimapán and Tamazunchale and, to some extent, in the Zongolica region (Carrillo-Martínez and Suter, 1982; Suter, 1984, 1987; Meneses-Rocha et al., 1997). At least one of these thrust faults has a large thermal anomaly associated with the overthrust section.

The stratigraphic column for this area is shown in Figure 6. A locally developed rift interval of Triassic-Jurassic age forms the base of the sedimentary succession. This interval crops out at several localities along the frontal portions of the fold belt (Figure 1) (e.g., Salvador, 1991; Padilla y Sánchez, 1982; Suter,

1990). These Early Mesozoic rocks have also been penetrated in the subsurface (e.g., Perez-1 well near Ciudad Mante, Salvador, 1987, 1991), and Triassic-Jurassic grabens are clearly imaged on fold, belt seismic data (J. Galicia, personal communication, 2000; Gray et al., unpublished work). Given this outcrop and subsurface distribution, we infer that this interval underlies most of the frontal structures of the Sierra Madre Oriental. The postrift section is quite variable. Evaporite deposits are regionally developed in units ranging in age from Late Jurassic through Aptian and are locally greater than 2 km thick (Carillo-Bravo, 1971; Valencia-Islas, 1993). Most of the Upper Jurassic through Santonian rocks are carbonates, ranging from platform interior and margin to deep basinal facies, depending on the specific area and age of the section. These platform sequences are capped by a succession of Upper Cretaceous shales, marls, and thin limestones (Pessagno, 1969; Salvador, 1991).

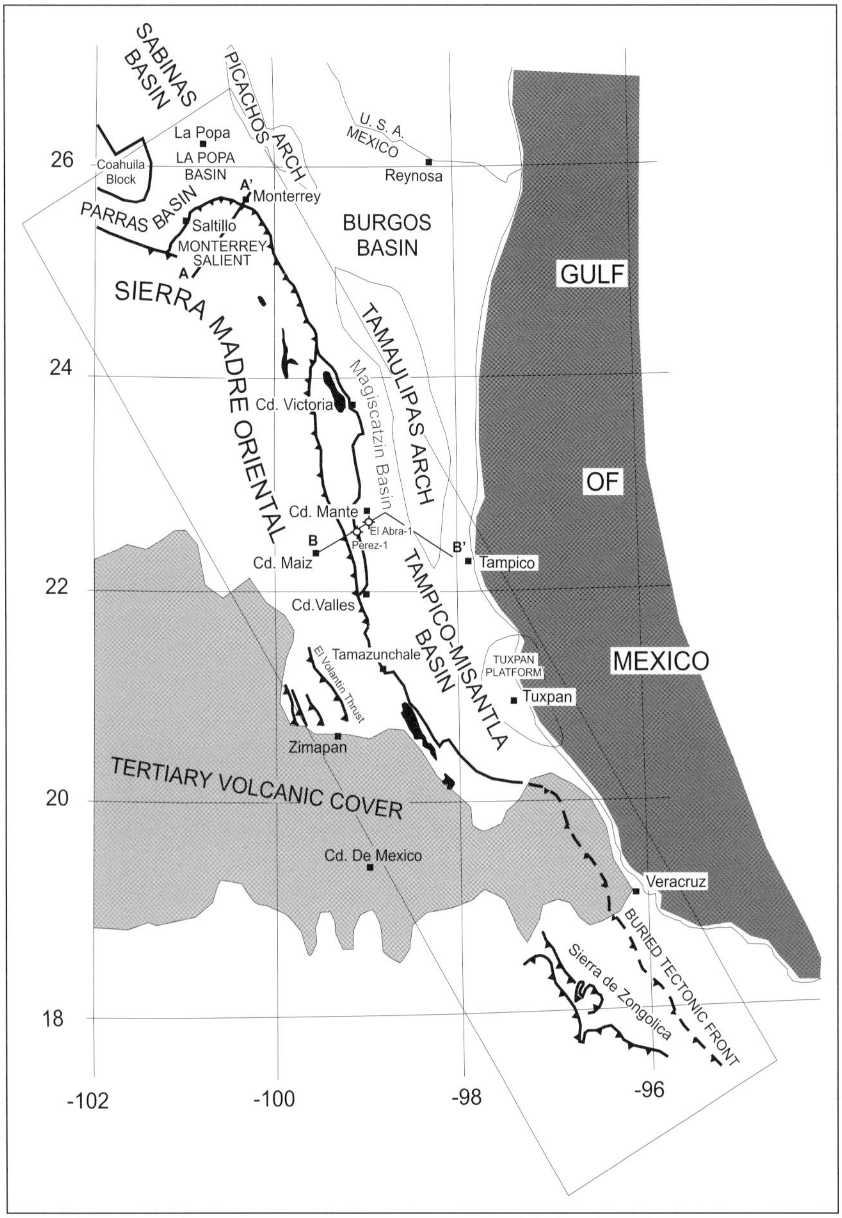

Hydrocarbon Distribution

Hydrocarbon exploration began early in the twentieth century with the discovery of oil in the Tampico-Misantla Basin. To date, no production has been established in the fold belt north of the Veracruz Basin, although gas and oil shows have been common. Giant oil fields are located along the southern nose of the Tamaulipas arch and in the Golden Lane and Poza Rica trends of the Tuxpan Platform. Large reserves of heavy oil are present in Paleocene sandstones in the Tampico-Misantla Basin. The only liquid production from within the fold belt occurs in the Veracruz Basin, where fold-belt structures are buried beneath upper Eocene and younger deposits. Gas production has been established in salt-cored structures of similar age in the Sabinas Basin (Gray et al., 1997). Oil seeps are relatively common in the fold belt, and large paleo-accumulations are now exposed in platform-margin carbonates along the front of the fold belt (Minero, 1991; Valencia-Islas, 1993; Yurewicz et al., personal communication, 2000).

Age of Deformation

The age of deformation in the Sierra Madre Oriental is coincident with the Laramide orogeny in northern North America. Unfortunately, the term *Laramide* has been used to connote both a particular style (and geographic area) of deformation (e.g., Dickinson and Snyder, 1977) and a time of deformation (Late Cretaceous to Eocene). To avoid confusion, the term *Laramide-style* is used in this paper to denote basement-involved structures, whereas *Laramide-age* is

FIGURE 1. Regional map of the Sierra Madre Oriental showing the major tectonic features. The rectangular box marks the area discussed in this paper. The thick black line shows the present topographic limit of the Sierra Madre Oriental fold belt. Where this boundary appears to be thrusted, the line is marked with barbs on the upthrown side. Where the frontal edge is unornamented, the boundary is a high-angle, basement-cored structure. Note that two fronts are shown along the boundary between Ciudad Victoria and Ciudad Valles. Where dashed, the eastern edge of the Sierra Madre Oriental is buried beneath Tertiary cover. Line A-A' is the location of the cross section shown in Figure 15. Black areas are exposures of Triassic-Jurassic rift sequences. Light shaded area is Tertiary volcanic cover. Structural features are after Padilla y Sánchez, 1982; Padilla y Sánchez et al., 1994; and Suter, 1990.

used to denote any style of deformation that occurs between the Cenomanian and Eocene (Schmidt and Perry, 1988).

The onset of Laramide-age deformation in the Sierra Madre Oriental is marked by an influx of clastics to the foreland region at about the Turonian-Coniacian boundary (de Cserna, 1956; Baker, 1970; McBride et al., 1974; Padilla y Sánchez, 1982; Eguiluz de Antuñano, 1984; Soegaard et al., 1997) (Figure 6). Clastic rocks in the Parras, La Popa, and Burgos Basins were derived from the north and west, although the highlands that provided most of these sediments are presently buried beneath thick Tertiary volcanic deposits of the Sierra Madre Occidental (Baker, 1970; McBride et al., 1974; Pérez-Crúz, 1992; Ye, 1977). Detritus derived from the Lower Cretaceous platforms was being shed into the Parras and La Popa Basins by the Maestrichtian (Baker, 1970; McBride et al., 1974), and into the Tampico-Misantla Basin by Paleocene-Eocene time (Suter, 1984; Bitter 1986, 1993).

The oldest directly dated structures are the Misión and Lobo-Ciéniga thrust faults at the eastern edge of the Valles–San Luis Potosí Platform, which are cut by a pluton 62.2 ± 1.5 Ma (Suter, 1984). The El Volantín thrust, along the western edge of this platform (Carrillo-Martinez and Suter, 1982), is dated by K/Ar techniques at 62 Ma (this paper).

The youngest structures are folds and faults that occur in Paleocene-Eocene rocks along the length of this orogen. In general, Upper Cretaceous and Paleocene rocks are conformable with the underlying Lower Cretaceous section, suggesting that the main phase of Laramide-age deformation is post-Paleocene. The youngest rocks involved in the deformation in the La Popa Basin are early Eocene (Vega-Vera et al., 1989), although their presence in salt-withdrawal synclines makes it difficult to resolve their relationship to Laramide-age deformation. In the subsurface of the Veracruz Basin, unfaulted upper Eocene strata unconformably overlie folded and faulted middle-Eocene rocks. This provides the tightest constraints on the end of Laramide-age deformation in the entire fold belt (Mossman and Viniegra, 1976; Cruz-Helú et al., 1977; Prost et al., 1995; Meneses-Rocha et al., 1997).

In summary, deformation appears to have begun in the western hinterland of the Sierra Madre Oriental as early as the Cenomanian. Detritus derived from Lower Cretaceous carbon-

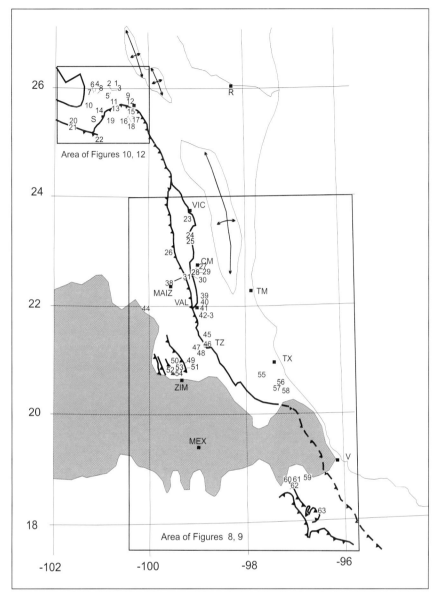

FIGURE 2. Map showing the data localities for this study. The localities are listed with the place name, latitude, and longitude in Table 1. The area of the larger box at the bottom of this diagram is shown in Figures 8 and 9. The areas of Figures 10 and 12 are shown in the smaller box near the top of the diagram. Abbreviations: R, Reynosa; S, Saltillo; M, Monterrey; VIC, Ciudad Victoria, CM, Ciudad Mante; TM, Tampico; VAL, Ciudad Valles; MAIZ, Ciudad del Maiz; TZ, Tamazunchale; TX, Tuxpan; MEX, Mexico City; V, Veracruz.

ates is present in clastic rocks as old as Maestrichtian. Active deformation in the Sierra Madre Oriental can be constrained only during the earliest Paleocene. Folding and faulting relationships in the La Popa, Tampico-Misantla, and Veracruz Basins indicate that deformation continued into the Eocene. In all locations, deformation appears to have been over by the middle Eocene, although this is tightly constrained only in the Veracruz Basin. This age of deformation is coincident with the timing of the Laramide orogeny in the United States (Gries, 1983; Steidtman et al., 1983; Kulik and Schmidt, 1988; Craddock et al., 1988; Cerveny, 1990).

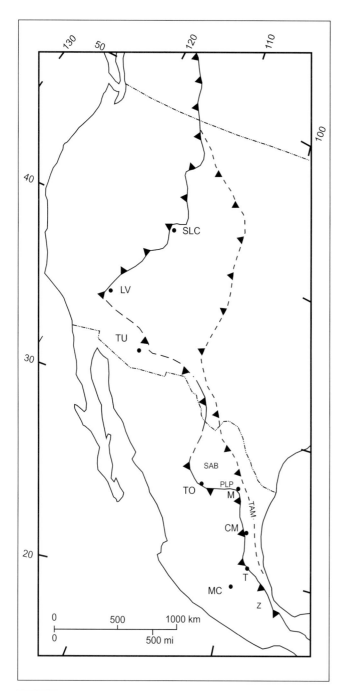

FIGURE 3. Map showing the relationship of the Sierra Madre Oriental to Sevier-Laramide system in North America. The solid line is the trace of the thin-skinned (Cordilleran) deformation front. This is shown as a dashed line in Arizona and New Mexico, where thin-skinned deformation is absent and the Laramide-age structures form a very complex zone (e.g., Drewes, 1988). The completely dashed line marks the eastern limit of basement-involved deformation of the same age. Abbreviations: CM, Ciudad Mante; LV, Las Vegas; M, Monterrey; MC, Mexico City; PLP, Parras–La Popa Basin; SAB, Sabinas Basin; SLC, Salt Lake City; T, Tamazunchale; TAM, Tamaulipas arch; TU, Tucson; TO, Torreon; Z, Sierra de Zongolica. After Coney (1978), Dickinson and Snyder (1977), Campa-Uranga (1985), Tardy et al. (1975), Drewes (1988), and Aranda-García (1991).

Analytical Methods

Several tools were used to investigate the thermal history of this large area. The integration of these constraints provides a more reliable interpretation of the magnitude of burial, uplift, and thermal stress than techniques that rely on traditional organic maturity indicators alone. As shown in Figure 7, K/Ar ages on diagenetic illite are used to constrain the prograde, or heating, portion of a burial history (e.g., Pevear, 1992, 1998). These techniques can also be used to directly date the timing of fault movement (e.g., Vrolijk and van der Pluijm, 1999). Aqueous fluid-inclusion homogenization temperatures are used to estimate the maximum burial temperatures. Apatite fission track and U-Th/He ages constrain the retrograde, or cooling, portion. The procedures used to perform and interpret these analyses are presented in the following sections.

Potassium Argon Dating

The technique of K/Ar dating of illite has recently been applied to a variety of problems such as determining an age of diagenesis and timing of fault movement (Pevear, 1998; Vrolijk and van der Pluijm, 1999). Diagenetic illite forms in sedimentary environments at 80°–120°C and has a closure temperature to argon diffusion of 250°–300°C (Pevear, 1998;

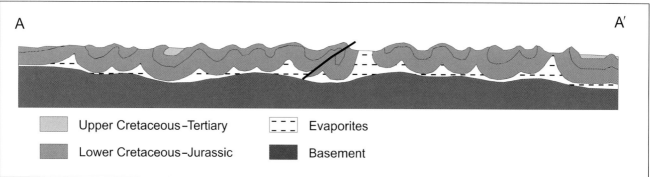

FIGURE 4. Northeast-southwest cross section drawn across the Monterrey salient. This section demonstrates the preponderance of folds and the lack of thrusts in this deformed belt. Section modified after de Cserna (1956). Location of section shown in Figure 1.

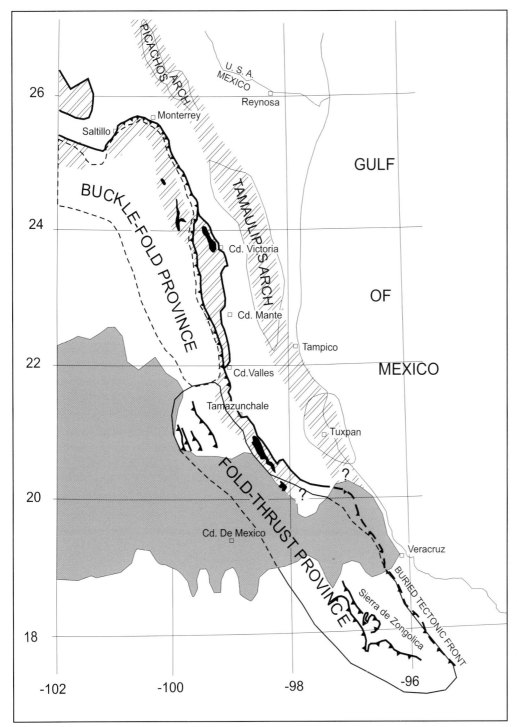

FIGURE 5. Schematic drawing of the structural provinces in the Sierra Madre Oriental. The predominant structural style is large-amplitude buckle folds. Thrust faults are shown with barbs on the upthrown side. Areas of basement-involved structure are shown with the diagonal shading pattern. Black shaded areas are exposures of Triassic-Jurassic rift sequences. Modified after de Cserna (1956), Guzmán and de Cserna (1963), and Padilla y Sánchez et al. (1994).

nique because of the large percentage of smectite in the rocks when deposited and the small amounts of the 2M (detrital) polytype.

Fluid-inclusion Analysis

The maximum burial temperatures for these samples were determined using homogenization temperatures of aqueous fluid inclusions. Aqueous fluid inclusions were used rather than more standard techniques, such as vitrinite reflectance, because most of the exposed rocks are platform carbonates that contain little or no vitrinite maceral. Fluid inclusions examined in this study usually occur as secondary inclusions in microfractures or crosscutting

Hunziker, 1987). If a sedimentary rock is heated above 80°C but stays below the closure temperature, the radiometric age of the diagenetic illite should equal the mean age of that diagenetic event over which the illite formed.

Our analyses follow the procedure outlined in Pevear (1992, 1998), in which the percentage of diagenetic illite is determined by the amount of 1M polytype. A plot of K/Ar age versus percentage of diagenetic (or detrital) illite in the sample then allows extrapolation to the age of diagenesis. Bentonites and carbonate rocks are particularly well suited to this tech-

veinlets. These relationships ensure that the inclusions reflect diagenetic conditions and are not inherited, although rare primary inclusions in authigenic cements were also observed.

The homogenization temperatures of aqueous fluid inclusions can be particularly good estimates of maximum temperature at peak burial. This technique consists of finding the highest homogenization temperature (Th) for aqueous, methane-bearing, fluid-inclusion assemblages. A fluid-inclusion assemblage is a petrographically associated group of inclusions that may have different sizes and shapes but have similar liq-

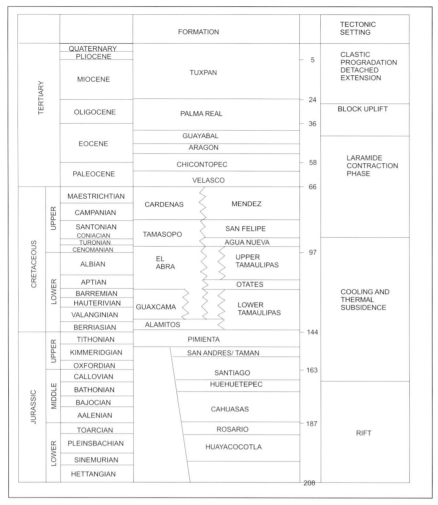

FIGURE 6. Stratigraphic column for the central portion of the study area. After Goldhammer et al. (1991) and Yurewicz et al., personal communication (2000).

uid/vapor ratios and homogenization temperatures (Goldstein and Reynolds, 1994).

This method assumes that methane is the principal dissolved gas phase in the inclusions and that the water is saturated or nearly saturated with methane at the trapping conditions. If these assumptions are true, the resulting homogenization temperatures are expected to be very close to the temperature at the time they were entrapped (Hanor, 1980). In sedimentary basins around the world, the assumption of methane presence in water has been determined empirically to be generally valid. As a result, the highest homogenization temperature measured from aqueous fluid-inclusion assemblages is commonly accurate to within 5°–10°C of the maximum burial temperature (Pottorf and Vityk, 1997).

There are some important limitations to maximum temperature interpretations from fluid-inclusion assemblage data. Fluid inclusions may not be trapped at all stages during the burial history of a rock, and therefore the maximum temperature in a sample may not be recorded. Similarly, a sufficient number of fluid-inclusion assemblages must be investigated to ensure that the highest-temperature assemblage in the sample is measured. For these reasons, we assume that data collected from individual localities represent

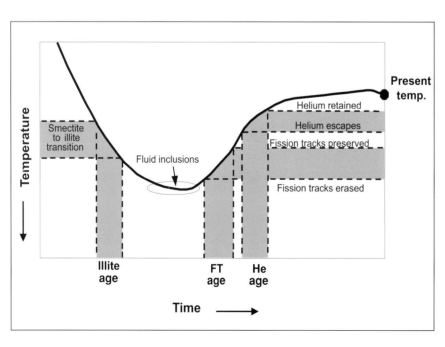

FIGURE 7. Schematic diagram showing the integrated thermochronometry approach used in this study. The dark curved line presents a hypothetical temperature history of a given bed through time. Potassium-argon dating on diagenetic illite provides a constraint on the heating (burial) portion of burial history. Fluid-inclusion homogenization temperatures or other paleothermometers constrain the maximum temperatures the bed experienced. The fission-track and helium ages on apatite constrain different parts of the cooling (uplift) history of the bed.

TABLE 2. Potassium/argon age data.

Locality	Formation	Stratigraphic age	Size fraction	Ave. ^{40*}Ar, ppm	^{40*}Ar/ total ^{40}Ar	Ave. % K	Age	Interpreted age
27	U. Tamaulipas	Albian-Aptian	coarse	0.01868	0.386	3.252	81.0 ± 2.1	51.3 ± 1.4
			meduim	0.01638	0.370	4.453	52.3 ± 1.4	
			fine	0.01619	0.376	4.585	50.3 ± 1.3	
39	U. Tamaulipas	Albian-Aptian	coarse	0.01994	0.258	5.848	48.5 ± 1.3	49.6 ± 1.3
			meduim	0.01891	0.264	5.412	49.7 ± 1.3	
			fine	0.01766	0.175	5.078	49.5 ± 1.3	
41	San Felipe	Maestrichtian	coarse	0.01680	0.632	4.588	52.1 ± 1.4	< 44
			meduim	0.01539	0.450	4.505	48.6 ± 1.3	
			fine	0.01319	0.362	4.288	43.8 ± 1.2	
49a	Soyatal	Maestrichtian	coarse	0.02243	0.644	5.709	55.8 ± 1.4	61.2 ± 1.6
			meduim	0.02301	0.342	5.372	60.8 ± 1.5	
			fine	0.02174	0.309	5.002	61.6 ± 1.6	
49b	fault gouge	N/A	coarse	0.009594	0.338	1.376	97.0 ± 2.5	62 ± 1.7
			meduim	0.01705	0.426	3.974	60.9 ± 1.6	
			fine	0.01923	0.321	4.319	63.1 ± 1.7	

λ_β = 4.962 x 10^{-10}/year

$(\lambda_\varepsilon + \lambda'_\varepsilon)$ = 0.582 x 10^{-10}/year

^{40}K/K = 1.193 x 10^{-4} g/g

minimum estimates of the burial temperature. However, combining data from many locations increases confidence that the true maximum temperatures for the region are captured.

Apatite Geochronological Data

Accessory apatite has long been recognized as a useful thermochronometer through applications of the fission-track method in sedimentary basins (Gleadow et al., 1983; Naeser et al., 1989). Recently, helium dating of apatite has been used to constrain the low-temperature thermal history of igneous and sedimentary rocks (Zeitler et al., 1987; Lippolt et al., 1994; Farley et al., 1996; Wolf et al., 1996, 1997; Warnock et al., 1997). By combining these two techniques, low-temperature fluctuations (less than 120°C) in a basin's thermal history may be resolved.

Apatite fission-track age analysis is predicated on the spontaneous fission of ^{238}U (typically found in apatite at concentrations of 5–100 ppm), which leaves behind a highly damaged trail in the crystal lattice. These so-called fission tracks form continuously, but they tend to shorten and anneal when subjected to elevated temperatures during time. In an apatite grain with a chlorine composition of 0.5 wt%, fission tracks will totally anneal above 120°C in five to ten million years. Little or no track shortening will be detected below about 70°C over the same period. For further discussion of the apatite fission-track technique, see Green et al. (1986), Laslett et al. (1987), Carlson (1990), Crowley et al. (1991), Carlson et al. (1999), Donelick et al. (1999), and Ketcham et al. (1999).

Apatite fission-track ages were calculated using the *weighted mean* (Mahon, 1996). The fit of individual grain ages to the overall weighted mean was tested using the reduced chi-squared test, also known as the mean square weighted deviate (MSWD). The expectation is that the MSWD will be 1.0 for completely random data sampled from the same population. When the MSWD is greater than 1.0, the fit is poorer than expected (Mahon, 1996).

Increased chlorine concentration in apatite is correlated with older fission-track ages (Gleadow and Duddy, 1981). Measurements of etch-pit diameter parallel to the c-axis, or *Dpar* (Burtner et al., 1994), indicate that this parameter correlates strongly with measured chlorine (Carlson et al., 1999). Grains with *Dpar* values of 1.7 μm contain approximately 0.5 wt% chlorine.

Apatite helium dating (U-Th/He dating) is predicated on the natural α decay of thorium and uranium to stable lead isotopes and helium. Helium does not bond to other atoms, so it is very susceptible to diffusion out of the crystal lattice at elevated temperatures. The partial retention temperature for helium diffusion in apatite is approximately 40° to 80°C during a few million years (Farley et al., 1996; Wolf et al., 1996, 1997; House et al., 1998, 1999). The helium ages are measured and corrected for alpha recoil loss using the compensation procedure of Farley et al. (1996).

We use closure temperature (Dodson, 1973; McDougall and Harrison, 1988) to relate the age of a sample to temperature for paleothermometers with vastly different kinetic reactions. Loss of daughter product (helium or fission tracks) continues to occur below the closure temperature. At a constant cooling rate of 10°C/m.y., the estimated closure temperatures for apatite fission-track annealing and helium diffusion are 120°C and 75°C, respectively, using the kinetics of Laslett et al. (1987) and Wolf et al. (1996).

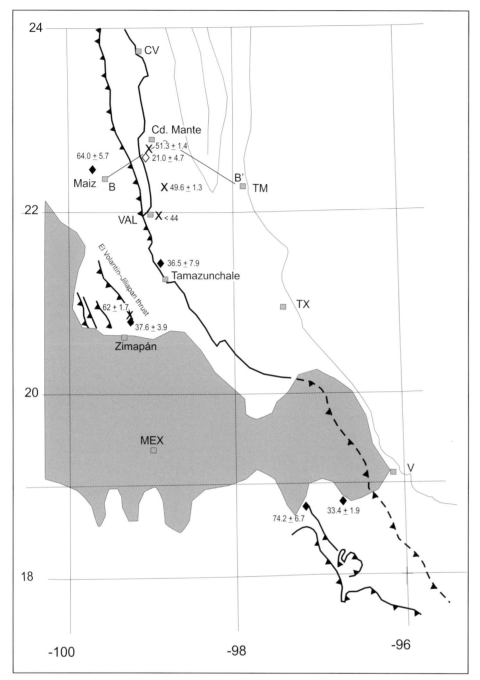

FIGURE 8. K/Ar-illite and apatite fission-track sample localities and results from the southern portion of the study area. Data are listed in Tables 2 and 4. Diamonds = apatite fission-track samples; X = K/Ar-illite samples. The K/Ar results show that diagenesis was occurring in this area at approximately 50 Ma. The K/Ar-illite age on fault gouge for the El Volantín thrust is also shown.

fractions of sample number 27 yield ages of 49–52 Ma. The consistency of these ages indicates that they are not contaminated by detrital clay or mica. Sample number 41 is from a marl in the San Felipe Formation, and it gives a similar age of 44 Ma. This heating overlaps in time with the waning stages of deformation, suggesting that the thermal event that resulted in clay diagenesis may be coupled with regional deformation.

Two additional samples were taken from the El Volantín thrust fault to try to determine the age of movement. The El Volantín thrust formed along the western margin of the Valles–San Luis Potosí Platform. The results from these samples are also shown in Table 2 and Figure 8 (samples number 49a and b). Medium- and fine-fraction illites from both fault gouge and footwall shales immediately below the thrust give 62-Ma ages, which we interpret to date the actual timing of movement (Pevear, 1998; Vrolijk and van der Pluijm, 1999).

Fluid Inclusions

Aqueous-inclusion homogenization temperatures were obtained from the 35 localities shown in Table 3 and Figures 9 and 10. Each locality includes analyses of 2 to 40 different samples. This data set also includes three additional aqueous inclusion homogenization temperatures from Valencia-Islas (1993). The presence of oil-bearing fluid inclusions are also noted (Figures 9 and 10). The API gravity of hydrocarbon-bearing inclusions has been determined, where possible, by using an Exxon-Mobil proprietary technique.

Results and Interpretation

Illite Age Analysis

Three samples of Upper Cretaceous rocks were analyzed to determine the diagenetic age of samples along the frontal portion of the fold belt. These results are presented in Table 2 and Figure 8. They indicate that these rocks were heated through the 80° to 120°C window at approximately 50 Ma. Samples number 27 and number 39 are particularly useful for this determination, because they are bentonites interbedded with the Tamaulipas Superior Formation along the axis of the thermal anomaly. These rocks were originally all smectite and are now 1M illite (mixed-layer illite/smectite with 85% illite layers). All size fractions of sample number 39 and the fine and medium

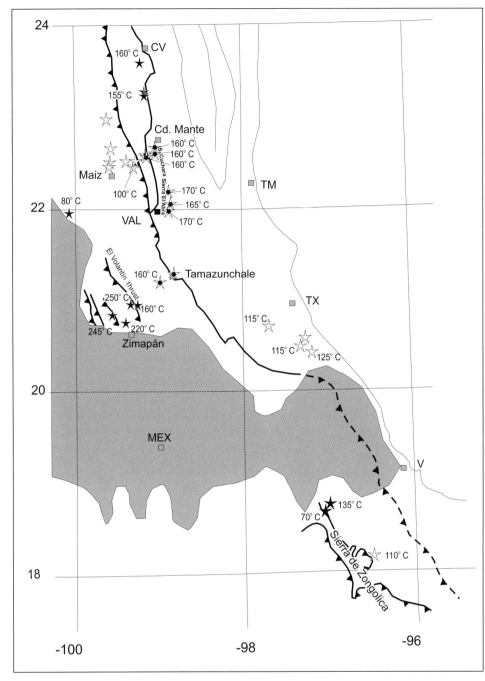

FIGURE 9. Fluid-inclusion sample localities and results from the south-central Sierra Madre Oriental and adjacent basins. Data are listed in Table 3. Sample localities are shown with a star. Open stars have hydrocarbon and aqueous inclusions; filled stars have only aqueous inclusions. Open stars with a black dot in the center are hydrocarbon fluid-inclusion localities that also contain pyrobitumen. Labeled stars have aqueous fluid-inclusion assemblages that provide consistent homogenization temperatures. Outlines of tectonic features are as in Figure 1. Note the consistent 55°–170°C temperatures distributed along the frontal edge of the fold belt. Also note that the El Volantín thrust west of Tamazunchale (TZ) brings hotter rocks over cooler rocks.

abruptly at the sole of the El Volantín thrust (Carrillo-Martínez and Suter, 1982; Suter, 1987) (Figure 9). The hanging-wall rocks west of this thrust have fluid-inclusion homogenization temperatures 60–90°C hotter than those of the footwall section. Temperatures as high as 250°C extend into the footwall rocks 30 m below the thrust. The maximum temperatures from rocks 500 m below the thrust are 160°C.

Homogenization temperatures from the central and southern Sierra Madre Oriental are shown in Figure 9. These samples range in age from Late Jurassic to Maestrichtian. From Ciudad Victoria to Tamazunchale, temperatures along the frontal portion of the fold belt are consistently between 155° and 170°C. Hydrocarbons present in this high-temperature region have largely been altered to pyrobitumen. West of Ciudad Mante, the maximum temperatures decrease markedly, although the temperature data are sparse. This westward decrease in temperature is also apparent in the Sierra de Zongolica (Figure 9), although overall, this region yields temperatures 35–60°C cooler than the Mante-Tamazunchale region.

In contrast with the Mante and Zongolica areas, temperatures increase west of Tamazunchale. This increase occurs

Core samples were also taken from four wells that penetrate Lower Cretaceous carbonate rocks of the Tuxpan Platform complex (Table 3; Figure 9) (Coogan et al., 1972). These samples give aqueous homogenization temperatures from 115° to 125°C. These are cooler than the frontal portion of the fold belt and similar to those in the Zongolica region.

The homogenization temperatures from the Monterrey salient area are shown in Figure 10. These samples range in age from Late Jurassic to Eocene. The temperatures vary from 140° to 250°C, with most temperatures between 150° and 200°C. Rocks in the Monterrey salient appear to have been slightly hotter than rocks in the Parras and La Popa Basins, with the exception of the westernmost sample from the Parras Basin. Temperatures in the La Popa Basin range from 140° to 155°C.

TABLE 3. Fluid-inclusion data. C = calcite, D = dolomite, F = fluorite, Q = quartz, Thmax(aq) = maximum homogenization temperature from aqueous fluid-inclusion assemblages, API = API gravity of liquid hydrocarbon inclusions.

sample	Thmax (aq)	Host	API⁰
1	155	C	
2	140	C	42–52
8		Q	42
9	150	Q	
10	166	C	
12	200	Q	
13	260	Q	
15	200	Q	
16	200	C	
17	200	Q	
18	165	C	
20	195	C	
22	175	C	
23	160	Q	
24	155	C	
25	155	C	
26		C	34–43
27	160	C	37–46
28		C	19–29, 32–40
29	160	C	28–46
31		C	37–46
32	160	Q	39–44
33	100	C	27–35
34		C	15–23
35		D	27–35
37		C	20–23
38		C	22–31
40	170	C	23–46
42	165	C	29–47
43	170	C, D	30–42
44	80	F	
46		C	38–47
47	160	C	36
48	160	C	
50	160	F	
51	250	Q	
52	245	C	
54	220	C	
55	115	C	16–33
56		C	16–29
57	115	C	18–33
58	125	C	23–33
61	135	C	
62	70	C	
63	110	Q	10–12, 24–27, 44–52

The consistency of maximum homogenization temperatures obtained from these regionally distributed samples suggests that these temperatures approximate the true maximum temperatures. These data are contoured in Figure 11. The temperature contours suggest a very elongated thermal anomaly, centered on the frontal edge of the fold belt, with a locally constrained steep western gradient and a gentle eastern gradient. The smoothness and regional extent of these contours indicate that these temperature anomalies result from a regional heating event and were not caused by intrusions or local hydrothermal circulation.

These data also suggest that the expected normal progression of increasing temperatures toward the interior of a deformed belt does not apply in this area. Rocks west of Ciudad del Maiz are 50° to 75°C cooler than rocks of the same age along the frontal edge of the fold belt at Ciudad Mante. In the Monterrey salient, the most interior sample of the fold belt is 25°C cooler than the frontal edge, and at Zongolica, the interior sample is 35°C cooler. The primary reason for this trend appears to be the lack of structural imbrication of this section (Figure 4). In the absence of thrust ramps and duplexes, hotter, deeper rocks are not brought to the surface. Apparently, both pre- and postdeformation burial of these rocks was minimal, because rocks exposed in the cores of the large folds have never experienced high temperatures. Mapped thrust faults do occur in the Tamazunchale-Zimapán region, where they coincide with large temperature inversions that bring hotter, older rocks over cooler, younger rocks (Figure 11).

The La Popa Basin contains early Eocene rocks that are some of the youngest strata preserved along the front of the Sierra Madre Oriental (Vega-Vera et al., 1989). Fluid-inclusion homogenization temperatures of 140°–155°C were obtained from these rocks, which indicates that the regional heating must postdate early Eocene time. All of the additional overburden required to achieve these temperatures has been eroded.

Hydrocarbon fluid inclusions are also abundant across much of the area (Figure 9). The main exceptions are the very hottest rocks found west of the El Volantín thrust and in the Monterrey salient. In general, the API gravity of the inclusions also decreases eastward and westward away from the high-temperature region (Table 3), consistent with a decrease in source-rock maturity associated with the lower temperatures.

Apatite Fission-track Dating

Results from apatite fission-track age analyses on the outcrop samples are listed in Table 4. Twenty-two samples of sedimentary rocks yielded sufficient apatite to be dated with the fission-track method. These samples range in depositional age from Early Cretaceous to Eocene, as do the fluid-inclusion samples. The average *Dpar* values for these samples range from 1.7 µm to 2.4 µm, which correspond to chlorine contents of 0.5 wt% to >1 wt% Cl, respectively. The ages from the southern portion of the Sierra Madre Oriental are presented in Figure 8, and those from the Monterrey area are presented in Figure 12.

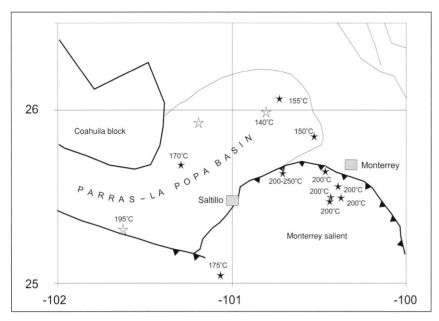

FIGURE 10. Fluid-inclusion sample localities and results from the Monterrey area. Data are listed in Table 3. Symbols are as in Figure 9.

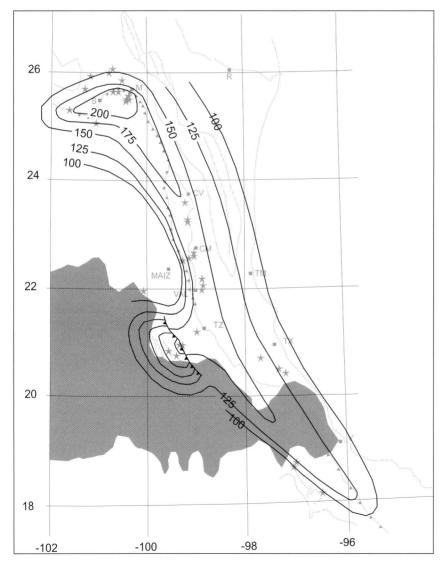

Every sample dated has a fission-track age considerably younger than the age of the enclosing strata from which it was sampled, indicating that these ages are all reset. This was expected from the high temperatures obtained from the fluid inclusions. The fission-track data by themselves also indicate that each sample must have experienced temperatures greater than 120°C, given the range of chlorine contents estimated from the *Dpar* measurements. There is no correlation between *Dpar* and age for these samples, but *Dpar* and average track length are moderately correlated ($r^2=0.28$). The implications of these results are discussed in the following paragraphs.

The fission-track ages can be grouped into three regions, reflecting provinces of similar cooling history. In the northern zone, fission-track ages less than 30 Ma extend from La Popa south to Ciudad Mante. Farther south along the axis of the high-temperature region, the fission-track ages are between 30 and 40 Ma. The third zone is in the fold belt to the west, where these ages increase to more than 60 Ma (Figure 13). In a general way, the age trends parallel the temperature trends. In the north, where the temperatures are hotter, the ages are youngest. In the south, temperatures are lower and ages are older. To the west, temperatures are lowest and ages are oldest, overlapping with the known ages of deposition and deformation in the fold belt. The two ages in the western province appear never to have been reset, and therefore have not been heated above the closure temperature for fission tracks. Where the two eastern-age provinces meet, the maximum burial temperatures (and hence

FIGURE 11. Contours of aqueous homogenization temperatures for the study area. Contour interval is 25°C. This map shows the two high-temperature regions centered over the Monterrey salient and west of the El Volantín thrust (see Figure 1 for locations). These contours also show the steep-sided nature of the temperature anomaly on the western flank and the gentle temperature gradient on the eastern flank. The overall geometry of these temperature contours suggests an origin resulting from burial beneath a foreland basin.

TABLE 4. Apatite age data. U-Th/He ages in parentheses are considered unreliable—sample 10 because of small grain size and sample 30 because it contains abundant monazite inclusions. For a discussion of these limitations of the technique, see Farley et al. (1996).

Locality	Weighted mean FT Age	MSWD	He age	Dpar (μm)	# tracks	Mean length +1 s. d. (μm)
2	19.6 ± 2.3	0.87	15.2 ± 0.8	2.38	123	13.82 ± 2.41
3	21.1 ± 4.3	2.76		2.27	148	13.80 ± 1.97
4	23.4 ± 2.5	0.68		2.1	205	13.24 ± 2.49
5	22.8 ± 2.5	0.57	20.1 ± 1.2	2.18	200	13.54 ± 2.41
6	23.3 ± 2.5	1.05		2.31	161	13.51 ± 2.36
7	24.1 ± 2.1	0.88	17.0 ± 0.8	2.33	201	13.88 ± 1.93
7	26.8 ± 3.1	0.63		2.32	200	13.99 ± 2.08
8	25.9 ± 2.3	0.63	21.9 ± 1.3	2.34	200	13.55 ± 2.07
10	24.6 ± 2.8	0.91	(11.4 ± 0.6)	2.22	153	13.73 ± 1.98
11	24.6 ± 2.7	0.62		2.1	182	14.02 ± 1.91
11	29.2 ± 2.8	0.99		2.18	137	14.21 ± 1.94
12	26.1 ± 2.0	0.89	18.1 ± 0.9	1.92	200	14.30 ± 1.42
14	21.7 ± 4.4	1.54		2.09	149	13.62 ± 2.24
19	27.3 ± 1.8	1.02	18.1 ± 1.1	1.97	203	13.82 ± 1.77
20	24.4 ± 2.1	1.1		2.12	200	13.98 ± 2.11
21	23.9 ± 3.7	1.24		2.23	79	14.07 ± 1.56
30	21.0 ± 4.7	0.38	(7.7 ± 0.5)	2.25	70	13.34 ± 2.00
36	64.0 ± 5.7	1.21		2.02	103	12.74 ± 1.99
45	36.5 ± 7.9	0.59		2.03	36	13.69 ± 1.59
53	37.6 ± 3.9	0.84		1.7	40	12.15 ± 1.43
59	74.2 ± 6.7	2.51		1.9	201	12.70 ± 1.75
60	33.4 ± 1.9	0.81		1.7	204	13.64 ± 1.55

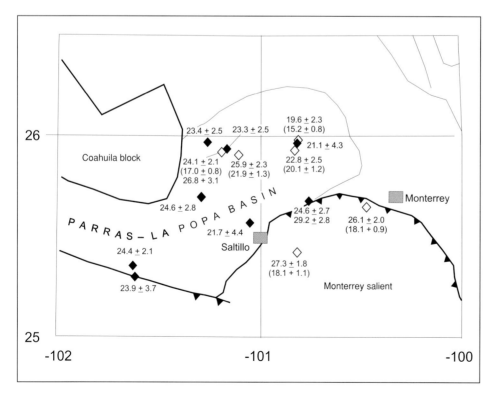

FIGURE 12. Apatite fission-track sample localities (diamond symbols) and data from the Monterrey area. Data are listed in Table 4. These data show very consistent ages over a composite stratigraphic interval more than 7 km thick. Filled diamonds are samples dated with apatite fission tracks only. Open diamonds are samples dated using both fission tracks and the U-Th/He technique on apatites. U-Th/He ages on the same apatite samples are shown in parentheses.

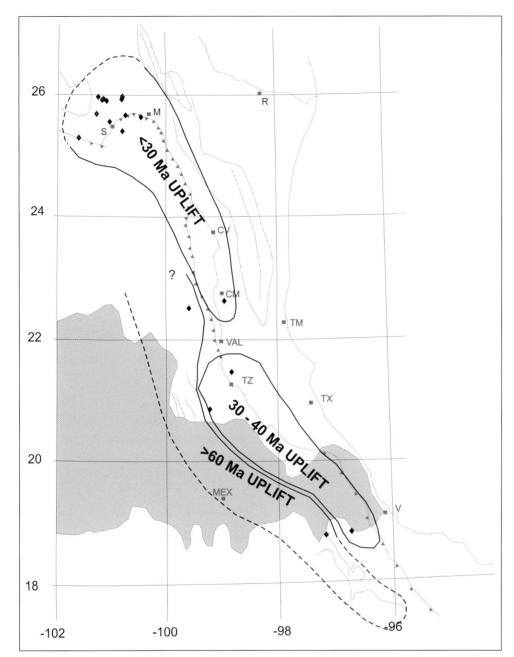

FIGURE 13. Map showing interpreted apatite fission-track age "provinces." Remarkably consistent cooling ages of 22–27 Ma (Miocene) characterize the northern province, which may extend northward into the Sabinas Basin. Cooling ages of 30–40 Ma (Oligocene) characterize the southeastern province, which suggests that it may have been uplifted earlier than the area to the north. Cooling ages of 60–80 Ma (Laramide) characterize the western province. These older ages are coincident with the cooler homogenization temperatures, suggesting that the foreland basin did not overlap the interior portions of the fold belt.

between *Dpar* and track lengths suggests that some continued track shortening occurred in grains with low chlorine contents. Together, these observations suggest that these rocks were cooled rapidly through high annealing temperature (i.e., >90°C) and were cooled more slowly through low annealing temperatures (i.e., <90°C).

interpreted amounts of overburden) are the same. This suggests that cooling (uplift) began earlier in the south.

There is abundant evidence from the fission-track data that the cooling (uplift) rates were relatively rapid everywhere along this basin. As mentioned earlier, there is no correlation between the *Dpar* values (chlorine content) and the calculated ages, suggesting that cooling from 130° to 90°C was rapid throughout most of the region. Slow cooling rates, coupled with the variation in annealing kinetics represented by these different chlorine contents, would have affected fission-track ages.

Further evidence of rapid cooling can be seen in the track-length distribution data (Table 4; Figure 14). There is essentially no difference in track-length distributions for samples from each of the three age provinces. The moderate correlation

Apatite Helium Dating

Eight of the original 22 apatite separates were chosen for further analysis using helium dating techniques (Table 4; Figure 12). All of the helium ages have been reset, as were the fission-track ages, and are younger than the depositional age of the strata from which they were taken. The apatite helium ages are all lower than the fission-track ages but do not represent a single population (average age 17.7 ± 0.9 Ma, P(χ^2)<0.001).

The helium ages range from 15 to 22 Ma, whereas the corresponding fission-track ages range from 20 to 27 Ma (Figure 12). As mentioned earlier, the closure temperature for helium dating (~75°C) is approximately 45°C cooler than fission-track closure temperatures (~120°C). This difference in closure temperatures can be used to determine a quantitative estimate of the cooling rate. A crossplot of apatite fission-track and helium

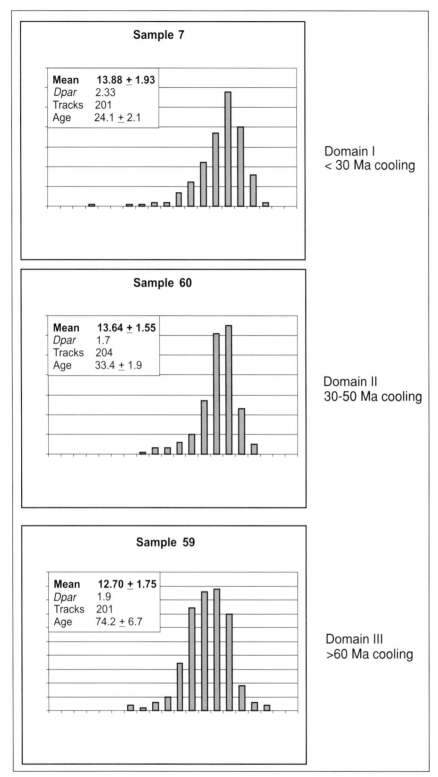

FIGURE 14. Track-length histograms for three representative samples, one from each fission-track age domain. Sample numbers are listed at the top of each diagram. Summary data (mean length, *Dpar*, number of tracks, age) for each sample are listed in the inset box. Each sample has more than 200 grains counted. The domains with young reset ages (I and II) have long mean track lengths (>13.5 μm) and small standard deviations. The sample from the western domain with older reset ages has a slightly lower mean track length at 12.7 μm, but still has a low standard deviation. These data suggest that all the samples cooled relatively rapidly.

ages, with isograds for various constant cooling-rate histories, is shown in Figure 15. This is a quick-look method for estimating the cooling-rate history when both apatite fission-track and helium age data are available.

Several conclusions can be drawn from Figure 15. Samples 8, 5, and 2 all appear to have cooled at about 10°C/my. Although sample 5 (Maestrichtian) is stratigraphically older than sample 2 (early Eocene), it has older apparent ages for both thermochronometers. In other words, the data suggest that sample 5 cooled earlier than sample 2. This is evidence that deformation occurred prior to the episode of cooling recorded by the apatite thermochronometers. Sample 7 clearly exhibits a slower cooling-rate history than the other samples in the Parras Basin. The apparent ages from the two samples in the Monterrey salient (samples 12 and 19) indicate an even slower paleo cooling rate, yet do not exhibit a correlation between *Dpar* and fission-track age. Samples 12 and 19 are from the same stratigraphic unit and are structurally equivalent. The fission-track ages are similar (26.4 ± 2.0 and 28.3 ± 1.9 Ma) and the helium ages are identical (18.1 ± 0.9 and 18.1 ± 1.1 Ma).

DISCUSSION

All of the data presented herein indicate that the frontal portion of the Sierra Madre Oriental was heated to temperatures high enough to reset apatite thermochronological systems and thermally degrade hydrocarbons reservoired in traps. These data also indicate that these paleotemperatures decreased rapidly to the west in places, and more gradually toward the east. The regional distribution and the smooth contours of this temperature anomaly make it unlikely to have been caused by local intrusion or hydrothermal circulation. Only one locality, number 50, contained minerals indicative of hydrothermal activity, including fluorite, barite, and hydrothermal clays. This locality lies directly below the El Volantín thrust, which brings hotter rocks up over cooler rocks and is west of the axis of the main thermal anomaly. Hydrothermal minerals are absent from all other localities; at the same time, these rocks retain relatively

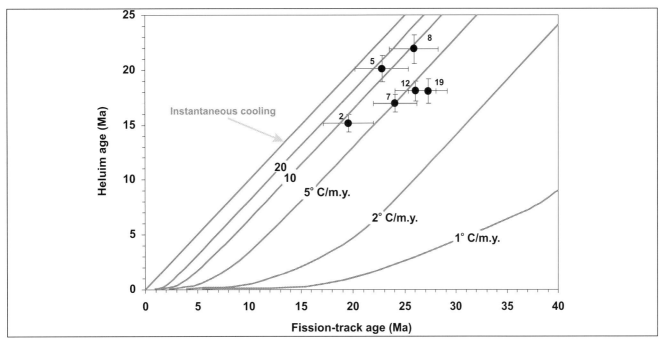

FIGURE 15. Graph comparing apatite fission-track and U-Th/He ages from the same samples. Sample localities are listed in Table 1. Dating results are shown in Table 4. The diagonal curves are modeled, constant cooling-rate isograds from instantaneous cooling to 1°C/my. This represents a quick-look method for estimating the cooling-rate history when both apatite fission-track and helium age data are available. The cooling rates for these Mexican samples between the two closure temperatures are 5° to 10°C/my.

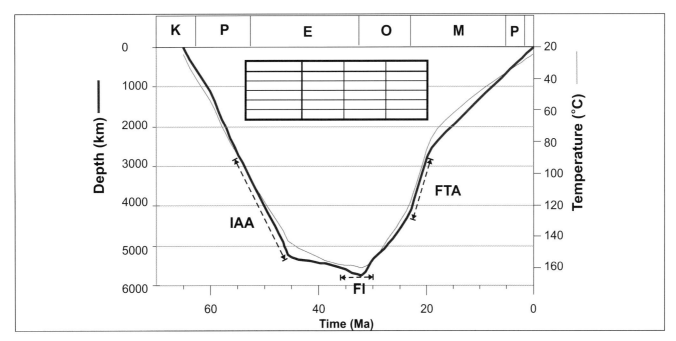

FIGURE 16. Burial-history diagram constrained by thermochronological and well data from the Ciudad Mante area. The site is based at the El Abra-1 well (see Figure 1 for the well location). Formation tops were taken from Valencia-Islas (1993). The modeled bed is the top of the El Abra Formation. The timing of burial is constrained by K/Ar ages on bentonites (labeled IAA) interbedded in the Upper Cretaceous Tamaulipas Superior Formation. These data indicate that the Tamaulipas Superior beds were undergoing burial diagenesis at approximately 50 Ma. The estimated temperature at that time was approximately 150°C, based on the kinetics of illite growth (Pevear, 1992). The maximum temperature of 160°C reached at this locality is estimated from the homogenization temperatures (labeled FI) on the El Abra Formation in the Sierra Cuchara. The age of uplift is constrained by apatite fission-track ages (labeled FTA) from the immediately overlying Mendez Formation. The estimated maximum burial at this locality exceeds 5.5 km. This entire section has now been eroded.

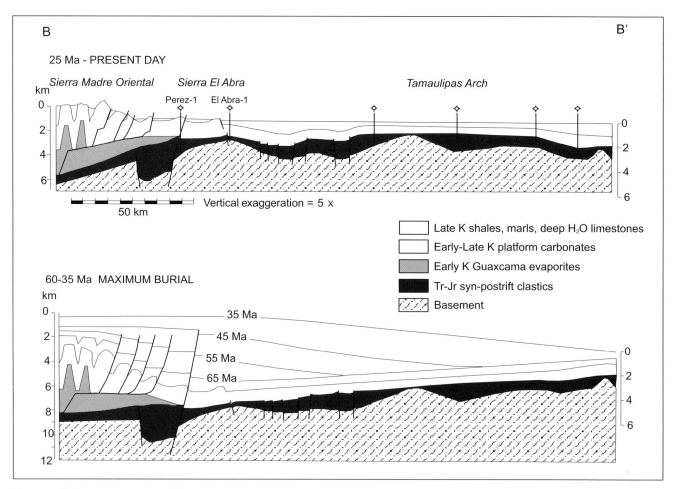

FIGURE 17. Reconstruction of the foreland basin at the time of maximum burial. Time lines in the foreland section are schematic only and are meant to mimic the depositional rates used in the burial-history diagram. Line of section is shown on Figures 1 and 10.

pristine depositional textures. Evidence for deep-seated intrusive igneous rocks is lacking along the axis of the thermal anomaly, and no intrusions of any kind have been mapped. Local intrusives are mapped in the region of Zimapán and to the west of the study area, but none of these bodies is of sufficient size to have caused more than a small, local thermal anomaly (e.g., Valencia-Islas, 1993).

We interpret these data instead to indicate that this thermal anomaly was caused by deep burial commencing with the Laramide-age deformation and continuing some time after the orogeny had ceased. A burial-history plot that honors these data at the location of the El Abra-1 well is shown in Figure 16. This figure follows the history of the top of the El Abra Limestone and is based on the stratigraphy encountered in the El Abra-1 well near Ciudad Mante (Figure 1). As stated in the section on analytical methods, the heating portion of the burial history is constrained by the K/Ar ages on illite, the maximum burial is constrained by the fluid-inclusion homogenization temperatures, and the uplift portion is constrained by the fission-track and helium ages on apatites. The data used to constrain this diagram are derived from outcrop samples less than 5 km from the well location.

In this model, rapid burial at the El Abra locality began at about 65 Ma. Maximum burial occurred at about 50 to 45 Ma, and by 40 Ma, the region was being uplifted and cooled. Note that our estimate of missing section at the well is greater than 5 km. Previous estimates of missing section at this location were less than 2 km (Valencia-Islas, 1996). Because of the large amount of missing section and the regional nature of the heating event, we infer that the agent responsible for this thermal anomaly was a sedimentary (foreland) basin deposited on top of the frontal portions of the fold belt. Remnants of this basin are preserved in the La Popa, Tampico-Misantla, and possibly Veracruz Basins.

This basin is inferred to have had the classic foreland basin shape, i.e., thick near the fold belt on the west and thinning toward the east (Figure 10). It was also actively subsiding during the formation of the fold belt, as shown by the Eocene K/Ar ages obtained on the Cretaceous bentonites. If the geothermal gradient during basin formation was constant along the strike of the basin, then burial was greater in the north and lesser in the south. If a conservative estimate of the geothermal gradient is assumed (30°C/km), then the estimated burial at Monterrey was approximately 7 km. In the La Popa, Mante,

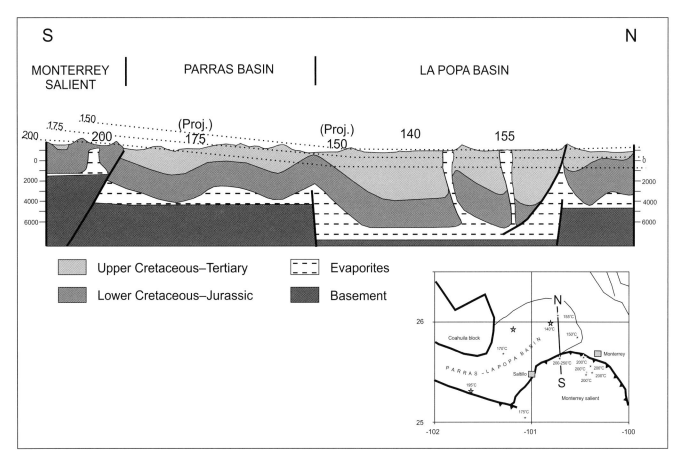

FIGURE 18. North-south schematic cross section from the La Popa Basin to the Monterrey salient. The geology along this section is modified after Giles and Lawton (1998) and Gray et al. (1997). The steps in the basement are taken from Gray and Johnson (1995) and Gray et al. (1997). Aqueous fluid-inclusion homogenization temperatures are shown at the surface locations where the samples were taken. Isotherm contours are projected into the subsurface with a constant gradient of approximately 30°C/km. This section shows that the 150°C isotherm coincides with the top of the presently exposed La Popa Basin. Projection of this isotherm to the south suggests that at most, ~2 km of Parras Basin strata may have overlain the rocks of the Monterrey salient prior to burial beneath the younger foreland basin. These relationships also indicate that large-scale deformation occurred prior to the late burial and that the Monterrey salient has been uplifted $\pm$ 2 km more than the basins to the north. The vertical scale is in meters subsea level. No vertical exaggeration is given.

and Tamazunchale areas, the estimated post–Early Eocene burial is 5–6 km. At the Tuxpan platform and in the Zongolica area, maximum burial would have been on the order of 4 km. Areas in the interior of the fold belt, such as Ciudad del Maiz and Zimapán, were uplifted and cooled during Laramide time and probably were never covered by foreland sediments. Lower geothermal gradients would increase the estimated missing section in these areas.

A cross section through the El Abra-1 well location is presented in Figure 17. This figure shows the present structural configuration (from Yurewicz et al., personal communication, 2000; Gray et al., unpublished work) and a restoration of the entire 5+ km of missing section. It appears from this reconstruction that the present fold-belt relief formed much of the topographic sill for the western margin of the basin. The depicted structural deformation of the foreland sequence is stylistic in this reconstruction, because the two events overlapped in time.

Estimates for the thickness of the basin fill from fluid-inclusion data do not allow for more than 2 km of Parras Basin strata to be present over the rocks of the Monterrey salient at the time this younger basin was forming (Figure 18). In addition, when seen in profile view, the fluid-inclusion isotherms clearly show that much, if not all, of the structural complexity of the region was present prior to the late burial. In particular, the maximum temperatures show no relationship to depth to basement or thickness of Late Cretaceous sedimentary cover. Instead, the present ground surface is much closer to an isothermal structure. In addition, the isothermal contours suggest that the Monterrey salient may have experienced as much as 2 km of additional uplift when compared with the basins immediately to the north. The present data are not dense enough to indicate whether this uplift is accommodated at the present mountain front or is distributed over a wider area.

We do not have access to temperature data from the four wells in the Poza Rica–Golden Lane trend. We estimate, however, that reservoir temperatures probably do not exceed 110°C at locality 58, and 100 °C at the other three localities. The fluid-inclusion temperatures are 15° to 25°C greater than our estimated reservoir temperatures today, suggesting that as much as a kilometer of section may be missing at these localities.

The timing and magnitude of uplift established through the use of these data agree very well with the regional relationships in the Gulf of Mexico province to the east. Major eastward progradation occurred there beginning in the middle Oligocene (Pérez-Crúz, 1992; Coogan et al., 1972). Sedimentation rates increased further in the Miocene, as evidenced by the establishment of growth faulting east of the Tamaulipas and Tuxpan arches and renewed subsidence in the Veracruz Basin (Cruz-Helú et al., 1977; López-Ramos, 1982; Pérez-Crúz, 1992). It is apparent that the source of this large sediment influx was the uplifting and eroding of this foreland basin immediately to the west. Nearly all traces of this formerly thick Tertiary basin are now eroded, leaving only the high-temperature imprint on the underlying Mesozoic rocks.

The thermal structure of this paleobasin suggests that the Magiscatzin and Tampico-Misantla Basins are not prospective for liquid hydrocarbons near this depositional axis. Reservoired liquid hydrocarbons above Upper Jurassic source rocks were cracked to lighter hydrocarbons (e.g., Minero, 1991; Yurewicz, et al., personal communication, 2000). It appears that the interior of the fold belt may have a higher potential for liquids than the frontal portion because of a lack of structural imbrication and minimal amounts of Paleocene-Eocene burial. This interpretation differs considerably from several recent studies that suggest the frontal portions of the fold belt and adjacent basins to the east are presently in the oil-generating window (e.g., Valencia-Islas, 1993, 1996; Holguin-Quiñones, 1991).

CONCLUSIONS

The relationships presented herein suggest the following sequence of events at the leading edge of the Sierra Madre Oriental and adjacent basins. The Laramide-age and -style deformation was important locally beginning at the Cretaceous-Tertiary boundary and extending into the Eocene. During the early Tertiary, a thick foreland basin began to develop along the present leading edge of the fold belt. The formation and inversion of this basin were diachronous along its length. Oil migrated into traps beneath the axis of this basin prior to maximum burial. Burial in this basin reached 7 km locally and at least 5 km in the central portion of the foreland area. Temperatures were between 125°C and 150°C by about 50 Ma. Maximum regional temperatures were about 170° C. Hydrocarbons present in these traps were cracked to gas and residual pyrobitumen. Burial-history modeling suggests that maximum burial was reached at about 35 Ma at Ciudad Mante.

The entire region began to cool very shortly after maximum burial. Cooling ages are 40–35 Ma in the south, where deposition was less, and average 24 Ma in the north, where deposi-

tion was greatest. Interior to the fold belt, rocks cooled during Laramide time, indicating that they were not affected by this younger foreland basin phase. These relationships suggest that much of the area covered by this foreland basin was heated beyond the oil window and is now prospective only for gas, whereas areas interior to the fold belt may remain locally prospective for oil.

ACKNOWLEDGMENTS

This paper has benefited from constructive reviews by Lori Summa, Tim Lawton, Gary Prost, and Max Suter. The authors would like to thank Mario Aranda, Samuel Eguiluz, Paul Enos, Jose Galicia, Kate Giles, Bob Goldhammer, Bob Goldstein, Tim Lawton, Randy Marrett, Javier Meneses-Rocha, Chris Soegaard, Max Suter, and Jim Wilson for introductions to different parts of the area and for freely discussing their ideas. Paul Enos and Randy Marrett also provided additional samples. We also thank Jim Reynolds at Fluid Inc.; Don Hall at Fluid Inclusion Technologies; Ray Donelick at Donelick Analytical, Inc.; Ken Farley at California Institute of Technology; Mike Koch at ExxonMobil Upstream Research Co.; and Geochron Laboratories for providing analyses and discussions of the data. We thank ExxonMobil Upstream Research Co. and ExxonMobil Exploration Co. for permission to publish this report.

REFERENCES CITED

Aranda-García, M., 1991, El segmento San Felipe del cinturón calbagado, Sierra Madre Oriental, estado de Durango, México: Boletín Asociación Mexicano Géologos Petroleros, v. 16, p. 18–36.

Baker III, R. A., 1970, Stratigraphy and sedimentology of the Cañon del Tule Formation (Upper Cretaceous), Parras Basin, Northeastern Mexico: Ph.D. dissertation, University of Texas at Austin, 322 p.

Bitter, M. R., 1986, Sedimentology and petrology of the Chicontopec Formation, Tampico-Misantla Basin, eastern Mexico: Master's thesis, Brigham Young University, Provo, Utah, 174 p.

Bitter, M. R., 1993, Sedimentation and provenance of Chicontopec sandstones with implications for uplift of the Sierra Madre Oriental and Texiutlán massif, east-central Mexico, in J. L. Penile and R. F. Perkins, eds., Mesozoic and Early Cenozoic development of the Gulf of Mexico and Caribbean region, a context for hydrocarbon exploration: Proceedings of the 13th Annual Research Conference, Gulf Coast Section, Society for Sedimentary Geology (SEPM) Foundation, p. 155–172.

Burtner, R. L., A. Nigrini, and R. A. Donelick, 1994, Thermochronology of Lower Cretaceous source rocks in the Idaho-Wyoming thrust belt: AAPG Bulletin, v. 78, p. 2649–2657.

Campa-Uranga, M. F., 1985, The Mexican thrust belt, in D. G. Howell, ed., Tectonostratigraphic terranes of the Circum-Pacific region: Circum-Pacific Council for Energy and Mineral Resources Earth Science Series, v. 1, p. 299–313.

Carlson, W. D., 1990, Mechanisms and kinetics of apatite fission-track annealing: American Mineralogist, v. 75, p. 1120–1139.

Carlson, W. D., R. A. Donelick, and R. A. Ketcham, 1999, Variability of apatite fission-track annealing kinetics: 1. Experimental results: American Mineralogist, v. 84, p. 1213–1223.

Carrillo-Bravo, J., 1971, La Plataforma Valles–San Luis Potosí: Boletín de la Asociación Mexicana Géologos Petroleros, v. 23, p, 1–113.

Carrillo-Martínez, M., and M. Suter, 1982, Tectonica de los alrededores de Zimapán, Hidalgo, in Excursion a la region de Zimapán y areas circundantes: Sociedad Geologica Mexicano, Convención Geologica Nacionál Sexto, p. 1–20.

Cerveny III, P. F., 1990, Fission-track thermochronology of the Wind River Range and other basement-cored uplifts in the Rocky Mountain foreland: Ph.D. dissertation, University of Wyoming, Laramie, 189 p.

Coney, P. J., 1978, Mesozoic-Cenozoic Cordilleran plate tectonics: Geological Society of America Memoir 151, p. 33–50.

Coogan, A. H, D. G. Bebout, and C. Maggio, 1972, Depositional environments and geologic history of the Golden Lane and Poza Rica Trend, Mexico: An alternative view: AAPG Bulletin, v. 56, p. 1419–1447.

Craddock, J. P., A. A. Kopania, and D. V. Wiltschko, 1988, Interaction between the northern Idaho-Wyoming thrust belt and bounding basement blocks, central western Wyoming, in C. J. Schmidt and W. J. Perry Jr., eds., Interaction of the Rocky Mountain foreland and the Cordilleran thrust belt: Geological Society of America Memoir 171, p. 333–352.

Crowley, K. D., M. Cameron, and R. L. Schaefer, 1991, Experimental studies of annealing etched fission tracks in fluorapatite: Geochemica Chosmochemica Acta, v. 55, no. 5, p. 1449–1465.

Cruz-Helú, P., V. R. Verdugo, and P. R. Barcenas, 1977, Origin and distribution of Tertiary conglomerates, Veracruz Basin, Mexico: AAPG Bulletin, v. 61, p. 207–226.

de Cserna, Z., 1956, Tectónica de la Sierra Madre Oriental de México, entre Torreón y Monterrey: Contributions, 20th International Geological Congress, 87 p.

de Cserna, Z., 1989, An outline of the geology of Mexico, in A. W. Bally, ed., The geology of North America—An overview: Geological Society of America, Decade of North American Geology, v. A, p. 233–264.

Dickinson, W. R., and W. S. Snyder, 1977, Plate tectonics of the Laramide orogeny: Geological Society of America Memoir 151, p. 355–366.

Dodson, M. H., 1973, Closure temperature in cooling geochronological and petrological systems: Contributions to Mineralogy and Petrology, v. 40, p. 259–274.

Donelick, R. A., R. A. Ketcham, W. D. Carlson, 1999, Variability of apatite fission-track annealing kinetics: II. Crystallographic orientation effects: American Mineralogist, v. 84, p. 1224–1234.

Drewes, H., 1988, Development of the foreland zone and adjacent terranes of the Cordilleran orogenic belt near the U. S.–Mexican border, in C. A. Schmidt and W. J. Perry Jr., eds., Interaction of the Rocky Mountain foreland and the Cordilleran thrust front: Geological Society of America Memoir 171, p. 447–463.

Eguiluz de Antuñano, S., 1984, Tectonica Cenozoica del norte de México: Boletín de la Asociación Mexicana Géologos Petroleros, v. 36, p. 43–62.

Eguiluz de Antuñano, S., 1991, Discordancia Cenomania sobre la plataforma de Coahuila: Boletín de la Asociación Mexicana Géologos Petroleros, v. 41, p. 1–17.

Farley, K. A., R. A. Wolf, and L. T. Silver, 1996, The effects of long alpha-stopping distances on (U-Th)/He ages: Geochimica Cosmochimica Acta, v. 60, p. 4223–4229.

Giles, K. A., and T. F. Lawton, 1998, La Popa Basin Salt Tectonics Seminar Guidebook: 90 p.

Gleadow, A. J. W., and I. R. Duddy, 1981, A natural long-term track annealing experiment for apatite: Nuclear Tracks and Radiation Measurements, v. 5, p. 169–174.

Gleadow, A. J. W., I. R. Duddy, and J. F. Lovering, 1983, Fission track analysis: A new tool for the evaluation of thermal histories and hydrocarbon potential: Australian Petroleum Explorationists Association Journal, v. 23, 93–102.

Goldhammer, R. K., P. J. Lehmann, R. G. Todd, J. L. Wilson, W. C. Ward, and C. R. Johnson, 1991, Sequence stratigraphy and cyclostratigraphy of the Mesozoic of the Sierra Madre Oriental, northeast Mexico, a field guidebook: Gulf Coast Section, Society for Sedimentary Geology (SEPM), 86 p.

Goldstein, R. H., and T. J. Reynolds, 1994, Systematics of fluid inclusions in diagenetic minerals: Society for Sedimentary Geology (SEPM) Short Course 31, 199 p.

Gray, G. G., and C. R. Johnson, 1995, Structural and tectonic evolution of the Sierra Madre Oriental, with emphasis on the Saltillo-Monterey corridor: A field trip guidebook: AAPG Annual Convention Field Trip Guidebook 10, 17 p.

Gray, G. G., S. Eguiluz, R. J. Chuchla, and D. A. Yurewicz, 1997, Structural evolution of the Saltillo-Monterrey corridor, Sierra Madre Oriental: Applications to exploration challenges in fold-thrust belts: AAPG/ Asociación Mexicana Géologos Petroleros International Research Symposium Guidebook, 52 p.

Green, P. F., I. R. Duddy, A. J. W. Gleadow, P. R. Tingate, and G. M. Laslett, 1986, Thermal annealing of fission tracks in apatite: 1. A qualitative description: Chemical Geology (Isotope Science Section), v. 59, p. 237–253.

Gries, R. R., 1983, North-south compression of Rocky Mountain foreland structures, in J. D. Lowell and R. R. Gries, eds., Rocky Mountain foreland basins and uplifts: Rocky Mountain Association of Geologists, p. 9–32.

Guzmán, E. J., and Z. de Cserna, 1963, Tectonic History of Mexico, in O. E. Childs and B. W. Beebe, eds., Backbone of the Americas: Tectonic history from pole to pole: AAPG Memoir 2, p. 113–129.

Hanor, J. S., 1980, Dissolved methane in sedimentary brines: Potential effect on the PVT properties of fluid inclusions: Economic Geology v. 75, p. 603–609.

Holguín-Quiñones, N. G., 1991, La geochímica del petróleo en México: Boletín de la Asociación Mexicana Géologos Petroleros, v. 41, p. 37–50.

House, M. A., B. P. Wernicke, and K. A. Farley, 1998, Dating topography of the Sierra Nevada, California, using apatite (U-Th)/He ages: Nature, v. 396, p. 66–69.

House, M. A., K. A. Farley, and B. P. Kohn, 1999, An empirical test of helium diffusion in apatite: Borehole data from the Otway Basin, Australia: Earth Planetary Science Letters, v. 170, p. 463-474.

Hunziker, J. C., 1987, Radiogenic isotopes in very low-grade metamorphism, in M. Fry, ed., Low temperature metamorphism: Blackie Academic & Professional Publishers, Glasgow, p. 200–226.

Ketcham, R. A., R. A. Donelick, and W. D. Carlson, 1999, Variability of apatite fission-track annealing kinetics: III. Extrapolation to geological time scales: American Mineralogist, v. 84, p. 1235–1255.

Kulik, D. M., and C. J. Schmidt, 1988. Region of overlap and styles of interaction of Cordilleran thrust belt and Rocky Mountain foreland, in C. A. Schmidt and W. J. Perry Jr., eds., Interaction of the Rocky Mountain foreland and the Cordilleran thrust front: Geological Society of America Memoir 171, p. 75–98.

Laslett, G. M., P. F. Green, I. R. Duddy, and A. J. W. Gleadow, 1987, Thermal annealing of fission tracks in apatite: 2. A quantitative analysis: Chemical Geology (Isotope Science Section), v. 65, p. 1–13.

Lippolt, H. J., M. Leitz, R. S. Wernicke, and B. Hagedorn, 1994, (U+Th)/He dating apatite: Experience with samples from different geochemical environments: Chemical Geology, v. 112, p. 179–191.

López-Ramos, E., 1982, Geologia de Mexico 3 (3rd ed.): Public Education Ministry, Mexico, D.F., Mexico, 446 p.

Mahon, K. I., 1996, The new "York" regression: Application of improved statistical method to geochemistry: International Geology Review, v. 38, p. 293–303.

McBride, E. F., A. E. Weidie, J. A. Wolleben, and R. C. Laudon, 1974, Stratigraphy and structure of the Parras and La Popa Basins, northeastern Mexico: Geological Society of America Bulletin, v. 84, p. 1603–1622.

McDougall, I., and T. M. Harrison, 1988, Geochronology and thermochronology by the ^{40}Ar/^{39}Ar Method: New York, Oxford University Press, 212 p.

Meneses-Rocha, J. J., D. Rodríguez Figueroa, J. Toríz Gama, J. Banda Hernández, R. Hernández De La Fuente, and V. Valdivieso Ramos, 1997, Geologic field trip to the Zongolica fold and thrust belt: AAPG/ Asociación Mexicano Géologos Petroleros International Research Symposium Guidebook, 103 p.

Minero, C. J., 1991, Sedimentation and diagenesis along open and island-protected windward carbonate platform margins of the Cretaceous El Abra Formation, Mexico: Sedimentary Geology, v. 71, p. 261–288.

Morelos-García, J. A., 1996, Geochemical evaluation of the southern Tampico-Misantla Basin, Mexico: Oil-oil and oil-source rock correlations: Ph.D. dissertation, University of Texas at Dallas, 635 p.

Mossman, R. W., and F. Viniegra, 1976, Complex fault structures in the Veracruz province, Mexico: AAPG Bulletin, v. 60, p. 379–388.

Naeser, N. D., C. W. Naeser, and T. H. McCulloh, 1989, The application of fission-track dating to the depositional and thermal history of rocks in sedimentary basins, in N. D. Naeser and T. H. McCulloh, eds., Thermal history of sedimentary basins: Methods and case histories: New York, Springer-Verlag, 319 p.

Padilla y Sánchez, R. J., 1982, Geologic evolution of the Sierra Madre Oriental between Linares, Conception del Oro, Saltillo, and Monterrey, Mexico: Ph.D. dissertation, University of Texas at Austin, 218 p.

Padilla y Sánchez, R. J., R.G. Martínez Serrano, and V. Tórres Rodríguez, 1994, Mapa Tectonica de Los Estados Unidos Mexicanos, Escala 1:2,000,000: Instituto Nacional de Estadistica, Geografia e Informatica, 2 sheets.

Perez-Cruz, G., 1992, Geologic evolution of the Burgos Basin, northeastern Mexico: Ph.D. dissertation, Rice University, Houston, Texas, 155 p.

Pessagno, E. A., 1969, Upper Cretaceous stratigraphy of the western Gulf Coast area of Mexico, Texas, and Arkansas: Geological Society of America Memoir 111, 139 p.

Pevear, D. R., 1992, Illite age analysis, a new tool for basin thermal history analysis, in Y. K. Kharaka and A. S. Maest, eds., Water-Rock Interaction: Rotterdam, A. A. Balkema Publishers, p. 1251–1254.

Pevear, D. R., 1998, Illite and hydrocarbon exploration: Proceedings, National Academy of Science, v. 96, p. 3440–3446.

Pottorf, R., and M. O. Vityk, 1997, Applications of fluid-inclusion technology to hydrocarbon systems: Pan American Conference on Research on Fluid Inclusions, Annual Meeting Abstracts, v. 7, p. 51.

Prost, G., R. Marrett, M. Aranda, S. Eguiluz, J. Galicia, and J. Banda, 1995, Deformation history of the Sierra Madre Oriental, Mexico, and associated hydrocarbon generation-preservation: First AAPG/ Asociación Mexicano Geólogo Petroleros Joint Research Conference Abstracts, p. 3.

Salvador, A., 1987, Late Triassic–Jurassic paleogeography and origin of the Gulf of Mexico Basin: AAPG Bulletin, v. 71, p. 419–451.

Salvador, A., 1991, Triassic-Jurassic, in A. Salvador, ed., The Gulf of Mexico Basin: Geological Society of America, Decade of North American Geology, v. J, p. 131–180.

Schmidt, C. A. and W. J. Perry Jr., 1988, preface, in C. A. Schmidt and W. J. Perry Jr., eds., Interaction of the Rocky Mountain foreland and the Cordilleran thrust front: Geological Society of America Memoir 171, p. ix–xi.

Soegaard, K., K. Giles, F. J. Vega, T. Lawton, A. T. Daniels, N. Halik, H. Ye, J. M. Garrison, L. A. Hunnicutt, K. A. McMillan, and K. A. Shannon, 1997, Structure, stratigraphy, and paleontology of Late Cretaceous–Early Tertiary Parras–La Popa foreland basin near Monterrey, Mexico: AAPG Annual Convention Field Trip Guidebook 10, 135 p.

Steidtman, J. R., L. C. McGee, and L. T. Middleton, 1983, Laramide sedimentation, folding, and faulting in the southern Wind River Range, Wyoming, in J. D. Lowell and R. R. Gries, eds., Rocky Mountain foreland basins and uplifts: Rocky Mountain Association of Geologists, p. 161–179.

Suter, M., 1984, Cordilleran deformation along the eastern edge of the Valles–San Luis Potosí carbonate platform, Sierra Madre Oriental fold-thrust belt, east-central Mexico: Geological Society of America Bulletin, v. 95, p. 1387–1397.

Suter, M., 1987, Structural traverse across the Sierra Madre Oriental fold-thrust belt in east-central Mexico: Geological Society of America Bulletin, v. 98, p. 249–264.

Suter, M., 1990, Geología de la hoja Tamazunchale, estados de Hidalgo, Querétaro, y San Luis Potosí: Carta Geológica de México, Serie de 1:100,000, núm. 22.

Tardy, M., J. F. Longoria, J. Martínez-Reyes, L. M. Mitra-S., A. M. Patiño, R. Padilla y Sánchez, and R. C. Ramírez, 1975, Observaciones generales sobre la estructura de la Sierra Madre Oriental: La Alloctina del conjunto Cadena Alta-Altiplano Central, entre Torreón, Coahuila y San Luis Potosí, S. L. P., México: Universidad Nacional Autonomia México Instituto Geología Revista, v. 1, p. 1–11.

Valencia-Islas, J. J., 1993, Évolution tectonique et historie thermique de la plateforme Valles San Luis Potosí (Méxique): Implications sur la distribution des hydrocarbures et des gites mineraux: Ph.D. dissertation, Université Paul Sabatier de Toulouse III, Toulouse, France, 179 p.

Valencia-Islas, J. J., 1996, Implicaciones de la historia térmica de la plataforma Valles–San Luis Potosí en la distribución de los hidrocarburos y yacimientos minerales: Boletín de la Asociación Mexicana Géologos Petroleros, v. 45, p. 1–19.

Vega-Vera, F. J., L. M. Mitra-Salazar, and E. Martinez, 1989, Contribución al conocimiento de la estratigrafía del grupo Difunta (Cretacio Superior–Terciario) en el noreste de México: Universidad Nacional Autonomia México Instituto Geología Revista, v. 8, p. 179–187.

Vrolijk, P. V., and van der Pluijm, B., 1999, Fault gouge: Journal of Structural Geology, v. 21, p. 1039–1048.

Warnock, A. C., P. K. Zeitler, R. A. Wolf, and S. C. Bergman, 1997, An evaluation of low temperature apatite U-Th/He thermochronometry: Geochimica Cosmochimica Acta, v. 61, p. 5371–5377.

Wolf, R. A., K. A. Farley, and L. T. Silver, 1996, Helium diffusion and low-temperature thermochronometry of apatite: Geochimica Cosmochimica Acta, v. 60, p. 4231–4240.

Wolf, R. A., K. A. Farley, and L. T. Silver, 1997, Assessment of (U-Th)/He thermochronometry: The low-temperature history of the San Jacinto mountains, California: Geology, v. 25, p. 65–68.

Ye, H., 1997, The arcuate Sierra Madre Oriental orogenic belt, NE Mexico: Tectonic infilling of a recess along the southwestern North American continental margin, in K. Soegaard, K. Giles, F. J. Vega, T. Lawton, A. T. Daniels, N. Halik, H. Ye, J. M. Garrison, L. Hunnicutt, K. A. McMillan, and K. A. Shannon, Structure, stratigraphy, and paleontology of Late Cretaceous–Early Tertiary Parras–La Popa Foreland Basin near Monterrey, Mexico: AAPG Annual Convention Field Trip Guidebook 10, p. 82–115.

Zeitler, P. K., A. L. Herczig, I. McDougall, and M. Honda, 1987, U-Th-He dating of apatite: A potential thermochronometer: Geochimica Cosmochimica Acta, v. 51, p. 2865–2868.

Meneses-Rocha, J. J., 2001, Tectonic evolution of the Ixtapa graben, an example of a strike-slip basin in southeastern Mexico: Implications for regional petroleum systems, *in* C. Bartolini, R. T. Buffler, and A. Cantú-Chapa, eds., The western Gulf of Mexico Basin: Tectonics, sedimentary basins, and petroleum systems: AAPG Memoir 75, p. 183–216.

8

Tectonic Evolution of the Ixtapa Graben, an Example of a Strike-slip Basin of Southeastern Mexico: Implications for Regional Petroleum Systems

Javier J. Meneses-Rocha
PEMEX Exploración y Producción, Villahermosa, Tabasco, Mexico

ABSTRACT

The Ixtapa graben is located in the center of the Strike-slip Fault province of the Sierra de Chiapas, Mexico. In this graben, rocks of middle Cretaceous (Albian-Cenomanian) to Pleistocene age represent a section in which successively younger beds lie to the southeast. This section is 15,365 m thick and represents marine, transitional, and continental environments with numerous vertical and lateral facies changes through the whole section and unconformities in the uppermost part. Along the flanks of the graben, beds are upturned and form positive flower structures.

Stratigraphic information indicates that from Albian to Maestrichtian, most of the Sierra de Chiapas underwent alternating phases of tectonic quiescence associated with the development of an extensive carbonate platform and phases of tectonic uplift that exposed parts of the platform and led to deposition of terrigenous clastics in shoreline and deep-water environments. During the Laramide orogeny, the Ixtapa graben was subject to gentle deformation that produced a disconformity between the Paleocene and the middle Eocene. From late Eocene to early Miocene, movement along the faults of the Strike-slip Fault province was predominantly vertical. This movement became sinistrally transcurrent at the beginning of the middle Miocene, when coarse- and fine-grained terrigenous clastics accumulated in the Ixtapa graben. This transtensional phase was followed by a transpressive episode related to a northeast-oriented maximum horizontal stress that gave rise to the Neogene fold belt of southeastern Mexico and concomitant recession of the shoreline to its present position. During the late Miocene–early Pliocene, 6770 m of coarse- and fine-grained terrigenous clastics was redeposited in the Ixtapa graben. The vertical succession of the lithofacies that make up this continental sequence reveals that its deposition was the response to normal block faulting of the basement caused by the sinistral shift of the main bounding faults. This transtensional phase in central Chiapas triggered a gravity tectonism realm in which the Macuspana and Comalcalco Basins started to develop. At the end of the Pliocene, a transpressive phase gave rise to the deformation of the flanks of the Ixtapa graben, and during the Quaternary, volcanic sediments were deposited in angular unconformity on the continental sediments. The development of the Ixtapa graben in the context of the evolution of the Strike-slip Fault province and surrounding areas suggests that the total sinistral shear across this province is approximately 70 km, and that probably no fault in this province has a displacement greater than 16 km. Thus, the tectonic evolution of the Ixtapa graben, in light of the regional structural and stratigraphic analysis of southeastern Mexico, helps to explain the coexistence of a shear zone, a fold belt, and an extensional realm in the southernmost rim of the Gulf of Mexico Basin from the Neogene to the present. It also provides important insights into the regional petroleum-systems assessment of the producing trends of this part of Mexico.

INTRODUCTION

The Ixtapa graben is located in the central part of the shear belt of the Sierra de Chiapas (Figures 1, 2, 3, and 4). This graben exposes the most complete middle Cretaceous (informal term used in Mexico for the Albian-Cenomanian stages)–Quaternary section of southeastern Mexico and displays structural features that are typical of transtensional and transpressional settings (Figure 4). These characteristics and the graben's key position in the regional structure of southeastern Mexico make this graben an important piece of evidence, not only to infer the Cretaceous, Tertiary, and Quaternary tectonic events in the southernmost rim of the Gulf of Mexico Basin, but also to constrain the time, nature, and amount of movement associated with the active Pacific-Caribbean margin throughout the western segment of the Motagua–Polochic–Cayman Trough transform (Figure 1).

The Ixtapa graben also provides some details of the tectonic and sedimentary processes that take place in a strike-slip basin situated on a wide, continental-plate boundary zone where a transform system meets a convergent margin. For petroleum exploration, the study of the evolution of the Ixtapa graben provides the regional structural framework to understand the genesis of two elements of the petroleum systems of the Mesozoic and Tertiary producing trends of southeastern Mexico—trap and timing.

This paper reconstructs the tectonic development of the Ixtapa graben on the basis of its stratigraphic, sedimentologic, and structural characteristics. In particular, the time and mode of deformation of this graben are related to the regional paleogeography of southeastern Mexico and to the tectonic evolution of the Strike-slip Fault province. Data from field mapping and measured sections, combined with local and regional structural and tectonics studies carried out by the author during several years of petroleum-exploration activities in southeastern Mexico, are the basis of this study.

PREVIOUS STUDIES

Early investigations of the geology of North America and the Caribbean considered the Sierra de Chiapas to have formed part of a region whose geological characteristics are very different from those of the rest of Mexico. Thus, Hill (1898, in Schuchert, 1935) pointed out that the north-trending Rocky Mountains ceased in "the great scarp of Mexico" to the north of Tehuantepec. This author observed that the structural lines of Central America, the Greater Antilles, the Caribbean Sea, and northern South America formed an east-striking structural belt, which he named the Antillean Orogenic System, trans-

verse to the dominant northern and southern continental mountain chains of North and South America. Boese (1905) thought that the Isthmus of Tehuantepec was the northernmost point where "the bridge between the two Americas rested," and that a similar point should exist at the southernmost end of Central America. Subsequently, Schuchert (1935) recognized that Central America could be divided into two regions with different tectonic histories: a northern region, which he named Nuclear Central America, and a younger, southern region. In the former, he included Chiapas, British Honduras (now Belize), Guatemala, Honduras, and northern Nicaragua, and in the second, he included southern Nicaragua, Costa Rica, and Panama. By following Hill's idea, Schuchert (1935) described the historical geology of Nuclear Central America in terms of the Antillean geosyncline, and he interpreted the geologic history of Mexico, north of the Isthmus of Tehuantepec, in terms of the Mexican geosyncline.

The ideas of these authors apparently gathered acceptance at that time; however, the summaries of the geologic history of Mexico published in subsequent years assumed that the Laramide events which affected northern and eastern Mexico extended southward into Chiapas, and that this orogeny produced the present structural pattern of the Sierra de Chiapas. Thus, Alvarez (1949) described a Mesozoic geosyncline extending from Chihuahua into Chiapas; de Cserna (1960) and Guzmán and de Cserna (1963) divided the tectonic evolution of Mexico into three geotectonic cycles, which were subsequently incorporated into the historical geology of Nuclear Central America by Dengo and Bohnenberger (1969).

Only those geologists who conducted fieldwork in Chiapas (e.g., Mullerried, 1949; Gutiérrez-Gil, 1956; Sánchez-Montes de Oca, 1979) observed that in this part of Mexico, an important tectonic event took place in the Neogene, and that this episode had formed the present structural pattern of the Sierra de Chiapas. These authors, however, did not agree on two subjects—the time and nature of the Laramide orogeny, and the time and nature of the Neogene diastrophism. On the one hand, Mullerried (1949) thought the Laramide orogeny had not affected the Sierra de Chiapas, because he observed that Cretaceous beds are gently tilted. On the basis of field observations, he proposed that the main orogeny had occurred in the late Miocene-Pliocene. During that time, folding and reverse faulting were the main deformation mechanisms. On the other hand, Gutierrez-Gil (1956) considered that in Chiapas, the Laramide orogeny lasted from Late Cretaceous to Eocene but with relatively gentle effects. For this author, folding and normal faulting were intense at the end of the Miocene "or during the Pliocene." In contrast to Mullerried and Gutierrez-Gil, Sanchez-Montes de Oca (1979) considered that in Chiapas, the Laramide orogeny extended from Late Cretaceous to Paleocene, producing block faulting related to left-lateral strike-slip faults, as well as intra-Cretaceous and Late Cretaceous–Paleocene unconformities. According to Sanchez-Montes de Oca (1979), this event was succeeded by a new tectonic pulse at the end of

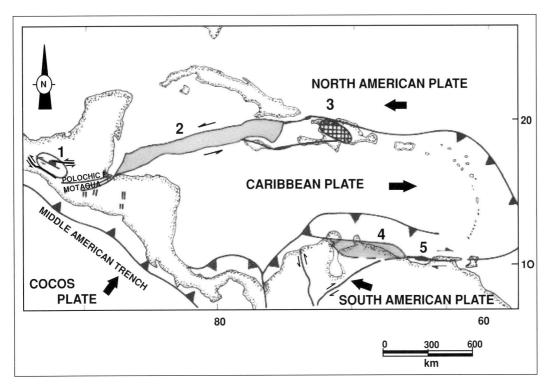

FIGURE 1. Present plate-tectonic setting of the study area: 1, Ixtapa graben and Strike-slip Fault province; 2, Cayman Trough; 3, Plate boundary zone at Hispaniola; 4, Falcon-Bonaire Basin; 5, Cariaco Basin. (Map adapted from Mann et al., 1984, and Jordan, 1975.)

the Oligocene, which gave rise to the filling of the Ixtapa graben and to an Oligocene and/or Eocene/early Miocene unconformity. For this author, the main displacement along the left-lateral strike-slip faults occurred during the middle and late Miocene. On the basis of Landsat images, Sanchez-Montes de Oca (1979) and other PEMEX geologists divided the Sierra de Chiapas into five tectonic provinces: (1) the Sierra homocline; (2) the Miramar province (box folds and reverse faults); (3) the Yaxchilan province (long, narrow anticlines limited by reverse faults; (4) the Simojovel province (anticlines and reverse faults); and (5) the Strike-slip Fault province.

Presently, the left-lateral strike-slip faults are recognized by most geologists familiar with Chiapas field geology; however, there is no consensus regarding their evolution. These faults form a left-stepping en-echelon array with the Motagua-Polochic fault system (Figures 1 and 2). This system and the Cayman Trough are considered the transform boundary between the North American and Caribbean Plates (Molnar and Sykes, 1969). The nature of the triple junction between these two plates and the Cocos plate is controversial. Some authors (e.g., Burkart, 1983) interpreted it as a simple trench-trench-transform, but others (e.g., Meneses-Rocha, 1985, and Guzmán-Speziale et al., 1989) suggested that plate-boundary deformation along the southern margin of the North American Plate has been diffuse and has spread over a wide area of southeastern Mexico.

Numerous plate-tectonic models have been proposed to explain the interaction among the North American, Caribbean, and Cocos Plates, and extensive literature exists about the Motagua–Polochic–Cayman Trough trend. Nevertheless, little is known about the relationship between this trend and the left-lateral strike-slip faults of Chiapas. Recently, Guzmán-

Speziale and Meneses-Rocha (2000) suggested that both fault systems might be connected through the Reverse-fault province, which acts as a stepover (fault jog) (Figures 1 and 2).

Fieldwork in the Ixtapa graben began in the early 1960s when graduate students of the University of Illinois, Urbana, investigated the biostratigraphy, paleoecology, and systematic paleobiology of certain invertebrate fossil groups in the Ixtapa area as the subjects of their master's theses and Ph.D. dissertations. The most detailed geologic report of the area is that of Heuer (1965), who mapped the Ixtapa-Soyalo area and reported the general characteristics of the Eocene-Pliocene section. The results of these studies were published in several short reports (e.g., Langenheim et al., 1965; Frost, 1971). This effort culminated in 1974 with Frost and Langenheim´s publication *Cenozoic Reef Biofacies*, in which they included a partial geologic map of the Chicoasen–Ixtapa–Soyalo area. Petróleos Mexicanos (PEMEX) began fieldwork in the Ixtapa graben in 1964. From that time until 1982, the year I began my study in the Strike-slip Fault province, only a few reconnaissance traverses were made, and the authors of these studies did not describe in detail the stratigraphy, structure, or geologic history of this graben. Details about the Mesozoic stratigraphy and paleogeography of the Sierra de Chiapas have been published previously by Castro-Mora et al. (1975), Quezada-Muñetón (1983, 1987), and Blair (1987, 1988).

REGIONAL GEOLOGY AND STRUCTURE

The present structures of southeastern Mexico lie between two areas that have remained relatively stable since the beginning of the Cretaceous—the Yucatán platform to the northeast

and the Chiapas massif to the southwest (Figure 2). These structures belong to two tectonic domains: a northwest-oriented
trend, here called the Neogene fold belt of southeastern Mexico, and a northeast-oriented trend comprised of three Neogene depocenters, here named the Gulf Coast Tertiary basins of
southeastern Mexico. The first domain, the Neogene fold belt,
is formed by the Sierra de Chiapas and the Reforma-Akal uplift,
whereas the second trend, the Gulf Coast Tertiary basins, contains, from west to east, the Isthmian Saline, the Comalcalco,
and the Macuspana Basins (Figure 2).

Yucatán Platform

The Yucatán platform forms the foreland of the Sierra de
Chiapas. It is a relatively stable area that includes the offshore
Campeche Bank and the onshore Yucatán Peninsula (Figure 2).
The former is defined by Ordoñez (1936) as an extensive,
almost flat, carbonate bank, whereas the Yucatán Peninsula is
a region of gentle structural relief; both are covered mostly by
Cretaceous and Tertiary carbonate rocks.

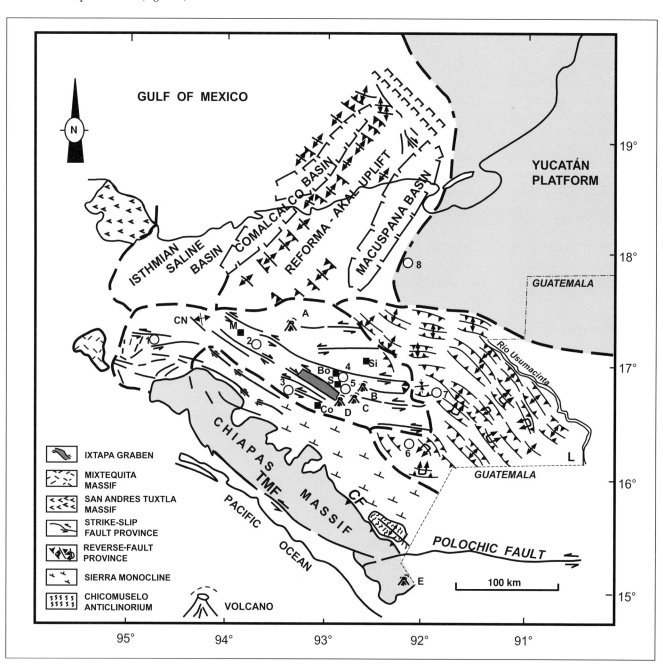

FIGURE 2. Generalized tectonic map of southeastern Mexico. 1, Sauzal-1 well; 2, Raudales-1 well; 3, Villa Allende-1 well;
4, Soyalo-1 well; 5, San Cristobal-1 well; 6, El Retiro-1 well; 7, Nazareth-51 well; 8, Cobo-301 well; A, Chichonal Volcano;
B, Tzontehuitz Volcano; C, Huitepec Volcano; D, Navenchauc Volcano; E, Tacana Volcano; Si, Simojovel syncline; CN,
Cerro Nanchital area; M, Malpaso area; Bo, Bochil area; S, Soyalo area; Co, Copoya area; L, Lacantun Area; TMF, Tonala-
Motozintla Fault; CF, Concordia Fault.

Chiapas Massif

The Chiapas massif is a mountain chain extending parallel to the Pacific coastal plain of Chiapas from the Isthmus of Tehuantepec to the Mexico-Guatemala border (Figure 2). Physiographically, it corresponds to the Sierra Madre province, which is characterized by a rugged topography with altitudes ranging from 25 to 3800 m (Tacana Volcano). The Chiapas massif is formed mainly by silicic plutonic (granite) and minor metamorphic rocks ranging in age from Precambrian to Pliocene. According to Muehlberger and Ritchie (1975), the southeastern end of the massif is crossed by the Polochic Fault, one of the three major faults crossing Nuclear Central America in a distributed shear boundary between the North American and Caribbean Plates. These authors also suggest that the limit between the massif and the Pacific coastal plain is fault controlled, because it is remarkably straight. This fault was studied in the field by J. M. García-Palomares (personal communication, 1979), who suggests that the fault is a left-lateral strike-slip fault of post-middle Miocene age. Paleomagnetic data obtained by Molina-Garza et al. (1992) from the batholitic complex and overlying red beds of the Todos Santos Formation indicate that the massif has not rotated or changed latitude with respect to North America after the Oxfordian.

NEOGENE FOLD BELT OF SOUTHEASTERN MEXICO

The Sierra de Chiapas

The Sierra de Chiapas extends from the Isthmus of Tehuantepec to the Mexico-Guatemala border and is formed mostly by Paleozoic, Mesozoic, and Cenozoic sedimentary rocks (Figures 1 and 2). It is bounded to the south by the Chiapas massif and to the north by the Yucatán platform. The structures exposed in this belt plunge toward the Reforma-Akal uplift and the Neogene Gulf Coast depocenters (i.e., the Isthmus Saline, Comalcalco, and Macuspana Basins).

The different structural styles exhibited by this structural belt and, to a lesser degree, the age of the rocks involved in its deformation allow us to identify four tectonic provinces: Strike-slip Fault province; Reverse-fault province; Sierra monocline; and Chicomuselo anticlinorium (Figure 2).

The structural term *Sierra de Chiapas* is equivalent to the *Pliegues frontales de Chiapas* of Alvarez (1958) and the *Sierra Madre Oriental de Chiapas* of Dengo (1968). Physiographically, this structural belt includes the Mountains of the North, the High Plains, the Mountains of the East, and the Central Depresión, as described by Mullerried (1949) and by Helbig (1964).

Although more than 40 exploratory wells have been drilled in the Sierra de Chiapas, it remains a frontier area. The only economically successful discoveries have been made in the foothills south of the giant fields of the Chiapas-Tabasco province (Cruz-Helú and Meneses-Rocha, 1999) more than 70, 150, and 250 km away from the uneconomic discoveries made

at Cerro Nanchital, Ocosingo, and Lacantun areas, respectively (Figure 2).

Strike-slip Fault Province

The Strike-slip Fault province (Sanchez-Montes de Oca, 1979) corresponds to the Mountains of the North and the High Plains physiographic provinces. The first province occupies the northern part of the Sierra de Chiapas and forms a valley-and-ridge topography with west-northwest trends and elevations of from 100 to 2000 m. The High Plains form the central part of the Sierra de Chiapas and comprise large, block-shaped mountains and valleys with elevations ranging from 1000 to 2860 m (Tzontehuitz Volcano). The Strike-slip Fault province is formed by a set of upthrown and downthrown blocks, bounded by left-lateral strike-slip faults (Figures 2 and 3). Northwest-trending en-echelon anticlines with middle–Upper Cretaceous and Paleogene rocks along their crests are present in most of the upthrown blocks, whereas strike-slip basins with Cenozoic rocks formed where the strike-slip faults are divergent or present double bends. Stratigraphically, this province is characterized by local unconformities and by thickness changes or lithologic variations across structural trends, indicating syndepositional tectonism.

Based on the fault orientation, this province can be subdivided into three regions—a western area where the faults have variable orientations, a central area where the faults trend northwest, and an eastern area where the faults trend in a western direction (Figure 3). The width of this shear belt is 130 km, and its length is 360 km. In the western and central areas, there is a detachment level in the Callovian salt deposits. The eastern part contains two detachment levels, one at the Callovian salt deposits and another in Lower Cretaceous anhydrites (Cobán Formation) (Figures 2, 6, and 7). However, basement involvement of some strike-slip faults is indicated by the presence of Pliocene intrusives (Damon and Montesinos, 1978) and Pliocene-Quaternary volcanoes at the ends of some faults or at zones of maximum-extension-produced fault stepping (Chichonal, Tzontehuitz, Huitepec, and Navenchauc Volcanoes; Figures 2, 3, and 6).

Four main features indicate sinistral-slip movement in most of the faults in this province: (1) the en-echelon pattern of the folds (Figures 2 and 3); (2) the sense of thrusting that occurs where the Tecpatán-Ocosingo and Tenejapa faults change in direction from east to nearly south (Figures 3 and 5); (3) the zones of compression and extension produced where en-echelon faults form right and left steps, respectively (Figure 4); and (4) horizontal slickensides.

There is no consensus regarding the amount of left-lateral displacement along the strike-slip faults. Viniegra (1971) mentioned displacements on the order of 10 km, but he gave no evidence for this estimate. Sanchez-Montes de Oca (1979) estimated displacements along each fault on the order of hundreds of kilometers, based on deformation of the synclines caused by faulting. However, Meneses-Rocha (1985) estimated

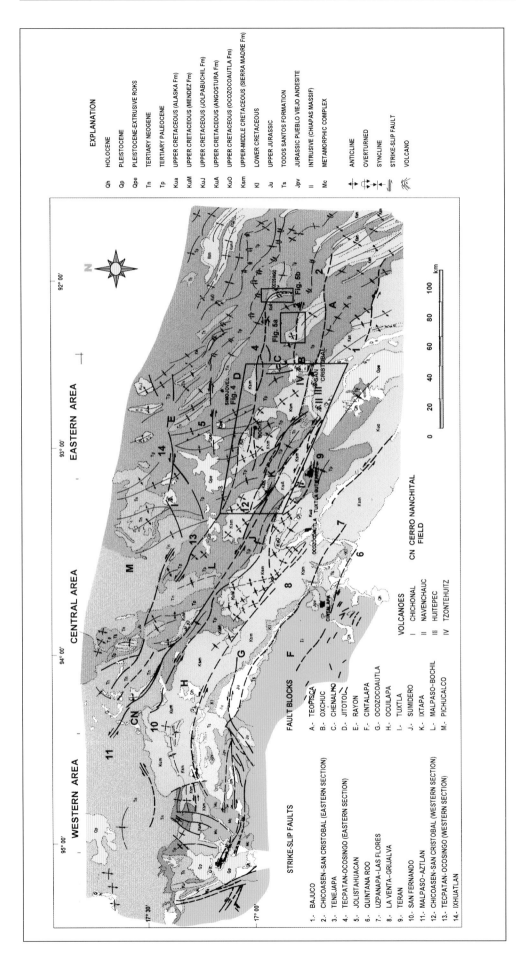

Figure 3. Simplified geologic map of the Strike-slip Fault province and of the northwestern part of the Sierra monocline, after Meneses-Rocha (1985, 1991). Locations of Figures 4 and 5 are outlined in solid black.

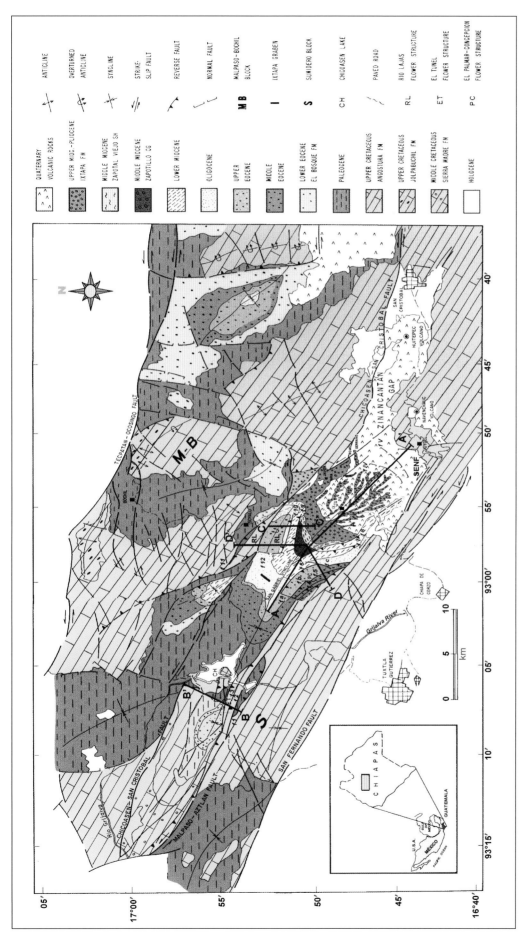

Figure 4. Geologic map of the Ixtapa graben and surrounding areas. Cross sections A-A′, B-B′, C-C′, and D-D′ are detailed in Figure 8.

4 and 5 km of left-lateral displacement along the Tecpatán-Ocosingo and Tenejapa faults, respectively (Figures 3 and 5). The first value was calculated by unfolding the Ocosingo anticline in a zone of compression formed where the fault changed direction from east to nearly south. The 5 km of left-lateral displacement along the Tenejapa fault was deduced from offsets of the axes of the Cancuc and Oxchuc anticlines, because both structures display similar stratigraphic units along their crests and flanks.

Reverse-fault Province

The Reverse-fault province is equivalent to the Yaxchilan and Miramar Tectonic provinces of Sanchez-Montes de Oca (1979). It occupies the eastern portion of the Sierra de Chiapas, extending from Ocosingo (near the Nazareth-51 well) to Río Usumacinta (Figure 2). Physiographically, this province corresponds to the Mountains of the East, which occupy the easternmost part of the Sierra de Chiapas known as Lacandona Jungle. The Mountains of the East consist of long, tightly folded mountains with northwest strike, separated by very narrow valleys. In general, these mountains decrease in elevation to the northeast from 2000 m to 500 m, as Río Usumacinta is approached. This province is an elongated structural trend, convex to the northeast, formed by long, narrow anticlines offset along their flanks by reverse faults that generally overthrust the intervening synclines. Box-shaped folds (detachment folds) or huge, asymmetric anticlines (fault propagation folds) without a common vergence are present in this province, as predicted by theoretical models of Davis and Engelder (1985) in areas where a fold-and-thrust belt rides atop a basal salt layer. Where the fold belt plunges toward the northwest and southeast, lower–middle Miocene beds crop out along the crests of the anticlines. Anticlines in the central portion of this province expose Upper Cretaceous or Paleocene–lower Eocene carbonates, whereas Tertiary terrigenous clastics crop out in the synclines. Throughout the province, jointing is widespread in the anticlines, striking approximately perpendicular to their axes. Normal faults divide the anticlines into upper and lower blocks. Other normal faults slightly displace the fold axes and the traces of some reverse faults. The folds of this province possess two principal detachment zones, one at the level of the Callovian salt deposits and another in the Upper Jurassic–middle Cretaceous anhydrites (Cobán Formation) (Figures 2, 6, and 7). Lack of angular unconformity in the sedimentary section of this province suggests that the only significant folding event is post-middle Miocene, because beds of that age form the youngest strata in the folds.

Sierra Monocline

The Sierra monocline corresponds physiographically to the Central Depresión, a huge plain broken by small hills through which the Río Grijalva flows northwest at an average altitude of 600 m. This province separates the Chiapas massif from the Strike-slip Fault and Reverse-fault provinces, and it can be subdivided into a northwestern region and a southeastern region

(Figure 2). The northwestern region is crossed by two northwest-trending faults (Quintana Roo and Uzpanapa–Las Flores Faults) that are vertical or dip very steeply to the southwest (Figure 3). The resulting structure resembles a huge staircase in which the upper step is the Cintalapa block and the lower step is the Ocozocoautla block (Figure 3). The Cintalapa block is a highstanding block, tilted southeastward. Granitic rocks of the Chiapas massif are exposed on its northwestern portion, and igneous, metamorphic, and sedimentary rocks ranging in age from Paleozoic to middle Cretaceous are exposed on its southeastern part. The Ocozocoautla block is a down-faulted and relatively undeformed block covered by a southeast-trending series of sedimentary rocks dipping approximately 15° to the northeast and ranging from Middle (?)–Late Jurassic to middle Cretaceous. This block is separated from the Strike-slip Fault province by the La Venta–Grijalva Fault. There is evidence for vertical motion along this fault and the Quintana Roo and Uzpanapa–Las Flores Faults during the Middle Jurassic.

The marked difference in deformation style between the two blocks adjacent to the La Venta–Grijalva Fault, as well as paleogeographic reconstructions, suggests that this fault coincides with the updip limit of the Callovian salt of Chiapas (Figures 3 and 6), implying that it must have evolved before and/or during deposition of salt. The en-echelon pattern displayed by the anticlines of the northern block is good evidence of left-lateral strike-slip movement of the La Venta–Grijalva Fault during the Tertiary.

The band of the Todos Santos Formation, which crops out on both sides of the Uzpanapa–Las Flores Fault, does not show any lateral displacement (Figure 3). In this region, the Todos Santos Formation is older than Oxfordian (i.e., pre-Late Jurassic).

Andesitic magma was intruded along the Uzpanapa–Las Flores Fault from the Callovian to the Oxfordian.

The Quintana Roo, Uzpanapa–Las Flores, and La Venta–Grijalva Faults follow the same northwest trend as two grabens filled with more than 1200 m of red beds in northwestern Guatemala (Burkart and Clemons, 1972).

In the southeastern region of the Sierra monocline, terrigenous clastics of the Todos Santos Formation and middle and Upper Cretaceous carbonates gently dip northeastward away from the Chiapas massif. On the basis of field data, Movarec (1983) and Burkart et al. (1989) proposed that the boundary in this region between the massif and the Mesozoic sedimentary strata is a northwest-oriented fault (the Concordia), which exhibits two periods of activity—an extensional phase of northeast-dipping, normal faulting in the Late Triassic–Early Jurassic, and a phase of high-angle, southwest-dipping, reverse faulting during the Laramide orogeny (Late Cretaceous–early Eocene).

Although no direct field evidence of high-angle reverse faulting has been reported in the northwestern region of the Sierra monocline (Cintalapa and Ocozocuautla blocks), structural data clearly suggest that this province evolved during two periods of deformation—a Late Triassic (?)–Middle Jurassic extensional phase, and a Laramide compressional phase.

Chicomuselo Anticlinorium

The Chicomuselo anticlinorium occupies about 2000 km² of the southeastern portion of the Sierra de Chiapas (Figure 2). The core is formed by Paleozoic rocks and the flanks by Mesozoic rocks. The Paleozoic rocks display two periods of deformation—a late Mississippian–early Pennsylvanian event responsible for metamorphosing Mississippian sediments, and a late Permian event that uplifted and folded the Permian as well as the Mississippian-Pennsylvanian rocks (Hernandez-García, 1973). This province is equivalent to the Comalapa anticlinorium of Burkart (1978, 1983) and Burkart et al. (1987), in which they also observed a Laramide deformation, because Campanian-Maestrichtian carbonates are part of the fold. Further study of this anticlinorium could provide important clues about the closure of the proto-Atlantic, as well as the position and orientation of the Yucatán block at the beginning of the Mesozoic.

Reforma-Akal Uplift

The Reforma-Akal uplift lies between the Comalcalco and Macuspana Basin, and corresponds onshore to the plunging nose of the Sierra de Chiapas (Figure 2). In this uplift, northeast compressional stresses, as well as flowage of a deeply buried horizon of the Callovian salt, gave rise to huge petroleum-bearing anticlines, faulted and often overturned, that comprise two of Mexico's most important producing provinces. They account for about 90% of the country's daily oil production. Onshore is the Chiapas-Tabasco petroleum province, where the main plays are in the Cretaceous–Upper Jurassic Kimmeridgian section, and where dolomitization and fracturing have enhanced reservoir quality. The offshore counterpart is known as the Sonda de Campeche, one of the richest petroleum provinces in the world. In this province, the fields are at depths of less than 100 m of water, and the plays are Kimmeridgian-Cretaceous dolomitized carbonates and Upper Cretaceous–Paleocene carbonate breccias, as well as Oxfordian sandstones associated with the normal faults crossing the northeastern part of the province (Angeles-Aquino et al., 1994). In both plays, halokinesis has played an important role in the resulting structural styles.

Gulf Coast Tertiary Basins of Southeastern Mexico

Isthmian Saline Basin

The Isthmian Saline Basin is the area of southeastern Mexico where the flowage of the Callovian salt (Isthmian Salt, equivalent in age to Louann Salt) was able to produce diapirs, domes, turtle structures, canopies, salt walls, and salt tongues. This basin lies in the northern part of the Isthmus of Tehuantepec, extending from the foothills of the Sierra de Chiapas into the continental shelf of the Gulf of Mexico (Figure 2). The western limit must lie somewhere between the San Andrés

Tuxtla massif (Figure 2) and the westernmost salt structures. Because of the abrupt termination of the occurrence of salt, it is believed that this limit could be a major fault (Viniegra, 1971; Sanchez-Ortiz, 1975; Meneses de Gyves, 1980). Eastward, the Isthmian Saline Basin is bounded by the Comalcalco Basin.

Salt mobilization probably started in the Early Cretaceous, as suggested by a Barremian-Campanian unconformity that can be observed in the southeastern margin of the basin (Cerro Pelón region). The onset of diapirism, however, began in the

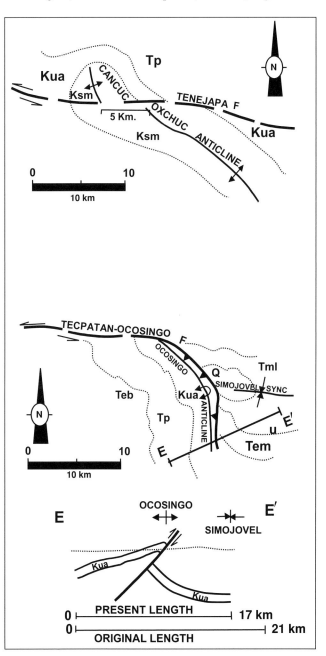

FIGURE 5. Maps and cross sections showing left-lateral displacements along the Tenejapa and Tecpatán-Ocosingo faults. See Figure 3 for location. Tp, Paleocene; Kua, Upper Cretaceous Angostura Formation; Ksm, middle Cretaceous Sierra Madre Formation; Tml, lower Miocene; Tem, middle–upper Eocene Lomut Formation; Teb, middle Eocene El Bosque Formation.

Oligocene and extended into the Quaternary. In the onshore part of the basin, the top of the salt masses is found at depths below sea level, ranging from 250 m in the southwest to 3000 m in the northeast.

Oil fields occur in Miocene sandstone wedges associated with elliptical and irregularly shaped salt masses. Subsalt play exploration in this petroleum province began in the mid-1970s, but because that activity then shifted toward the Mesozoic oil-bearing Chiapas-Tabasco area, this effort dropped to very low levels.

Comalcalco and Macuspana Basins

These basins are major zones of gravity tectonics lying between parallel listric down-to-basin normal faults that trend perpendicular to the northwest strike of the compressional structures within the Sierra de Chiapas and the Reforma-Akal uplift (Figure 2). These faults are synthetic and antithetic growth faults that sole out at the base of a middle Miocene shaly section. In these basins, the plays are Miocene-Pliocene sandstones related to rollover structures associated with secondary synthetic and antithetic normal faults. In the gas-prone Macuspana Basin, the faults are associated with middle Miocene shale diapirs and display inversion features. In the Comalcalco Basin, the detachment surface is a middle Miocene shaly horizon or, in some cases, the Callovian salt that flowed to produce this salt-withdrawal basin.

REGIONAL STRATIGRAPHY OF THE SIERRA DE CHIAPAS

Basement

Basement rocks in the Sierra de Chiapas are thought to be granitic and metamorphic, similar in age and composition to rocks found at the bottom of two wells—Cobo-301 in the Yucatán platform, and Villa Allende-1 in the Sierra de Chiapas—and to those rocks exposed in the Chiapas and Mixtequita massifs (Figure 2). Stratigraphic and structural information suggests that at least in the southern part of the Sierra de Chiapas, the basement formed a horst-and-graben topography during the early Mesozoic related to the rifting of the Gulf of Mexico Basin (Meneses-Rocha, 1985).

Jurassic

The oldest sedimentary rocks in the Sierra de Chiapas are nonmarine red beds corresponding to the Todos Santos Formation (Figure 7). This unit is well exposed on the Sierra monocline, mostly in a long, narrow band bordering the Chiapas massif (Figures 2 and 3). In the Strike-slip Fault province, this formation crops out along the crest of the Cerro Pelón anticline (Figure 3), and it has been drilled by four nonproductive wells—Sauzal-1, Raudales-1, Soyalo-1, and San Cristobal-1 (Figure 2). In the Reverse-fault province, the red beds were drilled by the nonproductive wells El Retiro-1 and Nazareth-51 (Figure 2).

From a regional view, the Todos Santos Formation must have accumulated contemporaneously with the Louann and Isthmian Salt deposits during the Middle Jurassic–Callovian (Salvador, 1987, 1991). In southeastern Mexico, these salt deposits are known as Isthmian Salt, and they exist in southern Veracruz, western and central Tabasco, and as far south as the neighborhood of Tuxtla Gutierrez and northern Guatemala (Figure 6). Northward, these deposits extend into the Gulf of Campeche, and they are connected to the Sigsbee knolls. In

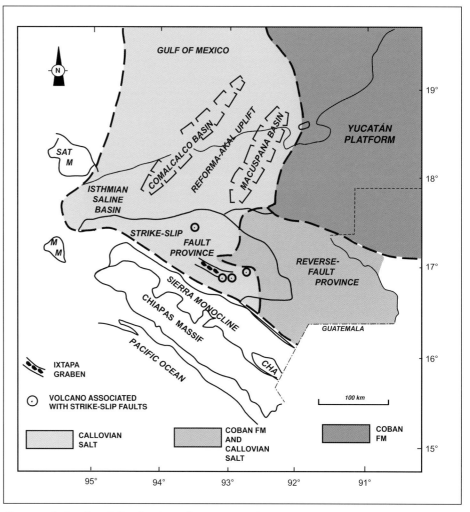

FIGURE 6. Regional distribution of evaporites that form zones of detachment in the Sierra de Chiapas. SATM, San Andrés Tuxtla Massif; MM, Mixtequita massif; CHA, Chicomuselo anticlinorium.

the Sierra de Chiapas, the limits of the salt are the northwest-trending edge of the Yucatán platform and the northwest-trending La Venta–Grijalva Fault, north of the Chiapas massif (Figures 2 and 6).

The stratigraphic position of the Todos Santos Formation and limited palinomorphic studies of this unit suggest that the red beds in the Sierra de Chiapas accumulated before, during, and after deposition of the salt. Support for a pre-Callovian age for the Todos Santos Formation is that in the southern part of the Reverse-fault province, the El Retiro-1 well drilled 52 m of red beds below 942 m of Callovian salt. Additional evidence is found in the area of La Concordia (Sierra monocline), where the lowest part of a section of red beds contains Bajocian-Bathonian palynomorphs (PEMEX internal geologic report). Information to suggest a Callovian age for the upper part of the Todos Santos Formation is the presence of the red beds below Oxfordian limestones and dolomites in the Raudales-1 well. A post-Callovian age for this formation is indicated by the presence of andesitic flows in the red beds in the northwestern part of the Sierra monocline (Figure 3). These andesites can be assigned to the Callovian-Hauterivian, based on seven isotopic dates reported by PEMEX geologists; however, the Kimmeridgian-Tithonian limestones contain fragments of andesites (Meneses-Rocha, 1977), which seems to favor a Callovian-Oxfordian age. Additional support for a post-Callovian age for the Todos Santos Formation is that the red beds in the Cerro Pelón anticline conformably underlie Kimmeridgian-Hauterivian limestones (Chinameca Formation), and in the wells Soyalo-1 and San Cristobal-1, the red beds are below Early Cretaceous strata.

The red beds deposited before the salt (i.e., pre-Callovian) probably accumulated as prograding alluvial fans and fluvial deposits in grabens or half grabens, bounded by active normal faults that formed during the early rifting phase of the formation of the Gulf of Mexico Basin. As stated earlier, these faults could have been the Quintana Roo, Uzpanapa–Las Flores, and La Venta–Grijalva Faults (Figure 3). Additional evidence to support a horst-and-graben topography during pre-Callovian time is that in the Villa Allende-1 well, the salt deposits overlie schists of late Mississippian age, and in the El Retiro-1 well, they overlie the Todos Santos Formation (Figure 2). In contrast to the presalt red beds, the deposition of the red beds during and after the salt probably was the product of erosion of the highstanding terrain that bounded the basin to the south.

During the Kimmeridgian-Tithonian, the Sierra de Chiapas was subject to a transgressive event that led to deposition of ooid-pellet sands (Zacatera group, Meneses-Rocha, 1977, 1990) in the western part of the Strike-slip Fault province. Limestones and evaporites formed in restricted and carbonate-shoal environments in the northwestern part of the Sierra monocline (San Ricardo Formation, Quezada-Muñeton, 1983; Blair, 1987). Northward, in the region of the Malpaso Fault (Figure 3), anhydrites and ooid limestones grade upward into shelfal and basinal facies (Malpaso Formation). Farther north, in the region of

Chinameca (Isthmus Saline Basin)–Cerro Pelón (foothills of the Sierra de Chiapas) (Figures 2 and 3), the Kimmeridgian-Tithonian consists of deep-water carbonates (Chinameca Formation). This lithostratigraphic unit is equivalent to the shaly limestones and calcareous shales that generated most of the hydrocarbons of the Chiapas-Tabasco and Sonda de Campeche provinces. During that time, red beds of the Todos Santos Formation still were accumulating on the southeastern part of the Sierra monocline and in the eastern parts of the Strike-slip Fault and the Reverse-fault provinces (Figures 2 and 7), as demonstrated by Nazareth-51 well, where a section of fine-grained red beds was drilled below a sequence of carbonates and evaporites of the Cobán Formation. Thus, it seems that a southeastward transgression and onlap took place across the region during the Late Jurassic.

Cretaceous

By Early Cretaceous, the sedimentologic conditions in the western part of the Sierra de Chiapas remained quite similar to those of the Late Jurassic (Figure 7). One significant change, however, took place in the northwestern part of the Sierra monocline, where the Lower Cretaceous reflects a progradation of terrigenous clastics (sandstone member of the San Ricardo Formation, Quezada-Muñeton, 1983). In contrast with the western part, the eastern part of the Sierra de Chiapas displays a major change in the pattern of sedimentation in the Early Cretaceous. Red beds of the Todos Santos Formation are succeeded by anhydrites and dolomites deposited in a restricted-evaporitic environment. These rocks are known as the Cobán Formation, and they reach a thickness of 450 m in the San Cristobal-1 well and 3620 m in the El Retiro-1 well, which could indicate that they were deposited across areas with different subsidence histories (i.e., different amounts of thinning), associated with Early and Middle Jurassic faults. According to Guzman-Vega and Mello (1999), this coastal sabkha environment was the setting of source-rock deposition that gave rise to the oils found in the Sierra de Chiapas and in the southern part of the Chiapas-Tabasco province (Artesa–Mundo Nuevo platform).

The middle Cretaceous (Albian-Cenomanian) is quite uniform throughout most of the Strike-slip Fault province and the Sierra monocline. It consists of a thick section of shelfal limestones and dolomites assigned to the Sierra Madre Formation. In some places, this unit ranges from Turonian to Santonian, equivalent in age to the deeper-water basinal limestones and cherts of the Jolpabuchil Formation, which are exposed in the northern portion of the central and eastern parts of the Strike-slip Fault province (Figure 7). In the Reverse-fault province, the Sierra Madre Formation ranges in age from Albian to Santonian and is mostly dolomitized (Figure 7).

In the Strike-slip Fault province, the Campanian is formed by four stratigraphic units that represent, from north to south, a basinal environment (Jolpabuchil and Méndez Formations), a carbonate shelf suitable for the development of rudists (Angostura Formation), and a siliciclastic shoreline (Ocozocoautla Formation). In contrast with the Campanian, the Maestrichtian is entirely calcareous in the southern region (Angostura Formation). In the westernmost part, the Méndez Formation is slightly metamorphosed (Alaska Formation). In the Reverse-fault province, the Maestrichtian consists of dolomites and limestones.

Surface and subsurface information indicates that in some locations of the Strike-slip Fault and Reverse-fault provinces, the Cretaceous stratigraphic sequence is continuous, although in others, an unconformity is indicated by the absence of strata ranging from Cenomanian to Campanian or from Hauterivian to Santonian (Figure 7).

Tertiary-Quaternary

In the Strike-slip Fault province, the Tertiary sequence consists mainly of terrigenous clastics, with limestone a minor component. Most of these rocks have been identified in PEMEX internal geologic reports with informal formation names. These sequences represent basinal, shelfal, and continental environments and are exposed mostly in structurally lower areas. They are characterized by local unconformities and by thickness changes or lithologic variations across structural trends. Pliocene-Quaternary rocks consist mainly of pyroclastic and andesitic rocks that are the product of the El Chichonal, Tzontehuitz, Huitepec, and Navenchauc Volcanoes, as well as of silicic intrusives in the southern part of the Pichucalco block (Santa Fe area) (Figures 2, 3, and 6).

In contrast with the Strike-slip Fault province, the Reverse-fault province presents a more uniform Tertiary section. Here, the Paleocene–lower Eocene is formed by dolomites and minor limestones. The middle and upper Eocene consist of an alternation of sandstones, dolomites, shales, and limestones. The lower Miocene unconformably lies (paraconformity) on the upper Eocene and consists of limestones (Macuspana Formation) and terrigenous clastics (Tulija Formation). In some places, the middle Miocene is present and consists of continental sandstones and shales.

THE IXTAPA GRABEN

General Description

The Ixtapa graben is located in the central part of the Strike-slip Fault province, bounded to the south by the Malpaso-Aztlán Fault and to the north by the Chicoasen–San Cristobal fault, which at its western end intersects at an acute angle a splay fault of the Malpaso-Aztlán Fault (Figure 3). The southeastern end of the graben is bounded by a mountain front formed by Cretaceous carbonate rocks intruded by the Navenchauc Volcano of Pliocene-Pleistocene age (Figure 4). The length of this graben is about 60 km, and its maximum width of about 10 km is in its southeastern end, from where it narrows to about 4 km to the northwest. In general, this graben is a southeast-plunging syncline in which successively younger beds dip toward the southeast. Along its limits, beds are upturned by reverse faults subparallel to the strike of the master faults. The Río Grijalva cuts across the center of the graben (Figures 4 and 9).

In map view, the southeastern part of the Ixtapa graben coalesces with a smaller, lozenge-shaped region 5 x 19 km in area, covered by gently dipping to horizontal volcaniclastic rocks that comprise the main record of the eruption of the Huitepec Volcano, at the center of the area (Figure 4). This region (here informally named Zinacantan area) is topographically higher than the Ixtapa graben and is bounded to the north by the eastern segment of the Chicoasen–San Cristobal Fault and to the south by the margin of the mountain front formed by Cretaceous carbonates.

Bounding Faults

Chicoasen–San Cristobal Fault

This fault limits the Ixtapa graben to the north with a trace characterized by prominent scarps. It extends for about 60 km with a N55°W strike, from the Chicoasen area to Colonia Romo Serrano, from where it follows an easterly trend for about 28 km in the study area (Figure 4).

In the study area, the Chicoasen–San Cristobal Fault is formed by two discontinuous strands, which produce an en-echelon pattern, and one wedge-shaped depression. The westernmost fault pattern is located north of Chicoasen Lake, where two strands of the Chicoasen–San Cristobal Fault overlap to the left 1 km and produce a depressed zone of crustal extension (Figures 4 and 9a). The width of this depressed zone is about 1.5 km, and the topographic relief is about 700 m below the surrounding mountains. The Río Grijalva flows northward along the axis of the depression and turns to the west at the end of the eastern strand of the Chicoasen–San Cristobal Fault. This strand cuts the southern side of a topographic high, which is a major outcrop of carbonate rocks. The western strand is marked by a fault-line scarp with the front face toward the downthrown side. In this region, short streams associated with small drainage areas originate in the highlands and flow transversely into the Río Grijalva. According to the theoretical models of pull-apart basin development proposed by Rodgers (1980), the Río Grijalva should follow a trend that corresponds to the strike of the deepest part of the basin and theoretically forms when the displacement of the master faults is twice their separation. Therefore, it is possible that the course of this stream in this region is controlled structurally.

The easternmost fault segment is located near Colonia Romo Serrano, where the northwest-trending segment of the Chicoasen–San Cristobal Fault forms an acute angle with the eastward segment, giving rise to the Romo Serrano sag pond (Figure 8b). In plan view, this depression is similar to the pitching grabens described by Lensen (1958).

The following features show that the Chicoasen–San Cristobal Fault has experienced left-lateral strike-slip motion: (1) horizontal and subhorizontal slickensides; (2) the tectonic landforms that form at the discontinuities of the fault (Figure 9); and (3) the en-echelon pattern and geometric relationship of the folds of the upthrown (northern) block of the fault and the fault itself (Figures 3 and 4).

Along its entire length, the upthrown block of the fault is on the north (Malpaso-Bochil block). This block exposes Cretaceous carbonate rocks along most of its surface. Toward the southeast, vertical displacement along the Chicoasen–San Cristobal increases; in that direction, the Ixtapa graben exposes successively younger strata and thus requires greater accommodation space (Figure 3). Physiographic features such as fault scarps, linear fault valleys, sag ponds, and hot springs indicate that the Chicoasen–San Cristobal Fault is still active.

Flower Structures

The central section of the Chicoasen–San Cristobal Fault cuts across a prominent hogback that parallels two prominent ridges. In cross section, these physiographic features form a "positive flower structure" (Río Lajas) made up of three upward-diverging fault-strands (faults 9, 10, and 11; Figures 4, 8c, and 8d). Upturned beds form the southern flank of the Río Lajas flower structure, because fault 9 juxtaposes lower Miocene beds against middle Miocene beds, and fault 10 juxtaposes Oligocene beds against lower Miocene beds (Figures 7c and d). The northern flank is made up of north-dipping Upper Cretaceous rocks that are juxtaposed against Paleocene beds by fault 11. This fault shows a reverse separation of about 250 m at the level of the base of the Upper Cretaceous. In map view, the axial trace of this flower structure, as well as the strike of the bedding, is parallel to the Chicoasen–San Cristobal Fault (Figure 4). Its northern flank is narrow, and its southern flank is overturned. This structure plunges to the east; to the west, it is truncated by a north-trending normal fault (fault 12), along which the downthrown block is to the west (Figure 4).

Malpaso-Aztlán Fault

This fault extends for about 54 km across rugged topography. In contrast with the Chicoasen–San Cristobal Fault, the Malpaso-Aztlán Fault displays an almost continuous trace. Again, the en-echelon arrangement of the folds located in the northern block (Figure 3) suggests that the Malpaso-Aztlán Fault underwent left-lateral motion. This is in agreement with fault-plane solutions determined by Guzmán-Speziale et al. (1989) in this region of Chiapas. Along its entire length, the upthrown block of the fault is to the south (Sumidero block). Because the Ixtapa graben exposes successively younger beds to the southeast, the vertical displacement along the Malpaso-Aztlán block increases in this direction.

Flower Structures

Along most of its length, the Malpaso-Aztlán Fault is nearly parallel to reverse faults that cut the Tertiary sequence of the Ixtapa graben. In cross section, this fault arrangement exhibits an upward-spreading pattern that can be described as a flower structure. Two main fault zones of this type develop along the southern flank of the Ixtapa graben—the El Tunel flower structure and the El Palmar–Concepción flower structure (Figures 4, 8b, and 8d).

The El Tunel flower structure is clearly distinct in the western section of the Malpaso-Aztlán Fault. It is made up of three diverging fault strands that in the field develop a large linear valley and two prominent scarps parallel to the Malpaso-Aztlán Fault (faults 1, 2, and 3 of Figure 8b). As shown in Figure 8b, the master wrench fault and the strand f_1 have a normal separation and bound a relatively downdropped syncline filled with Paleocene terrigenous clastics. Strand f_2 dips steeply toward the master fault and has a reverse separation. Strand f_3 also has a reverse separation, because it uplifts the Upper Cretaceous carbonates relative to the middle Eocene.

The El Palmar–Concepción flower structure is cut transversely by three faults trending north to northeast, which divide it into a western section (between fa and fb), a central section (between fb and c), and an eastern section (east of c) (Figure 4). The western and central sections are cut by four fault strands (f2, f4, f5, and f6) that are parallel to and dip steeply toward the Malpaso-Aztlán Fault. On both sections, hanging-wall transport is toward the northeast. Thus, faults 2, 4, and 5 carry Upper Cretaceous and middle Eocene beds over Upper Eocene beds (Figure 8d). In the eastern section, fault 6 carries upper Eocene beds over Oligocene and lower Miocene beds, whereas in the central section, this fault juxtaposes the upper Eocene against the lower Miocene. In both sections, reverse separation increases toward the northeast. The eastern section is made up of three fault strands (f6, f7, and f8) that dip toward the Malpaso-Aztlán Fault. Fault 6 juxtaposes almost-vertical upper Eocene beds against northeast-dipping Oligocene beds. Fault 7 carries Oligocene and lower Miocene over overturned lower Miocene beds. Fault 8 juxtaposes overturned lower Miocene beds against overturned middle Miocene beds. Thus, reverse separation increases toward the northeast. In the field, the fault slabs that make up this flower structure are clearly marked by prominent linear ridges, linear valleys, and hogbacks trending toward the northwest.

Southeastern Normal Fault

This fault limits the Ixtapa graben to the southeast. It appears as a fault-line scarp modified by mass wasting with a general height above the Ixtapa plain of about 500 m (SENF of Figure 4). The upthrown block is a mountain front formed of Cretaceous carbonates, and the hanging-wall block is formed of Pleistocene volcaniclastics that lie on the Ixtapa plain. Landslides on the front face probably represent recent activity. This fault parallels two tension fractures that cut the Pleistocene volcaniclastics (f iv and f v).

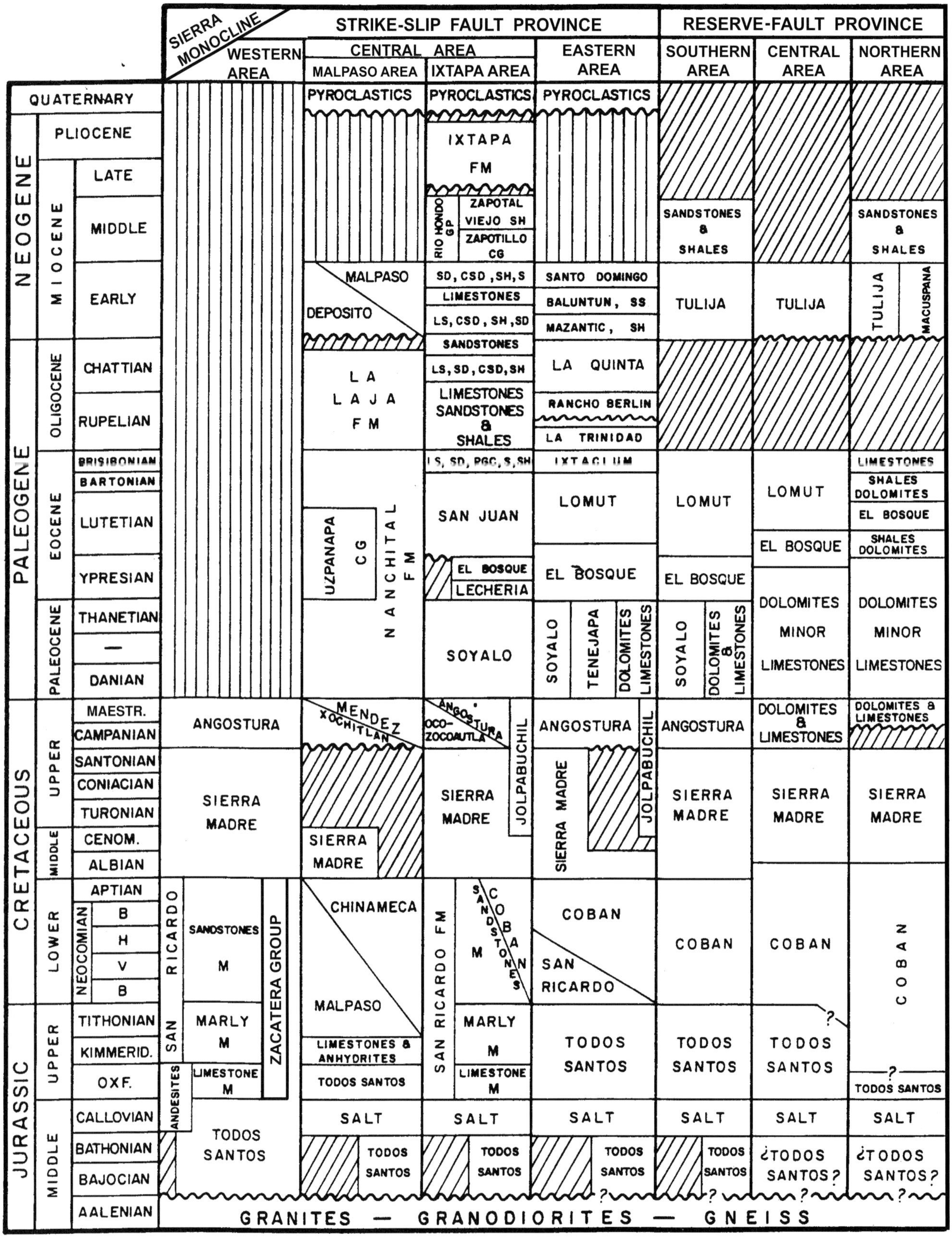

	SIERRA MONOCLINE	STRIKE-SLIP FAULT PROVINCE			RESERVE-FAULT PROVINCE		
		WESTERN AREA	CENTRAL AREA — MALPASO AREA	IXTAPA AREA	EASTERN AREA	SOUTHERN AREA	CENTRAL AREA	NORTHERN AREA

Time scale (left axis):

- QUATERNARY
- NEOGENE — PLIOCENE
- NEOGENE — MIOCENE: LATE / MIDDLE / EARLY
- PALEOGENE — OLIGOCENE: CHATTIAN / RUPELIAN
- PALEOGENE — EOCENE: PRIABONIAN / BARTONIAN / LUTETIAN / YPRESIAN
- PALEOGENE — PALEOCENE: THANETIAN / — / DANIAN
- CRETACEOUS — UPPER: MAESTR. / CAMPANIAN / SANTONIAN / CONIACIAN / TURONIAN
- CRETACEOUS — MIDDLE: CENOM. / ALBIAN
- CRETACEOUS — LOWER: APTIAN / NEOCOMIAN (B, H, V, B)
- JURASSIC — UPPER: TITHONIAN / KIMMERID. / OXF.
- JURASSIC — MIDDLE: CALLOVIAN / BATHONIAN / BAJOCIAN / AALENIAN

Chart units (selected labels, left to right):

- **Western Area:** SAN RICARDO — ZACATERA GROUP; SIERRA MADRE; SANDSTONES M; ANDESITES; MARLY M; LIMESTONE M; TODOS SANTOS
- **Malpaso Area:** PYROCLASTICS; MALPASO DEPOSITO; LA LAJA FM; UZPANAPA CG — NANCHITAL FM; ANGOSTURA; MENDEZ XOCHITLAN; SIERRA MADRE; CHINAMECA; MALPASO; LIMESTONES & ANHYDRITES; TODOS SANTOS; SALT; TODOS SANTOS
- **Ixtapa Area:** PYROCLASTICS; IXTAPA FM; RIO HONDO GP — ZAPOTAL VIEJO SH, ZAPOTILLO CG; SD, CSD, SH, S; LIMESTONES; LS, CSD, SH, SD; SANDSTONES; LS, SD, CSD, SH; LIMESTONES SANDSTONES & SHALES; LS, SD, PGC, S, SH; SAN JUAN; EL BOSQUE / LECHERIA; SOYALO; ANGOSTURA OCO-ZOCOAUTLA; SIERRA MADRE; SAN RICARDO FM — SANDSTONES M; MARLY M; LIMESTONE M; SALT; TODOS SANTOS
- **Eastern Area:** PYROCLASTICS; SANTO DOMINGO; BALUNTUN, SS; MAZANTIC, SH; LA QUINTA; RANCHO BERLIN; LA TRINIDAD; IXTAGUM; LOMUT; EL BOSQUE; SOYALO / TENEJAPA; DOLOMITES & LIMESTONES; ANGOSTURA; JOLPABUCHIL; SIERRA MADRE; COBAN; SAN RICARDO; TODOS SANTOS; SALT; TODOS SANTOS
- **Southern Area:** SANDSTONES & SHALES; TULIJA; LOMUT; EL BOSQUE; SOYALO; DOLOMITES & LIMESTONES; ANGOSTURA; SIERRA MADRE; COBAN; TODOS SANTOS; SALT; TODOS SANTOS
- **Central Area:** TULIJA; LOMUT; EL BOSQUE; DOLOMITES & LIMESTONES; MINOR LIMESTONES; DOLOMITES & LIMESTONES; SIERRA MADRE; COBAN; TODOS SANTOS; SALT; ¿TODOS SANTOS?
- **Northern Area:** SANDSTONES & SHALES; TULIJA; MACUSPANA; LIMESTONES; SHALES DOLOMITES; EL BOSQUE; SHALES DOLOMITES; DOLOMITES MINOR LIMESTONES; DOLOMITES & LIMESTONES; SIERRA MADRE; COBAN; TODOS SANTOS; SALT; ¿TODOS SANTOS?

GRANITES — GRANODIORITES — GNEISS

FIGURE 7. Regional Mesozoic-Cenozoic-Quaternary stratigraphy of the Sierra de Chiapas. This work follows the common practice in Mexico of using the term *middle Cretaceous* to include the Albian-Cenomanian stages.

Stratigraphy of the Ixtapa Graben

In the Ixtapa graben, successively younger beds dip to the southeast. They range, west to east, from middle Cretaceous to Pleistocene in age (Figures 4 and 8a). This measured sequence is 15,365 m thick and is divided into six units representing different depositional environments. The middle Eocene–early Pliocene sequence age dating was made on the basis of fossil content reported by Frost and Langenheim (1974) and by Daily and Durham (1966). The stratigraphy of the pre-Neogene sequence is summarized in Figure 10, the Neogene through Quaternary stratigraphy is shown in Figure 11, and a correlation with other stratigraphic units of the Sierra de Chiapas is shown in Figure 7.

Albian-Maestrichtian
(subtidal–middle shelf environment)

The Albian-Santonian crops out in the northwestern end of the Ixtapa graben, where it consists of dolomites and medium- to thick-bedded wackestones with scarce microfossils (Sierra Madre Formation). In this area, this unit was deposited in a subtidal environment and conformably underlies the Angostura Formation of Campanian-Maestrichtian age. The Angostura Formation consists of medium- to thick-bedded, light gray fossiliferous packstones with fragments of rudists, corals, algae, and gastropods that indicate a middle-shelf environment with zones suitable for the development of rudists. The complete thickness of the Cretaceous cannot be estimated in the graben area because of faulting. However, subsurface information indicates that the thickness is about 2000 m to the south and east of the graben.

Paleocene
(basinal-slope environment)

The Paleocene is represented by the Soyalo Formation, which conformably overlies the Angostura Formation and marks a phase of basin deepening. This unit consists of a lower unit, 450 m thick, composed of calcareous breccias, wackestone with planktonic foraminifers, and minor shales. The limestone beds are 30 to 150 cm thick and, in some intervals, display translational slides with partial development of clasts. The upper unit is 525 m thick and consists of thin-bedded siltstones and shales that in some levels form the matrix of large limestone clasts and carbonate breccias (debris flows) derived from an Upper Cretaceous platform.

Middle Eocene–Lower Miocene
(mixed siliciclastic–carbonate environment)

In the Ixtapa graben, an erosional contact with a basal pebble conglomerate marks a disconformity between the Paleocene and the middle Eocene. This disconformity is restricted to this graben and to the Copoya block located toward the south of the Ixtapa area. Northward of the Chicoasen–San Cristobal Fault, the Paleocene is overlain conformably by the lower Eocene. In the Ixtapa graben, the middle Eocene consists of a sequence of fossiliferous limestones whose middle part contains intercalated sandstones and calcareous sandstones. To the southeast, this sequence grades into sandstones, calcareous sandstones, pebble conglomerates, siltstones, and shales. The maximum thickness of this unit is 1350 m. These rocks were assigned to the San Juan Formation by Frost and Langenheim (1974).

The upper Eocene strata are similar to the middle Eocene in the Ixtapa graben; however, the higher proportion of carbonates, the lower proportion of pebble conglomerates, and the lesser thickness of the sequence (600 m) indicate a slower rate of sedimentation because of less subsidence of the graben area and less uplift of the southern clastic sources.

The Oligocene is 1495 m thick and consists of a lower unit (810 m) formed of fossiliferous limestones interbedded with thin beds of shales and sandstones; a middle unit (385 m) composed of an alternation of limestones, sandstones, calcareous sandstones, and shales, characterized by the presence of thin, coral biostromes; and an upper unit (300 m) that consists of sandstones.

The lower Miocene is 1500 m thick and consists of a lower unit (850 m) of limestones interbedded with calcareous sandstones, shales, and sandstones; a middle limestone unit (500 m); and an upper unit (150 m) composed of sandstones and siltstones.

Middle Miocene
(braided delta–brackish marine to
shallow-marine environments)

The middle Miocene is conformable with the lower Miocene and is made up of two formations that comprise the Río Hondo group—the Zapotillo Conglomerate below, and the Zapotal Viejo Shale above (Langenheim et al., 1965; Heuer, 1965). The Zapotillo Conglomerate is 800 m thick and marks the first input of coarse, terrigenous clastics into the Ixtapa graben. It consists of an alternation of conglomerates, sandstones, siltstones, and occasional limestones. The conglomerate beds are massive, with a generally sharp contact at the base. They are polymodal and clast to matrix supported. Clast size ranges from cobble to boulder; they are rounded to subrounded in shape and are composed of granites and gneisses, as well as quartzite and minor sandstone clasts. Imbrication is random, and normal grading is common. The sandstones are buff and yellowish, in beds ranging from 50 cm thick to massive. They consist of fine-grained, mature, and submature litharenite (volcanicarenite), and sublitharenite, occasionally calcareous and laminated. The gross composition of the sandstones and conglomerates, as well as the virtual absence of clasts of local derivation, suggests that this unit was deposited on a shallow-marine depression by a braided-delta system that flowed north from the Chiapas massif and onto a region underlain by rocks that provided abundant muddy and sandy material.

The Zapotal Viejo Shale is 690 m thick and consists of laminated siltstones that grade upward into fossiliferous shales

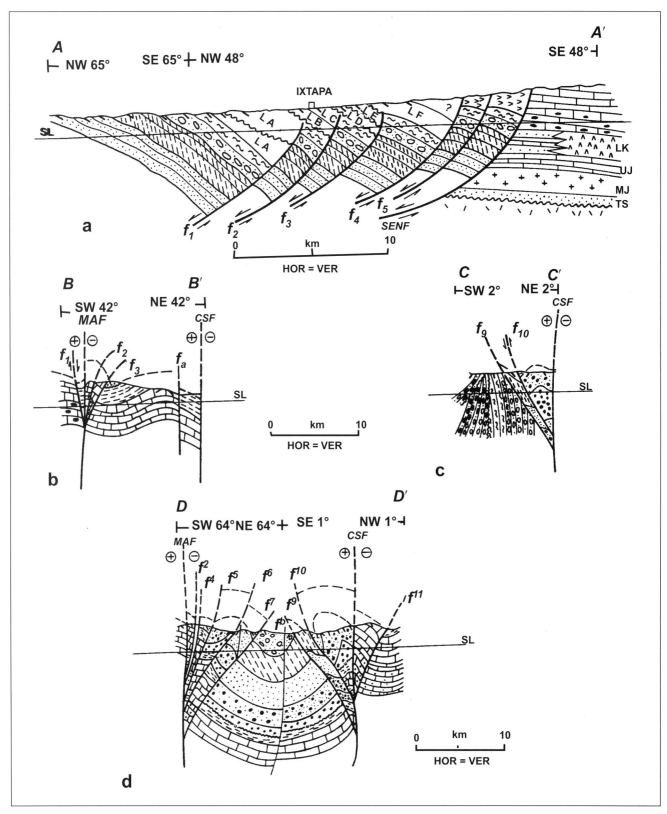

FIGURE 8 (a). Cross section along the axis of the southeastern part of the Ixtapa graben, showing internal unconformities within the Ixtapa Formation and block faulting caused by extension. LA, lithofacies association A; LB, lithofacies association B; LC, lithofacies association C; LD, lithofacies association D; LE, lithofacies association E; LF, lithofacies association F; LK, Lower Cretaceous (anhydrites, limestones, sandstones); UJ, Upper Jurassic; MJ, Middle Jurassic (salt); TS, Todos Santos Formation. See Figure 4 for location. (b), (c), (d). Cross sections showing flower structures that form along the flanks of the central part of the Ixtapa graben. Locations of cross sections are shown in Figure 4. Symbols are the same as in Figure 4.

with sideritic concretions, sparse desiccation cracks, and leaves. Based on the fossil content, Frost and Langenheim (1974) suggested that this unit was deposited in a mixed brackish-marine to shallow-marine environment.

Upper Miocene–Lower Pliocene (alluvial fan, fluvial, and lacustrine environments)

This unit is 6770 m thick and reflects a drastic change in sedimentation, because it consists entirely of continental conglomerates, sandstones, and siltstones. These rocks, which have been named the Ixtapa Formation by PEMEX geologists, unconformably overlie the middle Miocene sequence (Figures 4, 7a, and 10). This unconformity can be seen along the northern rim of the graben, where the Ixtapa Formation and the Río Hondo group form the overturned northern rim of the Río Lajas flower structure (Figures 4 and 8c). In this area, the Zapotal Viejo Shale dips 67° to the northwest, and the lowermost beds of the Ixtapa Formation are vertical. Rotating the overturned beds 90° clockwise would bring the lower beds of the Ixtapa Formation to their original position and would overturn the Río Hondo group. Thus, it can be presumed that this group was inclined about 23° at the time of deposition of the lower beds of the Ixtapa Formation.

Meneses-Rocha (1991) divided the Ixtapa Formation into six lithofacies associations, designated according to the lithofacies system of Miall (1978) and of Rust (1978). By considering the paleontological data reported by Daily and Durham (1966), the two lower lithofacies associations (A and B) can be assigned a late Miocene age, whereas the four upper lithofacies associations (C, D, E, and F) can be assigned to the latest Miocene and early Pliocene (Figure 10). The upper 970 m of this section cannot provide substantial insight into the sedi-

mentary history of the Ixtapa graben, because it is either poorly exposed or covered by Quaternary volcanic rocks (Figure 8a).

In the Ixtapa Formation, clasts of the conglomeratic beds were not studied in detail. The entire section has a similar composition, with only a small variation in lithofacies associations C and D, where limestone clasts constitute approximately 5%. In general, granite was the dominant clast type (50%), followed by gneisses (23%), sandstones (quartzite) (19%), and volcanic rocks (8%). Sandstone beds are mostly sublitharenites with subrounded grains. Rock fragments include volcanics and limestones. Detrital minerals include micas, magnetite, and hematite. Volcanic-rock fragments are especially abundant in lithofacies association C, which indicates contemporaneous volcanism.

The proportion of conglomerate beds decreases upward, from 76% (lithofacies association A) to 47% (lithofacies association F). Imbrication of clasts and tabular cross-beds is to the northwest. Bidirectional indicators, such as the long axes of ripple marks, also are toward the northwest.

The assemblage of clasts, sandstone composition, and sparse paleocurrent indicators suggests a source terrane located to the southeast and composed of acidic plutonic (granite), metamorphic (gneiss), and volcanic rocks. On the other hand, the roundness of the clasts and the abundance of resistant granitic clasts might indicate transportation over significant distances or fluvial gradients. In contrast, the great thickness of the conglomerates and the alluvial fan environment of lithofacies associations A and B suggest areas of adjacent high relief. Therefore, the roundness and the abundance of resistant clasts also might be interpreted in terms of recycling.

Field observations seem to suggest short transport of sedi-

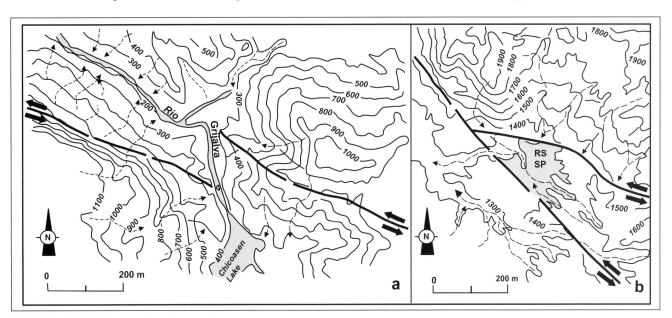

Figure 9 (a). Topographic map of the Chicoasen area showing the left-stepping overlap of two strands of the Chicoasen–San Cristobal Fault. Contour interval is 100 m. (b). Topographic map of the Romo Serrano area showing two segments of the Chicoasen–San Cristobal Fault that form a sag pond (RS SP). Contour interval is 100 m. Both maps are drawn from topographic map of DETENAL (1984).

ments from a southeastern source, rather than transport from the Chiapas massif, 50 km southwestward. The southeastern source, then, must have exposed either silicic plutonic, metamorphic, volcanic, and sedimentary rocks or conglomerates with clast composition similar to those of the Ixtapa Formation. The close match between the clast compositions of the Ixtapa Formation and those of the underlying Zapotillo Conglomerate seems to favor the second possibility. On this basis, it is proposed that after uplift, the Zapotillo Conglomerate and the overlying Zapotal Viejo Shale were eroded rapidly and provided the clasts, silt, and mud to make up the Ixtapa Formation. In addition, it is possible that southern terranes underlain by Mesozoic coarse-terrigenous clastics (e.g., Todos Santos and Ocozocoautla Formations) also have supplied a significant amount of debris.

Uppermost Pliocene–Pleistocene (alluvial [?]-lacustrine environment)

This sequence reaches a maximum thickness of 1200 m and rests with angular unconformity on the Ixtapa Formation. It consists of pyroclastic-fall deposits and pyroclastic deposits that probably were accumulated in an alluvial-lacustrine environment. These rocks form an extensive plain in the southeastern part of the graben and record the eruption of the Navenchauc and Huitepec Volcanoes. According to Damon and Montesinos (1978), the volcanic rocks exposed around the Navenchauc Volcano are hornblende andesites with middle to late Pleistocene K-Ar ages. The extrusive rocks of the Huitepec Volcano were not studied by these authors; however, hornblende andesites related to this volcano have been reported in unpublished PEMEX reports.

REGIONAL PALEOGEOGRAPHY AND TECTONIC DEVELOPMENT

The following tectonic events are inferred from the analysis of the stratigraphic and sedimentologic pattern of the Ixtapa graben, in light of the regional geology of southeastern Mexico. Numerical dates are after Palmer (1983).

Albian (regional tectonic quiescence) 113–97.5 Ma

During the Albian, much of central and southern Chiapas was a wide, shallow-water platform on which thick sequences of carbonates were deposited (Sierra Madre Formation) (Figure 12a). These deposits buried the Chiapas massif and cut off the source of terrigenous detritus that had reached the area during the Neocomian and Aptian (San Ricardo Formation). Evapor-

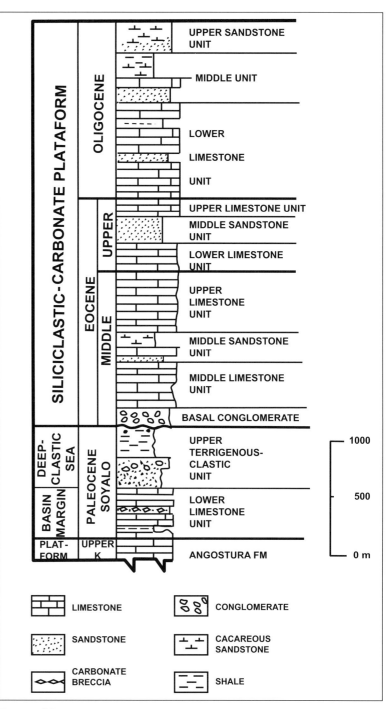

FIGURE 10. General stratigraphy of the Upper Cretaceous–Paleogene in the Ixtapa graben.

ites were restricted to an area that extended from northern Chiapas to the Yucatán Peninsula, and thin-bedded carbonates were deposited in a basinal environment that bordered the platform to the north and northwest (Jolpabuchil Formation). To the north, these deep-water deposits separated the extensive carbonate Sierra Madre platform from the Artesa–Mundo Nuevo isolated carbonate platform (Varela-Santamaria, 1995) that form the southern part of the Chiapas-Tabasco petroleum province. Well and seismic data indicate that by then, the

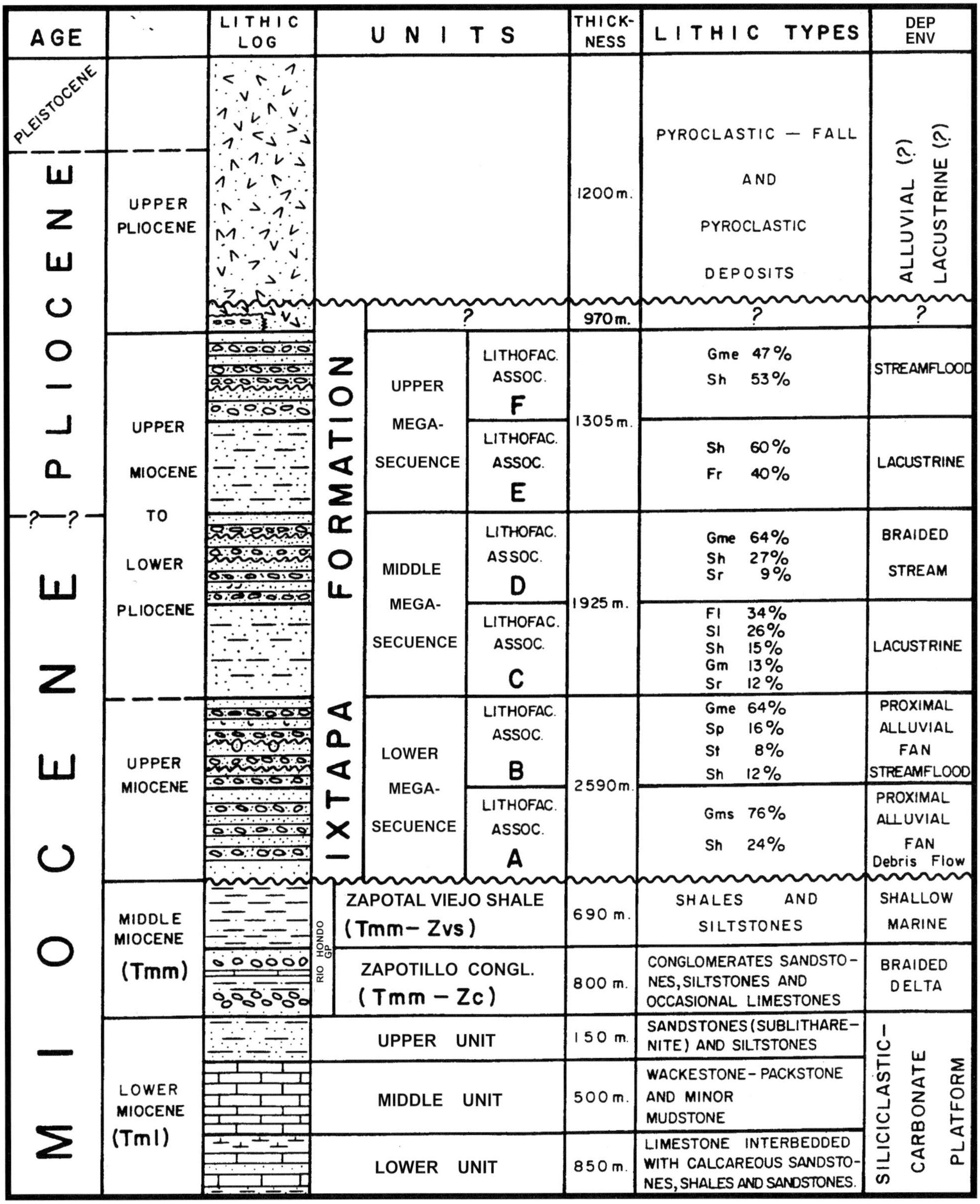

AGE		LITHIC LOG	UNITS			THICK-NESS	LITHIC TYPES	DEP ENV
PLEISTOCENE						1200 m.	PYROCLASTIC — FALL AND PYROCLASTIC DEPOSITS	ALLUVIAL (?) LACUSTRINE (?)
	UPPER PLIOCENE							
				?		970 m.	?	?
P L I O C E N E	UPPER MIOCENE TO LOWER PLIOCENE		I X T A P A F O R M A T I O N	UPPER MEGA-SECUENCE	LITHOFAC. ASSOC. F	1305 m.	Gme 47% Sh 53%	STREAMFLOOD
					LITHOFAC. ASSOC. E		Sh 60% Fr 40%	LACUSTRINE
				MIDDLE MEGA-SECUENCE	LITHOFAC. ASSOC. D	1925 m.	Gme 64% Sh 27% Sr 9%	BRAIDED STREAM
					LITHOFAC. ASSOC. C		Fl 34% Sl 26% Sh 15% Gm 13% Sr 12%	LACUSTRINE
M I O C E N E	UPPER MIOCENE			LOWER MEGA-SECUENCE	LITHOFAC. ASSOC. B	2590 m.	Gme 64% Sp 16% St 8% Sh 12%	PROXIMAL ALLUVIAL FAN STREAMFLOOD
					LITHOFAC. ASSOC. A		Gms 76% Sh 24%	PROXIMAL ALLUVIAL FAN Debris Flow
	MIDDLE MIOCENE (Tmm)		RIO HONDO GP	ZAPOTAL VIEJO SHALE (Tmm– Zvs)		690 m.	SHALES AND SILTSTONES	SHALLOW MARINE
				ZAPOTILLO CONGL. (Tmm – Zc)		800 m.	CONGLOMERATES SANDSTO-NES, SILTSTONES AND OCCASIONAL LIMESTONES	BRAIDED DELTA
	LOWER MIOCENE (Tml)			UPPER UNIT		150 m.	SANDSTONES (SUBLITHARE-NITE) AND SILTSTONES	SILICICLASTIC – CARBONATE PLATFORM
				MIDDLE UNIT		500 m.	WACKESTONE– PACKSTONE AND MINOR MUDSTONE	
				LOWER UNIT		850 m.	LIMESTONE INTERBEDDED WITH CALCAREOUS SANDSTO-NES, SHALES AND SANDSTONES.	

FIGURE 11. Lower Miocene–Pleistocene stratigraphy of the Ixtapa graben. Lithofacies code is from Miall (1978) and Rust (1978). Gms, massive or horizontally bedded conglomerate; Gme, massive or horizontally bedded conglomerate, commonly with basal erosional surfaces; Gm, massive, clast-supported cobble, pebble conglomerate; Sh, horizontally stratified sandstone; Sp, planar cross-laminated sandstone; St, trough cross-bedded sandstone; Sl, graded, finely laminated sandstone; Sr, ripple cross-laminated sandstone; Fl, laminated, very fine sandstone, siltstone, and mudstone; Fr, ripple cross-laminated siltstone.

granitic basement of the Chiapas-Yucatán block was exposed only in the present area of the Gulf of Tehuantepec (Sanchez-Barreda, 1981; Pedrazzini et al., 1982). The isotopic record of igneous activity in Chiapas indicates that magmatic activity ceased during the middle Cretaceous (Albian-Cenomanian). This lack of magmatism and the absence of any terrigenous input reflect a period of tectonic quiescence. The only tectonic disturbance took place in the region of Cerro Pelón, where early movements of the Middle Jurassic salt might account for the absence of the Albian and Cenomanian strata and the entire Lower Cretaceous sequence.

Cenomanian-Santonian (regional mild tectonism) 97.5–84 Ma

During the Cenomanian-Santonian, basinal deposition, including shaly sediments, extended farther to the east and southeast (Figure 12b). Areas of limited or no deposition extended into the carbonate platform. Evidence for this hiatus in deposition is the unconformities cutting out the Cenomanian and Santonian in some areas of the Sierra de Chiapas, as well as in the Artesa–Mundo Nuevo platform (Varela-Santamaria, 1995). The uneven distribution of these unconformities, as well as the beginning of coarse, terrigenous clastic deposition in the Gulf of Tehuantepec (Sanchez-Barreda, 1981; Pedrazzini et al., 1982), suggests that incipient tectonic movements were taking place by then that resulted in the growth of structural features and the partial uplift of the Chiapas massif. It is possible, however, that a drop in sea level during the mid-Cenomanian combined with these movements to produce the break in sedimentation and the development of karsted surfaces in the exposed areas. This process played an important role in the Albian reservoir rocks of the Artesa–Mundo Nuevo platform. These incipient tectonic instabilities could have been associated genetically with the inception of the collision of the Maya block against the Greater Antilles arc (Pindell and Dewey, 1982) or against the Honduras-Nicaragua block, also known as the Chortis block (Perfit and Heezen, 1978; Horne, 1989), along the Motagua Fault.

Campanian (onset of regional compressional regime) 84–74.5 Ma

Further uplift and emergence of the Chiapas massif took place during the Campanian, resulting in the accumulation of coarse, terrigenous detritus in a shoreline environment (Ocozocoautla Formation) (Figure 12c). Some parts of the platform remained exposed, and carbonate breccias with clasts derived from the platform were transported to the north (Xochitlan Formation, Quezada-Muñetón, 1987). According to Burkart et al. (1989): "the Ocozocoautla depositional basin was separated from an emerging island arc terrane by the Concordia fault zone and perhaps other high angle reverse fault systems." That the northern edge of the platform margin coincides with the trace of the Tecpatán-Ocosingo Fault could suggest that by this time, movement along this fault caused the necessary differen-

tial elevation to produce the erosion of the platform edge. It can be presumed that the Tecpatán-Ocosingo Fault and the high-angle reverse faults proposed by Movarec (1983) and by Burkart et al. (1989) (La Concordia fault zone) might have developed by reactivation of weakness lines produced during the normal block faulting that this region underwent during the rifting stage of the formation of the Gulf of Mexico Basin.

Rudist-bearing limestones remained common at the western margin of the platform (e.g., Ixtapa area), and lagoonal, intertidal, and supratidal facies extended into the eastern part. Basinward, carbonate breccia intertongued with two lithofacies—to the east, with the deep-water carbonates containing chert of the Jolpabuchil Formation, and to the west, with the basinal terrigenous fine clastics of the Méndez Formation. This last unit reflects the first great input of terrigenous clastics into a basinal environment during the Mesozoic, and it is equivalent in time and lithology to the flysch wedge deposited in northeastern and eastern Mexico during the early phase of the Laramide orogeny. This period of tectonism probably also affected the present area of the Gulf of Tehuantepec because, according to Pedrazzini et al. (1982), a paleontologic break is observed in this region between the lower Santonian and the Campanian.

The isotopic record of the Chiapas massif does not indicate magmatic activity during the Campanian, just as none was indicated during the Turonian-Santonian. However, evidence for magmatic activity during the Campanian could be the volcanic clasts observed in the Upper Cretaceous conglomerates drilled by the Salina Cruz No. 1 well and the calc-alkaline igneous clasts and volcanic ash reported by Burkart et al. (1989) in the Ocozocoautla Formation. This evidence is partly supported by the work of Carfantan (1981), who reported a phase of granitic intrusions (96 ± 10 Ma, 108 ± 10 Ma) and low-grade metamorphism (82.5 Ma) in the southern region of the Isthmus of Tehuantepec. In general, the Turonian-Santonian episode of regional mild tectonism and the Campanian episode of high-angle reverse faulting are contemporaneous but probably not genetically related to the last pulses of the Sevier orogeny of western North America (Armstrong, 1968).

In light of plate tectonics, the Cenomanian-Campanian tectonic events of Chiapas seem consistent with the closure of a subduction zone (arc-continent collision) in southern Guatemala. In addition to explaining an episode of high-angle reverse faulting in the southern part of southeastern Mexico, this regime may also account for the depositional pattern of the Ocozocoautla Formation and the synchronicity of the end of the compressional event and the emplacement of ultramaphic rocks along the Motagua fault zone of Guatemala (Dengo and Bohnenberger, 1969; Wilson, 1974; Williams, 1975; Rosenfeld, 1989). Thus, the regional geology of Chiapas may be compatible with the plate tectonic reconstructions that suggest a Late Cretaceous collision of the Maya block against the Greater Antilles arc (Pindell and Dewey, 1982) or against the Honduras-Nicaragua block (Perfit and Heezen, 1978), along the Motagua fault zone.

Maestrichtian (tectonic quiescence) 74.5–66.4 Ma

During the Maestrichtian, the previously exposed regions of the platform were covered by shallow-marine waters (Figure 12d). The Méndez and Jolpabuchil Formations continued to be deposited in a basinal environment that bordered the platform to the north and northwest. Rudist-bearing limestones (Angostura Formation) were common and formed scattered bioherms on the western part of the platform (e.g., Ixtapa area), and intertidal and supratidal facies extended southeastward. By that time, magmatic activity and the accumulation of a carbonate breccia at the margin of the platform started to decrease. This period of regional subsidence also affected the present area of the Gulf of Tehuantepec (Pedrazzini et al., 1982) and the area of the Concordia fault zone; in those areas, deep-water terrigenous clastics were accumulated (Quezada-Muñeton, 1987).

Paleocene (differential subsidence) 66.4–57.8 Ma

Sedimentation in the Sierra de Chiapas changed markedly during the Paleocene, when the long-established carbonate platform was flooded by flyschlike deposits (Soyalo Formation) and its central part underwent significant subsidence. To the east, a carbonate platform persisted (Figure 12e).

Debris flows seemed to have been confined to the present area of the Ixtapa graben at the end of the Paleocene, while turbidity-current deposits accumulated over the rest of the basinal area. This can indicate that debris flows were deposited in a local depression. This depression probably was formed by precursors to the present Ixtapa graben and the Achiote syncline; crustal instabilities along their margins were produced as the bounding highs started to emerge. It is possible that these unstable margins might have been controlled by vertical movements along the San Fernando and Chicoasen–San Cristobal Faults. Apparent differential subsidence also occurred along the central margin of the carbonate platform, as is indicated by a Paleocene basal breccia present in the Chenalho and Teopisca blocks, but in the intervening Oxchuc block, the breccia is not present (Figures 3 and 12e).

Even though no detailed paleogeographic study has been done on the turbiditic facies of the Soyalo Formation, the general lithologic aspect of this unit (litharenite) seems to indicate that it was produced mainly by the progressive erosion of Paleozoic sedimentary and metamorphic rocks exposed in southeastern Chiapas and western Guatemala (Santa Rosa and Chuacus groups, Dengo, 1968; Kesler et al., 1970, and Kesler, 1973). If this is the case, the Chortis block during that time was the southeastern extension of the continental margin of Chiapas.

Differential subsidence also took place in the Gulf of Tehuantepec. According to Sanchez-Barreda (1981), in the

northern part of this area, the "uplift of a possible basement block is recorded by the thinning of the Paleocene sediments in the direction of this feature." From a regional perspective, this phase of differential subsidence and flysch sedimentation in Chiapas marks the inception of a foreland basin that heralds the Laramide orogeny in southeastern Mexico. This event is contemporaneous with, although different in nature from, the onset of the Laramide orogeny in the North American Cordillera (Coney, 1972) and in the northern and central segments of the Sierra Madre Oriental (Tardy, 1980; Padilla y Sanchez, 1982; Suter, 1984).

Earliest Eocene (differential subsidence) 57.8–55 Ma

During the earliest Eocene, the paleogeography of Chiapas remained essentially the same as during the Paleocene, with the exception of the Soyalo and Copoya areas (Figures 2 and 12f). In those places, the distribution of the Lecheria Limestone indicates that a period of tectonic quiescence allowed the reestablishment of isolated carbonate banks on the older structural highs. The carbonate platform at that time extended westward and was dissected in that direction by the Ixtapa-Achiote depression, where turbidity currents continued to deposit sand-clay-sized material.

End of Early Eocene (compressional regime) 55–52 Ma

At the end of the early Eocene, progressive uplift of granitic intrusives (Chiapas massif), metamorphic terranes (Chuacus groups), and pre-Eocene sedimentary rocks located along the southern margin of Chiapas (Ixtapa-Copoya-Chicomuselo areas) was accompanied by deposition of continental and shoreline sediments (Figure 13a). Regression of the sea toward the northeast resulted in the continued offlap of terrigenous clastics rich in unstable minerals (micas, El Bosque Formation) across Paleocene and Cretaceous sediments and deposition of shoreline sequences (Lomut Formation) on the eastern part of the Sierra de Chiapas. These events mark an episode of tectonism dominated by vertical uplift in the hinterland, during which flyschlike deposits (Soyalo Formation) and carbonates were succeeded by molasselike deposits.

Evidence for vertical uplift in the hinterland consists of: (1) uplift of Cretaceous-Paleocene deposits "prior to deposition of the onlapping Eocene sequence" in the Gulf of Tehuantepec (Sanchez-Barreda, 1981); (2) monoclinal flexure (La Concordia fault system) that took place along the boundary between the Sierra monocline and the Chiapas massif (Movarec, 1983); and (3) the uneven distribution of an early Eocene disconformity across block-bounding faults of the south-central part of the Strike-slip Fault province. In the Malpaso-Bochil block (Soyalo area) and in the southeastern part of the Ocuilapa block, the Paleocene-Eocene sequence is continuous, whereas in the present area of the Ixtapa graben and in the Copoya block, the relation between these sequences is disconformable. In the Ixtapa

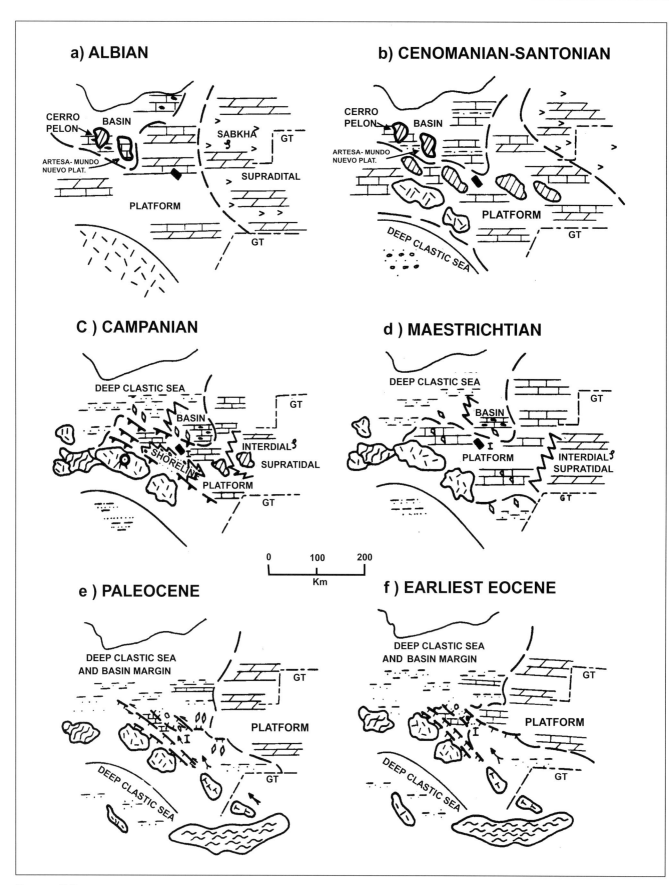

FIGURE 12. Paleotectonic maps of the Sierra de Chiapas, southeastern Mexico: (a) Albian; (b) Cenomanian-Santonian; (c) Campanian; (d) Maestrichtian; (e) Paleocene; and (f) earliest Eocene. Ixtapa graben location (I) and present Gulf of Mexico and Pacific Ocean borders are also shown. Symbols are the same as in Figure 13.

graben, the disconformity is marked by a quartz-bearing conglomerate at the base.

On the other hand, the transitional contact between Paleocene–early Eocene carbonates and the middle Eocene shoreline facies of the El Bosque Formation suggests tectonic quiescence in the northeastern part of the Sierra de Chiapas (Reverse-fault province). Northward, in the Reforma-Akal uplift and in the Campeche Sound, structural and stratigraphic evidence of the Laramide orogeny is lacking.

This period of tectonism represents the onset of the Laramide orogeny in Chiapas. In Chiapas, the most intense activity of this orogeny took place a little later than in the North America Cordillera (Coney, 1972) and the northern and central segments of the Sierra Madre Oriental (Tardy, 1980; Padilla y Sanchez, 1982; Suter, 1984), but earlier than in the Veracruz Basin (middle Eocene) of eastern Mexico (Mossman and Viniegra, 1976). The Laramide orogeny in southeastern Mexico presents the following characteristics: (1) It was not of regional extent; (2) it took place in an area where the Callovian salt deposits pinch out or are absent and where the Lower Cretaceous evaporites (Cobán Formation) did not accumulate (Figure 6); (3) it produced high-angle reverse faulting in the southern part of the Strike-slip Fault province and monoclinal flexure (Sierra monocline) between the sedimentary veneer and its basement (i.e., Chiapas massif) (Movarec, 1983); (4) the high-angle reverse faulting produced local disconformities; and (5) igneous activity was absent.

Middle Eocene (compressional regime) 52–40 Ma

During the middle Eocene, continued compression in the southern margin of the Sierra de Chiapas forced footwall blocks down, giving rise to a transgression of the sea from the northeast (Figure 13b). A deep, clastic sea to the northwest was bounded by a subsiding siliciclastic-carbonate platform farther to the southeast. A fluvial system farther to the southeast (El Bosque Formation) supplied terrigenous clastics to the platform. Along the southern margin of the deep, clastic sea, an unstable slope, probably resulting from reverse movement along the Tecpatán-Ocosingo and Malpaso-Aztlán Faults, continued to produce coarse-grained turbidites and conglomerate wedges (Uzpanapa Conglomerate) (Figure 13b). In the Gulf of Tehuantepec area, Eocene sediments onlap the older uplifted area, and uplift of a possible basement block took place in its northwestern part (Sanchez-Barreda, 1981). By middle Eocene, the present area of the Ixtapa graben occupied the southwestern part of the platform. In this area, the predominant lithology and microfauna of the middle Eocene section suggest the presence of a shallow-marine environment subject to high rates of terrigenous clastic influx that probably gave rise to low salinities. The nature of clastic influx into the Ixtapa area seems to indicate that the streams that brought these sediments flowed from the southeast, where the Chiapas massif and Mesozoic-Cenozoic sedimentary rocks were being eroded.

In the Ixtapa area, recycling and a multiple source for these sediments are evidenced by the roundness of pebbles and cobbles, the poor sorting, and the lack of correlation between roundness and size shown by the sandstone beds. The composition of some sandstone beds (calclithites), as well as the high rate of subsidence indicated by a sediment accumulation rate on the order of 161 m/my, suggests that intermittent tectonism in the source areas could account for the high rates of terrigenous clastic influx in the Ixtapa area.

Late Eocene (differential subsidence regime; block faulting) 40–36.6 Ma

After the Laramide orogeny, from late Eocene to early Miocene, the Strike-slip Fault province was subject to alternating up and down movements of fault blocks. During the late Eocene, east- and northwest-trending tectonic features began to develop, producing a zone of differential subsidence in the central part of this province (Figure 13c). The deep, clastic sea extended east to the present area of the Simojovel syncline, which may have undergone subsidence by downbuckling of the platform caused by vertical movement along the Tecpatán-Ocosingo and Ixhuatlan Faults. This is indicated by differences in lithology and depositional environments shown by the upper Eocene sequence of the Simojovel syncline in contrast with the equivalent sequences exposed to the north, east, and south of this depression. In the Simojovel syncline, the upper Eocene consists of shales and siltstones with a deep-marine fauna (Ixtaclum shale, Frost and Langenheim, 1974), whereas in the surrounding areas, the upper Eocene is made up of limestone and terrigenous clastics with a shallow-marine fauna.

By late Eocene, most of the deep, clastic sea occupied essentially the same area as during the middle Eocene, but the deposition of conglomerates (Uzpanapa Conglomerate) along its southern margin had ceased. In the southeastern part of the Sierra de Chiapas, the absence of upper Eocene–Oligocene rocks demonstrates that this region underwent a period of nondeposition or erosion. In the area now occupied by the Ixtapa graben, sediments continued to be deposited in a shallow-marine environment. In contrast with the middle Eocene, however, the rates of sedimentation and subsidence there, as well as the rate of uplift of the southeastern sources of terrigenous clastics, seem minor, as indicated by the lower proportion of pebble conglomerates and the reduced thickness of the upper Eocene. On the other hand, the content of a volcanic source during that time is indicated by the content of sanidine and microcline in the terrigenous fraction of some limestone beds.

Oligocene (differential subsidence regime; block faulting) 36.6–23.7 Ma

During the Oligocene, vertical movements in the central part of the Strike-slip Fault province were contemporaneous with a lowering of sea level, resulting in the narrowing of the platform (Figure 13d). This lowering of sea level did not affect the deep, clastic sea, because it still received sand- and silt-sized material and occupied essentially the same area as during the late Eocene. During the Oligocene, magmatic activity resumed in the Chiapas massif for the first time since Campanian times.

In the Simojovel syncline, Frost and Langenheim (1974) observed that during the early Oligocene, gentle folding probably disrupted sedimentation, because a "mild angular discordance separates the La Trinidad Sandstone from the suprajacent Oligocene Rancho Berlin Sandstone" (Figure 7). In addition, they proposed that the Pueblo Nuevo reef could have developed "on a local shoal area generated by faulting or folding along the axis of the central Chiapas seaway." Meneses-Rocha (1991), however, proposed that inversion, rather than folding, along the Ixhuatlan and Tecpatán-Ocosingo Faults could be a better explanation for the partial emergence of the Jitotol block where the Simojovel syncline is located.

In the Ixtapa graben area, shallow-water limestones were deposited over a wide area above the upper Eocene carbonate-siliciclastic platform, suggesting low relief and a brief period of tectonic quiescence in the southern upland areas. The subsequent emergence of these southern areas is indicated by the gradual increase in the accumulation of terrigenous clastics in the Ixtapa graben area. The composition of these sediments and their submature texture might indicate that faulting along the southern margin of the Ixtapa graben area promoted a high rate of erosion and deposition.

By the Oligocene, the lowering of the sea level exposed to erosion the easternmost part of the Sierra de Chiapas, as well as the present area of the Gulf of Tehuantepec (Sanchez-Barreda, 1981).

Early Miocene (differential subsidence regime; block faulting) 23.7–16.6 Ma

The extensional regime that prevailed at the end of the Oligocene continued into the early Miocene, when a rise of sea level caused the sea to onlap to the southeast over a platform region that had remained exposed during the Oligocene (Figure 13e). A narrow and shallow marine depression occupied the northeast part of the Reverse-fault province. In its early stage, the transgression seems to have been accompanied by block faulting, resulting in differences in lithology and depositional environments between the Simojovel syncline and the regions situated to the north (Macuspana area) and to the south (Ixtapa area) (Figure 13e). Whereas in the Simojovel syncline the lower Miocene section is shaly and contains planktonic foraminifers (Mazantic Shale of Frost and Langenheim, 1974), in the Ixtapa graben area, the equivalent section is composed of limestones and sandstones containing benthonic foraminifers and corals. In the Macuspana area to the north, the lower Miocene consists of fossiliferous limestones (Macuspana Limestone). In the southern margin of the deep, clastic sea, normal faulting along the Malpaso-Aztlán and La Venta–Grijalva Faults probably produced an unstable slope where coarse-grained terrigenous clastics accumulated (Malpaso Formation). This event was followed by a period of minor

transgressions and regressions related to the gradual uplift of the Chiapas massif and the sedimentary terranes located between the Ixtapa graben area and the massif. The positive elements supplied intermittently terrigenous clastics to the platform (Tulija Formation), whereas in regions away from the platform, a thick sequence of shales and siltstones was deposited at bathyal depths (Deposito Formation). The distribution of the lower Miocene in the Reverse-fault province might suggest that recurrent movement along two northwest-striking faults controlled the location of the depocenter. By that time, in the Gulf of Tehuantepec, proximal and distal turbidite deposits were accumulating (Sanchez-Barreda, 1981) contemporaneously with the emplacement of calc-alkaline intrusions along the southern margin of the Chiapas massif.

Localized unconformities and alternating up and down movements of fault blocks can be attributed to a continental transform plate boundary with alternating components of extension and compression (Norris et al., 1978). In this setting, the proposal of limited transcurrent motions in southeastern Mexico during the Paleocene–early Miocene is compatible with plate-tectonic reconstructions that suggest that the Honduras-Nicaragua block trailed the Caribbean Plate from the southern coast of Mexico to its present position along a transform fault (Malfait and Dinkelman, 1972; Pindell and Dewey, 1982; Wadge and Burke, 1983), and with the estimate made by Rosencrantz et al. (1988) for the initial opening of the Cayman Trough during the middle Eocene and the gradual resumption of magmatic activity in the Chiapas massif as the Chortis block moved eastward and exposed the Mexican margin to subduction of the Cocos Plate.

Middle Miocene (transtensile regime) 16.6–12.2 Ma

Inception of the Ixtapa Graben

The middle Miocene paleogeography of the Sierra de Chiapas area is difficult to reconstruct, because rocks of that age have been identified only in the Ixtapa graben, in the westernmost part of the Malpaso-Bochil block, and in the eastern and northeastern parts of the Sierra de Chiapas (Figures 2 and 7). In the Ixtapa graben, rocks of that age represent a shallow-marine environment (Zapotillo Conglomerate and Zapotal Viejo Shale). In the westernmost part of the Malpaso-Bochil block, the middle Miocene consists of shales and sandstones deposited in a deep-marine environment, and in the eastern and northeastern parts of the Sierra de Chiapas (Reverse-fault Province), it consists of continental deposits. In the northern part of the Isthmus of Tehuantepec (Isthmus Saline Basin), the middle Miocene represents a deep-marine environment but, in contrast with the lower Miocene, it shows an increase in sand-sized material. To the east, terrigenous clastics began to accumulate in the Comalcalco and Macuspana Basins, probably in deltaic environments, which drained the Yucatán platform (Figure 14a).

It thus seems probable that marine environments were displaced during the middle Miocene toward the northwest, and only a shallow seaway extended into the Ixtapa area (Figure 14a). In this area, the occurrence in the Zapotillo Conglomerate of clasts derived predominantly from a calc-alkaline intrusive source. Their coarseness and the composition of the matrix indicate that deposition of the Zapotillo Conglomerate was the result of a considerable tectonic uplift of the Chiapas massif and of the siliciclastic sources located southward. Furthermore, the almost total absence of carbonate clasts indicates that when the Zapotillo Conglomerate was deposited, the fault block that presently limits the Ixtapa graben to the north (Malpaso-Bochil block) was covered by siliciclastic Tertiary rocks and/or was under marine conditions, whereas the fault block bounding the graben to the south (Sumidero block) was exposed and thus contains preserved Tertiary terrigenous clastics. The Zapotillo Conglomerate in the Ixtapa graben is essentially conformable with the underlying strata; hence, it is proposed that only minor tilting, if any, occurred during the middle Miocene in the Ixtapa area.

The depositional system of the Zapotillo Conglomerate is here interpreted as a braided delta that prograded into a half graben that was under shallow-marine conditions. This coarse-grained delta was fed by rivers that flowed north from the Chiapas massif and worked their way through Mesozoic carbonate beds that underwent dissolution, through Mesozoic and Cenozoic siliciclastic beds that provided muddy and sandy sediment, and in some instances through first-cycle (Todos Santos Formation) and recycled pebbles and cobbles (Ocozocoautla Formation and some beds in the middle–upper Eocene sequence). Tectonic uplift of the hinterland might have been achieved by block faulting that extended northward, resulting in the inception of the Ixtapa graben. In this context, the subsequent transgression of the Zapotal Viejo Shale over the Zapotillo Conglomerate marks the foundering of the graben and the diminution of tectonism and uplift in the hinterland and along the Malpaso-Aztlán Fault. The early development of the Ixtapa graben can be associated with a strike-slip process because it is closely linked spatially and temporally to the following events: (1) the inception of left-lateral movements across the Polochic Fault of Guatemala and Chiapas (Burkart, 1978); (2) the older age assigned to the volcanic clasts that comprise the Colotenango Conglomerate of northwest Guatemala (± 12.3 Ma, Deaton and Burkart, 1984); and (3) an episode of dynamic metamorphism along the Tonalá-Motozintla fault that borders the southern margin of the Chiapas massif (J. M. García-Palomares, personal communication, 1990).

If we accept the older age assigned to the Colotenango beds (12.3 my) as the maximum age of sedimentation (Deaton and Burkart, 1984), then the Colotenango beds and the Zapotillo Conglomerate would record the initiation of activity along the transform segment of the North America–Caribbean Plate boundary. Thus, this activity seems to have propagated north-

ward from the Polochic Fault to the Malpaso-Aztlán Fault. Because it was related to the inception of the Ixtapa graben, this activity probably was characterized by divergent wrenching. In addition, the early development of the Ixtapa graben is contemporaneous with the development of an unconformity in the Gulf of Tehuantepec (Sanchez-Barreda, 1981) and to the inception of the Miocene Central American–Mexican volcanic arc (Damon and Montesinos, 1978).

End of Middle Miocene (transpressive regime) 12.2–11.2 Ma

At the end of the middle Miocene, a general uplift of the Sierra de Chiapas forced the shoreline to recede toward the north (Figure 14b). This period of emergence probably was related to an episode of transpressive wrenching along the faults bounding the Ixtapa graben, because in this depression, the Río Hondo group was tilted about 23° prior to deposition of the lowermost beds of the upper Miocene continental succession of the Ixtapa Formation. In terms of regional tectonics, the left-lateral motion along the strike-slip faults of Chiapas and Guatemala (Polochic Fault), combined with the subduction of the Cocos Plate beneath the North American Plate, produced a northeast-oriented, maximum horizontal compressive stress that gave rise to the folding and faulting of the Mesozoic-Cenozoic rocks of the Reforma-Akal uplift. There, seismic profiles indicate that the fold belt rode atop a salt layer at the end of the middle Miocene (Angeles-Aquino et al, 1994).

Late Miocene–Early Pliocene (transtensile regime) 11.2–3.4 Ma

Development of the Ixtapa Graben

In the Sierra de Chiapas, the only beds of late Miocene–early Pliocene age are in the Ixtapa graben, where they form a thick sequence of continental coarse- and fine-grained terrigenous clastics, indicating that that was the time of greatest tectonic activity in this graben (Figure 14c). This thick sequence has been called the Ixtapa Formation by PEMEX geologists. The thickness and persistence of coarse debris in this sequence require a high rate of subsidence in the graben and continued uplift and erosion of the source areas. This setting is commonly associated with strike-slip basins in which cyclicity of lithofacies is attributed to tectonic controls (Norris et al., 1978; Bluck, 1978; Crowell, 1982). In the Ixtapa graben, the best approach to understanding the nature of the tectonism that controlled sedimentation is through analysis of the vertical sequence of the Ixtapa Formation, because the progressive change in lithofacies must reflect repeated major episodes of uplift and subsidence.

Succession of Depositional Environments and Tectonic Interpretation

Six lithofacies associations make up the Ixtapa Formation along the El Camino Creek (Figure 4). These can be grouped

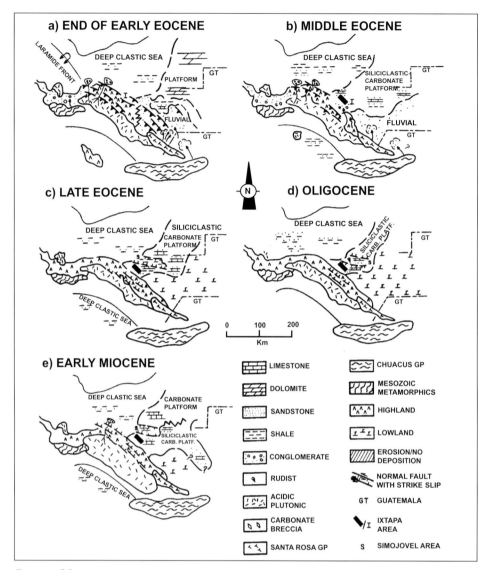

a) END OF EARLY EOCENE

b) MIDDLE EOCENE

c) LATE EOCENE

d) OLIGOCENE

e) EARLY MIOCENE

LIMESTONE
DOLOMITE
SANDSTONE
SHALE
CONGLOMERATE
RUDIST
ACIDIC PLUTONIC
CARBONATE BRECCIA
SANTA ROSA GP

CHUACUS GP
MESOZOIC METAMORPHICS
HIGHLAND
LOWLAND
EROSION/NO DEPOSITION
NORMAL FAULT WITH STRIKE SLIP
GT   GUATEMALA
IXTAPA AREA
S   SIMOJOVEL AREA

FIGURE 13. Paleotectonic maps of the Sierra de Chiapas, southeastern Mexico: (a) end of early Eocene; (b) middle Eocene; (c) late Eocene; (d) Oligocene; and (e) early Miocene. Present Gulf of Mexico and Pacific Ocean borders are also shown.

(Blair, 1987). In the Ixtapa graben, the proximal debris flows suggest an initial phase of tectonic subsidence probably achieved by faulting, whereas the incised-channel deposits indicate a change from active tectonism to a phase of tectonic quiescence, when erosion was greater than uplift (Figure 15, B–B′, C–C′). The normal fault (f_1) required to produce this tectonic subsidence would need to be parallel to the depositional strike of the conglomerates, that is, perpendicular to the strike-slip faults bounding the graben. Repeated vertical motions during tectonic subsidence are suggested by the internal coarsening-upward cycles observed in the proximal debris flows and by the internal unconformities that can be detected by observing the change in dip displayed by the beds of such deposits (Figure 8a).

Middle lacustrine–braided stream megasequence.—The middle megasequence is 1925 m thick and is also characterized by a coarsening-upward trend (Figure 11). This character results from the vertical succession of a lacustrine system below (lithofacies association C) followed by a fluvial system (lithofacies association D). Cyclic deposits of shorter duration are present in the lower lacustrine system.

into three megasequences (following Heward's 1978 definition) on the basis of their environment of deposition and the analysis of their vertical succession at the level of megasequences (Figure 11). The sediments that compose these megasequences were transported northwestward along the graben, and their vertical alternation resulted in cycles hundreds of meters thick.

Lower alluvial-fan megasequence.—This is an upward-coarsening megasequence 2590 m thick, composed of proximal debris flows (lithofacies association A), stratigraphically below incised-channel deposits (lithofacies association B) (Figure 11). The accumulation of a thick unit of proximal debris flows requires near-surface subsidence and relative uplift of the source area (Bluck, 1967; Steel and Wilson, 1975), whereas the deposition of incised-channel deposits over proximal debris flows can be explained in terms of gradual reduction of the relief in the source area during a phase of tectonic quiescence

The development of a lacustrine environment over an alluvial fan system is considered by Blair (1987) and Blair and Bilodeau (1989) as the response to tectonic subsidence that creates a topographic depression. According to Blair (1987), in ancient and modern examples of extensional basins, a lacustrine environment, rather than alluvial fan deposition, is the initial response to active tectonic subsidence, "due to different hydrologic controls on lake versus fan sedimentation." In the Ixtapa graben, the lacustrine system must have been the response to a renewed subsidence caused by the vertical motion along a new normal fault that might have been situated southeastward of the fault that generated the older megasequence (f_2, Figure 15 D–D′). The southeastward motion of the focus of faulting and a minor throw of the vertical movement are suggested by the thinning of the middle megasequence, the southeastward younging of its lithofacies associations, and its sediment dispersal.

Evidence for rapid sedimentation, probably related to a steep slope, is shown by soft sediment deformation, as suggested by mud and sand volcanoes that display graded finely laminated sandstone of the lithofacies association C. During the accumulation of this lithofacies association, hydrologic controls might have produced the contraction and expansion of the lake, as evidenced by the brief fluvial episodes represented by the cobble- and pebble-conglomerate beds. The abundance of tuffaceous material in lithofacies association C apparently heralds the inception of the "modern Chiapanecan volcanic arc" cited by Damon and Montesinos (1978). If this is the case, then the main source for the middle megasequence was the terrain situated around the Tzontehuitz Volcano because, according to Damon and Montesinos (1978), the augite andesites associated with this volcano yielded latest Pliocene K-Ar dates. A gradual reduction in the slope of fault 2 (f_2, Figure 15 E–E´) caused by an episode of tectonic quiescence, when erosion was greater than uplift, led to the progradation of a braided fluvial system (lithofacies association D) over the lacustrine system.

Upper lacustrine–streamflood megasequence.—This megasequence is 1305 m thick and also displays a coarsening-upward trend, which results from the transition of a lacustrine system (lithofacies association E) to streamflood deposits (lithofacies association F) (Figure 11). The origin of this megasequence is interpreted as that of the lower megasequence, i.e., the response to a phase of tectonic subsidence followed by an episode of relative tectonic quiescence (Figure 15, F-F´ and G-G´). However, the thinning of this megasequence, as compared with the underlying megasequence, the southeastward younging of the conglomerates, and their sediment dispersal, suggests that this sequence was generated by a normal fault with a lesser throw than that of the faults that generated the underlying megasequences, and also that this younger fault was situated farther southeastward (f_3). The southeastward migration of the faulting simultaneous with the migration of the depocenter might have been produced by extension caused by the sinistral shift of the Malpaso-Aztlán Fault, because this mechanism would explain the overlap and thinning of lithofacies in a southeasterly direction.

Repeated vertical movements along the three normal faults here postulated (f_1, f_2, f_3) might have produced the local unconformities that can be deduced from the changes in dip and strike observed within the Ixtapa Formation (Figure 8a). The oldest unconformity was produced by activity along the oldest fault, whereas the two younger unconformities seem to have been produced by intermittent motions along the second and third faults. As the Malpaso-Aztlán Fault propagated southeastward, the slip along the Chicoasen–San Cristobal Fault seems to have been dominantly oblique, with dip slip dominant over strike slip. This imbalance of dip slip and strike slip between the master faults could have produced the thickening of the Ixtapa Formation to the northeast, i.e., toward the Chicoasen–San Cristobal Fault.

Block-faulted basement caused by extension as a result of the sinistral shift of the main bounding fault has been proposed by Bluck (1978, 1980) for the Midland Valley (Scotland), where the strata overlap and thin away from the direction of strike-slip motion, and the paleocurrent orientations for both the coarse and fine sediments are almost the same. Local unconformities produced by vertical movements along parallel normal faults in a stepped arrangement have been described by Steel and Wilson (1975) in the Permian-Triassic North Minch basin of Scotland. In the case of the Ixtapa graben, this mechanism of syndepositional tectonic motion implies that the stratigraphic thickness of the late Miocene–Pliocene sequence does not reflect the true depth of the graben but is only the cumulative result of the banking of successively younger beds against at least three fault scarps. Thus, in this context it is analogous to the Ridge Basin described by Crowell (1982) in California. This mechanism also implies that the amount of lateral displacement along the Malpaso-Aztlán Fault during the late Miocene–Pliocene did not exceed 11 km, i.e., the length of the three depocenters where the Ixtapa Formation accumulated.

From a regional perspective, the main phase of growth of the Ixtapa graben (i.e., the main phase of left-lateral strike-slip faulting in Chiapas) is contemporaneous and perpendicular with the extensional regime that took place in the southern Gulf coastal plain that gave rise to the development of the northeast-oriented listric normal faults that limit the Macuspana and Comalcalco Basins. Although it is evident that the Macuspana and Comalcalco Basins were generated by local geologic characteristics, the northeast orientation of the listric normal faults that limit these two basins, relative to the orientation of wrenching in Chiapas, may suggest that these Neogene listric faults developed in the direction of theoretical "conjugate Riedel shears" or tension fractures. This event is also contemporaneous with the beginning of the main phase of halokinesis in the Isthmus Saline Basin.

Late Pliocene (transpressive regime) 3.4–1.6 Ma

Deformation of the Ixtapa Graben

At the end of the Pliocene, a transpressive tectonic phase resulted in the deformation of the Ixtapa graben. The Ixtapa Formation and older units were deformed by folding and tilting related to left-lateral motion along the master bounding faults (Figure 14d). Direct evidence for this transpressional motion are the Río Lajas, El Tunel, and El Palmar–Concepción flower structures, as well as the angular unconformity between the Ixtapa Formation and the overlying Quaternary strata. This period of transpressive wrenching became dominant in the rest of the Strike-slip Fault province, where a pattern of en-echelon folding related to sinistral slip became evident. The increase of compressive features of the strain ellipsoid of Chiapas probably was produced by the strengthening of a northeastward compressive component related to the subduction of the

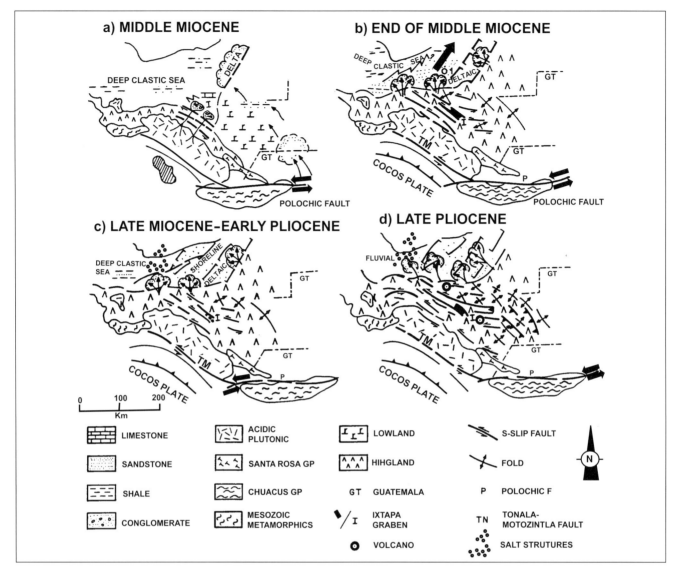

FIGURE 14. Paleotectonic maps of the Sierra de Chiapas, southeastern Mexico: (a) middle Miocene; (b) end of middle Miocene; (c) late Miocene–early Pliocene; and (d) end of Pliocene. Present Gulf of Mexico and Pacific Ocean borders are also shown.

Cocos Plate under the North American Plate. This final phase of compression, uplift, and erosion provided an additional input of terrigenous clastics into the petroleum-producing trends situated in the Gulf coastal plain of southeastern Mexico and gave rise to inversion features in the southern part of the Macuspana Basin.

If we compare the present structural pattern of the Strike-slip Fault province with the direction of theoretical features which could result in a strain ellipsoid produced by the east-west sinistral shear coupling between the Chiapas-Yucatán block and the Chortis block, along the Motagua-Polochic fault system, we can make the following observations (Figure 3): (1) the east-west orientation of strike-slip faults located in the eastern part of the province deviates approximately 15° from the theoretical orientation of riedel shears. This disparity might be justified if we accept that these faults formed by rejuvenation of old lines of weakness; (2) the northwest orientation of the

strike-slip faults of the central part of the province roughly coincides with the theoretical orientation of P faults (Tchalenko, 1970; Tchalenko and Ambraseys, 1970); and (3) the orientation of theoretical folding roughly corresponds to the average orientation of the folds observed in this province.

It thus seems probable that the development of this convergent phase of wrenching followed a sequential pattern, as has been observed in experimental models proposed by several authors (e.g., Tchalenko, 1970; Tchalenko and Ambraseys, 1970; Wilcox et al., 1973; Freund, 1974). In the early stages of movement, folds developed throughout most of the area. At a later stage, riedel shears started to appear in the eastern part of this province, displacing and rotating the axes of some folds. As deformation proceeded, P faults appeared in the central part, as a result of the resistance that riedel shears presented to any further movement. Contractional effects could have included the counterclockwise rotation of some en-echelon

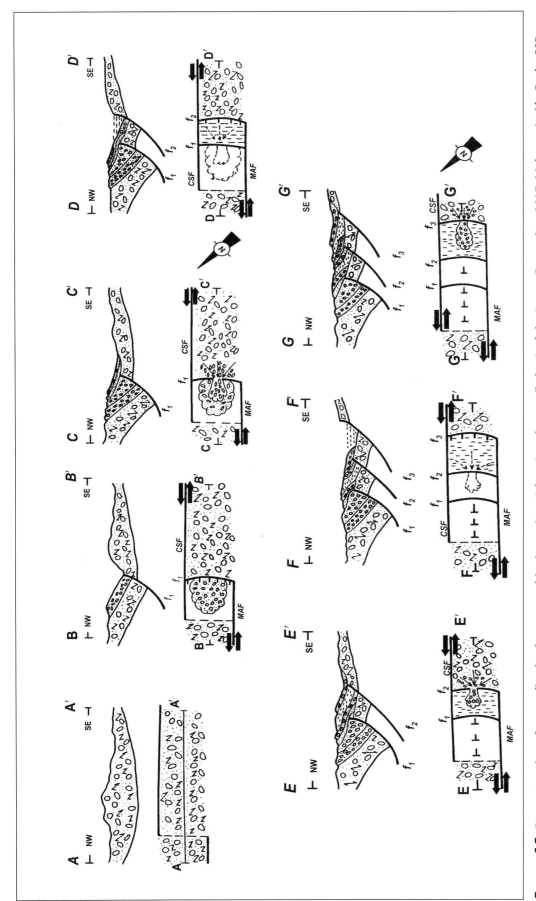

Figure 15. Reconstruction of generalized paleogeographies in sections (above) and maps (below) of the Ixtapa Formation. MAF, Malpaso-Aztlán Fault; CSF, Chicoasen–San Cristobal Fault. A-A′, initial situation with no overlap between MAF and CSF; B-B′, left-stepping produces a normal fault (f_1) that gives rise to a zone of tectonic subsidence where lithofacies association A accumulates; C-C′, a phase of tectonic quiescence in the zone of tension produces recession of the source area and the southeastward overlap of lithofacies association A by lithofacies association B (streamflood deposits); D-D′, further growth of the MAF produces a second normal fault (f_2), which generates a new depocenter. New hydrologic conditions led to deposition of a lacustrine system (lithofacies association C); E-E′, a gradual reduction in the slope of f_2 led to progradation of a braided fluvial system (lithofacies association D) over the lacustrine system; F-F′, in response to continued growth of the MAF, a third normal fault (f_3) is produced, giving rise to a new zone of tectonic subsidence. Lacustrine deposits (lithofacies association E) accumulate on the depression; G-G′, a phase of relative tectonic quiescence produces a gradual reduction of the slope of f_3, and the southeastward overlap of lacustrine deposits by a streamflood system (lithofacies association F).

folds (producing, in some cases, flank thrusting) and the development of flower structures along the margins of the Ixtapa graben. Extensional effects could have favored the activity of the El Chichonal and Tzontehuitz Volcanoes at the western ends of the Ixhuatlan and Tenejapa Faults.

In this scheme of deformation, it can be presumed that the variable orientations of the strike-slip faults of the western part of the province are a response to the final effect of the shear movement along P faults. According to Chinnery (1966), these faults might be considered "secondary faults." The existence of these horsetail structures and the diminution of en-echelon folds in this part of the Strike-slip Fault province seem to indicate that the wrench motion propagated westward during the late Miocene–Pliocene. The amount of left-lateral displacement during this phase of deformation along the Tecpatán-Ocosingo and Tenejapa faults was 4 and 5 km, respectively (Figure 5).

It is clear that those faults close to the Chiapas massif (Quintana Roo and Uzpanapa–Las Flores Faults), as well as the rest of the Sierra monocline, were not able to participate completely in this deformational sequence, because the massif could have formed a buttress, which restricted large movements. In addition, the fundamental causes for which the Reverse-fault Province displays a different tectonic style than the Strike-slip Fault province are: (1) the closeness of the former to the stable Yucatán platform; and (2) the presence in the former of an extremely thick level of salt and anhydrite in the Middle Jurassic Callovian and the Upper Jurassic–middle Cretaceous sequences, respectively (Figure 6).

Quaternary (transtensile regime)
1.6 Ma–Present

Extension of the Ixtapa Graben

After the latest major transpressive event in the Sierra de Chiapas, the Río Grijalva began to cut and dissect across the northwestern part of the Ixtapa graben. At the same time, volcanic sediments were deposited with angular unconformity upon the continental Ixtapa Formation (Figure 4). These sediments probably were deposited in an alluvial-lacustrine environment and reflect the activity of the Navenchauc and Huitepec Volcanoes, in the southeastern end of the graben.

During the last 1.6 my, strike-slip movement along the Malpaso-Aztlán Fault has resulted in the extension of the Ixtapa graben to the southeast and into the opening of a lozenge-shaped area (here informally named Zinacantan area) enclosed between the east-trending segment of the Chicoasen–San Cristobal Fault and the northern margin of the mountain front of Cretaceous carbonates (Figure 4).

A generalized outline of the development of the southeastern part of the Ixtapa graben is shown in Figure 16. In Figure 16b and 16c, we can observe that an offset of about 1 km along the Chicoasen–San Cristobal Fault (CSF), combined with the extension (5 km) of the Malpaso-Aztlán Fault (MAF) southeast-

ward, causes zone A to move away from B, producing tension and therefore subsidence (zone D). As this movement proceeds, zone A also moves away from C, so the reference points 1, 2, and 3 move to positions 1´, 2´, and 3´. Thus, a gap between these two blocks is produced.

From this model, it is possible to deduce the following: (1) the northwestern and northern margins of the mountain front (Figure 4) may be regarded as a fault-line scarp modified to some extent by mass wasting; (2) the locations of the Navenchauc and Huitepec Volcanoes were controlled by left-lateral strike-slip faulting, because they are located in the zones of maximum extension in the Ixtapa graben; (3) the rectangular drainage pattern of the southeastern part of the graben can be regarded as the result of the interception of streams that originated in the highland and flowed northwestward by two northeast-trending tension fractures that diverted their courses northeastward. These fractures are parallel to the normal faults that operated during the late Miocene–Pliocene (Figure 4); (4) the coalescence of the Ixtapa graben with the Zinacantan area will result in the formation of a wider graben; and (5) the amount of lateral displacement along the Malpaso-Aztlán Fault during the Quaternary is on the order of 5 km, i.e., the width of zone D (Figure 16c).

Evidence of recent left-lateral movements along the faults of the Strike-slip Fault province includes: (1) the solution of focal mechanisms obtained by Guzmán-Speziale et al. (1989); and (2) the almond-shaped depression formed where the northwest-trending segment of the Chicoasen–San Cristobal Fault intersects its east-trending segment (Romo Serrano area). In this area, it is possible to observe a small sag pond, associated with springs that emanate from the main fault scarp (Figure 9a).

Therefore, the amount of displacement across each fault of the Strike-slip Fault province is 4 to 5 km in the eastern part (Figure 5); 1 to 16 km in the central part, giving an average shear of 27 km across the eastern part (six faults); and 43 km across the central part (five faults). Thus, the total displacement across the province would be 70 km (Figure 3).

IMPLICATIONS FOR REGIONAL PETROLEUM SYSTEMS

The transpressive regime that is recorded in the Ixtapa graben at the end of the middle Miocene can be considered the main regional tectonic event leading to trap mechanisms in the Chiapas-Tabasco and Sonda de Campeche petroleum provinces, in that it produced a northeast-oriented compressive stress in a zone where a thick section of fine- and coarse-grained siliciclastics and carbonates rode atop a basal salt layer. Furthermore, the late Miocene–early Pliocene transtensile regime that affected the Ixtapa graben is associated with and may have increased the extensional phase that took place in the Macuspana and Comalcalco Basins, giving rise to the present petroleum traps in the gravity tectonic realm that is perpendicular to the compressional structures of southeastern Mexico. Finally, the late Pliocene transpressive event recorded at the Ixtapa graben eroded and breached some of the anticlines of the Sierra de Chiapas and produced inversion structures in the Macuspana Basin. All of these Neogene regional tectonic events also increased the input of coarse- and fine-grained siliciclastics to the petroleum provinces, leading to: (1) generation and migration of the Oxfordian, Tithonian, middle Cretaceous, and Tertiary oil and gas in all the producing trends; (2) deposition of reservoir and seal rocks and the increase of halokinesis in the Isthmian Saline and Comalcalco Basins; (3) deposition of reservoir, seal, and source rocks in the Macuspana Basin; and (4) formation of huge anticlines in the Sierra de Chiapas, whose petroleum potential waits for reassessment in light of new concepts and technologies.

SUMMARY

In the Strike-slip Fault province of Chiapas, stratigraphic information indicates that during the Albian-Santonian, the Ixtapa area remained on the carbonate platform. This platform then underwent gentle folding or block faulting that culminated in the Campanian with the emergence of the Chiapas massif and the high-angle thrusting along its northern margin (arc-continent collision). By Paleocene–early Eocene,

a regional transition from a phase of basin deepening to a phase of basin emergence gave rise to a disconformity in the Ixtapa area of the same age as the sedimentary fill in nearby blocks. This disconformity marks the effects of the Laramide orogeny. From late Eocene to early Miocene, movement along the faults bounding the Ixtapa graben was predominantly vertical. The movement of these faults was definitely transcurrent by the middle Miocene, giving rise to the opening of the Ixtapa graben. This episode was closely linked to the inception of sinistral slip along the Polochic Fault. This transtensional phase was followed by a regional transpressive episode that produced the tilting of the middle Miocene beds in the Ixtapa graben, and a northeast-oriented maximum horizontal stress in the Reforma-Akal uplift. From late Miocene to early Pliocene, a new transtensional phase gave rise to the redeposition of a thick sequence of continental coarse- and fine-grained terrigenous clastics in the Ixtapa graben. This phase is contemporaneous with the development of the extensional realm that prevailed in the Macuspana and Comalcalco Basins, as well as with the main phase of halokinesis in the Isthmian Saline Basin. At the end of the Pliocene, a regional transpressive phase gave rise to the deformation of the Ixtapa graben and to the erosion and breaching of some of the anticlines of the Sierra de Chiapas. During the Quaternary, the extension of the graben was contemporaneous with a volcanic phase, during which volcanic sediments were deposited

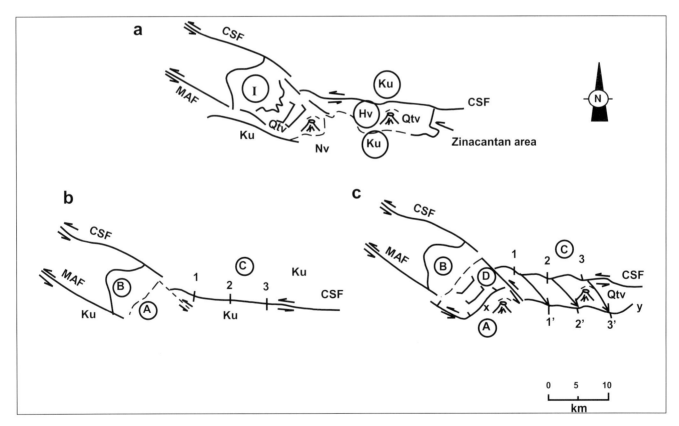

FIGURE 16. Map view showing development of the Ixtapa graben during the Quaternary. CSF, Chicoasen–San Cristobal Fault; MAF, Malpaso-Aztlán Fault; I, Ixtapa Formation; Ku, Upper Cretaceous; Qtv, Quaternary volcanics; NV, Navenchauc Volcano; HV, Huitepec Volcano. See text for discussion.

in angular unconformity on the continental sediments. The amount of shear displacement across each fault of the Strike-slip Fault province is 4 to 5 km in the eastern part and 1 to 16 km in the central part, giving an average shear displacement of 27 km across the eastern part (six faults) and 43 km across the central part (five faults). Thus, the total shear displacement across the province would be 70 km. This displacement was restricted to the area of the Sierra de Chiapas where the main zone of detachment is the Middle Jurassic Callovian Salt. However, some strike-slip faults extend into the basement, as is evidenced by the presence of volcanoes at the end of the traces of some faults and at the zones of maximum extension produced by the overlap of strike-slip faults. In this way, the shear belt of Chiapas constitutes a restraining bend located at the westernmost end of the North American–Caribbean transform boundary. Thus, plate-boundary deformation along the southern margin of the North American Plate has been diffuse and has spread over a wide area of southeastern Mexico. In this sense, it is analogous to the Neogene northern margin of the Caribbean Plate which, according to Burke et al. (1980) "is more than 200 km wide continental PBZ than a narrowly defined plate boundary." (Figure 1).

Combining fieldwork with regional stratigraphic, paleogeographic, and structural studies as well as with a systematic interpretation of modern seismic surveys is essential to the understanding of the genesis of petroleum systems, especially in those areas with a complex geologic evolution.

ACKNOWLEDGMENTS

Most of this paper stems from my master's and Ph.D. research at the University of Texas at Austin, conducted under the supervision of William R. Muehlberger, Amos Salvador, and John C. Maxwell, who shared with me their knowledge and experience and were invaluable guides and mentors throughout these studies. Dr. Muehlberger suggested the research on the Ixtapa graben and, along with Dr. Salvador, visited the study area and kindly provided endless hours of stimulating discussions and advice. David Lopez-Ticha, PEMEX geologist, suggested the study of the Strike-slip Fault province, along with Juan M. Quezada-Muñeton, Rafael Sánchez-Montes de Oca, Carlos Bortolotti-Andrade, and Juan M. Ham-Wong of PEMEX. Luis Sánchez-Barreda of Barreda & Associates, as well as Robert K. Goldhammer and Christopher A. Johnson of Exxon Exploration Company, contributed to the improvement of this work through helpful discussions in the field based on their knowledge and experience. Appreciation is also extended to Alberto Aquino-López and Miguel Varela-Santamaria of PEMEX for valuable assistance in the field and to Marco Guzmán-Speziale for valuable discussions. Gratitude is also expressed to PEMEX for providing subsurface data and financial and logistical support for fieldwork. My thanks are given also to Vicente Cruz-Ramos and Carlos Carrillo-Chavez for preparing the drawings. The original manuscript was improved by the constructive and incisive comments of Carlos A. Dengo, James L. Pindell, and Burke Burkart. The views and conclusions presented here are the sole responsibility of the author.

REFERENCES CITED

Alvarez, M., 1949, Unidades tectónicas de la República Mexicana: Sociedad Geológica Mexicana Boletín, Tomo 14, p. 1–22.

Alvarez, M., 1958, Tectónica profunda de Mexico: Boletín de la Asociación Mexicana de Geólogos Petroleros, v. 10, p. 163–182.

Armstrong, L. R., 1968, Sevier orogenic belt in Nevada and Utah: Geological Society of America Bulletin, v. 79, p. 429–458.

Anderson, T. H., R. J. Erdlac Jr., and M. Sandstrom, 1985, Late Cretaceous allochthonous and post-Cretaceous strike-slip displacement along the Cuilco-Chixoy-Polochic Fault: Tectonics, v. 4, p. 453–475.

Angeles-Aquino, F. J., J. Reyes-Nuñez, J. M. Quezada-Muñetón, and J. J. Meneses-Rocha, 1994, Tectonic evolution, structural styles and oil habitat of the Campeche Sound, Mexico: Transactions of the Gulf Association of Geological Societies, v. 44, p. 53–62.

Blair, T. C., 1987, Tectonic and hydrologic controls on cyclic alluvial fan, fluvial, and lacustrine rift-basin sedimentation, Jurassic–lowermost Cretaceous Todos Santos Formation, Chiapas, Mexico: Journal of Sedimentary Petrology, n. 57, p. 845–862.

Blair, T. C., 1988, Mixed siliciclastic-carbonate marine and continental syn-rift sedimentation, Todos Santos and San Ricardo Formations, western Chiapas, Mexico: Journal of Sedimentary Petrology, v. 58, p. 623–636.

Blair, T. C., and W. L. Bilodeau, 1989, Development of tectonic cyclothems in rift, pull-apart, and foreland basins: Sedimentary response to episodic tectonism: Geology, v. 16, p. 517–520.

Bluck, B. J., 1967, Deposition of some Upper Old Red Sandstone Conglomerates in the Clyde area: A study in the significance of bedding: Scottish Journal of Geology, v. 3, p. 139–167.

Bluck, B. J., 1978, Sedimentation in a late orogenic basin: The Old Red Sandstone of the Midland Valley of Scotland, in D. R. Bowes and B. E. Leake, eds., Crustal evolution in northwestern Britain and adjacent regions: Geological Journal Special Issue No. 10, p. 249–278.

Bluck, B. J., 1980, Evolution of a strike-slip fault-controlled basin, Upper Old Red Sandstone, Scotland, in P. F. Ballance and H. G. Reading, eds., Sedimentation in oblique-slip mobile zones: International Association of Sedimentologists Special Publication No. 4, p. 63–78.

Boese, E., 1905, Reseña acerca de la geología de Chiapas y Tabasco: Instituto Geológico de Mexico Boletín, v. 20, 113 p.

Burkart, B., 1978, Offset across the Polochic fault of Guatemala and Chiapas: Geology, v. 6, p. 328–332.

Burkart, B., 1983, Neogene North American–Caribbean Plate boundary: Offset along the Polochic fault: Tectonophysics, v. 99, p. 251–270.

Burkart, B., T. C. Blair, and D. Movarec, 1989, Late Cretaceous block-arc deposits of Chiapas, Mexico, and their relation to the orogen: Geological Society of America Abstracts with programs, v. 21, p. 5.

Burkart, B., and R. E. Clemons, 1972, Late Paleozoic orogeny in northwestern Guatemala, in Memorias 6.º Conferencia Geológica del Caribe, Margarita, Venezuela, p. 210–213.

Burkart, B., B. C. Deaton, C. Dengo, and G. Moreno, 1987, Tectonic wedges and offset Laramide structures along the Polochic fault of Guatemala and Chiapas, Mexico: Reaffirmation of large Neogene displacement: Tectonics, v. 6, p. 411–422.

Burke, K., J. Grippi, and A. M. C. Sengor, 1980, Neogene structures in Jamaica and the tectonic style of the Northern Caribbean Plate boundary zone: Journal of Geology, v. 88, p. 375–386.

Carfantan, J. C., 1981, Evolución estructural del Sureste de Mexico, paleogeografía e historia tectónica de las zonas internas mesozoicas: Universidad Nacional Autónoma de Mexico, Instituto de Geología Revista, v. 15, n. 2, p. 207–216.

Castro-Mora, J., C. Schalepfer, and E. Martinez, 1975, Estratigrafía y microfacies del Mesozoico de la Sierra Madre del Sur, Chiapas: Boletín de la Asociación Mexicana de Geólogos Petroleros, v. 27, no. 1–3, p. 1–103.

Chinnery, M. A., 1966, Secondary faulting: I. Theoretical aspects, II. Geological aspects: Canadian Journal of Earth Sciences, v. 3, p. 163–190.

Coney, P. J., 1972, Non-collision tectogenesis in western North America, *in* D. H. Tarling and S. K. Runcorn, eds., Implications of continental drift on earth sciences, v. 2: Academic Press, London and New York, p. 713–725.

Crowell, J. C., 1982, The tectonics of Ridge basin, Southern California, *in* J. C. Crowell and M. H. Link, eds., Geologic history of Ridge Basin, southern California: Pacific Section, Society of Sedimentary Geology (SEPM) Guidebook, p. 25–41.

Cruz-Helú, P., and J. J. Meneses-Rocha, 1999, PEMEX plots ambitious E&D spending increase: Oil & Gas Journal, June 15, p. 86–88.

Daily, F. K., and J. W. Durham, 1966, Miocene charophytes from Ixtapa, Chiapas, Mexico: Journal of Paleontology, v. 40, p. 1191–1199.

Damon, P. E., and E. Montesinos, 1978, Late Cenozoic volcanism and metallogenesis over an active Benoff zone of Chiapas, Mexico: Arizona Geological Society Digest, v. 11, p. 155–168.

Davis, D. M., and T. Engelder, 1985, The role of salt in fold and thrust belts: Tectonophysics, v. 119, p. 67– 88.

Deaton, B. C., and B. Burkart, 1984, Time of sinistral slip along the Polochic fault of Guatemala: Tectonophysics, v. 102, p. 297–313.

de Cserna, Z., 1960, Orogenesis in time and space in Mexico: Geologische Rundschau, v. 50, p. 595–605.

Dengo, G., 1968, Estructura geológica, historia tectónica y morfología de América Central: Mexico, D. F., Centro Regional de Ayuda Técnica, p. 52.

Dengo, G., and O. H. Bohnenberger, 1969, Structural development of northern Central America: AAPG Memoir 11, p. 203–212.

DETENAL, Cartas topográficas 1:50,000 de Chicoasén, Acala y Bochil: Dirección de Estudios del Territorio Nacional, Mexico, D.F.

Freund, R., 1974, Kinematics of transform and transcurrent faults: Tectonophysics, v. 21, p. 93–104.

Frost, S. H., 1971, Tertiary larger foraminiferal and coral succession, Northern Central America: Transactions Fifth Caribbean Geological Conference Bulletin, v. 5, p. 133–136.

Frost, S. H., and R. L. Langenheim, 1974, Cenozoic reef biofacies: Tertiary corals from Chiapas, Mexico: Northern Illinois Press, Urbana, Illinois, p. 338.

Gutierrez-Gil, R., 1956, Bosquejo geológico del Estado de Chiapas, in Geología del Mesozoico y Estratigrafía Pérmica del Estado de Chiapas: 20th International Geological Congress, Mexico, Excursión C-15, p. 9–32.

Guzman, E. J., and Z. de Cserna, 1963, Tectonic History of Mexico: AAPG Memoir, No. 2, p. 113–129.

Guzmán-Speziale, M., and J. J. Meneses-Rocha, 2000, The North America–Caribbean Plate boundary west of the Polochic fault system: A fault jog in Southeastern Mexico: Journal of South American Earth Sciences, v. 13, p. 459–468.

Guzmán-Speziale, M., W. D. Pennington, and T. Matumoto, 1989, The triple junction of the North America, Cocos, and Caribbean Plates: Seismicity and tectonics: Tectonics, v. 8, p. 981–999.

Guzman-Vega, M. A., and M. R. Mello, 1999, Origin of oil in the Sureste Basin, Mexico: AAPG Bulletin, v. 83, p. 1068–1095.

Helbig, K. M., 1964, La Cuenca Superior del Río Grijalva: Instituto de Ciencias y Artes de Chiapas, 247 p.

Heward, A. P., 1978, Alluvial fan sequence and megasequence models: With examples from Westphalian D–Stephanian B coal-fields, northern Spain, *in* A. D. Miall, ed., Fluvial sedimentation: Canadian Society of Petroleum Geologists Memoir 5, p. 59–76.

Hernández-García, R., 1973, Paleogeografía del Paleozoico de Chiapas, Mexico: Asociación Mexicana de Geólogos Petroleros Boletín, v. 25, p. 77–134.

Heuer, R. E., 1965, Geology of the Soyalo-Ixtapa Area, Chiapas, Mexico: Master's thesis, University of Illinois, Urbana, 104 p.

Horne, G. S., 1989, Pre-Cenozoic tectonic framework of the Chortis Block, northern Central America: Circum-Pacific Energy Council Meeting, San Jose, Costa Rica.

Jordan, T. H., 1975, The present-day motions of the Caribbean plate: Journal of Geophysical Research, v. 80, p. 4433–4439.

Kesler, S. E., 1973, Basement rock structural trends in southern Mexico: Geological Society of America Bulletin, v. 84, p. 1059-1064.

Kesler, S. E., W. L. Josey, and E. M. Collins, 1970, Basement rocks of western Nuclear Central America: The western Chuacus Group, Guatemala: Geological Society of America Bulletin, v. 81, p. 3307–3322.

Langenheim, R. L., S. H. Frost, and R. E. Heuer, 1965, Paleocene through Pliocene sequence in the Ixtapa-Soyalo region, Chiapas, Mexico: Geological Society of America Abstracts with Programs, Special Paper 115, p. 92.

Lensen, G. L., 1958, A method of graben and horst formation: Journal of Geology, v. 92, p. 513–530.

Malfait, B. J., and M. J. Dinkelman, 1972, Circum-Caribbean tectonic and igneous activity and the evolution of the Caribbean Plate: Geological Society of America Bulletin, v. 83, p. 251–272.

Mann, P., K. Burke, and T. Matumoto, 1984, Neotectonics of Hispaniola: Plate motion, sedimentation, and seismicity at a restraining bend: Earth and Planetary Science Letters, v. 70, p. 311–324.

Meneses de Gyves, J., 1980, Geología de la Sonda de Campéche: Boletín de la Asociación Mexicana de Geólogos Petroleros, v. 32, p. 1–26.

Meneses-Rocha, J. J., 1977, Estratigrafía, petrografía y diagénesis de rocas del Jurásico Superior en afloramientos del Istmo de Tehuantepec, Oaxaca: Tesis profesional, Facultad de Ingeniería, Universidad Nacional Autónoma de Mexico, 113 p.

Meneses-Rocha, J. J., 1985, Tectonic evolution of the Strike-slip Fault province of Chiapas, Mexico: Master's thesis, University of Texas at Austin, 315 p.

Meneses-Rocha, J. J., 1990, Marco Tectónico y Paleogeografía del Triásico Tardío-Jurásico en el Sureste de Mexico: Boletín de la Asociación Mexicana de Geólogos Petroleros, v. 39, p. 3–68.

Meneses-Rocha, J. J., 1991, Tectonic development of the Ixtapa graben, Chiapas, Mexico: Ph.D. dissertation, University of Texas at Austin, 308 p.

Miall, A. D., 1978., A review of the braided-river depositional environment: Earth-Science Reviews, v. 13, p. 1–62.

Molina-Garza, R. S., R. Van der Voo, and J. Urrutia-Fucugauchi, 1992, Paleomagnetism of the Chiapas massif, southern Mexico: Evidence for rotation of the Maya block and implications for the opening of the Gulf of Mexico: Geological Society of America Bulletin, v. 104, p. 1156–1168.

Molnar, P., and L. R. Sykes, 1969, Tectonics of the Caribbean and Middle America regions from focal mechanisms and seismicity: Geological Society of America Bulletin, v. 80, p. 1639–1684.

Mossman, R. W., and F. Viniegra, 1976, Complex fault structures in the Veracruz Province of Mexico: AAPG Bulletin, v. 60, p. 379–388.

Movarec, D., 1983, Study of the Concordia Fault System near Jerico, Chiapas, Mexico: Master's thesis, University of Texas at Arlington, 155 p.

Muehlberger, W., and A. W. Ritchie, 1975, Caribbean-American Plate boundary in Guatemala and southern Mexico as seen on Skylab IV orbital photography: Geology, v. 3, p. 232–235.

Mullerried, F. K. G., 1949, La orogénesis del sur y sureste de Mexico: Sociedad Geológica Mexicana Boletín, Tomo XIV, p. 73–100.

Norris, R. J., R. M. Carter, and J. M. Turnbull, 1978, Cenozoic sedimentation in basins adjacent to a major continental transform boundary in southern New Zealand: Journal of Geological Society of London, v. 135, p. 191–205.

Ordoñez, E., 1936, Principal physiographic provinces of Mexico: AAPG Bulletin, v. 20, p. 1277–1307.

Padilla y Sanchez, R., 1982, Geologic evolution of the Sierra Madre Oriental between Linares, Concepción del Oro, Saltillo and Monterrey, Mexico: Ph.D. dissertation, University of Texas at Austin, 217 p.

Palmer, A. W., 1983, The Decade of North American Geology: Geology, v. 11, p. 503–504.

Pedrazzini, C., N. Holguín, and R. Moreno, 1982, Evaluación Geológica-Geoquímica de la parte noroccidental del Golfo de Tehuantepec: Instituto Mexicano del Petróleo Revista, v. 14, p. 6–26.

Perfit, M. R., and B. C. Heezen, 1978, The geology and evolution of the Cayman Trench: Geological Society of America Bulletin, v. 89, p. 1155–1174.

Pindell, J., and J. F. Dewey, 1982, Permo-Triassic reconstruction of western Pangea and the evolution of the Gulf of Mexico/Caribbean region: Tectonica, v. 1, p. 179–211.

Quezada-Muñeton, J. M., 1983, Las Formaciones San Ricardo y Jerico del Jurásico Medio-Cretácico Inferior en el SE de Mexico: Boletín de la Asociación Mexicana de Geólogos Petroleros, v. 35, p. 37–64.

Quezada-Muñeton, J. M., 1987, El Cretácico Medio-Superior y el límite Cretácico Superior-Terciario Inferior en la Sierra de Chiapas: Boletín de la Asociación Mexicana de Geólogos Petroleros, v. 39, p. 3–98.

Rodgers, D. A., 1980, Analysis of pull-apart basin development produced by en echelon strike-slip faults, in P. F. Ballance and H. G. Reading, eds., Sedimentation in oblique-slip mobile zones: International Association of Sedimentologists Special Publication 4, p. 7–26.

Rosenfeld, J. H., 1989, The Santa Cruz ophiolite of Guatemala: A remnant of Early Cretaceous, pre-Caribbean oceanic crust: Geological Society of America Abstracts with Programs, v. 21, p. 39.

Rosenkrantz, E., M. I. Ross, and J. G. Sclater, 1988, Age and spreading history of the Cayman trough as determined from depth, heat flow, and magnetic anomalies: Journal of Geophysical Research, v. 93 (B3), p. 2141–2157.

Rust, B. R., 1978, Depositional models for braided alluvium, in A. D. Miall, ed., Fluvial Sedimentology: Canadian Society of Petroleum Geologists Memoir 5, p. 605–625.

Salvador, A., 1987, Late Triassic–Jurassic paleogeography and origin of the Gulf of Mexico Basin: AAPG Bulletin, v. 71, p. 419–451.

Salvador, A., 1991, Origin and development of the Gulf of Mexico Basin, in A. Salvador, ed., The Gulf of Mexico Basin: Geological Society of America, The geology of North America, v. J, p. 389–494.

Sanchez-Barreda, L. A., 1981, Geologic evolution of the continental margin of the Gulf of Tehuantepec in southwestern Mexico: Ph.D. dissertation, University of Texas at Austin, 191 p.

Sanchez-Montes de Oca, R., 1979, Geología petrolera de la Sierra de Chiapas: Boletín de la Asociación Mexicana de Geólogos Petroleros, v. 31, p. 67–97.

Sanchez- Ortiz, B., 1975, Aspecto sismológico de las estructuras salinas del Istmo: Asociación Mexicana de Geofisicos de Exploración, v. 16, p. 33–62.

Schuchert, C., 1935, Historical geology of the Antillean Caribbean Region: New York, John Wiley & Sons, 811 p.

Steel, R. J., and A. C. Wilson, 1975, Sedimentation and tectonism (Permo-Triassic) on the margin of the North Minch Basin, Lewis: Journal of Geological Society of London, v. 131, p. 183–202.

Suter, M., 1984, Cordilleran deformation along the eastern edge of the Valles–San Luis Potosí carbonate platform, Sierra Madre Oriental fold-thrust belt, east Central Mexico: Geological Society of America Bulletin, v. 95, p. 1387–1397.

Tardy, M., 1980, Contribution a l'étude geologique de la Sierra Madre Orientale du Mexique: Memoire de these de Doctorat d'Etat, Universite Pierre et Marie Curie de Paris, 445 p.

Tchalenko, J. S., 1970, Similarities between shear zones of different magnitudes: Geological Society of America Bulletin, v. 81, p. 1625–1640.

Tchalenko, J. S., and N. N. Ambraseys, 1970, Structural analysis of the Dascht-e-Bayas (Iran) earthquake fractures: Geological Society of America Bulletin, v. 81. p. 41–60.

Varela-Santamaria, M., 1995, Una plataforma aislada en el sureste de Mexico: Tesis de Maestria, División de Estudios de Postgrado, Facultad de Ingeniería, Universidad Nacional Autónoma de Mexico, 194 p.

Viniegra, O. F., 1971, Age and evolution of salt basins in southeastern Mexico: AAPG Bulletin, v. 55, p. 478–494.

Wadge, G., and K. Burke, 1983, Neogene Caribbean plate rotation and associated Central American tectonic evolution: Tectonics, v. 6, p. 633–643.

Wilcox, R. E., T. P. Harding, and D. R. Seely, 1973, Basic wrench tectonics: AAPG Bulletin, v. 58, p. 1348–1396.

Williams, M. D., 1975, Emplacement of Sierra de Santa Cruz, eastern Guatemala: AAPG Bulletin, v. 59, p. 1211–1216.

Wilson, H. H., 1974, Cretaceous sedimentation and orogeny in Nuclear Central America: AAPG Bulletin, v. 58, p. 1348–1396

Sedimentary
Basins

Lawton, T. F., F. J. Vega, K. A. Giles, and C. Rosales-Domínguez, 2001, Stratigraphy and origin of the La Popa Basin, Nuevo León and Coahuila, Mexico, *in* C. Bartolini, R. T. Buffler, and A. Cantú-Chapa, eds., The western Gulf of Mexico Basin: Tectonics, sedimentary basins, and petroleum systems: AAPG Memoir 75, p. 219–240.

9

Stratigraphy and Origin of the La Popa Basin, Nuevo León and Coahuila, Mexico

Timothy F. Lawton
Institute of Tectonic Studies, New Mexico State University, Las Cruces, New Mexico, U.S.A.

Francisco J. Vega
Instituto de Geología, Universidad Nacional Autónoma de Mexico, Mexico City, Mexico

Katherine A. Giles
Institute of Tectonic Studies, New Mexico State University, Las Cruces, New Mexico, U.S.A.

Carmen Rosales-Domínguez
Instituto Mexicano del Petróleo, Exploración-Geociencias, Mexico City, Mexico

ABSTRACT

Strata exposed in the La Popa Basin of northern Mexico range in age from Late Jurassic through middle Eocene and record the evolution of the region currently lying in the foreland of the Sierra Madre orogen from extensional to contractional tectonics. The facies distribution and geometry of the strata demonstrate that salt diapirism was active at least from late Aptian–middle Eocene. Jurassic rocks include evaporites of the Oxfordian-Kimmeridgian Minas Viejas Formation, which is exposed only in diapiric bodies in the basin. The evaporites contain three types of blocks transported to their present levels during diapirism: (1) mafic to intermediate metaigneous rocks with latest Jurassic $^{40}Ar/^{39}Ar$ cooling ages (146 Ma); (2) laminated and nodular gypsum; and (3) Kimmeridgian limestone (Zuloaga Limestone). This Upper Jurassic lithic assemblage represents different parts of a formerly thick (at least 1000 m) stratigraphic section deposited in an extensional, or pull-apart, basin. Post-Jurassic rocks of the basin range from Early Cretaceous to middle Eocene and record both carbonate and siliciclastic deposition in dominant deltaic, shallow-marine, and tidal settings, as well as subordinate basinal and coastal-plain environments. This exposed section is at least 6400 m thick. The Lower Cretaceous section is thin (~100 m) and locally consists of carbonate biostromes deposited on bathymetric highs adjacent to a diapiric salt wall. These carbonates are late Aptian–late Albian in age and much thinner than their regional correlatives. The lower part of the Upper Cretaceous is represented by the Indidura and Parras Formations, the former a basinal carbonate, the latter a prodeltaic or basinal shale that underlies the Difunta Group, a constructional continental-margin clastic wedge or embankment that filled the La Popa Basin. The Difunta Group in the La Popa Basin spans the Maestrichtian–middle Eocene. Lenticular carbonate beds as much as 350 m thick are interbedded with mud-rock intervals of the Difunta Group and represent deposition on bathymetric highs created by rising salt bodies.

INTRODUCTION

The La Popa Basin, northwest of Monterrey, Mexico, contains the most complete exposures of Mesozoic strata preserved in the foreland of the Sierra Madre Oriental. Recent advances in understanding of the biostratigraphy and structure of the basin (Vega, 1987; Vega and Perrilliat, 1989, 1995; Vega-Vera and Perrilliat, 1989, 1990; Vega-Vera et al., 1989; Giles and Lawton, 1999) indicate the need for a fresh review of the basin's stratigraphy. In addition, deformed panels of strata adjacent to salt piercement structures expose a much more complete stratigraphic section than was previously recognized. Revisions to the age of the Mesozoic and Paleogene stratigraphy permit a more accurate understanding of the tectonic history and development of sedimentary basins in this part of the Laramide foreland of northeastern Mexico (Figure 1). Relations among salt structures and stratigraphic units reveal that diapirism was continuous during deposition of all stratigraphic units exposed in the La Popa Basin, or at least from the late Aptian through middle Eocene. Moreover, blocks composed of thermally metamorphosed igneous rock are present in the diapirs and provide information about the basin's early development.

This paper serves as a brief overview of the stratigraphy of the La Popa Basin. It builds on and modifies the stratigraphic work of McBride et al. (1974) and Laudon (1975). In general, we employ the stratigraphic nomenclature of McBride et al. (1974), introducing new names only where they are essential to understanding basin stratigraphy. In addition, we briefly interpret the tectonic history recorded by the stratigraphy of the La Popa Basin.

GEOLOGIC SETTING OF THE LA POPA BASIN

The La Popa Basin is located in the foreland of the Sierra Madre Oriental fold belt (Figure 1). The basin has an unusual map pattern that results from the interference of laterally extensive, northwest-trending detachment folds and salt withdrawal basins whose dimensions are as yet not well understood (Figure 2a). The folds involve and expose rocks younger than Late Jurassic and thus are inferred to be detached in Jurassic evaporite (Gray et al., 1997; Marrett and Aranda-García, 1999). The Jurassic evaporite is exposed in the basin only in salt piercement structures that include the El Gordo, El Papalote, and La Popa diapirs and also as local exposures of gypsum along the La Popa weld (Figure 2a). The gypsum in all diapirs contains large metaigneous and carbonate blocks that were transported upward in the evaporite to present levels of exposure.

The entire foreland of the Sierra Madre Oriental was shortened during the Hidalgoan orogeny, which is approximately synchronous with the Laramide orogeny of the U. S. Cordillera (Guzmán and de Cserna, 1963). Prior to the development of Hidalgoan detachment folds, which subsequently have been

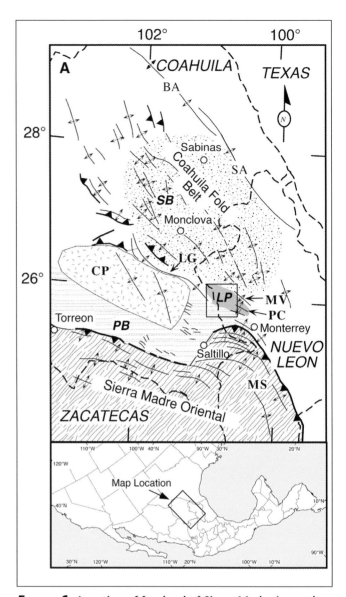

FIGURE 1. Location of foreland of Sierra Madre in northeastern Mexico (lower) and of the La Popa Basin in the foreland (upper). Explanation: BA, Burro arch; CP, Coahuila Platform; LG, La Gavia anticline; LP, La Popa Basin; MS, Monterrey salient of Sierra Madre fold belt; MV, detachment fold at Sierra Minas Viejas; PB, Parras Basin; PC, detachment fold at Potrero Chico; SA, Salado arch; SB, Sabinas Basin. Inset rectangle in the La Popa Basin is the area of Figure 2.

breached by erosion, the stratigraphy described in this paper was contiguous with that of the adjoining Sabinas and Parras Basins. Deepest burial was attained in the La Popa Basin near the end of the Hidalgoan event in the middle Eocene (50–45 Ma), followed by uplift that resulted in erosion of as much as 5–6 km of strata from the basin by the latest Eocene (~25 Ma; Gray et al., this volume). The basin forms a structural low between detachment folds that expose Lower Cretaceous limestone, with basement around the flanks of the basin at significantly shallower depth than in the basin center (Figure 2b).

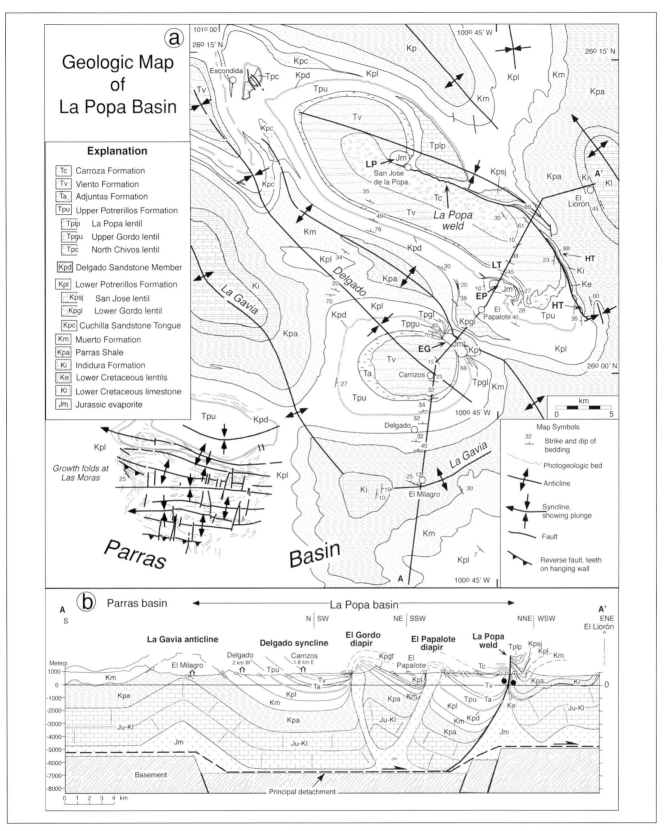

FIGURE 2. (a) Geologic map of the La Popa Basin and northwestern part of Parras Basin. Lower Potrerillos Formation includes Lower Siltstone, Lower Mudstone, and Middle Siltstone members; Upper Potrerillos Formation includes Upper Mudstone and Upper Sandstone members. BC, Boca La Carroza; EG, El Gordo diapir; EP, El Papalote diapir; HT, Cerro Huevo del Toro; LP, La Popa diapir; LT, Arroyo La Tinaja. (b) Cross section from northern Parras Basin across the La Popa Basin. Estimate of depth to basement is from sources in text. Both parts of the figure are modified from original geology by Dr. Hector Millán, University of Zaragoza, Zaragoza, Spain.

Contours on magnetic basement (Eguiluz de Antuñano, 1994) and stratigraphic reconstructions (Figure 2b) suggest that basement beneath the La Popa Basin lies at an elevation of -5500 to -7000 m. Paleogene strata in the Parras and Sabinas Basins largely were eroded during and after the mid-Tertiary uplift; therefore, the La Popa Basin contains the youngest foreland strata north of the Monterrey salient of the Sierra Madre fold belt (Figure 1).

STRATIGRAPHY

Rocks exposed in the La Popa Basin range in age from Late Jurassic through middle Eocene. The exposed rocks fall into two categories: (1) Upper Jurassic rocks present only in the diapirs; and (2) Lower Cretaceous through Eocene strata that comprise a continuously exposed stratigraphic section. Pre-Aptian Cretaceous rocks are not exposed at the surface but are present in the subsurface (Figure 3; Eguiluz de Antuñano, 1994). The descriptions here are restricted to rocks exposed at the surface. Although the stratigraphy is dominated by the Parras Shale and Difunta Group (McBride et al., 1974) of Late Cretaceous–middle Eocene age, recent biostratigraphic research indicates that older Cretaceous rocks are exposed locally. The stratigraphy of the La Popa Basin is dominantly siliciclastic, but lenticular carbonate units, termed *lentils* by McBride et al. (1974), are present locally. These lentils differ from correlative strata in the basin and adjoining region in that their facies record generally shallower water. These local shoals represent bathymetric highs associated in most cases with rising salt diapirs (McBride et al, 1974; Laudon, 1984).

Upper Jurassic Rocks

The diapirs consist of two main components—the diapiric evaporite itself and blocks included in the evaporite. The evaporite, consisting solely of gypsum, occupies diapiric stocks with dimensions on the order of 1–2.5 km across. The gypsum is pervasively deformed; it is foliated to mylonitic within several tens of meters of the diapiric margins, with foliation typically parallel to the adjacent contact with country rock. Although structural deformation has affected most of the gypsum, quarry exposures at the El Papalote diapir contain locally intact blocks of less deformed gypsum. The gypsum locally forms white, massive textures. Elsewhere, intact sections of gypsum contain chicken-wire or enterolithic textures consisting of slightly stretched nodules of gypsum in a mesh of dark-gray carbonate. The nodular evaporite is associated with meter-scale successions of gypsum and carbonate in laminations 1–2 mm thick. Evaporite fabrics resembling these are present in outcrop exposures of the Minas Viejas Formation in the Sierra Madre (Götte and Michalzik, 1992), suggesting that the evaporite blocks at the El Papalote diapir contain relict depositional fabrics. There is some subsurface evidence to suggest that much of the deformed gypsum probably represents a residuum, or caprock, remaining after the dissolution of halite from the

diapirs. For example, in Sierra Minas Viejas, 15 km southeast of the La Popa Basin (Figure 1), the Minas Viejas #1 well was spudded in gypsum on the crest of a detachment anticline and encountered 600 m (1967 ft) of gypsum, 900 m (2951 ft) of interbedded gypsum and halite, 2100 m (6885 ft) of halite, several meters of olivine basalt or dolerite, 520 m (1705 ft) of interbedded halite and black carbonaceous limestone, and finally finished in 370 m (1213 ft) of black carbonaceous limestone (Lopez-Ramos, 1982). The abundant gypsum and lack of halite in the upper part of the well may be interpreted as the result of halite dissolution. On the other hand, the succession encountered in the well may be a result of a long-term decrease in the salinity of the brines that deposited the Minas Viejas Formation. Such a salinity decrease might have come about as water of normal marine salinity had increasing access to the basin.

Blocks of nonevaporite lithologies are also present within the gypsum of the El Papalote diapir (Laudon, 1984; Garrison, 1998; Garrison and McMillan, 1999). The blocks, which include metaigneous and carbonate rocks, range in outcrop size from 1 m to as much as 200 m across. The igneous rocks are thermally metamorphosed to greenschist facies and include metaplutonic and metavolcanic rock types. Metaplutonic rocks include monzonite and biotite diorite with equigranular, phaneritic textures. Metavolcanic rocks include andesite and basalt with relict equigranular or porphyritic textures. Some blocks are vesicular. Vesicles are filled with secondary calcite and chlorite, and former igneous mineral assemblages in the rocks are altered to an assemblage of chlorite, epidote, actinolite, and albite. Chilled margins or concentric vesicle arrays suggestive of pillow-lava textures have not been observed in any igneous blocks. Trace-element characteristics of analyzed metavolcanic and metaplutonic rocks are similar, suggesting a common igneous origin by decompression partial melting of the asthenosphere during lithospheric extension, rather than by asthenospheric fluxing above a subducted slab (Garrison and McMillan, 1999). $^{40}Ar/^{39}Ar$ ages of 146 ± 1.6 Ma and 145.6 ±1.0 Ma on two metaplutonic samples probably represent metamorphic ages when the blocks cooled through 350° C in the latest Jurassic (Garrison and McMillan, 1999).

Carbonate blocks include several facies present in both the El Gordo and El Papalote diapirs (Figure 2a). The most common facies types are (1) black laminated dolostone; (2) bioclastic wackestone with interbedded intervals of gypsum and laminated gray carbonate mudstone; and (3) calcareous sandstone interbedded with fossiliferous, gray to grayish-brown sandy dolostone. A single block of facies type 2 in the northeastern corner of El Papalote diapir (Figure 4) has yielded *Nanogyra striata* (Smith) and *Buchia (Anaucella) concentrica* (Sowerby), a Kimmeridgian oyster and bivalve, respectively, known from the La Caja Formation (Buitron, 1984), as well as *Myophorella* sp. cf. *M. nodulosa* Bayle (Cox et al., 1969) and *Cymatoceras* (?) sp., an Oxfordian bivalve and nautiloid, respectively. A few badly preserved articulate brachiopods are also present in this block. A

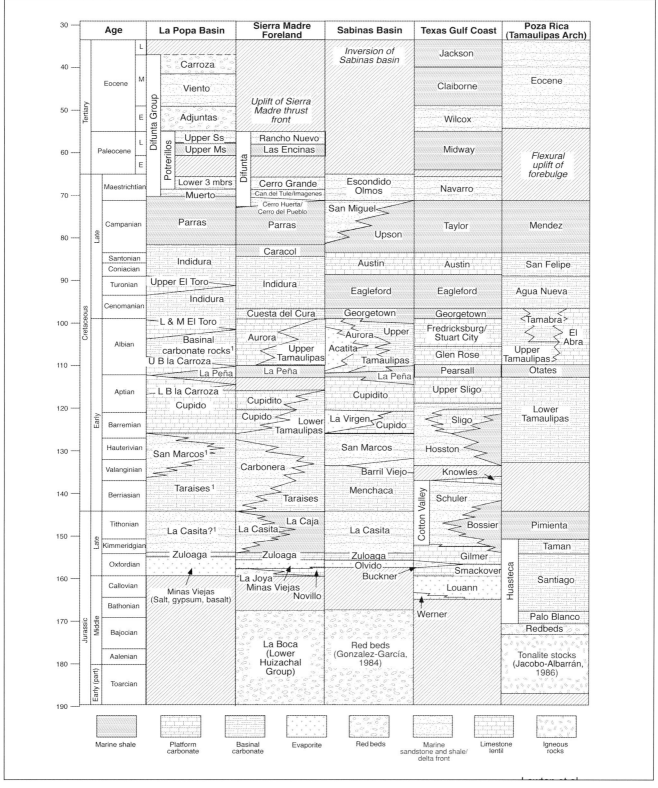

FIGURE 3. Correlation of stratigraphy of the La Popa Basin with surrounding regions. Intervals of chart with diagonal rule are hiatuses; descriptive phrases in hiatal intervals indicate interpretation of hiatus origin. Units indicated with superscript in La Popa Basin column are present only in the subsurface. Sources: La Popa Basin and Sierra Madre foreland— Eguiluz de Antuñano (1994), Fortunato and Ward (1982), Goldhammer (1999), Vega-Vera et al. (1989); Sabinas Basin— Cuevas-Leree (1985); Eguiluz, personal communication, 2000; Gonzalez-García (1984); Götte and Michalzik (1992); Lehman et al. (1998); Zwanziger (1992); Texas Gulf Coast—Goldhammer (1999), Imlay and Herman (1984), Scott (1984); Poza Rica—Cantú-Chapa (1979), Jacobo-Albarrán (1986), Torres et al. (1999), Viniegra and Castillo-Tejero (1970). Time scale is from Dickinson and Lawton (2001).

FIGURE 4. Block of Kimmeridgian limestone in the El Papalote diapir. View is toward the south. The horizontal width of the block (between upturned rims) is about 100 m.

large block of facies type 3 in the El Gordo diapir also contains *Nanogyra striata* (Smith), as well as *Codakia* sp. and articulate brachiopods.

The blocks of sulfate, carbonate, and igneous rocks incorporated in the diapiric gypsum comprise an assemblage of Upper Jurassic (Oxfordian-Kimmeridgian) lithologies that formed during lithospheric extension, asthenospheric melting, and initial subsidence of the La Popa Basin. The sulfates and carbonates were deposited in intertidal and supratidal environments that evolved from dominantly hypersaline to dominantly normal marine at the beginning of the Kimmeridgian. In the Sierra Madre, exposures of the laminated sulfate and carbonate facies have been interpreted as deposits of hypersaline, shallow subtidal to supratidal settings, and the nodular facies as the result of displacive diagenesis in carbonate muds (Götte and Michalzik, 1992). The massive gypsum has likewise been interpreted as deposits of shallow, subaqueous hypersaline settings (Götte and Michalzik, 1992). At Sierra Minas Viejas (Figure 1), the uppermost Minas Viejas Formation grades upsection through interbedded limestone, shale, and gypsum into the Zuloaga Limestone (Humphrey and Diaz, 1956; Longoria, 1984) and probably does so in the subsurface of the La Popa Basin as well, as indicated by the interbedded limestone and gypsum in the block at the El Papalote diapir.

In the Zuloaga Limestone of the Sierra Madre, similar facies to those present in the limestone blocks of the diapirs have been described and interpreted by Götte and Michalzik (1992): a bioclastic wackestone facies interbedded with gypsum and laminated carbonate mudstone represents lagoonal deposits that alternated between normal-marine and high salinities; black laminated carbonate mudstone represents deposits of a hypersaline, intertidal to shallow subtidal environment, and a

dolomite facies represents brine-reflux diagenesis. The carbonate facies, if interpreted as gradational with the upper part of the evaporite section, are consistent with development of increasingly normal marine conditions with time.

Mafic volcanic rocks have not been reported from exposed sections of the Minas Viejas Formation; however, intercalated halite and igneous rock are known from at least one subsurface example. As described above, the oil test at Sierra Minas Viejas encountered intervals of salt and basalt or diabase in what is probably the lower part of the Minas Viejas Formation (Lopez-Ramos, 1982). Based on the deep halite strata of the Minas Viejas well, Götte and Michalzik interpret an early episode of salt-pan deposition in the vicinity of the La Popa Basin (1992, their Figure 3). We infer that at least some of the mafic volcanic rocks were erupted onto and interbedded with the salt and subsequently were transported to present levels of exposure.

The limestone blocks at the El Papalote and El Gordo diapirs are Kimmeridgian in age. They are probably equivalent to the youngest Zuloaga Limestone rather than the La Casita and La Caja Formations, which are siliciclastic units where they are exposed in northeastern Mexico. The age of the Minas Viejas Formation is generally accepted as Oxfordian, and the age of the overlying Zuloaga is late Oxfordian–earliest Kimmeridgian (Figure 3; Götte and Michalzik, 1992). Because the limestone beds at the El Papalote diapir appear to be interbedded with gypsum, we also regard the uppermost part of the Minas Viejas Formation in the La Popa Basin as Kimmeridgian. This suggests that the Zuloaga Limestone in the basin is very thin and that evaporite deposition persisted longer in the La Popa Basin than in the evaporite ramp system in the present location of the Sierra Madre. Götte and Michalzik (1992) infer that in the Oxfordian, the Late Jurassic Mexican seaway south of the Coahuila block was separated from the Sabinas gulf, which occupied the Sabinas and La Popa Basins. Subsequent widespread transgression began to overtop basement blocks of the region, which resulted in increased marine conditions and deposition of the upper Oxfordian Zuloaga Limestone. Our data suggest that the La Popa and Sabinas Basins were not incorporated into the wider marine province until the Kimmeridgian.

Two incompletely resolved issues concern the metaigneous blocks in the diapirs: (1) the apparent cooling ages of the metaplutonic rocks are younger than the depositional ages of the salt and the carbonate blocks, yet the carbonate rocks are apparently unmetamorphosed; and (2) the metaplutonic rocks were somehow exhumed to be entrained in the evaporite, which suggests that they might be older than the evaporite.

These observations, in combination with the petrologic affinity of the metaplutonic and metavolcanic blocks, seem to refute our conclusion that the three diapiric block types are essentially all part of a genetically related Minas Viejas rift assemblage. Garrison and McMillan (1999) addressed the second problem by arguing that the metaplutonic blocks are intrusions into the older basement that were exhumed and exposed in horsts of the extensional terrane prior to final burial by evaporite deposits. This solution is not completely satisfying because there are no blocks in the diapirs of older basement into which the rift magmas intruded. The basement, exposed in the Coahuila block and encountered by oil wells in the Tamaulipas arch, consists of Grenville metamorphic rocks intruded by Permo-Triassic granites of arc affinity (Jacobo-Albarrán, 1986; McKee et al., 1988, 1999; Lopez, 1997; Torres et al., 1999). An observation that argues against a postsalt age for the metaplutonic blocks is the absence of other post-Kimmeridgian (e.g., Cretaceous, Tertiary) stratal blocks in the diapirs, indicating that the diapirs did not tend to sample postevaporite lithologies.

We conclude that the metaplutonic rocks were intruded into the lower part of the evaporite section contemporaneously with eruption of flows onto surface evaporites. Subsequently, the igneous rocks were metamorphosed, possibly by circulating hydrothermal fluids, near the end of the Jurassic, after deposition of the Kimmeridgian limestone. We suggest that the contrast in metamorphism between the igneous and carbonate blocks resulted simply from the original stratigraphic separation between them by the bulk of the Minas Viejas Formation. By analogy with the intercepts in the Minas Viejas #1 well, the mafic flows and intrusions were restricted to the lower part of the evaporite section. In contrast, the carbonate rocks were interbedded with the upper part of the evaporite section, which reaches a thickness of 600 m in the Sierra Madre (Götte and Michalzik, 1992) and was probably thicker in the deeply subsided La Popa Basin (Figure 2). High heat flow and circulating fluids would have had much greater impact on the deeper rocks, particularly after burial beneath the uppermost Jurassic section. The high thermal conductivity of the salt may have dissipated heat rapidly enough to prevent significant heating of the younger limestones. Finally, the apparent lack of pillows in the vesicular basalts in the diapir is consistent with major magmatism during the early development of the La Popa Basin while it was a salt pan, followed by a general absence of volcanism by the time the basin developed hypersaline, marginal-marine conditions.

In summary, during the Late Jurassic, the La Popa Basin was a pull-apart or rift basin. Evaporite deposition accompanied continental rifting, which began in the Middle Jurassic and continued into the Late Jurassic (Marton and Buffler, 1994; Dickinson and Lawton, 2001). Deposition began with a salt-pan phase accompanied by eruption and shallow intrusion of mafic magma. During this phase, the basin was internally drained and isolated from the world ocean. As rifting proceeded, halite deposition gradually gave way to gypsum deposition in a sabkha setting; widespread flooding of northeastern Mexico resulted in limestone deposition in the La Popa Basin by the early Kimmeridgian. The interpreted abrupt change in depth to basement at the edges of the La Popa Basin (Figure 2b) is the result of normal faults formed during crustal extension. These faults probably localized salt deposition while the La Popa Basin was isolated from the global ocean.

Lower Cretaceous Rocks

Lower Cretaceous strata in the La Popa Basin, although very thin by comparison with their regional counterparts, correlate with the Cupido, La Peña, Aurora, and Cuesta del Cura Formations (Figures 3 and 6). These strata are exposed only adjacent to the southeastern leg of the La Popa salt weld, in what we interpret as the footwall of the weld (Giles and Lawton, 1999), where steeply dipping beds face east and consist of dominant limestone interbedded with subordinate shale. The limestones range in age from late Aptian through Turonian (Figure 6). The exposed Lower Cretaceous section is approximately 100 m thick, an order of magnitude less than its typical regional thickness. We interpret the thinned section as the result of reduced accommodation adjacent to a diapiric salt wall that

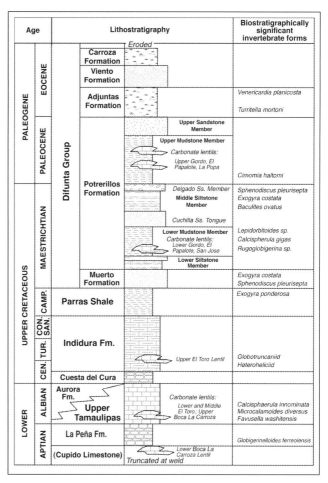

FIGURE 5. Cretaceous and Paleogene stratigraphy of the La Popa Basin.

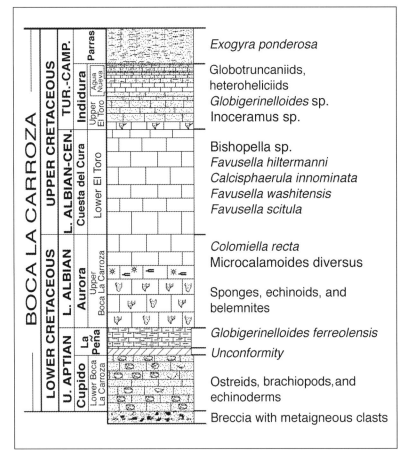

FIGURE 6. Lower Cretaceous stratigraphy exposed along the La Popa weld in the eastern part of the La Popa Basin. Local names defined here are given in the column adjacent to the graphic column; regional correlatives are in the adjoining column to the left. Biostratigraphically significant microfossils are listed to the right of the graphic column. Vertical scale is schematic.

was active during deposition of the Lower Cretaceous strata. These strata differ in facies significantly from time-equivalent basinal carbonates of the Cuesta del Cura, Upper Tamaulipas (Tamaulipas Superior), and Indidura Formations, and they are unlike shelfal facies of the partly equivalent Cupido and Aurora Formations. Named Lower and Upper Boca La Carroza lentils and Lower, Middle, and Upper El Toro lentils, they take their names from exposures at Boca La Carroza and Cerro Huevos del Toro, near the southeasternmost tip of the weld (Figure 2a).

Lower Boca La Carroza Lentil

At Boca La Carroza, the Lower Cretaceous strata consist of two lentils and an intervening shale-limestone unit that range in age from late to early Albian. The base of the limestone is structurally truncated at the weld. The basal lentil, about 1 m thick, consists of oyster (ostreoid) packstone with conspicuous orange-weathering igneous clasts. Although it has not yielded diagnostic forms, we tentatively correlate it with the Cupido

Limestone, based on its stratigraphic position (Figure 6). The igneous clasts were derived from exposed diapiric gypsum that lay adjacent to the limestone prior to weld development. Because this limestone differs from typical Cupido Limestone, we term it the Lower Boca La Carroza lentil (Figure 6).

La Peña Formation

The Lower Boca La Carroza lentil is overlain by the La Peña Formation, thinly bedded carbonate mudstone and shale with elongate black chert nodules (Figure 7). The La Peña is 11 m thick and overlies the Lower Boca La Carroza lentil on a sharp contact marked by a chert- and limestone-clast conglomerate that is 10–15 cm thick. Reddish-brown carbonate mudstone in the lowermost part of the formation contains abundant *Globigerinelloides ferreolensis* (Moullade) (Figure 8), recently reported from the upper Aptian La Peña Formation (Longoria et al., 1999). In the Sierra Madre Oriental, the La Peña is a similar thin-bedded cherty shale 66 m thick (Imlay, 1937; Longoria et al., 1999).

Upper Boca La Carroza Lentil

Above the shale, the Upper Boca La Carroza lentil consists of sponge-bearing, fossiliferous wackestone in its lower part (Figure 9) that grades upsection into thick-bedded carbonate mudstone. The upper lentil is about 90 m thick at Boca La Carroza. It contains *Colmiella recta* Bonet, a lower Aptian tintinnid. We correlate this lentil with the Aurora Limestone (Figures 3 and 6).

Lower El Toro Lentils

At the southeastern tip of the weld, the lower two of three El Toro lentils consist dominantly of thick-bedded carbonate mudstone with facies similar to the lentils at Boca La Carroza. Although massive, the limestone is divided by a recessive limestone interval into two prominent lentils that we term the Lower and Middle El Toro lentils. Those lentils are upper Albian in age (Figure 6). Key forms from the lower lentil include *Favusella washitensis* (Carsey) from the upper Albian, and *F. scitula* Michael and *Calcisphaerula innominata* Bonet from the Albian-Santonian. Fossils from the Middle El Toro lentil include *Microcalamoides diversus* Bonet, an Albian form (Figure 8).

Upper Cretaceous Rocks

The Upper Cretaceous section in the La Popa Basin includes the Indidura Formation, the Parras Shale, and the lower part of

FIGURE 7. The La Peña Formation at Boca La Carroza. Stratigraphic top is to the left. Contact with Lower Boca La Carroza lentil (arrow) is overlain by chert- and limestone-clast conglomerate. The La Peña consists of shale and chert nodules (c) elongate parallel to bedding. Scale bar in upper left is 15 cm long.

the Difunta Group (Figure 3). We describe the Difunta Group separately, in a section on Upper Cretaceous–Paleogene rocks.

Indidura Formation

The Indidura Formation is widely exposed on the east flank of the La Popa Basin on both sides of Mexico Highway 53 north of Mina. It is also present in the La Gavia anticline (Figure 2a) on the south flank of the La Popa Basin. In the basin, it is exposed in low ridges stratigraphically above the Lower Cretaceous lentils along the weld. In the La Gavia anticline west of the town of El Milagro, it forms light gray, ledgy exposures of rhythmically interbedded calcareous shale and argillaceous carbonate mudstone. Beds are 30 cm to 1 m thick and weather light gray to tan. Abundant large, thin-shelled inoceramid bivalves are present in the carbonate mudstone. In the upper part of the formation, tan to yellowish-gray tuff beds are present in the section. In the region, the Indidura ranges in age from middle Cenomanian to Santonian (Longoria et al., 1999). The upper tuffaceous part correlates with the Caracol Formation of the Sierra Madre, which is late Santonian–early Campanian (Longoria et al., 1999). The Caracol southwest of Saltillo in the Sierra de Parras, defined by Imlay (1937), consists of

interbedded shale and waxy tuffs deposited by turbidity currents. The Indidura likewise represents deep-water, basinal deposits.

Upper El Toro Lentil

The Upper El Toro lentil is stratigraphically in the Indidura Formation directly northeast of the weld. It is about 8 m thick and consists of wackestone with large, hemispherical sponges as much as 50 cm in diameter. It lies 100 m upsection of the Middle El Toro lentil and contains heteroheliciid (middle Albian–Maestrichtian) and globotruncaniid (Turonian–Maestrichtian) forms (Figure 8). We interpret its age as Turonian and correlate it with part of the Agua Nueva Member of the Indidura Formation (Figure 6).

Parras Shale

The Parras Shale lies gradationally on the Indidura Formation. It attains a thickness of 1500 m in the Parras Basin (Weidie, 1961; Weidie and Murray, 1967) and is locally of similar thickness in the La Popa Basin but thins to about 400 m adjacent to the La Popa weld (Figure 2b). The lower part of the Parras Shale consists of medium-gray to black fissile to subfissile

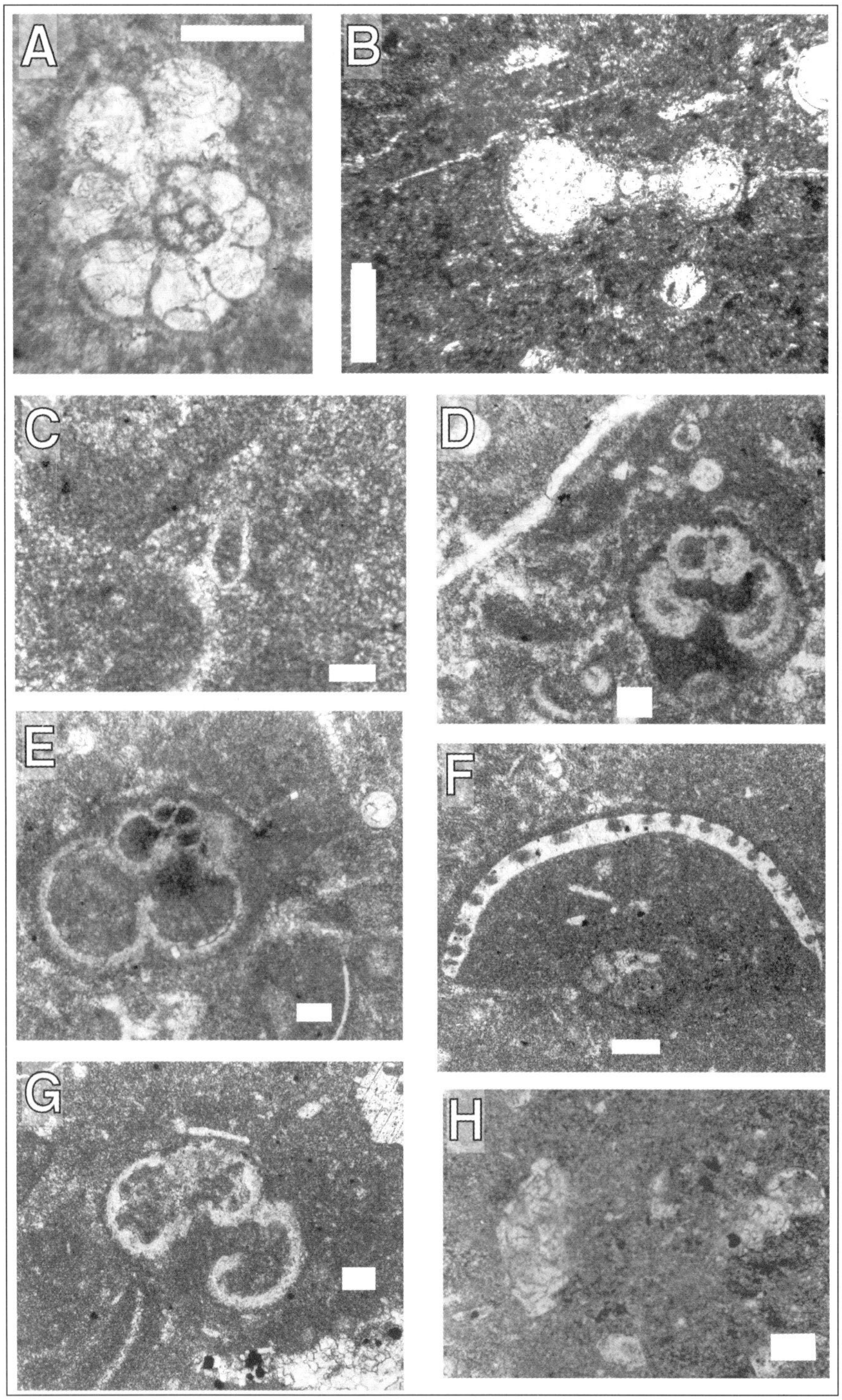

FIGURE 8. Selected microfossils from carbonate lentils near the La Popa weld. (a, b) Planktonic foraminifer, *Globigerinelloides ferreolensis* (Moullade), late Aptian, lower part of La Peña Formation at Boca La Carroza. (c) Tintinnid, *Colomiella recta* (Bonet), early Albian, Upper Boca La Carroza lentil. (d) Planktonic foraminifer, *Favusella washitensis* (Carsey), and calcisphaerulid, *Calcisphaerula innominata* Bonet (spherical objects), both middle Albian–Cenomanian, Lower El Toro lentil (middle Albian). (e) Planktonic foraminifer, *Favusella scitula* (Michael), late Albian, Lower El Toro lentil. (f) Pelagic ostracode (?), *Microcalamoides diversus* Bonet, early Albian, Middle El Toro lentil (lower–middle Albian). (g) Planktonic foraminifer, *Favusella* cf. *F. hiltermanni* (Loeblich and Tappan), Albian, Middle El Toro lentil (lower–middle Albian). (h) Heteroheliicid and small globotruncaniid, late Albian, Upper El Toro lentil (Turonian). Scale bars = 100 mm.

mudrock that becomes brownish-gray and silty upsection. At the top of the middle third of the formation, thin, very fine-grained sandstone beds 1–10 cm thick are present in the mudrock. The beds are continuous for tens to hundreds of meters in outcrop and have asymmetrical flute casts on their bases indicating east-southeast sediment dispersal. These beds are interpreted as basin-plain turbidites. The contact with the overlying Muerto Formation is gradational and marked by a succession of thin, very fine-grained sandstone beds with hummocky cross-stratification. The contact is designated at the base of the lowermost sandstone greater than 1 m thick (McBride et al., 1974). In the Parras Basin, the formation contains *Exogyra ponderosa*, a Campanian oyster (Weidie and Murray, 1967).

The Parras Shale represents prodelta deposits of a prograding clastic wedge (Weidie and Murray, 1967; McBride et al., 1973, 1975). Paleocurrent data from the overlying Difunta Group indicate that progradation was eastward in the Parras Basin and southeastward in the La Popa Basin (McBride et al., 1975). The vertical succession in the Parras Shale records a transition from basinal deposits through prodelta turbidites and into lower shoreface storm deposits.

Upper Cretaceous–Eocene Rocks (Difunta Group)

The interval of strata that encompasses the Difunta Group was defined as the Difunta Formation on the southern flank of the Parras Basin, southwest of Saltillo (Imlay, 1937). The Difunta Group was defined by Murray et al. (1962) to include all strata in the Parras Basin between the Parras Shale and the top of the Rancho Nuevo Formation, the uppermost exposed unit in the Parras Basin. The name was subsequently extended to the La Popa Basin to include all strata above the Parras Shale (McBride et al., 1974). This usage has persisted, although the upper formations of the Difunta Group in the La Popa Basin are now known to be younger than the Rancho Nuevo Formation (Figure 3; Vega and Perrilliat, 1989, 1995; Vega-Vera and Perrilliat, 1989; Vega-Vera et al., 1989).

Whereas the group is essentially all silici-clastic in the Parras Basin, it includes limestone lentils as much as 350 m thick in the La Popa Basin.

The Difunta Group was deposited in the foreland basin of the Sierra Madre Oriental by dispersal systems that transported sediment to the east in the Parras Basin and southeast in the La Popa Basin (McBride et al., 1974; McBride et al., 1975). Polarity of the depositional systems is based on facies distribution (Weidie and Murray, 1967; McBride et al., 1974) and paleocurrent data (McBride et al., 1975), although paleocurrent measurements also indicate that much of the detritus in the Parras Basin was derived from the south (Weidie and Murray, 1967). The part of the Parras Basin lying between Torreon and Saltillo (Figure 1) contains interbedded continental and shallow-marine deposits, whereas most of the La Popa Basin and the Parras Basin east of Saltillo lack continental strata. McBride et al. (1973, 1975) infer that separate river systems fed the Parras and La Popa Basins.

FIGURE 9. Sponge wackestone at base of Upper Boca La Carroza lentil.

Muerto Formation

The Muerto Formation represents the first major influx of coarse-grained detritus into the Parras and La Popa Basins. The formation has a type locality on the south flank of Delgado syncline (Figure 2a), where it is 570 m thick (McBride et al., 1974). It changes thickness dramatically in the vicinity of salt structures, thinning from 695 m to 67 m along a transect west of the El Gordo diapir beginning 4.5 km from the diapir and ending at the diapir (Weislogel, 2001). It pinches out adjacent to the La Popa weld both by lateral thinning and truncation beneath younger beds (Figure 10). The maximum thickness of the Muerto Formation reported from the La Popa Basin is 695 m near El Gordo diapir, but the formation appears to thicken westward.

The Muerto Formation consists of interbedded sandstone and subordinate siltstone and shale. It is resistant to weathering and forms rugged ridges throughout the La Popa Basin. Sandstones contain abundant hummocky-swaley cross-stratification and trough cross-beds in the lower part. Abundant *Ophiomorpha* and local ammonites and inoceramid fragments are present in these sandstones. The middle part of the Muerto contains abundant heterolithic strata, some of which are inclined to form accretion sets, wavy bedding, asymmetrical ripple cross-lamination, and meter-scale beds of very fine-grained sandstone packed with oysters. An upper sandstone-rich interval contains abundant cross-beds, interference ripples, and inclined heterolithic strata.

In general, the Muerto Formation represents a wave-dominated deltaic system (McBride et al., 1973, 1975) that consists of shoreface, foreshore, tidal, and lagoonal deposits that constitute an immense progradational clastic shoreline system (Weislogel, 2001). The formation contains abundant volcanic lithic grains and plagioclase, indicating a volcanic source (Tardy and Maury, 1973; McBride et al., 1975). The ammonite *Sphenodiscus pleurisepta* (Conrad) is found throughout the formation, indicating a Maestrichtian age (Figure 5).

Potrerillos Formation

The Potrerillos Formation is the thickest unit of the Difunta Group. It is generally concordant with the Muerto Formation but overlies the Muerto with marked angular discordance adjacent to the La Popa weld (Figure 10). Striking intraformational unconformities are also present in the Potrerillos Formation within 500 m of the El Papalote and El Gordo diapirs. These unconformities indicate that salt diapirism was active during deposition of the Potrerillos Formation (McBride et al., 1974; Laudon, 1984, 1996). McBride et al. (1974) defined five major members in the Potrerillos Formation. In ascending order, they are the Lower Siltstone, Mudstone (which we modify to Lower Mudstone), Middle Siltstone, Upper Mudstone, and Upper Sandstone Members. (Although not named after specific geographic localities, these members were capitalized and thus considered formal.) McBride et al. (1974) also defined a sandstone tongue, the Cuchilla Tongue, which lies stratigraphically near the middle of the Middle Siltstone Member in the western part of the La Popa Basin. We define an additional member, the Delgado Sandstone Member, which separates the Upper Siltstone and Upper Mudstone Members. It is an important marker at the top of the Cretaceous that can be traced throughout the La Popa Basin and takes its name from a village 7 km south-southwest of the El Gordo diapirs and the La Popa Weld (Figure 2). The Potrerillos Formation also contains conspicuous

FIGURE 10. View west of growth strata adjacent to the La Popa weld, which crosses the saddle at the black arrow. Explanation: Kp, Parras Shale; Km, Muerto Formation (delineated between black lines in lower right); Kpo, Potrerillos Formation; Tv, Viento Formation. Muerto thins away from the viewer and pinches out beneath angular unconformity (at white arrow), which causes Potrerillos to directly overlie Parras on the skyline. Rotated, intraformational normal fault (beneath the horizontal white arrow) is truncated beneath the Muerto Formation, although truncation is concealed behind the ridge with limestone talus.

limestone lentils adjacent to the El Papalote and El Gordo diapirs and the La Popa Weld. The Potrerillos Formation is 2300 m thick at its type locality 2 km east and southeast of the El Papalote diapir (McBride et al., 1974).

The Lower Siltstone Member consists predominantly of gray siltstone with subordinate intervals of extensively bioturbated, brown-weathering, very fine grained, calcite-cemented sandstone beds that are black on fresh surfaces. Beds are 1–2 m thick. The member contains *Sphenodiscus pleurisepta* (Conrad), *Exogyra costata*, and other Maestrichtian invertebrates (McBride et al., 1974; Vega-Vera and Perrilliat, 1990). It represents deposits of the delta platform, which occupies a position seaward of the delta front but above a break in slope that marks the upper extent of the prodelta region (McBride et al., 1975).

The Lower Mudstone Member consists of dark-gray to black shale and siltstone with thin, sharp-based beds of very fine-grained sandstone 1–20 cm thick. McBride et al. (1974) noted the "flyschlike" character of this unit, which is greater than 660 m thick at the type locality. The Mudstone Member represents turbidites of the delta platform (McBride et al., 1975) or prodelta region. The member records abrupt flooding of the La Popa Basin after deposition of the Lower Siltstone Member. At or near the base of the Lower Mudstone Member are three limestone lentils that are probably correlative. These include the Lower Gordo lentil (Figure 11), the lowermost lentil at the El Papalote diapir, and the San Jose lentil. Although laterally

variable in detail, these limestones are characterized by packstones and grainstones of red algal fragments and *Pseudorbitoides* forams. Large, red algal pisolites, or rhodoliths, as much as 10 cm in diameter are present in the lower lentil at the El Papalote diapir (Hunnicutt, 1998) and in the Lower Gordo lentil. Where measured, the Lower Gordo lentil is 60 m thick and the San Jose lentil is 38 m thick (McBride et al., 1974); directly north of the El Gordo diapir, the Lower Gordo lentil is about 100 m thick (Figure 11). The lower lentil at the El Papalote diapir attains a thickness of 11 m (Hunnicutt, 1998). Because it directly flanks the diapir, the lower lentil at the El Papalote diapir was formerly interpreted to represent exotic limestone transported to its present level by the diapir (Laudon, 1984, 1996). Our mapping indicates that the lentil may be traced laterally into the Potrerillos Formation. In addition, it contains the foraminifera *Lepidorbitoides* sp. and *Rugoglobigerina* sp., and *Calcisphaerula gigas*, confirming its Maestrichtian age.

The Middle Siltstone Member is dominantly light to medium olive-gray siltstone and sandy siltstone arranged in upward-coarsening successions, or parasequences, 5–30 m thick. Bedding is typically indistinct as a result of bioturbation, but hummocky cross-stratification is preserved locally in 15- to 60-cm sharp-topped beds of very fine grained, well-sorted sandstone. The parasequences are capped locally by lags of fine-grained sandstone with shell fragments, shark teeth, and oyster shells that represent winnowed transgressive deposits on flooding surfaces. One of these lags west of the El Papalote diapir is 3 m thick and contains articulated oysters as much as 25 cm in size, whole ammonites *(Sphenodiscus)*, bivalves, gastropods, and clasts of limestone and mafic metavolcanic rocks as much as 25 cm in diameter. Whole oysters are present in some thick siltstone intervals. Trace fossils include *Ophiomorpha*, *Thalassinoides*, and *Gyrolithes*.

The Middle Siltstone Member is 580 m thick at its type section east of the El Papalote diapir; the base is defined as that point at which siltstone exceeds 50% of the section (McBride et al., 1974). The upper part is gradational with amalgamated pebbly sandstone beds we define as the Delgado Sandstone Member, although near the diapirs, this contact is sharp and erosional. The Middle Siltstone Member was deposited on the delta platform (McBride et al., 1975) in shelfal and lower shoreface settings and possibly in lagoonal or bay environments. It is progradational with respect to the underlying Mudstone Member.

The Cuchilla Tongue is an eastward-thinning interval, composed dominantly of sandstone. It is present in the Middle Siltstone Member in the western part of the La Popa Basin, where it attains a thickness of 360 m (McBride et al., 1974). It pinches out in the central part of the basin (Figure 2a) and represents delta-plain, delta-front, and delta-platform deposits (McBride et al., 1974; McBride et al., 1975). A distributary-channel sandstone in the delta-plain deposits of the Cuchilla Tongue provides the only major nonmarine paleocurrent data in the La Popa Basin and indicates east-southeast sediment transport in the western third of the basin (McBride et al., 1975).

Delgado Sandstone Member

We here designate a persistent sandstone at the top of the Middle Siltstone Member as the Delgado Sandstone Member of the Potrerillos Formation. The type locality is a prominent exposure on the northeastern limb of the Delgado syncline, 500 m west of the El Gordo diapir (Figure 12). At this locality, the Delgado Member sharply overlies olive-gray siltstone and includes two upward-coarsening successions of fine-grained, hummocky-swaley sandstone beds 30–60 cm thick overlain by medium-grained sandstone with trough cross-beds. Chert pebbles, bivalves, and gastropods are locally concentrated as lags in scour troughs. Robust *Ophiomorpha* and clasts composed of *Sphenodiscus pleurisepta* (Conrad) filled with white siltstone and derived from the Middle Siltstone Member are present locally. The sandstone is overlain by 2 m of conglomerate with chert pebbles, white siltstone clasts, and oysters filled with white siltstone. *Sphenodiscus* and *Cimomia haltomi* (Aldrich), a Paleocene nautiloid known from the Rancho Nuevo Formation in the Parras Basin (McBride et al., 1974), are present as clasts in this conglomerate. We provisionally assign this thin Paleocene conglomerate to the basal part of the overlying Upper Mudstone Member of the Potrerillos Formation (Figure 12). The total thickness of the Delgado Sandstone Member at the type locality is 10 m.

Five hundred meters west of the El Papalote diapir, the Delgado Sandstone Member is similar to the occurrence west of the El Gordo diapir. It consists of 17 m of sandstone beds with hummocky cross-stratification and lags of chert pebbles and abundant metaigneous clasts derived from the diapir. It thins to a few beds of very fine grained sandstone on the western flank of the diapir (Figure 13). It is unconformably overlain by a thin conglomerate bed. East of both the El Gordo and El Papalote diapirs, the Delgado Member consists of interbedded siltstone and 15- to 50-cm beds of very fine grained sandstone with hummocky-swaley cross-stratification. This lower-shoreface interval is overlain unconformably by the Paleocene conglomerate.

FIGURE 11. View west across the El Gordo diapir of folded strata of the Muerto and Potrerillos Formations. The Potrerillos includes the Lower Gordo lentil, the type locality of the Delgado Sandstone Member, and the Upper Gordo lentil. The Lower Gordo lentil on the right skyline is about 100 m thick.

The Delgado Sandstone Member is latest Maestrichtian in age. It is interpreted to represent a shoreface system that capped the progradational Upper Siltstone Member and deepened from west to east. Upper and middle shoreface deposits are interpreted to be present on the west sides of the diapirs and lower shoreface deposits on their east sides. In the vicinity of the diapirs, the fossil evidence suggests that as much as five million years—from latest Maestrichtian to middle Paleocene—are missing at the unconformity between the Delgado Sandstone Member and the Upper Mudstone Member. Because the depositional systems below and above the unconformity are similar to lower and middle shoreface deposits, it is possible that very little time is missing at this unconformity elsewhere in the La Popa Basin.

The Upper Mudstone Member unconformably overlies the Delgado Sandstone Member. The basal part is a lenticular conglomerate that is not present everywhere. The conglomerate is thin west of both diapirs. West of the El Papalote diapir, it consists of 15–30 cm of conglomerate consisting of chert pebbles, oyster shells, bivalves, and cobbles of metaigneous rock, limestone, and orange calcareous siltstone. The conglomerate has a planar base and a matrix of fine- to medium-grained sandstone, and it ranges from clast to matrix supported. It has already been described west of the El Gordo diapir (Figure 12). Southeast of the El Papalote diapir, the conglomerate is 0–6.5 m thick and fills swales several tens of meters across. It rests on a sharp base and contains tightly packed oyster fragments and rounded clasts of very light gray sandy siltstone, some of which are oyster molds. One clast of very fine grained sandstone with a length of 1 m was observed. Clasts are present in diffuse, laminar, coarse-grained sandstone beds 30–40 cm thick. Tightly packed oysters locally make up 50% of the rock, and chert pebbles are present in the upper part. The conglomerate is overlain on a sharp contact by lenticular beds of brown-weathering, fine-grained sandstone with hummocky-swaley cross-stratification and convex-upward bed tops. East of the El Gordo diapir, the conglomerate consists of a thin oyster lag, locally with chert pebbles, overlain by glauconitic sandstone. The conglomerate rests with angular unconformity on the Delgado Sandstone Member.

Above the basal conglomerate are interbedded sandstone and siltstone that grade upsection into black, subfissile mudrock with large brown siderite concretions. Black shale is also present above the interval of siderite concretions, and it grades upsection into sandstone of the Upper Sandstone Member. *Cimomia haltomi* (Aldrich) is present at the base and near the top of the member (Vega-Vera et al., 1989; Vega and Perrilliat, 1995; Vega et al., 1999). *Venericardia (Baluchicardia) francescae* is present near the middle of the member (Vega and Perrilliat, 1995). All fossils recovered from the Upper Mudstone Member indicate a Paleocene age. The member is 400 m thick east of the El Papalote diapir (McBride et al., 1974); it thins to only 86 m on the east flank of the diapir.

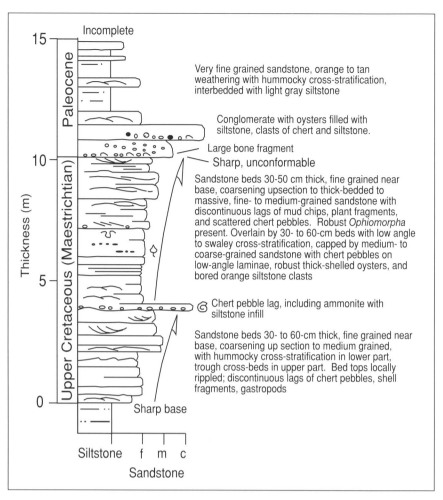

FIGURE 12. Measured section of the Delgado Sandstone Member of the Potrerillos Formation at type locality 500 m west of the El Gordo diapir.

The Upper Mudstone Member is interpreted to rest on a major sequence boundary and to constitute the transgressive systems tract and lower part of the highstand systems tract above that boundary. We interpret the oyster-rich basal conglomerate as a valley-fill deposit that rests on the sequence boundary and forms the base of the transgressive systems tract. The interval of siderite concretions is interpreted to represent the maximum flooding surface, and the shale above it represents the lower part of a highstand systems tract that also includes the overlying Upper Sandstone Member of the Potrerillos Formation.

The transgressive part of the Upper Mudstone Member contains the youngest limestone lentils in the La Popa Basin (Figure 5). These include five lentils at the El Papalote diapir (all except the lowermost lentil), the Upper Gordo lentil at the El Gordo diapir, and the La Popa lentil north of the La Popa weld. The lentils at the El Papalote diapir interfinger with the lower half of the Upper Mudstone Member (Figure 14). The lentils are thickest near the diapir, attaining individual thickness of as much as 70 m. They thin laterally and pinch out into the surrounding black shale within 1 km of the diapir. Individual lentils overlie unconformities that are strongly discordant

within a few tens of meters of the diapir and become concordant within a distance of 500 m from the diapir. The lentils are dominated by molluscan wackestone to packstone with abundant oysters and echinoid fragments. At their bases, they consist of conglomerate with clasts of older lentils and metavolcanic rocks. The lentils are onlapped by siliciclastic strata that range from thin-bedded turbidites to hummocky cross-stratified beds with fluid-escape structures. The turbidites alternate with beds of black shale, whereas overlying beds with hummocky cross-stratification are interbedded with olive-gray siltstone.

The carbonate lentils generally represent shallow-marine, subtidal rocks deposited adjacent to the El Papalote diapir. They grade laterally to correlative calciturbidites and debris flows derived from the carbonate accumulations. Unconformities and thinning adjacent to the diapir suggest that these turbidites were triggered by diapiric rise, probably during episodes of transgression and sediment starvation, such as are recorded by the correlative Upper Mudstone Member, that permitted development of diapiric topography. Intervening siliciclastics record shoaling and onlap, as increased rates of sedimentation overtook the constantly rising salt body.

FIGURE 13. Delgado Sandstone Member directly west of the El Papalote diapir. At this locality, the Delgado Sandstone Member consists of very fine grained sandstone with hummocky-swaley cross-stratification.

FIGURE 14. View northeast of El Papalote lentil 3, directly southeast of the El Papalote diapir. The lentil thins to the south (right) and interfingers with shale of the Upper Mudstone Member of the Potrerillos Formation. The lentil is seen twice in this view, because it is repeated by a normal fault that trends parallel to the plane of the photo. Explanation: Tpu, Upper Sandstone Member of Potrerillos Formation; Ta, Adjuntas Formation; Tv, Viento Formation. Arrow in the foreground indicates the Delgado Sandstone Member of the Potrerillos Formation.

At the El Gordo diapir, the Upper Gordo lentil correlates with the five upper carbonate lentils at the El Papalote diapir. Near the diapir, it consists of red algal, coral and sponge boundstone. Flanking the boundstone are more distal sponge-red algal packstones deposited as calciturbidites and fore-reef debris (Mercer and Giles, 1999). Apparently correlative thin beds of red algal-sponge grainstone form discontinuous exposures on the crest of the diapir. The Upper Gordo lentil attains a thickness of 50 m (McBride et al., 1974) and represents a bioherm or reef developed on the flank of the diapir, facing the open ocean. This reef shed debris into deeper water, and the thin-bedded grainstones on the diapir were probably swept from the reef onto the adjacent diapir by storms.

The La Popa lentil, lying north of the La Popa weld, creates the most prominent geographic feature in the La Popa Basin (Figure 15). The lentil is as much as 350 m thick, with a core of sponge and coral boundstone flanked by beds of strombolitic packstone with interbeds of allodapic grainstone (R. K. Goldhammer, personal communication, 2000).

The Upper Sandstone Member gradationally overlies the Upper Mudstone Member and forms the uppermost member of the Potrerillos Formation. It forms a prominent escarpment at its type section east of the El Papalote diapir, where it is 275 m thick (Figure 16; Shelley, 2001). Directly north of the El Papalote diapir, the member is only 125 m thick, indicating thinning adjacent to the diapir. The member consists mostly of fine-grained sandstone, ranging from very fine to rare, medium-grain sizes. In its lower part, it consists of thick beds with hummocky cross-stratification. Its upper part contains abundant planar-tabular cross-beds, locally with bidirectional foreset dips and lenticular pebbly beds that fill scours in the cross-stratified facies. In spite of their locally bidirectional nature, foreset dips are dominantly to the east. Lags contain chert pebbles and oyster fragments. *Rosselia* and *Asterosoma* are locally present in laminated sandstone intervals above flooding surfaces. The sandstone contains abundant plagioclase and potassium feldspar, silicic tuff, radiolarian chert and argillite, and intermediate volcanic grains.

The Upper Sandstone Member is interpreted to record progradation of a tidally influenced shoreface-shoreline complex. It contains an association of shoreface and ebb-tidal delta deposits (Shelley, 2001). The composition of the sandstone indicates a mixed volcanic and sedimentary source.

Adjuntas Formation

The Adjuntas Formation gradationally overlies the Potrerillos Formation. It is about 260 m thick at its type locality 3 km east-southeast of the El Papalote diapir (McBride et al., 1974). The formation consists of red siltstone and lenticular beds of fine- to medium-grained sandstone 1–2 m thick. In its lower part, it includes drab, olive-gray mud-rock intervals that alternate with broadly lenticular sandstone. This lower part contains

FIGURE 15. The La Popa lentil, 5 km east-southeast of the village of San Jose de la Popa. Main limestone cliff is 350 m high. Explanation: LP, La Popa lentil; SJ, San Jose lentil; Tc ,Carroza Formation. View is toward the northwest.

freshwater gastropods and other fossils that indicate an estuarine or marginal-marine setting (Vega and Perrilliat, 1992). The middle part contains horizons of calcareous nodules that we interpret as paleosols. The Adjuntas Formation was deposited in estuarine and coastal-plain settings and thus represents continued progradation above the Upper Sandstone Member of the Potrerillos Formation. It is early Eocene in age (Vega-Vera and Perrilliat, 1989; Vega-Vera et al., 1989) and correlates with the Wilcox Formation of the Texas Gulf Coast (Figure 3).

Viento Formation

The Viento Formation is a thick, sandstone-rich formation as much as 730 m thick at its type locality in Arroyo La Tinaja north of the El Papalote diapir (McBride et al., 1974). It overlies the Adjuntas Formation on a sharp contact (Figure 17). Although not extensively studied, the Viento appears to contain two dominant lithofacies associations. One consists of fine- to medium-grained sandstone and mudstone in horizontal beds and inclined, bundled bedsets with bimodal dip directions (Figure 18). We interpret these as bars deposited by tidal currents, probably in an outer sand-flat setting. The other facies association consists of medium-grained sandstone in lenticular channel-form beds interbedded with thinly bedded heterolithic strata (Figure 19). The channels contain lags of intraclasts, chert pebbles, and oyster shells and are interpreted as channel systems of the middle tidal flats. In Arroyo La Tinaja,

these channels are locally oriented generally northwest-southeast, parallel to the structural grain of the basin as defined by detachment folds and the La Popa weld. At Boca La Carroza, matrix-supported conglomerate is present in the Viento Formation near the La Popa weld. In addition to chert pebbles, the conglomerate there contains metaigneous clasts as much as 25 cm in diameter, indicating the presence of exposed evaporite with exotic blocks during Viento deposition. Sand-rich facies of the Viento Formation are concentrated adjacent to the weld, with sandstone decreasing with distance from the weld. The presence of exotic clasts and enrichment of sand adjacent to the weld indicate that diapirism was taking place during Viento deposition. Sand was preferentially deposited in the rapidly subsiding region near the salt wall, and metaigneous clasts were eroded from exposed salt of the wall. The Viento Formation remains undated but is likely middle Eocene (Lutetian; Vega-Vera et al., 1989). It is the record of a generally progradational, tidally influenced, coastal complex.

Carroza Formation

The youngest preserved formation of the Difunta Group, the Carroza Formation, is a red- to purple-weathering succession of mud-rich strata. It is present in the La Popa Basin only in the hanging wall of the La Popa weld. The formation is 620 m thick at its type locality about 5 km east-southeast of the village of San Jose de la Popa (McBride et al., 1974). In its lower

FIGURE 16. Upper part of Potrerillos Formation, 2 km southeast of the El Papalote diapir. Explanation: us, Upper Siltstone Member; d, Delgado Sandstone Member; um, Upper Mudstone Member; uss, Upper Sandstone Member. Contact of Upper Mudstone and Upper Sandstone members is at arrow. View is toward the north. Upper cliff is 160 m high.

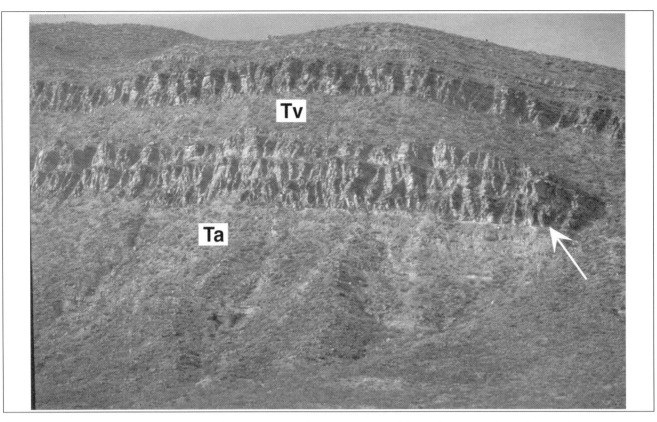

FIGURE 17. Adjuntas and Viento Formations on the southwestern flank of the Delgado syncline. Contact is at arrow. The hill is 120 m high.

FIGURE 18. Bundled heterolithic bedsets in the upper part of the Viento Formation, 10 km east-southeast of San Jose de la Popa. Note bimodal inclination directions in bedsets. Hammer handle is 30 cm long.

FIGURE 19. Channel-form sandstone bodies in the middle part of the Viento Formation, lower Arroyo La Tinaja. Dark gray beds are thinly bedded, mudstone-rich heterolithic strata.

part, it consists of thick red- to gray-weathering mudrock and siltstone with subordinate thin, tabular, fine-grained sandstone beds and thicker (1- to 2-m) beds with lateral-accretion bedsets. This part is interpreted as tidal-flat and lower coastal-plain deposits. The upper part of the section is predominantly a rich, reddish-brown weathering mudrock with lenticular channels of fine- to medium-grained sandstone. These rocks are interpreted as deposits of a low-gradient fluvial system and overbank complex. The Carroza represents continued progradation above the Viento Formation. The top of the Carroza is truncated everywhere by erosion. No direct age is yet available for the Carroza Formation, but Vega-Vera et al. (1989) inferred a middle Eocene (Lutetian-Bartonian) age.

DISCUSSION

The stratigraphy of the La Popa Basin records salt mobility from the late Aptian through middle Eocene. Early Cretaceous salt rise along what is now the La Popa weld created bathymetric highs that supported carbonate biostromes unlike correlative basinal or shelfal facies. Diapir-proximal thinning of Upper Cretaceous and Paleogene siliciclastic units and development of carbonate lentils demonstrate continued syndepositional salt rise and development of seafloor topography. This is also reflected by the presence of the metaigneous clasts in the

lentils, in the Upper Siltstone, Delgado, and Upper Sandstone Members of the Potrerillos Formation, and in the Viento Formation. These clasts record the exposure and erosion of the caprocks of the active diapirs.

Carbonate lentils typically developed during transgressive phases of deposition, as recognized by Laudon (1984). They are interbedded with mudstone members of the Potrerillos Formation and with Lower Cretaceous to Turonian basinal facies. Most lentils may be attributed to local bathymetric highs around the diapirs or along the weld during its diapiric history (Giles and Lawton, 1999); however, the massive, thick, laterally extensive La Popa lentil may have been situated on a larger-scale feature, such as a time-equivalent shelf break.

Rocks of the La Popa Basin also document the evolution of the basin through several tectonic stages. The Jurassic rocks exposed in the diapirs record an initial phase of basin development during lithospheric extension. Although the rock record is fragmentary because of evaporite deformation, the original Upper Jurassic stratigraphic section appears to have consisted, in ascending order, of asthenosphere-derived mafic rocks and halite, sabkha sulfates, and ramp carbonates. This succession records the early development of an isolated pull-apart basin that was subsequently integrated gradually into the broader marine realm of northeastern Mexico.

The Early Cretaceous–Santonian stratigraphy postdated rifting and probably records thermal subsidence following the breakup of Pangea and translation of Mexican crustal fragments into their current positions (e.g., Dickinson and Lawton, 2001). The Upper Cretaceous–Paleogene siliciclastics of the Parras Shale and Difunta Group are generally regarded as the fill of a foreland basin adjacent to the rising orogen of the Sierra Madre Oriental (McBride et al., 1974; McKee et al., 1975; Ye, 1997). Although the sedimentary component of the sandstone composition is compatible with derivation of detritus from the fold belt, the volcanic component (Tardy and Maury, 1973; McBride et al., 1975) is inconsistent with a sedimentary source terrane. The volcanic detritus in the basin indicates that a volcanic terrane was an important part of the source region. This may have been the Guerrero volcanogenic superterrane, an oceanic arc that collided with Mexico in the Aptian (Dickinson and Lawton, 2001) or the succeeding continental-margin arc that commenced in the Albian. The presence of thick submarine tuffs of early Late Cretaceous age at the top of the Indidura Formation and in its equivalent, the Caracol Formation, provides a case for the latter. These alternative sources will be resolved only by further detailed study of the provenance of Difunta Group sandstones.

Uplift and erosion of the La Popa Basin probably began in the early Tertiary. Hidalgoan detachment folds appear to affect all formations of the Difunta Group (Figure 2); therefore, shortening continued beyond the middle Eocene. Growth strata of Paleocene age onlap folds near Las Moras in the northern Parras Basin (Figure 2a); these folds are probably detached in the Parras Shale and so lack a salt-tectonic component. Although detachment folding was under way by the Paleocene, thermochonologic research suggests that major uplift and erosion of the basin were much later, in the Oligocene and early Miocene (Gray et al., this volume). Thus, uplift and erosion of the foreland of northeastern Mexico were in part unrelated to Hidalgoan deformation.

CONCLUSIONS

Rocks exposed in the La Popa Basin range in age from Late Jurassic to middle Eocene. Upper Jurassic rocks are present as blocks in diapiric structures in the basin and represent a rift assemblage of evaporites, thermally altered mafic igneous rocks, and intertidal and shallow-marine carbonates. Lower Cretaceous rocks are represented primarily by carbonate biostromes and subordinate shale equivalent to platform carbonates of the Cupido and Aurora Limestones and basinal shales of the La Peña Formation, respectively. Generally, deepwater sedimentation took place in the basin through the Santonian, with at least one additional biostrome deposited on a bathymetric high adjacent to a salt wall in the eastern part of the basin. This episode of deep-water carbonate sedimentation accompanied thermal subsidence of the basin following rifting. The siliciclastic successions of the Difunta Group were deposited in the basin by an eastward- to southeastward-prograding complex of deltaic shelf margins from the Campanian

to middle Eocene. In the La Popa Basin, the Difunta Group represents dominantly delta-front to prodelta deposits that filled a foreland basin adjacent to the Sierra Madre Oriental orogen. Growth-stratal geometries, stratal thinning, and facies distribution indicate that salt diapirism took place during the deposition of all strata exposed in the basin. Our preliminary understanding of growth strata on folds not associated with diapiric structures suggests that detachment folding did not affect the basin prior to the early Maestrichtian (Weislogel, 2001). Uplift and erosion of the basin to present levels of exposure began in the mid-Tertiary and appear to have postdated folding.

ACKNOWLEDGMENTS

We thank many individuals for discussions of La Popa geology in the field and the office. Among these are Gerardo Basurto, Dick Buffler, Samuel Eguiluz, Jennifer Garrison, Robert Goldhammer, Kyle Graff, Gary Gray, Joydeep Haldar, Steve Hall, Bruce Hart, Kevin Hon, Lela Hunnicutt, Philip Koch, Robert Laudon, Mary Beth McKee, James McKee, Nancy McMillan, David Mercer, Hector Millán, Jasper Peijs, Mark Rowan, Tim Seeley, Keith Shannon, David Shelley, Kris Soegaard, Ian Watson, Amy Weislogel, James Wilson, and Cindy Yeilding. This work was supported in part by the donors of the Petroleum Research Fund, administered by the American Chemical Society. Additional support came from the Institute of Tectonic Studies and the La Popa Basin Joint Industry Consortium at New Mexico State University. Reviews by Samuel Eguiluz, Gary Gray, and Robert Laudon improved the manuscript.

REFERENCES CITED

Buitron, B. E., 1984, Late Jurassic bivalves and gastropods from northern Zacatecas, Mexico, and their paleogeographic significance, in G. E. G. Westermann, ed., Jurassic-Cretaceous biochronology and paleogeography of North America: Geological Association of Canada, Special Paper 27, p. 89–98.

Cantú-Chapa, A., 1979, Biostratigrafía de la Serie Huasteca (Jurásico Medio) en el subsuelo de Poza Rica, Veracruz: Revista del Instituto Mexicano del Petróleo, v. 11, p. 14–24.

Cox, L. R., et al., 1969, Bivalvia, in R. C. Moore, ed., Treatise on invertebrate paleontology, Part N6(1): Geological Society of America and University of Kansas Press, Lawrence, p. 227–489.

Cuevas-Leree, J. A., 1985, Analysis of subsidence and thermal history in the Sabinas basin, northeastern Mexico: Master's thesis, University of Arizona, Tucson, 81 p.

Dickinson, W. R., and T. F. Lawton, 2001, Carboniferous to Cretaceous assembly and fragmentation of Mexico: Geological Society of America Bulletin, v. 113, p. 1142–1160.

Eguiluz de Antuñano, S., 1994, La formación Carbonera y sus implicaciónes tectónicas, estados de Coahuila y Nuevo León: Boletín de la Sociedad Geológica Mexicana, v. 50, p. 3–40.

Fortunato, K. S., and W. C. Ward, 1982, Upper Jurassic–Lower Cretaceous fan-delta complex, La Casita Formation of the Saltillo area, Coahuila, Mexico: Gulf Coast Association of Geological Societies Transactions, v. 32, p. 473–482.

Garrison, J. M., 1998, Implications of allochthogenic meta-igneous and carbonate blocks in El Papalote diapir, La Popa Basin, Nuevo León, Mexico: Master's thesis, New Mexico State University, Las Cruces, 144 p.

Garrison, J. M., and N. J. McMillan, 1999, Evidence for Jurassic continental rift magmatism in northeast Mexico: Allogenic meta-igneous blocks in El Papalote diapir, La Popa Basin, Nuevo León, Mexico, *in* C. Bartolini, J. L. Wilson, and T. F. Lawton, eds., Mesozoic sedimentary and tectonic history of north-central Mexico: Geological Society of America Special Paper 340, p. 319–332.

Giles, K. A., and T. F. Lawton, 1999, Attributes and evolution of an exhumed salt weld, La Popa Basin, northeastern Mexico: Geology, v. 27, p. 323–326.

Goldhammer, R. K., 1999, Mesozoic sequence stratigraphy and paleogeographic evolution of northeast Mexico, *in* C. Bartolini, J. L. Wilson, and T. F. Lawton, eds., Mesozoic sedimentary and tectonic history of north-central Mexico: Geological Society of America Special Paper 340, p. 1–58.

González-García, R., 1984, Petroleum exploration in the "Gulf of Sabinas"—a new gas province in northern Mexico, *in* J. L. Wilson, W. C. Ward, and J. Finneran, eds., A field guide to Upper Jurassic and Lower Cretaceous carbonate platform and basin systems, Monterrey-Saltillo area, northeast Mexico: Gulf Coast Section, Society for Sedimentary Geology (SEPM) Foundation, p. 64–76.

Götte, M., and D. Michalzik, 1992, Stratigraphic relations and facies sequences of an Upper Jurassic evaporitic ramp in the Sierra Madre Oriental (Mexico): Zentralblatt für Geologie und Paläontologie. Teil I, v. 1991, p. 1445–1466.

Gray, G. G., S. Eguiluz de Antuñano, R. J. Chuchla, and D. A. Yurewicz, 1997, Structural evolution of the Saltillo-Monterrey corridor, Sierra Madre Oriental: Applications to exploration challenges in fold-thrust belts—a field guidebook, privately published, 20 p.

Gray, G. G., R. J. Pottorf, D. A. Yurewicz, K. I. Mahon, D. R. Pevear, and R. J. Chuchla, 2001, Thermal and chronological record of syn- and post-Laramide burial and exhumation, Sierra Madre Oriental, Mexico, *in* C. Bartolini, R. T. Buffler, and A. Cantú Chapa, eds., this volume.

Guzmán, E. J., and Z. de Cserna, 1963, Tectonic history of Mexico, *in* O. E. Childs, and B. W. Beebe, eds., Backbone of the Americas—tectonic history from pole to pole: AAPG Memoir 2, p. 113–129.

Humphrey, W. E., and T. Diaz, 1956, Jurassic and Lower Cretaceous stratigraphy and tectonics of northeast Mexico: Unpublished report, Petroleros Mexicanos, 390 p.

Hunnicutt, L. A., 1998, Tectonostratigraphic interpretation of Upper Cretaceous to Lower Tertiary limestone lentils within the Potrerillos Formation surrounding El Papalote diapir, La Popa Basin, Nuevo Leon, Mexico: Master's thesis, New Mexico State University, Las Cruces, 181 p.

Imlay, R. W., 1937, Geology of the middle part of the Sierra de Parras, Coahuila, Mexico: Geological Society of America Bulletin, v. 48, p. 587–630.

Imlay, R. W., and G. Herman, 1984, Upper Jurassic ammonites from the subsurface of Texas, Louisiana, and Mississippi, *in* W. P. S. Ventress, D. G. Bebout, B. F. Perkins, and C. H. Moore, eds., The Jurassic of the Gulf Rim: Gulf Coast Section, Society for Sedimentary Geology (SEPM) Foundation, Proceedings of the Third Annual Research Conference, p. 149–170.

Jacobo-Albarrán, J., 1986, El basamento del distrito de Poza Rica y su implicación en la generación de hidrocarburos: Revista del Instituto Mexicano del Petróleo, v. 18, p. 5–24.

Laudon, R. C., 1975, Stratigraphy and petrology of the Difunta Group, La Popa and eastern Parras Basins, northeastern Mexico: Ph.D. dissertation, University of Texas, Austin, 294 p.

Laudon, R. C., 1984, Evaporite diapirs in the La Popa Basin, Nuevo León, Mexico: Geological Society of America Bulletin, v. 95, p. 1219–1225.

Laudon, R. C., 1996, Salt dome growth, thrust fault growth, and syndeformational stratigraphy, La Popa Basin, northern Mexico: Transactions of the Gulf Coast Association of Geological Societies, v. 46, p. 219–228.

Lehmann, C., D. A. Osleger, and I. P. Montanez, 1998, Controls on cyclostratigraphy of Lower Cretaceous carbonates and evaporites, Cupido and Coahuila platforms, northeastern Mexico: Journal of Sedimentary Research, v. 68, p. 1109–1130.

Longoria, J. F., 1984, Stratigraphic studies in the Jurassic of northeastern Mexico: Evidence for the origin of the Sabinas Basin, *in* W. P. S. Ventress, D. G. Bebout, B. F. Perkins, and C. H. Moore, eds., The Jurassic of the Gulf Rim: Gulf Coast Section, Society for Sedimentary Geology (SEPM) Foundation, Proceedings of the Third Annual Research Conference, p. 171–193.

Longoria, J. F., D. M. Clowes, and R. Monreal, 1999, The type Mesozoic succession of northern Mexico: Cañon La Casita, *in* C. Bartolini, J. L. Wilson, and T. F. Lawton, eds., Mesozoic sedimentary and tectonic history of north-central Mexico: Geological Society of America Special Paper 340, p. 287–318.

Lopez, R., 1997, High-Mg andesites from the Gila Bend Mountains, southwestern Arizona: Evidence for hydrous melting of lithosphere during Miocene extension and the Pre-Jurassic geotectonic evolution of the Coahuila terrane, northwestern Mexico: Grenville basement, a late Paleozoic arc, Triassic plutonism, and the events south of the Ouachita suture: Ph.D. dissertation, University of California, Santa Cruz, 147 p.

Lopez-Ramos, E., 1982, Geología de México, v. 2: Consejo Nacional de Ciencia y Technología, México, D. F., 454 p.

Marrett, R. J., and M. Aranda-García, 1999, Structure and kinematic development of the Sierra Madre Oriental fold-thrust belt, Mexico, *in* J. L. Wilson, W. C. Ward, and R. A. Marrett, eds., Stratigraphy and structure of the Jurassic and Cretaceous platform and basin systems of the Sierra Madre Oriental, Monterrey and Saltillo areas, northeastern Mexico—a field book and related papers: South Texas Geological Society, p. 69–98.

Marton, G., and R. T. Buffler, 1994, Jurassic reconstruction of the Gulf of Mexico basin: International Geology Review, v. 36, p. 545–586.

McBride, E. F., A. E. Weidie, and J. A. Wolleben, 1973, Deltaic and associated facies of Difunta Group (Late Cretaceous to Paleocene), Parras and La Popa Basins, Coahuila and Nuevo León, Mexico: Transactions of the Gulf Coast Association of Geological Societies, v. 23, p. 37–40.

McBride, E. F., A. L. Weidie Jr., and J. A. Wolleben, 1975, Deltaic and associated deposits of Difunta Group (Late Cretaceous to Paleocene), Parras and La Popa Basins, northeastern Mexico, *in* M. L. S. Broussard, ed., Deltas: Houston Geological Society, p. 485–522.

McBride, E. F., A. E. Weidie, J. A. Wolleben, and R. C. Laudon, 1974, Stratigraphy and structure of the Parras and La Popa Basins, northeastern Mexico: Geological Society of America Bulletin, v. 84, p. 1603–1622.

McKee, J. W., N. W. Jones, and T. H. Anderson, 1988, Las Delicias basin: A record of late Paleozoic arc volcanism in northeastern Mexico: Geology, v. 16, p. 34–40.

McKee, J. W., N. W. Jones, and T. H. Anderson, 1999, The late Paleozoic and early Mesozoic history of the Las Delicias terrane, Coahuila, Mexico, in C. Bartolini, J. L. Wilson, and T. F. Lawton, eds., Mesozoic sedimentary and tectonic history of north-central Mexico: Geological Society of America Special Paper 340, p. 161–189.

Mercer, D. W., and K. A. Giles, 1999, Growth strata analysis of the Upper Cretaceous to lower Paleogene Potrerillos Fm. adjacent to El Gordo salt diapir, Nuevo León, Mexico: Geological Society of America Abstracts with Programs, v. 31, p. 281.

Murray, G. E., A. E. Weidie Jr., D. R. Boyd, R. H. Forde, and P. D. Lewis Jr., 1962, Formational subdivisions of the Difunta Group, Parras Basin, Coahuila and Nuevo León, Mexico: AAPG Bulletin, v. 46, p. 374–383.

Scott, R. W., 1984, Significant fossils of the Knowles Limestone, Lower Cretaceous, Texas, in W. P. S. Ventress, D. G. Bebout, B. F. Perkins, and C. H. Moore, eds., The Jurassic of the Gulf Rim: Gulf Coast Section, Society for Sedimentary Geology (SEPM) Foundation, Proceedings of the Third Annual Research Conference, p. 333–346.

Shelley, D. C., 2001, Sedimentology, stratigraphy, and petrology of the Paleocene Upper Sandstone Member of the Potrerillos Formation, La Popa Basin, Mexico: Master's thesis, New Mexico State University, Las Cruces, 228 p.

Tardy, M., and R. Maury, 1973, Sobre la presencia de elementos de origen volcánico en las areniscas de los flyschs de edad cretácica superior de los estados de Coahuila y de Zacatecas, México: Boletín de la Sociedad Geológica Mexicana, v. 34, p. 5–12.

Torres, R., J. Ruiz, P. J. Patchett, and J. M. Grajales, 1999, A Permo-Triassic arc in eastern Mexico: Tectonic implications for reconstructions of southern North America, in C. Bartolini, J. L. Wilson, and T. F. Lawton, eds., Mesozoic sedimentary and tectonic history of north-central Mexico: Geological Society of America Special Paper 340, p. 191–196.

Vega, F. J., 1987, Contribuciones a la estratigrafía del Grupo Difunta en la cuenca de La Popa, Coahuila y Nuevo León: Segundo simposio de geología regional de México—Programa y resúmenes, Universidad Nacional Autónoma de México, Instituto de Geología, p. 29–30.

Vega, F. J., and M. C. Perrilliat, 1989, On a new species of Venecardia from the lower Eocene in northeastern Mexico (Difunta Group): New Orleans: Tulane University, Tulane Studies in Geology and Paleontology, v. 22, p. 101–106.

Vega, F. J., and M. C. Perrilliat, 1992, Freshwater gastropods from lower Eocene Difunta Group, northeastern Mexico: Journal of Paleontology, v. 66, p. 603–609.

Vega, F. J., and M. C. Perrilliat, 1995, On some Paleocene invertebrates from the Potrerillos Formation (Difunta Group), northeastern Mexico: Journal of Paleontology, v. 69, p. 862–869.

Vega, F. J., M. Perilliat, and L. M. Mitre-Salazar, 1999, Paleocene ostreids from the Las Encinas Formation (Parras Basin, Difunta Group), northeastern Mexico; stratigraphic implications, in C. Bartolini, J. L. Wilson, and T. F. Lawton, eds., Mesozoic sedimentary and tectonic history of north-central Mexico: Geological Society of America Special Paper 340, p. 105-110.

Vega-Vera, F. J., L. M. Mitre-Salazar, and E. Martínez-Hernández, 1989, Contribución al conocimiento de la estratigrafía del Grupo Difunta (Cretácico Superior–Terciario) en el noreste de México: Universidad Nacional Autonóma de México, Instituto de Geología, Revista, v. 8, p. 179–187.

Vega-Vera, F. J., and M. Perrilliat, 1989, La presencia del Eoceno marino en la cuenca de la Popa (Grupo Difunta), Nuevo León: Orogenia post-Ypresiana: Universidad Nacional Autonóma de México, Instituto de Geología, Revista, v. 8, p. 67–70.

Vega-Vera, F. J., and M. C. Perrilliat, 1990, Moluscos del Maastrichtiano de la Sierra el Antrisco, estado de Nuevo León: Universidad Nacional Autónoma de México, Paleontología Mexicana Número 55, 65 p.

Viniegra, F., and C. Castillo-Tejero, 1970, Golden Lane Fields, Veracruz, Mexico, in M. T. Halbouty, ed., Geology of giant petroleum fields: AAPG Memoir 14, p. 309–325.

Weidie, A. E. Jr., 1961, The stratigraphy and structure of the Parras Basin, Coahuila and Nuevo León, Mexico: Ph.D. dissertation, Louisiana State University, Baton Rouge, 73 p.

Weidie, A. E., and G. E. Murray, 1967, Geology of Parras Basin and adjacent areas of northeastern Mexico: AAPG Bulletin, v. 51, p. 678–695.

Weislogel, A. L., 2001, The influence of diapirism and foreland evolution on the depositional system, stratigraphy, and petrology of the Maastrichtian Muerto Formation, La Popa Basin, Mexico: Master's thesis, New Mexico State University, Las Cruces, 310 p.

Ye, H., 1997, Sequence stratigraphy of the Difunta Group in the Parras–La Popa foreland basin, and tectonic evolution of the Sierra Madre Oriental, NE Mexico: Ph.D. dissertation, University of Texas, Richardson, 197 p.

Zwanziger, J. A., 1992, New concepts in Mesozoic stratigraphy of Chihuahua, in P. C. Goodell, C. García-Gutiérrez, and I. Reyes-Cortés, eds., Energy resources of the Chihuahua Desert region: El Paso Geological Society, p. 77–124.

Eguiluz de Antuñano, S., 2001, Geologic evolution and gas resources of the Sabinas Basin in northeastern Mexico, *in* C. Bartolini, R. T. Buffler, and A. Cantú-Chapa, eds., The western Gulf of Mexico Basin: Tectonics, sedimentary basins, and petroleum systems: AAPG Memoir 75, p. 241–270.

10

Geologic Evolution and Gas Resources of the Sabinas Basin in Northeastern Mexico

Samuel Eguiluz de Antuñano

PEMEX Exploración, Reynosa, Tamaulipas, Mexico

ABSTRACT

The Sabinas Basin is located in northeastern Mexico in the states of Coahuila and Nuevo León. Basin fill is composed mainly of Mesozoic marine sediments deposited during long-term subsidence and folded during Late Cretaceous and Paleogene Laramide orogenesis. The origin of the basin is related to a rift associated with the opening of the Gulf of Mexico. More than 5000 m of sedimentary rocks was deposited in the Sabinas Basin. Three supersequences have been defined. The first represents synrift sediments and is composed of conglomerates and evaporites with associated basic igneous rocks. The following supersequence (144–96 m.y.) comprises several higher-frequency cycles represented by carbonate, evaporite, and coastal siliciclastic deposits of extensive platforms on a passive margin. The youngest supersequence (96–39.5 m.y.) consists mainly of regressive, terrigenous clastic facies deposited in a foreland setting. Subsidence was 40% to 70% greater during the initial rift stage than during subsequent depositional stages. Several lateral and vertical facies changes in the basin were controlled by the Coahuila and Tamaulipas basement blocks, as well as other, smaller blocks.

Laramide deformation is of the thin-skinned type. Four areas with distinct structural styles are recognized: (1) where Jurassic salt is the regional detachment level; (2) where salt diapirs formed; (3) where deformation was controlled by a basement high northeast of the basin; and (4) where the absence of Jurassic salt resulted in the development of fault-bend folding. Structural shortening calculated for different areas of the basin ranges from 16% to 26%. The critical-wedge model has been used to explain some of the deformational variations in the basin. The structures have natural fractures that provide permeability in the hydrocarbon reservoirs.

The Upper Jurassic La Casita Formation is considered to be the main hydrocarbon source. Total-organic-carbon (TOC) values are favorable, and although the formation's kerogens are mainly type III, they are characterized by a high transformation ratio.

In 23 years of exploitation, the Sabinas Basin has produced more than 350 bcf of dry gas. The average daily production rate per well ranges from 0.5 to 2.0 MMCFGD. Geostatistical modeling indicates that the La Gloria, La Casita, and Padilla-Virgen plays could contain total resources of more than 1000 bcf. Furthermore, a coal-degasification play may extend over an area exceeding 1000 km^2 and could contain additional potential resources amounting to 147 bcf of gas. Both of these resources could supply much of the local industry's methane needs for more than 20 years.

INTRODUCTION

This study aims to provide integrated surface and subsurface information for the Sabinas Basin, permitting a reconstruction of its geologic evolution and describing the hydrocarbon potential of its principal plays. The Sabinas Basin encompasses an area of about 37,000 km². It is in northeastern Mexico, in the central part of the state of Coahuila and the western part of the state of Nuevo León (Figure 1). The basin contains more than 5000 m of Mesozoic sedimentary rocks deposited in a marine environment. Its genesis is related to the opening of the proto–Gulf of Mexico.

The method used in this study consisted of integrating stratigraphic, lithologic, and paleontologic data, making correlations, and identifying the depositional environments of the hydrocarbon-bearing deposits. Thin sections were examined for each 5-m interval covering 3000- to 5000-m columns in more than 60 wells. In addition, outcrops were described and measured and samples were studied petrographically. Geophysical wells log were used to calibrate the subsurface samples. Using all this information, maps were made showing the distribution of facies, paleogeography, and isopachs for each of the formations present in the basin. Data were also included to identify the hydrocarbon-generation potential of various stratigraphic levels and to model the subsidence history of the basin.

With the support of preexisting maps and Landsat satellite images, structural analysis identified several structural styles in the basin. Statistical fracture, fault, fold, and lineament analyses were done in conjunction with seismic reflection and field data. Based on these analyses, structural sections were constructed and manually restored by line and area balancing. It is proposed that the deformation can be explained by the criti-

cal-wedge model, as applied to the Sierra Madre Oriental (Marrett and Aranda, 1999). The integration of the data mentioned here makes it possible to construct composite risk maps for the known productive trends and potential new trends in the basin. The evaluation of production data is encouraging for the existence of additional hydrocarbon potential in this basin.

The first geologic studies of the Sabinas Basin date to the early twentieth century, when Boese and Cavins (1927), Müllerried (1927), and Burckhardt (1930) made the first stratigraphic descriptions. Imlay (1940), with the support of Kane's measured sections and based on the previous works, introduced into the literature the names of the principal formations present in the basin. Alvarez (1949) defined the Sabinas Basin in structural terms, and Humphrey (1956) and Guzmán and De Cserna (1963), among others, described the basin in the classical framework of geosynclinal theory.

From the 1970s onward, Petróleos Mexicanos (PEMEX) carried out systematic geologic studies that by the mid-1970s had led to the discovery of several gas and condensate fields. This intense exploration activity ended in 1986, and exploration of the basin is currently deferred. During the years of petroleum exploration, geologic mapping of the entire basin was accomplished, and more than 3000 km of 2-D seismic data, as well as a limited amount of 3-D seismic, was obtained. In addition, the basin was totally covered by gravity and magnetic surveys, geological-geophysical integrations, and sedimentologic models for various stratigraphic levels. Geochemical, tectonic, and structural studies were also carried out, some of which have been published (Charleston, 1973; Márquez, 1979; González, 1984; Salvador, 1987, 1991; Cuevas, 1984).

More than 100 exploratory wells and 80 development wells have been drilled in the Sabinas Basin, but the characteristics of its hydrocarbon deposits are rarely mentioned in the literature. More than 350 bcf of dry gas has been produced during a period of 20 years from 11 fields; 96% of this production has come from three fields. Some of this information has been published previously (Linares and Montiel, 1987; González, 1984; Eguiluz, 1996, 1997).

ORIGIN OF THE SABINAS BASIN

The Sabinas Basin was initially bounded to the northeast and southwest by a series of high and low fault blocks (McKee et al., 1990; Eguiluz, 1994). The high blocks of the Tamaulipas arch and Coahuila peninsula are the main paleotectonic and paleogeographic basin limits that have persisted for a long time (Figure 2). The Sabinas Basin is in a depression formed by a series of subsided blocks that elevates to the northwest and separates the Sabinas Basin from the Chihuahua Basin (Eguiluz, 1984). High blocks of smaller dimension existed in the depression (La Mula and Monclova highs). Meanwhile, toward the southeast, other relatively low blocks formed the boundary separating the Sabinas Basin from the proto–Gulf of Mexico (Alfonso, 1978; Wilson, 1990), and communication with the

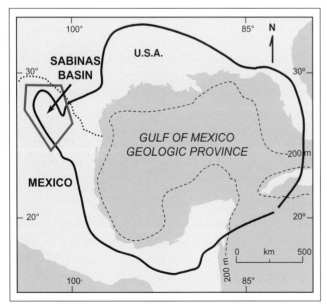

FIGURE 1. Location of the Sabinas Basin and the boundary of the Gulf of Mexico geologic province.

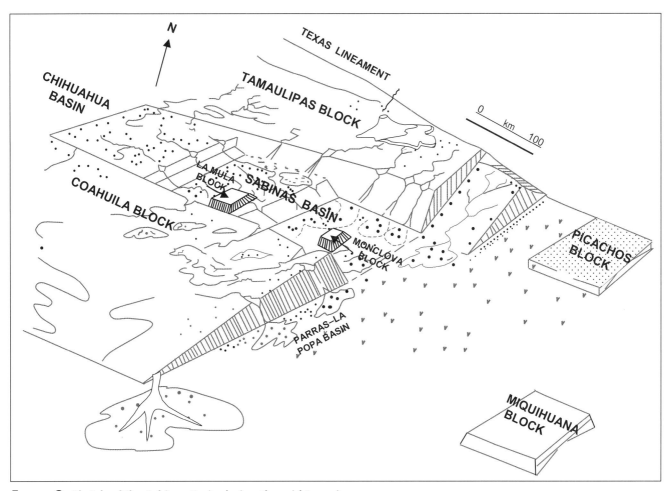

FIGURE 2. Sketch of the Sabinas Basin during the mid-Jurassic.

Central Mexican Basin was established through a trough of low blocks situated between the high blocks of Miquihuana and Coahuila.

In general terms, the lithologic composition of the blocks is different from one region to another. In the Coahuila block are Permian sedimentary rocks, with interspersed Pennsylvanian olistoliths and volcanic rocks that were deposited in the area of the Acatita–Las Delicias Valley in the state of Coahuila (McKee et al., 1988). The Tarahumara-1 well (Table 1 and Figure 3) penetrated siliciclastic sedimentary rocks with intercalated andesites and dacites. The age of this section is doubtful; two radiometric-age dates indicate 83–98 m.y. (Campanian-Albian), and a third reportedly places the age at 272 ± 22 m.y. (Early Permian). The Paila 1-A well penetrated ignimbrites dated by Rb/Sr at 236 ± 39 m.y. (Triassic), although an earlier study, which identified this rock as metavolcanic, determined their age to be 80.4 ± 3.2 m.y. (Campanian). The petrographic description of the rocks indicates alteration of its minerals, which probably explains the ambiguous age determinations. In the Las Delicias outcrop, the sedimentary rocks are intruded by granodiorite and tonalite that, according to the method utilized, vary in age from 203 ± 4 and 208 ± 8 to 224 ± 18 m.y. (Early Jurassic to Late Triassic). The Mayrán-1 well and Sierra

del Diablo outcrop, also situated on the Coahuila block, found metamorphosed rhyolitic volcanics with ages of 199 ± 20 and 197 m.y. (Early Jurassic). The Ceballos-1 well penetrated a schist, whose protolith could have been a sandy conglomerate, with volcaniclastics dated as 163 ± 8 m.y. (Late Callovian).

To the east of the Sabinas Basin, in the Tamaulipas block, schists and gneisses comprise a basement complex with radiometric ages ranging from 380 ± 30 m.y. in Cuatrociénegas-1 well to 215 ± 5 m.y. in the Carbón-1 well (Early Permian to Early Triassic). The Carmen outcrop and the cores of the Magvi-1, Yerbabuena-1, Cacanapo-101, Gerardo-1-A, Palau-1, Metatosa-101, and Nazca-1 wells correlate with the internal zone of the Marathon-Ouachita tectonic belt (Flawn et al., 1961). The exception is the Concordia-1 well, which penetrated a phyllite dated at 190 ± 7 m.y. (Early Jurassic) and which differentiates it from the ages and rocks types in the internal Ouachita zone, as described by Flawn et al. (1961). Its lithology situates it in the frontal zone of the Ouachita belt, although its age is within the range of younger igneous rocks described later in this paper.

In the marginal areas of the basin, intrusive or hypabyssal igneous rocks of calc-alkaline composition predominate. In the Potrero de La Mula outcrop (Jones et al., 1984) and in the Oro-

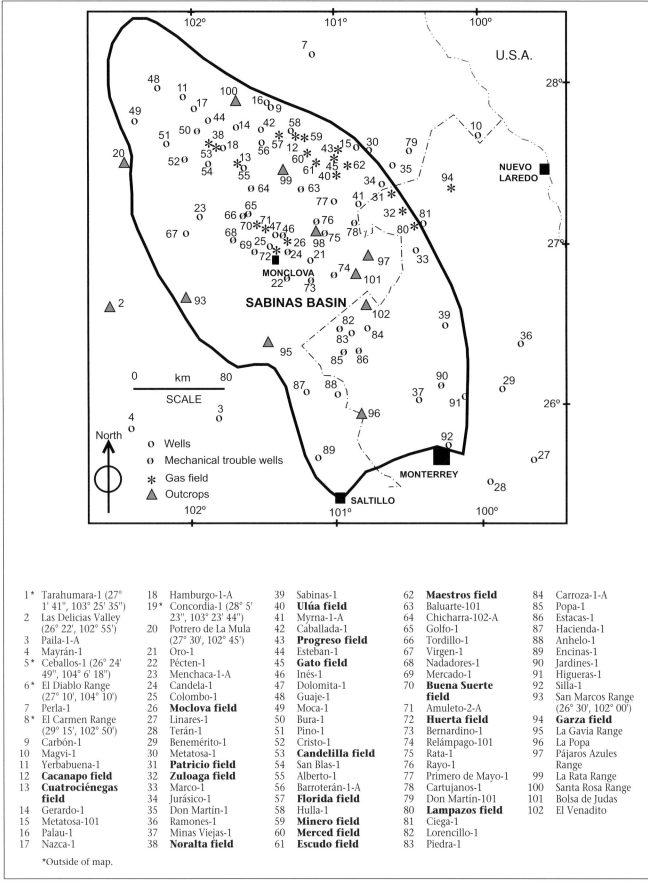

FIGURE 3. Location map of wells, field gas, and outcrops cited in text, and key to numbered sites on location map.

1*	Tarahumara-1 (27° 1' 41", 103° 25' 35")	18	Hamburgo-1-A	39	Sabinas-1	62	**Maestros field**	84	Carroza-1-A
2	Las Delicias Valley (26° 22', 102° 55')	19*	Concordia-1 (28° 5' 23", 103° 23' 44")	40	**Ulúa field**	63	Baluarte-101	85	Popa-1
3	Paila-1-A	20	Potrero de La Mula (27° 30', 102° 45')	41	Myrna-1-A	64	Chicharra-102-A	86	Estacas-1
4	Mayrán-1	21	Oro-1	42	Caballada-1	65	Golfo-1	87	Hacienda-1
5*	Ceballos-1 (26° 24' 49", 104° 6' 18")	22	Pécten-1	43	**Progreso field**	66	Tordillo-1	88	Anhelo-1
6*	El Diablo Range (27° 10', 104° 10')	23	Menchaca-1-A	44	Esteban-1	67	Virgen-1	89	Encinas-1
7	Perla-1	24	Candela-1	45	**Gato field**	68	Nadadores-1	90	Jardines-1
8*	El Carmen Range (29° 15', 102° 50')	25	Colombo-1	46	Inés-1	69	Mercado-1	91	Higueras-1
9	Carbón-1	26	**Moclova field**	47	Dolomita-1	70	**Buena Suerte field**	92	Silla-1
10	Magvi-1	27	Linares-1	48	Guaje-1	71	Amuleto-2-A	93	San Marcos Range (26° 30', 102° 00')
11	Yerbabuena-1	28	Terán-1	49	Moca-1	72	**Huerta field**	94	**Garza field**
12	**Cacanapo field**	29	Benemérito-1	50	Bura-1	73	Bernardino-1	95	La Gavia Range
13	**Cuatrociénegas field**	30	Metatosa-1	51	Pino-1	74	Relámpago-101	96	La Popa
14	Gerardo-1	31	**Patricio field**	52	Cristo-1	75	Rata-1	97	Pájaros Azules Range
15	Metatosa-101	32	**Zuloaga field**	53	**Candelilla field**	76	Rayo-1	99	La Rata Range
16	Palau-1	33	Marco-1	54	San Blas-1	77	Primero de Mayo-1	100	Santa Rosa Range
17	Nazca-1	34	Jurásico-1	55	Alberto-1	78	Cartujanos-1	101	Bolsa de Judas
		35	Don Martín-1	56	Barroterán-1-A	79	Don Martín-101	102	El Venadito
		36	Ramones-1	57	**Florida field**	80	**Lampazos field**		
		37	Minas Viejas-1	58	Hulla-1	81	Ciega-1		
		38	**Noralta field**	59	**Minero field**	82	Lorencillo-1		
				60	**Merced field**	83	Piedra-1		
				61	**Escudo field**				

*Outside of map.

1, Pecten-1, Colombo-1, Monclova-5, Menchaca-1, and Linares-1 wells (Figure 3 and Table 1), the oldest age obtained was 234 ± 8 m.y. The youngest was 160 ± 6 m.y. (Middle Triassic to early Oxfordian). The radiometric ages from the Terán-1 and Benemérito-1 wells are questionable because of rock alteration; their radiometric ages are 98 and 100.7 m.y. (Albian) and 138 ±9 m.y. (Berriasian), respectively. The emplacement of the youngest ages is not aceptable, because deeply seated plutonic rocks underlie and have a discordant contact with much older rocks (Benemérito-1 well) without any evidence of contact metamorphism. In these cases, the alteration of micas into chlorite may interfere with the true age of the intrusive rock because of stratigraphic relations and the minimum acceptable age for this magmatic stage, which may correspond to the cooling age of 160 ± 6 m.y. (early Oxfordian) found in the Pecten-1 well and others that could have had similar cooling ages but were emplaced earlier.

Andesitic and dacitic volcanics in the basement complex are limited to the area of the Patricio-1, Zuloaga-1, Marco-1, and Jurásico-1 A wells. A radiometric age of 48 ± 3 m.y. (middle Eocene) obtained from basement volcanic rocks in the Patricio-1 well is inconsistent with the regional geology in the area of those wells, because there is no other surface or subsurface evidence of Cenozoic igneous activity. However, the pre-Oxfordian conglomerates present in the subsurface reveal andesitic volcanic activity prior to their deposition, as observed in Metatosa-1 and Zuloaga-1 wells.

Although the basal sedimentary rocks of Mesozoic age are not exposed in the Sabinas Basin, well data (Inés-1, Caballada-1, Progreso-1, Esteban-1, Gato-1, Ulúa-1, Myrna-1 A, etc.) reveal that the oldest sedimentary rocks in the basin are conglomerates derived from elevated blocks of the preexisting basement complex (Figures 2, 3, and Table 1) during a period of active faulting (McKee et al., 1990). In general, the conglomerates are clast supported. Clasts are subrounded to subangular, with diameters exceeding 2 cm, and consist largely of aphanitic igneous rocks with some ferromagnesian minerals and, to a lesser extent, metamorphic rocks. The sandy fraction of the matrix, which is reddish in color and well cemented by quartz, typically has quartz-filled fractures. These red-bed conglomerates appear to be alluvial-fan deposits with thicknesses varying from a few meters to more than 1000 m (Figure 3 and Table 1). The greatest thicknesses in this conglomeratic unit are located along the margins of the basin, and it is thin or absent over the Tamaulipas and Coahuila blocks.

In conventional cores from the Don Martín-1 well and the

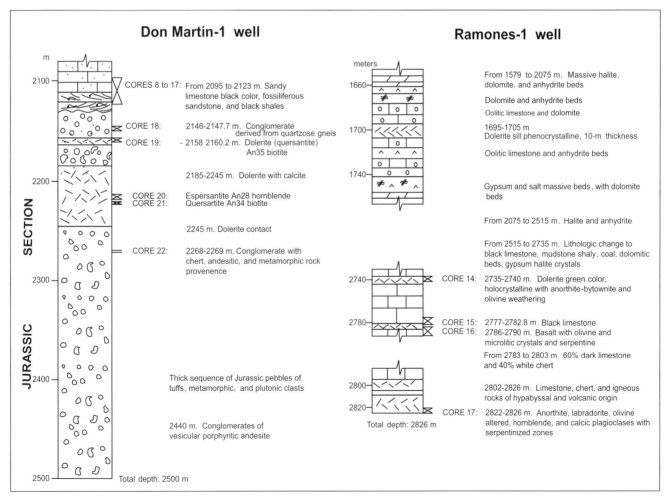

FIGURE 4. Stratigraphic columns of Callovian-Oxfordian age of Don Martín-1 and Ramones-1 wells.

deepest core of the Ramones-1 well, dolerites occur, intercalated with these basal conglomerates and evaporites (Figure 4). In addition, trachytes are intercalated with the Jurassic evaporites found in the Minas Viejas-1 well (Humphrey and Díaz, 1956). Exotic blocks of the same trachytes are entrained in the salt diapirs at La Popa (Garrison and McMillan, 1997). The same stratigraphic relations are observed in outcrop in the area of Galeana, Nuevo León, at Lomas de San Pablo, where the Huizachal group red beds (Triassic–Middle Jurassic) and the underlying evaporites are intruded by dolerite dikes, similar in appearance to those found in the subsurface of the Sabinas Basin. There are also reports of gabbros and dolerites intruding conglomerates of the basal complex on the Tamaulipas block in the Ramones-1 and Don Martín-1 wells. In all cases, these rocks are dark in color, with 50% to 60% mafic minerals content. A radiometric age of 75 ± 3.2 m.y. (Campanian) was obtained from the basal conglomerates at the Sabinas-1 well, but this age is considered to be anomalously young. The only other nearby igneous rocks of young age are Pleistocene basalts in the area of Múzquiz, Coahuila state. Farther afield, volcanic rocks of Coniacian age (Anacacho Formation) occur in Texas (Luttrell, 1977). It is noteworthy that the basic igneous rocks in the Sabinas Basin are closely associated with the pre-Oxfordian conglomerates and evaporites and do not intrude overlying rocks of Oxfordian or younger age. Therefore, the conglomerates are considered to be pre-Oxfordian.

Some authors (Charleston, 1973; Alfonso, 1978) have classified the Sabinas Basin as an intracratonic basin, either an aborted rift or aulacogen (Flores, 1981) or a transtensional pull-apart type basin (Santamaría et al., 1991; Longoria, 1984). The author considers the Sabinas Basin to be an intracontinental basin, peripheral to the cratonic edge that formed during the initial phase of evolution as a protorift (Salvador, 1991).

In the regional history of the basal complex, the following phases are recognized: The oldest, of late Paleozoic and Early Triassic age, began with marine siliciclastic sedimentation and andesitic vulcanism that were compressed during the Marathon Ouachita orogeny and intruded by granitic stocks of an intermittent, persistent magmatic arc. The second phase was the long-term vulcanism and plutonic activity that advanced from the west into eastern Mexico during the Late Triassic to Early Jurassic. This was succeeded by another compressive period (early Middle Jurassic?) that deformed the Zacatecas, Huayacocotla, and La Boca Formations (Eguiluz, 2000).

The third phase is the initial rift stage, which resulted in extensional deformation and fragmented, tilted horst blocks. The Coahuila block was tilted to the west and to the southwest, as documented by the deltaic deposits of Jurassic-Neocomian age that onlap its southwestern margin and the alluvial deposits flanking its northeastern margin (Eguiluz and Aranda, 1984). Meanwhile, the Tamaulipas block was gently warped and tilted to the northeast and southeast, as observed in seismic and well data. The margins of the Sabinas Basin may have had steep walls, as suggested by the presence of narrow belts of thick cobble and boulder conglomerates observed in the wells in (Table 1) and seismic sections (Eguiluz, 1984, 1994), which suggests rapid subsidence along a narrow basin margin.

Calc-alkaline magmatism occurred during the early and intermediate stages of rift development, succeeded by deposition of the continental conglomerates, then by evaporites (salt) during the initial marine incursion into the deeper parts of the basin (Figure 5), and finally by basaltic vulcanism during the more advanced and final rift stage. The rifting, basaltic magmatism, and first marine deposition are considered by the author to be associated with formation of the early Gulf of Mexico.

SEDIMENTARY EVOLUTION

There are three main depositional supersequences in the Sabinas Basin associated with the rift, drift, and foreland stages of its evolution (Figure 5). The first (supersequence 1) has not yet been subdivided into minor cycles, and data concerning its age are speculative. Supersequence 1 consists of conglomerates, salt interbedded with fine-grained red and green terrigenous rocks, anhydrite, and carbonates reaching a total thickness of more than 2500 m in the Minas Viejas-1 and Virgen-1 wells. From its stratigraphic position, this period of deposition probably occurred during the Callovian (?) and lower Oxfordian (Goldhammer et al., 1991). These rocks mostly occur in the central part of the basin (Figure 6). Toward the Tamaulipas Peninsula, the salt decreases in thickness or disappears, to be re-placed by anhydrite and carbonates whose combined thickness there is 500 m. These chemically precipitated stratigraphic units correspond to the Minas Viejas (Humphrey and Díaz, 1956) and Olvido (Heim, 1926) Formations. The Minas Viejas is composed mainly of salt, and the Olvido is of mixed lithology. Its lower two-thirds consist of anhydrite and interbedded carbonates, and its upper third is dominated by high-energy carbonates assigned to the next supersequence.

Updip toward the flanks of the basin, these evaporitic facies change to marginal-marine terrigenous rocks. The lower part of the La Gloria Formation (Imlay, 1936) may also be part of the supersequence 1 depositional cycle penetrated in the Escudo-1, Don Martín-1, Florida-101, Zuloaga-1, and Lampazos-1 wells. Similar facies changes occur toward the southwestern limit of the basin. In the adjacent areas of the Coahuila, Monclova, and La Mula basement blocks, rapid facies changes are noted from these thick marine carbonates and evaporites (Minas Viejas and Olvido Formations) to thin wedges of litoral sandstones and conglomerates of the La Gloria Formation, as noted in the Ines-1, Monclova-5, Nadadores-101, Mercado-1, and Golfo-1 wells. Farther updip, in the Myrna-1 A, Oro-1, and Pecten-1 wells, the La Gloria Formation was not deposited.

Deposition of cycle I of supersequence 1 occurred in sabkha and coastal marine environments during a period of continuous subsidence. In supersequence 1, numerous high-frequency depositional cycles probably occurred but have not yet been

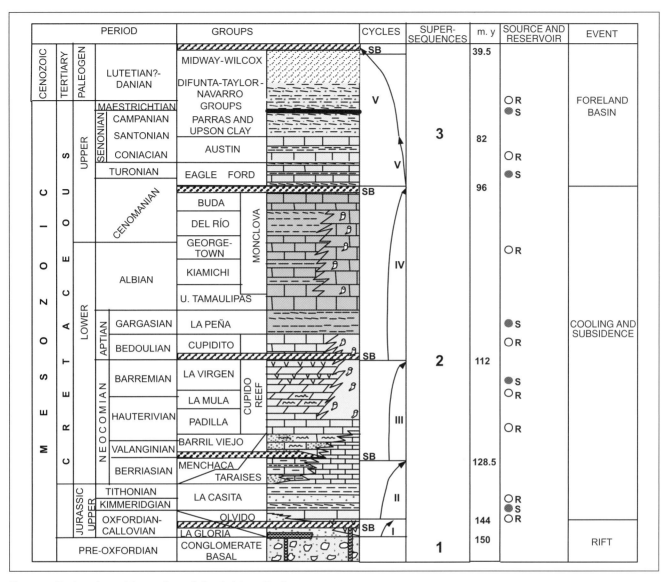

FIGURE 5. Stratigraphic section of the Sabinas Basin.

resolved. The whole set may be grouped in a transgressive-regressive cycle with an ill-defined upper-sequence limit, not yet observed in outcrop nor in the subsurface. This limit may possibly be represented by a breccia on top of the Olvido Formation evaporites, just beneath the first appearance of the high-energy carbonates.

Supersequence 2 represents the drift stage and comprises several depositional cycles (II, III, and IV); cycle I corresponds to Upper Oxfordian to Berriasian age, cycle II to Valanginian to Aptian age, and cycle III to Aptian to Cenomanian age. The base of cycle II (Figure 5) appears to be an erosional sequence boundary represented by a brecciated zone at the top of the Olvido Formation evaporites. The brecciated zone underlies the upper third of the Olvido Formation and consists predominantly of high-energy carbonates. The upper-cycle boundary is marked by a lithologic change and the absence of lower Valanginian fossils. Diagnostic fossils of this cycle include *Favreina* (upper Oxfordian), *Ataxioceras* sp., *Idoceras* sp., *Vir-*

gathosphinctes sp., *Hildoglochiceras* sp. ammonites; *Calpionella alpina, Crassicollaria parvula, Saccocoma* sp., microfossils and ammmonites of Kimmeridgian-Tithonian age; and *Stenosmellopsis* sp., *Calpionella eliptica, C. alpina, Calpionellopsis oblonga,* and *Calponellites darderi,* which are foraminifera of Berriasian age.

The upper third of the Olvido Formation consists of carbonates with evaporite nodules, calcarenites, and high-energy oolitic grainstones deposited on a complex carbonate platform. Deeper-marine environments are represented toward the top of the formation, where bioclastic wackestones and mudstones predominate and contain *Ataxiocecas* sp. ammonites of Lower Kimmeridgian age. These carbonate units mark a transgressive systems tract (TST) that grades updip into terrigenous coastal onlap facies of the La Gloria Formation (Figure 7). The thickness of the carbonates reaches a maximum of 400 m downdip.

The Olvido and La Gloria Formations are covered by black

TABLE 1. Radiometric age and petrologic studies of the basement complex of the Sabinas Basin.

Ref. map	Well/locality	No. core	Depth (m)	Rock type
1	Tarahumara-1 well	2	2868-2873	Andesite
	Tarahumara-1 well	3	2990-2998	Latite
2	Las Delicias Valley		outcrop	Granodiorite
	Las Delicias Valley		outcrop	Granodiorite
3	Paila-1A well	9	2418-2419	Ignimbrite
4	Mayrán-1A well	6	3039-3041	Perlitic rhyolite
5	Ceballos-1 well	32	5921-5930	Feldspathic schist
6	El Diablo Range		outcrop	Welded rhyolite tuff
7	Perla-1 well	1	1781-1788	Feldspathic gneiss
	Perla-1 well	2	1813-1815	Gneiss
8	El Cármen Range		outcrop	Schist
	El Cármen Range		outcrop	Schist
	El Cármen Range		outcrop	Schist
9	Carbón-1 well	cut chip	3365	Feldspathic schist
10	Magvi-1 well	3	3455-3458	Green schist
11	Yerbabuena-1 well	1	4563-4572	Muscovite schist
12	Cacanapo-101 well	5	4300-4301	Hornfels granoblastic of quartz, plagiclase, and sillimanite, grain oriented
13	Cuatrociénegas-1 well	4	4357-4357.6	Gneiss quartz feldspathic
14	Gerardo-1A well	1	3300-3303	Hornfels phelitic
15	Metatosa-101 well	2	2586-2589	Quartz feldspathic schist
16	Palau-1	2	2863-2863.5	Quartz feldspathic schist
17	Nazca-1 well	2	3095-3096	Metamorphic rock
18	Hamburgo-1 well	1	4592-4595	Dioritic hornfels, holocrystalline plagioclase, and amphibol
19	Concordia-1 well	22	4112-4116	Phillite pelitic
20	Potrero La Mula		outcrop	Granodiorite
	Potrero La Mula		outcrop	Granodiorite
21	Oro-1 well	1	3833-3835	Tonalite hornblende
22	Pécten-1 well	6	2216-2225	Granite
23	Menchaca-1A well	11	2755-2757	Granite
24	Candela-1 well	1	4284-4286	Metavolcanic aphanitic striped rock
25	Colombo-1 well	cut chip	4740	Diorite of quartz
26	Monclova-5 well	7	3930-3934	Tonalite chloritic porfidic with sodic plagioclase, potassic feldspar, and qua...
27	Linares-1 well	7	2628-2629.5	Granodiorite
28	Terán-1 well	1	1402-1409.5	Granodiorite
29	Benemérito-1 well	2	2580-2588	Granodiorite
30	Metatosa-1 well	6	2554-2559	Dacite with epidote, garnet, and chlorite
31	Patricio-1 well	7	3922-3926	Andesite phorfitic
32	Zuloaga-1 well	3	2596-2604	Andesitic tuff
33	Marco-1 well	2	2835-2840	Andesitic-type rock
34	Jurásico-1A well	2	3600-3603	Andesitic breccia, plagioclase sodic, oligoclase with vesicular aspect
35	Don Martín-1 well	19	2158-2160.2	Quersanite An 35 biotite
	Don Martín-1 well	20	2210.5-2212.6	Espersantite An 28 hornblende
	Don Martín-1 well	21	2218-2220	Buchite (dolerite dike)
36	Ramones-1 well	17	2822-2826	Gabro
37	Minas Viejas-1 well	11	3610-3619	Tracolite
38	Noralta-1A well			Volcanic and schists conglomerates in sandy matrix cemented by silex
39	Sabinas-1 well	13	2079-2083	Basalt
	Sabinas-1 well	14	2581-2590	Angular conglomerates-kind andesite
40	Ulúa-1 well	9	3712-3718	Conglomerates of igneous rocks into sandy matrix
41	Myrna 1A well	1	3746-3748	Igneous breccia
42	Caballada-1 well	8	1692-1692.6	Igneous and metamorphic conglomerates
	Caballada-1 well	9	1988-1988.6	Igneous and metamorphic conglomerate
43	Progreso-1 well	2	3222-3229	More 70-m quartz sandstone, reddish
44	Esteban-1 well	1	2985-2988	Igneous rock conglomerate
45	Gato-1	cut chip	2582-3913	Conglomerate and red sands
46	Inés-1 well	4	4070-4072	Conglomerate igneous clast
	Inés-1 well	5	4250-4252	Conglomerate igneous clast
47	Dolomita-1 well	2	4500-4501	Conglomerate igneous clast

Mineral test	Isotopic analysis		Age (m.y.)	Conglomerate thickness (m)	Reference
Whole rock	K/Ar	0.004964	83 ± 7	0	PEMEX file, reported 272 ± 22, study IMP
Sanidine	K/Ar	0.005879	98 ± 8	0	PEMEX file
Biotite	K/Ar		203 ± 4		Denison et al., 1970
Biotite	K/Ar		208 ± 8		McKee et al., 1990
Whole rock	Rb/Sr	0.7083	236 ± 39	0	Denison, PEMEX file, 1975, K/Ar-80.4 study IMP
Plagiclase	Rb/Sr	0.1194	199 ± 20	0	Denison, PEMEX file, 1975
Mica	K/Ar	0.0001167	168 ± 8	0	Denison, PEMEX file, 1979
	Rb/Sr		197		McKee et al., 1988
Whole rock	K/Ar		99.8	80	Mobil Oil, PEMEX file, 1964, quartzite, chloritic
Mica			240	0	Flawn et al., 1961
Muscovite	K/Ar		263 ± 5		Denison et al., 1969
Isochrone	Rb/Sr		275 ± 20		Denison et al., 1969
			215 ± 5	105	Denison, PEMEX file, 1975
			233 ± 9	88	in A.R. Sánchez, 1989
Mica	K/Ar		248 ± 20	25	Denison, PEMEX file
	K/Ar		381 ± 30	82	PEMEX file
				140+	PEMEX file
				8	PEMEX file
				20	PEMEX file
				30	PEMEX file
				16	PEMEX file
Whole rock	K/Ar	0.01166	190 ± 7	0	Krueger Enterprise, PEMEX file, 1981
Hornblende	K/Ar		206 ± 4	0	Denison et al., 1969
Hornblende	K/Ar		211 ± 8	0	Jones et al., 1984
			230 ± 11	0	Denison, PEMEX file, 1980
Feldspar	K/Ar		160 ± 6	518	Denison, PEMEX file; also report 225 ± 20 m.y.
Biotite	K/Ar	0.01003	164 ± 7	293	Denison, PEMEX file, 1980
	K/Ar	0.02099	329 ± 12	0	Kruegur Enterprise, PEMEX file, 1981
				85	Denison, PEMEX file, 1975
				210	PEMEX file, IMP report, 1978
Biotite	K/Ar	0.01456	234 ± 8	0	Krueger Enterprise, PEMEX file, 1975
Whole rock	K/Ar		98,100.7	0	Mobil Oil, PEMEX file, 1975
Chloritized biotite	K/Ar	0.008371	138 ± 9	0	Krueger Enterprise, PEMEX file, 1972
				0	Denison, PEMEX file
			48 ± 3	0	PEMEX file, IMP report
				281	PEMEX file, 1981
				0	PEMEX file
				0	PEMEX file, 1983
				370+	PEMEX file, 1964
					PEMEX file, 1964
					PEMEX file, 1964
				300+	PEMEX file
					PEMEX file
				31+	PEMEX file
Whole rock	K/Ar	0.004477	75 ± 3.2	1023+	Krueger enterprise, PEMEX file, 1972
					PEMEX file, 1972
				32+	PEMEX file
				0	PEMEX file
				520+	PEMEX file
					PEMEX file
				70+	PEMEX file
				126+	PEMEX file
				394+	PEMEX file
				719+	PEMEX file
					PEMEX file
				165+	PEMEX file

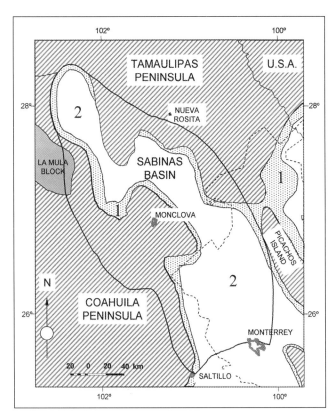

FIGURE 6. Callovian (?)–lower Oxfordian paleogeography of the Sabinas Basin. Striped areas indicate emerged lands. (1) Shoreface sands of the La Gloria Formation; and (2) sabkha facies (salt, anhydrite, and carbonates) of the Minas Viejas and Olvido Formations.

carbonaceous shales of the La Casita Formation (Imlay, 1937), whose basal member represents a maximum flooding surface (MFS) covering a broader area than the preceding cycle and containing *Idoceras* sp. ammonites of Kimmeridgian age. In the Sabinas Basin, the La Casita Formation consist of three members: a lower member of shale, a middle member of sandstone interbedded with carbonates, and an upper shale member. The shales, carbonates, and sandstones of the La Casita Formation record continuous transgression and indicate gradual deepening of the basin. During this time, the northwestern limit of the basin was flooding, and the La Mula block was surrounded by the sea. The Concordia-1 well confirms the extent of the Jurassic sea toward the border between the states of Chihuahua and Coahuila. The Sabinas and Chihuahua Basins probably were connected at this time, and the Coahuila and Tamaulipas blocks probably were separated by a marine channel (Figures 8 and 9).

Higher-frequency cycles occurred during the mid-Kimmeridgian and Tithonian, but in general, a high sea level was maintained. The thickness of the La Casita Formation in the Sabinas Basin ranges from 60 m updip to 800 m downdip. The base of this sedimentary cycle consists of organic-rich, carbonaceous shales deposited in an outer neritic environment. The distribution of the deeper facies and the greatest thickness of shales in the La Casita Formation coincide with the position

of the underlying pre-Kimmeridgian salt (Minas Viejas Formation), and updip, the La Casita Formation pinches out or changes to coastal sand facies (Figures 8, 9, and 10).

The intermediate sandstone member of the La Casita Formation coarsens toward the margins of the Coahuila and Tamaulipas blocks, changes facies, and thins toward the center of the basin (Figures 9 and 10) . The environments recognized in the sandstone are middle to inner neritic (Figures 8 and 9). Fossil plants have been encountered in well cores and indicate deltaic environments (Merced field, Cacanapo-101 and Escudo-1 wells).

Facies interpreted from well logs are indicative of channel and progradational deposits. The thickness of the middle sandy unit varies from 300 to 600 m. Sandstone composition varies from litharenites near the Tamaulipas block to arkoses with abundant microcline clasts near the La Mula and Monclova blocks (Figure 10). Grain size decreases away from the provenance area, changing from coarse/medium grained to fine/very fine grained. Texturally immature conglomerate predominates in the proximity of the Coahuila block. The La Casita sandstones are generally cemented by silica with silica overgrowths, and their average porosity is less than 5%.

The upper portion of the La Casita Formation consists of black calcareous shales and siltstones less than 50 m thick that may represent a high-frequency flooding zone. The Jurassic-Cretaceous boundary should be present in this unit, but it is

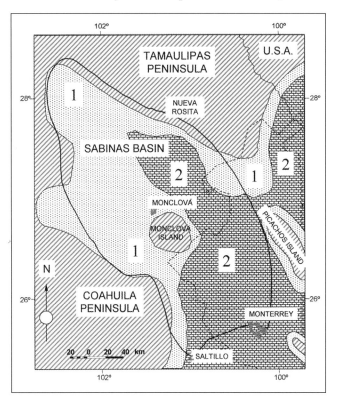

FIGURE 7. Upper Oxfordian–lower Kimmeridgian paleogeography of the Sabinas Basin. Striped areas indicate emerged lands. (1) Shoreface sands of the La Gloria Formation; (2) lagoonal carbonates of the Olvido Formation.

difficult to identify lithologically and is recognized only through biostratigraphic data. The areal distribution of the La Casita Formation varies with age, being limited in the Kimmeridgian and more widely distributed in the Tithonian (Figures 8 and 9). The organic-matter content of these rocks ranges from 0.5 to 2.0% TOC and is thermally overmature with vitrinite-reflectance values ranging from 2 to 4 (González and Holguin, 1992). It is considered to be the main gas source rock for this basin.

Clayey carbonates and terrigenous rocks of variable grain size were deposited during the Berriasian. Toward the Coahuila block, alluvial-fan conglomerates of the San Marcos Formation prevail (Imlay, 1940), and toward the Tamaulipas block, alluvial-plain facies composed of sandstone and conglomeratic red beds predominate (Hosston Formation). The La Mula block remained emerged, while the Monclova block was covered by the sea (Pecten-1 and Oro-1 wells). Downdip, toward the basin axis (Figures 11 and 15), there was a high-energy carbonate marine platform represented by the Menchaca Formation (Imlay, 1940) which grades toward the southeast and east into open platform and basinal facies represented by interbedded shales and carbonates of the Taraises Formation (Imlay, 1936). This is interpreted to represent a low-angle ramp of the type proposed by Ahr (1973).

A few coral-sponge patch reefs around the ramp edge are exposed at the Cañon de la Huasteca (Vokes, 1963). This cycle may represent a highstand systems tract (HST) and prograding

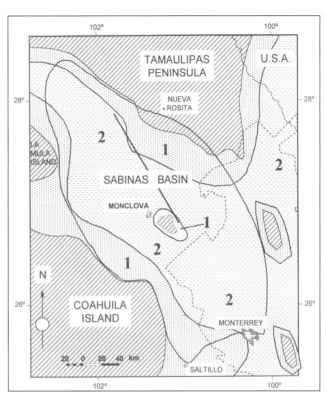

FIGURE 9. Tithonian paleogeography of the Sabinas Basin. Striped areas indicate emerged lands. (1) Shoreface sands and (2) deep-marine shale facies of the La Casita Formation. Cross section is shown in Figure 10.

platform. Updip from the basin axis at the top of this cycle (128.5 m.y.), there may exist a type 1 sequence boundary, represented by erosion and exposed in outcrops of the San Marcos Formation in the San Marcos Range. McKee et al. (1990) reported an unconformity here that could be the postulated sequence boundary. The lack of strata with lower Valanginian fauna in northeastern Mexico may denote nondeposit or erosion during this period.

The thickness of the Menchaca Formation in wells and outcrops varies from 250 to 350 m. The rock is compact and does not display any primary porosity. Low organic-carbon content in the fine-grained facies indicates that it was not a viable source rock. The age of the Menchaca Formation is considered to be Berriasian, based on the presence of foraminifera *Calpionella darderi, Stenosemellopsis* sp., *Calpionellopsis oblonga, Calpionella alpina,* and *C. elliptica* and ammonites *Berriasella* sp. and *Spiticeras* sp. (Imlay, 1940).

A new deposition cycle (III) followed deposition of the Menchaca Formation, with the accumulation of sandstones and finer-grained terrigenous rocks of the Barril Viejo Formation (Imlay, 1940). The source of these siliciclastic rocks was apparently the exposed Coahuila and Tamaulipas blocks. The lateral stratigraphic relationships are very similar to those of the Menchaca Limestone (Figures 11 and 15). Westward, toward the boundary of the Coahuila block, the formation is transitional

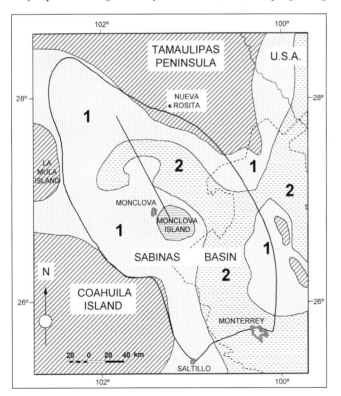

FIGURE 8. Mid-Kimmeridgian paleogeography of the Sabinas Basin. Striped areas indicate emerged lands. (1) Sandy facies and (2) deep-marine shale facies of the La Casita Formation. Cross section is shown in Figure 10.

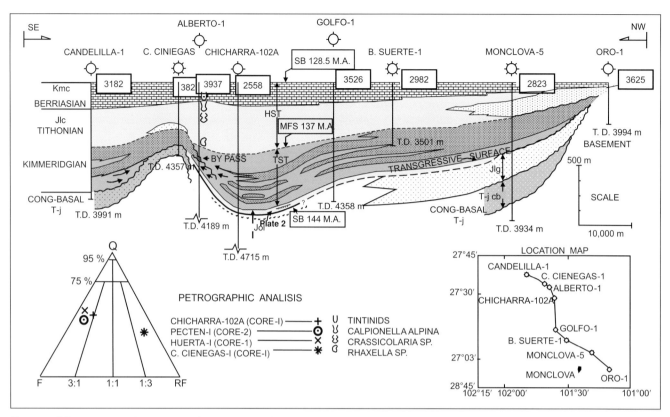

FIGURE 10. Stratigraphic section showing lateral changes between the Monclova and Tamaulipas blocks in the La Casita (Jlc) and the Menchaca (Kmc) Formations, the Olvido Formation (Jol), La Gloria Formation (Jlg), and the Triassic-Jurassic (T-j).

into San Marcos Formation conglomerates. Southeastward and eastward toward the Tamaulipas block, it grades into the Hosston Formation across an extensive ramp-type platform. Toward the ancestral Gulf of Mexico, it changes to clayey carbonates of the upper Taraises Formation. Toward the Chihuahua Basin to the northwest, it is transitional to continental fluviolacustrine sandstones of the Las Vigas Formation (Burrows, 1910), denoting regression from that direction. According to Burkhardt (1930), this regressive arkosic sandstone and siltstone unit was deposited in a warm, humid climate. However, according to Imlay (1940) and to McKee (1990), deposition of the Barril Viejo Formation was related to reactivation and uplift of the Coahuila block. Its facies corresponds to littoral deposits, as observed in outcrops of the La Gavia, Menchaca, and Obayos ranges. These change to alluvial facies toward the northeast and southwest—continental facies to the northwest and calcareous, pelitic platform facies to the east and southeast (Figure 12).

The thickness of the Barril Viejo Formation varies from 250 to 350 m. This unit has no appreciable porosity, because it is highly compacted and well cemented. A lower Hauterivian age is assigned to the Barril Viejo Formation, based on the presence of *Acanthodiscus magnificus* Imlay, *A. radiatus*, *A. bruguiere*, *Leopoldia crassicostata* Imlay, and *L. victoriensis*. Deposition of the Barril Viejo may have started with strata beneath the beds containing ammonites and could include the upper Valanginian, as suggested by the fauna content of the Taraises Forma-

tion in the El Fraile (Potrero Chico), Minas Viejas, and Pajaros Azules ranges and in the Barril Viejo Formation in the La Gavia range.

Subsidence of the Sabinas Basin continued during the Hauterivian and Barremian, and a gradual change in sedimentation occurred at that time. Carbonate deposition of the Padilla Formation (Imlay, 1940), with an average thickness of 150 m, occurred in an arcuate northeast-southwest-oriented belt (Figure 13). Dolomite predominates, with increasing content of interbedded siltstones and reddish shales toward the continental areas of the Coahuila and Tamaulipas blocks. Toward the northwestern part of the Sabinas Basin and the La Mula block, there is a belt of red-and-green fine-grained terrigenous rocks of the La Mula Formation (Imlay, 1940) that thins toward the open sea to the east and southeast of the basin (Figures 14 and 15). This unit thickens toward the Chihuahua Basin, where it is transitional to the Las Vigas Formation (Burrows, 1910). Both units change laterally into coarse siliciclastic rocks of the Hosston and San Marcos Formations, as did the Barril Viejo Formation (Figures 12 and 15). The hydrocarbon source potential of the Padilla Formation has not been studied, but it is tentatively considered to be of poor to nonsource quality. Maximum porosity in the Padilla dolomites is 8%, whereas it does not exceed 5% in the La Mula Formation sandstones.

The Padilla Formation is considered to be upper Hauterivian in age, based on the first appearance of *Chophatella decipi-*

ens. The Padilla consists of prograding facies deposited during a sea-level HST and represents the initial platform under the Cupido and La Virgen formations (Imlay, 1940). The La Virgen Formation consists of evaporites deposited in a lagoonal carbonate sabkha environment, whereas the Cupido Formation is a barrier reef complex (Murillo-Muñeton, 1999) that restricted the entry of normal salinity seawater into the lagoon (Figures 15 and 16); both formations were deposited as prograding systems during a sea-level highstand. The La Virgen Formation reaches thicknesses of 600 to 800 m and changes facies laterally into the terrigenous rocks of Hosston and San Marcos Formations. Toward the Chihuahua Basin, the evaporitic deposits (Cuchillo Formation; Burrows, 1910) have lateral continuity with the Sabinas Basin, whereas toward the east and southeast of the Cupido reef, the La Virgen is transitional into pelagic carbonates of the Lower Tamaulipas Formation (Figures 15 and 16). The La Virgen Formation contains five units—three dolomitic carbonate packages (Marquez et al., 1979) separated by two evaporite units; the dolomites generally have 6% to 8% porosity, and the carbonate-evaporitic facies have moderate source-rock potential.

Near the city of Monterrey, Goldhammer et al. (1991) identified an erosional contact at the top of the Cupido Formation, which contact has also been observed in several localities in the Sabinas Basin (ranges of Obayos, La Virgen, Menchaca,

etc.). This is interpreted to be a high-order sequence boundary. Once again, a TST appears (cycle IV) above this sequence boundary, comprising at least three major cycles in supersequence 2. Where the basal, high-energy carbonates transition upward into low-energy carbonates, these rocks were originally assigned to the Cupido Formation (Imlay, 1936; Humphrey, 1949) but are presently known as the Cupidito Limestone (Wilson, 1977). The lateral distribution of this marine unit TST extended over the Coahuila and Tamaulipas blocks, which had previously remained emerged (Figure 17). The Cupidito Limestone is a gradational development of the upper part of the Cupido barrier-reef facies at the eastern limit of an extensive platform that extended westward into the Chihuahua Basin. To the east, it is replaced by micritic carbonate basinal facies of the Lower Tamaulipas Formation (Figure 17).

The average thickness of the Cupidito Limestone is 250 m, thinning to 5 m toward the northwest at the boundary between the states of Coahuila and Chihuahua in the Concordia Range. The presence of *Salpingoporella* sp., *Acroporella* sp., and *Clypeina* sp. indicates that the Cupidito Limestone was deposited during the Barremian-Aptian as a TST.

Continuation of basin subsidence and increasing relative sea level is represented by the shales and clayey limestones of the La Peña Formation (Imlay, 1936; Humphrey, 1949), corresponding to an MFS that extended over the entire Sabinas

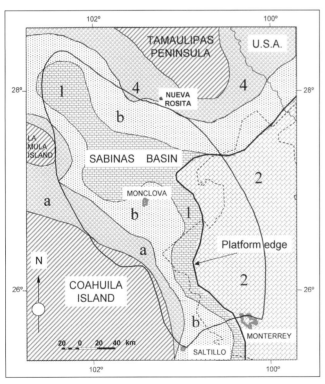

FIGURE 11. Berriasian paleogeography of the Sabinas Basin. Striped areas indicate emerged lands. (1) High-energy carbonates of the Menchaca Formation; (2) low-energy, open-shelf carbonates of the Taraises Formation; (3a) alluvial facies and (3b) shoreface or delta facies of the San Marcos Formation; (4) alluvial-plain facies of the Hosston Formation.

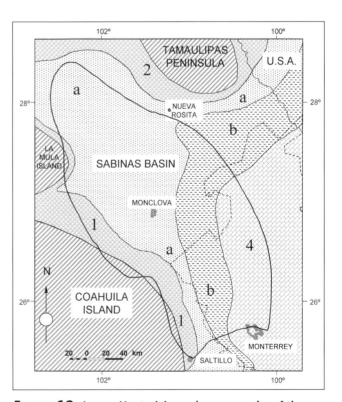

FIGURE 12. Lower Hauterivian paleogeography of the Sabinas Basin. Striped areas indicate emerged lands. (1) Alluvial facies of the San Marcos Formation; (2) alluvial-plain facies of the Hosston Formation; (3a) shoreface sands and (3b) silty limestone of the Barril Viejo Formation; (4) open-shelf limestone of the Taraises Formation.

Basin and adjacent Tamaulipas block. In contrast, on the Coahuila block it becomes a sandy facies (Figure 18) referred to as the Las Uvas Formation. This unit has an average thickness of 100 m to 150 m, and thins or is absent over the Cupido reef facies (Eguiluz, 1994). The source-rock potential of these rocks is poor, with TOC values typically less than 0.5%. The pelitic facies of the La Peña Formation correspond to a deep-basinal environment that developed in the upper Aptian, as indicated by the presence of *Nannoconus trutti, N. minutus, N. wasalli, N. elongatus, Hedbergella robesae,* and the ammonites *Dufreoyia justinae, Cheloniceras* sp., and *Ritydohoplites* sp.

Deposition of basinal micritic carbonates and shales began in the Albian and continued into the Cenomanian with three predominantly carbonate units (Upper Tamaulipas, Georgetown, and Buda Formations) interrupted by shale units ("Kiamichi" and Del Rio Formations), as shown in Figure 5. The base of the Upper Tamaulipas Formation (Muir, 1936) has an average thickness of 450 m and contains *Colomiella recta, C. mexicana,* and *Calpinollopsella maldonadoi,* indicating a lower Albian–middle Albian age.

Above the Upper Tamaulipas Formation is a tripartite unit composed of two shales and a clayey limestone that reaches a thickness of 150 m in the central and northwestern portions of

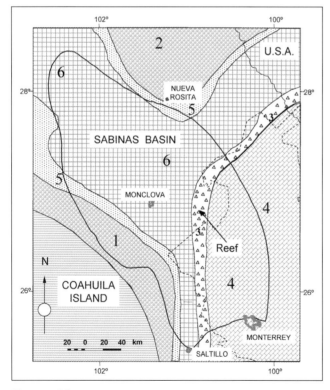

FIGURE 13. Upper Hauterivian paleogeography of the Sabinas Basin. Striped areas indicate emerged lands. (1) Alluvial facies of the San Marcos Formation; (2) alluvial-plain facies of the Hosston Formation; (3) reef facies of the Padilla Formation; (4) open-shelf limestone facies of the Taraises Formation; (5) shoreface sands of the San Marcos and Hosston Formations; (6) carbonate lagoonal facies of the Padilla Formation.

the Sabinas Basin, thinning to 10 m along a featheredge in the northeast and eastern parts of the basin. It has been incorrectly correlated with the Kiamichi Formation (Hill, *in* Sellards et al., 1932), but the facies and lithology are different from the type locality of this unit. It is more appropriately identified here as the Sombreretillo Formation (Bishop, 1970). The presence of *Microcalamoides diversus, Sacocoma* sp., and *Calciesphaerulla innominatax* indicates that these strata were deposited during the late to middle Albian and the early to late Albian.

Above the Sombreretillo Formation is an 80- to 150-m thick interval of micritic limestone with flint nodules representing open-platform to basinal conditions. This unit, containing *Phitomnella ovallis* and *Favusella* sp., is referred to as the Georgetown Formation of upper Albian to lowermost Cenomanian (Hill, *in* Sellards et al., 1932). Its distribution is very broad over the entire Sabinas Basin, losing its identity only toward the southeastern and eastern parts of the basin. The second, upper-clayey unit is the Del Río Formation (Hill and Vaughan, *in* Sellards et al., 1932), a fine-grained body of shale and sandstone containing *Bonetocardiella betica, B. conoidea,* and *Stomiosphaera sphaerica,* indicating an early Cenomanian age. Its thickness varies from 5 to 40 m, and it is thickest in the north and northeastern parts of the basin, pinching out toward the southeast. The upper carbonate unit, with a thickness of 10 to 50 m, corresponds to the Buda Formation (Hill, *in* Sellards et al., 1932). It consist of basinal calcareous mudstones with planktonic fauna. As with the Sombreretillo, Georgetown, and Del Río Formations, the Buda pinches out and changes facies near the southeastern and eastern edges of the Sabinas Basin.

The five previously described units change facies toward the southwest and northeast of the Coahuila and Tamaulipas blocks, which were buried by that time. During the Albian, restricted platforms developed that were characterized by shallow-marine environments and evaporite deposition. The Coahuila platform (Garza, 1973) and the Maverick platform (Smith, 1966) next to the Sabinas Basin are the principal areas of facies changes recognized as being associated with this period (Figure 19). A small, calcarenite shoal also developed adjacent to the La Rata Range. This deposition was influenced by salt growth in the structural zone of salt diapirs (Eguiluz, 1996). The total thickness of the Albian through Cenomanian section in this basin varies from 700 m in the northwest to 500 m in the southeast. The porosity of these rocks is very poor, because they consist mainly of very compact carbonates. No source-rock analyses are available, but their hydrocarbon-generating potential is considered to be poor or nil.

From the Aptian to the lower Cenomanian, extensive carbonate platforms prevailed on which carbonates and evaporites were deposited. The Coahuila and Maverick platforms were characterized by prograding cycles in the early Albian, which became accretionary toward the end of their existence in the late Albian and earliest Cenomanian (Eguiluz, 1990b; Marquez, 1979; Smith, 1966). The intermittent influx of terrigenous sediments from north-interrupted carbonate deposition and several minor flooding cycles are recognized in the central and northwestern parts of the Sabinas Basin. Accumulation rates of sediments on the carbonate banks (1200 m) and in the central part of the basin (885 m) differed by almost 30%, suggesting a platform model of an open-ramp type during a high sea-level cycle.

During most of the Late Cretaceous, the cycles of deposition in the Sabinas Basin (cycles V and V′) consist of overall regressive and prograding sequences, in contrast with the preceding cycles of the Jurassic to Aptian (150–112 m.y.) and the Aptian to Cenomanian (112–96 m.y.). Six formational units make up this supersequence 3. At its base is the Eagle Ford Formation (Roemer, *in* Sellards et al., 1932), followed by the Austin Limestone (Shumard, *in* Sellards et al., 1932); the Upson shale (Dumble, 1892); and the progressively coarser, terrigenous San Miguel, Olmos, and Escondido Formations (Dumble, 1892).

The Eagle Ford Formation consists of a series of thin black shales interbedded rhythmically with sandy limestone and carbonate-cemented sandstone. The Eagle Ford is approximately 300 m thick and has a wide distribution across the northwestern, northeastern, and central portions of the basin. There is a facies change toward the southeastern portion, where a greater carbonate component and thicker bedding have been named the San Felipe Formation (Figure 20). Westward, toward the area of the Coahuila platform, there is another facies change to calcareous shales, clayey limestone, and sandstones of the Indidura Formation (Kelly, 1936). The Eagle Ford Formation represents a TST deposited in a middle-neritic environment. A late Cenomanian to early Turonian age is indicated by the presence of *Inoceramus labiatius, Clavihedbergella simplex, Hedbergella amabilis, Rotalipora cushmani,* and *R. greenhornesis.* Fine-grained units within the Eagle Ford Formation have TOC values of 0.5% to 1% (González and Holguin, 1992). Potential reservoir facies commonly are highly compacted, with poor porosity.

The Austin Formation is a series of thick, interbedded, light-gray, clayey limestone and calcareous shales 200 to 300 m thick, with widespread, thin, interbedded tuffs (Figures 20 and 21). Facies changes in the Austin are similar to those described for the Eagle Ford Formation, representing similar depositional environments. The principal difference between the two formations is the higher content of calcium carbonate in the Austin, caused by a change of climate and/or shallowing of the platform during the prograding cycle, which developed at a high eustatic level (Figures 5 and 21). The presence of *Inoceramus onduluplicatus, Globotruncana laparenty, G. concavata, Hedbergella planispira,* and *Praeglobotruncana* sp. indicates a Coniacian-Santonian age for the Austin Formation. There are few source-rock analyses of this unit, and those that exist provide similar TOC values to the underlying Eagle Ford Formation (0.5% to 1.0%).

The Upson Formation consists of black laminated shale,

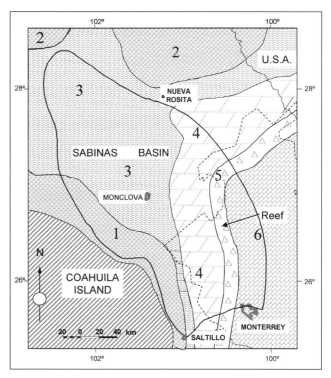

FIGURE 14. Upper Hauterivian–lower Barremian paleogeography of the Sabinas Basin. Striped areas indicate emerged lands. (1) Alluvial facies of the San Marcos Formation; (2) alluvial plain of the Hosston Formation; (3) silt, shales, and reddish sandstones of the La Mula Formation; (4) lagoonal carbonates of the Padilla Formation; (5) reef facies of the Cupido Formation; (6) open-shelf carbonates of the Lower Tamaulipas Formation. Depositional model is shown in Figure 15.

100 to 150 m thick, with the same areal distribution as the previous upper Cretaceous units. The Upson correlates with and is lithologically indistinguishable from the thicker Parras shale to the south (Imlay, 1936). Its depositional environment is interpreted to be a prodelta within a flooding cycle during initial development of the Laramide foreland basin.

The San Miguel Formation is a delta-front facies more than 400 m thick consisting of alternating shaly and sandy progradational units. Grain size is fine at its base and coarsening toward the top, where distributary channels have been identified on well logs. The San Miguel is dated as latest Santonian and early Campanian from interbedded marine strata containing *Globotruncana* sp. and *Heterohelicids* sp. This unit is widely distributed in the central and northeastern portions of the basin. According to available data, the San Miguel delta in the Nueva Rosita area might be older than the La Popa delta.

The Olmos Formation consists of thickly bedded, coarse- to medium-grained sandstones. Its well-log character is typical of distributary channels and bars. The presence of *Globotruncana elevata, G. rosseta,* and *G. stuatiformis* marks the last marine component of this cycle and suggests a late Campanian age. Distribution of these rocks is restricted to the central and northeastern portions of the Sabinas Basin (Figure 22).

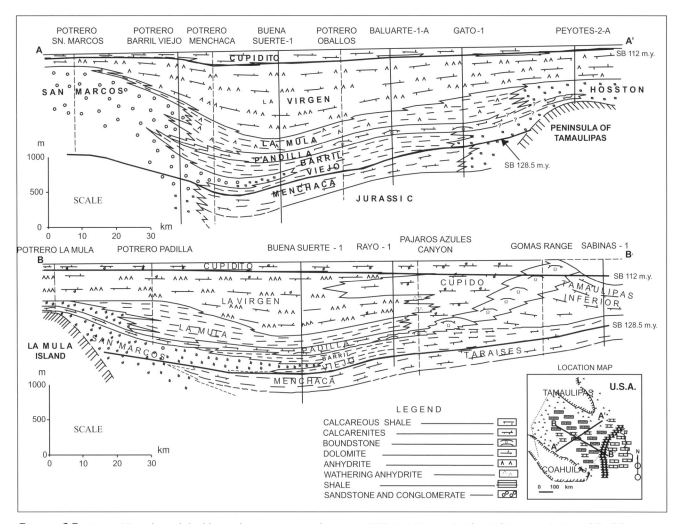

FIGURE 15. Depositional model of boundary sequences between 128.5–112 m.y. in the Sabinas Basin (modified from Marquez, 1979).

The Escondido Formation consists of conglomeratic sandstone with several cyclical shales and coal seams near its base, suggesting an oscillating floodplain. The Escondido has an average thickness of 800 m. Well logs indicate the presence of numerous channelized sandstone bodies characterized by cross-bedding and rapid lateral facies changes that impede correlations between wells. The Escondido Formation correlates with deltaic facies of the El Muerto Formation of the Difunta group found in the La Popa (Giles et al., 1999). From its stratigraphic position, the Escondido Formation is assigned a Maestrichtian age. By that time, it appears that the Nueva Rosita and La Popa deltas had emerged and prograded toward the Las Adjuntas syncline southeast of Monclova (Figure 22). The sequence comprising the Olmos through the Escondido Formations represents a major regressive cycle (HST).

No rocks younger than the Cretaceous are preserved in the Sabinas Basin. However, in the Rio Escondido Basin, east of the Sabinas Basin, there is a significant thickness (1000 m) of Paleocene deep-marine shales (Midway group) above the Escondido Formation. Midway equivalents may also have been deposited above the Escondido Formation in the Sabinas Basin

in the Nueva Rosita area. A burial/thermal-history analysis of the Sabinas Basin indicates that the Escondido was buried to a depth sufficient to mature the constituent organic matter to bituminous coal (at least 1500 m additional of sedimentary column) and the organic-rich Jurassic rocks into the dry-gas window (Figure 23).

The presence of Ypresian strata in the La Popa area (Vega-Vera and Perrilliat, 1989) also suggests that the Nueva Rosita region experienced post-Cretaceous marine flooding with prograding deltaic platform conditions into the post-Ypresian. This event apparently ended marine sedimentation in the Sabinas Basin. Continental sedimentation may have continued, thereby converting the Escondido Formation peats into coal. Another explanation for the high maturity of the Escondido coals could be an elevated geothermal gradient caused by igneous activity. However, there are only minor volcanics in the basin and no evidence of post-Escondido intrusive rocks.

The rates of deposition of the three megasequences varied considerably (Table 2), with the highest rate of deposition occurring during the synrift stage (cycle I). Assuming that 2500 m of marine sediments was deposited from the Callovian into

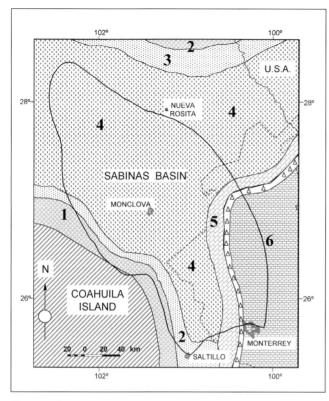

FIGURE 16. Upper Barremian paleogeography of the Sabinas Basin. Striped areas indicate emerged land. (1) Alluvial facies of the San Marcos Formation; (2) shoreface San Marcos; (3a) alluvial plain of the Hosston Formation; (3b) shoreface Hosston; (4) evaporitic sabkha of the La Virgen Formation; (5) postreef and reef complex facies of the Cupido Formation; (6) open-shelf carbonate facies of the Lower Tamaulipas Formation; (7) reefal facies of the Padilla Formation.

the Oxfordian (150–144 m.y.), an average sedimentation rate of 0.178 mm/y may be calculated. The next megasequence, comprising cycles II and III between the Oxfordian and Lower Cretaceous (144–128.5 and 128.5–112 m.y., respectively), was deposited on a passive margin controlled by subsidence caused by crustal cooling. Calculated sedimentation rates are 0.080 and 0.10 mm/y, at least 40% lower than those of the earlier cycle. Rates of sedimentation of the basinal facies are ~30% less than those of the contemporaneous shallow-water facies on the Coahuila and Maverick platforms (cycle IV, 112–96 m.y.). Calculated sedimentation rates in the foreland basin for the Upper Cretaceous to Paleogene sedimentary cycles (Table 2) were lower than those of the previous megasequence (cycle V, 96–82 m.y. = 0.039 mm/y, and cycle V′, 82–39.5 m.y. = 0.062 mm/y).

STRUCTURAL GEOLOGY

The Sabinas Basin initially developed on the margin of the North American craton during the early Mesozoic opening of the Gulf of Mexico. There are as yet no kinematic or geophysical data to discriminate between a transtensional (Longoria,

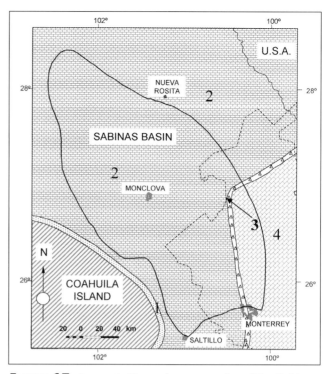

FIGURE 17. Lower Aptian paleogeography of the Sabinas Basin. Striped areas indicate emerged land. (1) Shoreface sands; (2) high-energy lagoonal facies of the Cupidito Formation; (3) Cupido reef; (4) open-shelf carbonate of the Lower Tamaulipas Formation. Depositional model is shown in Figure 15.

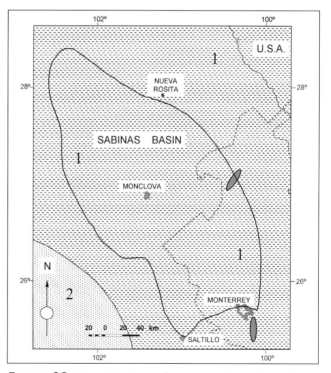

FIGURE 18. Upper Aptian paleogeography of the Sabinas Basin. (1) Shales and limy mudstone of basinal facies of the La Peña Formation; (2) shoreface sandstone of the Las Uvas Formation.

1984; Santamaría, 1991) or extensional model for its origin. The early emplacement of basic igneous rocks and the basin's asymmetrical transverse profiles (as observed on seismic reflection) suggest that the basin could have originated by rifting, possibly as part of an aulacogen (Dickinson and Yarborough, 1979).

Prolonged subsidence during the Upper Jurassic and Lower Cretaceous formed an Atlantic-type passive margin on the northeastern edge of the basin. The accumulation and deformation of foreland basin fill observed in the La Popa and Nueva Rosita areas are indicative of a final compressional stage of evolution for the Sabinas Basin. Laramide compressional deformation of the sedimentary cover during this last stage of the basin's history was less intense than that of the adjacent Sierra Madre Oriental and Chihuahua Basin. In contrast to the Sierra Madre Oriental, the Sabinas fold belt displays bidirectional vergence toward the paleotectonic edges of the Coahuila and Tamaulipas blocks, salt domes pierce the sedimentary column, and bifurcated folds occur with well-defined periclinal plunges and a lesser degree of shortening. The Sabinas Basin also lacks the Cenozoic basin-and-ranges-type deformation observed in the Chihuahua Basin to the west. In the Sabinas Basin, the Laramide, or *Hidalgoense* (de Cserna, 1960), structural framework is divided into three deformation levels: (1) Jurassic subevaporites, (2) Jurassic evaporites, and (3) Jurassic postevaporites.

The Sabinas Basin has regional detachment levels below and above the Jurassic evaporites that produced thin-skinned-type deformation. Regional detachment of the post-salt sedimentary section occurred wherever Jurassic salt is found. Where salt is absent, the basal complex is involved in the deformation (thrusted). Ductile lithologies such as evaporites (La Virgen Formation) or shales (Parras and Upson Formations) provided local detachment levels. Where the salt pinches out, regional detachment cuts upsection to younger stratigraphic levels and emerges at the basin edges.

Compressive structures predominate in the Sabinas Basin. Kinematic data (from folds, striations, and stylolites) indicate shortening parallel to the direction of tectonic transport. For the construction of the balanced sections presented herein (Figure 25), longitudinal linear balance was used for flexural sliding rocks, except in the case of evaporites, which were area balanced.

Four well-defined structural styles (Eguiluz, 1997) are identified in the Sabinas Basin (Figure 24). Where Jurassic salt is present, anticlinal folds are long and narrow, and their structural axes are consistently oriented approximately NW45°SE. Structures can be followed for tens of kilometers and generally have well-defined plunging noses. The folds are disharmonic and bifurcated. Narrow anticlines are separated by broad synclines and display vergence mainly to the southwest, although vergence to the northeast also occurs. A Jurassic salt-detachment level is estimated to exist at a depth of 2000 m.

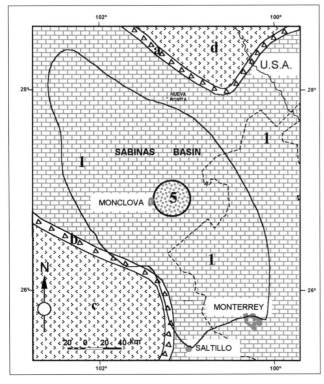

FIGURE 19. Albian paleogeography of the Sabinas Basin. (1) Basinal carbonate facies of the upper Tamaulipas–Georgetown formations; (2a) Stuart City reef facies; (2b) Viesca reef; (3) lagoonal evaporitic facies of the Acatita Formation; (4) lagoonal evaporitic facies of the Macknight Formation; (5) Monclova Formation calcarenites.

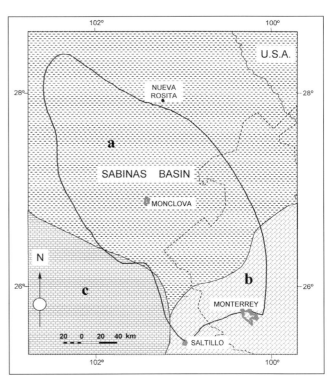

FIGURE 20. Upper Cenomanian–Turonian paleogeography of the Sabinas Basin. Open-shelf facies are identified by (a) shales and mudstones of the Eagle Ford Formation; (b) silty limestones of the San Felipe Formation; (c) shales and silty carbonates of the Indidura Formation.

Toward the basin margins where Jurassic evaporites are absent, faults are emergent, and very high-relief anticlinal structures occur. Anticlines there tend to have more deeply eroded cores and therefore expose older rocks. Radical changes are observed in the trend of the fold axes, with north-northwest and east-west orientations, e.g., the La Madera and La Fragua mountain ranges on the southwestern edge of the basin and in the Sierra Hermosa de Santa Rosa on the northeastern margin. Vergence of structures is toward the basin margins in the San Marcos, La Madera, La Fragua, and Hermosa de Santa Rosa anticlines. In this case, the basal complex is involved in the deformation. Schists in the basal complex may constitute deep detachment levels with emerging, duplex-type imbricated slopes (Figure 25, section III–III′, Monclova–Buena Suerte field).

A third structural style is observed on the Tamaulipas block where the Laramide structures are buried by younger rocks. Folding is relatively gentle in that part of the basin. Where the Jurassic evaporites pinch out, the detachment level emerges by ramping up to the northeast and with fold vergence in the same direction (Figure 25, section II–II′). At depth, the structures are more complex because of the presence of antithetic reverse faults, as seen in the Merced gas field (Vázquez and Eguiluz, 1997).

Finally, where salt is thickest in the southeastern and northeastern portions of the basin (La Popa area; Giles et al., 1999; Obayos, Eguiluz, 1997), a band of salt domes intrudes the anticlinal structures (San José de la Popa, Bolsa de Judas, northwestern Pájaros Azules, La Rata, Obayos, etc.) and box-type folds are common (Gray et al., 1997). This style is very distinctive and is restricted to a belt of complex, circular, dome-shaped structures (Giles et al., 1999) that are easily recognized on detailed geologic maps, seismic sections, and wells (Carroza-1, Primero de Mayo-1, etc.).

Kinematic data indicate shortening of the basin and tectonic transport in a NE40°SW direction. Area-balanced and restored sections show 26% shortening in the middle and southwestern portions of the basin (Figure 25, cross section I–I′). To the northeast, however, shortening does not exceed 16% on the Tamaulipas block (Figure 25, cross section II–II′). Structural shortening of the Sierra Madre Oriental salient is much greater, ranging from 30% to 35% (Marrett and Aranda, 1999).

The critical-wedge mechanical model (Davis et al., 1983; Dahlen et al., 1984) is as valid for the Sabinas Basin as it is for the Sierra Madre Oriental (Marrett and Aranda, 1999). This model explains how the horizontal contraction of material, detached from its substrate, tends to produce a wedge that grows vertically, caused by gravity. The geometric wedge develops and reflects a balance between the stress necessary to produce the wedge material's internal deformation and the stress necessary to cause movement along the basal detachment. A relatively strong basal detachment will tend to concentrate the shortening and produce a relatively steeply tapered wedge, whereas a weak basal detachment will tend to distribute the shortening over a more extensive zone and create a more gently tapering wedge.

Fracture systems are well developed in the structures of the Sabinas Basin. According to Nelson's classification (1985), the main fractures are of type 1 (ab) and type 2 (ac), forming sub-orthogonal grids that can be observed in outcrop, conventional cores, aerial photographs, and satellite images. Fracture density is greater in compacted, coarse-grained siliciclastic rocks and dolomites than in limestones and siliciclastic mudstones. Compacted limestones can be a good seal for hydrocarbons, because their degree of fracturing is minimal. The radius of curvature of the folds also influences the fracture density. Wells drilled in the hinge areas of folds of the Sabinas Basin commonly have encountered gas shows. Known gas reservoirs in the basin are generally found in structures above basement highs and near antithetic faults; nonetheless, this does not exclude other styles, in which brittle folded strata and fracture systems develop.

The latest deformation stage in the basin is poorly constrained, because the youngest preserved sedimentary rocks are of Maestrichtian age. Evidence from thermal maturity of coals presented in this study suggests that sedimentation probably continued during the Paleocene and lower Eocene (Vega-Vera and Perrilliat, 1989). Therefore, the main deformation and uplift could have been post-Ypresian (early Eocene) and pre-Rupelian (early Oligocene).

Immediately to the east of the Sabinas Basin, in the Burgos

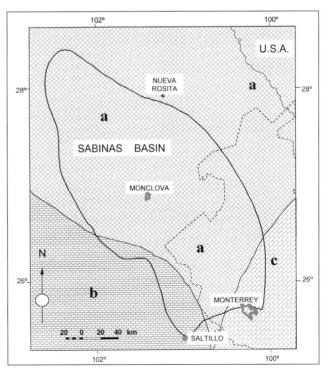

FIGURE 21. Coniacian paleogeography of the Sabinas Basin. Open-shelf facies: (a) Chalk and limestone wackestone of the Austin Formation; (b) limestone and shale of the Indidura Formation; (c) silty limestone of the San Felipe Formation.

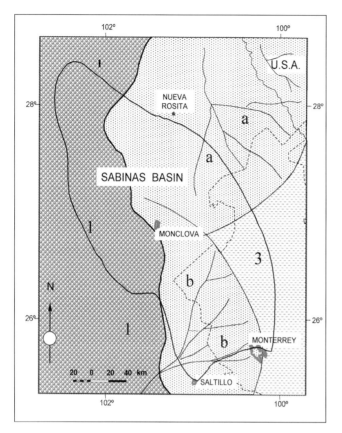

FIGURE 22. Campanian-Maestrichtian-Paleogene paleo-geography of the Sabinas Basin. (1) Alluvial-plain continental facies; (2a) Nueva Rosita deltaic facies complex; (2b) La Popa–Difunta delta facies complex (Taylor, Navarro, Midway, Wilcox, and Difunta groups); (3) prodelta shale facies (Parras-Mendez-Wilcox Formations).

Basin, the principal erosional unconformities were developed during the late Eocene. The 39.5-m.y. unconformity is a sequence boundary at the top of the Crockett–Cook Mountain Formations, with deep erosion of preexisting stratigraphic units. The late Eocene Yegua and Jackson Formations form a transgressive-regressive facies wedge deposited on this unconformity. Recent data (Gray et al., 1997) from fluid analyses, radiometric dating, and apatite fission-track analyses indicate that the frontal zone of the Sierra Madre Oriental was uplifted during this period. Igneous intrusions and mineralization in the Chihuahua and Sabinas Basins (Candela intrusive lineament) have been dated between 32 m.y. and 25 m.y. by Kennet et al. (1982). Therefore, the intrusions could have occurred after Laramide folding.

Basalts of Neogene age are present in the areas of Nueva Rosita and Ocampo, Coahuila state, and may represent recent extension. Geomechanical well-breakout data in the Sabinas Basin indicate that the present regional minimum-stress component (SHmin) is an active extension trending north-northwest–south-southeast and northwest-southeast (Zoback, 1989), possibly resulting from compression (SHmax) perpendicular to the present continental margin.

A HISTORY OF OIL AND GAS EXPLORATION

The earliest exploratory wells in the Sabinas Basin were drilled during the 1930s. The first two (San Marcos-1 and San Marcos-2) were drilled on the San Marcos and Agua Chiquita anticlines on the southern margin of the basin. The wells penetrated Neocomian clastics with no hydrocarbons shows. In the 1950s, PEMEX began its initial exploration of the basin. Toward the northeastern margin, PEMEX drilled several Peyote wells with gas shows. Subsequently, the Chupadero-1 was a dry hole, and the Don Martín-1 and -101 wells had gas shows which, when tested, were from saltwater-bearing strata. Several Garza wells were then drilled. The Garza-1 had gas shows at 2188 m in the La Casita Formation and established the first production of gas/condensate from 475-m depth in the Eagle Ford Formation. Estimated gas production was 84,753 cf/d through a $\frac{1}{4}$-inch choke. The Garza-101 well produced dry gas from 2322 to 2430 m and from 2360 to 2390 m in the La Gloria Formation at the rate of 32,036 cf/d and 14 kg/cm^2 via a $\frac{1}{4}$-inch choke. The reservoir at Garza-101 is a combination stratigraphic/anticlinal trap, oriented northwest-southeast, covering 24 km^2. Porosity ranges from 3% to 6% in fractured sandstones, with a net pay thickness of 20–40 m. Nonrisk-weighted ultimate reserves are calculated to be 106 bcf, but the field has not been exploited because of lack of nearby infrastructure.

During the 1960s, PEMEX carried out a regional evaluation of the Sabinas Basin and began an integrated exploration campaign. In 1968, the Anhelo-1 and Minas Viejas-1 wells were drilled and abandoned as dry tests. In 1969, the Gato-1 well had gas shows in the Cupidito and Taraises formations, and three production tests were run, two of which flowed hydrocarbons. Another test was run in the Upper Tamaulipas Formation. All of the tested intervals had been drilled with heavy muds, and the testing was performed in cased holes. As a result, the gas flowed with very little pressure.

In 1971 and 1972, the Guaje-1, Baluarte-1, and Sabinas-1 wells were drilled. The first of these had gas shows but was drilled off-structure, and the other two were dry. In 1972, drilling began on the Buena Suerte-1 well. This took 36 months of operation and resulted in gas shows at several stratigraphic levels, but the well was abandoned as wet. At the same time, the Buena Suerte-1-A, -2 and -2-A wells were drilled; the first two were abandoned before reaching their objectives because of mechanical problems. The Buena Suerte-1-A well found encouraging signs of gas in the Padilla and Virgen Formations. After many drilling problems and 18 months of operation, the Buena Suerte-2-A well was completed in 1975 as a dry gas producer from 2495 to 2515 m in fractured carbonates of the Padilla Formation. Production was estimated at 1.345 MMCFGD, flowing at 62 kg/cm^2 through a $\frac{1}{4}$-inch choke. The gas had a heat content of 984 Btu/ft^3. This field has an area of 13.4 km^2. The Padilla-Virgen reservoir had reserves of 69.7 bcf, of which 68.5 mmcf has been produced since 1997 with 10 development wells. The success of the Buena Suerte-2-A well opened the doors to a new, structurally complex Mesozoic gas province.

In 1976, the Monclova-1 well was completed as a gas pro-

ducer and initially flowed 3.6 MMCFGD dry gas at a pressure of 170 Kg/cm^2 through a $\frac{1}{4}$-inch choke, from 2120–2173 m, in dolomites of the basal La Virgen Formation. Matrix porosity is 4% to 7%, but reservoir quality is enhanced by natural fracturing. The La Virgen reservoirs encompass an area of 25 km^2. The

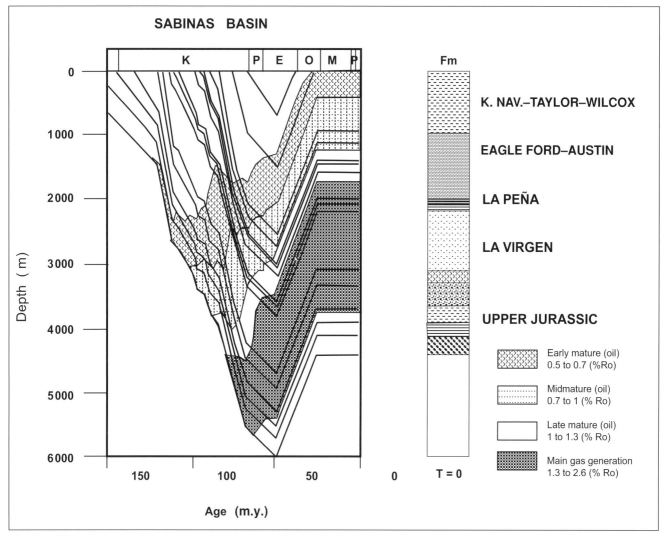

FIGURE 23. Evolution of the hydrocarbon-generation window in the Sabinas Basin.

TABLE 2. Sedimentation rates of the Sabinas Basin and Coahuila Platform sequences.

Age (m.y.)	Time (m.y.)	Thickness (m)	Sedimentation rate (mm/year)	Cycles	Event
150–144	14	2500	0.178	I	Synrift
144–128.5	15.5	1250	0.08	II	Passive margin
128.5–112	16.5	1650	0.1	III	Passive margin
112–96	16	885	0.055	IV	Sabinas basinal facies
112–96	16	1230	0.076	IV	Coahuila platform
96–82	14	550	0.039	V	Regressive facies
82–39.5	42.5	3200	0.062	V'	Foreland basin

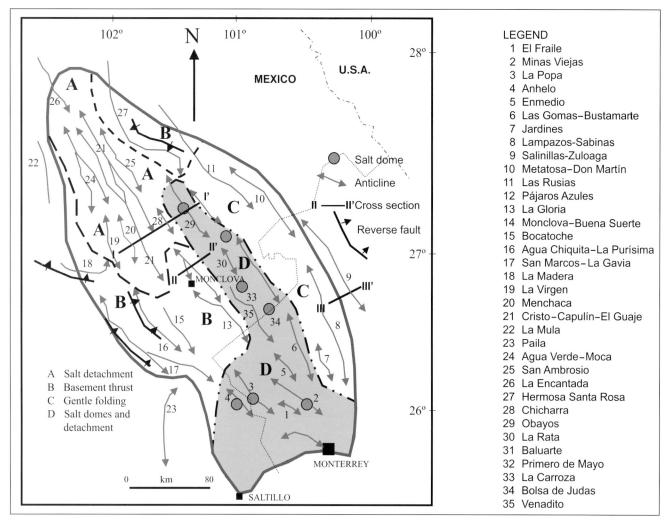

Figure 24. Structural map of the Sabinas Basin.

field also produced from fractured sandstones of the Upper Jurassic La Casita Formation and from carbonates of the Lower Cretaceous La Padilla and La Mula formations. In previous studies, the original reserves in those plays were estimated at 192.5 bcf, of which 164 bcf has been produced from 11 wells in 23 years of exploitation. Development of the Buena Suerte and Monclova fields proved that they comprise a single, compartmentalized structure that produces mainly from three reservoirs—Padilla, La Mula, and La Virgen—with only two wells producing from the La Casita reservoir.

In that same year (1976), further exploratory success was achieved with the Lampazos-1 well from a depth of 2400–2425 m in fractured sandstones of the Upper Jurassic La Gloria Formation with porosities as low as 5% to 8%. Dry gas was tested at the rate of 4.3 MMCFGD at a pressure of 180 kg/cm² through a ¼-inch choke. The Lampazos field has produced from the La Gloria, La Casita, Taraises, and Georgetown Formations. The best productive interval is in the Upper Jurassic La Gloria and La Casita. The area of the field is 11 km². Of the original reserves of 71.8 bcf, 52.6 bcf has been recovered in 23 years from 15 wells.

Intense exploratory activity between 1976 and 1985 resulted in additional successes, with the discovery of the Florida, Huerta, Zuloaga, Maestros, Gato, Escudo, and Patricio fields (Figure 26). Productive intervals in the La Casita, Padilla, and Austin Formations in these one-well fields have yielded a total of 7.7 bcf from reserves calculated to be 136 bcf. In 1986, the Merced-1 well was the last successful exploratory well drilled in the Sabinas Basin. Merced-1 flowed 4.4 MMCFGD at a pressure of 206 kg/cm² through a ¼-inch choke at 3765 to 3767 m from sandstones of the La Casita Formation, with matrix porosities as low as 3% to 5%. However, once again, natural fractures significantly enhance porosity/permeability in these reservoirs, which have produced 69.5 bcf in 12 years. Merced field reserves are estimated at 161 bcf (Vázquez and Eguiluz, 1997).

The Noralta, Ulúa, Candelilla, Cuatro Ciénegas, Cacanapo, Garza, Minero, and Progreso fields have not been developed. Some wells have gas containing variable amounts of CO_2 and H_2S (0.1%), and their aggregate reserves are estimated to be about 136 bcf .

A total of 101 exploratory wells has been drilled in the Sabi-

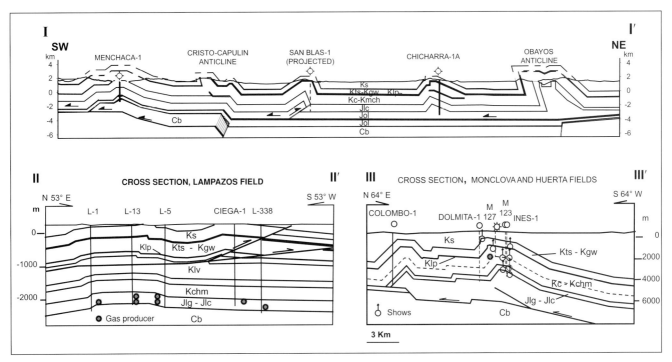

FIGURE 25. Structural sections of the Sabinas Basin. Section I–I': Cb, basement complex formed by Paleozoic schists underlying Triassic–mid-Jurassic red beds. Jol, limestone and evaporites of Callovian-Oxfordian and basal Kimmeridgian age (salt intrudes the core of the anticline and is in detachment zone). Jlc, thick section of shales and sandstones of Kimmeridgian-Tithonian age (La Casita Formation). Kc-Kchm, limestones, sandstones, and evaporites of Neocomian age (Menchaca, Barril Viejo, Padilla, La Mula, La Virgen, and Cupidito Formations). Klp, upper Aptian shales (La Peña Formation). Kts-Kgw, basinal limestones of Albian-Cenomanian age (Tamaulipas Superior, Sombreretillo, Georgetown, Del Río, and Buda Formations). Ks, shales, thin limestones, and sandstones of Cenomanian-Maestrichtian age (Eagle Ford, Austin, San Miguel Upson, and Taylor-Navarro-Difunta groups). Vertical and horizontal scales are the same; location is shown in Figure 24. Section II–II': Lampazos field is a gentle anticline with emergent upthrust in the top of evaporites of the La Virgen Formation. Section III–III': Monclova-Huerta fields are a fold-propagation fault with detachment in the basement complex (Cb).

nas Basin. Five were located in the Monclova–Buena Suerte field (Amuleto-1, -1-A, and -1-B; Ines-1; and Dolomita-1), and one on the Lampazos field (Ciega-1). Twenty-two wells were abandoned because of mechanical problems, and 19 resulted in discoveries, 11 of which have been commercial producers of dry gas. The overall exploratory success rate is 26%, with cumulative production until 1997 of 3362 bcf and remaining reserves of 404 bcf (Figure 26a). In 1999, the Minero well was successfully completed, and estimated reserves in the structure have increased to 55 bcf (Figure 26a).

Eighty-eight development wells have been drilled, with a success rate of 52%. Forty-six were producers, and as of 1997, 36 wells were producing a total of 23 MMCFGD. Average production rate per well is 1.67 MMCFGD for the Monclova–Buena Suerte field, 2.03 MMCFGD for the Merced field, and 0.50 MMCFGD for the Lampazos field. These rates compare favorably with the average daily output per well in the neighboring Burgos Basin, where the average daily production rates per well typically range from 0.5–0.7 MMCFGD (Echánove, 1986). Estimated initial production rates in the Sabinas Basin discovery wells normally were lower than the productivity registered by the field development wells. Development wells in

the Buena Suerte field initially produced as much as 27 MMCFGD (BS-104), 23.3 MMCFGD (BS-92), and 13.8 MMCFGD (BS-2A); in the Monclova field, 25.2 MMCFGD (M-3), 22.9 MMCFGD (M-127), and 8.3 MMCFGD (M-1); in the Lampazos field, 16 MMCFGD (L-71), 12.6 MMCFGD (L-31A), and 11.6 MMCFGD (L-1); and in the Merced field, 17.1 MMCFGD (Me-11) and 16.9 MMCFGD (Me-21). The maximum total production reached in the Sabinas Basin was approximatly 158 MMCFGD in 1979, with a rate of decline of 25% per year. In 1983, this stabilized at 35 MMCFGD. In 1991, the basin produced nearly 9 MMCFGD, but in 1995, with the development of the Merced field, gas production of the Sabinas Basin rose again to a maximum of 58 MMCFGD, with only eight new wells, afterward declining an average of 25% per year. Presently, the basin produces 12 MMCFGD (Figure 26b).

GAS RESOURCES IN THE BASIN

Based on geochemical analyses (González and Holguin, 1992), the Upper Jurassic La Casita Formation is considered to be the primary hydrocarbon source rock in the Sabinas Basin.

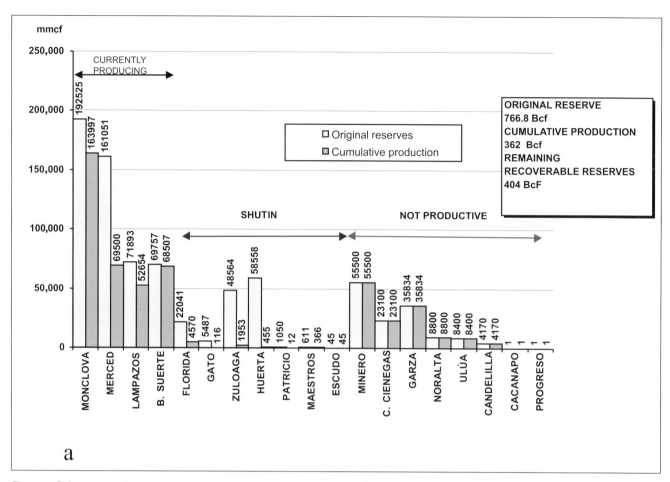

FIGURE 26a. Fields discovered and cumulative gas production in the Sabinas Basin.

Source-rock facies in the La Casita are black carbonaceous shales containing type III and minor type II organic material. TOC content averages 2%, with vitrinite-reflectance values ranging from 2 to 3, indicating that these rocks are currently in the dry-gas generation window. The density of the source rock, calculated from well logs, is $2.65g/cm^3$. Average net source-rock thickness is 70 m, distributed over an area of about 20,000 km^2. Using these values, net source-rock volume is calculated to be 7.42×10^{16} t. Assuming an initial standard hydrogen index for type III organic material of 250 mg HC/g TOC and using the measured present-day hydrogen index of 60 mg HC/g TOC from rocks of the basal the La Casita Formation, the resulting difference is 190 mg of expelled HC/g TOC. This value, multiplied by the hydrocarbon mass (7.42×10^{16}g) and by the conversion constant K (10^{-6}kg/mg), gives a value of hydrocarbon mass of 1.4098×10^{13}. In a conversion graph showing the relation of the hydrocarbon mass in kilograms and the gas equivalent in cubic feet, 930×10^{14} ft^3 of gas could have been generated by the La Casita Formation. Of that total volume, however, only a fraction of less than 1% could have been trapped (Allen and Allen, 1990).

Using a Monte Carlo statistical method, the average potential gas resources for the La Casita, La Gloria and Padilla–La Virgen plays are 1105.9, 830.3, and 646.4 bcf, respectively, or a total of 2500 bcf. Applying geologic risk factors of 0.5, 0.5, and 0.4 to the three plays, respectively, reduces the potential gas resources to 1422 bcf .

Cumulative production from the basin is about one-quarter of the estimated recoverable reserves. Opportunities for future exploitation include the Monclova–Buena Suerte field, where only two wells have been completed in the La Casita reservoir. Several wells that encountered play in the La Casita were abandoned because of drilling problems, mud invasion, and cased-hole testing. In addition, development wells were drilled at a regular spacing without regard to geologic factors, such as density and distribution of natural fractures. A review of the drilling history, production tests, and geologic analysis shows that the La Casita reservoir at Monclova–Buena Suerte field could yield about 91.7 bcf in an area of 37 km^2, and in the hanging wall of the structure, where the Huerta field is located (Figure 25, cross section III–III′), additional remaining reserves of Jurassic and Cretaceous plays of 58 bcf are estimated to exist.

Another opportunity for future development is the Lampazos field, which is separated from the Zuloaga field by a structural saddle. The designation of producer intervals in the Lampazos field has been inconsistent. Some of the wells that produce in the La Casita Formation have been reported as producing from the La Gloria Formation, and vice versa. In addi-

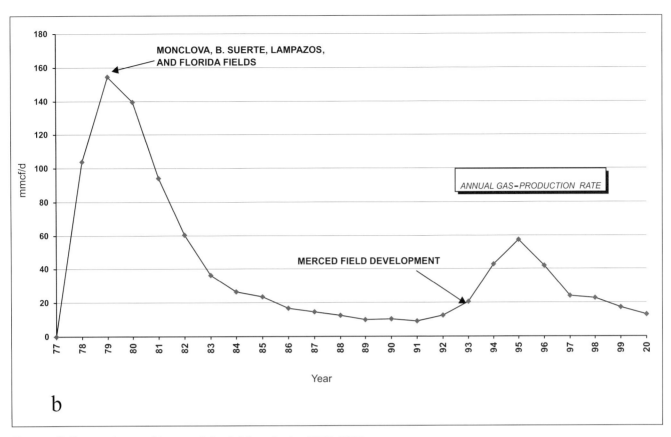

b

FIGURE 26b. Production history of the Sabinas Basin, 1977–2000.

tion, some zones that were reported as being invaded by water are gas-productive downdip, indicating that these zones may in fact be productive over a larger area. Preliminary analysis suggests that the Lampazos structure may have remaining reserves of 16.2 bcf for the La Casita reservoir and 2.9 bcf for the La Gloria reservoir, and the Zuloaga structure has 46 bcf of potential remaining reserves in the La Casita Formation.

The Cuatrociénegas, Candelilla, and Noralta fields are situated close to one another in the northeast of the basin. A 304.8-mm (12-in) diameter, 68-km pipeline is required to commercialize this gas. The gas in these three fields contains 15% to 40% CO_2. Cuatrociénegas field produces from the Cupidito Formation, and the others produce from the La Casita Formation. Pressure of 75 kg/cm^2 and a flow rate of 2.6 MMCFGD were recorded in the Cuatrociénegas well, and the Candelilla and Noralta wells recorded pressure of 224 kg/cm^2 and a flow rate of 4.8 MMCFGD (Linares and Montiel, 1987). Recoverable reserves in the Cupidito are estimated to be 23 bcf. For the La Casita play, recoverable reserves are estimated to be 4.1 and 8.8 bcf, respectively, for the Candelilla and Noralta structures. The Bura structure is adjacent to Noralta and Candelilla, with the three forming a single, compartmentalized structure covering an area of about 33 km^2, which may contain reserves of 22 bcf (9.1 bcf for the Bura structure).

The Garza field has not yet been developed. It lacks infrastructure and would require a 40-km pipeline to connect it to the Lampazos field. Garza encompasses an area of 23 km^2. The main La Casita Formation reservoir is located at 2100 m and has an average matrix porosity of 4% in fractured sandstones. Reservoir pressure is low, at 14 kg/cm^2. A preliminary calculation predicts reserves of 35.8 bcf of dry gas in this reservoir.

Although the Sabinas Basin has 2-D seismic coverage of 95% over the lower relief areas and most structures have already been tested, there are still nine untested structural leads with estimated median reserves ranging from 79.8 to 12 bcf; these leads may incorporate an additional 290 bcf with a median risk of 0.35 (Table 3). A review of the numerous exploratory wells that found gas but could not be tested or completed because of mechanical problems is considered necessary for the resumption of exploration in the area. Understanding of the area has been limited by off-structure wells, overbalanced drilling, testing of fractured reservoirs through casing, lack of hydrofrac completions, and fields developed on nonoptimal well spacing. Historic drilling problems in the Sabinas Basin consist of circulation loss, subsurface water flows, and natural hole deviations. These problems affect drilling times and have increased well costs by as much as 30%.

The success factors in the development of the Merced field were adequate 2-D seismic coverage, positioning of wells on the crest of the fold where natural fracture density is greatest, and the use of open-hole completions. In addition, the Merced field still has reserve potential. In spite of drilling with heavy

TABLE 3. Potential gas resources in the Sabinas Basin.

Reevaluating mechanically troubled wells on structures	Potential resources (mmcf)
1° de Mayo-1 (La Casita play)	29,000
Barroterán-1A (La Casita play)	34,000
Ulúa-1 (La Casita play)	78,740
Florida-1 (La Casita play)	94,300
Chicharra-102A (La Casita play)	35,500
Subtotal	**271,500**

Field	Original resources (mmcf)	Remaining resources (mmcf)
Monclova (Cretaceous play)	192,525	29,449
Buena Suerte (Cretaceous play)	69,757	1250
Montclova–Buena Suerte (La Casita play)	97,385	91,749
Huerta-Colombo (La Casita and Cretaceous plays)	58,558	58,103
Noralta-Candelilla-Bura (La Casita play)	22,013	
Lampazos:		
(La Casita play)	32,533	16,292
(La Gloria play)	37,360	2947
Zuloaga (La Casita play)	48,564	46,611
Garza (La Casita play)	35,843	
Cuatrociénegas (Cupidito play)	23,103	
Merced (La Casita play)	161,051	91,551
Ulúa (Cretaceos play)	8400	
Florida (Padilla play)	22,041	17,471
Minero (La Casita play)	55,000	
Subtotal		**499,782**

Leads	Potential resources (mmcf) (including risk factor)
Marea (La Virgen and La Casita plays)	79,800
Elipsoide (Padilla and La Casita plays)	35,600
Libertad (La Virgen and La Casita plays)	19,300
Cimarrón (Padilla and La Casita plays)	15,300
Nipón (Padilla and La Casita plays)	42,200
Estancias (Padilla play)	28,700
Pirinola (La Casita play)	12,000
Trinchón (Padilla and La Casita plays)	31,600
Takama (La Casita play)	25,600
Subtotal	**290,100**

Coal degas	Potential resources (mmcf)
Upper Cretaceous play	**147,000**

Resources in drilled structures, leads, and fields	=	**1,061,382 mmcf**
Methane-coal degas resources	=	**147,000 mmcf**
TOTAL RESOURCES		**1,208,382 mmcf**

mud (1.98 g/cm^2), there were several zones with gas shows above the developed reservoir. None of these zones has yet been tested. In 1998, the Minero-1 well was successfully drilled on the Merced structural trend and found reserves of 55.5 bcf in the La Casita Formation. The Padilla Formation had shows but was not tested.

Of the 15 drilled structures in the basin that were considered for additional hydrocarbon reserves, five are considered attractive for full reevaluation of their potential (Table 3). More than 271 bcf of additional reserves could be added from the La Casita play in the Primero de Mayo, Barroterán, Ulúa, Florida, and Chicharra structures.

Additional gas potential in the Sabinas Basin may also exist in methane resources of the Upper Cretaceous coal beds. According to the Consejo Nacional de Recursos Naturales (1994), 12,270 million tons of middle- to high-volatility bituminous coal exist near the surface in an area covering 6900 km^2. Coal seams range in thickness from 1.5 to 3.0 m. Analyses of samples and interpretation of records of 12 deep wells in the area of Nueva Rosita indicate that the coal seams occur at depths varying from the surface to 900 m. The coal is interbedded with sandstones and has a gross thickness of 30–70 m over an extensive exploitable area of 1000 km^2. Assuming a net coal thickness of 10 m, 14.7 x 10^9 mt of bituminous coal occur in this area (Eguiluz, 1998). Using the San Juan or Greater Green River Basins in the western United States as analogs, with a minimum of 100 cf of gas in situ per ton of coal (Tyler et al., 1997) and a 10% recovery factor, 147 bcf of low-pressure gas may be obtained by coal degasification in the Sabinas Basin (Table 3).

The Sabinas Basin has an infrastructure of pipelines capable of transporting gas to the steel-production facility of Altos Hornos in the city of Monclova. This facility requires more than 100 MMCFGD of dry gas to operate, and it consumes 36.5 bcf annually. To guarantee gas supplies for the next 20 years, a volume of 730 bcf is required. Efficient development of Sabinas Basin gas resources, using geologic principles and modern technology, could provide sustained production of 50 MMCFGD during 20 years.

CONCLUSIONS

The Sabinas Basin developed through several evolutionary phases. The first phase was a long period of Triassic–Lower Jurassic magmatic-arc activity followed by lithospheric rupture and rift development associated with the opening of the Gulf of Mexico. Basic lava flows associated with conglomerates and evaporites characterize the synrift basin fill.

The evolution of the Sabinas Basin resulted in three depositional supersequences. The oldest is the mid-Jurassic synrift sequence, which has yet to be subdivided into high-frequency cycles. The intermediate supersequence corresponds to subsidence during crustal cooling. It is subdivided into three higher-frequency cycles separated by two sequence boundaries. The

cycles are represented by carbonates, evaporites, and coastal siliciclastic rocks deposited over extensive platforms on a passive margin. The platforms were Upper Jurassic, Neocomian-Aptian, and Albian-Cenomanian in age. Many lateral and vertical facies changes in the basin were controlled by the Coahuila and Tamaulipas basement blocks and other blocks with smaller dimensions.

The youngest supersequence is represented by regressive terrigenous strata deposited in a foreland basin setting characterized by ramp-slope and deltaic environments related to the initial pulses of orogenic uplift to the west. The basin was episodically compressed between 96 Ma and 39.5 Ma, after the Ypresian marine conditions ceased in the Sabinas area and maximum folding occurred.

Subsidence rate during the synrift phase is calculated to have been 0.178 mm/y, from 0.55 to 0.1 mm/y during the cooling phase, and from 0.039 to 0.062 mm/y during foreland deposition. Therefore, subsidence during the initial stage was 40% to 70% more rapid than during the youngest recorded deposition.

The Laramide, or *Hidalgoense,* compressional deformation is generally thin skinned. In the Sabinas Basin, four areas display distinctive structures styles: (1) Where Jurassic salt exists in the subsurface, it acted as a regional detachment; (2) where Jurassic salt is thickest, salt diapirs occur; (3) where deformation was influenced by a basement high on the edge of the basin, folding was gentle; and (4) where Jurassic salt is absent, fault-bend folding developed.

Structural shortening calculated for different areas in the basin ranges from 16% to 26%, which is less than in the adjacent Sierra Madre Oriental or Chihuahua Basin. The critical-wedge model has been used to explain some of the varations in deformational geometry found in the basin.

The Upper Jurassic La Casita Formation is the main hydrocarbon source. Its TOC values are favorable, with mainly type III kerogen and evidence of having had a high transformation ratio. Potential resources generated for the La Casita Formation are calculated to be 930 x 10^{14} ft^3 of gas, of which only a fraction of 1% could have been trapped. Natural fractures provide permeability in the known hydrocarbon reservoirs. Coal-bed methane is also a potential resource, with the possibility of more than 147 bcf being present in the northeastern part of the basin.

In 23 years of exploitation, the Sabinas Basin has produced more than 350 bcf of dry gas. High average per-well daily production rates indicate a high probability of commercial feasibility. Geostatistical modeling indicates that identified plays could contain more than 1000 bcf of additional gas resources.

The development of discovered fields and the location of exploration wells have not been adequate. Operations have suffered from the use of heavy drilling muds, nonoptimal well testing, and frequent machanical problems during drilling. By building on previous knowledge and the application of new technology, the Sabinas Basin could provide additional production from previously developed and undeveloped fields, from problem wells that were abandoned in large structures, from new exploration leads, and from coal-bed methane resources. Together, these opportunities could provide additional reserves of more than 1000 bcf of dry gas.

ACKNOWLEDGMENTS

The data published here were compiled from several published papers authorized previously by PEMEX Exploration.

Thanks to Claudio Bartolini for his invitation to participate in this publication and his help in improving the content and readability of this article. Special thanks to Jon Blickwede and Joshua Rosenfeld for their constructive and careful reviews and translation of the original manuscript. Thanks to David Harris for his help in improving this article.

REFERENCES CITED

Ahr, W. M., 1973, The carbonate ramp: An alternative to the shelf model: Transactions of the Gulf Coast Association of Geological Sciences, v. 23, p. 221–225.

Allen, A. P., and R. J. Allen, 1990, Basin analysis: Oxford, U.K., Blackwell Scientific Publications, 451 p.

Alfonso, Z. J., 1978, Geología regional del sistema sedimentario Cupido: Boletín de la Asociación Mexicana de Geólogos Petroleros, v. 30, p. 1–55.

Alvarez Jr., M., 1949, Tectonics of Mexico: AAPG Bulletin, v. 33, p. 1319–1335.

Bishop, A. B., 1970, Stratigraphy of Sierra de Picachos and vicinity, Nuevo León, Mexico: AAPG Bulletin, v. 54, no. 7, p. 1245–1270.

Boese, E., and O. A. Cavins, 1927, The Cretaceous and Tertiary of southern Texas and northern Mexico: University of Texas at Austin, Bureau of Economic Geology Bulletin, no. 2748, p. 7.

Burckardt, C., 1930, Étude synthetique sur le Mésozoïque Mexicain: Mémoires de la Société Paléontologique Suisse, v. L, p. 1–280.

Burrows, H. R., 1910, Geology of northern Mexico: Sociedad Geológica Mexicana Boletín, v. 7, p. 85–103.

Charleston, S., 1973, Stratigraphy, tectonics and hydrocarbon potential of the Lower Cretaceous, Coahuila Series, Coahuila, Mexico: Ph.D. dissertation, University of Michigan, Ann Arbor, 268 p.

Consejo Nacional de Recursos Minerales, 1994, Inventario minero y exploración del carbón en el Estado de Coahuila: Secretaría de Energía, Minas e Industria Paraestatal, Subsecretaría de Minas, ISBN 9686710 426, 122 p.

Cuevas, L. J. A., 1984, Análisis de subsidencia e historia térmica en la Cuenca de Sabinas, Noreste de México: Boletín de la Asociación Mexicana de Geólogos Petroleros, v. 34, p. 56–100.

Dahlen, F. A., J. Suppe, and D. Davis, 1984, Mechanics of fold-and-thrust belts and accretionary wedges: Cohesive Coulomb theory: Journal of Geophysical Research, v. 89, p. 10087–10101.

Davis, D. M., and T. Engelder, 1985, The role of salt in fold-and-thrust belts: Tectonophysics, v. 119, p. 67–88.

de Cserna, Z., 1960, Orogenesis in time and space in Mexico: Geologishe Rundschau, v. 50, p. 595–605.

Dickinson, R. W., and H. Yarborough, 1979, eds., Plate tectonics and hydrocarbon accumulation: AAPG Continuing Education Course Note Series 1, 62 p.

Dumble, E. T., 1892, Notes on the geology of the valley of the middle Rio Grande: Geological Society of America Bulletin, v. 3, p. 219–230.

Echánove, E. O., 1986, Geología petrolera de la Cuenca de Burgos: Boletín de la Asociación Mexicana de Geólogos Petroleros, v. 38, p. 3–74.

Eguiluz de A., S., 1984, Tectónica cenozoica del norte de México: Boletín de la Asociación Mexicana de Geólogos Petroleros, v. 34, p. 41–62.

Eguiluz de A., S., 1990a, Distribución regional de facies albianas en el norte de México: Sociedad Geológica Mexicana, 10.° Convención Geológica Nacional, Resúmenes, p. 71–72.

Eguiluz de A., S., 1990b, Un hiato Aptiano en el Noreste de México: Revista de la Sociedad Mexicana de Paleontologia, v. 2, p. 57–68.

Eguiluz de A., S., 1991, Discordancia cenomaniana sobre la Plataforma de Coahuila: Boletín de la Asociación Mexicana de Geólogos Petroleros, v. 49, p. 1–17.

Eguiluz de A., S., 1994, La Formación Carbonera y sus implicaciones tectónicas, Estados de Coahuila y Nuevo León: Sociedad Geológica Mexicana Boletin, v. 50, p 3–39.

Eguiluz de A., S., 1996, Potencial gasífero de la Cuenca de Sabinas, Coahuila, México: Sociedad Geológica Mexicana, 13.° Convención Geológica Nacional, Resúmenes, p. 23–24.

Eguiluz de A., S., 1997, Potencial gasífero de las rocas mesozoicas en la Cuenca de Sabinas, Estado de Coahuila, México, in Oil and gas exploration and production in fold and thrust belts: AAPG/Asociación Mexicana de Geólogos Petroleros, Second Joint Hedberg Research Symposium Field Guide, p. 1–14.

Eguiluz de A., S., 1998, Recurso de gas metano contenido en el carbón de la Cuenca de Sabinas, Coahila: Primera Reunión Nacional de Ciencias de la Tierra, Simposia Regionales, Resúmenes, p. 163.

Eguiluz de A., S., 2000, Origen y evolución de una cuenca Triásico-Jurásica en el centro de México: Universidad Nacional Autónoma de México, Departamento de Geología, Universidad de Sonora, Publicaciones ocasionales no. 2, Cuarta reunión sobre la Geología del Noroeste de México y áreas adyacentes, Resúmenes, 27 p.

Eguiluz de A., S., and G. M. Aranda, 1984, Economic oil possibilities in clastic rocks of the Neocomian along the southern margin of the Coahuila Island: Gulf Coast Section, Society for Sedimentary Geology (SEPM) Foundation, Upper Jurassic and Lower Cretaceous carbonate platform and basin systems field trip, p. 43–51.

Flawn, T. P., A. Goldstein Jr., B. P. King, and E. C. Weaver, 1961, The Ouachita System: University of Texas at Austin, Bureau of Economic Geology Publication no. 6120, p. 420.

Flores, L. R., 1981, Análisis tectónico estructural del Golfo de Sabinas a partir de datos de subsuelo, superficie y estructural: Instituto Mexicano del Petróleo, proyecto C 1079, internal report.

Garrison, M. J., and N. J. McMillan, 1997, Origin and implications of allochthogenic blocks in the Papalote Evaporite Diapir, La Popa Basin, Mexico: AAPG, Structure, stratigraphy and paleontology of Late Cretaceous–Early Tertiary Parras–La Popa Foreland Basin near Monterrey, Mexico, Field Trip Guidebook no. 10, 135 p.

Garza, G. R., 1973, Modelo sedimentario del Albiano-Cenomaniano en la porción sureste de la Plataforma de Coahuila: Boletín de la Asociación Mexicana de Geólogos Petroleros, v. 25, p. 309–339.

Giles, A. K., F. T. Lawton, and J. F. Vega-Vera, 1999, Salt tectonics of Cretaceous-Paleogene La Popa Basin, Nuevo León, México: Asociación Mexicana de Geólogos Petroleros/AAPG Third Joint International Conference Field Trip, sponsored by Dallas Geological Society, 109 p.

Goldhammer, R. K., P. J. Lehmann, R. G. Todd, J. L. Wilson, W. C. Ward, and C. R. Johnson, 1991, Sequence stratigraphy and cyclostratigraphy of the Mesozoic of Sierra Madre Oriental, northeast Mexico: Gulf Coast Section, Society for Sedimentary Geology (SEPM) Field Guide, 85 p.

Gonzáles, G. R., 1984, Petroleum exploration in the Gulf of Sabinas—a new gas province in northern Mexico: Gulf Coast Section, Society for Sedimentary Geology (SEPM) Foundation Field Guide, Upper Jurassic and Lower Cretaceous carbonate platform and basin systems, p. 64–76.

González, G. R., and N. Holguin Q., 1992, Las rocas generadoras de México: Boletín de la Asociación Mexicana de Geólogos Petroleros, v. 42, p. 9-23.

Gray, G. G., S. Eguiluz de A., J. R. Chucla, and A. D. Yurewicz, 1997, Structural evolution of Saltillo-Monterrey Corridor, Sierra Madre Oriental: Applications to exploration challenges in fold-thrust belts, in Oil and gas exploration and production in fold and thrust belts: AAPG/Asociación Mexicana de Geólogos Petroleros, Second Joint Hedberg Research Symposium Field Trip, p. 20.

Guzmán, E. J., and Z. de Cserna, 1963, Tectonic history of Mexico, in Backbone of the Americas—Tectonic history from pole to pole: A symposium: AAPG Memoir 2, p. 113–129.

Heim, A., 1926, Notes on the Jurassic of Tamazunchale (Sierra Madre Oriental, Mexico): Eclogae Geologicae Helvetiae, v. 20, p. 84–87.

Humphrey, W. E., 1949, Geology of the Sierra de los Muertos area, Mexico: Geological Society of America Bulletin, v. 60, p. 80–176.

Humphrey, W. E., 1956, Tectonic framework of northeast Mexico: Transactions of the Gulf Coast Association of Geological Societies, v. 6, p. 25–35.

Humphrey, W. E., and T. Diaz, 1956, Jurassic and Lower Cretaceous stratigraphy and tectonics of northeast Mexico, PEMEX file, NE-M-799, 186 p.

Imlay, W. R., 1936, Evolution of the Coahuila Peninsula, Mexico, part IV, Geology of the western part of the Sierra de Parras: Geological Society of America Bulletin, v. 47, p. 1091–1152.

Imlay, W. R., 1937, Geology of the middle part of the Sierra de Parras, Mexico: Geological Society of America Bulletin, v. 48, p. 587–630.

Imlay, W. R., 1940, Neocomian faunas of northern Mexico: Geological Society of America Bulletin, v. 51, p. 117–190.

Jones, W. N., W. J. McKee, D. B. Márquez, J. Tovar, E. L. Long, and S. T. Laudon, 1984, The Mesozoic La Mula Island, Coahuila, Mexico: Geological Society of America Bulletin, v. 95, p. 1226–1241.

Kelly, A. W., 1936, Evolution of the Coahuila Peninsula, Mexico, part II, Geology of the mountains bordering the valleys of Acatita and Las Delicias: Geological Society of America Bulletin, v. 47, p. 1009–1039.

Kennet, F. C., C. T. Foster, and E. P. Damon, 1982, Cenozoic mineral deposits and subduction-related magmatic arcs in Mexico: Geological Society of America Bulletin, v. 93, p. 533–544.

Linares, F. M., and H. D. Montiel, 1987, Estudio potencial económico del Campo Noralta, Golfo de Sabinas, Distrito Frontera Noroeste: Ingeniería Petrolera, Revista no. 10, p. 38–42.

Longoria, F. J., 1984, Stratigraphic studies in the Jurassic of northeastern Mexico: Evidence for the origin of Sabinas Basin, in The Jurassic of the Gulf Rim: Gulf Coast Section, Society for Sedimentary Geology (SEPM) Foundation, Third Annual Research Conference Proceedings, p. 171–193.

Luttrell, E. P., 1977, Carbonate facies distribution and diagnoses associated with volcanic cones—Anacacho Limestone (Upper Cretaceous), Elaine Field, Dimmit County, Texas, in Cretaceous carbonates of Texas and Mexico: Applications to subsurface exploration: University of Texas at Austin, Bureau of Economic Geology, Report of Investigations no. 89, p. 260–285.

Márquez, B., 1979, Evaluación petrolera de sedimentos carbonatados del Cretácico en el Golfo de Sabinas, NE de México: Ingeniería Petrolera, v. 19, no. 8, p. 28–36.

Marrett, R., and G. M. Aranda, 1999, Structure and kinematic development of the Sierra Madre Oriental fold-thrust belt, Mexico: South Texas Geological Society Field Trip, Stratigraphy and structure of the Jurassic and Cretaceous platform and basin systems of the Sierra Madre Oriental, p. 69–98.

McKee, W. J., W. N. Jones, and H. T. Anderson, 1988, Las Delicias Basin, a record of Late Paleozoic arc volcanism in northeastern Mexico: Geology, v. 16, p. 37–40.

McKee, W. J., W. N. Jones, and L. E. Long, 1990, Stratigraphy and provenance of strata along the San Marcos Fault, Central Coahuila, Mexico: Geological Society of America Bulletin, v. 102, p. 593–614.

Muir, J. M., 1936, Geology of the Tampico Region, Mexico: AAPG Special Publication, 280 p.

Mülleried, F., 1927, Informe preliminar acerca de la geología y zonas petrolíferas de una parte de la región carbonífera de Coahuila y Nuevo León: Instituto Geológico de México, Folleto de Divulgación, no. 26, p. 3–21.

Murillo-Muñeton, G., 1999, Stratigraphic architecture, platform evolution, and mud-mound development in the Lower Cupido Formation (Lower Cretaceous), northeastern Mexico: Ph.D. dissertation, Texas A&M University, College Station, 53 p.

Nelson, A. R., 1985, Geologic analysis of naturally fractured reservoirs: Houston, Gulf Publishing Company, 320 p.

Salvador, A., 1987, Late Triassic–Jurassic paleogeography and origin of Gulf of Mexico Basin: AAPG Bulletin, v. 71, p. 419–451.

Salvador, A., 1991, Origin and development of the Gulf of Mexico Basin, in The Gulf of Mexico Basin: Geological Society of America, The Geology of North America, v. J, p. 389–444.

Sánchez, A. R., 1989, Contribuciones a la interpretación tectónica de Coahuila utilizando datos Geofísicos: Master's thesis, Universidad Nacionál Autónoma de México, Departamento de Estudios de Postgrado, Facultad de Ingeniería, 56 p.

Santamaría, O. D., A. F. Ortuño, T. Adatte, U. A. Ortíz, R. A. Riba, and N. S. Franco, 1991, Evolución geodinámica de la Cuenca de Sabinas y sus implicaciones petroleras, Estado de Coahuila: Instituto Mexicano del Petróleo internal report.

Sellards, E. H., W. S. Adquins, and F. B. Plummer, 1932, The Geology of Texas, v. I: The University of Texas at Austin, Bulletin No. 3232, 996 p.

Smith, I. C., 1966, Physical stratigraphy and facies analysis, Lower Cretaceous formations, northern Coahuila, Mexico: Ph.D. dissertation, University of Michigan, Ann Arbor, 157 p.

Tyler, R., A. R. Scott, W. R. Kaiser, and R. G. McMurray, 1977, The application of a coalbed methane producibility in defining coalbed methane exploration fairways and sweet spots: University of Texas at Austin, Examples from the San Juan, Sand Wash, and Piceance Basins: Bureau of Economic Geology and Gas Research Institute, Report of Investigations no. 244, 52 p.

Vázquez, R. A., and S. Eguiluz de A., 1997, Development of the Merced Field, Golfo de Sabinas, Northeast Mexico, in Oil and gas exploration and production in fold and thrust belts: AAPG/Asociación Mexicana de Geólogos Petroleros, Second Joint Hedberg Research Symposium, 4 p.

Vokes, E. H., 1963, Geology of the Cañon de la Huasteca area in the Sierra Madre Oriental, Nuevo León, Mexico: Tulane University, New Orleans, Louisiana, Tulane Studies in Geology, v. 1, p. 125–148.

Vega-Vera, F. J., and M. C. Perrilliat, 1989, On a new species of Venericardia from the Lower Eocene in Northeastern Mexico (Difunta Group): Tulane University, New Orleans, Louisiana, Tulane Studies in Geology and Paleontology, v. 22, p. 101–106.

Wilson, J. L., 1990, Basement structural controls on Mesozoic carbonate facies in Northeastern Mexico—A review, in M. E. Tucker, J. L. Wilson, P. D. Crevello, J. R. Sarg, and J. F. Read, eds., Carbonate platforms, facies, sequences and evolution: International Association of Sedimentologists Special Publication 9, p. 235–255.

Wilson, J. L., and G. Pialli, 1977, A Lower Cretaceous shelf margin in Northern Mexico, in Cretaceous carbonates of Texas and Mexico: Applications to subsurface exploration: The University of Texas at Austin, Bureau of Economic Geology, Report of Investigations no. 89, p. 286–298.

Zoback, M. L., 1989, Global patterns of intraplate stresses: A status on the world stress map project of the International Lithosphere Program: Nature, v. 341, p. 291–298.

Prost, G., and M. Aranda, 2001, Tectonics and hydrocarbon systems
of the Veracruz Basin, Mexico, *in* C. Bartolini, R. T. Buffler, and
A. Cantú-Chapa, eds., The western Gulf of Mexico Basin:
Tectonics, sedimentary basins, and petroleum systems: AAPG
Memoir 75, p. 271–291.

11

Tectonics and Hydrocarbon Systems of the Veracruz Basin, Mexico

Gary Prost

Gulf Canada
Calgary, Alberta, Canada

Mario Aranda

PEMEX Exploración y Producción
Mexico City, Mexico

ABSTRACT

The Veracruz Basin occurs along the southwest margin of the Gulf of Mexico. It is bounded on the north by the Trans-Mexican volcanic belt, on the west by the Sierra Zongolica fold-thrust belt, and on the south by the Saline Basin, and it is separated from the Gulf of Mexico by structural highs associated with the Los Tuxtlas and Anegada Volcanoes.

The Veracruz Basin lies on transitional crust weakened by a Triassic-Jurassic transform margin along which Yucatán (the Maya block) moved south during the opening of the Gulf of Mexico. During the latest Jurassic and Cretaceous, this area comprised a passive margin with a carbonate platform. Laramide east-directed thrusting in the Sierra Zongolica initiated clastic deposition and crustal loading that probably led to development of a foreland basin along the eastern limit of thrusting. Eocene to mid-Miocene basin formation may have been a result of continued crustal downwarping in the foreland of the Zongolica thrust belt or of initiation of left-lateral extension, based on analogy to the Tehuacán Basin to the west. Initiation of the Cocos-Nazca spreading center during the mid-Miocene led to oblique convergence between the Farallon and North American Plates and caused left-lateral movement along the Motagua-Polochic system, contractional deformation in the Chiapas fold-thrust belt, and structural inversion and possible right-lateral transpression in the Veracruz Basin. The basin continued to deepen as it deformed internally, collecting as much as 12 km of Tertiary sediments. Folds in the Catemaco area of the southernmost Veracruz Basin rotate, without tear or lateral faulting, from northwest-southeast to northeast-southwest. The Catemaco area is considered to be on the northwest limb of a regional orocline extending 400 km from San Andres Tuxtla, Veracruz, to Tenosique, Tabasco, and characterized by Neogene growth-thrust folds.

Marine carbonates of the Jurassic Tepexilotla Formation and the Cretaceous Orizaba, Maltrata, Guzmantla, and Méndez/Atoyac Formations may have generated both oil and gas, and Tertiary source rocks produced biogenic gas. Reservoirs exist in Cretaceous carbonate and siliciclastic units and in Tertiary sandstones. Migration most likely occurred along the top Upper Jurassic Tepexilotla Formation sandstone, in the top mid-Cretaceous Orizaba Formation carbonate breccias, and in sandstones throughout the Tertiary section. Interbedded shales and tight carbonates form seals. Modeling suggests that oils found in the basin generated and began migration in the Late Cretaceous about 80 Ma and that generation continues today. Thermogenic gas generation may have begun about 48 Ma, and biogenic gas is probably generating today.

INTRODUCTION

The Veracruz Basin is of interest to petroleum geologists as a possible analog to the prolific Tampico Basin to the north and the Isthmus Saline–Reforma Basin to the south (Figure 1). Early work (Benavides, 1956; Mossman and Viniegra, 1976) focused on production from thrusted Cretaceous units along the western margins of the basin. Production in the Tertiary was addressed by Benavides (1956) and Cruz Helú et al. (1977). A petroleum-systems approach (Alvarado, 1980) examined source rocks, traps, and reservoirs to describe various plays in the basin and areas for future exploration. Buffler (1991) reported that ultimate recovery in the Veracruz Basin, as of December 31, 1987, was 120 million barrels of oil, 10 million barrels of condensate, and 1.09 tcf of gas. PEMEX (2000) reports that the Veracruz Basin had 473 million barrels of original oil in place and 2.5 tcf of gas. Reserves for the basin and some typical fields are given in Table I.

The Veracruz Basin has been discussed in the context of Gulf of Mexico regional tectonics and development of northern Central America (Moore and Castillo, 1974; Suter, 1991; Vásquez Meneses et al., 1992; Ortuño Arzate, 1992; Johnson and Barros, 1993; Ortega-Gutiérrez et al., 1994; Burkart, 1994; Marton and Buffler, 1994; and Reed, 1994, 1995). Aranda-García (1999) and Aranda-García and Marrett (1999) suggest a contractional origin for two structures in the southern Veracruz Basin. The southern part of the basin is considered a part of the southern Mexico Neogene (Chiapas) thrust belt (Figure 2). Rojas (1999) interpreted the Veracruz Basin as a doubly vergent compressional basin buttressed between two basement highs.

Hydrocarbon source rocks and their environments were described by Holguin et al. (1995), Guzmán et al., (1996), and Guzmán-Vega and Mello (1999). The origin of Tertiary biogenic gas was discussed by Dahl (1994). The present study attempts to integrate recent field observations, seismic data interpretation, geochemical sampling, plate-tectonic concepts, and basin modeling to provide an understanding of the structural setting and petroleum systems in the Veracruz Basin.

FIGURE 1. Index map showing the location of the Veracruz Basin with respect to the Gulf of Mexico and surrounding basins.

STRUCTURAL EVOLUTION
Pre-Laramide Tectonics

Prior to the Late Jurassic, the basement in the Sierra Zongolica–Veracruz region consisted of a series of accreted terranes that included, from west to east, the Oaxaca, Cuicateco, and Maya metamorphic complexes and tectono-stratigraphic terranes (Delgado-Argote et al., 1992). During the Late Jurassic, the Gulf of Mexico opened by southward movement of the Yucatán Block along a transform margin that probably coincided with the continental slope (Moore and Castillo, 1974). This zone, called the Western Main Transform (WMT) by Marton and Buffler (1994), may have been responsible for the transtension that generated a series of northwest-trending horsts along the Gulf coast of Mexico that during the Creta-

TABLE 1. Reserves for the Veracruz Basin and some typical fields in millions of barrels (mmbbl) and billions of cubic feet (bcf). From PEMEX (2000).

Field	Reservoir age	Original oil in place (mmbbl)	Original gas in place (bcf)	Produced oil (mmbbl)	Produced gas (bcf)
Veracruz Basin	All	473.3	2500	443.9	578
Angostura	Cretaceous	116.1	19.6	115	19.2
Cocuite	Tertiary	0	234.2	0	61.2
Cópite	Cretaceous	27.3	433.3	22.4	315.5
Mata Pionche	Cretaceous	250.2	671.7	240.2	437.7
Mecayucan	Cretaceous	48.2	391.5	39	186.3
Mirador	Tertiary	0	133.5	0	1.6

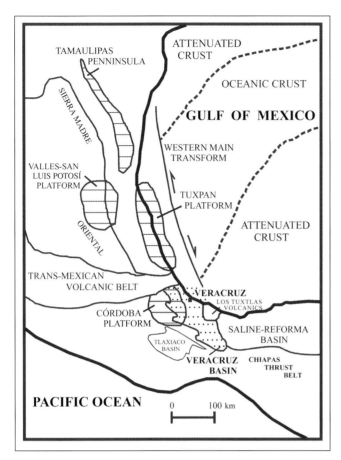

FIGURE 2. Tectonic map of eastern Mexico showing the Western Main Transform, Tamaulipas peninsula, Tuxpan platform, Valles–San Luis Potosí platform, Córdoba platform, Trans-Mexican volcanic belt, and Los Tuxtlas high.

ceous formed a series of carbonate platforms, including the Córdoba Platform (Sedlock et al., 1993; Ortega-Gutiérrez et al., 1994; Figure 2). The present basin would evolve between two structural zones in this extended, transitional crust. Some strands of this faulted margin may exist west of the basin in the present Sierra Zongolica. The main transform boundary may be coincident with the eastern margin of the basin and the volcano-tectonic Anegada high and the Los Tuxtlas fault. The western margin of the basin is defined by a structural hinge (deep fault?) that extends north-northwest from the Novillero field to past the Vibora well.

During the Cretaceous, the Veracruz region was in a passive-margin setting with basinal deposits both east and west of the Córdoba carbonate platform, centered roughly on the town of Córdoba. Some seismic data suggest that the platform had a simple ramp margin on the east, with the top Cretaceous dipping gently eastward; as yet, there are no seismic data or well evidence of reef development. Because platform carbonates are encountered in the easternmost wells penetrating the Cretaceous, it seems likely that the platform environment continued some distance farther east. In addition, palynology from the Mata Espino-1 well indicates that salt in the well is Albian-

Aptian, i.e., an intraplatform evaporite, rather than Jurassic, as in other basins around the Gulf of Mexico. It thus appears likely that the platform extended at least halfway across the present basin, and that the eastern margin (with or without a reef) either foundered during the early Tertiary or has been displaced by Tertiary strike-slip faulting.

Laramide Orogeny

The Laramide orogeny began in this area during the Late Cretaceous (perhaps as early as Turonian, 90 Ma) with deposition of clastics of the Tecamalucan Formation. This event, considered to be a result of eastward subduction of the Pacific Plate beneath the west coast of Mexico, generated east-directed thrusts and folds in the Sierra Zongolica (Johnson and Barros, 1993; Ortega-Gutiérrez et al., 1994) and thrusts and folds in the intracratonic Tlaxiaco Basin, now involved in the western Zongolica segment of the Sierra Madre Oriental (López, 1985; Figure 2). Seismic data across the Córdoba platform margin show a complex of stacked thrusts that obliterate most original platform relationships. Deformation moved from west to east and caused deposition of Paleocene–lower Eocene formations east of the sierra. The easternmost extent of thrusting is unknown because it lies buried beneath several thousand meters of Tertiary sediments of the Veracruz Basin.

Laramide thrust loading and sedimentation derived from the eroding thrust belt probably caused a foreland basin to begin to subside isostatically to compensate for the increased load. One cannot say whether a "typical" foreland basin developed, because thrusting is toward a continental margin (rather than a continental interior), and the foreland extent of thrusting has not been determined. The Anegada–Los Tuxtlas area shows minor structural uplift prior to mid-Miocene (15 Ma), based on thinning of units seen in seismic data and paleontological correlations. This may represent a peripheral bulge associated with flexural loading caused by thrusting.

A basement high probably existed north of the Veracruz Basin during the Late Cretaceous to Paleogene. Thinning in the Paleocene Chicontepec Formation indicates that such a high existed at the Macizo de Teziutlán/Santa Ana high (Viniegra, 1966) during the Paleocene. Benavides (1956) reported outcrops of both pre–Upper Jurassic and Paleogene-Neogene intrusions and volcanic rocks in this area.

Eocene to Mid-Miocene Deformation

Crustal downwarping, strike-slip pull-apart, and back-arc spreading may all be invoked to provide accommodation space for 5 to 7 km of Eocene-Oligocene sediments. Flexural downwarping of the crust has been used to explain the deposition of more than 8 km of sediments in the Parras Basin in the foreland of the Sierra Madre Oriental of northern Mexico. By analogy, it is possible that similar forces were at work in the Veracruz Basin. In the Veracruz Basin, however, the leading edge of the thrust belt has itself subsided in the basin, suggesting that other forces are at work here. We have found no evidence for obvious major strike-slip faults in the basin. Back-arc spreading has been used to explain the Los Tuxtlas thermal bulge during the early Tertiary (Reed, 1995). Available seismic lines suggest that a gentle high existed at Los Tuxtlas at the start of the Oligocene, but back-arc spreading is not considered to be a major factor in basin formation, because there is no known volcanism of this age, and vitrinite data suggest that the basin is relatively cool (R. Roman-Ramos, personal communication, 1995).

Two major fault systems are proposed to assist in basin formation during the Eocene. The first, trending approximately N20°W and extending from near the Víbora well in the north to the Cocuite field, then projecting south of the Novillero field, is here called the Novillero-Víbora (NV) trend (Figures 3, 4, 5, and 6). The master fault has not been seen at the surface or in seismic data, possibly because of its depth, but isopach maps showing abrupt thickness increases in the interval between the base Miocene and mid-Miocene make a case for the existence of a fault on the west side of the main depocenter. The apparent en-echelon alignment of folds along the NV trend suggests a component of left-lateral strike-slip. Although left-lateral kinematic indicators (fault-plane orientations, striations, lineations, and sense of movement) exist on north-northwest to north-northeast faults along the mountain front of the Sierra Zongolica (Figure 7), no direct evidence exists for a strike-slip fault in the Veracruz Basin. An Eocene–early Miocene age is assigned to this trend, based on analogous strike-slip faulting and graben formation in the Tehuacán Basin on the west side of the Sierra Zongolica (Figure 3).

Deformation in the Tehuacán Valley is a result of transten-sion between the north-south Azumbilla sinistral-normal fault bounding the west side of the valley and down-to-the-west normal faults bounding the east side of the valley. The valley fill is Eocene-Oligocene. Pre-Miocene basin formation in eastern Mexico could have been a result of southeast-directed shortening and northeast-southwest extension related to motion vectors between the Guadalupe (Farallon) and North American Plates between 25 and 15 Ma (Mammerickx and Klitgord, 1982). Late Eocene–Oligocene north-south folds in the Veracruz Basin are the result of local compression along northnorthwest faults. Growth strata on the Novillero anticline are dated as Oligocene to lower Miocene (Rojas, 1999). Some structures, such as those at Loma Bonita and Ixhuapan, show several episodes of growth up to the Holocene, based on seismic lines and paleontological work. The existence of structures through time is critical if, as modeling suggests, hydrocarbons were generated and migrated between 90 Ma and the present.

The second major fault system, named the Los Tuxtlas–Anegada fault, extends N50°W through the Los Tuxtlas volcanic field and projects offshore N40°W along the eastern margin of the Anegada high (Jacobo A. et al., 1992). This fault has been mapped in the field and on satellite images. Offshore seismic lines show an apparent coincidence with the faulted eastern margin of the Anegada high. This alignment may be the present-day expression of the deep WMT, which was active during the Late Jurassic opening of the Gulf of Mexico.

Mid-Miocene to Recent Deformation

The initiation of the Cocos-Nazca spreading center between 15 and 12.5 Ma (mid-Miocene) changed the plate movement vector of the Cocos Plate to northeast (Figure 4; Mammerickx

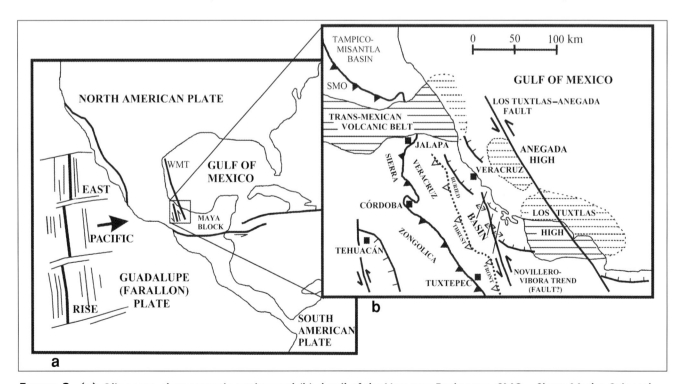

FIGURE 3. (a) Oligocene plate-tectonic setting and (b) detail of the Veracruz Basin area. SMO = Sierra Madre Oriental.

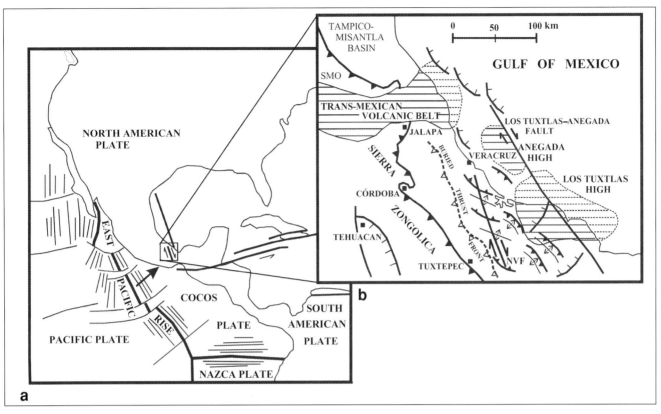

FIGURE 4. (a) Pliocene plate-tectonic setting and (b) detail of the Veracruz Basin area. SMO = Sierra Madre Oriental, NVF = Novillero-Víbora fault trend.

FIGURE 5. Structural map of the Veracruz Basin showing distribution of oil and gas fields and location of the Sardinero and Catemaco anticlines. Modified after Alvarado, 1980, and Aranda-García, 1999.

and Klitgord, 1982; Drummond, 1986; Ingle, 1995). The Veracruz Basin continued to subside as a result of northeast-directed compression and to deform internally, preserving complex growth sequences on fold flanks. Kooi and Cloetingh (1989) have demonstrated that postrift compression can cause significant flexural downbending in the center of a basin. This was accompanied by major thermal uplift of the Los Tuxtlas and Anegada highs to the east (Jacobo A. et al., 1992). Structural uplift in the Los Tuxtlas area can also be seen as a mid-Miocene unconformity in a series of break-thrust detachment folds in seismic data (Figure 8; Aranda-García, 1999).

Deformation of the upper crust in southeastern Mexico is represented mainly by regionally distributed contractional growth folds. This is ultimately the result of oblique convergence of the Farallon and North American Plates. Syntectonic beds date structures of the Veracruz Basin and elsewhere in southern Mexico at 3.4 to 15.5 Ma (Aranda-García, 1999). We use the term *southern Mexico Neogene belt* to refer to the coeval regional contractional deformation that includes structures from Veracruz to Chiapas.

FIGURE 6. Geologic map of the Vera-cruz Basin showing distribution of sur-face units, fields, and seismic lines men-tioned in this paper. Modified after Alvarado, 1980. Q = Quaternary deposits, T = Tertiary deposits, K = Cre-taceous units, J = Jurassic units, Met = Jurassic metamorphics, Ig = Tertiary volcanics.

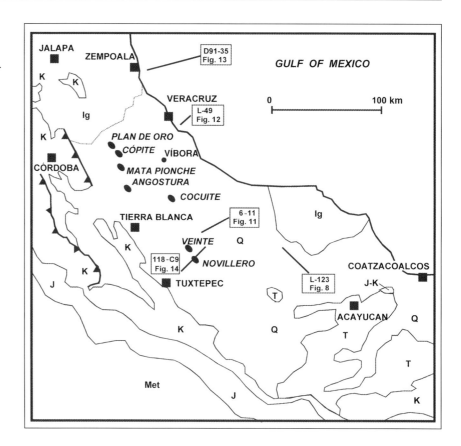

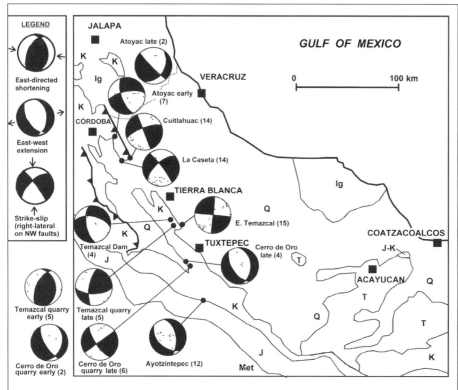

FIGURE 7. Outcrop locations for kinematic measurements along the western flank of the Veracruz Basin. At four locations, measure-ments are separated into "early" and "late" sets, based on crosscut-ting relationships, thus avoiding simultaneous shortening and ex-tension in the same direction. Fault-slip data are displayed as pseudo fault-plane solutions, simi-lar to earthquake fault-plane solu-tions, by using a lower hemisphere equal-area stereonet (method of Marrett and Allmendinger, 1990). The number of measurements at each outcrop is shown in parenthe-ses next to the location name. Thrust faults are indicated at Te-mazcal quarry (early) and Cerro de Oro (early) but are not shown on existing maps. Normal faulting is supported by measurements at Ayotzintepec, Cerro de Oro quarry (early), and Cerro de Oro quarry (late). Left-lateral movements are indicated at Atoyac (early), Temaz-cal dam, and Temazcal quarry. Right-lateral displacement is shown at Atoyac (late), Cuitlahuac, La Caseta, and Cerro de Oro quarry (late).

One might expect a thickening of the crust in this tectonic setting (Kooi, 1991), and regional gravity maps support this for the Veracruz Basin. Whereas gravity highs related to near-surface mantle material encircle the Gulf of Mexico, a gravity low (–1100 mGal) exists beneath the Veracruz Basin (Viniegra, 1956; Drummond, 1984). This is likely the result of the great thickness of Tertiary sediments.

Out-of-the-basin thrusting occurred between the Sierra Zongolica and the Anegada high as a result of northeast-directed compression, which is consistent with present-day tectonic-plate movement directions (Drummond, 1984). There is evidence for both steepening of onlapping sediments with depth on the backlimb of shallow thrusts (which suggests that vertical movements were dominant) and for folded thrusts with rotated limbs showing growth sedimentation (Figure 9; Poblet et al., 1997). One clearly listric thrust has been mapped along the west flank of the Anegada high, and Aranda-García (1999) reports that the Sardinero and Catemaco anticlines are fault-bend folds developed by thrusting (Figures 5 and 8).

With the change in shortening direction came renewed movement in a reverse and dextral sense along the NV trend and the Los Tuxtlas–Anegada fault. A change from left-lateral to right-lateral offset along northwest-oriented trench-parallel faults in southern Mexico at 10 Ma was noted by Johnson and Barros (1993). Present-day focal-plane solutions to earthquakes indicate dextral offset for northwest-striking fault planes (Figure 10; Dziewonski et al., 1981; Dziewonski and Woodhouse, 1983; Drummond, 1984). The east-northeast shortening since mid-Miocene also caused northeast-directed thrusting in Chiapas and left-lateral transpression along the Motagua-Polochic transform margin (Meneses-Rocha, 1985; Pindell and Barrett, 1990; Suter, 1991; Burkart, 1994; García-Molina, 1994).

Burkart and Scotese (1990) and Burkart (1994) have defined the Orizaba fault zone as a system of right-lateral faults that extend from near Orizaba and the western edge of the Veracruz Basin to the Polochic-Motagua fault zone near the Mexico-Guatemala border (Figure 10). They propose that the Orizaba fault system has been active throughout the Neogene, that it has 130 to 160 km of dextral offset, and that it is associated with extension across the Veracruz Basin. We feel that at least since mid-Miocene, the NV trend may be a part of this previously defined system, although we see evidence for compression (out-of-the-basin thrusting) since the mid-Miocene. Burkart (personal communication, 2000) suggests that the Chiapas massif may have once been adjacent to the San Andres Tuxtlas volcanics, and that 130–160 km of offset is indicated. Although we see evidence for right-lateral faulting in the Veracruz Basin, we have no basis for estimating the amount of lateral offset.

On the order of 4 to 5 km of post–middle Miocene deposition occurred in the deepest parts of the Veracruz Basin, based on depth-converted seismic data. Folds developed in a right-lateral en-echelon pattern above the NV trend, forming the Vibora, Cocuite, Mirador, and Loma Bonita structures, among

others. East-dipping thrusts associated with these folds dip 20–30° toward the proposed master fault.

East-directed thrusted folds developed along at least four approximately north-south-striking fronts, with structures getting younger to the east (Figure 5). Modeling suggests that liquid hydrocarbons were and are being generated in the basin center (further detailed in following sections). Yet in the absence of remigration of liquid hydrocarbons up nearby faults, these structures have little liquids potential because of their Neogene-Quaternary age. They may have greater potential for biogenic and/or remigrated thermal gas.

The post–mid Miocene shortening direction is N20–40°E, based on outcrop kinematic measurements (joint and fault orientations, lineations) and fold orientations measured on Cretaceous units along the east flank of the Sierra Zongolica (Figure 7). This is consistent with normal faults trending north-northeast in the eastern third of the Sierra Zongolica. The near-vertical open fractures (N10-30°E) of this event, as seen in outcrops and cores, enhance the permeability of and, through secondary dissolution, improve the porosity of the Cretaceous carbonate reservoirs in the buried thrusts. Crosscutting relationships show that this is the youngest event in the sierra. The northeast-shortening event probably generated the west-northwest-trending folds in the southern Veracruz Basin. These folds contain Miocene rocks at the surface at San Juan Evanjelista; thus, the event is younger than 23 Ma (start of the Miocene). Fold axes, perhaps influenced by older trends, suggest a shortening direction of N40–60°E in western and southern parts of the basin. This latest compression event appears to continue to the present, because seismic lines show evidence of thinning in young sediments on the flanks of near-surface structures along the south- eastern margin of the basin (Figure 11). Regional seismic lines display progradational geometries in beds younger than 3.4 Ma, indicating that the available accommodation space was reduced because of recent structural growth.

Folds in the Catemaco area of the southernmost Veracruz Basin rotate, without tear or lateral faulting, from northwest-southeast to northeast-southwest. This area is considered to be part of the northwest limb of a regional orocline extending 400 km from San Andres Tuxtla, Veracruz, to Tenosique, Tabasco. The orocline is a salient of the southern Mexico Neogene fold-thrust belt and is characterized by growth-thrust folds (Aranda-García, 1999; Aranda-García and Marrett, 1999).

The Veracruz Basin is asymmetric in cross section, with gravity showing the deepest part along the western margin between Tuxtepec and Cuitlahuac (Viniegra, 1956). In map view, the basin is roughly triangular, with an apex in the north near the Víbora-1 well and widening toward the south. Structural deformation, based on east-west seismic lines across the basin, varies from extension in the north to shortening in the south. In the southern part of the basin, shortening is about 13% at the mid-Miocene level. The Sardinero anticline shows 15% shortening during development of the fault-bend fold and 2% shortening in the deeper buckle detachment (Aranda-García, 1999; Figure

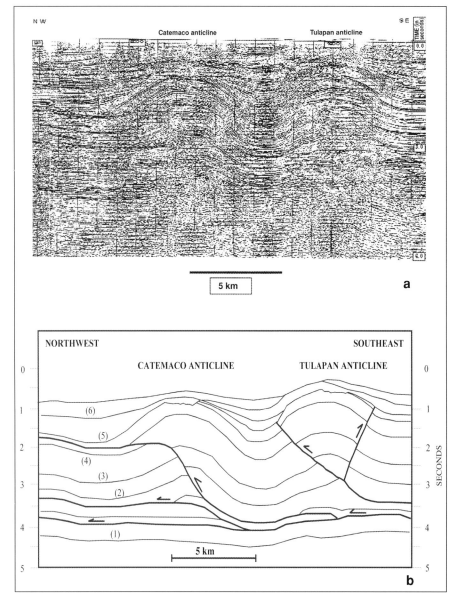

FIGURE 8. PEMEX seismic line 123 of the Catemaco and Tulapan anticlines. (a) Uninterpreted line. (b) Interpreted line. Reflectors 1, 2, 3, 4, and 5 correspond to prekinematic layers. Reflector 6 is the top of the growth strata, which indicates an end to deformation at 5.5 Ma (upper Miocene). Onlapping reflectors between horizons 5 and 6 correspond to growth units on both anticlines. From Aranda (1999).

12). The shortening decreases to 3% in the center of the basin, and there is 2% extension in the northern part of the basin. The areal association of shortening and an eastern uplift suggests that compression occurred simultaneously with uplift of the Tuxtlas-Anegada highs and exists only where these highs serve as buttresses to constrain eastward movement of sediments in the basin (Rojas, 1999). Where the buttresses do not exist, sediments are free to move east under the influence of gravity. In the northern Veracruz Basin, seismic data show basement tilted east-southeast toward the Gulf of Mexico, and extensional growth faulting is the dominant structural style east of the present coastline (Figure 13).

Structural Plays

Four structural plays exist in the Tertiary of the Veracruz Basin and one in the buried Laramide thrust belt. Recent studies indicate an important stratigraphic component for some plays in the basin (e.g., Novillero-Veinte field):

1) Footwall structures associated with fault-related folds produce at Novillero and Veinte fields (Figures 5, 6, and 14). Hydrocarbons probably migrated up along faults from a deep source.

2) West-directed hanging-wall, fault-related folds along the same trends as in (1) above, including Loma Bonita, Mirador, Zafiro, Cocuite, and Víbora, form the second type of structural play (Figures 5, 6, and 14).

3) East-directed fault-related folds, such as those at Acula, Morillo, Macuile, and Anegada, are along the east side of the basin (Figures 5, 6, and 11).

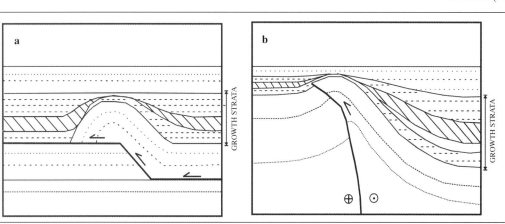

FIGURE 9. Schematic diagram of growth strata one might expect to be associated with thrusting (a) and convergent strike-slip (b).

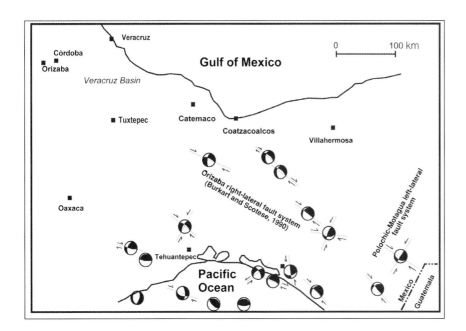

FIGURE 10. Regional earthquake fault-plane solutions displayed as the lower hemisphere of an equal-area stereonet. These show that the northwest faults of the Orizaba fault zone tend to have right-lateral strike-slip displacement. From the Harvard Central Moment Tensor Group database.

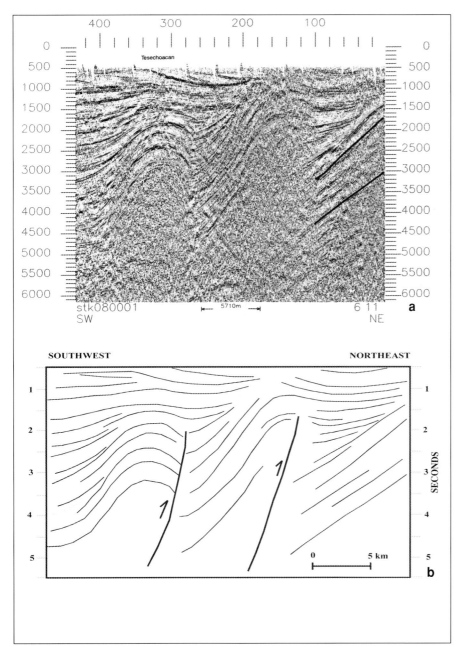

FIGURE 11. PEMEX seismic line 6-11. (a) Uninterpreted line. (b) Interpretation showing thinning of youngest sediments on shallow structures in the southeastern part of the Veracruz Basin.

FIGURE 12. PEMEX seismic line 49. (a) Uninterpreted line. (b) Interpreted line. Reflector 1 corresponds approximately to the lower Eocene at 50 Ma. Reflector 2 is in the upper Oligocene at 27 Ma. The onlapped surface (3) is above the middle Miocene at 13 Ma. Surface 4, at the base of a series of clinoforms, corresponds to upper Pliocene at 3.4 Ma. Reflectors between horizons 3 and 4 are synkinematic growth strata and above 4 are postkinematic layers. From Aranda (1999).

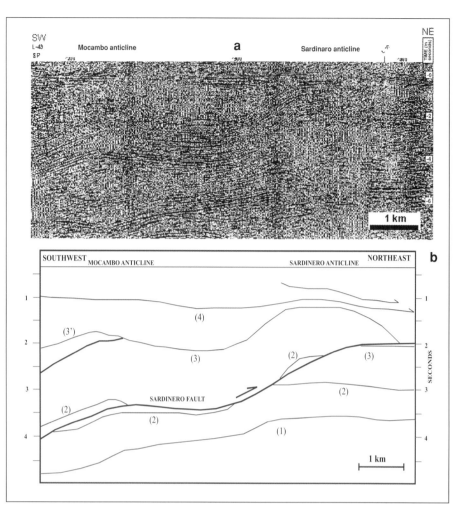

4) The updip corner of tilted fault blocks, such as are seen offshore and onshore north of Veracruz, and rollover into growth faults that form large structures offshore from Zempoala are in the northern part of the basin (Figures 5, 6, and 13).

5) The primary structural play in the Laramide thrust belt beneath the westernmost Veracruz Basin is a four-way, hanging-wall, stacked, thrusted fold (e.g., Angostura, Mata Pionche fields, Figures 5 and 6).

GEOCHEMISTRY

Upper Jurassic deep-marine source rocks analogous to those in the Tampico-Misantla Basin to the north may be present east of the Córdoba Platform along the eastern and offshore portions of the present Veracruz Basin. The central and western basin contained primarily platform deposits through the Mesozoic, with Jurassic source rocks thin to absent. However, a Lower–middle Cretaceous evaporite-carbonate intraplatform assemblage could contain source intervals (Holguin et al., 1995).

Natural gas in the Veracruz Basin is interpreted to be of bacterial, thermal, and mixed origins (Holguin et al., 1995). Bacterial gases generally range from –5 to –75 per mil $d^{13}C$ methane. Thermogenic gases exhibit $d^{13}C$ methane of -45 or heavier (more positive). Gases with intermediate $d^{13}C$ methane values would be mixtures of bacterial and thermal gases. Tertiary dry gas, based on $d^{13}C$ values of –65 to –67, are found in at least two fields (Holguin et al., 1995).

Gases of bacterial (biogenic) origin are found only in Tertiary reservoirs and primarily in the southern part of the basin (Dahl, 1994). Tertiary reservoirs, however, may contain thermal oil and gas as indicated by the Tertiary biomarker oleanane (Holguin et al., 1995). Thermogenic and mixed gases in Tertiary reservoirs indicate that deep faults most likely pro-

vided conduits for upward migration of hydrocarbons from Mesozoic source rocks (Guzmán-Vega and Mello, 1999). The implication is that liquid accumulations may exist below Tertiary gas fields.

All oils were derived from carbonate source rocks based on $d^{13}C$ values between –24 and –28‰, prystane/phytane less than 1, C_{35} hopanes greater than or equal to C_{34} hopanes, and the presence of hexahydrobenzohopanes (Guzmán et al., 1996). Saturate and aromatic isotopic compositions suggest that the oils were derived from a marine source. Two principal families are present in the Veracruz Basin. The first is a Jurassic–Lower Cretaceous marine carbonate family similar to Tampico-Misantla basin oils. The second oil group in the Veracruz Basin is an Upper Cretaceous marine carbonate-evaporite family similar to oils seen in the Sureste Basin (Saline, Chiapas-Tabasco subbasins). These oils are recognized by $d^{13}C$ values between –23 and –24, low sterane/hopane ratios, high C_{24}-tetracyclic/C_{26}-tricyclic terpane ratios, and C_{33} hopanes greater than C_{34} hopanes (Guzmán et al., 1996; Guzmán-Vega and Mello, 1999).

SOURCE ROCK

Five units are considered candidates for hydrocarbon source rocks in the Veracruz Basin (Figure 15). These include the Upper Cretaceous Méndez/Atoyac, Guzmantla, and Maltrata

Formations; the middle Cretaceous Orizaba Formation; and the Upper Jurassic Tepexilotla Formation (Alvarado, 1980). All contain type II (oil-prone) marine-algal kerogen. In addition, the Miocene La Laja and Oligocene Horcones Formations reportedly have high organic content, although no correlation to oil has been found (Holguin et al., 1995).

Samples from the Tithonian Tepexilotla Formation have total-organic-carbon (TOC) values that average 2% and S_2 values of about 5 mg/g (Holguin et al., 1995). S_2 represents the remaining oil-generative capacity in the rock. This source rock falls in the marine carbonate oil family.

The Aptian to Cenomanian Orizaba Formation contains thin source horizons with algal, sulfur-rich organic matter. TOC ranges from 1.4% to 6.4%, HI is greater than 500, and S_2 is greater than 5.0 mg/g. Organic matter consists of type II oil-prone kerogen (Holguin et al., 1995).

The richest, most widely distributed, and best-known source rocks are in the Turonian Maltrata Formation. As much as 80 m of shaly marine limestones have TOC contents greater than 2% and HI greater than 500. Organic matter consists of type II oil-prone kerogen, and S_2 is about 5 mg/g (Holguin et al., 1995). The Maltrata is in the marine carbonate source family.

The Turonian to Campanian Guzmantla Formation limestone has zones with 1% TOC and S_2 of as much as 8.5 mg/g. The organic matter is type II oil-prone kerogen, and biomarkers suggest that this sample correlates to the marine carbonate-evaporite oil family.

The Maestrichtian Méndez/Atoyac Formation contains potential source-rock intervals based on a weathered sample with 2.8% TOC. Carbon isotopes and biomarkers suggest they fall in the marine carbonate oil family.

Oligocene and Miocene shales are found to contain greater than 2% TOC, but this source consists primarily of type III (gas-prone) kerogen with S_2 less than 1 mg/g. These deltaic facies probably generated mainly gas.

RESERVOIR UNITS

Primary reservoir units in the Tertiary include sandstones and conglomerates in the middle Miocene Encanto and Depósito and in the lower Miocene La Laja Formations (Cruz Helú et al., 1977; Figure 15). The Encanto produces at the Macuile field; the Depósito produces at the Coapa, Cocuite, and Veinte fields; and the La Laja produces at the Mirador and Novillero fields (Alvarado, 1980).

The principal reservoirs in the Meso-zoic folded belt beneath the western Veracruz Basin include the Upper Cretaceous Méndez/Atoyac Formation limestone, the Upper Cretaceous Guzmantla Formation limestone, and the middle Cretaceous Orizaba Formation dolomite. Porosity occurs as vugs or in breccias in the carbonates. Fracturing is often essential for economic production. The Méndez/Atoyac Formation produces from the Angostura field; the Guzmantla Formation produces from the Miralejos, Higueras, Cópite, and Rincon Pacheco

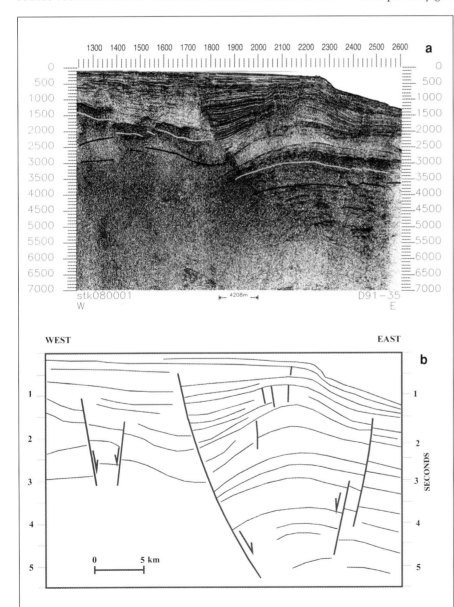

FIGURE 13. PEMEX seismic line D91-35. (a) Uninterpreted line. (b) Interpretation showing growth faulting in the northern part of the Veracruz Basin.

fields; and the Orizaba Formation produces from the Mecayucan and Remudadero fields (Alvarado, 1980).

GENERATION AND MIGRATION

Hydrocarbon generation and expulsion were modeled using Platte River Associates' BasinMod 1-D program. This program applies kinetics equations to a stratigraphic section, heat-flow history from a well or basin, and kinetic parameters based on kerogen type and source-rock richness to model the maturity, transformation ratio, and time of generation of source rock (Figure 16). A major uncertainty in these models is the amount of rock missing because of periods of erosion. The thickness of a unit also changes, depending on where it is measured in the basin; this too will affect the depth of burial and timing of generation calculated in the model. Kinetic parameters include kerogen type (known to be type II, marine), an estimated historical heat flow, present-day heat flow (from bottom-hole temperatures), and program-default values for activation energies.

This model used a generalized stratigraphic section from near the basin center based on published work (e.g., Cruz Helú et al., 1977). The present geothermal gradient of 20°C/km is taken from an estimate by R. Roman-Ramos (PEMEX, personal communication, 1995) derived from bottom-hole temperatures. This gradient was used to generate a present-day heat flow of 25 mW/m². Whereas an average continental heat flow is 58 mW/m² (Platte River Associates, 2000), the low Veracruz Basin heat flow can be explained by rapid basin subsidence and the inability of the sedimentary fill to equilibrate to a normal geothermal gradient. Estimated historical heat flow (Figure 17) is reasonable for passive margins and rifted basins (Platte River Associates, 2000). Past high heat flows are related to Jurassic regional metamorphism observed in the western Sierra Zongolica.

Modeling suggests that an Upper Jurassic Tepexilotla source rock would have finished generating oil and gas prior to 78 Ma. Because most basin structures did not exist at that time, any expulsed oil or gas was probably lost. In the basin center, this unit entered the gas-preservation window (vitrinite reflectance of 1.3–2.6% Ro) at about 80 Ma. Any oil that might have been reservoired in the Tepexilotla at that time would have cracked to gas in situ.

Kerogen in the Orizaba Formation would have begun the transformation to oil at about 80 Ma (Late Cretaceous) and would have ended at about 45 Ma (middle Eocene). Thermogenic gas would have been generated beginning at about 48 Ma (Figure 18). This oil and gas may be trapped in structures of Laramide age. In the basin center, the Orizaba is presently in the gas-preservation window.

Source intervals in the Upper Cretaceous Maltrata Formation shales started to generate oil at about 55 Ma (early Eocene) and started to produce gas at about 40 Ma (late Eocene). Modeling suggests that the Guzmantla carbonate source would

have begun to generate oil at 50 Ma (middle Eocene) and gas at about 35 Ma (early Oligocene). Laramide-age folds were fully developed and could have trapped those oils. Folds and traps associated with Eocene to mid-Miocene deformation were developing at that time and could have contained those oils. The Maltrata Formation is presently in the late-mature oil-to-gas-preservation window in the basin center (Figure 19). The Guzmantla Formation is in the late-mature oil-preservation window (1.0–1.3% Ro).

Source rocks in Upper Cretaceous Méndez Formation shales began to generate oil at about 45 Ma and are presently in the late-generation stage (Figure 20). Those oils would have migrated into Laramide, early Tertiary, or mid-Miocene to Holocene folds. Modeling suggests that the Méndez is presently in the midmature oil-preservation window (0.7–1.0% Ro) in the basin center.

Biogenic gas generated from organic-rich intervals in the Tertiary were not modeled. Generation of bacterial gas does not require deep burial and maturation. This gas is probably evolving at the present time.

The above timing is based on a composite stratigraphic section for the deepest part of the basin. Ideally, detailed stratigraphic information would be available from many parts of the basin. A thick, basin-center model will give early generation and migration, whereas one would expect hydrocarbons to generate later in parts of the basin where the section is thinner. The overall trend anticipated for the Veracruz Basin is for generation to occur earlier in the western, deeper parts of the basin and later along the east side, where basin fill is thinner and younger. The Anegada and Los Tuxtlas volcanics are not considered significant sources of heat flow in the eastern Veracruz Basin.

A late Tertiary timing of migration is supported by observations from the Angostura field. Undegraded oil from the Angostura-17 well is produced from the Upper Cretaceous Guzmantla to Méndez section. Cross sections show that the Cretaceous reservoir was breached during the lower Miocene (Figure 21). Any hydrocarbons in the reservoir at that time would have been biodegraded, water washed, or eroded. Therefore, oil migration into the Upper Cretaceous reservoirs would have occurred long enough after the mid-Miocene unconformity that the oils were not lost or degraded.

Migration most likely occurred along sandstones near the top of the Upper Jurassic Tepexilotla Formation and along carbonate breccias at the top of the mid-Cretaceous Orizaba Formation. Tight carbonates form seals at the top of Cretaceous reservoirs. Cretaceous reservoirs occur as shallow-buried thrusted folds throughout the western third of the basin. Reservoirs consist of fractured, brecciated, and/or vuggy limestones. Oils most likely migrated updip from the basin center, then vertically along faults until trapped by tight carbonates or interbedded shales (Figure 22). It is possible that oil and gas migrated updip from the west into Laramide thrusted folds. Because these folds began to subside into the basin in the early

Tertiary, hydrocarbons would have migrated during Laramide deformation and prior to basin subsidence. No production has been established in the Sierra Zongolica west of the Veracruz Basin.

Migration into Tertiary traps probably occurred along faults connecting Upper Cretaceous source intervals with sandstones throughout the Tertiary section (Figure 23). Deepening of the basin and associated tilting and remigration may have occurred throughout the Tertiary. Interbedded shales form top seals in the Tertiary section. Reservoir sandstones all lie in or above the oil-preservation window; therefore, any oil trapped in these units would be preserved as liquids. Most oil in Tertiary reservoirs is found in Miocene formations, implying that (1) older source rocks are less mature (generated later) than modeled, because they occur in less deeply buried parts of the basin; (2) secondary migration has taken place; or (3) much of the oil comes from Méndez Formation or younger sources.

CONCLUSIONS

The Veracruz Basin lies along a Jurassic transform that defines the western margin of the Gulf of Mexico. During the Cretaceous, the basin was situated on a passive margin occupied by the Córdoba carbonate platform. East-directed thrusting from Late Cretaceous to the Eocene caused folding in the western to central part of the basin. The eastern extent of this folding event is unknown. Laramide folds with Upper Cretaceous reservoirs produce hydrocarbons along the western margin of the basin. Thrust loading probably initiated a foreland basin that may be the proto-Veracruz Basin. Eocene to mid-Miocene extension with a left-lateral component is proposed to explain 5 to 7 km of early Tertiary sediments and development of en-echelon folds along the western basin bounding the Novillero-Vibora trend. A change of Pacific Plate motions between 12 and 15 Ma led to late Miocene to Holocene contraction and flexure of the upper crust. As the basin continued to deepen, another 4 to 5 km of basin fill was deposited simultaneously with out-of-the-basin east- and west-directed thrusting. Folds that developed in the Tertiary section of the western and northern Veracruz Basin contain oil and primarily thermogenic gas; in the east-

ern and southern Veracruz Basin, they contain primarily biogenic gas or lack hydrocarbons.

Five potential source intervals, based on TOC and S_2 values, include the Upper Jurassic Tepexilotla Formation shales, middle Cretaceous Orizaba Formation limestone, Upper Cretaceous Maltrata Formation limy shale, Guzmantla Formation limestone, and Méndez Formation shale. Oligocene-Miocene deltaic shales may also source oil and gas in the basin. Oils are derived from type II kerogen and carbonate source rock, whereas gas is derived biogenically, from overmature oil or from type III kerogen. Reservoirs include Upper Cretaceous fractured, brecciated, or vuggy limestones and Tertiary sandstones. Seals are either tight carbonates or interbedded shales.

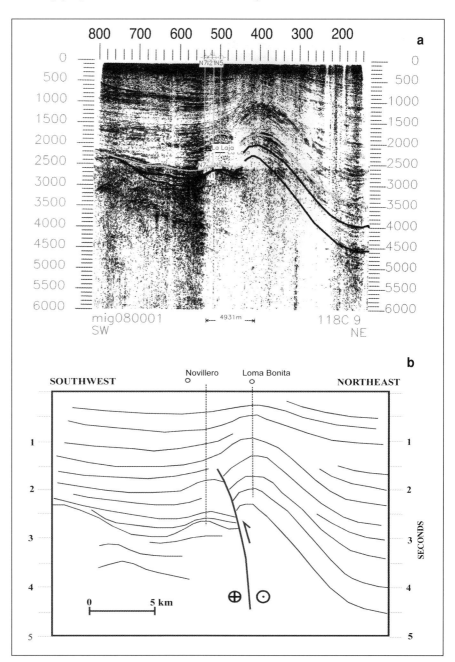

FIGURE 14. PEMEX seismic line 118-C9. (a) Uninterpreted line. (b) Interpretation showing west-directed thrusting in the Tertiary section, Veracruz Basin.

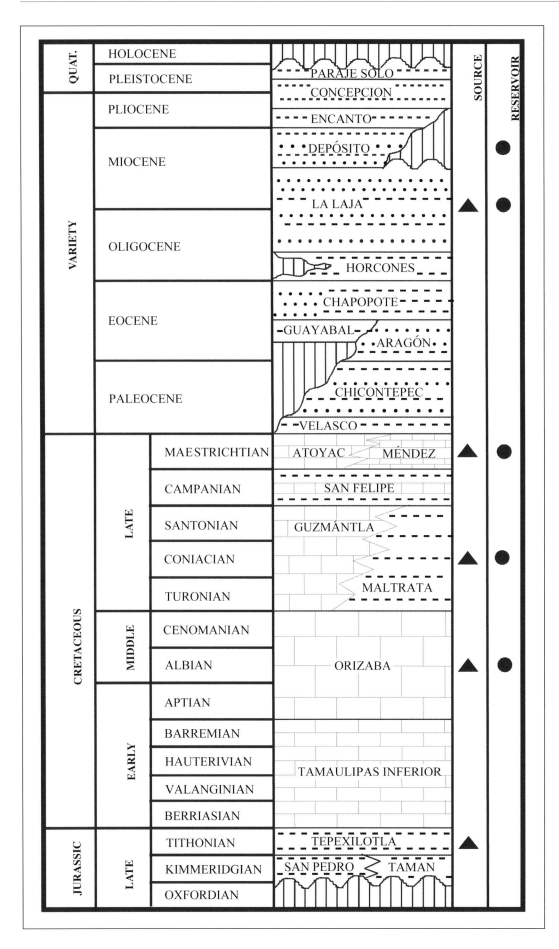

Figure 15. Veracruz Basin stratigraphic column showing reservoir and source-rock intervals.

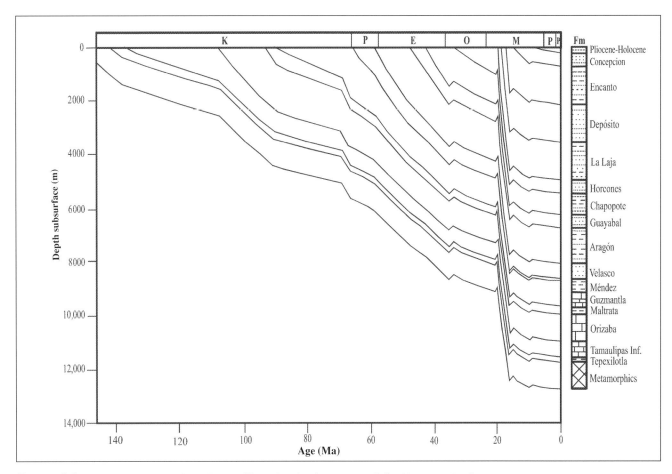

FIGURE 16. Subsidence curve for a "typical" section in the center of the Veracruz Basin.

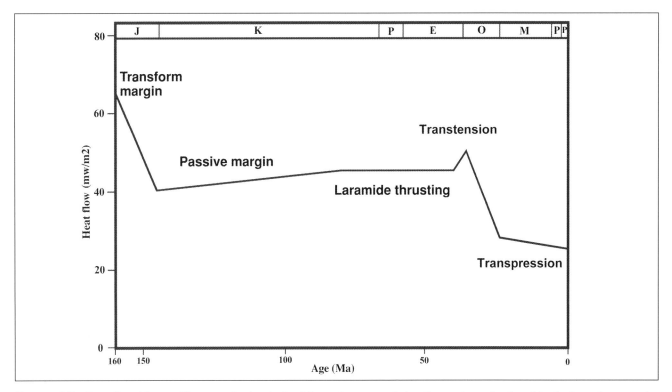

FIGURE 17. Heat-flow history, Veracruz Basin, derived from the present-day thermal gradient and typical passive-margin and rift-setting heat flows.

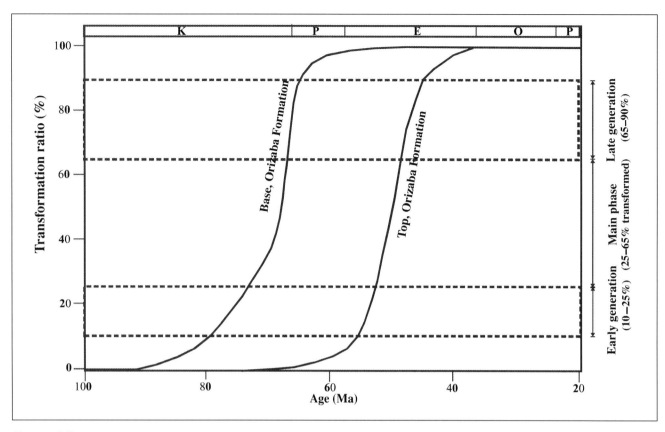

FIGURE 18. Transformation ratio for kerogens in the Orizaba Formation through time.

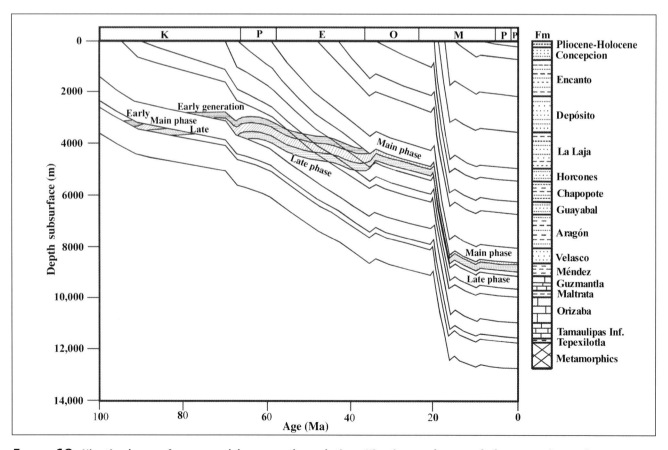

FIGURE 19. Kinetic phases of source-rock kerogens through time. Kinetics are shown only for source intervals.

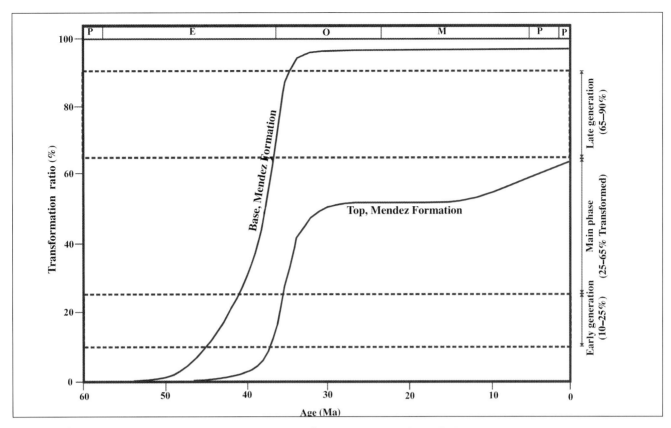

FIGURE 20. Transformation ratio of kerogens in the Méndez Formation through time.

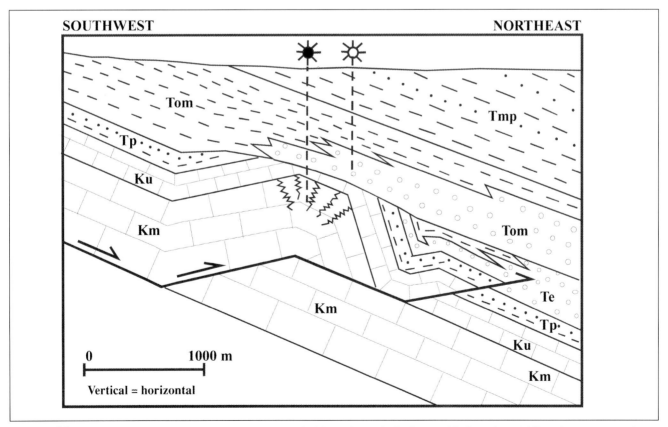

FIGURE 21. Schematic cross section of the Angostura field. Undegraded oil at and below the middle Miocene unconformity shows that migration occurred post-lower Miocene. Km = mid-Cretaceous, Ku = Upper Cretaceous, Tp = Paleocene, Te = Eocene, Tom = Oligocene-Miocene, Tmp = Miocene-Pliocene.

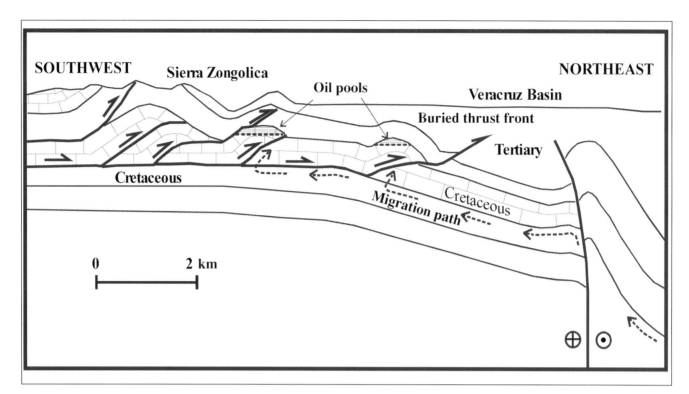

FIGURE 22. Schematic diagram of possible migration routes into Cretaceous reservoirs and traps.

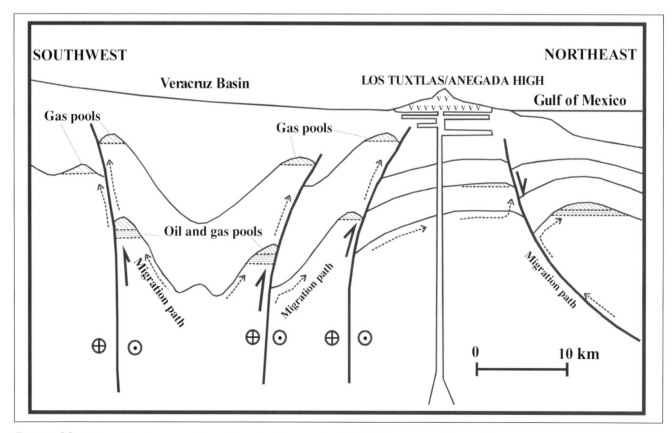

FIGURE 23. Schematic diagram of possible migration routes into Tertiary reservoirs and traps.

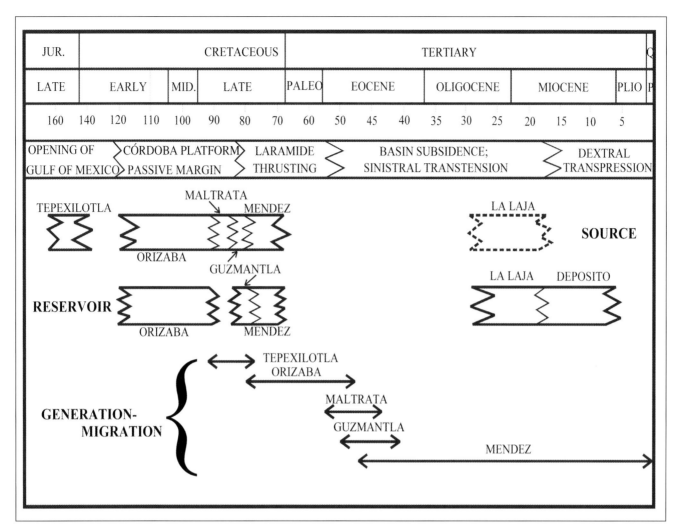

FIGURE 24. Hydrocarbon-system event chart.

Modeling of a stratigraphic section in the basin center suggests that generation and migration of liquid hydrocarbons may have begun at about 92 Ma in the Tepexilotla Formation and continue today in the Méndez Formation (Figures 20 and 24). Onset of generation would be younger for thinner stratigraphic sections. Gas generation in the Orizaba Formation may have begun at about 48 Ma, in the Maltrata at about 40 Ma, and in the Guzmantla at about 35 Ma. The Méndez Formation began to generate at about 45 Ma and continues to generate liquid hydrocarbons today. Biogenic gas is probably generating today. Most Cretaceous reservoirs in the western part of the basin are at fairly shallow depths and are presently in the oil preservation window. Tertiary reservoirs are all in the oil-preservation window.

ACKNOWLEDGMENTS

We thank PEMEX for permission to publish this paper. Much of the work reported here is the result of studies involving dedicated workers. Among these are PEMEX employees Javier Banda, Juan Toriz Gama, Miguel Espinoza Nava, Rafael Quintos, Javier Meneses-Rocha, and Hector Vizcarra; our colleagues at the Instituto Mexicano del Petróleo (IMP) — Jorge Jacobo Albaran, Salvador Ortuño Arzate, Alberto Rocha, and René Tellez; and Peter Bentham, Chip Carney, Rich Gibson, David Grass, and Ronald Nelson. We thank Josh Rosenfeld and Richard Chuchla for critical reviews of the manuscript.

REFERENCES CITED

Aranda-García, M., 1999, Evolution of Neogene contractional growth structures, southern Gulf of Mexico: Master's thesis, University of Texas at Austin, 169 p.

Aranda-García, M., and R. Marrett, 1999, Evolution of Neogene contractional growth structures, southern Gulf of Mexico: Thrust Tectonics 99 (abs.): Royal Holloway College, University of London, p. 125.

Alvarado, J. G., 1980, Perspectivas petroleras del distrito de exploración Córdoba, Veracruz: Ingeniería Petrolera, v. 20, no. 7, p. 5–18.

Benavides, G. L., 1956, Notas sobre la geologia petrolera de México, *in* E. J. Guzmán, ed., Simposio sobre los yacimientos de petróleo y gas: America del Norte, v. 3, 20.° Congreso Geológico Internacional, p. 351–562.

Buffler, R. T., 1991, Seismic stratigraphy of the deep Gulf of Mexico basin and adjacent margins, *in* A. Salvador, ed., The Gulf of Mexico Basin: Geological Society of America, The geology of North America, v. J, p. 353–387.

Burkart, B., 1994, Northern Central America, *in* S. K. Donovan and T. A. Jackson, eds., Caribbean geology: An introduction: Kingston, Jamaica, University of West Indies (UWI) Publisher's Association, p. 65-284.

Burkart, B., and C. R. Scotese, 1990, The Orizaba fault zone: Link between the Mexican Volcanic Belt and strike-slip faults of northern Central America: EOS Transactions of the American Geophysical Union, v. 71, p. 1559.

Cruz Helú, P., R. Verdugo V., and R. Barcenas P., 1977, Origin and distribution of Tertiary conglomerates, Veracruz Basin, Mexico: AAPG Bulletin, v. 61, p. 207–226.

Dahl, J. E., 1994, Orígen y ocurrencia de gas biógeno en la cuenca de Veracruz: El Trilobites de la Asociación Mexicana de Geólogos Petroleros Delegación Poza Rica, p. 9.

Delgado-Argote, L., M. López-Martínez, D. York, and C. M. Hall, 1992, Geologic framework and geochronology of ultramafic complexes of southern Mexico: Canadian Journal of Earth Sciences, v. 29, p. 1590–1604.

Drummond, K. J., 1984, Geodynamic map of the circum-Pacific region, northeast quadrant: Circum-Pacific Council for Energy and Mineral Resources, 1:10,000,000.

Drummond, K. J., 1986, Plate tectonic map of the circum-Pacific region, northeast quadrant: Circum-Pacific Council for Energy and Mineral Resources, 1:10,000,000.

Dziewonski, A. M., T.-A. Chou, and J. H. Woodhouse, 1981, Determination of earthquake source parameters from waveform data for studies of global and regional seismicity: Journal of Geophysical Research, v. 86, p. 2825–2851.

Dziewonski, A. M., and J. H. Woodhouse, 1983, An experiment in the systematic study of global seismicity: Centroid-moment tensor solutions for 201 moderate and large earthquakes of 1981: Journal of Geophysical Research, v. 88, p. 3247–3271.

García-Molina, G., 1994, Structural evolution of southeast Mexico (Chiapas-Tabasco-Campeche) offshore and onshore: Ph.D. dissertation, Rice University, Houston, Texas, 161 p.

Guzmán, M., M. Mello, and N. Holguin, 1996, Geochemical features of oils from western and southern Gulf of Mexico: Reconstruction of the depositional environment of their source rocks (abs.): AAPG Bulletin, v. 80, p. 1298–1299.

Guzmán-Vega, M. A., and M. R. Mello, 1999, Origin of oil in the Sureste Basin, Mexico: AAPG Bulletin, v. 83, p. 1068–1095.

Holguin, N., F. Galindo, E. Serrano, A. Romero, A. Sosa, G. Martínez, R. Roman, 1995, The HC generative subsystems of productive Mexican basins: Fourth Brazilian Geophysical Society International Congress/First Latin American Geophysical Union Conference, Rio de Janeiro, Brazil, Expanded Abstracts, v. 1, p. 73–75.

Ingle Jr., J. C., 1995, Formation and deformation of Neogene basins around the Pacific (abs.): AAPG Bulletin, v. 79, no. 4, p. 588.

Jacobo A., J., M. Garduño, F. Innocenti, Manetti, G. Pasquare, and S. Tonarini, 1992, Datos sobre el vulcanismo neogénico-reciente del complejo volcánico de Los Tuxtlas, Edo. de Veracruz, México: Evolución petrológica y geovulcanológica: 11.º Convención Geológica Nacional, Veracruz, Libro de Resúmenes, p. 97–98.

Johnson, C. A., and J. A. Barros, 1993, Tertiary tectonics and margin truncation in southern Mexico: Gulf Coast Section, Society for Sedimentary Geology (SEPM) Foundation, 13th Annual Research Conference Proceedings, p. 181–191.

Kooi, H. 1991, Tectonic modeling of extensional basins: The role of lithospheric flexure, intraplate stress and relative sea level change: Academisch Proefschrift, Vrije Universiteit, Drukkerij Elinkwijk B.V., Utrecht, Netherlands, p. 123–145.

Kooi, H. and S. Cloetingh, 1989, Some consequences of compressional tectonics for extensional models of basin subsidence: Geologische Rundschau, v. 78, p. 183–195.

Lopez, T. D., 1985, Revisión de la estratigrafía y potencial petrolero de la cuenca de Tlaxiaco: Boletín de la Asociación Mexicana de Geólogos Petroleros, v. 37, no. 1. p. 49–92.

Mammerickx, J., and K. D. Klitgord, 1982, Northern East Pacific Rise: Evolution from 25 m.y. B.P. to the present: Journal of Geophysical Research, v. 87, no. B8, p. 6751–6759.

Marrett, R. A., and R. W. Allmendinger, 1990, Kinematic analysis of fault slip data: Journal of Structural Geology, v. 12, p. 973–986.

Marton, G., and R. T. Buffler, 1994, Jurassic reconstruction of the Gulf of Mexico Basin: International Geology Review, v. 36, p. 545–586.

Meneses-Rocha, J., 1985, Tectonic evolution of the strike-slip fault province of Chiapas, Mexico: Master's thesis, University of Texas at Austin, 315 p.

Moore, G. W., and L. D. Castillo, 1974, Tectonic evolution of the southern Gulf of Mexico: Geological Society of America Bulletin, v. 85, p. 607–618.

Mossman, R. W., and F. Viniegra, 1976, Complex fault structures in Veracruz province of Mexico: AAPG Bulletin, v. 60, p. 379–388.

Ortega-Gutiérrez, F., R. L. Sedlock, and R. C. Speed, 1994, Phanerozoic tectonic evolution of Mexico, *in* R. C. Speed, ed., Phanerozoic evolution of North American continent—Ocean transitions: Geological Society of America, Decade of North American Geology, Continent-Ocean Transect Volume, p. 265–306.

Ortuño Arzate, S., J. P. Xavier, and J. Delfaud, 1992, Analisis tectónico estructural de la cuenca de Zongolica a partir de imagenes de satellite Landsat MSS: Revista del Instituto Méxicano del Petróleo, v. 24, no. 1, p. 11–45.

PEMEX Exploración y Producción, 2000, Las reservas de hidrocarburos de México, p. 82.

Pindell, J. L., and S. F. Barrett, 1990, Geological evolution of the Caribbean region; a plate-tectonic perspective, *in* G. Dengo and J. E. Case, eds., The Caribbean Region: Geological Society of America, The geology of North America, v. H, p. 405–432.

Platte River Associates Inc., 2000, User manual for BasinMod 1-D, petroleum systems software: Online user help, support@platte.com, Boulder, Colorado.

Poblet, J., K. McClay, F. Storti, and J. A. Muñoz, 1997, Geometries of syntectonic sediments associated with single-layer detachment folds, *in* D. J. Anastasio, E. A. Erslev, D. M. Fisher, and J. P. Evans, eds., Fault-related folding: Journal of Structural Geology, v. 19, nos. 3–4, p. 369–381.

Reed, J. M., 1994, Probable Cretaceous-to-Recent rifting in the Gulf of Mexico Basin, part 1: Journal of Petroleum Geology, v. 17, p. 429–444.

Reed, J. M., 1995, Probable Cretaceous-to-Recent rifting in the Gulf of Mexico Basin, part 2: Journal of Petroleum Geology, v. 18, p. 49–74.

Rojas, L., 1999, Tectonic evolution of the Zongolica fold thrust belt and the Veracruz Basin: Master's thesis, Royal Holloway College, University of London, 72 p.

Sedlock, R. L., F. Ortega-Gutiérrez, and R. C. Speed, 1993, Tectono-stratigraphic terranes and tectonic evolution of Mexico: Geological Society of America Special Paper 278, 146 p.

Suter, M., 1991, State of stress and active deformation in Mexico and western Central America, *in* D. B. Slemmons, E. R. Engdahl, M. D. Zoback, and D. D. Blackwell, eds., Neotectonics of North America: Geological Society of America, Decade Map Volume 1, p. 401–421.

Vazquez Meneses, M. E., P. E. Villaseñor R., R. S. Quiñones, and M. A.

Islas C., 1992, Neotectónica del sureste de México: Revista del Instituto Méxicano del Petróleo, v. 24, no. 3, p. 12–37.

Viniegra, F., 1956, Isogamas, Cuenca de Veracruz, *in*: E. J. Guzmán, ed., Simposio sobre los yacimientos de petróleo y gas: Tomo III, América del Norte, 20.º Congreso Geológico Internacional, 1:500,000.

Vinicgra, F., 1966, Palcogcografía y tectónica del Mcsozóico cn la provincia de la Sierra Madre y Macizo de Teziutlán: Boletín de la Asociación Mexicana de Geólogos Petroleros, v. 18, p. 145–171.

Stratigraphy

Bartolini, C., H. Lang, A. Cantú-Chapa, and R. Barboza-Gudiño, 2001, The Triassic Zacatecas Formation in central Mexico: Paleotectonic, paleogeographic, and paleobiogeographic implications, in C. Bartolini, R. T. Buffler, and A. Cantú-Chapa, eds., The western Gulf of Mexico Basin: Tectonics, sedimentary basins, and petroleum systems: AAPG Memoir 75, p. 295–315.

12

The Triassic Zacatecas Formation in Central Mexico: Paleotectonic, Paleogeographic, and Paleobiogeographic Implications

Claudio Bartolini
International Geological Consultant
Houston, Texas, U.S.A.

Harold Lang
Jet Propulsion Laboratory, California Institute
of Technology, Pasadena, California, U.S.A.

Abelardo Cantú-Chapa
Instituto Politécnico Nacional
Mexico City, Mexico

Rafael Barboza-Gudiño
Instituto de Geología, Universidad Autónoma de
San Luis Potosí, San Luis Potosí, Mexico

ABSTRACT

Middle to Late Triassic turbidite sequences are exposed in the states of Zacatecas and San Luís Potosí in central Mexico. These strata, assigned mostly to the Zacatecas Formation, accumulated in continental slope, toe-of-slope, and basin-plain environments along the passive continental margin of western Pangea. Strata of the Zacatecas Formation are age equivalent to rocks of the Antimonio Formation and Barranca Group in Sonora, the La Boca Formation in Tamaulipas and Nuevo León, and unnamed strata in Baja California. Based on their age, the Zacatecas turbidites correlate with a drop in sea level during the Permian-Triassic assembly of Pangea. The Triassic paleogeographic setting of Mexico is complex and poorly understood, because only dispersed Triassic outcrops exist across Mexico. However, the biogeographic affinities of the faunas from the Zacatecas Formation in central Mexico with those from equivalent strata in Baja California and Sonora suggest that these three regions were connected through the eastern Pacific, and that the Atlantic Ocean did not exist during the Ladinian-Carnian. The Zacatecas sequences underwent three periods of compressive deformation: one during their obduction onto the continental margin at some time during the latest Triassic–earliest Jurassic (?); a second during the Middle to Late Jurassic (Oxfordian) (?), apparently related to transpression; and a third during the Late Cretaceous to Tertiary Laramide orogeny.

INTRODUCTION

The Zacatecas Formation in the states of Zacatecas and San Luís Potosí in central Mexico is a succession of turbidites commonly described as a flysch sequence, containing pillow lavas and minor limestone that accumulated on the passive continental margin of western Pangea. The formation contains Middle to Late Triassic marine index fossils, especially ammonites and lamellibranchs (Burckhardt and Scalia, 1905; Burckhardt, 1930; Cantú-Chapa, 1969; Gallo et al., 1993). López-Ramos (1980) also reported Late Triassic (early Carnian) faunas.

Carrillo-Bravo (1971) first used the name Zacatecas Formation informally for sandstone and shale exposed west of the town of Charcas because of lithologic similarities with the rocks reported by Burckhardt (1930) and Burckhardt and Scalia (1905) near the city of Zacatecas. More recently, strata formerly assigned to the Zacatecas Formation have been named La Ballena Formation (Silva-Romo, 1993) at the Sierra de Salinas and El Ahogado and El Bote Formations (Monod and Calvet, 1992) west of the city of Zacatecas.

Because the Triassic sequences in Charcas, Sierra de Salinas, and the city of Zacatecas are lithologically indistinguishable, of the same age, contain the same faunas, and display the same depositional facies, we use the name Zacatecas Formation in all three localities.

Strata of the Zacatecas Formation provide critical tectono-stratigraphic information because they originated during the latest stage of the consolidation of the supercontinent Pangea during the Permian-Triassic. Although our work focuses on the Zacatecas Formation in Zacatecas and San Luís Potosí states, we also incorporated in our analysis the Triassic sequences of the Antimonio Formation and Barranca Group in Sonora. Our goal is to establish the Triassic tectono-depositional setting and to provide a paleogeographic analysis and comprehensive regional biogeographic correlation of the Zacatecas Formation.

LOCATION OF THE OUTCROPS

Rocks that we assign to the Zacatecas Formation are exposed in three localities in the states of Zacatecas and San Luís Potosí, central Mexico: (1) immediately west-northwest of Zacatecas city, (2) at the Sierra de Salinas, and (3) in the Charcas area (Figure 1).

Zacatecas City

Previous Work

At the turn of the last century, Burckhardt and Scalia (1905, 1906) studied the stratigraphy and mapped outcrops along the Arroyo Calavera ó Pimienta, west of Zacatecas. They discovered faunas that allowed assignment of these rocks to the Late Triassic (Carnian). Burckhardt (1930) subsequently found ammonites and lamellibranchs that confirmed this age assignment. McGehee (1976) and Ranson et al. (1982) used petrographic and petrologic criteria to document the tectonic nature of contacts in the Zacatecas Formation.

Recently, Monod and Calvet (1992) subdivided rocks of the Zacatecas Formation into three new formations. In ascending order, these are the La Pimienta Formation, El Bote Flysch Formation, and El Ahogado Formation. Monod and Calvet claimed that the pillow lavas in this sequence are correlative with the Fresnillo Formation (Cantú-Chapa, 1974) or Chilitos Formation (de Cserna, 1976) in the Fresnillo area of Zacatecas state. Cantú-Chapa (1974) first dated these strata as late Valanginian–early Hauterivian with the ammonoids *Distoloceras* aff. *parritense* Imlay, *Distoloceras laticostatum* Imlay, *Distoloceras* aff. *nodosum* Imlay, *Mexicanoceras* aff. *neohispanicum* Bose, *Taraisites* aff. *neoleonense* Cantu, and *Taraisites* sp. Early Cretaceous radiolarians were identified from the Plateros Formation in Fresnillo (Dávila-Alcocer, 1981), and Early Cretaceous (Valanginian) radiolarians are contained in chert in pillow lavas in other localities of Zacatecas (Yta, 1992). Thus, the Zacatecas and Fresnillo (or Chilitos) Formations are not age equivalent. The limestones in the sequence confuse the

U. S. A.

Gulf of California

Gulf of Mexico

N

LOCALITIES
1. Charcas Area (Zacatecas Formation)
2. Sierra de Salinas (Zacatecas Formation)
3. West City of Zacatecas (Zacatecas Formation)
4. Sierra de Teyra (Taray Formation)
5. Central Sonora (Barranca Group)
5a. Central Sonora (Unnamed Units)
6. Northwest Sonora (El Antimonio Formation)
7. Baja California Sur (Unnamed Unit)
8. Michoacan (Arteaga Complex)
9. Baja California Norte (De Indio Formation)
10. Tamaulipas (La Boca Formation)

0   400
km

FIGURE 1. Location map of Mexico showing the Triassic localities for this study.

interpretation even more. Monod and Calvet (1992) suggested that limestones may be Cretaceous (?), Maldonado-Koerdell (1948) claim they are Jurassic, and McGehee (1976) believes they are Triassic.

Stratigraphy

According to Monod and Calvet (1992), the El Bote Formation consists of black phyllite and light brown quartzite showing trace fossils. The overlying El Ahogado consists of fossiliferous (lamellibranchs) black slate intercalated with quartzite.

West of the city of Zacatecas, the combined thickness of the El Bote and El Ahogado Formations exceeds 180 m.

Age

Burckhardt and Scalia (1905) discovered and described Upper Triassic fossils at Arroyo Pimienta west of Zacatecas city. They identified the ammonoids *Juvavites* (*Anatomites*) sp. Mojsvari, *Clionites* sp., and *Trachyceras* sp. Fossils collected by Burckhardt (1930) at Puente del Ahogado in the vicinity of Arroyo Calavera ó Pimienta include the lamellibranchs *Halobia* cfr. *austriaca* Mojs., *Palaeoneilo zacatecana*, *Palaeoneilo longa* Burckhardt, *Palaeoneilo Broilii* Burckhardt, *Palaeoneilo triangularis* Burckhardt, *Palaeoneilo burkarti* Burckhardt, *Palaeoneilo frechi* Burckhardt, *Palaeoneilo villadae* Burckhardt, *Palaeoneilo mexicana* Burckhardt, *Palaeoneilo cordobae* Burckhardt, *Palaeoneilo boesei* Burckhardt, *Palaeoneilo inflata* Burckhardt, *Palaeoneilo humboldti* Burckhardt, *Palaeoneilo circularis* Burckhardt, *Palaeoneilo cordiformis* Burckhardt, *Palaeoneilo quadrata* Burckhardt, *Palaeoneilo waitzi* Burckhardt, *Palaeoneilo ledaeformis* Burckhardt, *Palaeoneilo costata* Burckhardt, and *Palaeoneilo ordoñezi* Burckhardt. He also identified the ammonoids *Trachyceras smithi* Burckhardt sp., *Protrachyceras* sp., *Clionites* sp., and *Juvavites* (*Anatomites*) *mojsvari* Burckhardt. These faunas suggest a Late Triassic, early Carnian age (Burckhardt, 1930).

Fossils reported from La Colina, south of Arroyo Calavera, include: *Avicula hofmanni* Bittner, *Avicula hofmanni* var. *pseudopterinea* Frech, *Cassianella* (*Burckhardita*) *boesei* Frech, *Cassianella* (*Burckhardita*) *aguilerae* Frech, *Cassianella* n. sp. aff. *decussata* Mstr., and *Palaeoneilo* sp. These faunas indicate an early Carnian age (Burckhardt, 1930).

Cuevas-Pérez (1985) collected one sample from a calcareous horizon in the Zacatecas strata that contained the conodonts *Neogondolella polygnathiformis* Budurov and Stefanov, *Neogondolella* sp., *Epigondolella primitia* Mosher, and *Epigondolella* ex. aff. *parva* Kosur. These faunas indicate a late Carnian age (zone of *macrolobatus*).

Sierra de Salinas, Zacatecas

Previous Work

Chávez-Aguirre (1968) first described the Zacatecas Formation as a sequence of conglomerate, quartzarenite, and green slate exposed near the Rancho La Ballena and as a schistose sequence consisting of schist and sericitic phyllite in other parts of the Sierra de Salinas. Chávez-Aguirre (1968) and López-Ramos (1980) provided paleontologic data from the Zacatecas strata. The Zacatecas Formation was described in a Mexican Geological Society field-trip guidebook (1982) as a series of gray sandstones, shales, and thin-bedded red siltstones and green phyllites with intercalations of gray, recrystallized limestone. In a regional study, Labarthe-Hernández et al. (1982) provide stratigraphic and structural descriptions of these Triassic rocks. Cuevas-Pérez (1985) provided regional stratigraphic and paleogeographic interpretations of the Zacatecas in central Mexico. The stratigraphy, depositional environments, sediment provenance, and structural deformation of the Zacatecas rocks at the Sierra de Salinas were studied in detail by Silva-Romo (1993). An integrated tectonic analysis incorporating regional stratigraphic, structural geology, and sediment provenance analyses for the Guerrero terrane also provided limited isotopic and sediment source analysis for the Zacatecas strata in the Sierra de Salinas (Centeno-García et al., 1993; Centeno-García, 1994; Centeno-García and Silva-Romo, 1997).

Stratigraphy

In the Sierra de Salinas, the Zacatecas Formation is intensely deformed and metamorphosed to chlorite and sericite phyllites that display a generally dark color and shiny luster. Most outcrops are on the south side of the mountain; the Zuloaga Limestone occupies the northern flank above the Zacatecas Formation. At the Cañada Comanja, between the Cerro Grande and Cerro La Leona, the Zacatecas Formation consists of unmetamorphosed dark green and black sandstone and shale. The green beds weather to light green, brownish-orange, and reddish-orange. The black beds weather to light and dark gray. The sandstone is fine- to very fine-grained and primarily thin bedded, with some medium- and thick-bedded sandstone. The shale is fissile. Abundant primary sedimentary structures include convolute bedding, load casts, pillow-and-ball structures, and mullion structures. Other common primary sedimentary structures are flute casts, cross-bedding, rill marks, ripple marks, horizontal parallel laminations, and graded bedding. Fossils found in these strata were identified as *Sirenites* sp. of probable Triassic age (Wolfgang Stinnesbeck, personal communication, 1994). The Zacatecas strata are overlain by the Nazas Formation and the Zuloaga Limestone.

Silva-Romo (1993) provided a tentative structural thickness of 2500 m for the strata. Our measured thickness of the Zacatecas Formation northeast of La Ballena and west of Cerro Grande is 319 m, undoubtedly an incomplete stratigraphic thickness.

Age

Chávez-Aguirre (1968) found fragments of *Sirenites* sp. at several localities in the Sierra de Salinas, and Silva-Romo (1993) collected *Clionites* sp., *Palaeoneilo* sp., and *Halobia* sp. in the vicinity of La Ballena village. These faunas are similar to the Late Triassic faunas described by Burckhardt (1930) near the city of Zacatecas. At the Arroyo de La Huerta in the La Ballena locality, exposures of the Zacatecas Formation yielded ammonite impressions of the family Beyrichitidae Spath that ranges in age from Early to Middle Triassic (Gallo et al., 1993; Gómez-Luna et al., 1997). Tristán-González and Tórres-Hernández (1994) collected ammonites near La Ballena village that were identified as *Parairachiceras* sp., *Pseudolococers* sp., and *Metadinarites* sp. and suggest a Middle Triassic (Anisian) age. A collection of poorly preserved fossils from the same outcrops was identified as *Halobia* sp. and *Paleoneilos* sp. of possible Carnian age (López-Ramos, 1980). Chávez-Aguirre (1968) collected one ammonite specimen northeast of Rancho La Ballena identified as *Sirenites* sp. of Carnian to Norian age (Late Triassic).

Sediment Provenance

Thin sections from sandstone collected from the Zacatecas Formation in the Sierra de Salinas have been studied to determine sedimentary provenance (Silva-Romo, 1993; Silva-Romo et al., 1993; Centeno-García, et al., 1993; Centeno-García, 1994). These reports show that the composition of the sandstone (quartz, chert, feldspar, and minor volcanic grains) suggests source areas in the North American craton. According to the classification of Dickinson (1985), the sandstones of the Zacatecas Formation in the Sierra de Salinas were derived from continental blocks or from an uplifted metamorphic and plutonic complex in an orogenic belt (Centeno-Garcia et al., 1993). Sediment transport was to the southwest. Centeno-García and Silva-Romo (1997) suggest that Sm-Nd isotopic ratios indicate sediment sources for the Triassic basins in the Grenvillian belt that extends from Chihuahua to Oaxaca.

Our study of 10 thin sections from sandstones collected at the Sierra de Salinas shows that the sandstone is fine to very fine grained, moderate to well sorted, and commonly contains subangular to subrounded grains cemented by calcite. Quartz grains constitute more than 80% of the rock, feldspars plus lithic fragments range from 20% to 25%, and heavy minerals are less than 1%. Based on a simple comparison with Krynine's genetic classification of quartz types (Folk, 1980), quartz grains are of plutonic, vein, and gneiss origin. Feldspars were derived from both igneous and metamorphic rocks, as suggested by concentric growth. Metamorphic rock grains are schistose, recrystallized, and stretched. Volcanic grains are commonly porphyritic, and sedimentary grains are all chert. Metamorphic grains are the predominant rock grains. Heavy minerals include rounded zircon, tourmaline, micas, magnetite, epidote, hematite, and hornblende. Based on mineralogy, these sandstones are subarkoses (Folk, 1980). The general composition and heavy mineral content suggest that the sandstones were derived from crystalline and sedimentary sources.

Charcas Area, San Luis Potosí
Previous Work

Carrillo-Bravo (1971) described the clastic sequence exposed in the vicinity of Charcas and, based on lithologic similarities and stratigraphic relations, proposed the name Zacate-

cas Formation for these strata. Upper Triassic marine faunas collected in the Charcas area were assigned a Late Triassic age (Cantu-Chapa, 1969). Martínez (1972) described the Zacatecas northwest of Charcas as an alternating sequence of siltstone, shale, and sandstone. Butler (1972) described these same rocks as metamorphosed sandstone, quartz arenite, argillite, and conglomerate of possible Paleozoic age. Extensive geologic studies of the Charcas mountains included mapping, structural geology, and stratigraphy of the Zacatecas Formation, and similar studies of the overlying Jurassic-Tertiary rocks have been carried out recently (Tristán-González and Tórres-Hernández, 1992; Tristán-González and Tórres-Hernández, 1994). Small, scattered outcrops of the Zacatecas Formation were reported at Presa de Santa Gertrudis, north of the town of Charcas (Tristán-González et al., 1995). Barboza-Gudiño et al. (1998) provide a large-scale tectonic scenario of the Triassic-Jurassic sequences of Charcas.

Barboza-Gudiño (1993) conducted field studies showing that Zacatecas outcrops in Charcas extend to the east, to the Sierra de Catorce.

Stratigraphy

The Zacatecas Formation in the Charcas area consists of beds of dark gray, green, and grayish-green shale, sandstone, conglomerate, and siltstone. Weathered rocks are normally yellow, brownish red, or grayish yellow. The shale is fissile and contains concretions; the sandstone and siltstone vary from thin to thick bedded. Some sandstone beds pinch out locally to form lensoid bodies, but in general they have a well-defined stratification. Shale is predominant locally in the section, but overall the section is predominantly sandstone. Primary sedimentary structures in these strata include ripple marks, load casts, laminations, and sole marks.

Cuevas-Pérez (1997) measured an incomplete stratigraphic section of 240 m along the Arroyo Ojo de Leño, near Charcas. Martínez (1972) estimated that the thickness of the Zacatecas Formation in Charcas is more than 250 m. In the same area, Butler (1972) determined a tentative thickness of more than 2000 m. A well drilled by PEMEX 25 km east of the town of Charcas cut 4640 m of the Zacatecas Formation (López-Infanzón, 1986); however, this thickness may reflect only duplication by overthrusting.

Age

Cantú-Chapa (1969) identified the ammonite *Juvavites* sp. from outcrops of the Zacatecas Formation northwest of Charcas, in San Luís Potosí. This specimen is similar to the ammonoids described by Burckhardt (1930) in Puente del Ahogado, in Zacatecas. Therefore, its age is also considered to be early Carnian. Gallo et al. (1993) and Gómez-Luna et al. (1997) confirmed the early Carnian age of Zacatecas strata in the Charcas area based on the ammonite *Anatomites* aff. *herbichi* Mojsijovics and the protobelemnoidea *Aulacoceras* sp.

Stratigraphic Relationships of the Zacatecas Formation

To date, no rocks older than Triassic have been reported from the localities described above (Figure 2). As for the upper contact, west of the city of Zacatecas, the Zacatecas Formation is in fault contact with Lower Cretaceous pillow lava and limestone of probable Cretaceous (?) age (Monod and Calvet, 1992). In the Charcas area, San Luís Potosí, and Sierra de Salinas, Zacatecas volcanic-sedimentary rocks of the Nazas Formation rest unconformably on the Zacatecas Formation, but where the Nazas is absent, either because of erosion or nondeposition, the Zuloaga Limestone (Oxfordian) covers the Zacatecas unconformably (Silva-Romo et al., 1993; Tristán-González and Tórres-Hernández, 1994; Bartolini, 1998). Locally, the Zacatecas Formation is also overlain by continental clastics that are considered part of the La Joya Formation (Barboza-Gudiño et al., 1997).

Depositional Setting

There is a consensus that sediments of the Zacatecas Formation were deposited in deep-water marine environments in the states of Zacatecas and San Luís Potosí (Monod and Calvet, 1992; Silva-Romo, 1993; Silva-Romo et al., 1993; Centeno-García et al., 1993; Gallo et al., 1993; Centeno-Garcia, 1994; Tristán-González and Tórres-Hernández, 1994; Barboza-Gudiño et al., 1998; Bartolini, 1998). Silva-Romo et al. (1993) proposed that the Zacatecas Formation in the Sierra de Salinas represents middle and outer submarine-fan facies, according to the classification of Mutti and Lucchi (1972). In the same locality, Centeno-García (1994) and Bartolini (1998) identified convoluted bedding that suggests deposition in continental slope and/or toe-of-slope settings. Monod and Calvet (1992) also suggested slope depositional environments for the El Ahogado and El Bote Formations (formerly Zacatecas Formation) in the vicinity of Zacatecas city.

Turbidite deposition along the passive continental margin of western Pangea in Mexico was apparently in response to a global sea-level lowstand during the Permian and Triassic (Vail et al., 1977). Turbidite deposition occurred adjacent to shelf margins, particularly in slope, toe-of-slope, and basin-plain settings. The deep-marine, poorly reworked, fine- to very fine-grained sand may represent bypass turbidite deposits. Thus, the Triassic turbidite sediments of the Zacatecas Formation

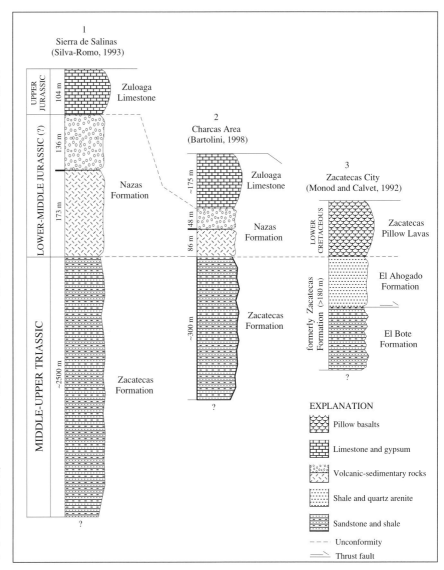

Figure 2. Stratigraphic relationships between the Zacatecas Formation and overlying formations.

were most likely deposited in type A basins, according to the classification of Mutti and Normark (1987) and Mutti (1992). Type A basins are formed on oceanic crust with large, long-lived sediment source(s) and little or no tectonic activity. Indeed, the stratigraphy, paleontology, sedimentology, and absence of igneous activity during the sedimentation of the Zacatecas Formation bear a remarkable similarity to turbidites deposited in relatively deep-marine settings (Mutti, 1985). One problem with this interpretation is the failure to identify rocks that underlie the Zacatecas Formation; however, it is reasonable to assume that the Zacatecas strata were underlain originally either by thinned continental or transitional crust, depending on how far inside the basin the turbidites were deposited. The Zacatecas turbidites match fairly well the characteristics of a sand/mud-rich depositional system of deep-water basin margins (Reading and Richards, 1994). In this model, high- and low-density turbidites form lobes of interbedded sands and muds that are controlled chiefly by point-source feeder systems.

The main types of turbidite deposits differ from each other mainly in terms of where the sand is deposited in the system. The turbidites of the Zacatecas Formation best fit type I depositional systems of Mutti (1985) and Mutti and Normark (1987), which are characterized by sand deposited predominantly in nonchannelized lobes, as indicated by the tabular geometry and lateral extent of the sandstones in distal, relatively flat basinal areas. The convoluted bedding of Zacatecas strata record the downslope flow of unconsolidated sediments on an inclined surface, and although turbidites commonly dominate in the distal parts of the basin and convoluted bedding is more typical of slope and toe-of-slope settings, either type of deposit can be formed (Shanmugam et al., 1995). Whether this depositional surface was part of the continental rise or slope is uncertain.

Based on facies associations, bedding characteristics, and sedimentologic features of the sedimentary deposits in Sierra de Salinas, Charcas, and west of the city of Zacatecas, we believe that the depositional setting of the Zacatecas is comparable to the basin-floor fan model (Shanmugam et al., 1995), in which very complex and laterally extensive sand-rich submarine fans consist of lobe-sheet-dominated turbidites. The Triassic turbidites of the Zacatecas represent facies B, C, and D in the classification of Mutti and Normark (1987), Mutti (1992), and Reading and Richards (1994). Facies B, C, and D of type I are normally characterized by remarkable lateral continuity and tabular geometry over distances of tens of kilometers. Facies B contains conglomerate and conglomeratic sandstone, and facies C is medium- to fine-grained sandstone with minor shale. The sandstone beds are bounded by even, parallel surfaces having good lateral continuity. Small-shale clasts (Tristán-González and Tórres-Hernández, 1994), either scattered or concentrated in distinct lenses and levels, are common in this sandstone. In Sierra de Salinas, facies C and D are more common than facies B. Facies D (distal facies), on the other hand, is comprised of fine and very fine sandstone, siltstone, and increasing amounts of shale that has lateral continuity. Internal structures are thin-current laminae, which can be either parallel undulating or convoluted. The turbidite sequences exposed along the San Rafael anticline in Charcas show a combination of facies C and D and minor facies B with thin conglomerate beds. In the vicinity of Zacatecas city, Monod and Calvet (1992) suggested that the El Bote and Ahogado Formations record slope deposits. Despite intense deformation that obscures the original sedimentologic features of these outcrops, local rhythmic interbedding of sandstone and shale generally resemble the fan facies recognized in the other two areas; therefore, middle- and outer-fan facies should not be ruled out.

The distribution and relative predominance of turbidite facies types in the three areas represent, in general, the original sedimentological record; however, because repetition of the section is caused by folding and thrust faulting, the predominance of a specific lithology should be interpreted with caution.

Paleogeography

Where Are the Triassic Platform Facies in Central Mexico?

The presence of Upper Triassic deep-marine strata in several localities in central Mexico (Figure 3) leads to an important question. Where were the shallow-water facies? Has the shelf record not been found because it is too narrow or missing because of erosion? In general, during lowstand conditions, shelves are exposed and eroded (or not preserved), but the deep-marine record is well represented in the basins (Vail et al., 1977)

The only reported Triassic (?) shallow-marine facies that may be equivalent to the Zacatecas Formation are exposed at Cerro La Cruz (Mezquital area), 8 km northeast of the town of Aramberri, Nuevo León (Figure 3; Bartolini, 1998), described by Meiburg et al. (1987) and DeLeón-Gómez (1988). We measured 115 m of strata at Cerro La Cruz that consist of sandstone, shale, and limestone. Thin sections from limestone samples were studied by Dr. Wolfgang Stinnesbeck (Universität Karlsruhe, Germany), who concluded that most allochems (>90%) are grains composed partly or entirely of micrite. These grains are elliptical in cross section, averaging about 0.08 to 0.1 mm in diameter. They are interpreted to be fecal pellets. Some larger coprolites are also present. They reach diameters of 0.15 to 0.18 mm and show complex sievelike internal structures with pierced longitudinal canals. Canals are rounded in cross section and are arranged to form three or more curved series parallel to the plane of bilateral symmetry. These bilaterally symmetrical groups of longitudinal canals are typical of anomuran coprolites. The ichnofossils are favreine-form species and may belong to *Favreina* or *Parafavreina*. These fecal pellets, known to be at least as old as Triassic, are attributed generally to fossil thalassinid crustaceans such as *Callinassa* (Brönnimann, 1972; Förster and von Hillebrandt, 1984), although the form genus *Thoronetia* may be the product of galatheid decapods (Brönnimann, 1972; Schweigert et al., 1997) and *Helicerina ruttei* (Schweigert et al. 1997) of brachyurid crabs. Additional bioclastic grains include fragments of echinoderms, brachiopods, rare gastropods, and ostracodes. This rock is grain-supported with sparite cement and thus is a pelsparite or biopelsparite in the classification of Folk (1959, 1962) or a pelletal grainstone in the classification of Dunham (1962). This sediment type with abundant fecal pellets (e.g., *Favreina*) typically characterizes shallow subtidal and intertidal coastal areas with restricted water circulation in lagoons, coastal ponds, and tidal flats (Wilson, 1975; Flügel, 1982), but it may also occur in supratidal or nonmarine environments. So far, however, only three species of internally structured microcoprolites have been described from nonmarine environments (Schweigert et al., 1997). These belong to the ichogenus *Helicerina* and seem to be unrelated to the specimens that we found. The presence of echinoderms and brachiopods also indicates marine conditions, because both groups do not exist in brackish environments (Flügel, 1982).

FIGURE 3. Triassic sedimentary and sedimentary-volcanic facies exposed in Mexico.

1994; Stewart et al., 1997). Also, since the beginning of the last century (Dumble, 1900a,b), Triassic transitional facies of the Barranca Group have been documented in detail throughout central Sonora (King, 1939; Wilson and Rocha, 1949; Alencaster-de Cserna, 1961a, b; Silva-Pineda, 1961; Weber, 1980; Weber et al., 1980; Potter and Cojan, 1985; Stewart and Roldán-Quintana, 1991).

Recently, the tectonic and paleogeographic setting of the southeastern margin of North America in northwestern Mexico has been described as either a disrupted continental margin affected by a large-scale Middle Jurassic transpressive structure or as a continuous Precambrian to Jurassic paleogeographic belt that wrapped around the southeastern corner of the continent. The existence of Triassic platform facies in Sonora can be explained by both of these two hypotheses.

Stanley and González-León (1995) highlighted the striking similarities in lithology, stratigraphy, thickness, and depositional environments of the Luning Formation of west-central Nevada and the Antimonio Formation of Sonora. They conclude that the biogeographic affinities and the faunal links between the two regions are clear. They suggest that the simplest way to explain these similarities is by left-lateral displacement of the Luning and Antimonio along the Mojave-Sonora megashear (Silver and Anderson, 1974). This requires lateral displacement of 1000 km, the present distance between Luning Formation outcrops in Nevada and El Antimonio Formation outcrops in northwestern Sonora.

A second explanation, originally conceived by Kay (1951), is the existence of a continuous continental margin with Paleozoic belts that wrap around the southwestern corner of North America. In this model, neither the Triassic transitional facies (Barranca Group) nor the shallow-marine facies (Antimonio Formation) that overlie the Paleozoic strata are expected to be truncated structurally. Likewise, Stewart et al. (1990), Poole (1993), and Poole et al. (1995) describe the continuity of the Paleozoic belts in the western United States. Neither hypothesis, however, provides a logical explanation for the lack of Triassic deep-marine facies in Sonora.

The exact orientation, size, and configuration of the Late Triassic Zacatecas basin are unknown. However, shallow-marine facies in Nuevo León may record a coeval platform located to the east-northeast of the present location of deep-marine Zacatecas strata. Unfortunately, because most pre-Cretaceous rocks lie beneath a fold belt, it is not feasible to document the extent of the platform facies. On the other hand, the broad, northwest-trending distribution of Triassic rocks across Mexico may indicate either the expression of the fold belt or the original trend of the continental margin with a basin facies in the southwest and a platform facies in the northeast.

Where Are the Triassic Deep-marine Facies in Sonora?

Triassic platform facies (Figure 3) have long been recognized in northwestern Sonora (Keller, 1928; Burckhardt, 1930; Alencaster-de Cserna, 1961a, b; Gónzalez-León, 1980; Stanley et al., 1994; Stanley and González-León, 1995; Lucas et al., 1997) and most recently in central Sonora (Lucas and González-León,

Paleobiogeography

Triassic Marine Faunas of Mexico and their Paleobiogeographic Relation to the Eastern Pacific

Late Triassic marine strata exist in Charcas, Sierra de Salinas, and the vicinity of Zacatecas city (Burckhardt and Scalia, 1905; Burckhardt, 1930; Cantú-Chapa, 1969; Gallo et al., 1993). Triassic marine rocks have also been identified in Sonora (Burckhardt, 1930; González-León, 1980, 1997a, b; Estep et al., 1997a, b; Lucas et al., 1997; McRoberts, 1997; Stanley, 1997; Stanley and González-León, 1997) and Baja California (Jones et al., 1976; Finch and Abbot, 1977; Gastil et al., 1981). All of these studies report Triassic ammonoids (ceratites) and lamellibranchs (*Monotis-Halobia-Palaeoneilo*). Stanley and González-León (1997) and Damborenea and González-León (1997) report Triassic corals and lamellibranchs from Sonora.

West of the city of Zacatecas, shale of the Zacatecas Formation contains *Trachyceras* sp., *Prototrachyceras* sp., *Clionites* sp., and *Anatomites* sp. associated with *Halobia* sp. and *Palaeoneilo* sp. (Burckhardt, 1930). These ammonoids have an Anisian to Norian age distribution and are reported in central Europe (Alps and Balkans) and Asia and in Alaska, Nevada, and California in western North America (Arkell et al., 1957; Westermann, 1973; Tozer, 1982).

The occurrence of these ammonoids in central Mexico (Zacatecas city and Charcas) raises a paleozoographic problem, because localities in central Mexico are geographically isolated from areas of marine sedimentation in the eastern Pacific region (Cantú-Chapa, 1994). Still, the *ceratites* of central Mexico show a general biogeographic affinity with western North America (Table 1), in particular with California (*Prototrachyceras*) and Nevada (*Clionites* and *Anatomites*).

From a biogeographic point of view, it is noteworthy that some of the same genera also have been reported in Sonora, northwestern Mexico (González-León, 1997a; Estep et al., 1997b; Lucas et al., 1997; McRoberts, 1997; Stanley, 1997; Stanley et al., 1994). These studies suggest marine communication between these regions, which today are separated by hundreds of kilometers.

The presence of Carnian-Norian *Juvavites* in the Charcas area adds another paleogeographic element to the picture (Cantú-Chapa, 1969). This cosmopolitan genus is known from Europe (Alps-Sicily), Asia (Himalayas), southeast Asia (Indonesia), Alaska, British Columbia, and California (Arkell et al., 1957). Apparently, *Juvavites* ranged from central Mexico to the eastern Pacific (Table 1). In addition, Gallo et al. (1993) found in central Mexico the Triassic ammonite *Anatomites* aff. *herbichi* Mojsijovics, which has the same geographic range as *Juvavites*.

REGIONAL CORRELATIONS

Based on these data, we are able to correlate the Triassic strata from central, eastern, and northern Mexico, as shown in

TABLE 1. Comparative stratigraphic distribution of ceratites genera in central Mexico and the eastern Pacific of western North America (Arkel et al., 1957; Estep et al.; 1997; Lucas et al., 1997; Tozer, 1982; Westermann, 1973). Fossils are shown as a group, not in stratigraphic order.

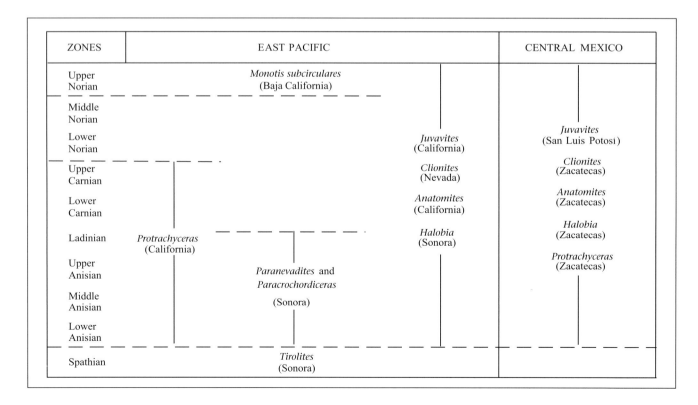

Table 2. Continental, shallow-marine, deep-marine, and intraoceanic arc successions are included in this correlation.

Central Mexico

Lithostratigraphic and chronostratigraphic correlation among the Upper Triassic Zacatecas Formation in Charcas, San Luís Potosí, Sierra de Salinas, Zacatecas, and west of the city of Zacatecas is established with some measure of certainty. Another unit that may correlate with the Zacatecas Formation is the Taray Formation (Cordoba, 1964), but its age has not yet been determined paleontologically. Taray exposures exist only on the western flank of the Sierra El Solitario de Teyra in northern Zacatecas state.

Taray Formation

The Taray Formation was named by Cordoba (1964) for a deformed sequence of shale, sandstone, limestone, dolomite, and chert of unknown age exposed on the western flank of the Sierra El Solitario de Teyra in northern Zacatecas. López-Infanzón (1986) correlated the formation with the Middle to Late Triassic Zacatecas Formation. Anderson et al. (1990) proposed a probable Jurassic age for the Taray and considered these rocks a mélange. Bartolini (1998) did not correlate the Taray with the Zacatecas Formation because of the absence of fossils in the Taray and the lithologic differences between the two formations.

Stratigraphy

The thickness of the Taray Formation is unresolved because of structural complexity and poor exposure. The Taray Formation is predominantly black shale and minor sandstone, chert, and dolostone. The shale weathers yellowish to greenish-yellow, is laminated, and is fissile in the unmetamorphosed exposures. Typically, it is extremely contorted and strained from dynamic metamorphism. Thin- to very thin-bedded chert is a minor component. In general, Taray rocks are partially recrystallized and intensely fractured and oxidized. Locally, the formation contains shale partings. Chert beds are folded on northeast-trending axes and show no signs of foliation. Sandstone in the Taray is dark green, medium-grained, arkosic, and thin to thick bedded. Thin dolostone beds are rarely found. Locally, intraformational conglomerate occurs, which contains chalky-white (dolostone?) clasts. Rare sandy, dark-green conglomerates with flattened clasts also are present.

TABLE 2. Stratigraphic correlation of Triassic strata in Mexico.

	AGE	TAMAULIPAS	ZACATECAS SAN LUIS POTOSÍ	CENTRAL SONORA	NORTHWEST SONORA	BAJA CALIFORNIA
LATE TRIASSIC	Late Norian					UNNAMED FORMATION (Finch and Abbot, 1974) (Jones et al., 1976)
	Middle Norian			S. SANTA TERESA (Stewart et al.,1997)		
	Early Norian		ZACATECAS FORMATION (Burckhardt, 1930) (Cantú-Chapa, 1969) (Gallo et al., 1993)		EL ANTIMONIO FORMATION (Burckhardt, 1930) (Alencaster, 1961) (González-León, 1980) (Stanley et al., 1994) (Stanley and González-León, 1995) (Lucas et al., 1997) (Estep et al., 1997a)	
	Late Carnian	LA BOCA FORMATION (Mixon et al., 1959) (Weber, 1997)		S. LA FLOJERA (Lucas and González-León, 1994) BARRANCA GROUP (Alencaster, 1961) (Silva-Pineda, 1961) (Weber, 1980) (Weber et al., 1980)		
	Early Carnian					
MIDDLE TRIASSIC	Ladinian-Anisian					
EARLY TRIASSIC	Smithian					De INDIO FORMATION (Gastil et al., 1981) (Buch and Delattre, 1993)

Overall, the Taray Formation is intensely deformed. Except for chert beds, the folding displays two foliation sets well exposed in the Arroyo El Taray. In general, one foliation strikes N70°E and dips 59°SE; the second foliation strikes N57°W with a dip of 46°SW. Folds are normally isoclinal, and the fold axes strike northeast. The northwest-striking foliation affects both the Taray and the Nazas Formations, whereas the northeast-striking foliation is developed only in the Taray Formation. That the Nazas is affected only by the northwest-striking foliation suggests that the northeast-trending foliation predates the Nazas Formation.

The Nazas Formation rests unconformably on the Taray. This same unconformable relationship was observed also between the Zacatecas and the Nazas Formations at the La Ballena locality in the Sierra de Salinas (Silva-Romo et al., 1993; Bartolini, 1998) and at Charcas (Tristán-González and Tórres-Hernández, 1992; Tristán-González and Tórres-Hernández, 1994; Bartolini, 1998).

Age of the Taray Formation

Rogers et al. (1961) assigned a Permian (?) age for the Taray Formation, based on field relations. Cordoba (1964) assigned a tentative Early Devonian to Early Pennsylvanian age for the Taray, based on its lithologic similarity to the Tesnus Formation and Caballos Novaculite in the Marathon region of Texas. The Taray was assigned a Late Triassic age (López-Infanzón, 1986), based on correlation with the Zacatecas Formation's position below the Nazas Formation. More recently, the Taray Formation was considered Jurassic in age (Anderson et al., 1991), possibly the deep-water equivalent of the Oxfordian Zuloaga Limestone. Our samples of Taray black shale were processed for palynomorphs and for conodonts, but the rock residues were barren (Cornell, University of Texas at El Paso; Ethington, University of Missouri, written communication, 1996). There are no fossils reported from the Taray Formation, so its age is uncertain. Its stratigraphic position below the Nazas Formation suggests a pre-Jurassic age.

Depositional Environments

Sedimentologic and lithologic characteristics of the Taray Formation indicate that it was probably deposited in relatively deep marine environments. Black shale is considered an indicator of deep, quiet, anoxic marine waters. The chert and dolostone beds are consistent with pelagic sedimentation in deep-water settings. Accumulation of these pelagic sediments must have been interrupted episodically by influx of coarse clastic sediment and intraformational gravels.

Eastern Mexico
La Boca Formation (Tamaulipas)

At the Novillo Canyon in Tamaulipas state (Figure 1), continental red beds of the La Boca Formation were originally assigned a Late Triassic age on the basis of their fossil plant content (Mixon et al., 1959). A revision of the fossil flora from strata of the La Boca at the same Novillo Canyon locality (Weber, 1997) identified the species *Ctenophyllum braunianum*, *Chiropteris* sp. (Kurr?), *Laurozamites yaqui*, and questionable fragments of *Podozamites* sp. According to Weber, this flora is Carnian in age and is found in the Santa Clara Formation in Sonora and in other Carnian formations elsewhere in North America.

Northwestern Mexico
Barranca Group (Central Sonora)

Sedimentary transitional sequences of Late Triassic age occur in central Sonora (Figure 1). These strata were first studied and named the Barranca Division by Dumble (1900a, b) and were later called the Barranca Formation by King (1939). Wilson and Rocha (1949) divided these strata into three subdivisions, which were later renamed (from older to younger) the Arrayanes, Santa Clara, and Coyotes Formations, which together form the Barranca Group (Alencaster-de Cserna, 1961a), nomenclature that is used today.

The fossil flora and fauna of the Barranca Group have been studied in detail by Silva-Pineda (1961), Alencaster-de Cserna (1961b), Weber (1980), and Weber et al. (1980). Identification of approximately 7000 fossil plant species supports a Late Triassic (Carnian) age for the Santa Clara Formation (Weber, 1997). Potter and Cojan (1985), Cojan and Potter (1991), and Stewart and Roldán-Quintana (1991) documented marine-delta and alluvial-fan deposits for Barranca strata in the Sierra de San Javier, central Sonora. Petrographic data also exist for the Barranca Group (Stewart and Roldán-Quintana, 1991), as well as detailed petrologic, sediment-provenance, and depositional environmental data for the Santa Clara Formation (Cojan and Potter, 1991). The absence of volcanic detritus in the section is notable (Stewart and Roldán-Quintana, 1991), especially in comparison with the Nazas Formation to the east, where the main clastic components are volcanic rock fragments. According to Stewart and Roldán-Quintana (1991), sedimentary rocks of the Barranca Group were deposited in pull-apart basins associated with strike-slip faults. The paleogeographic and tectonic framework of the Barranca Group with respect to other Upper Triassic rocks in central Mexico is still poorly understood.

El Antimonio Formation (Northwestern Sonora)

The first Late Triassic (Carnian) ammonites in northern Mexico were reported by Keller (1928) and Burckhardt (1930) from the type locality of El Antimonio Formation in northwestern Sonora (Figure 1). Alencaster-de Cserna (1961a) studied the Antimonio Formation stratigraphy, paleontology, and environments of deposition. Twenty years later, González-León (1980) collected fossils, redefined the stratigraphy, and assigned the name Antimonio Formation to these strata, which range in age from Late Triassic to Early Jurassic. Ever since, the Antimonio Formation has been assigned a Late Triassic–Early Jurassic age (Stanley et al., 1994; Lucas and González-León, 1994; Stanley and González-León, 1995).

An extensive taxonomic study of the Upper Triassic faunas from the Antimonio Formation was made by Stanley et al. (1994). Tectonic, paleogeographic, and biogeographic settings for the Antimonio Formation were provided by Stanley and González-León (1995). At Sierra La Flojera, 5 km southeast of Hermosillo, Lucas and González-León (1994) discovered an 80-m-thick marine section containing the ammonoid *Hannaoceras nodifer* (Hyatt and Smith) of late Carnian age. They proposed that these rocks correlate in part with the Antimonio Formation. At Sierra de Santa Teresa, 20 km southeast of Hermosillo, limestone beds in a marine sequence yielded the sponges *Cinnabaria expansa*, *Fania* sp., and *Nevadathalamia cylindrica*, as well as the corals *Retiophyllia* sp., *Astraeomorpha sonorensis*, and *Chondrocoenia waltheri* of Norian age (Stewart et al., 1997). More recently, Lucas et al. (1997) documented an Early Triassic age (Spathian) based on the ammonoids *Tirolites* sp. Mojsijovics and *Pseudosageceras diener* and conodont faunas from a 150-m section in the lower Antimonio Formation. In the Sierra El Alamo Muerto, Estep et al. (1997a) identified the Middle Triassic (Anisian) ammonoids *Paranevadites* cf. *furlongi* (Smith) and cf. *Paracrochordiceras* sp. from a 180-m section of the Antimonio Formation.

Baja California Sur

Jones et al. (1976) reported the presence of Upper Triassic lamellibranchs of the genus *Monotis* in a thick volcanic-sedimentary sequence at Punta Hipolito in western Baja California Sur state (Figure 1). The litharenite, chert, limestone, andesite breccia, tuff, and pillow basalt sequence contains *Monotis subcircularis* and *Halobia* sp. of Late Triassic age (Finch and Abbot, 1977).

Baja California Norte

Triassic marine strata near El Volcan, Baja California Norte state (Figure 1), were first reported by Gastil et al. (1981). They discovered that this marine clastic and carbonate sequence contains the conodonts *Neospathodus bicuspidatus* (Muller), *Xaniognathus* cf., *X. elongatus* (Sweet), *Ellisonia triassica* (Muller), and the ammonoid *Meekoceras* sp., all of which indicate an Early Triassic (Smithian) age. More recently, Buch and Delattre (1993) used the name De Indio Formation for more than 300 m of this same Lower Triassic shallow-marine, fossiliferous sandstone, carbonaceous argillite, limestone, and conglomerate sequence.

Discussion

Structural Deformation

Multiple lines of evidence suggest that the Triassic Zacatecas Formation and equivalent strata have undergone several stages of compressive deformation (Table 3). Various authors do not agree about the specific structural style, timing, and tectonic setting of the deformational events. The following is a summary of published descriptions of structures and microstructures in the Zacatecas and overlying Nazas and Zuloaga Formations.

In the vicinity of Zacatecas, Monod and Calvet (1992) described deformation in the Zacatecas Formation as folds, one lineation, and one foliation with northwest-southeast orientations, ascribed to the Late Cretaceous–Tertiary Laramide orogeny. Silva-Romo (1993) measured, at the Sierra de Salinas, a northeast-southwest-trending foliation with a general dip to the southeast, asymmetric folds with axes oriented northwest-southeast, a crenulation lineation oriented north-north-west–south-southeast, a north-northeast–south-southwest lineation, and both northwest-southeast and northeast-southwest-oriented noncoaxial drag folds. Tristán-González and Tórrez-Hernández (1994) reported north-south-oriented, east-northeast-vergent folds, and one northeast-southwest-striking, northwest-dipping foliation in both the Zacatecas and Nazas Formations in the San Rafael area, near Charcas. Centeno-García and Silva-Romo (1997) reported two phases of deformations of the Zacatecas Formation at Sierra de Salinas. They described an older deformation with fold axes oriented northeast, a southwest-northeast foliation, and a second phase related to the Laramide orogeny with axial cleavage and folds with a general northwest trend. Also in the Sierra de Salinas, Bartolini (1998) described folds with north-south axes in the Zacatecas Formation. Because all folding described above does not affect the overlying Oxfordian Zuloaga Limestone, it is pre-Laramide, but because the Nazas Formation is missing, the age of these folds relative to the Nazas is unknown. Our measurements in three outcrops in the Sierra de Salinas show a predominantly southwest-northeast foliation, almost subparallel to bedding, and a second foliation parallel to Laramide fold axes. The Zacatecas Formation is strongly folded and exhibits one foliation and one crenulation, possibly the result of two pre-Laramide events. The northwest-southeast regional foliation and the northwest-southeast folds that affect the Nazas throughout northern and central Mexico are provisionally bracketed by the Middle to Late (Oxfordian) Jurassic, because it affects only rocks as young as Oxfordian. This deformation may be linked to transpressive activity along the arc (Bartolini, 1998), but we lack compelling evidence. Finally, Laramide deformation throughout Mexico (as described by numerous authors) is generally characterized by north-north-west–south-southeast-trending folds and thrust faults with an east-northeastward vergence.

In most of northern and central Mexico, Laramide deformation commonly overprints older deformations. It is imperative to pursue further detailed structural analyses to establish the distinctive structural signature of accretionary, transpressive, and Laramide processes in this area.

TABLE 3. Tentative timing of tectonic events, their structural styles, and formations affected, central Mexico (after Cuevas-Pérez, 1985; Monod and Calvet, 1992; Silva-Romo, 1993; Silva-Romo et al., 1993; Centeno-García et al., 1993; Tristán-González and Tórres-Hernández, 1994; Barboza-Gudiño et al., 1998; Bartolini, 1998).

	FIRST EVENT	SECOND EVENT	THIRD EVENT
REGION AFFECTED	Zacatecas and San Luis Potosí	North-central Mexico	Mexico
FORMATIONS AFFECTED	Zacatecas and Taray Formations	Zacatecas, Taray, and Nazas Formations Base of La Gloria and Zuloaga Formations	Paleozoic-Cretaceous Formations
AGE	Latest Triassic–earliest Jurassic (?) (Prior to the emplacement of the subaerial volcanic arc [Nazas Formation])	Middle-Late Jurassic (?) (During Nazas emplacement and until the deposition of Zuloaga and La Gloria Formations [Early Oxfordian])	Late Cretaceous–Early Tertiary (Coniacian to Eocene)
STRUCTURAL FEATURES	NE-SW-striking foliation with dips to the northwest NE-SW-striking lineation Crenulation foliation	Northwest-striking Foliation NW-SE-oriented folds NW-SE Crenulation foliation	NNW-SSE-trending folds and thrust faults with an east-northeast vergence N-S-oriented foliation with dips to the WSW
TECTONIC PROCESS	Subduction-related obduction and/or accretionary processes (?)	Large-scale NW-trending left-lateral-transpressive continental structures	E-NE low-angle subduction of Farallon oceanic plate foreland fold-thrust belt

TECTONIC SETTING

The Mojave-Sonora Megashear

The Mojave-Sonora megashear (Figure 3) is a proposed left-lateral transform fault responsible for the disruption of the southeastern corner of the North America craton (Silver and Anderson, 1974). According to this hypothesis, a fragment of crust from California and Nevada, which includes Precambrian basement, Paleozoic miogeoclinal and eugeoclinal rocks, and Triassic strata, was displaced as much as 800 km to the southeast during the Middle to Late Jurassic (Silver and Anderson, 1974; Anderson and Silver, 1979). Thus, Triassic marine strata of the Antimonio Formation in northwest Sonora and fluvio-deltaic rocks of the Barranca Group in central Sonora are part of the proposed displaced fragment, also known as the Caborca block.

Lucas and González-León (1994) agree with displacement along the Mojave-Sonora megashear in the Mid-Jurassic and suggest also that the Middle Triassic marine strata exposed in central Sonora are part of the Antimonio terrane, an allochthonous block above the Caborca terrane (Stanley and González-León, 1995). The Antimonio terrane supposedly is not indigenous to Sonora but was once part of Nevada (González-León, 1989; Stanley and González-León, 1995). The allochthonous nature of the Caborca block (which contains the Antimonio terrane) has been questioned by Stewart et al. (1990), Poole

(1993), and Poole et al. (1995), who point out the undisrupted nature of Paleozoic miogeoclinal and eugeoclinal belts in the western United States and Baja California (Gastil et al., 1991). More recently, Molina-Garza and Geissman (1999) carried out detailed paleomagnetic studies in Neoproterozoic to Cretaceous rocks on both sides of the proposed trace of the Mojave-Sonora megashear in Sonora, but paleomagnetic results are inconsistent with the proposed southeastward displacement of the Caborca block along the Mojave-Sonora megashear. In addition, gravity evidence for the existence of continental-scale, Late Jurassic transform faults is inconclusive (Bartolini and Mickus, this volume).

In central Mexico, outcrops of the Zacatecas Formation are older than and lie south of the Mojave-Sonora megashear. Consequently, it is asserted that these rocks too were displaced southeastward about 1000 km at some time during the Middle to Late Jurassic (Silver and Anderson, 1974; Anderson and Silver, 1979). Timing of displacement along the megashear was later proposed to have occurred 150 Ma (Anderson et al., 1982). If 1000 km of left-lateral displacement is restored, the Triassic rocks of the Antimonio Formation in Sonora would be next to the Luning Formation in Nevada (Stanley and González-León, 1995), and the Zacatecas Formation in central Mexico would restore somewhere in Sonora. Biogeographically, this resolves the isolated position of the Zacatecas faunas by placing them adjacent to the eastern Pacific region. On the other hand, if Sonora is indigenous to northwestern Mexico, a transpressive structure south of the Mojave-Sonora megashear would be required to displace the Zacatecas Formation from Sonora to its present position, 1000 km southeast of the megashear in central Mexico. This sort of speculation has resulted in a proliferation of numerous scenarios involving tectono-stratigraphic terranes to explain the presence of Triassic marine rocks in central Mexico.

Tectono-stratigraphic Terranes

In central Mexico, Triassic deep-marine successions of the Zacatecas Formation in the states of Zacatecas and San Luís Potosí are considered fragments of upper levels of oceanic crust that have been accreted onto cratonic Mexico during the Campanian-Eocene Laramide orogeny (Campa and Coney, 1983). More recently, accretion of the same terranes has been assigned a Middle to Late Jurassic age (Centeno-García et al., 1993; Sedlock et al., 1993; Silva-Romo, 1993; Silva-Romo et al., 1993; Centeno-García, 1994; Centeno-García and Silva-Romo, 1997).

Originally, Triassic rocks of the Zacatecas Formation west of the city of Zacatecas were considered part of the Guerrero terrane, and the Zacatecas strata in Sierra de Salinas and Charcas were considered part of the Sierra Madre terrane (Figure 4; Campa and Coney, 1983). More recently, Sedlock et al. (1993)

revised the terrane nomenclature and proposed 17 tectono-stratigraphic terranes in Mexico and Central America. The Tepehuano terrane, which includes rocks of the Zacatecas and Taray Formations, encompasses the Parral and Sombrerete subterranes, as well as parts of the Sierra Madre, Guerrero, and Cortes terranes. The Tepehuano terrane is bounded on the north by the Mojave-Sonora megashear and includes all outcrops of the Zacatecas Formation in Zacatecas and San Luís Potosí.

Centeno-García (1994), on the other hand, subdivided the Guerrero terrane into three subterranes: the Zihuatanejo-Huetamo, the Teloloapan, and the Zacatecas. The Guerrero terrane encompasses the Triassic Zacatecas Formation west of Zacatecas city, and the Sierra Madre terrane comprises Triassic strata of La Ballena Formation (Silva-Romo, 1993) in the Sierra de Salinas and Charcas. We concur with Monod and Calvet (1992), who assign an early Cretaceous age for pillow lava and limestone in the vicinity of Zacatecas city, which may be correlative with the Chilitos Formation (de Cserna, 1976). We maintain the name Zacatecas Formation for Triassic turbidites in all three areas—Zacatecas city, Sierra de Salinas and Charcas—and possibly in Sierra de Catorce (Barboza-Gudiño, 1993). But there is no reason to use the name La Ballena Formation (Silva-Romo, 1993) or El Ahogado and El Bote Formations (Monod and Calvet, 1992) for rocks that formed in the same

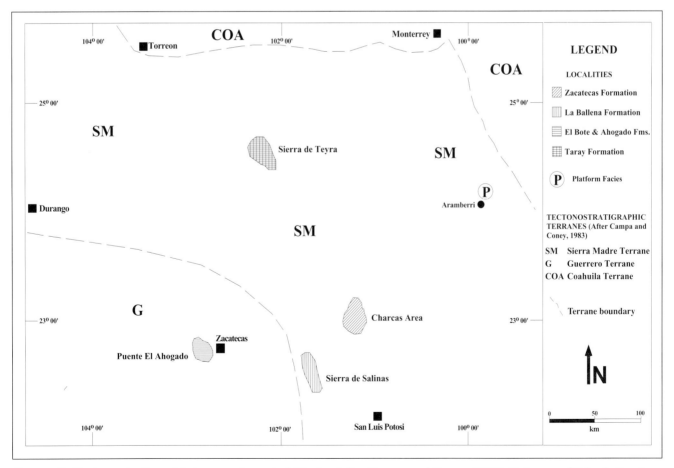

FIGURE 4. Tectono-stratigraphic terranes in central Mexico (after Campa and Coney, 1983). Outcrops of Triassic Zacatecas Formation in Zacatecas and San Luis Potosí are also shown.

basin, are of the same age, exhibit identical lithologies and fauna, show similar environments of deposition, and are compositionally similar.

According to Centeno-García (1994), the Arteaga complex, yet another terrane, is an ocean-floor assemblage exposed near the Pacific coast of Michoacán state, which constitutes most of the basement of the Guerrero terrane. The complex is made up of several formations and unnamed lithologic units that consist of clastic rocks (shale and sandstone), volcanic flows, pillow lavas, metaintrusives (gabbro-diorite-granite), dikes, chert, metasediments, and erratic blocks of limestone and chert. The entire complex is assigned a tentative Late Triassic–Early Jurassic age (Centeno-García, 1994).

As far as we are concerned, the age of the Arteaga complex is unknown. A Late Triassic (Ladinian-Carnian) age, based on unidentified radiolarians, was reported in an abstract (Campa et al., 1982), but no taxonomic, collection locality, or stratigraphic documentation was ever published. K-Ar ages from metamorphic rocks of the complex (Grajales-Nishimura and López-Infanzón, 1983) show three groups of ages: Early to Middle Jurassic, middle Cretaceous, and Eocene. These ages apparently represent thermal/metamorphic events, not the original age of the rocks. Thus, the age of the Arteaga complex remains undocumented. Conversely, the Triassic age of the Zacatecas Formation is firmly established at all localities in central Mexico (Burckhardt and Scalia, 1905; Burckhardt, 1930; Cantú-Chapa, 1969; Gallo et al., 1993). It appears unacceptable to link the origin and evolution of the Triassic Zacatecas Formation with rocks of the Arteaga complex, whose age and geologic setting are unknown.

Based on data provided in our lithostratigraphic and biostratigraphic documentation of the Zacatecas Formation, we believe that these strata were deposited adjacent to the craton. The Arteaga complex was deposited in a basin floored by oceanic crust, according to Centeno-García (1994). Because of shortening during Middle Jurassic compression plus additional shortening during the Late Cretaceous–Tertiary Laramide orogeny (40%; Suter, 1987), the original distance between the Arteaga Basin and the Zacatecas Basin must have been at least 1000 km. The possibility of the same source areas for both the Arteaga complex and Zacatecas Formation sediments was addressed by Centeno-García (1994) and Centeno-García et al. (1993), who compared sediment composition, Sm-Nd-isotopic ratios, and trace-element concentrations between the Arteaga complex in Michoacán and the Zacatecas Formation in central Mexico. They concluded that because these regions are very far from each other, their original paleogeographic relationship is uncertain.

According to Centeno-García and Silva-Romo (1997), the Arteaga complex was accreted onto nuclear Mexico at some time during the Middle to Late Jurassic. We believe that this timing of accretion is inconsistent with the unconformable stratigraphic position of the Early to Middle Jurassic Nazas Formation above the Late Triassic Zacatecas Formation every-

where in Zacatecas and San Luís Potosí (Silva-Romo, 1993; Jones et al., 1995; Barboza-Gudiño et al., 1997; Bartolini, 1998). Our field data indicate that the tectonic emplacement of the Zacatecas Formation occurred prior to the deposition of the Early to Middle Jurassic Nazas Formation, probably at some time in the latest Triassic to earliest Jurassic. The unconformable position of the Oxfordian Zuloaga Limestone on both the Zacatecas and Nazas Formations precludes accretion of the Zacatecas Formation after Late Jurassic (Oxfordian). In attempting to date the tectonic emplacement of the Zacatecas Formation as Middle to Late Jurassic, Centeno-García and Silva-Romo (1997) stated that the age of the Nazas Formation (volcanic arc) is Late Jurassic, based on only one 158 Ma U-Pb date from the Caopas Formation in northern Zacatecas (Jones et al., 1995). But this age is questionable, because previous studies of the Nazas Formation in northern and central Mexico have documented the age of the Cordilleran magmatic arc, of which the Nazas is part, with Rb-Sr, Ar-Ar, and K-Ar data. All of these data consistently show a Late Triassic to Middle Jurassic age (Fries and Rincón-Orta, 1965; Denison et al., 1971; Pantoja-Alor, 1972; Halpern et al., 1974; Damon et al., 1984; López-Infanzón, 1986; Grajales et al., 1992; Bartolini and Spell, 1997; Bartolini, 1998). Field evidence is also consistent with these abundant isotopic ages, showing that by Late Jurassic time, the volcanic arc was already inactive, as proved by the presence of overlying Oxfordian marine rocks of the Zuloaga Limestone and the La Gloria Formation throughout northern and central Mexico (Bartolini, 1998).

The exact timing of the several structural events that affect Zacatecas strata is a complex issue. Stratigraphic, paleontologic, and geochronologic information, as well as field relations, indicates that the Zacatecas Formation was deformed during three compressive tectonic events: (1) during the latest Triassic–earliest Jurassic (?), when northeast-southwest-trending structures were formed; (2) during the Middle–Late Jurassic (Oxfordian), when northwest-southeast-trending structures were formed; and (3) during the Late Cretaceous–Early Tertiary (Coniacian-Eocene), when another set of northwest-southeast-trending structures were formed. We find it unacceptable to use Jurassic K-Ar dates in the Arteaga complex on the Pacific coast of Michoacán (Grajales-Nishimura and López-Infanzón, 1983) to explain the structural evolution of the Triassic Zacatecas Formation in central Mexico.

If the Triassic western margin of Pangea was a passive continental margin, we believe that an associated platform, a continental slope and rise, and a basin floor must have existed. The tectonic juxtaposition of the Zacatecas deep-marine facies upon coeval platform facies is consistent with this idea and is similar to the tectonic juxtaposition of Paleozoic basinal facies upon contemporaneous platform facies in Nevada (Roberts, 1951). Unfortunately, the only area where shallow-marine successions of possible Triassic (?) age occur is in the state of Nuevo León, but there are no outcrops of Zacatecas Formation in that area. In any case, our conceptual hypothesis differs from that of Centeno-García and Silva-Romo (1997), who assume the tectonic juxtaposition of Zacatecas deep-marine facies directly on the craton. Neither of these two alternatives can be proved because no pre-Triassic rocks have been found exposed in any of the areas where the Zacatecas crops out in central Mexico.

There are other problems with the proposed east-northeastward direction of Zacatecas obduction, because the general trend of Zacatecas structures and structural fabrics is northeast-southwest. What was the mechanism of exhumation?

Clearly, the tectono-depositional framework and structural evolution of the Zacatecas Formation require more geologic data from central Mexico. A reliable explanation of the geologic history of the Zacatecas and San Luís Potosí regions is not possible using information from the Arteaga complex in Michoacán state, especially because the tectonic setting of the Arteaga complex itself is highly speculative. No faunas have been identified in the complex, and the age or ages of its informal formations is not documented.

CONCLUSIONS

Preliminary Assessment of Tectono-stratigraphic Evolution of the Zacatecas Formation

The following events (Figure 5) are speculative, because of the isolated and incomplete nature of the exposures, lack of pre-Triassic outcrops, polyphase structural deformation, and ubiquitous regional metamorphism:

1) deposition during the Middle to Late Triassic of the deep-marine strata of the Zacatecas Formation on slope, toe-of-slope, and basin-plain sites along the western Pangea passive continental margin (presently in the states of Zacatecas and San Luís Potosí). Turbidity currents deposited submarine fans where classical turbidite sequences are well represented. Predominantly clastic sedimentary piles, especially sandstone, shale, conglomerate, and rare limestone, were deposited, but tuffaceous layers reported from the stratigraphic section west of the city of Zacatecas (Monod and Calvet, 1992) record periods with low rates of turbidite deposition and deposition of airfall deposits from volcanic centers.

2) northeast-eastward tectonic emplacement of the Zacatecas Formation clastic marine strata onto the platform facies at some time during the latest Triassic or earliest Jurassic, producing complex structural deformation of the Zacatecas Formation at Charcas (Tristán-González and Tórres-Hernández, 1992; Tristán-González and Tórres-Hernández, 1994; Barboza-Gudiño, et al., 1998; Bartolini, 1998) and at Sierra de Salinas (Silva-Romo et al., 1993; Centeno-García, 1994). The tectonic transport of the Zacatecas Formation was possibly triggered by the development, along western North America, of a convergent continental margin, perhaps with a low-angle subducting slab, as early as the Late Triassic.

3) construction of an Early to Middle Jurassic (?) subaerial volcanic arc (Nazas Formation) above the allochthonous Zacatecas Formation, which was exposed at the surface. The age of this continental magmatic arc is constrained by the lower contact of the Nazas Formation with the Middle to Late Triassic Zacatecas Formation and by the upper contact of the Nazas Formation with the Late Jurassic (Oxfordian) Zuloaga Limestone. Thus, the continental volcanic arc in the states of Zacatecas and San Luís Potosí spanned the Early and Middle Jurassic. The age of this continental arc farther north in the western United States ranges from Early Triassic to Late Jurassic (Armstrong and Suppe, 1973; Miller, 1978; Palmer, 1983; Dilles and Wright 1988; Tosdal et al., 1989; Asmeron et al., 1990; Lipman, 1992). The age of the magmatic arc in northwestern Mexico ranges from Late Triassic to Late Jurassic (Damon et al., 1991), and in north-central Mexico, the arc ranges from Late Triassic to Middle Jurassic (Fries et al., 1965; Denison et al., 1969; Denison et al, 1971; Pantoja-Alor, 1972; Halpern et al., 1974; Damon et al., 1984; López-Infanzón, 1986; Jones et al., 1990, 1995; Anderson et al., 1991; Grajales et al., 1992; Bartolini and Spell, 1997; McKee et al., 1997; Barboza-Gudiño et al., 1998; Bartolini, 1998).

4) A second deformation event that affected both the Zacatecas and the Nazas Formations began at some time in the Middle Jurassic and apparently ended in the earliest Oxfordian, as suggested by the presence of foliation at the base of the Oxfordian La Gloria Formation in Durango and in the Zuloaga Limestone in San Luís Potosí (Bartolini, 1998). This foliation may possibly be related to transpressive activity during the Middle to Late Jurassic (Silver and Anderson, 1974; Anderson and Silver, 1979; Longoria, 1985). This relationship is inconclusive; the existence of transpressive structures still is debated.

5) an increased rate of erosion in the arc at some time during the Jurassic, as suggested by the presence of thick clastic deposits in upper stratigraphic levels in the Nazas arc sections. The exact timing of continental clastic sedimentation is unknown, but it occurred prior to the marine transgression in the Oxfordian.

6) early Oxfordian marine incursion in central Mexico, promoted by loss of relief in the magmatic arc caused by intraarc extension (Bartolini, 1998) and by rifting along the Gulf of Mexico region (Salvador, 1991a, b; Marton and Buffler, 1994). Marine limestone and evaporite deposits of the Zuloaga Formation were deposited on both the Zacatecas and Nazas Formations. By the Late Jurassic (Early Oxfordian), volcanism in the continental Nazas arc must have ceased (Bartolini, 1998).

(7) A third tectonic event coincides with the Laramide orogeny, which began in Coniacian time (de Cserna, 1989; Lang and Frerichs, 1998). This regional episode of compressive deformation is responsible for the evolution of the Mesozoic fold-and-thrust belt (de Cserna, 1956; Padilla y Sánchez, 1982; Suter, 1987). Structures characteristic of this event are northwest-trending folds and thrust faults with east-northeast vergence.

ACKNOWLEDGMENTS

We acknowledge NASA and Exxon Exploration Company for providing funds to Claudio Bartolini. Servicios de Explo-

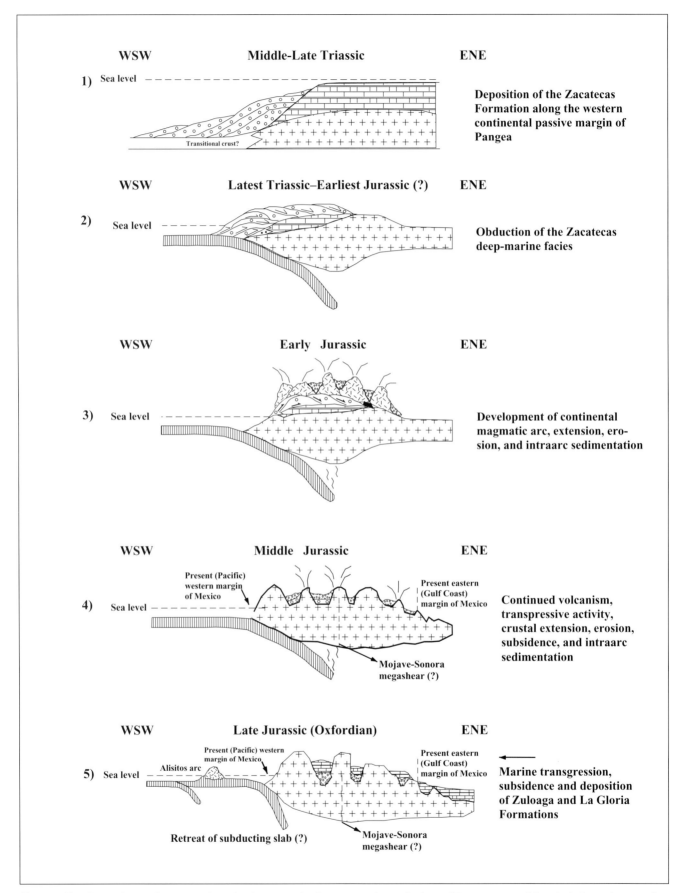

FIGURE 5. Chronology of proposed geologic events in the Mesozoic evolution of Zacatecas and San Luis Potosí, central Mexico.

ración Peñoles, Torreón office, supported Bartolini's fieldwork in central Mexico. Comisión Federal de Electricidad allowed the authors to consult materials from its library. Rafael Barboza's field studies were funded by Consejo Nacional de Ciencia y Tecnología. The authors are grateful to Chris Johnson and Nimio Tristan for their constructive review of the manuscript.

REFERENCES CITED

Alencaster-de Cserna, G., 1961a, Estratigrafía del Triásico Superior de la parte central del Estado de Sonora: Parte 1, Paleontología del Triásico Superior de Sonora: Instituto de Geología, Universidad Nacional Autonóma de México, Paleontología Mexicana II, parte 2, 18 p.

Alencaster-de Cserna, G., 1961b, Fauna fósil de la Formación Santa Clara (Cárnico) del Estado de Sonora: Parte III, Paleontología del Triásico Superior de Sonora: Instituto de Geología, Universidad Nacional Autonóma de México, Paleontología Mexicana II, parte 2, 38 p.

Anderson, T. H., J. W. McKee, and N. W. Jones, 1990, Jurassic (?) melange in north central Mexico: Geological Society of America Abstracts with Programs, v. 22, no. 3, p. 3.

Anderson, T. H., J. W. McKee, and N. W. Jones, 1991, A northwest-trending Jurassic old nappe, northernmost Zacatecas, Mexico: Tectonics, v. 10, p. 383–401.

Anderson, T. H., and L. T. Silver, 1979, The role of the Mojave-Sonora megashear in the tectonic evolution of northern Sonora, Mexico, *in* T. H. Anderson and J. Roldán-Quintana, eds., Geology of northern Sonora: Geological Society of America Annual Meeting Guidebook, p. 1–22.

Anderson, T. H., V. A. Schmidt, K. K. Cohen, and L. T. Silver, 1982, Timing of activity along the Mojave-Sonora megashear in Mexico (abs.): EOS, American Geophysical Union Transactions, v. 63, p. 915–916.

Arkell, W. J., W. M. Furnish, B. Kummel, A. K. Miller, R. C. Moore, O. H. Schindewolf, P. C. Sylvester-Bradley, and C. W. Wright, 1957, Treatise on invertebrate paleontology, pt. K, *in* R. C. Moore, ed., Mollusca 4, Cephalopoda Ammonoidea: Kansas City, Geological Society of America and University of Kansas Press, 490 p.

Armstrong, R. L., and Suppe, J., 1973, Potassium-Argon geochronometry of Mesozoic igneous rocks in Nevada, Utah, and southern California: Geological Society of America Bulletin, v. 84, p. 1375–1392.

Asmeron, Y., R. E. Zartman, P. E. Damon, and M. Shafiqullah, 1990, Zircon U-Th-Pb and whole-rock Rb-Sr age patterns of lower Mesozoic igneous rocks in the Santa Rita Mountains, southeastern Arizona: Implications for Mesozoic magmatism in the southern Cordillera: Geological Society of America Bulletin, v. 102, p. 961–968.

Barboza-Gudiño, J. R., 1993, Geología de la Sierra de Catorce, San Luís Potosí, Mexico: Actas, Facultad de Ciencias de la Tierra, Universidad Autónoma de Nuevo León, p. 9–18.

Barboza-Gudiño, J. R., J. R. Tórrez-Hernández, and M. Tristán-González, 1997, Some pre-Oxfordian red beds and related stratigraphic units in the southern and northeastern Central Plateau, Mexico: Geological Society of America Abstracts with Programs, v. 29, no. 2, p. 2.

Barboza-Gudiño, J. R., M. Tristán-González, and J. R. Tórrez-Hernández, 1998, The Late Triassic–Early Jurassic active continental margin of western North America in northeastern Mexico: Geofísica International, v. 37, no. 4, p. 283–292.

Bartolini, C., 1998, Stratigraphy, geochemistry, geochronology and tectonic setting of the Mesozoic Nazas Formation, north-central Mexico: Ph.D. dissertation, University of Texas at El Paso, 557 p.

Bartolini, C., and T. Spell, 1997, An Early Jurassic age (Ar-Ar) for the Nazas Formation at the Cañada Villa Juarez, northeastern Durango, Mexico: Geological Society of America Abstracts with Programs, v. 29, no. 2, p. 3.

Brönnimann, P., 1972, Remarks on the classification of fossil Anomuran coprolites: Paläontologische Zeitschrift, v. 46, p. 99–103.

Buch, I. P., and M. P. Delattre, 1993, Permian and Lower Triassic stratigraphy along the 30th parallel, eastern Baja California Norte, Mexico, *in* R. G. Gastil and R. H. Miller, eds., The prebatholithic stratigraphy of Peninsular California: Geological Society of America Special Paper 279, v. 279, p. 77–90.

Burckhardt, C., 1930, Synthétique sur le Mésozoique méxicain: Society of Paleontology Suisse Memoir, v. 49–50, p. 1–280.

Burckhardt, C., and S. Scalia, 1905, La faune marine du Trias supérieur de Zacatecas, Méxique: Boletín del Instituto de Geología, Universidad Nacional Autonóma de Mexico, v. 21, 44 p.

Burckhardt, C., and S. Scalia, 1906, Géologie des environs de Zacatecas: 10th International Geological Congress, Mexico, excursion guidebook, v. 16, 26 p.

Butler, J., 1972, Geology of the Charcas mineral district, San Luís Potosí: Master's thesis, Colorado School of Mines, Golden, 135 p.

Campa, M. F., and P. Coney, 1983, Tectono-stratigraphic terranes and mineral resource distributions in Mexico: Canadian Journal of Earth Sciences, v. 20, p. 1040–1051.

Campa, M. F., J. Ramirez, and C. Bloome, 1982, La secuencia volcanico-sedimentaria metamorfizada del Triásico (Ladiniano-Cárnico) de la región de Tumbiscatio, Michoacán: Sociedad Geológica Mexicana, 6.° Convención Nacional, Resúmenes, p. 48.

Cantú-Chapa, A., 1969, Una nueva localidad del Triásico Superior marino en Mexico: Instituto Mexicano del Petróleo, Revista, v. 1, p. 71–72.

Cantú-Chapa, C. M., 1974, Una nueva localidad del Cretácico Inferior en México: Revista del Instituto Mexicano del Petróleo, v. 6, no. 4, p. 51–54.

Cantú-Chapa, A., 1994, Mexico, márgen occidental de la Pangea según evidencias biogeográficas del Pérmico al Jurásico Inferior: Revista Mexicana del Petróleo, v. 36, no. 345, p. 28–35.

Carrillo-Bravo, J., 1971, La Plataforma Valles–San Luís Potosí: Boletín de la Asociación Mexicana de Geólogos Petroleros, v. 23, p. 1–102.

Centeno-García, E., 1994, Tectonic evolution of the Guerrero Terrane, western Mexico: Ph.D. dissertation, University of Arizona, Tucson, 220 p.

Centeno-García, E., P. J. Coney, J. Ruíz, J. Patchett, and F. Ortega-Gutiérrez, 1993, Tectonic significance of the sediments of the Guerrero Terrane from petrographic, trace element, and Nd-isotopic studies, *in* F. Ortega-Gutiérrez, P. Coney, E. Centeno-García, and A. Gómez-Caballero, eds., A First Circum-Pacific and Circum-Atlantic Terrane Conference: Instituto de Geología, Universidad Nacional Autónoma de Mexico, p. 30–33.

Centeno-García, E., and G. Silva-Romo, 1997, Petrogenesis and tectonic evolution of central Mexico during Triassic-Jurassic time: Revista Mexicana de Ciencias Geológicas, Universidad Nacional Autónoma de Mexico, Instituto de Geología, v. 14, no. 2, p. 244–260.

Chávez-Aguirre, R., 1968, Bósquejo geológico de la Sierra Peñon Blanco, Zacatecas: Tesis Profesional, Facultad de Ingeniería, Universidad Nacional Autónoma de México, 67 p.

Cojan, I., and P. E. Potter, 1991, Depositional environment, petrology, and provenance of the Santa Clara Formation, Upper Triassic Barranca Group, eastern Sonora, Mexico, in E. Pérez-Segura and C. Jacques-Ayala, eds., Studies of Sonoran geology: Geological Society of America Special Paper 254, p. 37–50.

Córdoba, D. A., 1964, Geology of the Apizolaya Quadrangle (east half), northern Zacatecas, Mexico: Master's thesis, University of Texas at Austin, 111 p.

Cuevas-Pérez, E., 1985, Geologie des Alteren Mesozoikums in Zacatecas und San Luís Potosi, Mexiko: Ph.D. dissertation, Hessen Marbur/Lahn, Germany, 182 p.

Cuevas-Pérez, E., 1997, Late Triassic (Carnian-Norian) paleogeography of central Mexico, in C. M. González-León and G. D. Stanley Jr., eds., Publicaciones Ocasionales No. 1. Estación Regional del Noroeste, Instituto de Geología Universidad Nacional Autónoma de Mexico, U.S.-Mexico Cooperative Research: International Workshop on the Geology of Sonora Memoir, p. 13–16.

Damborenea, S. E., and C. M. González-León, 1997, Late Triassic and Early Jurassic bivalves from Sonora, Mexico: Revista Mexicana de Ciencias Geológicas, Universidad Nacional Autónoma de Mexico, Instituto de Geología, v. 14, no. 2, p. 178–201.

Damon, P. E., M. Shafiqullah, K. DeJong, and J. Roldan-Quintana, 1991, Chronology of Mesozoic magmatism in Sonora and northwest Mexico: Geological Society of America Abstracts with Programs, v. 23, no. 5, p. 127.

Damon, P. E., M. Shafiqullah, and J. Roldan-Quintana, 1984, The Cordilleran Jurassic arc from Chiapas (southern Mexico) to Arizona: Geological Society of America Abstracts with Programs, v. 17, p. 140.

Dávila-Alcocer, V. M., 1981, Radiolarios del Cretácico Inferior de la Formación Plateros del Distrito Minero de Fresnillo, Zacatecas: Universidad Nacional Autónoma de México, Instituto de Geología, v. 5, p. 119–120.

de Cserna, Z., 1956, Tectónica de la Sierra Madre Oriental de Mexico entre Torreón and Monterrey: Contribución del Instituto Nacional para la investigación de recursos minerales de México: 20.° Congreso Geológico Internacional, 87 p.

de Cserna, Z., 1976, Geology of the Fresnillo area, Zacatecas, Mexico: Geological Society of America Bulletin, v. 87, p. 1191–1199.

de Cserna, Z., 1989, An outline of the geology of Mexico, in A. W. Bally and A. R. Palmer, eds., The Geology of North America—An overview, v. A: Geological Society of America, p. 233–264.

DeLeón-Gómez, H., 1988, Geologische Kartierung (1:10,000) des Gebietes El Porvenir am Ostrand des Aramberri Uplifts, Sierra Madre Oriental, Mexico: Diplomakartierung, Fachbereich Geowissenschaften der Technischen Universitat Clausthal, 64 s.

Denison, R. E., W. H. Burke, E. A. Hetherington, and J. B. Otto, 1971, Basement rocks framework of parts of Texas, southern New Mexico, and northern Mexico, in K. O. Seeward and D. A. Sundeen, eds., The geologic framework of the Chihuahua tectonic belt: West Texas Geological Society Symposium, p. 3–14.

Denison, R. E., G. S. Kenny, W. H. Burke, and E. A. Hetherington, 1969, Isotopic ages of igneous and metamorphic boulders from the Haymond Formation (Pennsylvanian), Marathon Basin, Texas, and their significance: Geological Society of America Bulletin, v. 80, p. 245–256.

Dickinson, W. R., 1985, Interpreting provenance relations from detrital modes of sandstones, in G. G. Zuffa, ed., Provenance of Arenites: Dorodrecht, Netherlands, Reidel Publishing Company, p. 333–361.

Dilles, J. H., and J. E. Wright, 1988, The chronology of early Mesozoic arc magmatism in the Yerington district of western Nevada and its regional implications: Geological Society of America Bulletin, v. 100, p. 644–652.

Dumble, E. T., 1900a, Triassic coal and coke of Sonora, Mexico: Geological Society of America Bulletin, v. 11, p. 10–14.

Dumble, E. T., 1900b, Notes on the geology of Sonora, Mexico: American Institute of Mining Engineers, Transactions, v. 29, p. 122–152.

Dunham, R. J., 1962, Classification of carbonate rocks according to depositional texture, in W. E. Ham, ed., Classification of carbonate rocks: AAPG Memoir 1, p. 108–121.

Estep, J. W., S. G. Lucas, and C. González-León, 1997a, Middle Triassic ammonites from Sonora, Mexico: Revista Mexicana de Ciencias Geológicas, Universidad Nacional Autónoma de Mexico, Instituto de Geología, v. 14, no. 2, p. 155–159.

Estep, J. W., S. G. Lucas, and C. González-León, 1997b, Late Triassic (Late Carnian) ammonoids at El Antimonio, Sonora, Mexico, in C. M. González-León and G. D. Stanley Jr., eds., Publicaciones Ocasionales No. 1. Estación Regional del Noroeste, Instituto de Geología Universidad Nacional Autónoma de Mexico, U.S.-Mexico Cooperative Research: International Workshop on the Geology of Sonora Memoir, p. 16–18.

Finch, J. W., and P. J. Abbot, 1977, Petrology of a Triassic marine section, Vizcaino Peninsula, Baja California Sur, Mexico: Sedimentary Geology, v. 19, p. 253–273.

Flügel, E., 1982, Microfacies analysis of limestones: Heidelberg, Springer-Verlag, 633 p.

Folk, R. L., 1959, Practical petrographic classification of limestones: AAPG Bulletin, v. 43, p. 1–38.

Folk, R. L., 1962, Spectral subdivision of limestone types, in W. E. Ham, ed., Classification of carbonate rocks: AAPG Memoir 1, p. 62–84,

Folk, R. L., 1980, Petrology of sedimentary rocks: Austin, Texas, Hemphill Publishing Company, 182 p.

Förster, R., and A. von Hillebrandt, 1984, Das Kimmeridge des Profeta-Jura in Nordchile mit einer Mecochirus-Favreina-Vergesellschaftung (Crustacea, Decapoda-Ichnogenus): Mitteilungen Bayerische Staatssammlung Paläontologie und historische Geologie, v. 24, p. 67–84.

Fries, J. C., and C. Rincon-Orta, 1965, Nuevas aportaciones geocronologicas y tecnicas empleadas en el laboratorio de geocronometria: Revista del Instituto de Geologia, Universidad Nacional Autónoma de México, v. 73, part 2, p. 57–133

Gallo, P. I, L. Gómez, M. E. Contreras, and P. E. Cedillo, 1993, Hallazgos paleontológicos del Triásico marino en la región central de México: Revista de la Sociedad Mexicana de Paleontología, v. 6, no. 1, p. 1–9.

Gastil, R. G., et al., 1991, The relation between the Paleozoic strata on opposite sides of the Gulf of California, in E. Pérez-Segura and C. Jacques-Ayala, eds., Studies of Sonoran geology: Geological Society of America Special Paper 254, p. 7–18.

Gastil, R. G., R. H. Miller, B. R. Wardlaw, N. J. Silberling, and D. V. Le Mone, 1981, Lower Triassic strata near El Volcán, Baja California, Mexico: Geological Society of America Abstracts with Programs, v. 13, p. 57.

Gómez-Luna, M. E., P. E. Cedillo, M. B. Contreras, P. I. Gallo, and C. A. Martínez, C.A., 1997, El Triásico marino en la meseta central de Mexico: Implicaciones paleogeográficas: Segunda convención sobre la evolución geológica de México y recursos asociados, Simposia y Coloquio, extended abstracts, p. 67–71.

González-León, C., 1980, La Formación Antimonio (Triásico Superior–Jurásico Inferior) en la Sierra del Alamo, Estado de Sonora: Universidad Nacional Autónoma de Mexico, Revista del Instituto de Geología, v. 4, p. 13–18.

González-León, C., 1989, Evolución de terrenos Mesozoicos en el noroeste de Mexico: Universidad de Sonora, Boletín del Departamento de Geología, v. 6, no. 1–2, p. 39–54.

González-León, C., 1997a, The Triassic-Jurassic boundary in the Antimonio Formation from new evidences, in C. M. González-León and G. D. Stanley Jr., eds., Publicaciones Ocasionales No. 1. Estación Regional del Noroeste, Instituto de Geología Universidad Nacional Autónoma de México, U.S.-Mexico Cooperative Research: International Workshop on the Geology of Sonora Memoir, p. 33–36.

González-León, C., 1997b, Sequence stratigraphy and paleogeographic setting of the Antimonio Formation (Late Permian–Early Jurassic), Sonora, Mexico: Revista Mexicana de Ciencias Geológicas, Universidad Nacional Autónoma de México, Instituto de Geología, v. 14, no. 2, p. 136–148.

Grajales, J. M., D. J. Terrell, and P. E. Damon, 1992, Evidencias de la prolongación del árco magmático Cordillerano del Triásico Tardío-Jurásico en Chihuahua, Durango y Coahuila: Boletín de La Asociación Mexicana de Geólogos Petroleros, v. 42, no. 2, p. 1–18.

Grajales-Nishimura, J. M., and M. López-Infanzón, 1983, Estudio petrogenético de las rocas ígneas y metamórficas en el prospecto Tomatlán-Guerrero-Jalisco: Instituto Mexicano del Petróleo, Internal Report, Proyecto C-1160, 69 p.

Halpern, M., J. G. Guerrero, and C. M. Ruiz, 1974, Rb-Sr dates of igneous and metamorphic rocks from southeastern and central Mexico: A progress report: Resumenes, Reunion Anual de la Union Geofisica Mexicana, p. 30

Jones, D. L., M. C. Blake Jr., and C. Rangin, 1976, The four Jurassic belts of northern California and their significance to the geology of the southern California borderland, in Aspects of the geologic history of the California continental borderland: Pacific Section, AAPG Miscellaneous Publication 24, p. 343–362.

Jones, N. W., J. W. McKee, T. H. Anderson, and L. T. Silver, 1990, Nazas Formation: A remnant of the Jurassic arc of western North America in north-central Mexico: Geological Society of America Abstract with Programs, v. 22, no. 7, p. A327.

Jones, N. W., J. W. McKee, T. H. Anderson, and L. T. Silver, 1995, Jurassic volcanic rocks in northeastern Mexico: A possible remnant of a Cordilleran magmatic arc, in C. Jacques-Ayala, C. González-León, and J. Roldán-Quintana, eds., Studies on the Mesozoic of Sonora and adjacent areas: Geological Society of America Special Paper 301, p. 179–190.

Kay, M., 1951, North American geosynclines: Geological Society of America Memoir 48, 143 p.

Keller, W. T., 1928, Stratigraphische Beobachtungen in Sonora (nordwest Mexico): Ecologae Helvetiae, v. 21, p. 327–335.

King, E. R., 1939, Geologic reconnaissance in northern Sierra Madre Occidental de Mexico: Geological Society of America Bulletin, v. 50, p. 1625–1722.

Labarthe-Hernández, G., M. Tristán-González, and A. Aguillón-Reyes, 1982, Estudio geológico minero del área de Peñon Blanco, Estados de San Luís Potosí y Zacatecas: Universidad Autónoma de San Luís Potosí, Instituto de Geología y Metalurgía, Folleto Técnico 76, 63 p.

Lang, H. R., and W. E. Frerichs, 1998, New planktic foraminiferal data documenting Coniacian age for Laramide orogeny onset and paleo-oceanography in southern Mexico: The Journal of Geology, v. 106, p. 635–640.

Lipman, P. W., 1992, Magmatism in the Cordilleran United States; progress and problems, in B. C. Burchfiel, P. W. Lipman, and M. L. Zoback, eds., The cordilleran orogen: Conterminous U.S.: Geological Society of America, The Geology of North America, v. G-3, p. 481–514.

Longoria, J. F., 1985, Tectonic transpression in the Sierra Madre Oriental, northeastern Mexico: An alternative model: Geology, v. 13, p. 453–456.

López-Infanzón, M., 1986, Estudio petrogenético de las rocas ígneas de las Formaciones Huizachal y Nazas: Boletín de la Sociedad Geológica Mexicana, Tomo 47, v. 2, p. 1–32.

López-Ramos, E., 1980, Geología de Mexico, Tomo I, 2da Edición: Public Education Ministry, Mexico City, Mexico, 454 p.

Lucas, S. G., J. W. Estep, C. González-León, R. K. Paull, N. J. Silberling, M. B. Steiner, and J. E. Marzlof, 1997, Early Triassic ammonites and conodonts from Sonora, northwestern Mexico: Neues Jahrbuch für Geologie und Paläontologie-Abhandlungen., v. 9, p. 562–574.

Lucas, S. G., and C. González-León, 1994, Marine Upper Triassic strata at Sierra La Flojera, Sonora, Mexico: Neues Jarbruch für Geologie und Paläontologie Mh., v. 1, p. 34–40.

Lucas, S. G., B. S. Kues, J. W. Estep, and C. González-León, 1997, Permian-Triassic boundary at El Antimonio, Sonora, Mexico: Revista Mexicana de Ciencias Geológicas, Universidad Nacional Autónoma de México, Instituto de Geología, v. 14, no. 2, p. 149–154.

Maldonado-Koerdell, M., 1948, Nuevos datos geológicos y paleontológicos sobre el Triásico de Zacatecas: Análes de la Escuela Nacional de Ciencias Geológicas, v. 5, p. 291–306.

Martínez, P. J., 1972, Exploración geológica del área El Estribo–San Francisco, San Luís Potosí: Boletín de la Asociación Mexicana de Geólogos Petroleros, v. 24, nos. 7–9, p. 327–402.

Marton, G., and R. T. Buffler, 1994, Jurassic reconstruction of the Gulf of Mexico Basin: International Geology Review, v. 36, p. 545–586.

McGehee, R. V., 1976, Las rocas metamórficas del Arroyo La Pimienta, Zacatecas, Zacatecas: Boletín de la Sociedad Geológica Mexicana, v. 37, p. 1–10.

McKee, J. W., N. W. Jones, and T. H. Anderson, 1997, Is "Nazas" a junior synonym of "Huizachal," and does it matter?: Geological Society of America Abstracts with Programs, v. 29, no. 2, p. 39.

McRoberts, C. A., 1997, Late Triassic (Norian-Rhaetian) bivalves from the Antimonio Formation, northwestern Sonora, Mexico: Revista Mexicana de Ciencias Geológicas, Universidad Nacional Autónoma de México, Instituto de Geología, v. 14, no. 2, p. 167–177.

Meiburg, P., J. R. Chapa-Guerrero, I. Grotehusmann, T. Kustusch, P. Lentzy, H. DeLeón-Gómez, and M. A. Mansilla-Terán, 1987, El basámento pre-Cretácico de Aramberri-estructura clave para comprender el decollement de la cubierta Jurásica-Cretácica de la Sierra Madre Oriental, México: Actas, Facultad de Ciencias de la Tierra, Universidad Autónoma de Nuevo León, v. 2, p. 15–22.

Miller, C. F., 1978, An Early Mesozoic alkalic magmatic belt in western North America, in D. G. Howell and K. A. McDougall, eds., Mesozoic paleogeography of the western United States: Pacific Section, Society for Sedimentary Geology (SEPM), Pacific Coast Paleogeography Symposium 2, p. 163–172.

Mixon, R. B., G. E. Murray, and T. E. Diaz, 1959, Age and correlation of Huizachal Group (Mesozoic), state of Tamaulipas, Mexico: AAPG Bulletin, v. 43, p. 757–771.

Molina-Garza, R. S., and J. W. Geissman, 1999, Paleomagnetic data from the Caborca terrane, Mexico: Implications for Cordilleran tectonics and the Mojave-Sonora megashear hypothesis: Tectonics, v. 18, no. 2, p. 293–325.

Monod, O., and P. H. Calvet, 1992, Structural and stratigraphic reinterpretation of the Triassic units near Zacatecas (Zacatecas) central Mexico: Evidence of a Laramide nappe pail : Zentrallblatt für Geologie und Paläontologie, Teil I, H. 6, p. 1533–1544.

Mutti, E., 1985, Turbidite systems and their relations to depositional sequences, in G. G. Zuffa, ed., Provenance of arenites, NATO-ASI Series: Dorodrecht, Netherlands, Reidel Publishing Company, p. 65–93.

Mutti, E., 1992, Turbidite sandstones: Agip S.p.A., Donato Milanese, 275 p.

Mutti, E., and R. F. Lucchi, 1972, Le torbiditi dell'Appenino Settentrionale: Introduzioni all'analisi di facies: Geological Society of Italy Memoir, v. 11, p. 161–199.

Mutti, E., and W. R. Normark, 1987, Comparing examples of modern and ancient turbidite systems, problems and concepts, in J. K. Leggett and G. G. Zuffa, eds., Marine Clastic Sedimentology Case Studies: London, Graham and Trotman Ltd., p. 1–38.

Padilla y Sánchez, R. J., 1982, Geologic evolution of the Sierra Madre Oriental between Linares, Concepcion del Oro, Saltillo and Monterrey, Mexico: Ph.D. dissertation, University of Texas at Austin, 217 p.

Palmer, A. R., 1983, The decade of North American Geology 1983 geologic time scale: Geology, v. 11, p. 503–504.

Pantoja-Alor, J., 1972, Datos geológicos y estratigráficos de la Formacion Nazas: Memoria de la Sociedad Geológica Mexicana, Segunda Convención Nacional, Mazatlán, Sinaloa, México, p. 25–32; 194–196.

Poole, F. G., 1993, Ordovician eugeoclinal rocks on Turner Island in the Gulf of California, Sonora, Mexico: Universidad Nacional Autónoma de Mexico, Instituto de Geología, III Simposio de la Geología de Sonora y Areas Adyacentes, Hermosillo, Sonora, Mexico, Abstracts, p. 103.

Poole, F. G., J. H. Stewart, W. B. N. Berry, A. G. Harris, R. J. Repetski, R. J. Madrid, K. B. Ketner, C. Carter, and J. M. Morales-Ramirez, 1995, Ordovician ocean-basin rocks of Sonora, Mexico, in J. D. Cooper, M. L. Droser, and S. C. Finney, eds., Ordovician Odyssey: Society for Sedimentary Geology (SEPM), 7th International Symposium on the Ordovician System, v. 77, p. 267–275.

Potter, P. E., and I. Cojan, 1985, Description and interpretation of the type section of the Barranca Group east of Rancho La Barranca, Municipio de San Javier, Sonora, in R. Weber, ed., Simposio sobre floras del Triásico Tardío, su fitogeografía y paleoecología: III Congreso Latinoamericano de Paleontología, Universidad Nacional Autónoma de México, Instituto de Geología, Memoria, p. 101–105.

Ranson, W. A., L. A. Fernández, W. B. Simmons, and S. Enciso de la Vega, 1982, Petrology of the metamorphic rocks of Zacatecas, Zacatecas, Mexico: Boletín de la Sociedad Geológica Mexicana, v. 1, p. 37–59.

Reading, H. G., and M. Richards, 1994, Turbidite systems in deep-water basin margins classified by grain size and feeder system: AAPG Bulletin, v. 78, no. 5, p. 792–822.

Roberts, R. J., 1951, Geology of the Antler Peak Quadrangle, Nevada: U. S. Geological Survey Geologic Quadrangle Map GQ-10.

Rogers, C. L., Z. de Cserna, R. Van Vloten, E. Tavera, and R. Ojeda, 1961, Reconocimiento geológico y depósito de fosfatos del norte de Zacatecas y áreas adyacentes en Coahuila, Nuevo León y San Luís Potosí: Consejo de Recursos Naturales No Renovables, Boletín 56, 322 p.

Salvador, A., 1991a, Triassic-Jurassic, in A. Salvador, ed., The Gulf of Mexico basin: Geological Society of America, The geology of North America, v. J, p. 131–180.

Salvador, A., 1991b, Origin and development of the Gulf of Mexico basin, in A. Salvador, ed., The Gulf of Mexico basin: Geological Society of America, The geology of North America, v. J, p. 389–444.

Schweigert, G., D. B. Seegis, A. Fels, and R. R. Leinfelder, 1997, New internally structured decapod microcoprolites from Germany (Late Triassic/Early Miocene), Southern Spain (Early/Middle Jurassic) and Portugal (Late Jurassic): Taxonomy, paleoecology and evolutionary implications: Paläontologische Zeitschrift: v. 71 (1/2), p. 51–69.

Sedlock, R. L., F. Ortega-Gutiérrez, and R. C. Speed, 1993, Tectono-stratigraphic terranes and tectonic evolution of Mexico: Geological Society of America Special Paper 278, 143 p.

Sexta Convencion Geologica Nacional, 1982, Libreto Guia, Excursion Geologica a la Cuence Mesozoica del Centro de Mexico, Estados de San Luís Postosíi and Zacatecas: Sociedad Geologica Mexicana, México City, México.

Shanmugam, G., R. B. Bloch, S. M. Mitchell, W. J. Beamish, R. J. Hodgkinson, J. E. Damuth, T. Straume, S. E. Syvertsen, and K. E. Shields, 1995, Basin-floor fans in the North Sea: Sequence stratigraphic models vs. sedimentary facies: AAPG Bulletin, v. 79, no. 4, p. 477–512.

Sílva-Pineda, A., 1961, Flora fósil de la Formación Santa Clara (Cárnico) del Estado de Sonora, Mexico: Instituto de Geología, Universidad Nacional Autónoma de México, Paleontología Mexicana II, parte 2, 118 p.

Silva-Romo, G., 1993, Estudio de la estratigrafía y estructuras tectónicas de la Sierra de Salinas, Estados de San Luís Potosí, Mexico: Tesis de Maestría, Universidad Nacional Autónoma de México, 140 p.

Silva-Romo, G., J. Arellaño-Gil, and C. C. Mendoza-Rosales, 1993, El papel de la secuencia marina Triásica en la evolución Jurásica del norte de Mexico, in F. Ortega-Gutiérrez, P. Coney, E. Centeno-García, and A. Gómez-Caballero, eds., First Circum-Pacific and Circum-Atlantic Terrane Conference: Instituto de Geología, Universidad Nacional Autónoma de México, p. 139–143.

Silver, L. T., and T. H. Anderson, 1974, Possible left-lateral early to middle Mesozoic disruption of the southwestern North American craton margin: Geological Society of America Abstracts with Programs, v. 6, p. 955

Stanley Jr., G. D., 1997, Upper Triassic fossils from the Antimonio Formation, Sonora and their implications for paleoecology and paleogeography, in C. M. González-León and G. D. Stanley Jr., eds., Publicaciones Ocasionales No. 1. Estación Regional del Noroeste, Instituto de Geología Universidad Nacional Autónoma de Mexico, U.S.-Mexico Cooperative Research: International Workshop on the Geology of Sonora Memoir, p. 62–65.

Stanley Jr., G. D., and C. González-León, 1995, Paleogeographic and tectonic implications of Triassic fossils and strata from the Antimonio Formation, northwestern Sonora, in C. Jacques-Ayala, C. González-León, and J. Roldán-Quintana, eds., Studies on the Mesozoic of Sonora and adjacent areas: Geological Society of America Special Paper 301, p. 1–16.

Stanley Jr., G. D., and C. González-León, C., 1997, New Late Triassic scleractinian corals from the Antimonio Formation, northwestern Sonora, Mexico: Revista Mexicana de Ciencias Geológicas, Universidad Nacional Autónoma de México, Instituto de Geología, v. 14, no. 2, p. 202–207.

Stanley Jr., G. D., C. González-León, M. R. Sandy, B. Senowbari-Daryan, P. Doyle, M. Tamura, and D. II. Erwin, 1994, Upper Triassic invertebrates from the Antimonio Formation, Sonora, Mexico: Palaontological Society Memoir 36, 33 p.

Stewart, J. H., R. Amaya-Martínez, R. G. Stamm, B. R. Wardlaw, G. D. Stanley Jr., and C. H. Stevens, 1997, Stratigraphy and regional significance of Mississippian to Jurassic rocks in Sierra Santa Teresa, Sonora, Mexico: Revista Mexicana de Ciencias Geológicas, Universidad Nacional Autónoma de Mexico, Instituto de Geología, v. 14, no. 2, p. 115–135.

Stewart, J. H., F. G. Poole, K. B. Ketner, J. Roldán-Quintana, and R. Amaya-Martínez, 1990, Tectonics and stratigraphy of the Paleozoic and Triassic southern margin of North America, Sonora, Mexico: Arizona Geological Survey, Special Paper 7, p. 183-202.

Stewart, J. H., and J. Roldán-Quintana, 1991, Upper Triassic Barranca Group: Nonmarine and shallow-marine rift-basin deposits of northwestern Mexico, in E. Pérez-Segura, and C. Jacques-Ayala, eds., Studies of Sonoran Geology: Geological Society of America Special Paper 254, p. 19–35.

Suter, M., 1987, Structural traverse across the Sierra Madre Oriental fold-thrust belt in east-central Mexico: Geological Society of America Bulletin, v. 98., p. 249–264.

Tosdal, R. M., G. B. Haxel, and J. E. Wright, 1989, Jurassic geology of the Sonoran Desert region, southern Arizona, southeastern California, and northernmost Sonora: Construction of a continental margin magmatic arc, in J. P. Jenney and S. J. Reynolds, eds., Geologic evolution of Arizona: Arizona Geological Society Digest 17, p. 397–434.

Tozer, E. T., 1980, Triassic Ammonoidea: Geographic and stratigraphic distribution, in M. R. House, and J. R. Senior, the Ammonoidea: London, Academic Press, p. 397–431.

Tozer, E. T., 1982, Marine Triassic faunas of North America: Their significance for assessing plate and terrane movements: Geologische Rundschau, v. 71, no. 3, p. 1077–1104.

Tristán-González, M., and J. R. Tórres-Hernández, 1992, Cartografía Geológica 1:50,000 de la Hoja Charcas, Estado de San Luís Potosí: Universidad Autónoma de San Luís Potosí, Instituto de Geología y Metalurgía, Folleto Técnico 115, 94 p.

Tristán-González, M., and J. R. Tórres-Hernández, 1994, Geología de la Sierra de Charcas, Estado de San Luís Potosí, Mexico: Revista Mexicana de Ciencias Geológicas, Universidad Nacional Autónoma de México, Instituto de Geología, v. 11, no. 2, p. 117–138.

Tristán-González, M., J. R. Tórres-Hernández, and J. L. Mata-Segura, 1995, Geología de la hoja Presa de Santa Gertrudis, San Luís Potosí: Universidad Autónoma de San Luís Potosí, Instituto de Geología, Folleto Técnico no. 122, 50 p.

Vail, P. R., R. M. Mitchum Jr., and S. Thompson III, 1977, Seismic stratigraphy and global changes of sea level, Part 4: Global cycles of relative changes of sea level, in C. E. Payton, ed., Seismic stratigraphy applications to hydrocarbon exploration: AAPG Memoir 26, p. 83–97.

Weber, R., 1980, Megafósiles de Conífera del Triásico Tardío y del Cretácico Tardío de Mexico y consideraciones generales sobre las Coníferas Mesozoicas de Mexico: Universidad Nacional Autónoma de Mexico, Revista, v. 4, no. 2, p. 111–124.

Weber, R., 1997, How old is the Triassic flora of Sonora and Tamaulipas? and news on Leonardian floras in Puebla and Hidalgo, Mexico: Revista Mexicana de Ciencias Geológicas, Universidad Nacional Autónoma de Mexico, Instituto de Geología, v. 14, no. 2, p. 225–243.

Weber, R., A. Zambrano-García, and F. Amozurrutía-Silva, 1980, Nuevas contribuciones al conocimiento de la tafloflora de la Formación Santa Clara (Triásico Tardío) de Sonora: Revista de la Universidad Nacional Autónoma de México, v. 4, no. 2, p. 125–137.

Westermann, G. E. G., 1973, The Late Triassic bivalve Monotis, in A. Hallam, ed., Atlas of paleobiogeography: London, Elsevier Scientific Publishing Company, p. 251-258,

Wilson, J .L., 1975, Carbonate facies in geologic history: Heidelberg, Springer-Verlag, 471 p.

Wilson, I. F., and V. S. Rocha, 1949, Coal deposits of the Santa Clara District near Tonichi, Sonora, Mexico: U. S. Geological Survey Bulletin 962-A, p. 1–80.

Yta, M., 1992, Étude géodynamique métallogénétique d'un secteur de la "Faja de Plata," Méxique: La zone de Zacatecas-Francisco I Madero-Saucito: Thèse de doctorat, Université d'Orléans, France, 266 p.

Blickwede, J. F., 2001, The Nazas Formation: A detailed look at the early Mesozoic convergent margin along the western rim of the Gulf of Mexico Basin, *in* C. Bartolini, R. T. Buffler, and A. Cantú-Chapa, eds., The western Gulf of Mexico Basin: Tectonics, sedimentary basins, and petroleum systems: AAPG Memoir 75, p. 317–342.

13

The Nazas Formation: A Detailed Look at the Early Mesozoic Convergent Margin along the Western Rim of the Gulf of Mexico Basin

Jon F. Blickwede

IHS Energy Group, Houston, Texas, U.S.A.

DEDICATION

I dedicate this paper to the family of Don Domingo Juárez of Caopas, Zacatecas, as well as to the memory of Manuel López Infanzón, geologist of the Instituto Mexicano del Petróleo.

ABSTRACT

The lower Mesozoic Nazas Formation, where it crops out in the Sierra de San Julián in northern Zacatecas state, Mexico, consists of slightly metamorphosed volcanic rocks and unfossiliferous continental red beds measuring a combined total thickness of approximately 1 km. The age of the Nazas remains equivocal, although limited radiometric dates from the Nazas and adjacent units indicate that the formation is probably Triassic and/or Early to Middle Jurassic in age.

The Nazas Formation, as defined in the study area, consists of two members: (1) a thick (average 946-m) lower volcanic member and (2) a thin (average 146-m) upper red-bed member. The lower volcanic member consists of interstratified ash-fall tuffs, ash-flow tuffs, lava flows, and lahars. More than two-thirds of the volcanic rocks are andesitic and dacitic pyroclastic rocks. Rhyolites and latites also occur. The lower volcanic member is thought to have been deposited as a composite cone complex. The upper red-bed member contains channel-fill, sheet-flood, and debris-flow deposits, representing prograding medial to proximal alluvial-fan facies. These sediments are compositionally and texturally immature volcarenites, volcanic lithic arkoses, and arkoses; all detritus was derived from the lower volcanic member. The alluvial fan(s) probably developed in an active graben system that formed immediately after the deposition of the lower volcanic member. Graben formation was likely a result of crustal collapse above a depleted magma chamber.

The Nazas overlies more highly metamorphosed volcanic rocks of the Rodeo Formation; the Rodeo may be part of the same volcanic succession as the Nazas, which would call for their combination into a single formational unit. The Nazas is unconformably overlain by shallow-marine carbonates of the Upper Jurassic (Oxfordian) Zuloaga Formation, equivalent to the Smackover Formation of the U.S. Gulf Coast. The Nazas Formation has often been correlated with the Huizachal Group of northeastern Mexico, and thus by inference to other passive-margin, rift-related, lower Mesozoic red beds of the Gulf of Mexico Basin, such as the Eagle Mills Formation of the U.S. Gulf Coast. It may be at least partially age equivalent, but the Nazas possesses certain characteristics, most notably the volume and composition of its constituent volcanic rocks, which suggest that it was deposited in a convergent-margin setting. As such, the Sierra de San Julián is interpreted to represent part of a transitional belt where the western Gulf of Mexico passive margin is superposed on a Pacific-margin magmatic arc. Among other unsolved problems in the region is the abruptness of change from a convergent- to passive-margin tectonostratigraphic regime at the Nazas/ Zuloaga contact.

INTRODUCTION

Lower Mesozoic continental red beds and associated volcanic rocks are known from scattered localities in northern Mexico, from Durango state southward to Veracruz state (Figure 1). Exposures occur mainly in dissected Laramide folds of the Sierra Madre Oriental and the Central Meseta. To the east, in the states of Nuevo León, Tamaulipas, Veracruz, and Hidalgo, the red beds generally have been designated the Huizachal Group, with its component La Boca (older) and La Joya

(younger) Formations. To the west, in San Luis Potosí, Coahuila, Zacatecas, and Durango states, the lower Mesozoic red beds and associated volcanics have generally been referred to as the Nazas Formation. This original distinction in stratigraphic nomenclature was a result of geographic separation and was not necessarily meant to imply that the Huizachal Group and Nazas Formation are noncorrelative. However, as detailed in this paper, there are indeed some significant differences between the Nazas Formation and the Huizachal Group that support their maintenance as separate stratigraphic units. The recognition of these different characteristics also aids in understanding the early history of the western Gulf of Mexico Basin.

Apart from some unpublished work performed by the state oil company Petróleos Mexicanos (PEMEX) and the Instituto Méxicano del Petroleo (IMP), all of the detailed stratigraphic and petrologic investigations of the early Mesozoic red beds in northern Mexico have focused on the Huizachal Group in Tamaulipas and Nuevo León (Mixon et al., 1959; Mixon, 1963; Carrillo Bravo, 1961; Corpstein, 1974; Allen, 1976; Lazzeri, 1979; Stilwell, 1980; Götte, 1986, 1990; Meiburg et al., 1987; Michalzik, 1988, 1991). The Nazas Formation, on the other hand, has been described only in a general manner in the published literature (de Cserna, 1956; Rogers et al., 1957, 1961;

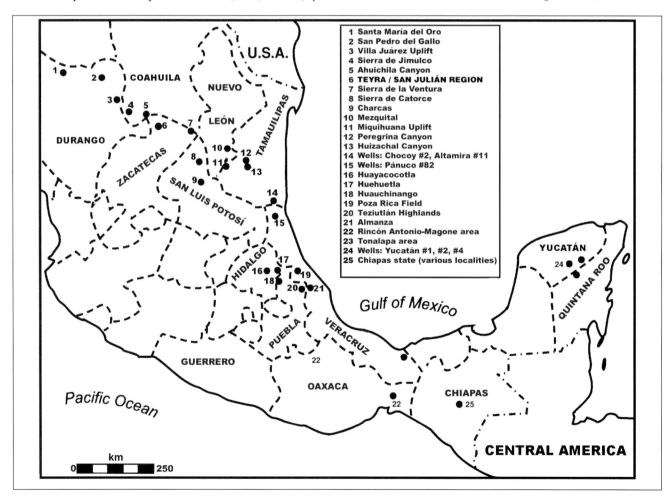

FIGURE 1. Distribution of lower Mesozoic red beds in Mexico (modified from Imlay et al., 1948).

Córdoba, 1964, 1965; Pantoja Alor, 1963, 1972; Belcher, 1979; Tardy, 1980; Blickwede, 1981b; López Infanzón, 1986; Aranda García, 1991; Anderson et al., 1991; Grajales Nishimura et al., 1992; Jones et al., 1995; Bartolini, 1997; Bartolini and Spell, 1997; Barboza Gudiño et al., 1997; Bartolini, 1998).

The primary purpose of this paper is to present a detailed description of the lithostratigraphy and petrology of the Nazas where it crops out in the Sierra de San Julián of northern Zacatecas state, which is the most complete exposure of the formation. Although the type section of the Nazas Formation was originally designated at the Villa Juárez uplift (Pantoja Alor, 1963, 1972; location 3 in Figure 1) about 200 km west of the Sierra de San Julián, I recommend that it be redesignated as the Piedras Blancas section described in this report. The data on the Nazas sections presented here were collected during fieldwork carried out in

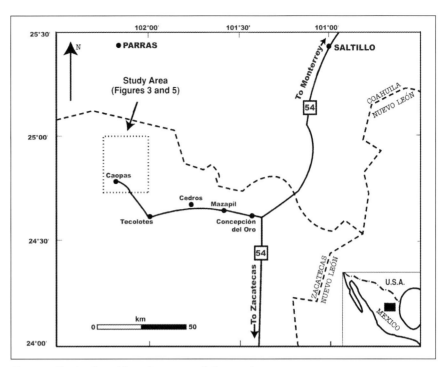

FIGURE 2. Regional location map of the study area.

1980; this information was incorporated in a master's thesis at the University of New Orleans (Blickwede, 1981a). Based on these observations, an interpretation is made of the petrogenesis, environment of deposition, and tectonostratigraphic regime represented by the Nazas volcanic and sedimentary rocks in the Sierra de San Julián and their relationship to the western Gulf of Mexico Basin as a whole. The final section of this paper focuses on some controversial issues concerning the general stratigraphy of the pre–Upper Jurassic Oxfordian section in the Teyra–San Julián region.

The Sierra de San Julián is located in northern Zacatecas state, Republic of Mexico, about 75 km west of the mining town of Concepción del Oro (Figure 2). The range trends northwest for approximately 40 km between latitude 24°45'N, longitude 102°03'W, and latitude 25°N, longitude 102°15'W. Topographic map coverage is provided by sheet G-13-D-59 (scale 1:50,000) of the Comisión de Estudios del Territorio Nacional (CETENAL). The most detailed geologic map available is the Hoja Apizolaya 13R-1(9) (scale 1:100,000) of Córdoba (1965), which covers the Sierra de San Julián and the Sierra de Teyra to the southwest.

Although the Nazas Formation crops out almost continuously along the Sierra de San Julián, the best exposures are found in the central part of the range from Cerro San Julián to Cerro San Antonio. The locations of three measured sections are shown in Figure 3.

Physiographic and Geologic Setting

The Teyra–San Julián region is situated in a transition zone between two major physiographic provinces: the Cross Ranges

of the Sierra Madre Oriental ("Sector Transversal de Parras") and the Central Meseta (Figure 4). The Cross Ranges are characterized by roughly east-west-trending, doubly plunging, tightly spaced Laramide (Late Cretaceous–early Tertiary) fold ridges, which are asymmetric to the north (Humphrey, 1956; de Cserna, 1956; Weidie et al., 1967, 1970). The Central Meseta is a broad, high plateau in which generally northwest-oriented mountain ranges, often intruded and tilted by igneous plutons, are separated by wide bolsons.

The generalized lithostratigraphy of the Teyra–San Julián region is shown, along with a surface geologic map, in Figure 5. Because of Laramide uplift, only a minor volume of Cenozoic rocks is preserved in the region, mostly thin Tertiary-Holocene alluvial deposits found in localized topographic lows. However, more than 5 km of Mesozoic sedimentary, igneous, and metamorphic rocks is exposed in the mountain ranges of the region. The Upper Jurassic Oxfordian through Upper Cretaceous units display the classic western Gulf of Mexico Basin marine-carbonate to terrigenous-clastic stratigraphic succession, starting with the Oxfordian Zuloaga Formation limestones and ending with the Campanian-Maestrichtian Parras Shale. In contrast, beneath the Oxfordian rocks are four major volcanic and sedimentary units that are not typical of the classic western Gulf lower Mesozoic stratigraphic succession, as will be discussed later in this paper. From youngest to oldest, these are the Nazas, Rodeo, and Taray Formations; the fourth unit, the Caopas Formation, has an equivocal stratigraphic relationship with the other units. The ages of the pre-Oxfordian units remain controversial because of the lack of any observed autochthonous fossils and inconclusive radiometric dates. The wide variance in radiometric dates is probably related to multiple thermal events that have affected the region.

The most recent and comprehensive published work on the structure and tectonics of the Teyra–San Julián region is that of Anderson et al. (1991), who presented evidence for at least three distinct Mesozoic-Cenozoic deformation events in the area. The first and most important of these was a northeast-directed compressional event (Oxfordian) that affected the lowermost Zuloaga Formation and older units and produced a large, northwest-trending fold nappe (the San Julián uplift,

which is delimited by the Sierra de San Julián to the northeast and the Sierra de Teyra to the southwest) and associated smaller-scale structures. Most of the metamorphism of the pre-Oxfordian rocks is attributed by the authors to this event. There is evidence of a later, Laramide (Late Cretaceous to early Tertiary) north-directed gravity-glide compressional event, which produced the large east-west-oriented folds of the Cross Ranges. The detachment of the Laramide structures in the Teyra–San Julián region was along an evaporite bed in the lower Zuloaga Formation, and as such, this deformation did not affect the underlying units, including the Nazas. Finally,

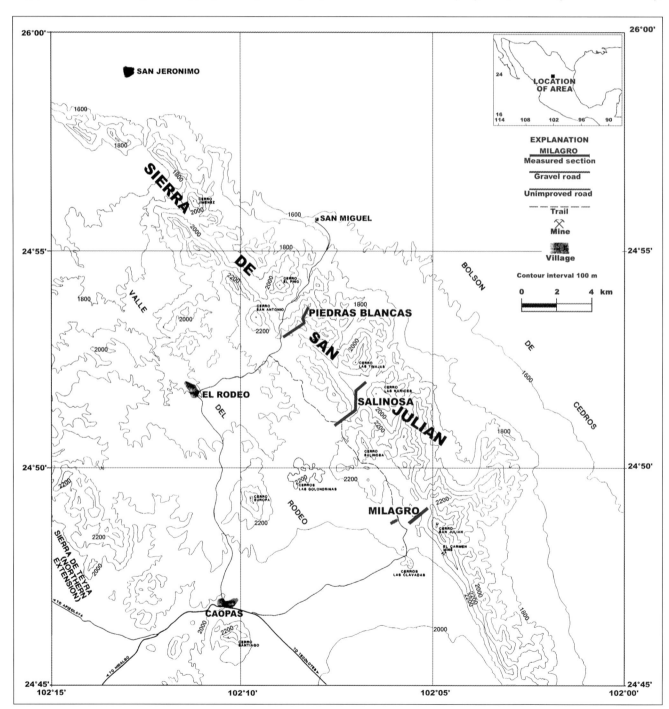

FIGURE 3. General topography of the Teyra–San Julián region, location of measured sections, and access routes.

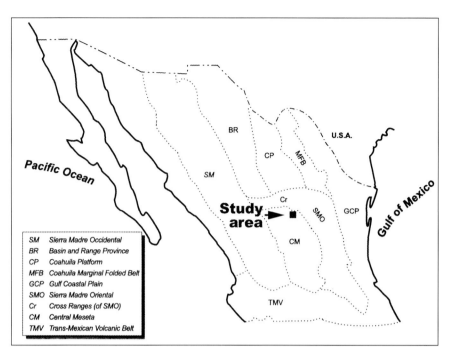

FIGURE 4. Major physiographic provinces of northern Mexico.

uplift in the late Tertiary steepened the axial planes of some Oxfordian nappe-related folds, tilted peripheral strata, and rotated the axes of Laramide folds such that they have since been truncated by erosion.

Other Studies of the Nazas Formation in the Teyra–San Julián Region

Rogers et al. (1957) and Rogers et al. (1961) were the first to describe and map the lower Mesozoic red beds in the Teyra–San Julián region, which they referred to as the Huizachal Formation because of the gross similarities with the Huizachal red beds in eastern Mexico. They observed that the unit consists of a series of unfossiliferous, predominantly red siltstones, sandstones, and conglomerates separated by angular unconformities from the overlying Oxfordian limestones of the Zuloaga Formation and underlying phyllites and schists of the Rodeo Formation. Rogers and colleagues also noted the presence of rhyolites in association with the red beds. They considered the red beds and associated volcanics to be late Callovian or early Oxfordian.

Córdoba (1964, 1965) produced a 1:100,000-scale geologic map of the Teyra–San Julián area incorporating the map of Rogers et al. (1961) and differentiating the pre-Oxfordian rocks. An abridged version of Córdoba's map is shown in Figure 5. Córdoba (1964) also outlined the general geology of the area and was the first to use the informal designation *Nazas Formation* for the pre-Oxfordian red beds in the Teyra–San Julián region, correlating the unit with the red beds and volcanic rocks that had been described in northern Durango state (Kellum, 1936; Pantoja Alor, 1963). Córdoba (1964) described a 155-m-thick incomplete exposure of the Nazas in the Cerros Comales area of the Sierra de Teyra, where he noted a preponderance of red and green siltstones, shales, and conglomerates. I made a cursory examination of these outcrops in the Cerros

Comales and believe that what Córdoba (1964) described as siltstones and shales are more likely ash-fall tuffs.

Belcher (1979) performed a general description of three sections of upper Nazas sediments (the "upper red-bed member" of this report) in the Sierra de San Julián as part of a regional study concerning the stratigraphy and paleomagnetic aspects of the lower Mesozoic red beds of northern Mexico. Two of Belcher's sections (localities 14 and 15) were diligently sought but not found, and his third section (locality 16) has been revised herein (upper red-bed member, Piedras Blancas section). Belcher was the first to recognize the volumetric predominance of volcanic rocks over sediments in the Nazas of the Teyra–San Julián region.

López Infanzón (1986) examined the Nazas Formation in the Sierras de San Julián, Teyra, and Candelaria and observed bounding unconformities with the overlying Zuloaga Formation (at least at the San Julián and Candelaria outcrops) and underlying Taray Formation (in Arroyo Palo Blanco of the Sierra de Teyra). López Infanzón proposed that the metasedimentary rocks originally designated as lower Rodeo Formation by Rogers et al. (1963) be assigned to the Taray and the volcanic rocks of the upper Rodeo Formation be considered part of the Nazas.

Anderson et al. (1991), during their structural analysis of the region, examined the Nazas in the Sierra de San Julián and argued that the red-bed sedimentary rocks at the top of the Nazas should be separated from the volcanic rocks and given formational status as the La Joya Formation.

Most recently, Bartolini (1998) generally described various stratigraphic sections, obtained additional radiometric dates, and carried out surface mapping of the Nazas Formation in the Teyra–San Julián region as part of a regional study of the unit in north-central Mexico.

A number of these other authors' observations and inter-

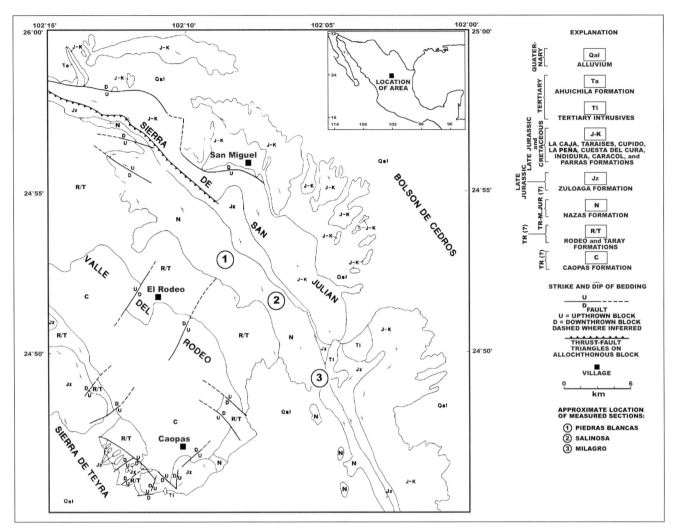

FIGURE 5. Geologic map of the Teyra–San Julián region, Zacatecas state (modified from Córdoba, 1965).

pretations of the Nazas Formation, as well as the pre-Oxfordian geology of the Teyra–San Julián region in general, are addressed in the section at the end of this paper.

LITHOSTRATIGRAPHY OF THE NAZAS FORMATION, SIERRA DE SAN JULIÁN

Three Nazas sections in the central part of the Sierra de San Julián were selected for measurement, sampling, and description. From north to south, the sections are designated Piedras Blancas, Salinosa, and Milagro and are located approximately 5 km apart (Figures 3, 5, and 6). Piedras Blancas and Salinosa are complete sections. The lowest portion of the Nazas is not exposed at Milagro, but the section was chosen because the upper red-bed member is well exposed. Each subdivision of the Nazas members is assigned an informal name based on locality and proximity to the formational base or the base of the exposed portion of the formation. For example, the lowest subdivision at Piedras Blancas is designated PBa, the next highest subdivisions are designated PBb, PBc, and so on. Units in different sections that share similar designations (e.g., PBa, Sa, Ma) are not necessarily correlative. Field descriptions of the Nazas subdivisions were made using the stratification terminology of McKee and Weir (1953) and the rock-color chart of Goddard et al. (1970). The rock colors cited in this report are those observed on dry, unweathered surfaces.

Based on the fieldwork and subsequent analyses of thin sections made from outcrop samples, the following overall observations can be made. The Nazas Formation of the Sierra de San Julián averages 1132 m in thickness and consists of mainly intermediate volcanic rocks and volcaniclastic sedimentary rocks. Pyroclastic rocks are the most abundant lithologic type, accounting for about 65% (45% ash-fall tuffs, 20% ash-flow tuffs) of the formation. Lava flows make up 10% of the formation, and volcaniclastic sediments 25%. Two members are recognized: (1) a thick (average 946-m) volcanic member that contains minor (less than 10%), poorly sorted conglomeratic units and (2) a thin (average 146-m) upper red-bed member. The basal contact of the Nazas Formation with the underlying Rodeo Formation is difficult to define, and this problem is discussed in more detail later in this paper. Although the upper contact of the Nazas with the Zuloaga in the Sierra de San Julián is covered and was not directly observed, the two units appear to have an unconformable relationship. Apart from the abrupt change in depositional environments from continental/volcanogenic to open-marine/nonvolcanogenic across the contact, the upper 20–30 m of the Nazas is deeply weathered, and the difference in the measured dips of the beds above and below the contact varies from 20° to 60°.

The two Nazas members can be traced easily from section to section, but correlation of individual units is much more difficult, particularly in the lower volcanic member. The distribution of lava flows and ash-flow tuffs was controlled by local topography at the time of eruption; thus, lateral correlation problems are to be expected. Ash-fall tuffs, on the other hand, tend to form a continuous blanket on preexisting topography and are commonly correlative over great distances. It is unclear why many of the Nazas ash-fall tuffs seem to be laterally discontinuous; perhaps these deposits were redistributed by postdepositional surface runoff.

Systematic chemical analysis of each volcanic unit, beyond the scope of this study, is necessary for unit correlation and a detailed interpretation of the petrogenesis and eruptive history of the lower volcanic member. In the upper red-bed member, PBn, So, and Mi are lithologically equivalent, as are PBm and Mg.

Piedras Blancas Section

The Nazas exposure at Piedras Blancas Pass, located about 6 km northeast of the village of El Rodeo along the road to San Miguel (Figure 3), is the most easily accessible and complete in the entire Teyra–San Julián region. Details of how to access this and the other measured sections are described in Blickwede (1981a).

Lower Volcanic Member

Unit PBa (0–244 m above the defined base of the formation): The basal unit (PBa) of the Nazas at Piedras Blancas is a 244-m-thick series of grayish-red-purple and brownish-gray andesitic and dacitic crystal-lithic ash-fall tuffs that resemble massive siltstones. Scanning electron-microscope images of the unit revealed abundant devitrified volcanic glass shards. The entire unit displays a pronounced slaty cleavage with northwest orientation. The first meter of PBa is a dusky red dacitic lithic-vitric ash-fall lapillistone; angular accidental ejecta as much as 5 cm in diameter, derived from the underlying Rodeo andesite, are abundant at the contact but rapidly diminish in quantity upward. The lapillistone grades into the overlying tuffs. At 86–87 m, there is a laminated tuffaceous bed, in which many sand-sized grains are oriented with their long axes perpendicular to bedding. The genesis of this interval is not known. The long axes of the vertically oriented grains are parallel to the slaty cleavage plane. Thus, perhaps all grains were originally oriented subparallel to bedding, as would be expected in an epiclastic (reworked) tuff, and were later reoriented by deformation.

Unit PBb (244–279 m): Unit PBb is a grayish-yellow-green vitric ash-fall tuff of unknown composition. The upper 1 m of the unit is weathered to light brownish gray.

Unit PBc (279–352 m): Unit PBc is a light greenish-gray to light brownish-gray vitric ash-flow tuff of unknown composition. It forms a resistant ridge that caps the first hill traversed in the section. Abundant ellipsoidal cavities (1 mm–0.5 m in long dimension), subparallel to the flow texture, characterize this unit. Some of these may have formed by dissolution of the collapsed pumice fragments. Alternatively, Mackin (1960) and Chapin and Lowell (1979) have suggested that such cavities form by concentration of gases along laminar-flow planes during deposition and compaction of an ash-flow tuff. The boundary between PBc and PBd is located in a covered interval from 352 m to 364 m.

Unit PBd (364–438 m): PBd is a series of grayish-yellow-green and grayish-red vitric and vitric-crystal ash-fall tuffs. Mineralogic composition is indeterminate because of a pervasive silicification affecting the entire unit and total replacement of feldspars by calcite and sericite. However, ghosts of randomly oriented shards are visible in thin section.

Unit PBe (438–468 m): Unit PBe is the oldest Nazas sedimentary unit in the Piedras Blancas section. The dominant lithology is pale reddish-brown to dark reddish-brown clayey, matrix-supported, granular pebble conglomerate. A 20-cm-thick vitric ash-fall tuff occurs at 461 m, indicating at least two conglomerate depositional units. The conglomerates are texturally immature; clay matrix makes up about 40% by volume, and clasts are subangular-subrounded. No grading, stratification, or preferred clast orientation was observed. Although silicification has obscured clast mineralogy, clast color and texture suggest that the conglomerate is heterolithologic and volcanogenic. Most of the clasts display shard ghosts in thin section.

Unit PBf (468–535 m): Unit PBf is another series of vitric and vitric-crystal ash-fall tuffs, similar to unit PBd. Pale red and grayish red are the most common colors, with thin zones of mottled blackish red that probably indicate horizons that were subjected to prolonged subaerial exposure. Individual fall units show fining upward of ejecta and highly oxidized tops; at least six of these fall units were observed in PBf, ranging from less than 1 m to 11 m in thickness. A planar-bedded, tuffaceous epiclastic interval occurs at 494–495 m. Unlike the interval at 86–87 m in PBa, all sand-sized grains are oriented with their long axes subparallel to the bedding plane.

Unit PBg (535–638 m): Unit PBg is a resistant sequence of very dusky red andesite or latite. It is not known if PBg represents one or many individual lava-flow units.

Unit PBh (638–670 m): PBh consists of at least nine grayish-red to blackish-red vitric-crystal ash-fall tuff units averaging 3 m in thickness. The upper 20 m of PBh is distinguished by a network of 5- to 10-cm-wide grayish-yellow-green reduced zones along fractures.

Unit Pbi (670–695 m): PBi is a resistant vitric ash-flow tuff.

Its overall color is pale pink to pale red-purple with pale green collapsed pumice fragments.

Unit PBj (695–787 m): PBj is the uppermost series of ash-fall tuffs in the Piedras Blancas section. These vitric-crystal ash-fall tuffs are grayish red and grayish red-purple with grayish-yellow-green reduced zones along fractures as in PBh. At least 17 individual fall units were recognized.

Unit PBk (787–830 m): PBk is a resistant, pale red-purple to grayish-red-purple rhyolite lava flow that caps the highest ridge traversed in the Piedras Blancas section.

Unit PBl (830–847 m): PBl is the youngest volcanic unit exposed at Piedras Blancas. It is a dark reddish-brown crystal-vitric ash-flow tuff of trachytic, latitic, or andesitic composition.

Upper Red-bed Member

The upper 197 m of the Nazas Formation at Piedras Blancas is composed entirely of red-bed sedimentary rocks. Two units, PBm and PBn, have been defined.

Unit PBm (852–901 m): PBm is characterized by a dark reddish-brown clayey, sandy siltstone with scattered channel-fill conglomerates and sandstones. The bulk of the siltstone is apparently massive and contains randomly oriented rip-up clasts (less than 1 cm long) of laminated, highly oxidized, fine siltstone. Sand-sized grains are randomly distributed and show no preferred orientation. Clay content approaches 50%. Some planar-bedded siltstones occur just above and adjacent to the coarse, channel-fill deposits. The channel-fill lenses display scoured basal contacts with the massive siltstones and grade upward into thin, planar-bedded siltstones. Load structures occur at the base of the channels. Channels increase in dimension from the bottom to the top of the unit. Near the base, they may be less than 1 m wide and 20 cm thick, but most channels above 870 m are at least 60 m wide and 5 m thick. Grain size increases in direct proportion to channel dimension. Granule conglomerates without lag predominate in the lowermost channels, whereas pebble conglomerates with a distinct cobble lag are common toward the top of PBm.

Unit PBn (901–1044 m): PBn crops out in three resistant ridges at the top of the Piedras Blancas section. From 901–996 m, 87 individual grayish-red conglomeratic beds are exposed, each averaging about a meter in thickness. Each bed displays crude normal grading; seven of these actually grade to a thin (5- to 10-cm) planar-bedded siltstone. Conglomerate clasts are not imbricated, but at the base of each unit, many are oriented subparallel to the bedding plane; above this, the clasts are oriented more randomly. Contacts between the grading units lack any evidence of scour. All beds are planar in form. The 901–996-m conglomerates are poorly sorted, are grain and matrix supported, and contain about 15% detrital clay. Clasts are subangular-subrounded. From 996 to 1021 m, two thick (7-m and 18-m) nongraded and unstratified boulder-conglomerate beds crop out. Although many of the larger, tabular clasts are oriented subparallel to bedding, most are randomly oriented in a sandy matrix. Clay content is somewhat higher than in the

901- to 996-m interval. The top 23 m of PBn, below the contact with the Zuloaga Formation, is highly altered, and as such, sedimentary structures are obscured. The texture, however, suggests that conglomerates such as those at 996–1021 m directly underlie the Zuloaga.

Salinosa Section

The Nazas Formation is moderately well exposed in a northeast-trending canyon that begins at Puerto Salinosa and divides Cerro Las Narices and Cerro Las Tinajas downstream. A 1149-m-thick section was measured in this canyon. The best access is by a trail from Puerto Milagro, 5 km to the southeast (Figure 3).

Lower Volcanic Member

Unit Sa (0–43 m): The basal Nazas unit at Salinosa, Sa, is a sequence of grayish-red to blackish-red dacitic crystal ash-fall tuffs. Two dacitic crystal-lithic lapillistone intervals appear at the base of the formation (0–2 m) and at 21–30 m. Lapilli in the basal interval are angular fragments of the upper Rodeo andesite. A laminated tuffaceous bed occurs at 5–7 m, similar to PBa 86–87 m. A northwest-oriented slaty cleavage is well developed. Apart from the difference in thickness, there is an overall resemblance between Sa and PBa.

Unit Sb (43–149 m): Sb is a very dusky red lithic-vitric ash-flow tuff of unknown composition. A welded texture with collapsed pumice fragments is visible in thin section.

Unit Sc (149–384 m): Sc comprises a series of grayish-red to blackish-red andesitic crystal ash-fall tuffs. The pyroclastic sequence is interrupted by an epiclastic conglomerate interval at 257–263 m. The conglomerate is grayish-red, nonstratified, matrix supported, and poorly sorted. Grain size ranges from clay (about 20% of total volume) to boulders. Clasts are subrounded-angular. Although the conglomerate is heterolithologic, all clasts are volcanic rock fragments; many were apparently derived from the underlying crystal ash-fall tuffs.

Unit Sd (384–389 m): Sd is a yellowish-gray andesitic lava flow.

Unit Se (389–425 m): Se closely resembles the conglomerate at 257–263 m in Sc. However, there are two depositional units here (20 m and 15 m thick) separated by a 1-m-thick, planar-bedded coarse sandstone which fines upward to a coarse siltstone.

Unit Sf (442–522 m): Sf consists of blackish-red andesitic crystal ash-fall tuffs, similar to the tuffs in unit Sc.

Unit Sg (534–634 m): A change in composition occurs in the unit Sg, which is composed mainly of medium light-gray dacitic vitric-crystal ash-fall tuffs.

Unit Sh (634–691 m): Sh is a sedimentary unit dominated by nonstratified clayey siltstones. The only trace fossils observed in the Nazas Formation in the study area, randomly oriented burrows (Scoyenia?), are found in Sh. The lower 8 m of Sh consists of greenish-gray, massive, clayey, sandy siltstone beds alternating with dark greenish-gray, planar-bedded, very fine to fine siltstones. The massive siltstones average about 1 m thick, and the planar-bedded siltstones are less than 5 cm thick. From 642 to 643 m, laminated, clayey, very fine to fine siltstone is exposed; it contains alternating light and dark-green laminae (less than 1 mm to 5 mm thick) that resemble varves. At 643–645 m is another bed of massive, clayey, sandy siltstone, grayish green at the base and grading to very dusky red-purple at the top. Randomly oriented burrows occur in the oxidized red-purple portion. At 645–649 m is a planar cross-bedded and planar-bedded, greenish-gray, silty, very fine to fine sandstone. Individual beds are 5–10 cm thick. From 649–691 m, four massive, clayey, sandy, granular, pebbly silt-stone beds are exposed; their thicknesses are, from base to top, 11 m, 16 m, 12 m, and 3 m. The upper 0.5 m of each is a pla-nar-bedded coarse sandstone. Color in each grades from gray-ish green to very dusky red-purple; the red-purple portions contain abundant burrows.

Unit Si (691–707 m): Si is a pale grayish-red to blackish-red dacitic crystal ash-fall tuff.

Unit Sj (707–780 m): Sj is a pale red-purple to grayish-red-purple lava flow, probably a rhyolite. It is similar in appearance to unit PBk.

Unit Sk (780–814 m): Most of Sk is covered by recent allu-vium; however, from 785 to 796 m, a grayish-red to blackish-red andesitic crystal-vitric ash-fall tuff is exposed, and float in the covered interval indicates an upward continuation of the same lithology.

Unit Sl (814–823 m): S1 is a massive, dusky-brown, clayey, sandy, granular pebble conglomerate that grades to a 1-m-thick, planar-bedded, granular, fine to very coarse sandstone. The conglomerate is both matrix and grain supported and con-tains about 20% clay. Clasts are subrounded to very angular.

Unit Sm (823–1067 m): Sm is a thick sequence of grayish-red, blackish-red, and yellowish-gray to light olive-gray ash-fall tuffs and lapillistones. Composition is andesitic-dacitic, and there is a high percentage of lithic and crystal components.

Unit Sn (1067–1108 m): Sn is the uppermost volcanic unit in the Salinosa section. It is a grayish-red-purple dacitic or rhy-olitic vitric-crystal ash-flow tuff. Devitrified, welded glass shards and collapsed pumice fragments are visible in thin section.

Upper Red-bed Member

Unit So (1108–1149 m): Unit So is a conglomeratic red-bed unit that is virtually identical to the beds at 901–996 m in unit PBn. It is highly weathered near the contact with the Zuloaga Formation.

Milagro Section

A 1084-m-thick incomplete exposure of the Nazas Forma-tion crops out along the northeast-trending arroyo that leads up to the Milagro Pass, about 2 km northwest of the El Carmen barite mine (Figure 3). The lowest 156 m was measured where the formation crops out in a hogback immediately northwest of the Cerros Las Clavadas. Most of the succeeding 500 m of

section is covered by alluvium in the valley between the hog-backs and the main part of the range.

Lower Volcanic Member

Unit Ma (0–83 m): The lowest exposed unit at Milagro is a nonresistant, yellowish-gray, clayey, silty, granular, very fine to very coarse sandstone. No stratification was observed, and clay content exceeds 40%. Clasts are subrounded to angular. The upper portion of Ma is a darker, olive-gray color and more highly weathered, indicating prolonged subaerial exposure. Slaty cleavage is well developed.

Unit Mb (83–156 m): Mb is a highly resistant, medium-gray, vitric-crystal ash-flow tuff that forms the above-men-tioned hogback. Its mineralogic composition is unknown. Abundant ellipsoidal cavities occur throughout the unit. Mb is similar in appearance to unit PBc.

Unit Mc (498–548 m and 565–588 m): Mc is a very dusky red vitric-crystal ash-fall tuff, which crops out in the valley between the hogbacks and Milagro Pass. Randomly distrib-uted, very dark red patches of intense iron-oxide stain give Mc the appearance of a conglomerate in hand specimen. The unit also contains scattered calcite veinlets less than 1 mm thick.

Unit Md (678–708 m): Md is a dusky-brown latitic or andesitic lava flow.

Unit Me (708–797 m): Me is a series of andesitic or latitic vitric-crystal ash-flow tuffs. Colors range from moderate red to dusky red to brownish gray–brownish black. Scattered acciden-tal lapilli derived from the underlying unit Md were observed from 708 to 784 m. The contact zone with the overlying ash-flow unit Mf is pale greenish yellow and appears to have been "baked."

Unit Mf (797–879 m): Mf is composed of dacitic or rhy-olitic crystal-vitric and crystal ash-flow tuffs. Colors are moder-ate red and dusky red to very dusky red-purple in the top 11 m of the unit.

Upper Red-bed Member

Unit Mg (879–898 m): Mg consists of dusky red to moder-ate red clayey, fine to medium, sandy siltstones with scattered channel-fill conglomerates and sandstones of the same color. Most siltstones are massive, with randomly distributed, angu-lar clasts of sand and cobble size. However, the siltstones adja-cent to and immediately overlying the coarse channel-fill lenses are characterized by low-angle ripple lamination, planar bedding, and occasional scour-and-fill structures. Clay content is about 30–40%. The channel-fill lenses display scoured con-tacts with the underlying massive siltstones and rapidly grade upward into the stratified siltstones. Some load structures at the base of the channels were observed. The channel fill is a poorly sorted, clayey, granular, fine to very coarse sandstone with about 10% detrital clay matrix. The thickness of the chan-nels ranges from 30 cm to 1 m; lateral dimensions are un-known because of the narrow width of exposure in the arroyo bed. Mg closely resembles unit PBm at Piedras Blancas.

Unit Mh (922–979 m): After a 24-m interval of recent alluvial cover, Mh is a section of moderately well sorted, planar-bedded, coarse sandstones to granule conglomerates and predominantly ripple-laminated siltstones. The coarse intervals range from 2 to 12 m in thickness, and the siltstones from 0.1 to 2 m. All beds are planar in form. The lower contacts of the coarse intervals are sharp (erosional) but horizontal, without notable scour; scattered load structures were observed. The coarse intervals grade into the overlying siltstones. Mh is dusky red to moderate red and grayish red.

Unit Mi (979–1084 m): Mi, the uppermost Nazas unit at Milagro, is a sequence of crudely fining-upward conglomerate beds. The unit is virtually identical to the uppermost units at Piedras Blancas (PBn 901–996 m) and Salinosa (So). Mi is deeply weathered from about 1060 m to the top of the exposure.

PETROLOGY OF THE NAZAS FORMATION, SIERRA DE SAN JULIÁN

A total of 160 outcrop samples were obtained and analyzed from the Nazas Formation in the Piedras Blancas, Salinosa, and Milagro measured sections in the Sierra de San Julián—120 samples from volcanic units and 40 from sedimentary units.

Volcanic Rocks

For the volcanic units, petrographic data were gathered on phenocryst mineralogy, mechanical composition, and alteration of phenocrysts and matrix. The classification systems used for the Nazas volcanic rocks are as follows: mineralogic composition of volcanic rock types, International Union of Geological Sciences (IUGS) Subcommission on Systematics of Igneous Rocks (1979); mechanical composition of pyroclastic rocks, Cook (1965); granulometric classification of pyroclastic rocks, Fisher (1961).

The mineralogic composition of volcanic rocks was determined on the basis of phenocryst mineralogy. Plagioclase grains twinned according to the albite or albite-Carlsbad laws were examined with a petrographic microscope fitted with a three-axis universal stage; for some samples, the anorthite content was determined by the Slemmons (1962) method and Rittmann (1929) zone method. A few relatively fresh feldspars from various units in the formation were also analyzed by x-ray diffraction. Although not enough data were gathered on feldspar composition to allow precise classification of each volcanic unit, a limited compositional range was defined for those samples containing a significant number of phenocrysts. It should be noted that although the mineralogic composition of volcanic rocks was determined on the basis of phenocryst mineralogy, the prefix *pheno-* is not used in this report.

Quantitative compositional data on 24 volcanic rock samples are presented in Table 1. Precise mineralogic compositional classification was not achieved for most of the volcanic rocks; this was a result of highly altered feldspars that did not stain and the very fine-grained nature of the matrix. Thus, all

feldspars other than those twinned according to the albite or albite-Carlsbad laws were counted as unknown feldspars. In certain samples, more than 65% of all feldspars are albite or albite-Carlsbad twins, and the sample can be positively identified as an andesite or dacite, depending on the percentage of quartz. In many of the intermediate volcanic units, however, the quartz content is quite variable; thus, these have been classified as andesite-dacite. It is unlikely that any of the plagioclase-rich volcanic rocks are basaltic, because the maximum anorthite content of plagioclase crystals examined is 43% (andesine), with a mean value of An_{27} (calcic oligoclase). Volcanic rocks containing less than 65% diagnostically twinned plagioclase have been assigned a range of possible compositions (e.g., andesite-latite). The only exceptions to this are the lava flow units PBk and Sj, for which x-ray diffraction analysis indicated sanadine as the predominant feldspar, and these could be defined as rhyolites.

For the most part, the ash-fall tuffs are too fine grained to make a quantitative optical determination of mineralogic composition. A visual estimate of unidentified feldspar, plagioclase, and quartz percentages was performed and, in conjunction with x-ray diffraction data, a compositional range was defined for crystal-rich ash-fall tuffs. Ash-fall tuffs predominantly or entirely composed of altered glass are classified only by mechanical composition. Similarly, most of the ash-flow tuffs are characterized by a high vitric component and could not be min- eralogically classified by the methods used in this study.

Figure 7 shows the approximate relative volumetric abundance of mineralogic composition groups represented in the Nazas volcanic rocks of the Sierra de San Julián, using the IUGS (1979) classification system. The following general observations can be made: (1) Of the volcanic rocks which were compositionally defined, 70% are andesites and dacites; (2) all compositionally defined ash-fall tuffs are andesites and dacites; (3) ash-flow tuffs (at least those with a relatively high crystal component) are more silicic than ash-fall tuffs; and (4) lava flows are bimodal andesite-latite/rhyolite.

The Nazas volcanic rocks are in general highly altered, especially the pyroclastic units. Original glass has been devitrified to a mixture of quartz and feldspar, the latter commonly being further altered to clay. Feldspar phenocrysts are vacuolized and partially or totally replaced by calcite and sericite. The matrix is partially altered to sericite or illite and, in some cases, is entirely silicified. Ferromagnesian minerals (pyroxene, hornblende, biotite) have been replaced by hematite, which imparts the reddish color to many of the pyroclastic rocks.

Figure 8 shows the approximate relative abundance of the mechanical composition groups represented in the Nazas pyroclastic rocks, using the classification system of Cook (1965). Overall, the vitric component is dominant in the Nazas tuffs, although many ash-fall tuffs contain abundant crystals (mainly fragments thereof) of feldspar and quartz, and the ash-flow tuffs contain a high percentage of lithic fragments.

Each volcanic rock type (ash-fall tuffs, ash-flow tuffs, and

TABLE 1. Mineralogic composition of Nazas Formation volcanic rocks, Sierra de San Julián.

Sample	Unit	% unidentified feldspar	Essential phenocrysts % plagioclase/ An content	% quartz	Varietal and accessory phenocrysts Type	%
*R44	PBa	30	70	25		
R47	PBa	33	67	5		
*R21	PBg	50	50	5	hornblende and biotite	< 5
R32	PBk	58: x-ray diffraction indicates sanidine	42	42		
*P4	Sa	25	75	25		
*P8	Sc	20	80	5		
*P11	Sc	15	85/An_{29}, An_{43}	5		
*P12	Sc	15	85	< 5		
*P13	Sd	20	80/An_{24}	5		
P15	Sf	22	78/An_{35}	< 5		
*P16	Sg	30	70	40	pyroxene	< 5
*P19	Si	30	70	30		
*P20	Sj	60: x-ray diffraction indicates sanidine	40	40		
*P21	Sk	30	70	< 5		
*P25	Sm	20	80/An_{24}	10		
*P30	Sn	85	15	30		
M6	Mc	71	29/An_{20}	20	pyroxene	< 10
M7	Md	57	43/An_{22}	15	pyroxene	3
M8	Me	85	15/An_{14}	10		
*M9	Me	80	20	10		
M10	Mf	82	18	36	pyroxene	6
M11	Mf	7	93/An_{18}	24	pyroxene	5
M13	Mf	68	32/An_{21}	30	pyroxene	2
M14	Mf	83	17/An_{14}	43		

* Indicates visual estimate; feldspar percentages are recalculated to 100%.

lava flows) is characterized by certain diagnostic microtextures. The pyroclastic texture of the tuffs is largely obscured by post-depositional vapor-phase mineralization and devitrification, but in all cases, at least some shard "ghosts" were observed. Both ash-fall and ash-flow tuffs contain a mixture of devitrified glass shards and fragmented phenocrysts (i.e., hypocrystalline texture), but ash-flow tuffs were distinguished on the basis of a subparallel orientation of the shards and collapsed pumice fragments (welded texture). In general, ash-fall tuffs are better sorted than ash-flow tuffs, and in certain cases, they reveal internal stratification produced by individual fall units of varying grain size and/or color.

The lava-flow units are holocrystalline and aphanitic-porphyritic, with some showing flow alignment in thin section. Phenocrysts are mainly unfragmented and euhedral or anhedral. Quartz in both the lava flows and tuffs is water-clear and resorbed, displaying embayments and rounded terminations.

Sedimentary Rocks

Thin sections of sedimentary rock samples were point-counted by the standard stage-grid-system technique. Data were gathered on framework grain composition, size, roundness, contacts, and alteration, as well as the percentage of matrix and cement. Cumulative grain-size curves were constructed for most of the point-counted sediment samples, corrected as per Friedman (1958), and used to determine mean grain size, standard deviation, and sorting. Grain-size classes are those of Wentworth (1922). The sediments were classified mineralogically and texturally according to the system of Folk (1974).

Texture

Quantitative grain-size analysis was performed on six sandstone samples (R34, R36, and R37 from unit PBm; M17 from unit Mg; M20c and M22 from unit Mh) and sandy portions of four conglomerate samples (R9a from unit PBe; R42c from unit

PBn; P18 from unit Sh; M1 from unit Ma), and the resulting data are presented in Table 2.

Combining the Nazas sedimentary units from all three measured sections, the formation comprises approximately 50% conglomerates, 30% sandstones, and 20% siltstones. Although clay-sized detritus is abundant in most of the units, mud rocks are entirely lacking.

The sorting values presented in Table 2 are not absolute, because the clay and coarse gravel fractions, which make up a significant proportion of the conglomerates, could not be accounted for in thin-section analysis. Thus, the standard deviation data represent minimum sorting values, and the classes reflect a maximum degree of sorting. Nevertheless, the sorting value of one sample can be compared with another, which

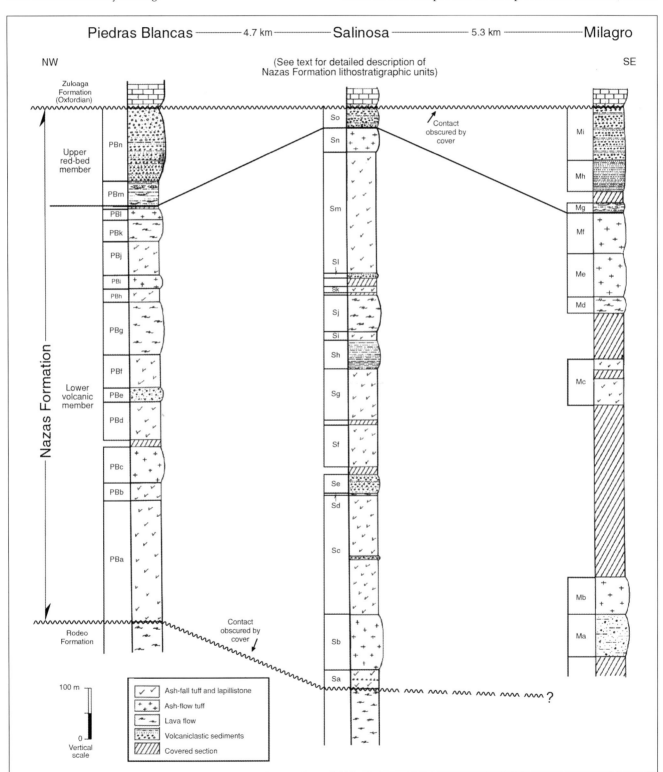

FIGURE 6. Generalized stratigraphic sections, Nazas Formation, Sierra de San Julián, Zacatecas state, Mexico.

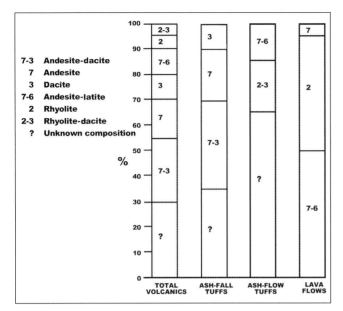

FIGURE 7. Mineralogic composition groups of the Nazas Formation volcanic rocks at Sierra de San Julián, Zacatecas state, and their approximate relative volumetric abundance.

may aid in differentiating types of sedimentary deposits (e.g., a water-laid sandstone should be better sorted than the sand fraction of a debris-flow conglomerate). Field observation revealed that sediments of the Nazas Formation in the Sierra de San Julián are moderately to extremely poorly sorted. The most poorly sorted units are those interstratified with the volcanic rocks in the lower volcanic member. Channel-fill sandstones of PBm and Mg and the conglomerates of PBn, So, and Mi are

intermediate in degree of sorting. The planar-bedded sandstones and associated siltstones in unit Mh are the best-sorted sediments in the entire study area, although these water-laid deposits are still only moderately sorted.

The roundness of each point-counted framework grain was estimated by comparison with the roundness chart of Powers (1953). An analysis of these data is shown in Table 3. As with sorting, the roundness of the detrital grains in the Nazas sediments reflects their low textural maturity. Overall, only 6% of the grains are rounded, and most of these grains are relatively soft volcanic rock fragments (VRFs). No well-rounded grains were observed in thin section, although some of the cobbles and boulders in the conglomerates of units PBn, So, and Mi fall into this category. Most grains are subrounded to angular. Even among VRFs, which are the least mechanically resistant of the three main framework types, 34% are classed as subangular.

The data on quartz grain roundness are potentially misleading. Most quartz, which is classed as subrounded or rounded (comprising 19% of all quartz grains), is virtually nonabraded. Such grains are unfragmented volcanic quartz with rounded terminations caused by the resorption of SiO_2 prior to or during extrusion of the parent volcanic rock.

Four types of grain contacts were defined by Taylor (1950): tangential, long, concavo-convex, and sutured. Long and concavo-convex contacts are the most common types in the Nazas Formation sediments because of the high percentage of volcanic rock fragments. These relatively soft grains deformed on compaction, such that their boundaries conform to the edges of adjacent grains. The higher-strength feldspar and quartz grains display tangential (point) contacts where they are in juxtaposition with each other. Sutured contacts (interpenetra-

TABLE 2. Mean grain size (Mz), standard deviation(s), and sorting class of Nazas Formation sandstones and conglomerate sand fractions.

Sample	Unit	Mz/size-class	s/size-class range	Sorting class
R9a	PBe	-0.6∅ / vcs	1.54∅ / cs-g	p
R34	PBm	0.64∅ / cs	1.07∅ / ms-vcs	p
R36	PBm	0.16∅ / cs	0.77∅ / cs-vcs	m
R37	PBm	1.52∅ / ms	0.71∅ / ms-cs	m
R42c	PBn	1.25∅ / ms	1.07∅ / fs-cs	p
P18	Sh	0.68∅ / cs	1.70∅ / fs-g	p
M1	Ma	1.0∅ / cs	1.25∅ / fs-vcs	p
M17	Mg	0.66∅ / cs	1.0∅ / ms-vcs	m-p
M20c	Mh	0.54∅ / cs	0.73∅ / ms-vcs	m
M22	Mh	0.32∅ / cs	1.0∅ / ms-vcs	m-p

fs = fine sand
ms = medium sand
cs = coarse sand
vcs = very coarse sand
g = granule
p = poorly sorted
m = moderately sorted
Calculations and sorting classes as per Folk (1974).
Size classes as per Wentworth (1922).

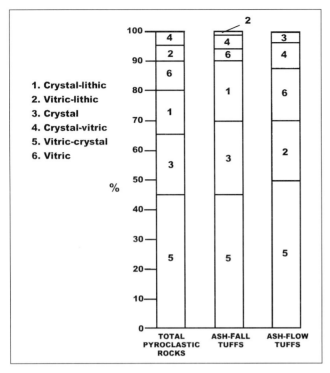

FIGURE 8. Mechanical composition groups of Nazas Formation pyroclastic rocks at Sierra de San Julián, Zacatecas state, and their approximate relative volumetric abundance.

tion of grains by pressure solution) were not observed. In those sediments containing a high percentage of clay matrix, there are fewer grain-to-grain contacts.

Primary porosity has been eliminated in all Nazas sedimentary units, mostly because of the high percentage of easily deformable VRFs. A minimal amount (less than 2%) of secondary porosity has developed in some of the sediments by intragranular dissolution of feldspar grains.

Composition and Classification

Framework grain composition of 10 Nazas sandstone, conglomerate sand-fraction, and siltstone samples is shown in Table 4, and the classification of these same samples is presented in Figure 9. Volcanic quartz, feldspar, and volcanic rock fragments (VRFs) make up the majority of the framework grains; only a few hematite pseudomorphs after pyroxene and/or hornblende were observed.

Quartz averages 23% of total framework grains, with a minimum of 10% (sample M17, unit Mg) and a maximum of 60% (sample M16, unit Mg). Almost all quartz is monocrystalline, water-clear (lacking vacuoles and microlites), and exhibiting straight or slightly undulose extinction.

Many quartz grains are unfragmented and display a hexagonal-bipyramidal shape, embayments, and rounded terminations. The above characteristics are typical of quartz derived from volcanic rocks. Rarely, there occur medium to very coarse sand-sized quartz grains that are finely polycrystalline (composed of five or more subcrystals). Only three such grains were observed, comprising far less than 1% of all quartz grains counted. The polycrystalline quartz was possibly derived from metamorphic rocks; however, there are no other indications of such a source. More likely, these grains were derived from certain altered tuffs in the lower volcanic member where the silicified matrix is made up of an interlocking mosaic of quartz crystals.

Feldspar averages 27% of all framework grains, with a minimum of 3% (sample R42c, unit PBn) and a maximum of 54% (sample M20c, unit Mh). The feldspar grains in the Nazas sedimentary units exhibit approximately the same degree and type of alteration as the feldspar phenocrysts in the penecontemporaneous volcanic unit. Thus, similar problems arose in attempting to identify feldspar types. However, albite and albite-Carlsbad twins are abundant, indicating a predominance of plagioclase over potassium feldspar. Two measurements of anorthite content were obtained from plagioclase grains in unit Mh, yielding values of An_{15} and An_{24}.

Rock fragments are the most abundant of the three framework constituents. They comprise an average of 50% of all framework grains, with a minimum of 2% (sample M16, unit Mg) and a maximum of 85% (sample R42c, unit PBn). Virtually all rock fragments in the sediments of the Nazas Formation are of volcanic origin (VRFs). The exceptions are rare intraclasts of penecontemporaneous siltstones found in a few granule conglomerate units. The VRFs were apparently derived from the Nazas volcanic rocks, although no attempt was made to

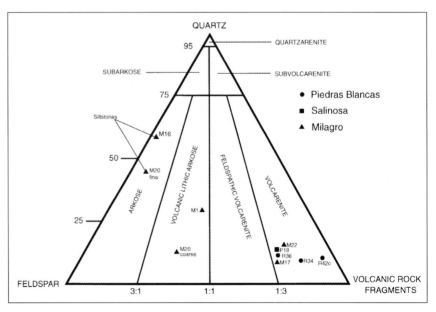

FIGURE 9. Classification of Nazas Formation sandstones, conglomerate sand fractions, and siltstones, Sierra de San Julián, Zacatecas state (classification system of Folk, 1974).

link individual VRFs to specific volcanic units. The majority of VRFs are fragments of pyroclastic rock and display the same types of alteration that characterize the Nazas volcanic rocks. Many VRFs have been highly deformed by compaction, such that recognition of grain boundaries is difficult. In some cases, authigenic clay rims delineate grain boundaries; more commonly, variations in texture and color in the crushed VRF pseudomatrix were used to distinguish individual grains.

For each of the 10 samples analyzed, the relative percentage of quartz, feldspar, and volcanic rock fragments was plotted on

TABLE 3. Framework grain roundness of Nazas Formation terrigenous sedimentary rocks.

	Very angular	*Angular*	*Subangular*	*Subrounded*	*Rounded*
Overall averages	%	%	%	%	%
All sediments	2	14	38	40	6
Quartz	2	43	36	18	1
Feldspar	3	28	53	15	1
VRFs	-	2	34	59	5
Data from individual samples					
R9a (unit PBe)					
Total frameworks	-	2	33	44	21
R34 (unit PBm)					
Total frameworks	-	10	44	43	3
Quartz	-	56	33	11	-
Feldspar	-	38	62	-	-
VRFs		3	43	52	2
R35 (unit PBm)					
Total frameworks	*	*	-	-	-
R36 (unit PBm)					
Total frameworks	-	11	43	43	3
Quartz	-	60	10	20	10
Feldspar (totally altered to clay)					
VRFs	-	-	52	47	1
R42c (unit PBn)					
Total frameworks	-	*	*	-	-
P14 (unit Se)					
Total frameworks	*	*	*	-	-
P18 (unit Sh)					
Total frameworks	-	2	40	56	2
Quartz	-	38	50	12	-
Feldspar	-	10	70	20	-
VRFs	-	-	32	66	2
P22 (unit S1)					
Total frameworks	-	*	*	*	-
M1 (unit Ma)					
Total frameworks	3	36	36	22	3
Quartz	6	43	45	6	-
Feldspar	3	51	34	11	-
VRFs	-	10	27	53	10
M17 (unit Mg)					
Total frameworks	10	30	33	26	1
Quartz	10	43	30	17	-
Feldspar	19	44	25	13	-
VRFs	-	2	44	52	2
M20c (unit Mh)					
Total frameworks	-	14	47	33	6
Quartz	-	22	33	45	-
Feldspar	-	18	68	13	-
VRFs	-	-	18	64	18
M20f (unit Mh)					
Total frameworks	-	*	*	-	-
M22 (unit Mh)					
Total frameworks	-	7	25	67	1
Quartz	-	38	50	12	-
Feldspar	-	8	58	33	1
VRFs	-	-	17	81	2

indicates predominant shape(s) represented

TABLE 4. Framework grain composition of Nazas Formation sandstones and conglomerate sand fractions, expressed in percentage of total framework grains.

Sample	R34	R36	R42c	P18	M1	M16	M17	M20c	M20f	M22
Unit	PBm	PBm	PBn	Sh	Ma	Mg	Mg	Mh	Mh	Mh
Quartz	11%	13%	12%	15%	30%	60%	10%	14%	45%	17%
Feldspar	11%	29%	3%	19%	37%	38%	13%	54%	50%	14%
Volcanic rock fragments	78%	58%	85%	66%	33%	2%	77%	32%	5%	69%

a Folk (1974) composition triangle (Figure 9). Six are volcarenites with 10–17% quartz, two are volcanic lithic arkoses with 14% and 30% quartz, and two are arkoses with 45% and 60% quartz. The variation in composition is primarily controlled by grain size and does not reflect any significant variability in provenance. The relatively quartz- and feldspar-rich arkoses (samples M16 and M20f) are both siltstones that occur in the same units as the VRF-rich, quartz-poor sandstones M17 and M20c; the volcarenitic sandstones grade into the arkosic siltstones. It appears that, as the volcanic detritus broke down to grain sizes finer than sand, VRFs completely disaggregated to yield clay-sized material, free quartz, and free feldspar. Arkose is the resulting composition. For sand-sized material, the amount of energy in the depositional environment also exerted a significant control on composition. Sample M20c (unit Mh) is just as coarse as many of the volcarenites, but it is a volcanic lithic arkose. However, M20c was deposited in the highest-energy flow regime of the entire formation, as indicated by its planar-bedded structure and relatively high degree of sorting. Relatively soft VRFs would have been mechanically unstable in the high-energy environment and were thus selectively reduced in relative abundance. The high feldspar and quartz content of sample M1 (unit Ma), which is also coarse, cannot be explained in the same manner as for sample M20c. It was deposited in a low-energy environment, as shown by its relatively poor sorting and lack of stratification. Much of the matrix of M1, however, is probably a deformed VRF pseudomatrix, and its actual composition is probably a feldspathic volcarenite or volcarenite.

Clay-sized intergranular material was counted during thin-section analysis, but authigenic and detrital components were not quantitatively differentiated. Authigenic clay, sericite or illite, probably makes up less than 5% of the nonframework constituents in all sediments analyzed. It occurs as thin (less than 10 μm) rims partially surrounding some framework grains, small pore fillings, and as a patchy replacement of detrital matrix, feldspars, and VRFs. A first-order gray mineral, probably a zeolite, was also observed as a minor pore fill. Hematite occurs as a replacement of both detrital matrix and VRFs, imparting the reddish color to most of the Nazas sediments. Detrital clay matrix, occurring as pore fill, comprises most of the nonframework composition. Sandstones from the upper red-bed member contain an average of 10–20% matrix by volume. Poorly sorted, matrix-supported conglomeratic units in the lower volcanic member contain as much as 40% matrix.

These estimates are probably somewhat high, because certain compacted VRFs are indistinguishable from true matrix. In any case, all Nazas sediments examined contain more than 5% detrital clay and thus are texturally immature.

ORIGIN OF THE LOWER VOLCANIC MEMBER

The rocks of the lower volcanic member of the Nazas Formation, which are predominantly intermediate-silicic pyroclastic rocks with interbedded lava flows and poorly sorted volcaniclastic sediments, are common constituents of composite cones such as the volcanoes of the Cascade Range in the northwestern United States. In particular, the presence of immature volcanogenic sediments suggests that the early Mesozoic landscape in the Teyra–San Julián region was characterized by the high relief of a composite volcano terrain rather than the low relief of a Huizachal Group or Eagle Mills Formation–type rift valley. Extensive hematite staining and lack of textures developed during subaqueous extrusion (e.g., hyaloclastites and pillow lavas) indicate that the lower volcanic member was formed in a subaerial environment.

Eruptive activity was mostly explosive, with some minor effusive eruptions also occurring. Vulcanian, Peléan, and Plinian eruptions took place, at present typical of circum-Pacific magmatic-arc volcanism (Macdonald, 1972). Glowing avalanches (depositing ash-flow tuffs and related ash-fall tuffs), Vulcanian explosion clouds (depositing ash-fall tuffs), and lava flows were all produced.

The location of central-vent and/or fissure sources and the relative position of the Nazas volcanic rocks at San Julián have not been determined. Nevertheless, three features suggest that the source vent or vents were located nearby: (1) the presence of rhyolitic, andesitic, and latitic lavas, which are fairly viscous and usually short traveled; (2) the massive, nongraded nature of most of the sediments, which typifies the proximal portions of the sediment apron surrounding a volcano (Parsons, 1969); and (3) the abundance of crystal fragments in the ash-fall tuffs, which most often are deposited near the volcanic center (Hay, 1959). In any case, the units of the lower volcanic member are considered part of a composite cone complex in a magmatic-arc setting. In support of this interpretation, Y and Rb values from the Nazas volcanic rocks, as well as from igneous rocks in the Rodeo and Caopas Formations, are consistent with those of volcanic-arc and collision granites (Jones et al., 1995).

The sedimentary rocks in the lower volcanic member are interpreted as lahars (single depositional units) and lahar assemblages (multiple depositional units), which are debris flows deposited on the flanks of a volcanic edifice. The lahars of the lower volcanic member are similar in composition, texture, and structure to certain conglomeratic beds in the upper red-bed member (e.g., PBn). However, they are inferred to pertain to a different depositional environment because of their interstratification with volcanic rocks and the fact that only one depositional mode is represented.

Lahars tend to follow preexisting streambeds where available and may obstruct later runoff, resulting in areas of ponded water. Hay (1959) described laminated siltstones and claystones that were deposited in small lakes formed by mudflows damming the Rabacca River on the island of Saint Vincent, West Indies, during the 1902 eruption of Mount Soufrière. The laminated siltstones at 642–643 m in unit Sh are interpreted to be such ponded-water deposits; this "varved" interval is interbedded with lahar deposits.

PROVENANCE AND ENVIRONMENT OF DEPOSITION OF THE UPPER RED-BED MEMBER

The sedimentary rocks of the upper red-bed member of the Nazas Formation are evidently first-cycle sediments derived from erosion of the penecontemporaneous Nazas volcanic rocks. This interpretation is supported by the following characteristics: (1) All detritus is attributable to a volcanic source; (2) volcanic rock fragments are similar to the Nazas volcanic rocks in texture and color (most are pyroclastic rock fragments); (3) as in the Nazas volcanic rocks, plagioclase is the predominant feldspar; (4) the anorthite content of plagioclase is comparable in both the volcanic and sedimentary rocks; (5) the alteration of feldspars is the same type (calcite and sericite replacement) and approximately the same degree in the volcanic and sedimentary rocks; and (6) the sedimentary rocks are compositionally and texturally immature, implying a short transport distance.

An interpretation of the depositional environment of the upper red-bed member is based on the following characteristics:

1) predominance of coarse grain sizes
2) textural and compositional immaturity
3) ubiquitous red coloration
4) lack of fossils
5) approximately unidirectional paleocurrents, as revealed in cross-bedding
6) predominance of sheetlike beds

These characteristics, considered together, are indicative of an alluvial-fan environment of deposition. Five types of deposits that characterize alluvial-fan deposits of dry regions were defined by Bull (1972). Four of these are represented in the upper red-bed member, each of which has a distinct degree of sorting, as shown in the cumulative grain-size frequency curves of Figure 10.

Water-laid Deposits

Sheetflood Deposits (Unit Mh)

Sheetfloods are relatively low viscosity flows that expand outward from the downstream end of channels incised in the proximal fan. The formerly channelled flow thus deteriorates into low-sinuosity, shallow, braided channels and bars which repeatedly divide and rejoin. Sheetflood deposits are typically planar-bedded sheets of sand and granules that contain an average of about 5% detrital clay. The coarse beds are commonly overlain by thin interfluve siltstones or mudstones (Bull, 1963, 1972).

The relatively well sorted, sheetlike sandstone and siltstone beds of unit Mh are interpreted as sheetflood deposits of a mid-fan environment. Some of the siltstone beds in Mh are thicker (as much as 2 m) than would be expected for an interfluve deposit and may in fact be mudflows. In accord with the interpreted mode of deposition, Mh represents the best-sorted (moderate to moderately well) sediments in the red-bed member (Figure 10).

Channel-fill Deposits (coarse lenses in units PBm and Mg)

Channels incised in the fan surface may head in the adjacent highlands or on the fan itself (Denny, 1967). The former are mainly agents of fan deposition with minor fanhead erosion, whereas the latter act to remove debris from the fan. Channels are most common on the proximal fan. The deposits which backfill fan channels tend to be coarser grained and more poorly sorted than sheetflood sediments (Bull, 1972).

The coarse lenses in units PBm and Mg are demonstrably stream channel deposits. The channel fill is composed of moderately to poorly sorted (Figure 10) sandstone and conglomerate, containing 10–20% clay matrix, that display crude normal grading to thin, stratified siltstones. The channels were scoured into very poorly sorted, massive, clayey, sandy siltstones that are probably mudflow deposits. Channels increase in dimension upward through the section.

Debris-flow Deposits

Debris flows are of high density and viscosity in comparison with sheetfloods and streamflows. They occur when a flow contains sufficient sediment of a wide size range, such that sediment entrainment becomes irreversible (the flow cannot selectively deposit any size fraction), and the flow acts more like a plastic mass than a Newtonian fluid (Bull, 1972). Flow is laminar; hence the base of a debris-flow unit forms an abrupt but nonscoured contact with the underlying material (Hooke, 1967). Mudflows are a variety of debris flow in which clay, silt, and sand sizes predominate (Varnes, 1958). Debris flows are most common on the proximal fan, near the fan apex (Bull, 1972).

Two types of debris-flow deposits are distinguished in the upper red-bed member of the Nazas in the Sierra de San Julián, based on the amount of entrained detrital clay:

High-viscosity Debris-flow Deposits (PBn 996–1021 m and massive siltstones of PBm and Mg)

High-viscosity debris-flow deposits lack stratification or grading, contain clasts that are randomly oriented, and have a high detrital clay content (Bull, 1972; Hooke, 1967). Bull (1964) performed grain-size analyses on 50 Holocene high-viscosity debris-flow samples, which yielded a mean clay content of 31%.

In the upper red-bed member of the Nazas, the portions of units designated as high-viscosity debris flows are poorly to

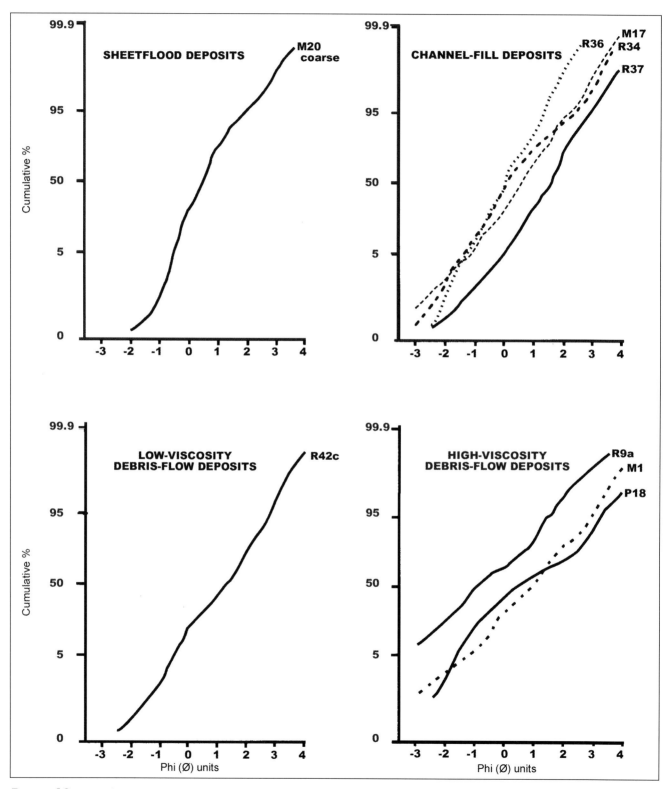

FIGURE 10. Cumulative grain-size frequency curves (probability ordinate), Nazas Formation terrigenous sedimentary rocks, Sierra de San Julián, Zacatecas state.

extremely poorly sorted and contain as much as 40% detrital clay. The massive, clayey siltstones of PBm and Mg are interpreted as viscous mudflows. The boulder beds of PBn 996–1021 m are viscous debris flows. These consist of thick, massive beds of randomly oriented sand and gravel clasts set in a silty matrix.

The lahars of the lower volcanic member are also high-viscosity debris flows but are not part of an alluvial fan complex.

Low-viscosity Debris-flow Deposits (PBn 901–996 m, So, Mi)

Low-viscosity (fluid) debris-flow deposits exhibit crude normal grading, subhorizontal orientation of tabular gravel clasts at and near the base of the flow, and a relatively low clay content (Bull, 1963, 1972). Bull (1963) estimated an average clay content of 17% for Holocene fluid debris flows in western Fresno County, California. Most of the clay occurs as a thick film around sand and gravel grains.

The portion of unit PBn from 901–996 m, unit So, and unit Mi are interpreted as being fluid debris-flow deposits. These beds are the most abundant type of deposit in the upper red-bed member. They consist of moderately to poorly sorted, crudely normal-graded, granule to cobble conglomerates (0.3 to 4 m thick) that are sometimes overlain by thin (5- to 10-cm-thick) planar-bedded siltstones. Conglomerate clasts display random orientation, except at the base of each bed, where they are oriented subparallel to the bedding plane. Clast imbrication is lacking. The conglomerate beds contain approximately 15% clay matrix.

The siltstones on top of the fluid debris-flow units are interpreted as being deposited by streamfloods, which are known to follow mudflows on modern alluvial fans (Sharp and Nobles, 1953). Steel (1974) has described similar debris-flow conglomerate/planar-bedded sandstone couplets in the Permo-Triassic New Red Sandstone of western Scotland.

Depositional Setting

The upper red-bed member is a generally coarsening-upward sequence composed of midfan to proximal-fan, stream-channel, and sheetflood deposits that are overlain by proximal-fan, debris-flow deposits. The overall sequence is thus interpreted to be progradational. The cause of this progradation is dependent on the process or processes that provided the geomorphic relief necessary for fan development, which is discussed below.

The prime requisite of alluvial-fan formation is the juxtaposition of highland and lowland (Denny, 1967), which may occur via three general geologic processes: (1) the differential erosion of adjacent rock of varying resistance, (2) volcanic edifice construction, or (3) block faulting.

Alluvial fans that accumulate by differential erosion of adjacent rock types of varying resistance should be mainly aggradational, because the highland source (the more resistant rock body) is not rejuvenated. A coarsening-upward sequence not related to fan progradation could conceivably be produced in such a setting by climatic change; a trend toward greater aridity would reduce vegetative cover and inhibit weathering, thereby optimizing the conditions for rapid erosion and deposition of coarser debris. The Nazas fan deposits, however, do not contain a record of such a climatic change, and the bedrock in the depositional basin (the lower volcanic member) was the same lithologic type as the highland source.

The volcaniclastic apron which accumulates at the base of a volcanic edifice is composed of coalescing alluvial fans ("volcanic molasse" of Mazarovich, 1972). A coarsening-upward sequence may occur during construction of a volcanic pile. Eruptive activity continually renews the supply of detritus in the fan catchment, and the entire cone complex progrades radially away from the volcanic center. Such a prograding apron, however, would be expected to contain interstratified volcanic rocks. Although the upper red-bed member is only partially preserved, more than 200 m of sediments was deposited without any volcanic interbeds, and it appears that volcanism had ceased, or at least greatly diminished, before the onset of fan deposition.

Walton (1979) suggested that a vertical increase in grain size of volcaniclastic apron sediments also may be caused by a cessation of volcanic activity. If fan deposition develops during a period of explosive volcanism that produces mainly fine, nonwelded pyroclastic debris, the apron sediments should be of similar fine-grain size as long as the volcano remains active. As soon as volcanism stops, erosion would remove the cover of fine pyroclastics and expose the nonfragmented, shallow, intrusive bodies that served as conduits for the material erupted; these would presumably supply coarser debris. In this model, VRF composition should change from predominantly pyroclastic rock detritus in the lower part of the fan sequence to mainly nonpyroclastic volcanic detritus in the upper, coarse part of the sequence. Such a VRF compositional change was not observed in the Nazas.

Although the upper and lower Nazas members were not observed in fault contact, block faulting of the lower volcanic member appears to be the most likely source of the upper red-bed member. In this model, the alluvial fan sequence was deposited in active grabens or half grabens which formed in the preexisting volcanic terrain after the cessation of volcanism (Figure 11).

Graben formation is a well-known adjunct of arc volcanism. Small radial grabens (sector grabens) may form on the flanks of composite volcanoes by stretching of the cone during an active volcanic phase or by removal of magma from the underlying chamber during volcanic quiescence or extinction (Macdonald, 1972). Large graben systems, or "volcano-tectonic depressions," are common in areas of recent explosive volcanic activity. In El Salvador and Nicaragua, large grabens have developed on the crests of broad, elongate arches after the eruption of voluminous dacitic and rhyolitic pyroclastic rocks

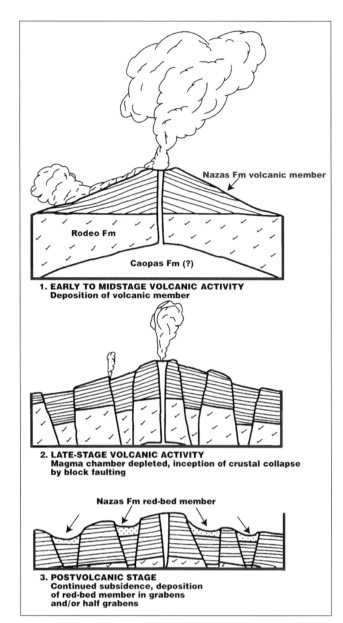

FIGURE 11. Environments of deposition and proposed volcano-tectonic evolution of the Nazas Formation.

(McBirney and Williams, 1965). Other examples of volcano-tectonic depressions are the Taupo Basin, New Zealand; Toba Basin, Sumatra; and the Ata depression on Kyushu, Japan (Macdonald, 1972). All are associated with cessation of explosive volcanism. As with the small, inactive-phase sector grabens, volcano-tectonic depressions are believed to form as magma is extruded and/or withdrawn from the underlying chamber(s), causing collapse of the overlying crust.

TECTONO-STRATIGRAPHIC REGIME OF THE NAZAS FORMATION

Despite their gross similarities, the Nazas Formation represents a distinctly different tectono-stratigraphic regime than other lower Mesozoic red beds in northeastern Mexico (Huiz-

achal Group, comprising the La Boca and La Joya Formations) and the U.S. Gulf Coast (Eagle Mills Formation). Instead of the assemblage of multiprovenance continental red beds and secondary mafic igneous rocks typical of the proto–Gulf of Mexico rifted margin, the section immediately underlying the western Gulf passive-margin prism in the Teyra–San Julián region and wherever the Nazas occurs is composed of voluminous intermediate-silicic explosive volcanic rocks and derived volcanogenic continental red beds typical of convergent-margin arc terranes. This distinction between the Nazas and other red beds in the western Gulf of Mexico province has been previously pointed out by several workers (Blickwede, 1981a, 1981b; Jones et al., 1990, 1995; Maher et al., 1991; Grajales Nishimura et al., 1992; Bartolini, 1997; Barboza Gudiño et al., 1999) and appears to be widely accepted by now.

However, a related phenomenon that remains unexplained is the abrupt transition from Pacific convergent-margin to Gulf of Mexico passive-margin settings (at the Nazas/Zuloaga contact) in the Teyra–San Julián region as well as other areas where the Nazas Formation has been described. Why is there no evidence of a "confluence" of the two tectonic regimes, such as a transitional interval of interbedded Pacific-margin volcanogenic rocks and Gulf-margin marine sedimentary rocks in the Late Jurassic?

One possible explanation is that the Nazas was deposited in a geographic location far from its position during the Oxfordian when it was transgressed by the western Gulf of Mexico seas and reached this later position by tectonic transport. Anderson et al. (1991) and Jones et al. (1995) presented a case for the Oxfordian compressional deformation event in the Teyra–San Julián region being related to movement along the postulated (but as yet unproved) Mojave-Sonora megashear, which according to them would have tectonically translated the Nazas and underlying section eastward for a distance of more than 1000 km from its original position. The megashear hypothesis is compelling, but I question the lack of direct evidence for the megashear as well as the speed at which the translation must have taken place if the associated deformation was confined to the Oxfordian. Even allowing for displacement during the entire span of the Oxfordian, about seven million years, tectonic transport along the megashear would have had to proceed at a pace of at least an average of 15 cm/yr to move the section now exposed in the Teyra–San Julián region from its postulated original position in the vicinity of what is now northern Sonora state. Although feasible, this rate of translation seems anomalously high for a transcurrent plate margin. Alternatively, the inception of seafloor spreading in the central Gulf of Mexico during the Oxfordian (?) could conceivably have caused a change in the convergence of the Pacific oceanic plates and Mexico, perhaps steepening the angle of subduction and rapidly shifting the axis of arc magmatism to the west.

REVIEW OF PRE-OXFORDIAN STRATIGRAPHY IN THE TEYRA–SAN JULIÁN REGION AND PROPOSAL OF "SAN JULIÁN COMPLEX"

The Nazas Formation is one of four major pre-Oxfordian stratigraphic units in the Teyra–San Julián region that appears to be related to a convergent margin: the Taray, Rodeo, and Nazas volcanic and terrigenous sedimentary units and the Caopas igneous pluton. All four units have been subjected to varying degrees of metamorphism, perhaps related to the Upper Jurassic deformation event described by Anderson et al. (1991). The following is a review of these pre-Oxfordian units and their stratigraphic interrelationships.

Taray Formation

Based on field observations, the oldest rocks exposed in the Teyra–San Julián region are a series of metamorposed conglomerates, sandstones, shales, radiolarites, and volcanic rocks referred to as the Taray Formation (originally named by Córdoba, 1964). The siliciclastic sediments in the Taray are heterogeneous in composition and have a varied provenance. Although no autochthonous fossils have yet been found in the Taray and no radiometric dates have been obtained from its volcanic rocks, this apparently marine unit has been tentatively correlated with the Upper Triassic Zacatecas Formation, which crops out approximately 200 km to the south around the city of Zacatecas (López Infanzón, 1986). The Taray and Zacatecas Formations, along with similar units in the states of Guerrero and Michoacán, are thought to be part of an accretionary prism that was welded onto the Pacific margin of Mexico during the early Mesozoic (López Infanzón, 1986; Anderson et al., 1990).

Rodeo and Nazas Formations

Above the Taray Formation and below the Upper Jurassic Oxfordian limestones of the Zuloaga Formation is a thick section of silicic-intermediate volcanic rocks and volcaniclastic sedimentary rocks referred to as the Rodeo and Nazas Formations, which together comprise the magmatic arc discussed previously. The age and stratigraphic relationship of the Rodeo and Nazas are still matters of debate because of limited exposure and a dearth of reliable radiometric dates from the units. López Infanzón (1986) proposed that the metasedimentary rocks originally designated as lower Rodeo Formation by Rogers et al. (1963) be assigned to the Taray, and the volcanic rocks of the upper Rodeo Formation be considered part of the Nazas. I tend to agree with the first proposal of López Infanzón regarding the Taray and would agree with the concept of treating the Nazas and Rodeo volcanic rocks as a single unit if additional radiometric dates support the concept. In addition, an apparent metamorphic-grade differential across the interpreted Rodeo/Nazas contact zone, described below, needs to be

confirmed or discounted before revising the stratigraphic nomenclature.

Rogers et al. (1961) originally proposed the name Rodeo Formation for the series of green, gray, and purple sericite and chlorite phyllites with a few metaconglomerate interbeds that crop out in and around the village of El Rodeo in the foothills of the Sierra de San Julián. Overlying the phyllite, along the road leading northeastward to the Piedras Blancas section, is a thick section of greenish-gray andesite, which consists of an altered chloritic matrix containing stretched plagioclase phenocrysts. The upper 60 m of the andesite, just below the contact with the Nazas Formation as defined in this study, is altered to dark purple. Córdoba (1964) measured and described a 996-m-thick section of metasediments located 2.5 km northeast of Cerro Europa, apparently equivalent to the section cropping out in El Rodeo village. López Infanzón (1986) reinterpreted Córdoba's section as belonging to the Taray Formation and apparently considered only the above-mentioned andesite (and perhaps other, age-equivalent volcanic units) to belong to the Rodeo Formation.

As mentioned above, the contact between the Rodeo Formation and the overlying Nazas Formation is equivocal, if indeed they are separate formations. In the central Sierra de Teyra, the formational contact was described as a pronounced angular unconformity by Córdoba (1964), but this locality apparently has not been found by later workers. I was unable to find the exact Rodeo-Nazas contact in any outcrops in the Sierra de San Julián, but I have tentatively placed the contact in a section of alluvial cover above the top of the purple altered andesite sequence mentioned above. Two phenomena suggest that a formational contact may occur somewhere in this covered interval: (1) the highly altered nature of the upper andesite, suggesting prolonged subaerial exposure, and, more important, (2) the apparent deformational-grade differential across the horizon. In thin sections, the phenocrysts in the tuffs above the inferred contact are not stretched as they are in the andesite. In addition, Anderson et al. (1991) observed a "widespread, northeast-striking foliation and other ill-defined northeast-trending structures" that were found only in the Rodeo and Caopas Formations, possibly reflecting a deformation event that predates the deposition of the Nazas. If one accepts López Infanzón's (1986) redefinition of the Rodeo Formation, then the only other possible horizon for a Rodeo-Nazas boundary is at the base of the andesite, but this would effectively combine the two units.

The absolute age of the Rodeo and Nazas Formations is not firmly established, but both units are tentatively thought to be Triassic and/or Early to Middle Jurassic. Of a total of 14 radiometric dates obtained from the two units (Denison et al., 1971; Pantoja Alor, 1972; López Infanzón, 1986; Bartolini, 1997; Bartolini, personal communication, 1998), only three indicate a pre-Oxfordian age and are considered potentially valid. A rhyolite sample from the Nazas exposure in the Villa Juárez uplift of Durango state, about 200 km west of the San Julián area, gave a lead-alpha date of 230 ± 20 my (Permian Guadalu-pian–Triassic Norian) (Pantoja Alor, 1972). More recently, Bar-tolini and Spell (1997) reported an $^{40}Ar/ ^{39}Ar$ date of 195.3 ± 5.5 Ma (Early Jurassic Sinemurian-Toarcian) from plagioclase in a rhyolite in the same Nazas section at Villa Juárez. Finally, a K-Ar date of 183 ± 8 Ma (Toarcian-Bathonian, Early to Middle Jurassic) was obtained by López Infanzón (1986) from a vol-canic unit in the Rodeo Formation near the Piedras Blancas section of this study (probably the andesite below my inferred contact with the Nazas), leading him to consider the overlying Nazas as Middle Jurassic in age. Nairn (1976) interpreted paleo-magnetic data acquired from the Nazas in the Villa Juárez area to indicate a Triassic age. Reinterpretation of the same data set by Cohen et al. (1986) showed that the data were also compat-ible with an Early Jurassic pole position.

Also currently at issue is whether to consider the red-bed alluvial sediments just below the Oxfordian Zuloaga Forma-tion in the Teyra–San Julián region as part of the Nazas ("upper red-bed member" of this paper and Blickwede, 1981a, 1981b) or as a separate formation (La Joya Formation of Anderson et al., 1991, and Jones et al., 1995). The use of the name La Joya for the red beds in the Teyra–San Julián area is misleading. As López Infanzón (1986) has pointed out, the upper member of the Nazas and the Huizachal Group (La Joya and La Boca For-mations) represent two petrogenetically distinct continental sedimentary cycles in the Mesozoic of northern Mexico. Un-like the Nazas red beds, the Huizachal Group is not associated with a significant volume of volcanic rocks and was derived from the erosion of a wide variety of rock types. Furthermore, the Huizachal Group is widely held to have been deposited in basins related to passive-margin (Gulf of Mexico) tectonics, and the Nazas is more typical of units deposited on a conver-gent (Pacific) margin. Finally, the La Joya Formation is transi-tional into the overlying Oxfordian carbonates and evaporites (Götte, 1986, 1990; Michalzik, 1988), and the Nazas upper red-bed member, as described in this report, is not.

Regardless of the name it is assigned, should the upper red-bed member be raised to formation status? Anderson et al. (1991) state that this reclassification is justified because the upper red-bed member is lithologically distinct and readily mappable, as well as having inferred differences in age and ori-gin when compared with the lower volcanic member. Al-though I agree that the upper red-bed member is mappable in its own right, I believe it important to maintain a stratigraphic nomenclature that conveys its close relationship with the lower volcanic member, i.e., either as members of the same for-mation or as formations within the same group.

In support of keeping the two units as part of the same for-mation, I would first argue that they are not lithologically dis-tinct. As shown in my petrologic analysis of the Nazas, the lower volcanic member in effect cannibalized itself to produce the upper red-bed member; all detritus in the upper red-bed member is volcanogenic, and all clasts in the conglomerates are potentially traceable to the lower member. This apparently holds true for Nazas sedimentary units wherever they have been described in northern Mexico. In addition, there is no reason to infer a significant difference in the age of the two Nazas members if one accepts the evolutionary model pre-sented in this paper. In my model, which was discussed in the section on provenance and environments of deposition, the development of the normal fault blocks in which the upper red-bed member is thought to have been deposited was caused by the depletion and collapse of the magma chambers which "fed" the lower volcanic member; as such, the faulting would have immediately followed the cessation of volcanism. Accordingly, I perceive the origin of the lower and upper mem-bers of the Nazas as intimately linked, with the existence of the upper member dependent on the existence of the lower mem-ber. Both are related to the same episode of arc volcanism in the region.

Caopas Formation

The fourth unit that has been described in the pre-Oxfor-dian section of the Teyra–San Julián region is the Caopas For-mation (de Cserna, 1956; Rogers et al., 1961; Córdoba, 1964, 1965; Denison et al., 1969; López-Infanzón, 1986; Anderson et al., 1991; Jones et al., 1995), a metamorphosed, hypabyssal igneous pluton that crops out in the core of the San Julián uplift (Figure 5). Based on recent petrographic and chemical analyses (López Infanzón, 1986; Jones et al., 1995) the pro-tolith of the Caopas Formation was apparently a granitic or dioritic porphyry. The poor exposure and structural complex-ity of the unit precludes the possibility of determining its true thickness in outcrop, but a minimum of 500 m was estimated by Córdoba (1964).

Neither the age, extrusive equivalents, nor country rock of the Caopas plutonic body is definitely known. Although con-tacts with the Rodeo (Lower and Middle Jurassic?) and Zuloaga Formations (Upper Jurassic Oxfordian) have been observed in a number of outcrops, they are either normal faults or shallow-dipping zones that display intense shearing (Anderson et al., 1991). In places where the Caopas-Rodeo contact is character-ized by ductile normal faulting, Anderson et al. (1991) ob-served intercalations of mylonitized Caopas porphyry with feldspathic volcanic beds of the Rodeo and stated that this phenomenon implies the existence of a transition between these units. I question whether this apparent intercalation of the Caopas and Rodeo could instead be related to deformation in the fault zone.

The Caopas Formation was originally interpreted to predate the Rodeo and Nazas Formations (de Cserna, 1956; Rogers et al., 1961; Córdoba, 1964, 1965; Blickwede, 1981a) and was thought to be as old as Precambrian, based on the high meta-morphic grade where it was examined and lithologic similarity to some Paleozoic and Precambrian high-grade metamorphic rocks in western Texas. Early attempts at radiometric dating of the Caopas (Rb-Sr method) yielded a broad scattering of dates, ranging from Early Permian to Early Cretaceous (Fries and

Rincón Orta, 1965; Denison et al., 1969; Denison et al., 1971), and many of the dates obtained were interpreted to be "reset" by thermal events postdating the emplacement of the igneous body. More recently, the Caopas has been shown to be variably metamorphosed, and in places it displays only a weak foliation (López Infanzón, 1986; Anderson et al., 1991), unlike the metamorphic rocks with which it was initially compared. These and other workers (Jones et al., 1995) have interpreted the Caopas to be as young as early Mesozoic in age and perhaps to be related genetically to the Rodeo and/or Nazas volcanic rocks. Jones et al. (1995) believe that the Caopas porphyry is the plutonic equivalent of the Nazas Formation volcanic rocks. In support of this interpretation, they cite the chemical similarity of a sample from the Caopas to samples of volcanic rocks from the Nazas Formation. Indeed, their data seem to show a general chemical kinship between not only the Caopas and Nazas, but the Rodeo Formation as well. As a caveat, Jones et al. (1995) point out that all their samples from the pre-Oxfordian section showed evidence of significant hydrothermal alteration, and as such, any conclusions based on mobile-element data (most of the chemical components analyzed) are suspect.

Jones et al. (1995) obtained a U-Pb date of 158 ± 4 Ma (Late Jurassic Oxfordian-Kimmeridgian) from zircons in a relatively undeformed sample of the Caopas Formation. This date must also be regarded as questionable because of the regional deformation and thermal alteration that occurred during the Late Jurassic (Anderson et al., 1991), which may have "reset" the radiometric "clocks." In addition, the Oxfordian-Kimmeridgian U-Pb date would make the Caopas pluton equivalent to the Zuloaga and/or La Caja Formations, which contain no volcanic rocks.

Observations made by Córdoba (1964, p. 18, 21) on an outcrop section 2.5 km northeast of Cerro Europa, near the Sierra de San Julián, support the interpretation originally held by workers in the 1950–1970s that the Caopas Formation predates the Rodeo and Nazas. In this section, interpreted by Córdoba to be part of the Rodeo Formation, he described a 116-m-thick conglomerate containing subangular to rounded pebbles of reworked Caopas Formation metamorphic rock. López Infanzón (1986) also examined this section near Cerro Europa, and although he did not address the existence of reworked Caopas in the conglomerate, he believed this entire section to be a part of the Taray Formation (Late Triassic?), which would require that the Caopas Formation be older still. Further study of this key outcrop seems necessary to help confirm the age of the Caopas pluton and its relationship to the other pre-Oxfordian units.

Proposal of "San Julián Complex"

Until a more comprehensive set of reliable radiometric dates or other age-constraining data are obtained from the pre-Oxfordian section in the San Julián Uplift and the stratigraphic relationship of its constituent units becomes more clear, I propose that this entire group of rocks be grouped together as the "San Julián Complex." This grouping also makes sense from the standpoint that the Taray, Rodeo, Nazas, and Caopas Formations all appear to be genetically related to the same convergent-margin phase in the geologic history of the region, just prior to the establishment of the western Gulf of Mexico passive margin.

CONCLUSIONS

1) The Nazas Formation in the Sierra de San Julián is composed of two members: (a) a thick (average 946-m) lower volcanic member and (b) a thin (average 146-m) upper red-bed member.

2) The total formation comprises 45% ash-fall tuffs, 20% ash-flow tuffs, 10% lava flows, and 25% volcaniclastic sediments. Of the sedimentary rocks, 50% are conglomerates, 25% are sandstone, and 25% are siltstones. Mudrocks are entirely lacking.

3) More than two-thirds of the volcanic rocks are andesites and dacites. Rhyolites and latites also occur.

4) Sedimentary rocks are compositionally and texturally immature volcarenites, volcanic lithic arkoses, and arkoses. Grain size is the primary determinant of composition.

5) Volcanic rocks, especially ash-fall tuffs, are highly altered. Common alteration products are hematite (from hornblende, biotite, and pyroxene), calcite and sericite/illite (as a patchy replacement of feldspar phenocrysts and matrix), and quartz (silicification of matrix or entire rock). Sedimentary rocks also exhibit extensive hematite, calcite, and sericite/illite alteration.

6) The lower volcanic member comprises a composite-cone complex. The source vent or vents were located nearby. Sedimentary units interstratified with the volcanic rocks are lahars.

7) The upper red-bed member contains low- and high-viscosity debris-flow deposits, representing prograding medial to proximal alluvial-fan facies. The detritus that formed the fan(s) was derived entirely from the lower volcanic member.

8) The upper red-bed member was deposited in a graben (or graben system) that formed in the lower volcanic member because of the cessation of volcanic activity and consequent crustal collapse.

9) The Nazas Formation is probably Triassic and/or Early to Middle Jurassic in age, and along with the underlying Rodeo and Caopas Formations, it comprises a part of a Pacific-margin magmatic arc along the western margin of North America. Thus, the Nazas represents a distinctly different tectono-stratigraphic regime from other circum–Gulf of Mexico lower Mesozoic continental red beds, such as the Huizachal Group and Eagle Mills Formation, that were deposited in rift basins associated with the opening of the Gulf of Mexico. The Nazas-Rodeo-Caopas arc terrane forms the foundation for a large portion of the western Gulf of Mexico passive margin. Still unresolved is the curiously abrupt nature of the transition from convergent margin to passive margin in the region.

10) Additional fieldwork and laboratory work are required to understand the stratigraphic relationships of the pre-Oxfordian units in northern Mexico. In the meantime, it is proposed that the Taray, Rodeo, Nazas, and Caopas Formations be combined as the "San Julián Complex," a series of units that were all part of a convergent margin which forms a significant part of the updip rim of the western Gulf of Mexico Basin.

ACKNOWLEDGMENTS

I would like to express my gratitude to the members of my thesis committee at the University of New Orleans, Al Weidie, Bill Ward, and Lou Fernández, for their invaluable guidance during the course of my master's of science project, which was the basis for this paper. I would also like to thank Claudio Bartolini and Dick Buffler for their encouragement to prepare this paper for publication. Nancy MacMillan, Leta Smith, and Bill Ward reviewed the manuscript and provided important suggestions for improvements. IHS Energy Group/Petroconsultants, S.A., provided support in the word processing of the manuscript and drafting of the graphic materials; specifically, Shereen Qian and Pablo Cassina of the Geneva office carried out this work. Financial support for the fieldwork and laboratory analyses was provided in part by the Shell Graduate Fund, University of New Orleans.

REFERENCES CITED

Allen, W. W., 1976, Petrology of the Middle Jurassic (?) La Joya Formation, Sierra Madre Oriental, southwestern Tamaulipas, Mexico: Master's thesis, Texas A&M University, College Station, 189 p.

Anderson, T. H., J. W. McKee, and N. W. Jones, 1990, Jurassic (?) mélange in north-central Mexico: Geological Society of America Abstracts with Programs, v. 22, no. 3, p. 3.

Anderson, T. H., J. W. McKee, and N. W. Jones, 1991, A northwest trending, Jurassic fold nappe, northernmost Zacatecas, Mexico: Tectonics, v. 10, p. 383–401.

Aranda García, M., 1991, El segmento San Felipe del cinturón cabalgado, Sierra Madre Oriental, Estado de Durango, Mexico: Boletín de la Asociación Mexicana de Geólogos Petroleros, v. 41, p. 18–36.

Barboza Gudiño, J. R., J. R. Torres Hernández, and M. Tristán Gonzalez, 1997, Some pre-Oxfordian red beds and related stratigraphic units in the southern and northeastern Central Plateau, Mexico: Geological Society of America Abstracts with Programs, v. 29, no. 2, p. 2.

Barboza Gudiño, J. R., M. Tristán Gonzalez, and J. R. Torres Hernández, 1999, Tectonic setting of pre-Oxfordian units from central and northeastern Mexico: A review: Geological Society of America Special Paper 340, p. 197–210.

Bartolini, C., 1997, The Nazas Formation of northern and central Mexico: Mesozoic volcanic-sedimentary arc sequences, not red beds: Geological Society of America Abstracts with Programs, South-Central/Rocky Mountain Sections, v. 29, no. 2, p. 3.

Bartolini, C., 1998, Stratigraphy, geochronology, geochemistry and tectonic setting of the Mesozoic Nazas Formation, north-central Mexico: Ph.D. dissertation, University of Texas at El Paso, 558 p.

Bartolini, C., and T. Spell, 1997, An Early Jurassic age (^{40}Ar/^{39}Ar) for the Nazas Formation at the Cañada Villa Juarez, northeastern Durango, Mexico: Geological Society of America Abstracts with Programs, South-Central/Rocky Mountain Sections, v. 29, no. 2, p. 3.

Belcher, R. C., 1979, Depositional environments, paleomagnetism, and tectonic significance of Huizachal red beds (lower Mesozoic), northeastern Mexico: Ph.D. dissertation, University of Texas at Austin, 276 p.

Blickwede, J. F., 1981a, Stratigraphy and petrology of Triassic (?) "Nazas Formation," Sierra de San Julián, Zacatecas, Mexico: Master's thesis, University of New Orleans, 100 p.

Blickwede, J. F., 1981b, Stratigraphy and petrology of Triassic (?) "Nazas Formation," Sierra de San Julián, Zacatecas, Mexico: AAPG Bulletin, v. 65, p. 1012.

Bull, W. B., 1963, Alluvial fan deposits in western Fresno County, California: Journal of Geology, v. 71, p. 243–251.

Bull, W. B., 1964, Alluvial fans and near-surface subsidence in western Fresno County, California: U.S. Geological Survey Professional Paper 437-A, 70 p.

Bull, W. B., 1972, Recognition of alluvial fan deposits in the stratigraphic record, in J. K. Rigby and W. K. Hamblin, eds., Recognition of Ancient Sedimentary Environments: Society for Sedimentary Geology (SEPM) Special Publication 16, p. 63–83.

Carrillo Bravo, J., 1961, Geología del anticlinorio Huizachal-Peregrina al noroeste de Ciudad Victoria, Tamaulipas: Boletín de la Asociación Mexicana de Geólogos Petroleros, v. 13, p. 1–98.

Chapin, C. E., and G. R. Lowell, 1979, Primary and secondary flow structures in ash-flow tuffs of the Gribbes Run paleovalley, central Colorado, in C. E. Chapin and W. E. Elston, eds., Ash-flow tuffs: Geological Society of America Special Paper 180, p. 137–154.

Cohen, K. K., T. H. Anderson, and V. A. Schmidt, 1986, A paleomagnetic test of the proposed Mojave-Sonora megashear in northwestern Mexico: Tectonophysics, v. 131, p. 23–51.

Cook, E. F., 1965, Stratigraphy of Tertiary volcanic rocks in eastern Nevada: Nevada Bureau of Mines Report 11, 60 p.

Córdoba, D. A., 1964, Geology of the Apizolaya quadrangle (east half), northern Zacatecas, Mexico: Master's thesis, University of Texas at Austin, 111 p.

Córdoba, D. A., 1965, Hoja Apizolaya 13R-1(9), Con resumen de la geología de la Hoja Apizolaya, Estados de Zacatecas y Durango: Universidad Nacional de México, Instituto de Geología.

Corpstein, P., 1974, Historical geology of Huizachal-Peregrina anticlinorium and northeastern Mexico, in Geology of Huizachal-Peregrina anticlinorium: Pan American Geological Society Guidebook, p. 1–9.

de Cserna, Z., 1956, Tectónica de la Sierra Madre Oriental de México, entre Torreón y Monterrey: 20th International Geological Congress, Mexico, 87 p.

Denison, R. E., G. S. Kenny, W. H. Burke Jr., and E. A. Hetherington, 1969, Isotopic ages of igneous and metamorphic boulders from the Haymond Formation (Pennsylvanian), Marathon Basin, Texas, and their significance: Geological Society of America Bulletin, v. 80, p. 245–246.

Denison, R. E., W. H. Burke Jr., E. A. Hetherington, and J. B. Otto, 1971, Basement rock framework of parts of Texas, southern New Mexico, and northern Mexico, in The geologic framework of the Chihuahua tectonic belt: West Texas Geological Society Publication 71-59, p. 3–14.

Denny, C. S., 1967, Fans and pediments: American Journal of Science, v. 265, p. 81–105.

Fisher, R. V., 1961, Proposed classification of volcaniclastic sediments and rocks: Geological Society of America Bulletin, v. 72, p. 1409–1414.

Folk, R. L., 1974, Petrology of sedimentary rocks: Austin, Texas, Hemphill's Publishing Company, 170 p.

Friedman, G. M., 1958, Determination of sieve-size distribution from thin-section data for sedimentary petrological studies: Journal of Geology, v. 70, p. 737–756.

Fries Jr., C., and C. Rincón Orta, 1965, Nuevas aportaciones geocronológicas y técnicas empleadas en el Laboratorio de Geocronometría: Universidad Nacional Autónoma de México, Instituto de Geología Boletín, v. 73, p. 57–133.

Goddard, E. N., P. D. Trask, R. K. De Ford, O. N. Rove, J. T. Singewald, and R. M. Overbeck, 1970, Rock-color chart: Geological Society of America, Publication No. RCC001, 16 p.

Götte, M., 1986, Beitrag zur fazies analyse, tektonik und mineralisation der Huizachal-und Minas Viejas Formation (Ob. Trias-Ob. Jura) im Raum Galeana, Sierra Madre Oriental/Mexiko: Diplomarbeit, Technische Hochschule Darmstadt, 120 p.

Götte, M., 1990, Halotektonische deformationsprozesse in sulfatgesteinen der Minas Viejas-Formation (ober Jura) in der Sierra Madre Oriental, nordost-Mexiko: Ph.D. dissertation, Technische Hochschule Darmstadt, 270 p.

Grajales Nishimura, J. M., D. J. Terell, and P. E. Damon, 1992, Evidencias de la prolongación del arco magmático Cordillerano del Triásico Tardío–Jurásico en Chihuahua, Durango, y Coahuila: Boletín de la Asociación Mexicana de Geólogos Petroleros, v. 42, n. 2, p. 1–18.

Hay, R. L., 1959, Formation of the crystal-rich glowing avalanche deposits of St. Vincent, B.W.I.: Journal of Geology, v. 67, p. 540–562.

Hooke, R. L., 1967, Processes on arid-region alluvial fans: Journal of Geology, v. 75, p. 438–460.

Humphrey, W. E., 1956, Tectonic framework of northeast Mexico: Gulf Coast Association of Geological Societies, Transactions, v. 6, p. 25–35.

Imlay, R. W., E. Cepeda, M. Alvarez, and T. Díaz, 1948, Stratigraphic relations of certain Jurassic formations in eastern Mexico: AAPG Bulletin, v. 32, p. 1750–1761.

International Union of Geological Sciences Subcommission on the Systematics of Igneous Rocks, 1979, Classification and nomenclature of volcanic rocks, lamprophyres, carbonatites, and melilitic rocks: Geology, v. 7, p. 331–335.

Jones, N. W., J. W. McKee, T. H. Anderson, and L. T. Silver, 1990, Nazas Formation: A remnant of the Jurassic arc of western North America in north-central Mexico: Geological Society of America Abstracts with Programs, v. 22, no. 7, p. A327.

Jones, N. W., J. W. McKee, T. H. Anderson, and L. T. Silver, 1995, Jurassic volcanic rocks in northeastern Mexico: A possible remnant of a Cordilleran magmatic arc, in C. Jacques-Ayala, C. M. González-León, and J. Roldán-Quintana, eds., Studies on the Mesozoic of Sonora and adjacent areas: Geological Society of America Special Paper 301, p. 179–190.

Kellum, L. B., 1936, Geology of the mountains west of the Laguna District: Geological Society of America Bulletin, v. 47, p. 1039–1063.

Lazzeri, J. J., 1979, Stratigraphy and petrology of the Middle Jurassic La Joya Formation, Miquihuana, Aramberri-Mezquital, and Real de Catorce areas, Mexico: Master's thesis, University of New Orleans, 109 p.

López Infanzón, M., 1986, Estudio petrogenético de las rocas ígneas en las formaciones Huizachal y Nazas: Sociedad Geológica Mexicana Boletín, v. 47, no. 2, p. 1–42.

Macdonald, G. A., 1972, Volcanoes: Englewood Cliffs, N. J., Prentice-Hall, Inc., 510 p.

Mackin, J. H., 1960, Structural significance of Tertiary volcanic rocks in southwestern Utah: American Journal of Science, v. 258, p. 81–131.

Maher, D. J., N. W. Jones, J.W. McKee, and T. H. Anderson, 1991, Volcanic rocks at Sierra de Catorce, San Luis Potosí, Mexico: A new piece for the Jurassic arc puzzle: Geological Society of America Abstracts with Programs, v. 23, no. 5, p. A133.

Mazarovich, O. A., 1972, Geotectonic conditions for the formation of molasse: Geotectonics, no. 4, p. 14–21.

McBirney, A. R., and H. Williams, 1965, Volcanic history of Nicaragua: University of California Berkeley Publications in Geological Sciences, v. 55, p. 1–65.

McKee, E. D., and G. W. Weir, 1953, Terminology for stratification and cross-stratification in sedimentary rocks: Geological Society of America Bulletin, v. 64, p. 381–390.

Meiburg, P., J. R. Chapa-Guerrero, I. Grotehusmann, T. Kustusch, P. Lentzy, H. de León-Gómez, and M. A. Mansilla-Teran, 1987, El basámento precretácico de Aramberri: Estructura clave para comprender el décollement de la cubierta Jurasica/Cretacica de la Sierra Madre Oriental, México?: Actas de la Facultad de Ciencias de la Tierra, Universidad Autónoma de Linares, Nuevo León, v. 2, p. 15–22.

Michalzik, D., 1988, Trias bis tiefste unter-kreide de nördostlichen Sierra Madre Oriental, Mexiko—Fazielle entwicklung eines passiven kontinentalrandes: Ph.D. dissertation, Technische Hochschule Darmstadt, 247 p.

Michalzik, D., 1991, Facies sequence of Triassic-Jurassic red beds in the Sierra Madre Oriental (NE Mexico) and its relation to the early opening of the Gulf of Mexico: Sedimentary Geology, v. 71, p. 243–259.

Mixon, R. B., 1963, Geology of the Huizachal red beds, Ciudad Victoria area, southwestern Tamaulipas, in Peregrina Canyon and Sierra de El Abra: Corpus Christi Geological Society Guidebook, 107 p.

Mixon, R. B., G. E. Murray, and T. Díaz, 1959, Age and correlation of Huizachal Group (Mesozoic), state of Tamaulipas, Mexico: AAPG Bulletin, v. 43, p. 757–771.

Nairn, A. E. M., 1976, A paleomagnetic study of certain Mesozoic formations in northern Mexico: Physics of the earth and planetary interiors, v. 13, p. 47–56.

Pantoja Alor, J., 1963, Hoja San Pedro del Gallo 13R-K(3): Universidad Nacional Autónoma de México, Instituto de Geología, Carta Geológica de México, serie 1:100,000 (map with text).

Pantoja Alor, J., 1972, La Formación Nazas del Levantamiento de Villa Juárez, Estado de Durango: Sociedad Geológica Mexicana, Memórias de la Segunda Convención Nacional, p. 25–31, 194–196.

Parsons, W. H., 1969, Criteria for the recognition of volcanic breccias: A review, in L. H. Larsen, M. Prinz, and V. Manson, eds., Igneous and metamorphic geology: Geological Society of America Memoir 115, p. 263–304.

Powers, M. C., 1953, A new roundness scale for sedimentary particles: Journal of Sedimentary Petrology, v. 23, p. 117–119.

Rittmann, A., 1929, Die Zonenmethode: Schweizerischen Mineralogie und Petrologie Mitteilungen, v. 9, p. 1–46.

Rogers, C. L., Z. de Cserna, E. Tavera, and S. Ulloa, 1957, Geología general y depósitos de fosfatos del Distrito de Concepción del Oro, Estado de Zacatecas: Instituto Nacional para la Investigación de Recursos Minerales, Boletín 38, 129 p.

Rogers, C. L., Z. de Cserna, R. Van Vloten, E. Tavera, and J. Ojeda, 1961, Reconocimiento geológico y depósitos de fosfatos del norte de Zacatecas y areas adyacentes en Coahuila, Nuevo León, y San Luis Potosí: Consejo de Recursos Naturales No Renovables, Boletín 56, 322 p.

Rogers, C. L., R. Van Vloten, J. Ojeda, E. Tavera, and Z. de Cserna, 1963, Plutonic rocks of northern Zacatecas and adjacent areas, Mexico: U.S. Geological Survey Professional Paper 475-C, p. C7–C10.

Sharp, R. P., and L. H. Nobles, 1953, Mudflow in 1941 at Wrightwood, southern California: Geological Society of America Bulletin, v. 64, p. 547–560.

Slemmons, D. B., 1962, Determination of volcanic and plutonic plagioclases using a three- or four-axis universal stage: Geological Society of America Special Paper 69, 61 p.

Steel, R. J., 1974, New Red Sandstone flood and piedmont sedimentation in the Hebridean Province, Scotland: Journal of Sedimentary Petrology, v. 44, p. 336–357.

Stilwell, R., 1980, Petrology and stratigraphy of the Triassic La Boca Formation, Sierra Madre Oriental, northeastern Mexico: Master's thesis, University of New Orleans, 151 p.

Tardy, M., 1980, Contribution à l'étude géologique de la Sierra Madre Oriental du Mexique: Ph.D. thesis, Université de Pierre et Marie Curie, Paris, France, 459 p.

Taylor, J. M., 1950, Pore-space reduction in sandstones: AAPG Bulletin, v. 34, p. 701–716.

Torres, R., J. Ruíz, P. J. Patchett, and J. M. Grajales Nishimura, 1999, A Permo-Triassic continental arc in eastern Mexico: Tectonic implications for reconstructions of southern North America: Geological Society of America Special Paper 340, p. 191–196.

Varnes, D. J., 1958, Landslide types and processes, in C. B. Eckel, ed., Landslides and engineering practice: Highway Research Board Special Report 29, 232 p.

Walton, A. W, 1979, Volcanic sediment apron in the Tascotal Formation (Oligocene?), Trans-Pecos Texas: Journal of Sedimentary Petrology, v. 49, p. 303–314.

Weidie, A. E., and G. E. Murray, 1967, Geology of Parras Basin and adjacent areas of northeastern Mexico: AAPG Bulletin, v. 51, p. 678–695.

Weidie, A. E., J. A. Wolleben, and E. F. McBride, 1970, Regional geologic framework of northeastern Mexico: Corpus Christi Geological Society Guidebook, p. 5–16.

Wentworth, C. K., 1922, A scale of grade and class terms for clastic sediments: Journal of Geology, v. 30, p. 377–392.

Angeles-Aquino, F., and A. Cantú-Chapa, 2001, Subsurface Upper
Jurassic stratigraphy in the Campeche Shelf, Gulf of Mexico, *in* C.
Bartolini, R. T. Buffler, and A. Cantú-Chapa, eds., The western Gulf
of Mexico Basin: Tectonics, sedimentary basins, and petroleum
systems: AAPG Memoir 75, p. 343–352.

14

Subsurface Upper Jurassic Stratigraphy in the Campeche Shelf, Gulf of Mexico

Francisco Angeles-Aquino
Instituto Politécnico Nacional–Petróleos Mexicanos, Mexico City, Mexico

Abelardo Cantú-Chapa
Instituto Politécnico Nacional, Mexico City, Mexico

ABSTRACT

This sedimentary-stratigraphic study of the Upper Jurassic rocks of the Campeche shelf, Gulf of Mexico, uses material from 50 exploratory wells to describe and formally designate three lithostratigraphic units in the shelf. The Ek-Balam upper Oxfordian Group consists of three members: (1) lumpy peloidal limestones at the base; (2) massive sand, mudstone, and bentonitic shales with organic material in the middle section; and (3) calcareous sandstones with anhydrite at the top. One of the best oil reservoirs in the Campeche shelf, the Ek-Balam Field produces from the sandy member of this group. The Akimpech Formation of Kimmeridgian age consists of four members: (1) bentonitic mudstones, (2) dolomitized limestones, (3) algal shales, and (4) oolitic limestones. The oolitic and dolomitized limestones are reservoir rocks from the Upper Jurassic in the Campeche offshore region; the algal shales and mudstones are possible source rocks. The Tithonian Edzna Formation consists of shales and clayey limestones with abundant organic material; it is the main source rock in this region. The Campeche shelf is an important oil-producing area located offshore in the southeastern Gulf of Mexico. Formal designation of Jurassic lithostratigraphic units will serve as an aid to oil exploration in this region.

INTRODUCTION

The Campeche shelf, as defined in this study, is located in territorial waters of Mexico in the southeastern Gulf of Mexico, between longitude 91°40'W–94°00'W and latitude 18°30'N–20°00'N. Physiographically, the shelf forms part of the continental platform to a depth of 500 m offshore from the states of Tabasco and Campeche. It has a surface area of about 15,800 km². Campeche shelf is an important oil-producing area with very important reservoirs and source rocks in Jurassic strata (Figure 1).

This study is supported by several unpublished stratigraphic studies by F. Angeles-Aquino, based on data taken from oil wells in the Campeche shelf. Several stratigraphic and sedimentologic studies have been carried out in the Campeche shelf (Angeles-Aquino, 1988, 1996a, b; Angeles-Aquino et al., 1994; Araujo et al., 1986; Basañez-Loyola and Brito-Arias, 1988; Cantú-Chapa, 1977, 1982, 1989, 1994; Flores-Varga, 1982; Landeros-Flores and Neri León, 1984; Lugo-Rivera et al., 1976; Ornelas-Sánchez et al., 1993). Geologic studies have also been carried out in the deep zone of the Gulf of Mexico (Buffler, 1991).

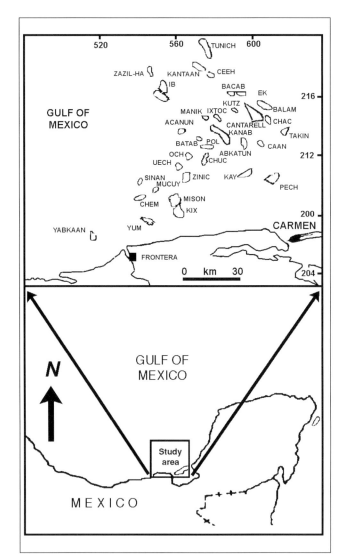

FIGURE 1. Location plan for the Campeche shelf, showing the more important producing fields.

STRATIGRAPHY

This stratigraphic study is based on material from 50 exploratory wells that have penetrated Upper Jurassic rocks. Analyzed were 4500 samples of cuttings, 30 cores, and all mechanical logs of the wells under study. With this material, eight lithofacies are defined and informally designated as A, B, C, D, E, F, G, and H; detailed descriptions follow. These stratigraphic units in the maritime region of Campeche are summarized in Figure 2.

Based on the wealth of information from exploration wells in offshore Campeche, stratigraphic nomenclature is formally proposed to subdivide the subsurface rocks from the Upper Jurassic. The impetus comes from oil operations that need to identify producing and source units in the region and to establish correlations with other areas, thus making it possible to expand oil exploration. The proposed lithostratigraphic

units correlate with units that crop out along the coast of the Gulf of Mexico and in the southeastern region of Mexico (Figure 3).

Oxfordian

Oxfordian rocks are represented by the La Gloria, Zuloaga, and Olvido Formations in northeastern Mexico; the Todos Santos Formation in southeastern Mexico in the states of Chiapas, Oaxaca, and Yucatán (Imlay, 1980); and the Santiago Formation in the eastern Sierra Madre in the states of Hidalgo and Veracruz, as well as in the subsurface of the Poza Rica oil district in eastern Mexico (Cantú-Chapa, 1971, 1992).

Ek-Balam Group

It is here proposed to formally designate as the Ek-Balam Group a sequence of rocks consisting of massive sand beds, bentonitic mudstone, and bentonitic limestones with pelletoids. (*Ek-Balam* means "Star-Tiger" in Mayan).

These rocks are important reservoirs and source rocks of Oxfordian hydrocarbons in the region. It is therefore necessary to refer to them constantly, because of their stratigraphic and economic importance. Ek-Balam is the most important producing field in Oxfordian rocks. The reservoir rocks are thick sand bodies, widely distributed throughout the region and easily identifiable in well logs. The term *group* is used because this unit consists of several lithologically well-defined units of significant thickness. Further subdivision of this unit into formations is not feasible.

This unit is in the subsurface of the offshore region of Campeche in the Gulf of Mexico. Its section type is in the Caan-1 well (Figure 1). Its lower contact has not been reached in oil drillings, and its upper contact is concordant with unit B of the Akimpech Formation. At the top of this unit, dense bodies of calcareous sandstone with anhydrite make a distinctive, divergent log signature in gamma-ray and lithodensity logs (Figure 2).

Caan-1 well is proposed as the type locality, its upper contact being located at a depth of 4985 m. The rocks belonging to this unit have penetrated between 57 m (Cantarell-2239 well) and 440 m (Caan-1 well), although the unit has never been completely drilled. Its best development is in the Ek-Balam field, where the sand bodies are very thick (± 231 m) and its expression in the gamma-ray and lithodensity well logs is very clear. The curves are displaced in opposite directions, indicating a clay-free porous unit.

The sand bodies are distributed mainly in the eastern section of the area, covering approximately 2500 km^2. These rocks are composed of fine to thick detrital clastics, including sandstones and laminated bentonitic mudstones (Figure 4). The Ek-Balam Group may be divided into three members for better study:

1) The lower member consists of peloidal wackestone to packstone, olive gray in color, clayey and sandy, with quartz

cement and intercalated evaporite (Ek-Balam and Batab fields). It grades east into bentonitic mudstone of lumpy appearance (Cantarell-2239 well, Caan-1 well, Chac-1 well and Cantarell-91 well). This member is characteristically developed in the Caan-1 well at 5370 m.

2) The middle member consists of a rhythmic alternation of calcareous sandstones, mudstones, and bentonitic shales. It has thick bodies of sand that grade up from conglomeratic sandstone or poorly consolidated sandy conglomerates at the base (Ek-101 well, 4539–4565 m; Balam-1 well, 4400–4428 m). There are also salt intrusions, as in the Ek and Balam fields.

3) The upper member is characterized by sandy limestones that grade into calcareous sandstones with anhydrite. The top of this unit is marked by anhydrite layers that range from 5 to 200 m thick (Chac-1 well, 4540 m; Bacab-21 well, 4336 m). Accessory minerals are pyrite, detrital quartz, anhydrite, and gypsum (Figure 4).

Salt in the Ek-Balam, Batab, and Malob fields is not in a consistent stratigraphic position in the Upper Jurassic sequence, although it is observed to be interstratified in some wells (Balam-1 and 101, Bacab-1 and 21). Seismic profiles and the studied wells suggest that the salt should underlie the base of the Ek-Balam Group.

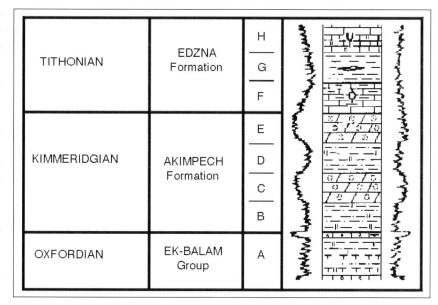

FIGURE 2. A stratigraphic table indicating the distribution and expression in the well logs of the eight lithofacies that make up the Upper Jurassic in Campeche: (A) sands, mudstone, and claystone; (B) terrigenous, mudstone, and claystone; (C) dolomitized limestones; (D) muds and algal shales; (E) oolitic dolomitized limestones; (F) clayey limestones with radiolarians; (G) calcareous shales with saccocoma; and (H) clayey limestones with tintinides.

Age

Fragments of ammonites obtained from the Balam-101 well, core 3, were classified as *Ochetoceras* and *Discosphinctes* (see Cantú-Chapa, this volume, chapter 1), which are characteristic of the upper Oxfordian. These ammonites indicate the top of this stage and support the correlation proposed here with other localities in the central-northern region of Mexico. The Jaguar Formation in western Cuba would also be correlative. The age assigned to the Ek-Balam Group is upper Oxfordian.

A microfossil identified as *Protoglobigerina oxfordiana* was found in core 16 of the Chac-1 well (Lugo-Rivera et al., 1976). The fossil was identified later as *Caucasella oxfordiana* by Longoria-Treviño (personal comunication, 1983). Ornelas-Sánchez et al. (1993) consider the same specimen to be *Globuligerina oxfordiana* and place it in the Oxfordian.

Stratigraphic Relations

In the eastern portion of the Campeche shelf, the upper Oxfordian rocks underlie the terrigenous rocks of unit B of the Kimmeridgian with apparent concordance, as stated above. The Caan 1-well has penetrated farthest into the Ek-Balam rocks, reaching peloidal carbonates.

Oxfordian rocks have been drilled only in the eastern portion of the area under study. The following wells encountered

the upper contact at the indicated depth: Balam-101 well (4690 m); Bacab-2 well (4735 m); Caan-1 well (4985 m); Cantarell-91 well (2971 m); Cantarell-2239 well (4720 m); Chac-1 well (4540 m); Ek-3 well (4275 m); Ek-31 well (4920 m); Ek-101 well (4395 m); and Ek-Balam-1 well (4872 m)

It is inferred that the Ek-Balam Group extends toward the west with a change of facies into lumpy peloidal limestones, which are observed sporadically in the Caan and Bacab fields (Figure 4).

Kimmeridgian

Rocks of this age are represented in Mexico by the La Casita Formation in the north, by the Taman Formation and San Andrés Member in the central-western region, and by the Chinameca and Todos Santos Formations in the south (Cantú-Chapa, 1971, 1992; Imlay, 1980) (Figure 3).

Akimpech Formation

It is here proposed to formally designate as the Akimpech Formation a sequence of carbonates and terrigenous rocks characterized by oolitic and partially dolomitized limestones, algal shales, and bentonitic mudstones, which are present throughout the Campeche shelf. All the wells included in this study reached Kimmeridgian rocks.

The Akimpech Formation is the most important reservoir rock of the Upper Jurassic of the region. It is widely distributed and mappable throughout the area (± 4500 km^2). Characteristics in the gamma-ray and lithodensity logs mark them as a sequence of dense carbonate banks.

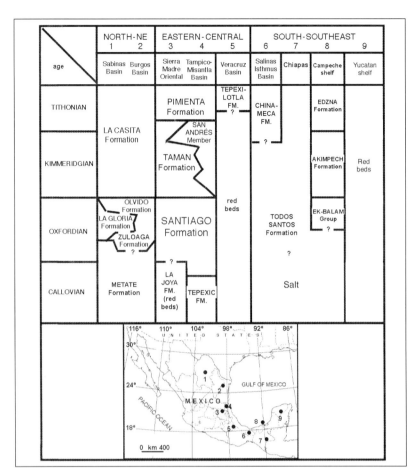

FIGURE 3. Location map and stratigraphic correlation of the Upper Jurassic units proposed for the Campeche shelf, with the northeast, center, and southeast of Mexico.

This formation is best exposed at the Uech-21 well at 4890 m. Locally shaly, less indurated rocks are evident at the Kokay-1A well at 5043 m; they have a thickness of 69 m at the algal unit. The Caan-1 well is designated as the Akimpech Formation type section, being the well that penetrates the entire interval. This name, Akimpech Formation, replaces such informal names as the Campeche Group and the Akimpech Group (Angeles-Aquino, 1988, 1996a).

Increased oil exploration in the Campeche shelf has enabled better definition of this unit. It is widely distributed, from the Cantarell field, in the eastern portion, to the Uech field, in the western portion, to the Zinic-1 well in the south-central portion. The name Akimpech ("land of the tick") proposed here is the original Mayan name for the city and state of Campeche.

In the offshore region of Campeche, 50 wells have reached the stratigraphic level of the Akimpech Formation and 18 wells have passed completely through. The thicknesses penetrated are highly variable, from 63 m (Mucuy-1 well) to 1272 m (Ku-407 well); the average thickness is 454 m.

The Akimpech Formation type locality is the Caan-1 well, where the upper and lower contacts are located at 4400 and 4824 m, respectively, with a thickness of 424 m. Its lower con-

tact is defined by well-log character because of the contrast between the calcareous sandstones and Oxfordian anhydrite that underlies bentonitic mudstones of unit B of the Akimpech Formation. This expression is best seen in the Chac-1 well (4540 m).

The upper contact of the Akimpech Formation is abrupt because of the dense layers of oolitic banks found at the top of this formation, in contrast to the clayey limestones of the overlying Edzna Formation. The contact is marked in the gamma-ray and lithodensity logs by two opposite curves (Figure 2). Its best expression is seen in the Uech-21 well (4890 m).

The Akimpech Formation is widely distributed throughout the area and constitutes an alternating sequence of carbonates and siliciclastic rocks, which can be divided into four informal units designated B, C, D, and E.

Lower Terrigenous Member B

The lower terrigenous member B has been reached by 22 of the wells in the offshore Campeche shelf. Its thickness varies from 75 m (Cantarell-2239 well) to 408 m (Zinic-1 well); it is more common in the eastern portion.

Member B consists of mudstone and bentonitic sandy shales, sporadically and thinly interbedded with sandstones and bentonitic microdolomite with anhydrite. This type of material appears to change laterally into carbonates toward the western portion (Figure 5).

Age

Cantú-Chapa (1977) identified the ammonites *Nebrodites* and *Taramelliceras* from the lower Kimmeridgian in core 14 of the Chac-1 well; these fossils are common in rocks of this age from central-northern Mexico (San Pedro del Gallo and Mazapil; Burckhardt, 1930). This core also provided some dacycladcea associated with dinoflagellates.

Stratigraphic Relations

Terrigenous member B conformably overlies the Oxfordian Ek-Balam Group and conformably underlies calcareous member C of the Akimpech Formation.

Lower Calcaerous Member C

This member has been encountered in 24 wells studied in this work in the offshore region of Campeche, of which 22 crossed the entire thickness. Thickness varies from 37 to 267 m. It is found mainly in the central portion of the Campeche shelf.

Carbonate member C consists of microcrystalline to mesocrystalline dolomites, packstone with incipient dolomitization, isolated interbeds of mudstone, and olive-gray sandy shales (Figure 6).

Age

This member overlies member B conformably. Member B contains *Nebrodites* and *Taramelliceras* ammonites from the lower Kimmeridgian, which were found in the Chac-1 well (Cantú-Chapa, 1977). The *Rhaxella sorbyana* has been found in the Kokay-1A well at a depth of 5290 m, as well as in the Chac-1 and Caan-1 wells. It characterizes the Kimmeridgian (Landeros-Flores and Neri-León, 1984). Ornelas-Sánchez et al. (1993) include this microfossil from member C in the *Pseudocyclammina lituus–Acicularia elongata elongata* biozone of the Kimmeridgian.

Stratigraphic Relations

Calcareous member C overlies terrigenous member B concordantly, and likewise concordantly underlies algal member D of the Akimpech Formation. It is distributed in the central and eastern portions of the Campeche shelf and grades into terrigenous rocks toward the east (Figure 6).

Upper Terrigenous Member D

Member D is recognized in 21 wells studied in this work, of which 14 crossed complete intervals. Its thickness varies from 23 m (Uech-21 well) to 387 m (Tunich-1 well). It is best developed in the eastern portion of the area. Its western extent is delineated by occurrences in Kokay-1A well, Uech-21 well, Chem-1A well, and Taratunich-201 well (Figure 7).

Terrigenous member D consists of claystones, mudstones, and sandy shales interbedded with carbonates and abundant algal material (Kokay-1A well). Reddish mudstone sediments are interbedded with marine terrigenous rocks. A continental origin is inferred for the red mudstones in the eastern shelf (Caan-1 well) (Figure 7).

Age

The upper terrigenous member D conformably overlies unit C, which is included in the *Pseudocyclammina lituus–Acicularia elongata elongata* biozone of the Kimmeridgian (Ornelas-Sánchez et al., 1993). Unit D also is probably of Kimmeridigian age.

Stratigraphic Relations

Member D lies conformably between members C and E, both carbonates from the Kimmeridgian. Unit D thins and disappears towards the west.

Upper Calcareous Member E

All 50 wells that were studied reached unit E, with 20 passing completely though it. Nearly all of the oil production in

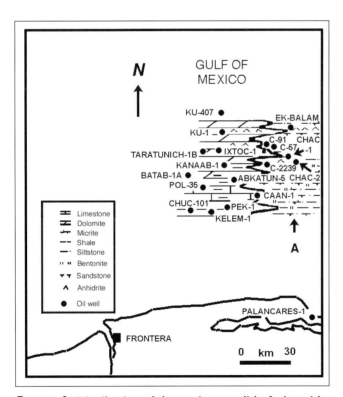

FIGURE 4. Distribution of the terrigenous lithofacies with continental influence from the Ek-Balam Group of the upper Oxfordian (A). The lithofacies grades into carbonates with a marine influence toward the western portion in the Campeche shelf.

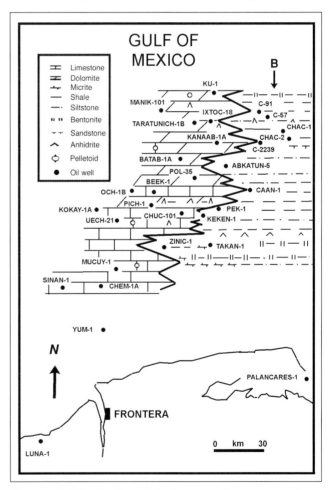

FIGURE 5. Distribution of the lower terrigenous member B of the Akimpech Formation (lower Kimmeridgian), located in the eastern portion of the area. The member changes into carbonates toward the west of the Campeche shelf.

the Campeche shelf is concentrated in this unit. Its thickness varies from 52 m (Zinic-1 well, 5142 m) to 373 m (Uech-21 well, 4890 m), with an average thickness of 225 m.

Member E consists of mesocrystaline (Uech-1 well) and microcrystaline dolomites (Chac-1 well). Petrographic studies carried out by Angeles-Aquino (1988, 1996a) indicate that these rocks are packstones, ooid and peloid grainstones, and pelletoid mudstones or wackestones, locally dolomitic. Oolitic lithoclastics form banks that have not been affected by the dolomitization seen at the other wells (Ki-101 well, 5250 m; Och-1B well, 4585 m; and Zinic-1 well, 5592 m) (Figure 8).

Age

No characteristic fossils have been found in member E. Ornelas-Sánchez et al. (1993) places it in the *Pseudocyclammina lituus–Acicularia elongata elongata* biozone of the Kimmeridgian (Balam-1 well, Caan-1 well, Cantarell-91 well, Chac-1 well, Chem-1A well, Pek-1 well, Pich-1 well, and Tunich-1 well).

Stratigraphic Relations

Member E overlies member D and underlies member F from the Tithonian. The upper limit is apparently concordant with

clayey limestone of the Tithonian. Member E extends throughout the area studied; its best development is in the western portion, where it forms part of the series of oolitic bars (Och-1B well, Uech-21 well, Ki-101 well, and Mucuy-1 well) (Figure 8, letter a). Toward the east, member E thins but forms possible interior bars. Farther east, member E becomes less pure, with increased terrigenous content and abundant micrite; this is interpreted as lagoonal facies (Zinic-1 well, Pich-1 well, and Chuc-101 well) (Figure 8, letter b).

Sedimentology and Diagenesis

It is inferred that ooids of member E originated in very shallow water in a restricted marine environment under high-energy conditions, giving rise to the formation of oolitic banks which restricted the circulation of water and allowed evaporation, thus enhancing dolomitization. In some cases, the energy conditions were sufficient for the development and growth of oolitic banks that line up in the form of linear ridges (Puerto Ceiba-1 well, Luna-1 well, Sinan-1 well, Uech-1 well, Och-1B well, Batab-1A well, Ku-407 well, Tunich-1 well, and Kambul-1 well).

A sedimentary model of the latest Kimmeridgian deposition

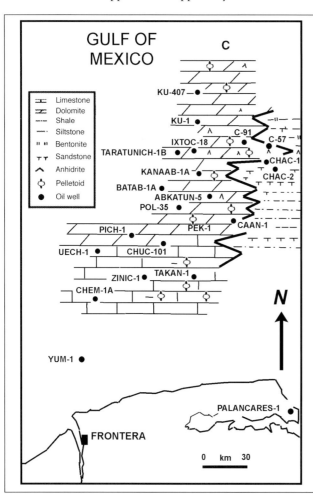

FIGURE 6. Lithofacies distribution of lower carbonate member C of the Akimpech Formation (lower Kimmeridgian), located in the central portion of the Campeche shelf.

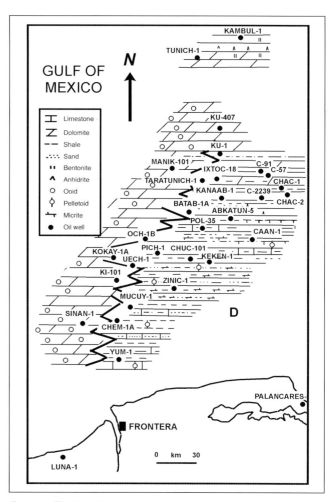

FIGURE 7. Distribution of upper terrigenous member D of the Akimpech Formation (lower Kimmeridgian), west of the Campeche shelf.

includes frontal oolitic bars interrupted by tidal channels that transported material, mostly ooids, to form small tidal deltas, which evolved into interior oolite bars in the lagoon (Figure 9).

Deposits of member E on the high-energy Campeche shelf allow us to recognize four zones of carbonate deposition. Changes in energy across these zones generated coarse-grained sediments in the bars and fine-grained sediments in the lagoon zone and seaward of the bars. These zones migrated landward with the transgression and tend to open toward the continent as a result of the influence of a terrigenous sedimentation, whose facies will be described below.

1) **Micritic bentonitic carbonate facies.** Located in the western portion of the Campeche shelf are bentonitic lime mudstones with scarce ooids and pelletoids (Kokay-1A well).

2) **Oolitic carbonate facies.** In the central-western portion of the Campeche shelf, grainstones with bioclastic grains and oolotic lime formed in a series of oolitic bars (Uech-1 and 21 wells, Och-1 well, Mucuy-1 well, and Pol-72 well).

3) **Evaporitic carbonated facies.** The central-eastern portion of the Campeche shelf contains mudstones and wackestones, with a few peloid packstones, oolites, evaporites, and mudstone rocks (Zinic-1 well, Chuc-101 well, Pich-1 well, Pek-1 well, and Abkatun-5 well).

4) **Terrigenous carbonated facies.** In the western portion

of the Campeche shelf, carbonates predominate, with terrigenous mudstone, fine-grained sands, and anhydrites (Caan-1 well, Cantarell-57 well, Chac-1 well, Chac-2 well, and the Ek-Balam field).

Tithonian

Rocks from the Tithonian are represented in northern and eastern Mexico (Burgos Basin, Sabinas Basin, Tampico-Misantla Basin, and the Sierra Madre Oriental) by the La Casita and Pimienta Formations (Cantú-Chapa, 1971, 1992, this volume, chapter 1; Imlay, 1980), in the eastern-central zone (Veracruz Basin) by the Tepexilotla Formation, and in the south and southeast of the country (Salinas Basin in the Isthmus, the Chiapas Mountains, and the Yucatán shelf) by the Chinameca and Todos Santos Formations (Imlay, 1980) (Figure 3).

Edzna Formation

Angeles-Aquino (1988) and Cantú-Chapa (1982, 1989) considered the Tithonian rocks of the Campeche shelf equivalent to the Pimienta Formation in eastern Mexico. According to recent well information, the Campeche rocks contain considerably more carbonates than the Pimienta Formation, which has more terrigenous rocks. We therefore propose the name Edzna Formation for the Tithonian rocks in the subsurface in the Campeche region.

The development of the Edzna Formation is shown in the Batab-1A well (top at 4400 m), Pol-72 well (top at 4285 m), and Chuc-101 well (top at 3893 m), where it reaches thicknesses of 295 m, 189 m, and 290 m, respectively. The Caan-1 well (4303 m) is nevertheless considered as the locality type, because this well penetrates the most complete lithologic sequence in the region.

The Edzna Formation rocks can be divided into three members: F, G, and H. The first was deposited irregularly over Kimmeridgian rocks, where carbonate rocks predominate on the E horizon. The second member is more uniformly and widely distributed; it is predominantly claystone. The third member is the most widely distributed of the three and is principally bentonitic chalk.

Lower Member F

This unit has been reached by nearly all the exploratory wells in the Campeche shelf, as well as by wells located outside the study area (Puerto Ceiba, Luna, Xicalango, and Palancares wells in the southeastern onshore region of Mexico) (Figure 10). Thickness varies from 20 m (Caan-1 well) in the east to 110 m (Batab-2 and 3 wells) in the west.

This member consists of clayey mudstone, light gray to dark brown in color, with abundant organic material and occasional thin intervals of dark-gray or black shale. These sediments are distributed uniformly in the Campeche shelf.

Age

Radiolarians and saccocomas are common in the Edzna Formation. Their stratigraphic position and correlation are well

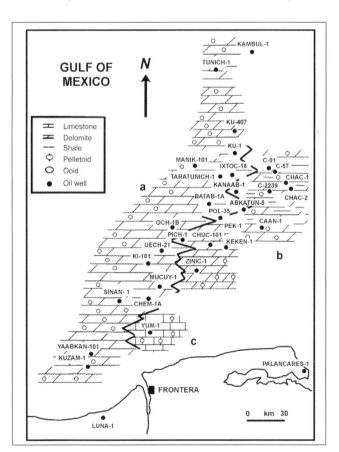

FIGURE 8. Distribution of the upper calcareous member E of the Akimpech Formation (Kimmeridgian). Mesocrystalline (a) and microcrystalline dolomites (b). Evaporitic carbonate facies (c). Campeche shelf.

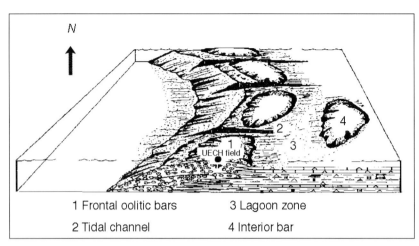

1 Frontal oolitic bars   3 Lagoon zone
2 Tidal channel   4 Interior bar

FIGURE 9. Sedimentary model of the Kimmeridgian rocks, showing shallow environment of high and low energy in the open and restricted waters of the Campeche shelf. Shown is the location of the Uech field, one of the most representative in this stage.

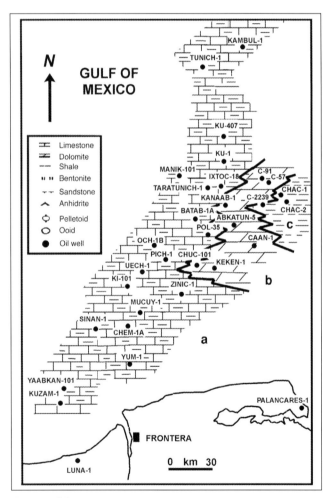

FIGURE 10. Distribution of the carbonated lithofacies of member F of the Edzna Formation (Tithonian). The central part of the area (a and b) is occupied by clayey carbonates that are dolomitized (c) and tend to become terrigenous toward the east because of the continental influence of the Campeche shelf.

marked in the composite records of the wells under study and in the transversal sections in the Campeche shelf that are built up from these, illustrating their continuity. In effect, this horizon can be seen in the section, limited by oolitic limestones from the Kimmeridgian at the base and by shales at the top.

Ornelas Sánchez et al. (1993) locate the biozone with *Saccocoma arachnoidea* and *Eothryx alpina* as either lower or middle Tithonian, although we cannot establish a more precise age.

Stratigraphic Relations

Member F overlies member E (of Kimmeridgian age) and underlies member G (Tithonian). The distribution of member F extends throughout the area of the study. The lithofacies of this carbonate unit alternate with clayey carbonates and gradually become dolomitized toward the east. According to information provided from the wells, the lithofacies would tend to change into clayey terrigenous rocks with carbonate influence toward the continent.

Middle Member G

This member has been reached by all the wells involved in this study, with the exception of the Taratunich-1B well. Its thickness varies from 39 m (Caan-1 well) to 171 m (Yum-1 well).

In member G, calcareous sandy shales of dark gray to black predominate, interspersed with dark-colored clayey limestones. It is the main source unit in the Campeche shelf (Figure 11). This shale unit is easily recognized in all the geophysical records comprising this work. The gamma-ray and lithodensity logs tend to converge to form a funnel, which denotes a high clay content. Its best expression is found in Batab-1A well (4259 m), Pol-72 well (4342 m), and Chuc-101 well (3939 m).

Age

Cantú-Chapa (1982) identified the ammonites *Suarites* in core 7 of the Tunich-1 well and assigned it an upper Tithonian age. This sample was cut into Member G. Ornelas Sánchez et al. (1993) placed member G rocks studied here within the *Saccocoma arachnoidea* and *Eotryx alpina* biofacies of the lower-middle Tithonian. In this member, *Saccocoma* sp., *Eotrix alpina*, *Lombardia arachnoidea*, *L. angulosa*, *L. filamentosa*, and calcified and pyritized radiolarians have been identified.

Stratigraphic Relations

Claystone member G conformably overlies calcareous member F in the Batab-1A well at 4410 m, Pol-72 well at 4506 m, and Chuc-101 well at 4043 m. It conformably underlies the upper carbonate member in the Batab-1A well at 4259 m, Pol-

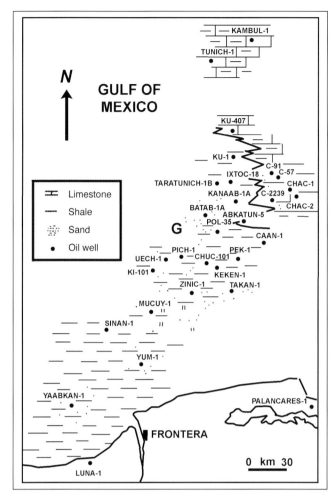

FIGURE 11. Distribution of lithofacies terrigenous member G of the Edzna Formation (Tithonian). The lithofacies tends to be carbonated toward the east. This horizon is considered the principal generator rock for hydrocarbons in the Campeche shelf.

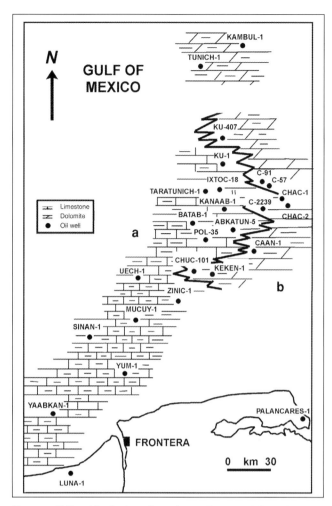

FIGURE 12. Lithofacies plan of member H of the Edzna Formation (Tithonian): calcareous-clayey (a) and clayey dolomites (b) in the Campeche shelf.

72 well at 4342 m, and Chuc-101 well at 3939 m. Member G is thicker in the western portion than in the eastern part of the Campeche shelf (Figure 11).

Upper Member H

This unit has been reached by all the wells involved in this study. It varies in thickness from 26 m (Abkatun-5 well) to 83 m (Chem-1A well) and is generally thinner in the east and thicker in the west. Its expression in the lithodensity and gamma-ray logs reflects a dense body that also shows variations in density, giving it a typically funicular form.

The rocks are mainly clayey and bentonitic lime mudstones, with a chalky appearance and dolomitization toward the east (Figure 12). Its best expression is in Mucuy-1 well (5339 m), Uech-1 well (4605 m), and Yum-1 well (4543 m), where it reaches thicknesses of 77 m, 63 m, and 66 m, respectively.

Age

Calpionellids (*Calpionella alpina, Crasicollaria massutiniana*), saccocomas, and calcified radiolarians predominate. Cantú-Chapa (1982) identified ammonites as *Durangites* sp. from core 8 of the Chac-1 well and *Salinites* sp. and *Protancyloceras* sp. from core 14 of the Chac-2 well, placing them in the uppermost Tithonian. Ornelas Sánchez et al. (1993) place the uppermost rocks in the *Crasicollarias* biozone, also corresponding to the upper Tithonian.

Stratigraphic Relations

Member H corresponds to the top of the Upper Jurassic (Tithonian). It overlies the G member concordantly and underlies the Lower Cretaceous units transitionally, as seen in the Mucuy-1 well, Uech-1 well, and Yum-1 well. The lateral distribution of member H is quite uniform and wide, covering the whole area of the study.

CONCLUSIONS

The stratigraphic sequence of the Upper Jurassic in the Campeche shelf is composed principally of carbonates and terrigenous rocks, having a wide distribution throughout the area. Underlying these deposits are thick bodies of salt and evaporites that have intruded locally into part of the Upper Jurassic sequence. They are considered pre–Upper Jurassic in age.

The sedimentary sequence of the Upper Jurassic has been subdivided into eight members: member A from the Oxfordian; members B, C, D, and E from the Kimmeridgian; and members F, G, and H from the Tithonian. In this work, we formally propose the following names: the Ek-Balam Group for the Oxfordian rocks, the Akimpech Formation for the Kimmeridgian rocks, and the Edzna Formation for the Tithonian rocks.

In the Upper Jurassic part of the Campeche shelf, conditions exist for deposition from open and restricted seawaters, which corresponds to a transgressive sequence.

ACKNOWLEDGMENTS

The authors are grateful to Paul Enos and Robert Goldhammer for their suggestions, which helped to improve the final product.

REFERENCES CITED

Angeles-Aquino, F. J., 1988, Estudio estratigráfico sedimentológico del Jurásico Superior en la Sonda de Campeche, México: Revista Ingeniería Petrolera, v. 28, no. 1, p. 45–55.

Angeles-Aquino, F. J., 1996a, Estratigrafía de las rocas del Jurásico Superior del subsuelo de la Sonda de Campeche, Golfo de México: Tesis de Maestría, Escuela Superior de Ingeniería y Arquitectura, Sección de Estudios de Posgrado e Investigación, Instituto Politécnico Nacional, México City, México, 93 p.

Angeles-Aquino, F. J., 1996b, Unidades generadoras del Jurásico Superior en la Sonda de Campeche México: Memorias del 5.º Congreso Latinoamericano de Geoquímica Orgánica, Cancún, México, p. 144–146.

Angeles-Aquino, F. J., J. Reyes-Núñes, J. M. Quezada-Muñetón, and J. J. Meneses-Rocha, 1994, Tectonic evolution, structural styles and oil habitat in the Campeche Sound, Mexico: Transactions of the Gulf Coast Association of Geological Societies, v. 44, p. 53–62.

Araujo-Mendieta, J., M. A. Basañez-Loyola, and M. A. D. Márquez, 1986, Estudio estratigráfico sedimentológico de las rocas del Jurásico en el prospecto "Área Marina de Campeche": Proyecto C-5008, Instituto Mexicano del Petróleo, internal report.

Buffler, R. T., 1991, Seismic stratigraphy of the deep Gulf of Mexico basin and adjacent margins, in A. Salvador, ed., The Gulf of Mexico Basin: Geological Society of America, The geology of North America, v. J., p. 353–357.

Basañez-Loyola, M. A., M. Brito-Arias, 1988, Estudio diagenético del Jurásico Superior en pozos de la Zona Marina de Campeche: Proyecto C-3043, Instituto Mexicano del Petróleo, internal report.

Burckhardt, C., 1930, Étude synthétique sur le Mésozoïque mexicain: Mémoire de la Société Paléontologique Suisse, v. 49–50, 280 p.

Cantú-Chapa, A., 1971, La Serie Huasteca (Jurásico Medio–Superior) del Centro Este de México: Revista del Instituto Mexicano del Petróleo, v. 3, no. 2, p. 17–40.

Cantú-Chapa, A., 1977, Las amonitas del pozo Chac-1, Norte de Campeche (Golfo de México): Revista del Instituto Mexicano del Petróleo, v. 9, no. 2, p. 8–39.

Cantú-Chapa, A., 1982, The Jurassic-Cretaceous Boundary in the subsurface of eastern Mexico: Journal of Petroleum Geology, v. 4, no. 3, p. 311–318.

Cantú-Chapa, A., 1989, Precisiones sobre el límite Jurásico-Cretácico en el subsuelo del Este de México: Revista de la Sociedad Mexicana de Paleontología, v. 2, no. 1, p. 26–69.

Cantú-Chapa, A., 1992, The Jurassic Huasteca Series in the subsurface of Poza Rica, eastern Mexico: Journal of Petroleum Geology, v. 15, no. 3, p. 259–282.

Cantú-Chapa, A., 1994, Informe bioestratigráfico: Instituto Politécnico Nacional, Escuela Superior de Ingeniería y Arquitectura, Sección de Estudios de Posgrado e Investigación, internal report.

Cantú-Chapa, A., 2001, Mexico as the western margin of the Pangea based on biogeographic evidence from the Permian to the Lower Jurassic: This volume, chapter 1.

Flores-Vargas, A., 1982, Paleosedimentación y diagénesis de las rocas carbonatadas productoras en el área Chac, de la Sonda de Campeche, México: Boletín de la Asociación de Ingenieros Petroleros de México, v. 22, p. 18–24.

Imlay, R. W., 1980, Jurassic paleobiogeography of the conterminous United States in its continental setting: U. S. Geological Survey, Professional Paper v. 1062, 134 p.

Landeros-Flores, R., and L. Neri-León, 1984. Apéndices paleontológicos de los pozos: Taratunich-1B, Chuc-101, Ku-407, Kambul-1, Mucuy-1, Ixtoc-18, Abkatun-5, Yum-1, Pich-1, Yaabkan-1, Caan-1, Cantarell-2239, Zinic-1 y Pol-79: Departamento de Paleontología, Superintendencia de Exploración, Zona Marina, Petróleos Mexicanos, internal report.

Lugo-Rivera, E., A. Díaz-Puebla, L. Neri-León, A. del Alto-Ramirez, J. Maldonado-Maldonado, and M. Ponce-Ramirez, 1976, Informe final del pozo Chac No. 1: Petróleos Mexicanos, Superitendencia General de Exploración, Departamento de Operación Geológica, Zona Sur, internal report.

Ornelas-Sánchez, M., N. Aguilera-Franco, S. Franco-Navarrete, M. Granados-Martínez, and R. Bello-Montoya, 1993, Biozonificación y análisis de facies del Jurásico Superior en pozos del Área Marina de Campeche: Revista de la Sociedad Mexicana de Paleontología, v. 6, no. 1, p. 11–47.

Cantú-Chapa, A., 2001, The Taraises Formation (Berriasian-Valanginian) in northeastern Mexico: Subsurface and outcrop studies, *in* C. Bartolini, R. T. Buffler, and A. Cantú-Chapa, eds., The western Gulf of Mexico Basin: Tectonics, sedimentary basins, and petroleum systems: AAPG Memoir 75, p. 353–370.

15

The Taraises Formation (Berriasian-Valanginian) in Northeastern Mexico: Subsurface and Outcrop Studies

Abelardo Cantú-Chapa
Instituto Politécnico Nacional, Mexico City, Mexico

ABSTRACT

The Berriasian-Valanginian Taraises Formation in the subsurface of northeastern Mexico consists of an intercalation of argillaceous limestone and shale. This formation is divided into two members, based on lithologic characteristics defined from electric and radioactivity logs from oil wells in northeastern Mexico. In addition, two small ammonite groups of the subfamily Olcostephaninae (upper Valanginian) collected at Potrero de García, Nuevo León, northeastern Mexico, are described: (1) *Capeloites neoleonense* n. sp. and (2) *Garcites potrerensis* n. gen. and sp. and *G. cavernensis* n. gen. and sp. The presence of *Dichotomites (Dichotomites)* sp. in the same beds supports an late Valanginian age for these strata. One specimen of *Capeloites* obtained from a core of the La Laja-8 well, south of Tampico City, eastern Mexico, designates an late Valanginian age for the interval drilled (2556–2562 m) in the so-called Lower Tamaulipas Formation in this area. Global geographic distribution of *Capeloites* is restricted to France, Spain, and Peru; its presence in Mexico supports a biogeographic link between Europe and America.

INTRODUCTION

The Taraises Formation was first described in northern Mexico by Imlay (1938) as "a limestone and marl facies with a large cephalopod fauna of Neocomian (Berriasian to lower Hauterivian) age. . . ." He divided this formation into two members. The lower member consists of a compact limestone separated by marly beds. The upper member consists of thin- to medium-bedded limestone and marl.

The contact between these two members is apparently conformable. The lower contact of the Taraises Formation with the underlying Upper Jurassic La Casita Formation is transitional. Its upper contact with the limestone of the Cupido Formation (Hauterivian–lower Aptian) is abrupt. The Taraises Formation represents open-water deposits of the Neocomian sea (Imlay, 1944), into which clay was introduced, probably from the Coahuila block.

Fossils of Berriasian to early Hauterivian age were studied by Burckhardt (1930) and by Böse (1923) in northern Mexico, and some of them were later considered as having been collected from the Taraises Formation. Berriasian to lower Hauterivian ammonites of this formation include the following genera: *Bochianites, Spiticeras, Olcostephanus, Valanginites, Mexicanoceras, Berriasella, Subthurmannia, Karakaschiceras, Neo-*

comites, Kilianella, Distoloceras, and *Acanthodiscus* (Imlay, 1937, 1938, 1940, 1944; Cantú-Chapa, 1966, 1972, 1992; Contreras, 1977; Peña-Muñoz, 1964; Young, 1988).

THE TARAISES FORMATION IN NORTHEASTERN MEXICO

In the present study, the Taraises Formation is described from the subsurface of northeastern Mexico using well logs and ammonite biostratigraphy (Figure 1a). New upper Valanginian ammonites of the Taraises Formation are described from outcrops in the Potrero de García, northeastern Mexico (Figure1b). A small specimen of the ammonite *Capeloites* sp. from the La Laja-8 well (core 11, 2556–2562 m) south of Tampico City is also described. The age of this latter specimen is late Valanginian; the core containing the ammonite was cut in the Lower Tamaulipas Formation in this area (Figure 2).

Neocomian ammonites from two oil wells in northeastern Mexico were originally described by the author without using the corresponding well logs. In the Anahuac-1 well, ammonites were incorrectly determined as upper Tithonian. In the Calichoso-1 well, ammonites were assigned to the upper Valanginian (Cantú-Chapa, 1963, 1972). Logs are used to establish the exact stratigraphic position of the fossils, and they are also correlated with logs and specimens of nearby wells, such as the Anahuac-4 and -5, Cadena-3, San Javier-2, Pesquerías-1, and Los Herreras-2 (Figure 1).

The top of the La Casita Formation (Upper Jurassic) and the bases of the Taraises and Cupido Formations (Lower Cretaceous) are determined with radioactivity logs (Figure 3). They are also correlated with electric logs in sections corresponding to the same formations drilled in the Anahuac-2 and Pesquerías-1 wells. Important biostratigraphic ammonite data are revised from wells where only electric logs are available (Figure 4).

The transitional contact between the Taraises and the La Casita Formations is placed at the decrease in value of the gamma and neutron curves (Figure 3). The spontaneous potential (SP) curve changes in value (south part of the section) in the Cadena-3, Los Herreras-2, and Pesquerías-1 wells. The SP curve is flat in the northern part of the section (Figure 4). This contact corresponds to the Jurassic-Cretaceous boundary in northeastern Mexico (Figures 3 and 4), and it was dated with

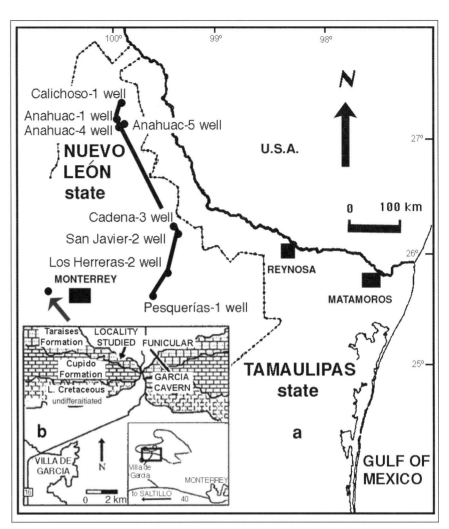

FIGURE 1. (a) Location map showing the oil wells studied. (b) Lower Cretaceous rocks in the Potrero de García, Villa de García, 20 km northwest of Monterrey, northeastern Mexico.

the late Tithonian ammonites *Proniceras, Salinites, Substeueroceras, Kossmatia,* and *Salinites* in the upper part of the La Casita Formation; specimens were found in several wells of the same area studied here (Cantú-Chapa, 1999). The Taraises–La Casita Formations contact is present at different depths from 3250 m (San Javier-2 well) to 1850 m (Pesquería-1 well) as a result of the Paras fault system in the studied area (Cantú-Chapa, 1989).

The flat SP curve does not allow for a precise determination of the Taraises and Cupido Formations; however, the contact is established utilizing gamma-ray well logs. Radioactivity and resistivity logs decrease in value at the contact (Figures 3 and 4).

Members

The Taraises Formation consists of interbedded limestone, shaly limestone, and shale in the subsurface of northeastern Mexico. It is divided into two members of approximately equal thickness. The lower member consists of approximately 140 m of a dense limestone with a thin, shaly bed near the base. The gamma-ray curve stays uniformly toward the left through this

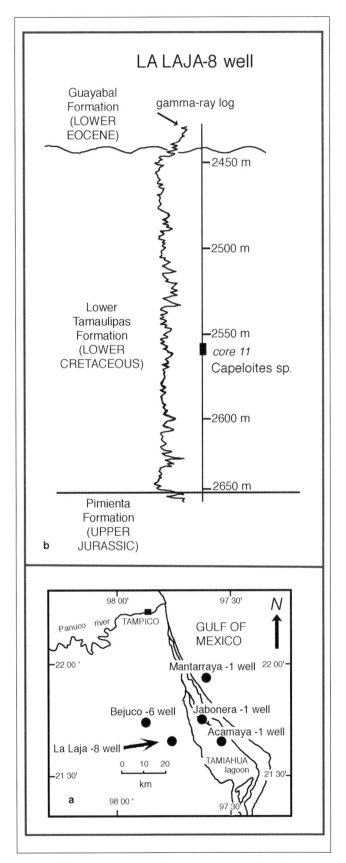

FIGURE 2. Location map (a) and gamma-ray log (b) showing the Lower Tamaulipas Formation (Lower Cretaceous) in La Laja-8 well, south of Tampico, eastern Mexico (Cantú-Chapa, 2001).

member, but it changes as it cuts across the transitional contact with the La Casita Formation (Figure 3). The SP curve is flat in the lower member and shows a change decreasing in value as it crosses the contact between the La Casita and Taraises Formations (Cadena-3, Los Herreras-2, and Pesquerías-1 wells; Figure 4). The resistivity and radioactivity logs display important value changes, increasing at the contact between the members, thus permitting characterization of the lower member (Figures 3 and 4).

No Berriasian ammonites were found in the cores studied. Possible age-equivalent rocks are inferred by their stratigraphic position. The upper Tithonian (Cantú-Chapa, 1999) and upper Valanginian ammonites in the wells establish the age of the lower member (Figures 3 and 4).

The upper member consists of about 130 m of shaly limestone. Gamma-ray and neutron curves are shifted to the center when recording this lithology in the Anahuac-4 and -5, San Javier-2, and Pesquerías-1 wells. A negative departure is observed in the two microresistivity curves through the upper member. The contact between the Taraises and Cupido Formations is abrupt, as shown by the gamma-ray and neutron curves (Figure 3). The resistivity curve presents an important change in value in the upper member of the Taraises Formation. Ammonites were found in the upper member (Figures 3 and 4).

Age Revision of the Core Number 12 from the Anahuac-1 Well

The stratigraphic study of the Taraises Formation in the Anahuac-1 well shows that the ammonites from core number 12 (2499–2503 m) were erroneously attributed to the upper Tithonian (Cantú-Chapa, 1963). Ammonites found in this core must be restudied. The exact position of core number 12 is here established in a stratigraphic cross section (Figure 4). The comparison of the radioactivity and electric logs of the Anahuac-2 well and the electric logs of the Anahuac-1 well shows that core number 12 of the latter well was taken from the upper member of the Taraises Formation. Also, the systematic position of core number 12 ammonites from the Anahuac-1 well are herein modified.

Protancyloceras anahuacensis Cantú-Chapa, 1963, was proposed as the name for a heteromorph ammonite with a smooth ventral line and simple ribs; it was found in this well and assigned to the upper Tithonian by Cantú-Chapa (1963, p. 28, Plate I, Figures 2-4) (Figure 5.1).

Later, *Bejucoceras* Cantú-Chapa (1976) was proposed as the name for a heteromorph ammonite with characteristics similar to those of *P. anahuasensis*. *Bejucoceras* was found associated with *Leopoldia victoriensis* Imlay in the Bejuco-6 well, (cores 3–5, 1904–1920 m) in eastern Mexico (Figure 2). Both fossils were assigned to the lower Hauterivian by Cantú-Chapa (1976); their age was established by *L. victoriensis* after Imlay (1937). However, the genus *Leopoldia* was restudied, and some forms were considered as belonging to *Karakaschiceras*. The radial

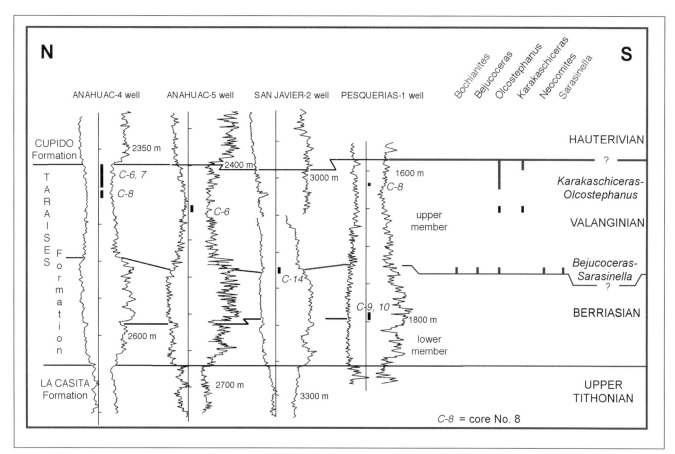

FIGURE 3. Cross section of the Taraises Formation (Lower Cretaceous) in oil wells in northeastern Mexico, according to radioactivity logs. Stratigraphic datum is top of Tithonian.

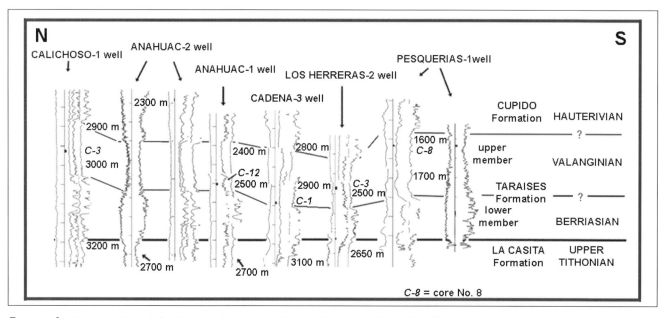

FIGURE 4. Cross section of the Taraises Formation (Lower Cretaceous) in oil wells in northeastern Mexico, according to radioactivity and electric logs. Stratigraphic datum is top of Tithonian.

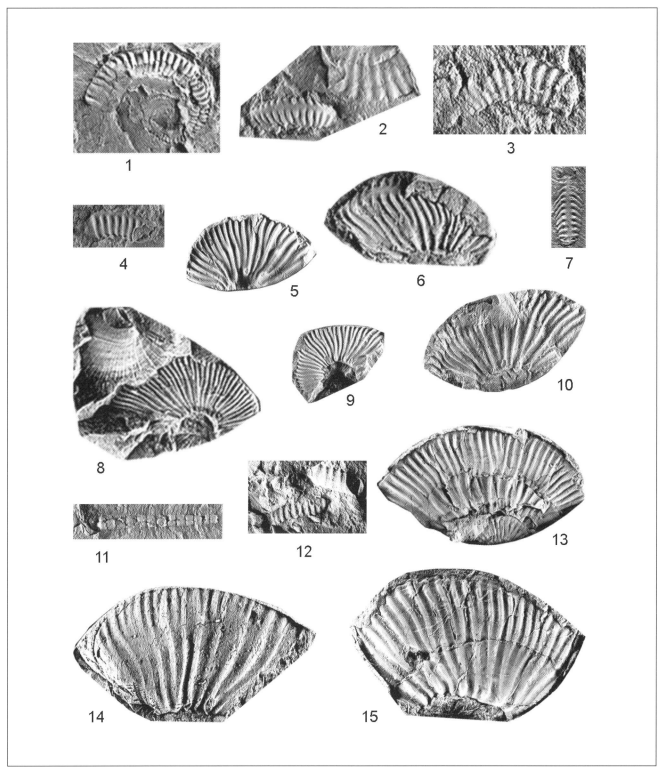

FIGURE 5. 1. *Bejucoceras simplecostatum* Cantú-Chapa, Anahuac-1 well, core 12 (2499–2503 m), Escuela Superior de Ingenieria y Arquitectura (ESIA)-1032; 2. *Bejucoceras* sp., Cadena-3 well, core 1 (2940–2946 m), ESIA-1033; 3, 4. *Bejucoceras* sp., San Javier-2 well, core 14 (3120–3124 m), ESIA-1034, ESIA-1035; 5, 8, 10. *Neocomites* sp., San Javier-2 well, core 14 (3120–3124 m), 5.8 contains a fragmented pelecypod, ESIA-1036, ESIA-1037, ESIA-1038; 6. Neocomitinae, Anahuac-1 well, core 12 (2499–2503 m), ESIA-1039; 7. Neocomitinae, ventral region, San Javier-2 well, core 14 (3120–3124 m), ESIA-1040; 9. *Neocomites* sp., Cadena-3 well, core 1 (2940–2946 m), ESIA-1041; 11, 12. *Bochianites* sp., Anahuac-5 well, core 6 (2452–2463 m), Los Herreras-2 well, core 3 (2479–2484 m), ESIA-1042, ESIA-1043; 13. Neocomitinae, Los Herreras-2 well, core 3 (2479–2484 m), ESIA-1058; 14, 15. cf. *Sarasinella* sp., Anahuac-1 well, core 12 (2499–2503 m), ESIA-1044, ESIA-1045. Age: Taraises Formation upper member (upper Valanginian), northeastern Mexico. All specimens are x 2.

direction of the ventrolateral tubercle characterizes this last genus (Thieuloy, 1971), which is now dated by the top of the lower Valanginian to the base of the upper Valanginian (Kemper et al., 1981).

The Mexican *Bejucoceras* was recently considered to be *Aegocrioceras* Spath (1924) of the lower Hauterivian from Europe by Wright et al. (1996). This last genus has simple ribs that cross transversely the ventral region (Kemper, 1992; Rawson, 1975). The genera differ by their ventral ribbing and age, which justifies separating them. *Bejucoceras* Cantú-Chapa, 1976, must be recognized as a Valanginian genus, as its smooth midventral band distinguishes it from the lower Hauterivian *Aegocrioceras* (Kemper et al., 1981). *Bejucoceras simplecostatum* Cantú-Chapa (1976, p. 65) must be conserved as the type species of *Bejucoceras*. *P. anahuasensis* (Cantú-Chapa, 1963) of the Anahuac-1 well (core 12), is here considered as a synonym of *B. simplecostatum* Cantú-Chapa, 1976.

Bejucoceras (Figure 5.2) is present in the upper member of the Taraises Formation (as it is found in cores from the three oil-wells studied here). It is associated with *Neocomites* sp. (Figure 5.9) and *Karakaschiceras* sp. (Figure 6.6) in the Cadena-3 well (core 1, 2940–2946 m). *Bejucoceras* sp. (Figure 5.3) is also associated with *Neocomites* sp. (Figure 5.5 and 5.8) and *Olcostephanus* sp. (Figure 6.10) in the San Javier-2 well (core 14, 3120–3124 m); a pelecypod fragment is present in the last core (Figure 5.8). *B. simplecostatum* Cantú-Chapa (Figure 5.1) was found associated with *Bochianites* sp. (Cantú-Chapa, 1963), and cf. *Sarasinella* sp. in the Anahuac-1 well (core 12, 2499–2503 m) (Figures 5.14 and 5.15).

Other Ammonites of the Taraises Formation from Oil Wells in Northeastern Mexico

Three ammonite groups were found in the upper member of the Taraises Formation in oil wells of northeastern Mexico: heteromorphs (*Bejucoceras* and *Bochianites*), neocomitins, and olcostephanins. They allow the Taraises Formation to be dated as Valanginian.

Heteromorph Ammonites

Bejucoceras has been studied previously. *Bochianites* sp. (Cantú-Chapa, 1963, Plate I, Figure 6, p. 29) was found in Anahuac-1 well (core 12), in the Anahuac-5 well (core 6, 2452–2463 m) (Figure 5.11), and the Los Herreras-2 well (core 3, 2479–2484 m) (Figure 5.12). *Bochianites* was studied by Young (1988) in the Taraises Formation in northeastern Mexico.

Neocomitins

Fragmented and involute ammonites are here assigned to *Neocomites* sp. because of their fine, sigmoidal ribs irregularly dividing at the umbilical bullae and in the middle of the flanks (Figure 5.5 and 5.10). They were found in the upper member of the Taraises Formation (Figures 3 and 4) and are associated with *Karakaschicheras* sp. in the San Javier-2 well (core 14). *Neocomites* sp. was also found in the Cadena-3 well (core 1) (Figure

5.9). Our specimens resemble *N. acuticostatus* Imlay (1937, p. 580, Plate 10, Figure 2) with their fine ribbing and irregular division of ribs.

Fragmented specimens of neocomitins were found from the base to the top of the upper member of the Taraises Formation in the Cadena-3 well (core 1) (Figure 6.1 and 6.4) and in the Anahuac-4 well (core 6, 2372–2378 m) (Figure 6.2). Their fine, sigmoidal ribs finishing in ventrolateral bullae and the narrow, flattened, and smooth venter allow identification of these specimens as neocomitins. Two tabulate, smooth ventral regions probably belonging to neocomitins were found in the upper member of the San Javier-2 well (core 14, 3120–3124 m) and Anahuac-4 well (core 6, 2372–2378 m) (Figures 5.7 and 6.5).

Karakaschiceras. Some neocomitins in Mexico were described as *Leopoldia* (Cantú-Chapa, 1976; Contreras, 1977; Imlay, 1938; Peña-Muñoz, 1964) or as *Karakaschiceras* Young (1988). One of them, from Samalayuca, Chihuahua, northern Mexico, was erroneously assigned to *Karakaschiceras* (Young, 1988, Plate 1, Figure 3) (compare Young, 1969); it represents a lower Kimmeridgian *Idoceras* (Cantú-Chapa, 1970).

Two compressed and involute specimens with attenuated ribbing in the middle part of the flanks and a smooth siphonal band were obtained from the studied oil wells. The ribs terminate in a small, ventrolateral radial node. Specimens appear to be somewhat similar to *K. victoriensis* (Imlay, 1938, p. 581, Plate 12, Figures 1, 2) in their lateral and ventrolateral ribbing and whorl section. However, the fragmented preservation does not permit their specific identification, although the ornamentation and presence of the siphonal smooth band permit a generic identification. They were found at the base and top of the upper member of the Taraises Formation in the Anahuac-4 well (core 6, 2372–2378 m) (Figure 6.3) and the Cadena-3 well (core 1, 2940–2946 m) (Figure 6.6).

Sarasinella. Two fragmented and slightly evolute specimens have slightly sigmoidal ribs that are bifurcated at the middle of the sides, some of them starting at a pair of umbilical bullae. The ventral side is not visible (Figure 5.14 and 5.15). The specimens resemble *Sarasinella trezanensis* Lory *in* Baumberger (1923, p. 307, Plate VIII, Figures 2–4) in their lateral ornamentation. Both specimens were found at the base of the upper member of the Taraises Formation in the Anahuac-1 well (core 12, 2499–2503 m). The specimen of Figure 5.15 was determined as *Substeueroceras*? *imlayi* Cantú-Chapa (1963, p. 40, Plate III, Figure 3); it is here assigned as cf. *Sarasinella* sp. because of its ribbing.

One fragmented and crushed specimen was obtained from the Los Herreras-2 well (core 3, 2479–2484 m). It is evolute, and its ribs are bifurcated at the middle of the flank and originate in the umbilical area. Its state of preservation does not allow observation of the umbilical bullae, as only one rib shows this structure (Figure 5.13). This specimen resembles cf. *Sarasinella* sp. of the Anahuac-1 well (Figure 5.15); it is here considered a neocomitin. It was found in the same core associated with a *Bochianites* sp. (Figure 5.12).

FIGURE 6. 1, 4. Neocomitinae, Cadena-3 well, core 1 (2940–2946 m), ESIA-1046, ESIA-1047; 2. Neocomitinae, Anahuac-4 well, core 6 (2372–2378 m), ESIA-1048; 3. *Karakaschiceras* sp., Anahuac-4 well, core 6 (2372–2378 m), ESIA-1049; 5. Neocomitinae, ventral region Anahuac-4 well, core 6 (2372–2378 m), ESIA-1050; 6. *Karakaschiceras* sp., Cadena-3 well, core 1 (2940–2946 m), ESIA-1051; 7. *Olcostephanus* sp., Pesquerías-1 well, core 8 (1615–1619 m), ESIA-1052; 8, 9. *Taraisites* sp., Calichoso-1 well, core 3 (2960–2967 m), ESIA-1053, ESIA-1054; 10. *Olcostephanus* sp., San Javier-2 well, core 14 (3120–3124 m), ESIA-1055; 11. *Olcostephanus* sp., Calichoso-1 well, core 3 (2960–2967 m). ESIA-1056; 12. *Olcostephanus* sp., Anahuac-5 well, core 6 (2452–2463 m), ESIA-1057. Age: Taraises Formation upper member (upper Valanginian), northeastern Mexico. All specimens are x 1.

One fragmented and evolute specimen from the Anahuac-1 well (core 12) was previously determined to be *Protacanthodiscus* sp. (Cantú-Chapa, 1963, p. 41. Plate III, Figure 1) (Figure 5.6). Some ribs are irregularly trifurcate at the middle of the flanks from a tubercle, while others are simple and intercalated; all finish at the ventrolateral edge, and a smooth siphonal band is present. This ammonite should not be identified as *Protacanthodiscus*. Its stratigraphic position is in a core from the Lower Cretaceous Taraises Formation. Constrictions were not observed that could identify it as *Kilianella*. It must, therefore, be considered as a neocomitin.

Taraisites and Olcostephanus and Concepts of Sexual Dimorphism in Systematic Paleontology of Ammonites

These two genera are present in the Calichoso-1 (core 3, 2960–2967 m), San Javier-2 (core 14, 3120–3124 m), and Pesquerías-1 (core 8, 1615–1619 m) wells herein studied (Figures 3, 4, and 6.7–6.12). Their systematic position is discussed as follows:

The genus *Taraisites* was proposed by Cantú-Chapa (1966), with its type species *Taraisites bosei* Cantú Chapa, 1966, a species from north-central Mexico described as *Astieria* aff. *Baini* Sharpe, 1856, by Böse (1923, Plate II, Figures 3–6). According to Cantú-Chapa (1966), attribution of this specimen to the genus *Rogersites* (*Olcostephanus* Neumayr, 1875) was erroneous (Imlay, 1937); its features do not agree with the definition of the latter genus. It differs from its type species by its ribbing and whorl section.

Taraisites is a form with two or three secondary and sharp ribs coming from the periumbilical tubercles; they are convex on the flanks and widely spaced on the ventral region. Its whorl section is compressed. Some species have a coiled scaphitoid. *Olcostephanus* is an inflated and involute form, with a dense secondary ribbing coming from the periumbilical tubercles.

Taraisites was not accepted by Riccardi et al. (1971), based on the following terms: "No proper consideration was given to growth stage and morphologic variation which were presumably known to Imlay (1938). Cantú-Chapa omitted the critical evidence that the small paratypes differ from the inner whorls of the large holotypes. . . ." Riccardi et al. (1971) discarded *Taraisites* because of subjective considerations, assuming it to be a different genus from *Olcostephanus*. To this author, the first genus represents a microconch of *Olcostephanus* because of supposed sexual dimorphism. When peristomal lappets and parabolic constrictions are preserved, specimens are implicitly considered as microconch forms.

The arguments of Riccardi et al. (1971) about dimorphic forms in *Olcostephanus* were taken by Cooper (1981), who divided it into four categories based on the dimorphic concepts: macroconchs, microconchs, immature macroconchs, and juvenile forms. Cooper described the first two categories, according to a detailed world study of *Olcostephanus* species. They were not referred to stratigraphic sections.

The classification of *Olcostephanus* species proposed by Cooper (1981) is based on some morphologic characteristics when they happened to be preserved in the form of parabolic constriction and peristomal lappets, thus allowing him to consider them as microconch specimens of the same species. The dimorphic characters in a given species represent a myth in the systematic of ammonites implemented with subjective bases, as indicated by several authors who favor the following concepts:

- "the majority of specimens appears to be macroconchs, specimens . . . are probably microconchs. . . . The dimorphic relationship of the microconch is, therefore, somewhat questionable. . . .¨ (Verma and Westermann, 1973, p. 212, 220).
- "Some macroconchs do not have their equivalent microconchs. . . . There are not enough dimorphic couples established to use them in the systematics of species. . . ." (Enay, 1966, p. 581).
- "The reality of the sexual dimorphism is currently admitted by numerous researchers . . . but . . . in the perisphinctid ammonites it is always difficult to recognize the microconchs if their lappets are not preserved. . . . (Cecca, 1986).

Researchers have expressed different opinions about how to establish the systematic of ammonite dimorphism:

- Place all microconchs specimens in a subgenus (Atrops, 1982).
- Consider certain heteromorph genera of different ages as microconchs because of their small size (Rawson, 1975, p. 138).
- Point out that "the identification of two dimorphic specimens is very difficult, below a certain diameter where the characters are the same. . . ." (Cecca, 1986).
- Some difficulties exist in separating specimens into microconchs and macroconchs (Brochwicz-Lewiñski and Rozak, 1976).

The Mexican olcostephanid specimens studied by Imlay (1937) are conserved at the University of Michigan under different numbers. Imlay never indicated that small forms correspond to the first whorls of shell from the same specimen, as was suggested by Riccardi et al. (1971), who determined *Taraisites* to be a synonym of *Olcostephanus*.

Olcostephanid specimens of similar sizes studied here have two sorts of ribbing: (1) A dense ribbing arising from periumbilical tubercles identifies *Olcostephanus*, from the Pesquerías-1 well (core 8, 1615–1619 m) (Figure 6.7), and (2) two or three coarse ribs beginning from a periumbilical tubercle identify *Taraisites huizachense* Cantú-Chapa (1972, p. 88, Plate 1, Figure 6), from the Calichoso-1 well (core 3, 2960–2967 m) (Figure 6.8).

Two large specimens of *Olcostephanus* were found in the Calichoso-1 well (core 3, 2960–2967 m) and in the Anahuac-5 well (core 6, 2452–2463 m) (Figure 6.11 and 6.12). One of them, which is crushed, shows the last umbilical tubercles, from where some fine ribs arise (Figure 6.11). The other one is involute and shows 15 umbilical ribs that terminate in acute and radially elongated tubercles (Figure 6.12). From the tubercles, bundles of five or six very fine ribs arise. This last specimen is somewhat similar to *Olcostephanus bakeri* (Imlay, 1937, p. 560, Plate 70, Figure 1). Our specimen differs by its smaller number of umbilical tubercles and by its finer ribs.

AGE OF THE TARAISES FORMATION IN THE SUBSURFACE OF NORTHEASTERN MEXICO

No ammonites were found in the lower member of the Taraises Formation in the subsurface of northeastern Mexico. The age of this member can be inferred by its stratigraphic position, because it overlies the Tithonian La Casita Formation (Cantú-Chapa, 1999).

However, two groups of ammonites were found in different oil wells from base to top in the upper member of the Taraises Formation. The *Bejucoceras-Sarasinella* association was found in the first 10 m of the upper member in the Anahuac-1 well (core 12, 2499–2503 m), San Javier-2 well (core 14, 3120–3124 m), and Cadena-3 well (core 1, 2940–2946 m) (Figures 3 and 4).

The second group, consisting of *Karakaschiceras* and *Olcostephanus,* was found in the last 60 m before the top of the upper member in the Calichoso-1 well (core 3, 2960–2967 m), Pesquerías-1 well (core 8, 1615–1619 m), Anahuac-4 well (core 6, 2372–2378 m), and Anahuac-5 well (core 6, 2452–2463 m) (Figures 3 and 4).

From subsurface data, there is no biostratigraphic evidence to separate the Berriasian from the lower Valanginian base or to indicate the top of the upper Valanginian deposits in the Taraises Formation. The last two stratigraphic levels are defined by the European *otopeta* and *callidiscus* zones.

The *Bejucoceras-Sarasinella* association can be assigned to the lower Valanginian (*pertransiens* zone) after European chronostratigraphy and stratigraphic position. *Sarasinella* is found in the lower and upper Valanginian rocks of southeastern France (Bulot and Thieuloy, 1993; Bulot et al., 1992; Thieuloy and Bulot, 1992). *Neocomites* shows a similar stratigraphic distribution; it is also known from the base of the lower Hauterivian strata in France and Spain (Company, 1987; Bulot and Thieuloy, 1993).

The *Karakaschiceras-Olcostephanus* association is found at the top of the upper member of the Taraises Formation. *Karakaschiceras* is known only at the lower–upper Valanginian boundary in Europe. Kutek et al. (1989) studied two groups of Valanginian *Karakaschiceras* species in central Poland, but they are separated by a hiatus. A similar distribution of this fossil is observed in the Valanginian rocks in southeastern France

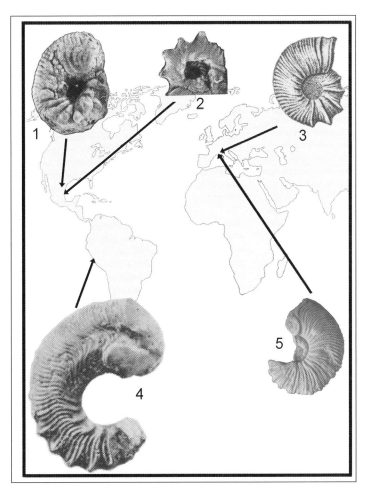

FIGURE 7. Biogeographic distribution of *Capeloites* (1, 2, 3–5) of the upper Valanginian from France, Peru, and Mexico; after Lissón (1937), Thieuloy (1969), Matheron (1878), and this work. All specimens are x 1.

(Bulot et al., 1992). This fossil could be assigned to the upper Valanginan base (European *verrucosum* zone), and it overlies the *Bejucoceras-Sarasinella* association in the upper member of the Taraises Formation in the subsurface. *Olcostephanus* has a wide stratigraphic distribution; it is known from the Valanginian to the lower Hauterivian (Autran, 1989; Bulot et al., 1992).

It is herein considered that the upper member of the Taraises Formation in the subsurface of northeastern Mexico represents the boundary defined by the early–late Valanginian age (*pertransiens-verrucosum* zones), as is well defined with the two ammonite associations *Bejucoceras-Sarasinella* and *Karakaschiceras-Olcostephanus.*

NEW TARAISES FORMATION AMMONITES (UPPER VALANGINIAN) FROM THE POTRERO DE GARCÍA, NORTHEASTERN MEXICO

Dwarf ammonites from calcareous shales in the Taraises Formation at the Potrero de García were, until now, unknown. This locality is found on the southern flank of the El Fraile anticlinal, 20 km northwest of Monterrey city, Nuevo León, in northeastern Mexico (Wilson and Ward, 1993) (Figure 1b). The ammonites permit the addition of important information to the regional biostratigraphy; we are able to update their international biogeographic distribution and to propose new systematic units.

Ammonites were collected by the writer and his family from the calcareous shales in the lower part of the upper member of the Taraises Formation; some of them are eroded. This member was dated as lower Hauterivian (Imlay, 1944) or upper Valanginian (Young, 1988). Ammonites are associated with dwarf echinoids, all preserved as pyritized internal molds.

ONE AMMONITE OF THE LA LAJA-8 WELL, EASTERN MEXICO

The La Laja-8 well, core 11 (2556–2562 m) provided a fragmented and negative ammonite mold from the micritic limestone of the Lower Tamaulipas Formation (Berriasian–lower Aptian) (Imlay, 1944). The La Laja-8 well was drilled south of the city of Tampico, eastern Mexico (Figure 2).

The position of the core in the compact, brown limestone of Lower Tamaulipas Formation (Lower Cretaceous) was established by the gamma-ray log, which shows a constant straight line. This formation is unconformably overlain by the Guayabal Formation shale (middle Eocene) (Figure 2). The La Laja-8 well is located in a syncline, which represents a major submarine erosional surface forming the Bejuco paleocanyon (Cantú-Chapa, 1987, 2001). The only negative mold of a small ammonite from the La Laja-8 well included in this study was provided by the Department of Subsurface Geology, PEMEX northern region in Tampico.

SYSTEMATIC PALEONTOLOGY

Subfamily OLCOSTEPHANINAE Haug, 1910
Genus *Capeloites* Lissón, 1937
Type species: *Capeloites larozai* Lissón, 1937, by original designation of Lissón (1937, p. 3, Plate I, Figure a).

Discussion. *Capeloites* was proposed by Lissón (1937) to characterize the dimorphic ornamentation present on the last whorl of small specimens collected by him in the El Cascajal, a locality near Lima, Peru. This genus was considered as a synonym of *Olcostephanus* by Arkell et al. (1957), although there is no similarity between them. Recently, Wright et al. (1996) included these two genera in the subfamily Olcostephaninae.

Worldwide distribution of *Capeloites* is known in France, Spain, and Peru (Autran, 1989; Lissón, 1937; Matheron, 1878; Thieuloy, 1969; Wilke, 1988) (Figure 7); its presence in Mexico adds a biogeographic link between these two hemispheres. Autran (1989) found *Capeloites* in the upper Valanginian in southeastern France, and Wilke (1988) collected it in the lower Hauterivian beds in Spain.

Capeloites neoleonense n. sp.
(Figure 8.2, 8.2a, 8.3, 8.3a-c)

Diagnosis. Involute, compressed whorl section, tabulated to rounded venter. Three sorts of ornamentation; primary and close-together ribs arise on the umbilical margin in the last whorl. Prominent, rectiradiate, and large primary ribs end in lateral tubercles; coarse and distant ribs cross the tabulated venter in the adapical part. Lateral ribbing and tubercles produce bundles of two or three fine, dense, and sigmoidal secondary ribs that cross transversely the rounded venter in the middle part of the last whorl. Periumbilical, fine, rectiradiate primary ribs originate fine, secondary ribs crossing slightly forward of the rounded venter in the adoral part.

Discussion. The proposed *Capeloites neoleonense* n. sp. differs from *C. larozai* (Lissón, 1937), and *C. perelegans* (Matheron, *in* Thieuloy, 1969) in having a rectiradiate, close-together, coarse ribbing on the flanks; they are distant on the venter of the adapical part. *C. neoleonense* is also considered comparable to *A. perelegans* Matheron (1878), because of the similar coarse, rectiradiate, and prominent primary ribs in the adapical half of both species; however, there are no secondary ribs in the Mexican ammonite (Figure 7).

Capeloites hispanicus (Mallada, *in* Autran, 1989, p. 154, Plate 3, Figure 11; Plate 10, Figure 2) from France has three sorts of ornamentation and does not bear ventral tubercles in the adapical part of the last whorl as does *C. neoleonense*. However, the fine and dense secondary ribs are convex in the adoral part of the French specimens and radial in the Mexican ammonite (Figure 7).

Holotype. Specimen ESIA-1011 (Figures 8.2, 8.2a), two fragmented paratypes, ESIA-1012 (Figures 8.3, 3a-c), ESIA-1059 from the Potrero de García, northeastern Mexico (Figure 1b).

Etymology. The name *neoleonense* is taken from Nuevo León state, where the Potrero de García is located and where the specimens were collected.

Occurrence. The Potrero de García, northeastern Mexico, upper part of the lower member of the Taraises Formation (Lower Cretaceous). Measurements are in Table 1.

Capeloites sp. A (Figure 8.5, 8.5a-c.)

Material. One specimen (ESIA-1014); same locality as *C. neoleonense* n. sp. (Figure 1a).

Description. Dwarf and dimorphic shell with rounded sides and venter, whorl section depressed; the lateral tubercles produce two sorts of ribbing. Coarse ribs alternate with two weakly secondary ribs on the ventral region in the adapical part. Bundles of three or four fine, sigmoidal, and dense ribs arising from a midlateral tubercle cross transversely the broad venter in the adoral part.

Discussion. The proposed *Capeloites* sp. A differs from *Capeloites larozai* (Lissón, 1937), *C. perelegans* (Matheron, *in* Thieuloy, 1969), and *C. neoleonense* n. sp. by its midlateral and rounded tubercles that produce two sorts of ribbing and by its wide venter in the adoral part.

TABLE 1. Whorl measurements (in mm) of *Capeloites* species from Peru, France, and Mexico; after Lissón (1937), Thieuloy (1969), Autran (1989), and this work. D = whorl diameter, UD = umbilical diameter, U = UD/D.

Specimen	D	UD	U
PERU			
larozai	46	12.0	0.26
FRANCE			
perelegans	25	06.9	0.27
hispanicus	28.5	10.5	0.36
	25	09.5	0.38
MEXICO (Potrero de García)			
neoleonense n.sp.			
holotype, ESIA-1011	27	07.0	0.25
paratype, ESIA-1012	23	06.0	0.26
paratype, ESIA-1058	25	06.0	0.24
Capeloites sp. A			
ESIA-1014	23	-	-
Capeloites sp.	14	06.0	0.42
La Laja 8 well, ESIA-1016			

Occurrence. Same locality and formation as the previously described specimens.

Capeloites sp. (Figure 8.1)

Material. From the La Laja-8 well, southern Tampico District, eastern Mexico, a negative and small ammonite mold was obtained from core 11 (2556–2562 m) (ESIA-1016) (Figure 2).

Description. Dwarf, semievolute shell, dimorphic ornamentation in the last whorl; inclined umbilical wall, sides - almost smooth; small periumbilical bullae, inclined slightly forward, and five prominent ventral spines, longitudinally elongated, very distant from one another in the adapical half. Ribs weakly sigmoid, inclined forward on the flank, ending in ventral bullae in the adoral half; a peristomal constriction is present.

Discussion. This specimen may be recognized by the prominence of the ventral spines, very distant from one another in the adapical part, and by its forward inclination of the weak sigmoid ribs near the peristome; this ornamentation contrasts with that of all species of *Capeloites* previously described. Percentage of the umbilicus/whorl diameter (U) varies from *larozoi* and *perelegans* of Peru and France (U = 0.26– 0.27) to *hispanicus* of France (U = 0.36–0.38) and *Capeloites* sp. of the La Laja-8 well of Mexico (U = 0.42) (after Autran, 1989; Lissón, 1937; Thieuloy, 1969; and this work) (Table 1).

Garcites n. gen.

Type species. Garcites potrerensis n. gen. and sp.

Diagnosis. Dwarf, involute, sphaerocone, moderately compressed to coronate whorl section; sides slightly subparallel to rounded; deep umbilicus with inclined wall; strong, rectiradiate, cuneiform, prominent, and very close-together primary ribs ending in lateral or ventrolateral tubercles, narrow intercostal spaces; rounded or tabulated venter crossed by distant or close-together, high, and simple ribs arising freely or joined to the tubercles. No suture line is observed. The name

alludes to the Villa de García, where the specimens were collected (Figure 1b).

Discussion. Garcites n. gen. differs from the upper Valanginian olcostephanins, *Valanginites* (Kilian, 1910), and *Dobrodgeiceras* (Nikolov, 1963) by its ventral ribs separated by very large or moderate spaces, which arise from or in front of rounded and close-together lateral or ventrolateral tubercles. In the last two genera, bundles of three or four ribs arise from a tubercle. *Garcites* n. gen. differs from the last two genera mentioned by its fewer, prominent, large, cuneiform, and rectiradiate primary ribs terminating in tubercles and by its rounded or tabulated venter, not as broad as that of *Dobrodgeiceras* from Peru and *Valanginites* from Europe (Riccardi and Westermann, 1970; Kemper et al., 1981) (Figure 9).

Garcites is also distinguished from *Dobrodgeiceras* by having fewer large primary ribs (approximately 11–14 vs. 18–20 per whorl), and secondary ribs (approximately 12–14 vs. 52 per whorl) (Figure 9.1, 9.2, 9.4, 9.5). *Garcites* presents a simple sculpture, while all previously described Olcostephaninae genera are ornamented with primary ribs terminated in umbilical tubercles, from which arise bundles of some secondary ribs; they cross densely and transversely the ventral region without interruption, with the exception of *Mexicanoceras* (Imlay, 1938).

The absence of midventral tubercles in some *Dobrodgeiceras* specimens from Peru is not considered a special morphologic character that would join them with *Valanginites*; therefore, they persist as different genera (Riccardi and Westermann, 1970). To these authors, species of *Dobrodgeiceras* are separated by the tubercle position; they are lateral in the European specimens and ventrolateral in the Peruvian fossils (Figure 9.4, 9.5).

The secondary ribs arising from ventrolateral tubercles zigzag irregularly on the venter in the Peruvian specimens, but the European forms possess a straight ribbing across the venter (Figure 9.4a, 9.5a). The morphologic features of the two groups of upper Valanginian *Dobrodgeiceras* species are different. The

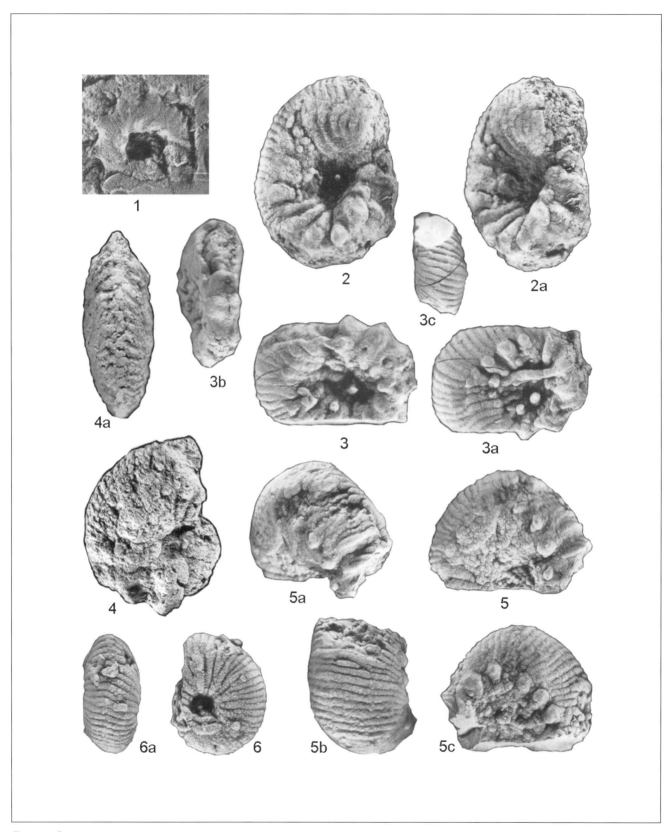

FIGURE 8. 1. *Capeloites* sp., latex replica of a fragmented specimen (ESIA-1016); 2, 2a, 3, 3a-c. *Capeloites neoleonense* n. sp.; 2, 2a, holotype (ESIA-1011); 3, 3a-c, paratype (ESIA-1012); 4, 4a. cf. *Karakaschiceras* sp. (ESIA-1031); 5, 5a-c. *Capeloites* sp. A (ESIA-1014). 6, 6a. *Dichotomites* (*Dichotomites*) sp., (ESIA-1028). Locality and age: All specimens except 8.1 are from the upper part of the lower member of the Taraises Formation (upper Valanginian), Potrero de García, northwest of Monterrey city, northeastern Mexico. Figure 8.1 is from the La Laja-8 well, core 11 (2556–2562 m), Tampico district, eastern Mexico. All specimens are x 2.

Peruvian forms are larger than the European ones (Figure 9.4a, 9.5a).

Garcites potrerensis n. gen. and sp.
(Figure 10.1, 10.1a-b, 10.3, 10.3a)

Diagnosis. Dwarf, involute, sides flattened. Whorl section slightly wider than high, small and deep umbilicus with inclined wall. Eleven rectiradiate, low, wide, and cuneiform ribs, very close together, commencing on the umbilical shoulder and ending in ventrolateral tubercles; very narrow intercostal spaces on the sides; wide and slightly rounded to tabulated venter crossed by very distant, high, and slightly forward ventral ribs; they arise independently and are intercalated between two ventrolateral tubercles.

Discussion. The proposed *Garcites potrerensis* n. gen. and sp. differs from *Olcostephanus beticus* (Mallada, *in* Wilke, 1988, Plate 11, Figures 5a, b) of the upper Valanginian from southeastern Spain by its involute form, its rectiradiate and cuneiform ribs ending in ventrolateral tubercles, and its very distant and high ventral ribs. Percentage of the umbilicus/whorl diameter (U) varies from very involute (U = 0.12–0.13) in *D. wilfride ventrotuberculatum* (Nikolov, *in* Thieuloy and Gazay, 1967) and *D. broggianum* (Lissón, *in* Riccardi and Westermann,1970), from France and Peru, respectively, to relative evolute (U = 0.27–0.30) in *G. potrerensis* (Table 2).

Etymology. The name *potrerensis* is taken from the Potrero de García locality, where the specimens were collected. Holotype, specimen ESIA-1017 (Figure 10.1, 10.1a-b); paratypes, specimen ESIA-1018 (Figures 10.3, 10.3a), ESIA-1060 from the Potrero de García, northeastern Mexico (Figure 1b).

Garcites cavernensis n. gen. and sp.
(Figure 10.6, 10.6a, 10.7, 10.7a)

Diagnosis. Dwarf, involute, depressed, and coronate shell. Small and deep umbilicus with inclined wall. Wide and rounded venter, and inclined sides toward the umbilical area. Fourteen large, rectiradiate, cunciform, and close-together primary ribs commencing on the umbilical wall, ending in ventrolateral tubercles and separated by narrow intercostal spaces. Coarse ventral ribs arise in front of each tubercle, pass straight across the rounded and wide venter, and are separated by spaces as large as the width of the ribs.

Discussion. The proposed *Garcites cavernensis* n. gen. and sp. differs from *G. potrerensis* by its ventral ribbing—the intercostal spaces are as large as the ribs. Ventral ribs are separated by very large spaces in *G. potrerensis*.

Etymology. It is herein proposed that *G. cavernensis* should keep this name, derived from the locality where it was collected near García Cavern, Potrero de García, northeastern Mexico.

Occurrence. Same locality and formation as the previously described specimens. Holotype: specimen ESIA-1025 (Figures 10.6, 10.6a); two paratypes—specimen ESIA-1026 (Figures 10.7, 10.7a) and ESIA-1061; same locality as the previously described specimens (Figure 1b). Measurements are in Table 2.

Garcites sp. A (Figure 10.2, 10.2a, 10.2b, 10.4, 10.4a–b)

Material. Two elliptically coiled specimens (ESIA-1020, ESIA-1021), same locality as the previously described speciman (Figure 1b); their adapical halves are flattened without sides and primary ribs, as if they were crushed, probably by posthumous compression.

TABLE 2. Whorl measurements (in mm) of *Garcites* n. gen., *Dichotomites* sp., and *Karakaschiceras* sp. from Mexico; *Dobrodgeiceras* from Peru and France (Riccardi and Westermann, 1970; Thieuloy and Garzay, 1967), and *Dichotomites* from France (Thieuloy, 1977). D = whorl diameter, UD = umbilical diameter, U = UD/D.

Specimen	*D*	*UD*	*U*
Garcites potrerensis n. gen. and sp. holotype, ESIA-1017 (holotype)	14.8	4	0.27
ESIA-1018	23.6	7	0.30
Garcites cavernensis n. gen. and sp. ESIA-1025 (holotype)	18.7	3.6	0.19
ESIA-1026	15.7	3.6	0.22
Garcites sp. A. ESIA-1020	19.7	4.3	0.22
EISA-1021	20.8	5.1	0.24
Dobrodgeiceras broggianum (Lissón)	34.6	4.5	0.13
FRANCE *D. wilfridi ventrotuberculatum* (Nikolov)	25.6	3.2	0.12
MEXICO *Dichotomites* (*Dichotomites*) sp. ESIA-1028	17.3	2.6	0.15
cf. *Karakaschiceras* sp ESIA-1031	23.0	3.6	0.15
FRANCE: *Dichotomites* (*D.*) *vergunnorum* Thieuloy	26.2	7.8	0.30

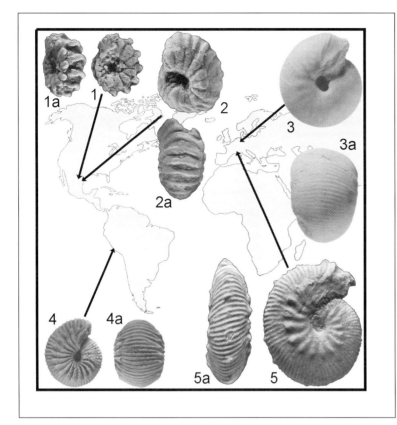

FIGURE 9. Biogeographic distribution of *Garcites* n. gen. (1–2), *Valanginites* (3), and *Dobrodgeiceras* (4–5) of the upper Valanginian from France, England, Peru, and Mexico, after Kemper et al. (1981), Riccardi and Westermann (1970), Thieuloy and Gazay (1967), and this work. All specimens are x 1.

Description. Adoral part with depressed whorl section, sides flattened; rectiradiate, large, cuneiform, and close-together primary ribs ending in ventrolateral tubercles; slightly rounded venter crossed by distant ribs borne freely in front of the tubercles. Adapical part flattened with only coarse, right, and ventral ribs.

Discussion. The proposed *Garcites* sp. A has similar ornamentation on the adoral part to *Garcites potrerensis*, but it differs by its crushed adapical part that conserves only ventral ribs.

Occurrence. Same locality and formation as the previously described specimen. Measurements are in Table 2.

Garcites sp. (Figures 10.5, 10.5a)

Description. Involute, dwarf shell, deep umbilicus, rounded venter, and compressed whorl section; rectiradiate, large, and close-together primary ribs separated by narrow intercostal spaces on the umbilical area; they finish in lateral tubercles; ribs are separated by wider interspaces on the venter.

Discussion. The proposed *Garcites* sp. is similar to *Garcites potrerensis* n. gen. and sp. in its ventral ribbing separated by large spaces, but it differs in its compressed whorl section (Figure 10.5, 10.5a).

Material. One eroded specimen (ESIA-1022); same locality

as the previously described specimens (Figure 1b). Measurements are in Table 2.

Family POLYPTYCHITIDAE Wedekind, 1918
Genus *Dichotomites* Koenen, 1909

Dichotomites (Dichotomites) sp. (Figure 8.6, 8.6a)

Material. One specimen (ESIA-1028) from the Potrero de García, northeastern Mexico (Figure 1b); concretional structures are added to the shell. Two small, fragmented specimens are covered by calcareous material at the umbilical part (ESIA-1029, ESIA-1030).

Description. Small, involute, and compressed shell; narrow, deep umbilicus with vertical umbilical well; suboval whorl section slightly higher than wide; slightly flat sides and rounded venter; fine and rectiradiate ribs freely start on the rounded umbilical shoulder; some of them are irregularly bidichotomous at the middle of the sides, alternating irregularly with bifurcate ribs; fine intercostal spaces; 27 secondary ribs in the adoral half of the last whorl cross slightly forward on the ventral region.

Discussion. *Dichotomites (Dichotomites)* sp. might be a juvenile form; because of a difference in size, it could not be compared with the two large species from the upper Valanginian of eastern and northeastern Mexico, *Dichotomites (Dichotomites) compressiusculus* (Imlay, 1938) and *Dichotomites (Dichotomites) mantarraiae* Cantú-Chapa (1990).

Thieuloy (1977) described small specimens of *Dichotomites (Dichotomites) vergunnorum* Thieuloy (1977, Plate 7, Figures 2–8, p. 418) from the upper Valanginian of southeastern France. They are more evolute than our specimen and have a greater number of ribs in the external half of the last whorl, 42 in the French specimen versus 27 in the Mexican *Dichotomites (Dichotomites)* sp. Measurements are in Table 2.

Family NEOCOMITIDAE Salfeld, 1921
Genus *Karakaschiceras* Thieuloy, 1971
Type species. *Hoplites biassalensis* Karakasch, 1889, p. 435, Plate 1, Figures 4–5, by original designation of Thieuloy (1971, p. 229).

cf. Karakaschiceras sp. (Figure 8.4, 8.4a)

Material. An eroded specimen (ESIA-1031) from the Potrero de García, northeastern Mexico (Figure 1b). The ornamentation is preserved only in the outer part of the sides and on the ventral region. Measurements are in Table 2.

Description. Compressed, involute, small shell; whorl section much higher than wide, subparallel sides, narrow and tabulated venter; slightly convex fine ribs in the middle, external sides; intercostal spaces narrower than ribs.

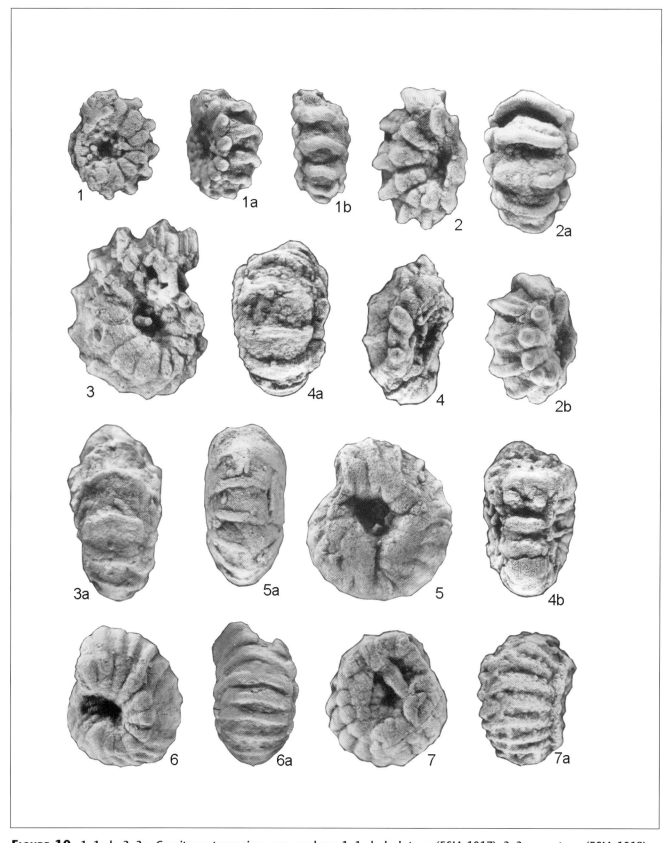

FIGURE 10. 1, 1a-b, 3, 3a. *Garcites potrerensis* n. gen. and sp.; 1, 1a-b, holotype (ESIA-1017); 3, 3a, paratype (ESIA-1018). 2, 2a, 2b, 4, 4a-b. *Garcites* sp. A (ESIA-1020, ESIA-1021). 5, 5a. *Garcites* sp. (ESIA-1022). 6, 6a, 7, 7a. *Garcites cavernensis* n. gen. and sp.; 6, 6a, holotype (ESIA-1025); 7, 7a, paratype (ESIA-1026). Locality and age: All specimens are from the upper part of the lower member of the Taraises Formation (upper Valanginian), from the Potrero de García, northwest of Monterrey City, northeastern Mexico. All specimens are x 2.

Discussion. This eroded and involute specimen is preserved in calcareous concretions in the middle, internal part of the sides; its ribbing is preserved only in the outer part of the sides and in the ventral region of the adoral half of the whorl. It is assigned to *Karakaschiceras* sp. by its tabulated venter and by its radial ribs, which slightly thin along the midventral region. Some species of *Karakaschiceras* from northeastern Mexico were studied by Imlay (1938) and by Young (1988); they are better preserved than the specimen described above.

Occurrence. Same locality and formation as the previously described specimens.

AGE OF THE AMMONITES FROM THE POTRERO DE GARCÍA

Capeloites and *Garcites* are predominant in the ammonite association collected from the lower part of the upper member of the Taraises Formation, approximately 100 m below the upper contact with the Cupido Formation (upper Hauterivian–lower Aptian; Imlay, 1944), from the Potrero de García, 20 km northwest of Monterrey, northeastern Mexico (Figure 1b). *Capeloites* was found in the upper Valanginian from southeastern France (Autran, 1989) and in the lower Hauterivian from Spain (Wilke, 1988); the exact age of this genus is still uncertain in Peru (Lissón, 1937). *Garcites* is a closely related form of *Dobrodgeiceras*, which was found at the base of the upper Valanginian, zone of *verrucosum*, in France (Thieuloy and Gazay, 1967; Thieuloy, 1977). The age of *Dobrodgeiceras* specimens from Peru was documented only as "upper Valanginian" by Riccardi and Westermann (1970).

Ammonites from the Potrero de García are associated with three small specimens of *Dichotomites* (*Dichotomites*) and one eroded specimen of *Karakaschiceras*. The age of *Dichotomites* is restricted to the upper Valanginian base in Germany (Kemper, 1978; Kemper et al., 1981) or to the zone of *trinodosum* (upper Valanginian), in the French small specimens studied by Thieuloy (1977). The age of *Karakaschiceras* is presently being discussed. It is placed at the lower–upper Valanginian boundary (*campylotoxus-verrucosum* zones) in France, Germany, and Poland (Bulot and Thieuloy, 1993; Kemper et al., 1981; Kutek et al., 1989) or at the upper Valanginian base (Kemper, 1992). Because of the cf. *Karakaschiceras-Dichotomites-Garcites* association, the Mexican ammonites here described could belong to the base of the upper Valanginian, as it is known in the Mediterranean area (Hoedemaeker et al., 1993).

AGE OF THE *CAPELOITES* FROM THE LA LAJA-8 WELL

This ammonite, found in the La Laja-8 well, core 11 (2556–2562 m), Tampico district, eastern Mexico, was obtained from limestone of the Lower Tamaulipas Formation (Berriasian–lower Aptian; Imlay, 1944) (Cantú-Chapa, 2001). The specimen can be correlated with ammonites reported here from the base of the upper Valanginian in Potrero de Garcia.

CONCLUSIONS

The Taraises Formation consists of open-water deposits that contain Berriasellids and Olcostephanids ammonites of Valanginian age. In the subsurface of northeastern Mexico, this formation is easily recognized with the aid of electric and radioactivity logs. The Taraises Formation is formed by two members: a lower member consisting of unfossiliferous compact limestone, which could be Berriasian in age, based on its stratigraphic position above the Late Tithonian La Casita Formation; and the upper member, consisting of fossiliferous marly limestone that contains Valanginian ammonites. Ammonites from several oil wells permitted the identification of the lower–upper Valanginian age boundary.

The new upper Valaginian ammonites found in outcrops of the Taraises Formation at Potrero de García, northeastern Mexico, not only represent important biostratigraphic information but also suggest a paleobiogeographic communication between Europe and America (Peru and Mexico). The upper Valanginian *Capeloites* sp. ammonite from the La Laja 8-well in the Lower Tamaulipas Formation is also an important biostratigraphic and biogeographic worldwide element.

REPOSITORY

Specimens described herein are housed in the Department of Geology at the Escuela Superior de Ingeniería y Arquitectura (ESIA), Instituto Politécnico Nacional (IPN), Mexico D. F. 07738, Mexico.

ACKNOWLEDGMENTS

Thanks to my wife, Françoise, and my sons Xochitl, Tonatiuh, and Yaocalli for helping me collect the small specimens from the Potrero de García locality. I also thank Hector Amezcua for his assistance with photographs. I thank Claudio Bartolini for translating the manuscript to English. I express my gratitude to Ryszard Myczynski and James Lee Wilson for providing constructive reviews of the manuscript. I thank Jaap Klein for making literature available.

REFERENCES CITED

Arkell, W. J., B. Kummel, and C. W. Wright, 1957, Mesozoic Ammonoidea, *in* R. C. Moore, ed., Treatise on invertebrate paleontology, part L. mollusca 4, cephalopoda, ammonoidea: Geological Society of America and University of Kansas Press, York and Lawrence, p. L80–L490.

Atrops, F., 1982, La sous-famille des Ataxioceratinae (Ammonitina) dans le Kimméridgien inférieur du Sud-Est de la France. Systématique, Evolution, Chronostratigraphie des genres (*Orthosphinctes* et *Ataxioceras*): Document du Laboratoire Géologique, Lyon 83, 463 p.

Autran, G., 1989, L'évolution de la marge nord-est provençale (Arc de Castelane) du Valanginien Moyen à l'Hauterivien à travers l'analyse biostratigraphique des séries de la région de Peyroules:

Séries condensées, discontinuités et indices d´une tectogenèse distenssive: Paléobiologie,Thèse Doctorat, Faculté des Sciences et Techniques de l'Université de Nice, 232 p.

Baumberger, E., 1923, Beschreibung zweier Valangienammoniten, nebst Bemerkungen über die Fauna des Gemsmätlihorizontes von Suizi im Justistal: Eclogae Geologicae Helvetiae, v. 18, no. 2, p. 303–313.

Böse, E., 1923, Algunas faunas cretácicas de Zacatecas, Durango y Guerrero: Instituto de Geología de México, 42, 219 p.

Brochwicz-Lewiñski, W., and Z. Rozak, 1976, Some difficulties in recognition of sexual dimorphism in Jurassic perisphinctids (Ammonoidea): Acta Palaeontologica Polonica, v. 21, no. 1, p. 97–114.

Bulot, L., and J-P. Thieuloy, 1993, Implications chronostratigraphiques de la révision de l´échelle biostratigraphique du Valanginien supérieur et de l´Hauterivien du Sud-Est de la France: Compte-Rendus de l' Académie des Sciences, Paris, T. 317, série II, p. 386–394.

Bulot, L., J.-P. Thieuloy, E. Blanc, and J. Klein, 1992, Le cadre stratigraphique du Valanginien supérieur et de l´Hauterivien du Sud-Est de la France: Définition des biochronozones et caractérisation de nouveaux biohorizons: Géologie Alpine, t. 68, p. 13–56.

Burckhardt, C., 1930, Étude synthétique sur le Mésozoïque mexicain: Mémoire de la Société Paléontologique Suisse, v. 49–50, 280 p.

Cantú-Chapa, A., 1963, Étude biostratigrafique des ammonites du Centre et de l´Est du Mexique: Société Géologique de France, Mémoire 99, no. 5, 103 p.

Cantú-Chapa, A., 1966, Se propone una nueva subdivisión de la Familia Olcostephanidae (Ammonoidea), del Cretácico Inferior (Taraisitinae subfam. nov. y Taraisites gen. nov.): Ingeniería Petrolera, v. 6, no. 12, p. 15–17.

Cantú-Chapa, A., 1970, El Kimeridgiano Inferior de Samalayuca, Chihuahua: Revista del Instituto Mexicano del Petróleo, v. 2, no. 3, p. 40–44.

Cantú-Chapa, A., 1972, Amonitas del Valanginiano Superior del pozo Calichoso no. 1, Norte de México: Revista del Instituto Mexicano del Petróleo, v. 4, no. 3, p. 88–89.

Cantú-Chapa, A., 1976, El contacto Jurásico-Cretácico, la estratigrafía del Neocomiano, el hiato Hauteriviano Superior–Eoceno Inferior y las amonitas del pozo Bejuco 6 (Centro-Este de México): Boletín de la Sociedad Geológica Mexicana, v. 37, p. 60–83.

Cantú-Chapa, A., 1987, The Bejuco Paleocanyon (Cretaceous-Paleocene) in the Tampico District, Mexico: Journal of Petroleum Geology, v. 10, no. 2, p. 207–218.

Cantú-Chapa, A., 1989, La Peña Formation (Aptian): A condensed limestone-shale sequence from the subsurface of NE of Mexico: Journal of Petroleum Geology, v. 12, no. 1, p. 69–83.

Cantú-Chapa, A., 1990, *Dichotomites* (*Dichotomites*) *mantarraiae* sp. nov., ammonita del Valanginiano Superior del pozo Mantarraya 1, Golfo de México: Revista de la Sociedad Mexicana de Paleontología, v. 2, no. 2, p 43–45.

Cantú-Chapa, A., 1992, *Novoleonites tovarensis* gen. nov. sp. nov. (Ammonoideo Neocomitino) del Valanginiano Superior del NE de México: Revista de la Sociedad Mexicana de Paleontologia, v. 5, no. 1, 37–43.

Cantú-Chapa, A., 1999, Confrontation of stratigraphic methods to define the Jurassic-Cretaceous boundary in eastern Mexico subsurface: Geological Society of America, Special Paper 340, p. 93–103.

Cantú-Chapa, A., 2001, Paleocanyons in the subsurface of eastern Mexico: Facts and uncertainties. This volume, chapter 19.

Cecca, F., 1986, Le genre *Richterella* Avram (Ammonitina, Périsphinctidés) dans le Tithonique inférieur de la bordure ardéchoise (Sud-Est de la France): Dimorphisme et variabilité: Geobios, no. 19, p. 33–44.

Company, M., 1987, Los ammonites del Valanginiense del sector oriental de las Cordilleras Béticas (SE de España): Thesis, Universidad de Granada, Spain, 295 p.

Contreras, M. B., 1977, Bioestratigrafía de las formaciones Taraises y La Peña (Cretácico Inferior), de la Goleta, Coahuila y Minillas, Nuevo León: Revista del Instituto Mexicano del Petróleo, v. 9, no. 1, p. 8–29.

Cooper, M. R., 1981, Revision of the Late Valaginian Cephalopoda from the Sundays River Formation of South Africa, with special reference to the genus *Olcostephanus:* Annals of the South African Museum, v. 83, pt. 7, 366 p.

Enay, R., 1966, L´Oxfordien dans la moitié sud du Jura français: Étude stratigraphique Nouvelles Archives du Muséum d´Histoire Naturelle de Lyon, fasc. 8, T. II, p. 331–624.

Hoedemaeker, P. J., and 15 others, 1993. Ammonite zonation for the Lower Cretaceous of the Mediterranean region; basis for the stratigraphic correlations within IGCP-Project 262: Revista Española de Paleontología, v. 8, no. 1, p. 117–120.

Imlay, R. W., 1937, Lower Neocomian fossils from the Miquihuana Region, Mexico: Journal of Paleontology, vol. 11, no. 7, p. 552–574.

Imlay, R. W., 1938, Ammonites of the Taraises Formation of northern Mexico: Bulletin of the Geological Society of America, v. 49, no. 4, p. 539–602.

Imlay, R. W., 1940, Neocomian faunas of northern Mexico: Bulletin of the Geological Society of America, v. 51, p. 117–190.

Imlay, R. W., 1944, Cretaceous formations of Central America and Mexico: AAPG Bulletin, v. 28, p. 1077–1195.

Karakasch, N., 1889, Über einige Neokomablagerungen in der Krim: Sitzungsberichte der kaiserlichen Akademie der Wissenschaften in Wien, Mathematisch-Naturwissenshaftliche Klasse (für 1889), p. 428–438.

Kemper, E., 1978, Einige neue, biostratigraphisch bedeutsame Arten der Ammoniten-Gattung *Dichotomites* (NW-Deutschland, Obervalangin): Geologisches Jahrbuch A45, p. 183–253.

Kemper, E., 1992, Die tiefe Unterkreide im Vechte-Dinkel-Gebiet (westliches Niedersächisisches Becken): Het Staringmonument te Losser, p. 7–95.

Kemper, E., P. F. Rawson, and J.-P. Thieuloy, 1981, Ammonites of Tethysan ancestry in the early Lower Cretaceous of north-west Europe: Palaeontology, 24, p. 251–311.

Kilian, W., 1910, Erste Abteilung: Unterkreide (Palaeocretacicum). Lieferung 2: Das bathyale Palaeocretacicum im südostlichen Frankreich; Valendis-Stufe: Hauterive-Stufe; Barreme-Stufe: Apt-Stufe, *in* Fritz Frech, ed., Lethaea Geognostica II, Das Mesozoicum, 3 (Kreide): E. Schweizerbart's sche Verlagsbuchhandlung, Stuttgart, p. 169–288.

Kutek, J., R. Marcinowski, and J. Wiedmann, 1989, The Wawal Section, Central Poland—An important link between Boreal and Tethyan Valanginian, *in* J. Wiedmann, ed., Cretaceous of the Western Tethys, Proceedings 3rd International Cretaceous Symposium, Tübingen, 1987: Stuttgart, E. Schweizerbart'sche Verlagsbuchhandlung, p. 717–754.

Lissón, C. I., 1937, Dos Amonites del Perú: Boletín de Minas, Lima, Talleres Gráficos San Miguel, p. 3–9.

Matheron, P., 1878–1880, Recherches paléontologiques dans le Midi de la France: Marseille, 12 p.

Neumayr, M., 1875, Die Ammoniten der Kreide und die Systematik der Ammonitiden: Zeitschrift der Deutschen Geologischen Gesellschaft, v. 27, p. 854–942.

Nikolov, T., 1963, New name for a Valanginian Ammonites genus: Geological Magazine, v. 100, no. 1, p. 94.

Peña-Muñoz, M. G., 1964, Ammonitas del Jurásico Superior y del Cretácico Inferior del extremo oriental del estado de Durango, México: Paleontología Mexicana, no. 20, 33 p.

Rawson, P. F., 1975, Lower cretaceous ammonites from north-east England: The Hauterivian Heteromorph Aegocrioceras: Bulletin of the British Museum, (Natural History) Geology, v. 26, no. 4. p. 131–283.

Riccardi, A. C., and G. E. G. Westermann, 1970, The Valanginian *Dobrodgeiceras* Nikolov (Ammonitina) from Peru: Journal of Paleontology, v. 44, no. 5, p. 888–892.

Riccardi, A. C., G. E. G. Westermann, and R. Levy, 1971, The Lower Cretaceous Ammonitina *Olcostephanus*, *Leopoldia*, and *Favrella* from west-central Argentina: Palaentographica, 136, A, p. 83–121.

Spath, L. F., 1924, On the ammonites of the Speeton Clay and the subdivisions of the Neocomian: Geological Magazine, v. 61, p. 73–89.

Thieuloy, J.-P., 1969, Sur la présence du genre Capeloites Lisson (Ammonoidea) dans le Néocomien des Basses-Alpes et la signification des espèces migratrices transatlantiques: Comptes-Rendus sommaire des Séances de la Société Géologique de France, Fascicule 7, p. 256–258.

Thieuloy, J.-P., 1971, Réflexions sur le genre *Lyticoceras* Hyatt, 1900 (Ammnoidea): Comptes-Rendus des Séances de l'Académie des Sciences, Paris, series D272, p. 2297–3000.

Thieuloy, J.-P., 1977, Les ammonites boréales des formations néo-comiennes du sud-est français (province subméditerranéenne): Geobios 10, p. 395–461.

Thieuloy, J.-P., and L. Bulot, 1992, Ammonites du Crétacé Inférieur du Sud-Est de la France: 1. Nouvelles espèces à valeur stratigraphique pour le Valanginien et l'Hauterivien: Géologie Alpine, t. 68, p. 85–103.

Thieuloy, J.-P., and M. Gazay, 1967, Le genre *Dobrodgeiceras* Nikolov en Haute Provence: Travaux du laboratoire de Géologie, Faculté des Sciences, Lyon, Nouvelle Série 14, p. 69–78.

Verma, H. M, and G. E. G. Westermann, 1973, The Tithonian (Jurassic) ammonite fauna and stratigraphy of Sierra Catorce, San Luis Potosí, Mexico: Bulletin of American Paleontology, v. 63, no. 277, p. 105–320.

Wilke, H.-G., 1988, Stratigraphie und Sedimentologie der Kreide im Nordwesten der Provinz Alicante (SE-Spanien): Berliner geowissenschaft 95, 72 p.

Wilson, J. L., and W. C. Ward, 1993, Early Cretaceous Carbonate platforms of northeastern and east-central Mexico, in J. A. Tonisimo, R. W. Scott, and J. P. Masse, eds., Cretaceous carbonate platforms; AAPG Memoir 56, p. 35–49.

Wright, C. W., J. H. Calloman, and M. K. Howarth, 1996, Cretaceous Ammonoidea, in R. L. Kaesler, ed., Treatise on invertebrate paleontology, part L, mollusca 4: The Geological Society of America and the University of Kansas, Lawrence, 362 p.

Young, K., 1969, Ammonites zones of Northern Chihuahua, in Guidebook of the Border Region: New Mexico Geological Society, 20th Field Conference, p. 97–101.

Young, K., 1988, *Karakaschiceras* and the Late Valanginian of Northern Mexico and Texas, in J. Wiedmann and J. Kullmann, eds., Cephalopods—present and past: Stuttgart, E. Schwizerbart'sche Verlagsbuchhandlung, p. 621–632.

16

Adatte, T., W. Stinnesbeck, H. Hubberten, J. Remane, and J. G. López-Oliva, 2001, Correlation of a Valanginian stable isotopic excursion in northeastern Mexico with the European Tethys, *in* C. Bartolini, R. T. Buffler, and A. Cantú-Chapa, eds., The western Gulf of Mexico Basin: Tectonics, sedimentary basins, and petroleum systems: AAPG Memoir 75, p. 371–388.

Correlation of a Valanginian Stable Isotopic Excursion in Northeastern Mexico with the European Tethys

Thierry Adatte
*Institut de Géologie, University of Neuchâtel
Neuchatel, Switzerland*

Wolfgang Stinnesbeck
*Geologisches Institut, Universität Karlsruhe
Karlsruhe, Germany*

Hans Hubberten
*Alfred Wegener Institut für Polar- und
Meeresforschung, Potsdam, Germany*

Jürgen Remane
*Institut de Géologie, University of Neuchâtel
Neuchatel, Switzerland*

José Guadalupe López-Oliva
*Facultad de Ciencias de la Tierra
Universidad Autónoma de Nuevo León
Linares, Mexico*

ABSTRACT

In the Sierra Madre Oriental of northeastern Mexico, two sections (La Huasteca and San Lucas) spanning Berriasian to lower Hauterivian rocks were analysed and are correlated by mean of calpionellid and ammonite ocurrences, microfacies, and stable isotopes (bulk rock). A major isotopic excursion (approximately 3‰) of both δ^{13}C and δ^{18}O was recognized in an interval of pelagic mudstone corresponding to the upper Valanginian. A similar δ^{13}C excursion was also observed in coeval strata of the southern Italian Alps and Appennines (Weissert and Channell., 1985; Weissert et al., 1989; Weissert and Lini, 1991; Lini et al., 1992); the northern Tethys margin (Föllmi et al. 1994); the Gulf of Mexico (Patton et al., 1984); and the North Atlantic (Robertson and Bliefnick, 1983) and Pacific (Douglas and Savin, 1973) Oceans. The δ^{13}C shift is independent of changes in microfacies, contents in organic matter, and mineralogical composition of the sediment. This stable isotopic pattern was also identified in the Vocontian basin in France (Hennig et al., 1999), and calibrated to the *Campylotoxus-Verrucosum* zones of the early/late Valanginian. Integration of our biostratigraphic and isotopic data indicates the presence, at San Lucas, of a complete Valanginian sequence in terms of European ammonite and calpionellid zones, whereas at La Huasteca some of the zones may be absent.

This late Valanginian δ^{13}C excursion is generally interpreted to be the first episode of Cretaceous greenhouse conditions. It reflects a major change in the global carbon budget that could have resulted from increased tectonic/volcanic activity, such as the Paraná continental flood basalts leading to increased atmospheric CO_2 and, consequently, greenhouse conditions.

Oxygen-isotopes data in the Mexican sections trend toward higher values, but with a significant temporal lag, as compared with the δ^{13}C curve. This indicates seawater and/or climatic cooling, which can be linked either to negative feedback (e.g., increase in weathering) leading to a decrease in atmospheric CO_2 or to intense oceanic spreading and a subsequent global sea-level rise.

INTRODUCTION

The San Lucas and La Huasteca sections in the Sierra Madre Oriental of northeastern Mexico (Figure 1) span the late Tithonian, Berriasian, and Valanginian (Figure 2). This time interval corresponds to the deposition of the upper La Casita, Taraises, and lowermost Cupido Formations. The lithology of the La Casita indicates a period of major clastic influx (Figure 3a) in northeastern Mexico (Imlay, 1936; Smith, 1981; Michalzik, 1988). Most of the detrital material is derived from the emerged Coahuila block (Fortunato, 1982; Goldhammer et al., 1991). The transition to the Taraises is characterized by a decrease of siliciclastic supply and an increase of carbonate production in shelf and pelagic environments (Michalzik, 1988; Michalzik and Schuman, 1994). The Taraises and lowermost Cupido consist of pelagic carbonates (Figure 3a, b), and the Cupido represents a late Hauterivian to Aptian platform system (Wilson and Pialli, 1977).

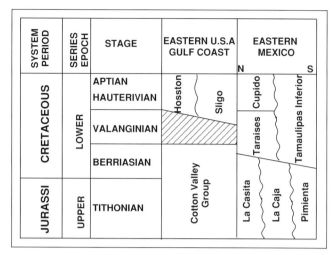

FIGURE 2. Table of uppermost Jurassic and lowermost Cretaceous formations of northeastern Mexico. Wiggly lines indicate gradation and transition of lithofacies.

Adatte et al. (1992, 1994, 1996a, b) have reported on a significant late Tithonian to Berriasian excursion of stable isotopes in Mexico. This excursion corresponded to the opening of the Mexican Basin toward the Tethyan domain during the sea-level cycle LZB-1 of Haq et al. (1988) and the Be3–Be4 interval of the Hardenbohl curve (Hardenbohl et al., 1998). The authors describe a major shift (aproximately 2‰) toward higher values of $\delta^{13}C$ (0–1‰) and $\delta^{18}O$ (> –6‰) in sections from northeastern and central Mexico and interpret these changes as the result of increased oceanic productivity and relative cooling during a transgressive interval. Carbon and oxygen data from these sections correlate well and are independent of lithology, thickness of sequences, and diagenetic overprint.

In the present paper, we extend our studies to the Valanginian of northeastern Mexico. Lini et al. (1992) reported a significant positive excursion of carbon isotopes in marly limestones of the upper Valanginian of southern Italy. They correlate this shift to coeval sections in the northern Tethys margin in the Swiss Alps (Föllmi et al., 1994); the Gulf of Mexico (Patton et al., 1984); and the North Atlantic (Robertson and Bliefnick, 1983; Cotillon and Rio, 1974) and Pacific (Douglas and Savin, 1973) Oceans and interpret this event as the first episode of Cretaceous greenhouse earth conditions.

We report on a positive excursion of $\delta^{13}C$ in the upper Taraises Formation, which correlates well with a similar isotopic excursion in the Vocontian basin (France). In that area, ammonites place it in the early/late Valanginian *Campylotoxus-Verrucosum* zones (Hennig et al., 1999). This positive shift reflects a major change in the global carbon budget, probably related to increased tectonic/volcanic activity (Paraná continental flood basalts) that increased atmospheric CO_2 and,

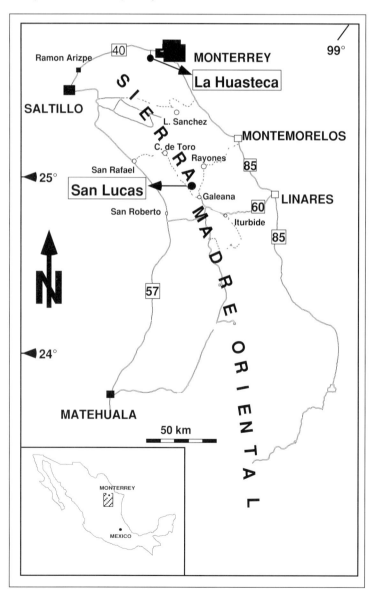

FIGURE 1. Location of the sections studied.

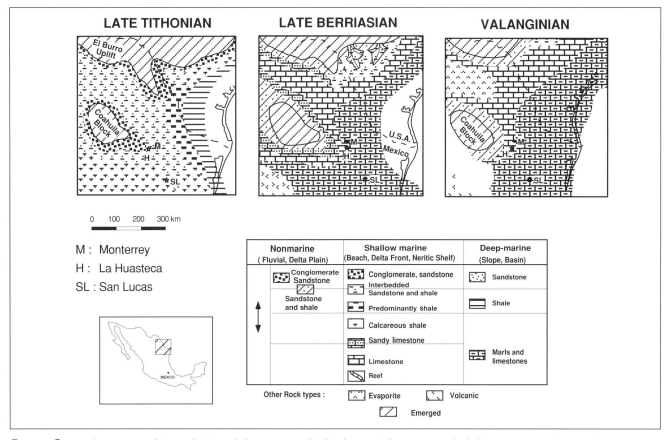

FIGURE 3a. Paleogeographic evolution of the area studied (after Smith, 1981; Michalzik, 1988; McFarlan and Menes, 1991; Salvador, 1991, modified).

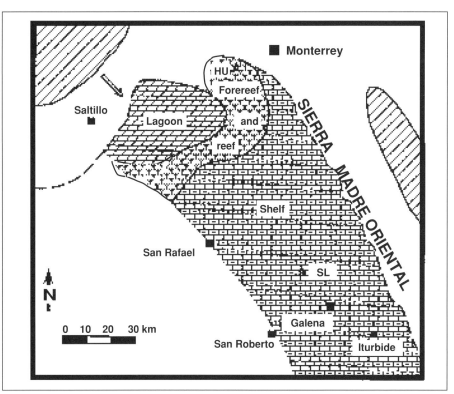

FIGURE 3b. Paleogeographic setting of the area studied (after Michalzik, 1988, modified), during the late Berriasian: installation of reef environment in northeastern Mexico.

therefore, greenhouse conditions. This $\delta^{13}C$ excursion is consequently interpreted generally as the first episode of a Cretaceous greenhouse condition.

LOCATION OF THE SECTIONS STUDIED HEREIN

Rancho San Lucas is located in the Sierra Madre Oriental approximately 25 km north of Galeana, Nuevo León, and is reached from there by a dirt road (Figure 1). The section is located on the eastern side of the valley in a small canyon.

The La Huasteca section is located approximately 20 km west of Monterrey, Nuevo León, in the La Huasteca canyon south of the suburb of Santa Catarina (Figure 1). It is reached by a dirt road that crosses the canyon. We sampled a section located southeast of the high cliffs at the canyon entrance, on the eastern side of the dry riverbed.

METHODS

Whole-rock and insoluble residue analyses were carried out for all the samples by X-ray diffraction (XRD) at the Geological Institute of the University of Neuchâtel. The samples were prepared following the procedure of Kubler (1987) and Adatte et al. (1996c). Two sample preparation methods were applied. Random powder of the bulk sample was used for characterization of the whole-rock mineralogy. Nearly 20 g of each rock sample was ground with a "jaw" crusher to obtain small chips (1 to 5 mm) of rock. Approximately 5 g were dried at a temperature of 110°C and then ground again to a homogenous powder with particle sizes < 40 μm. Of this powder, 800 mg (20 bars) were pressed in a powder holder covered with blotting paper, then analysed by XRD. Whole-rock compositions were determined by XRD (SCINTAG XRD 2000 Diffractometer) based on methods by Ferrero (1966) and Kübler (1983). The method for the semiquantitative analysis of the bulk-rock mineralogy (obtained by XRD patterns of random powder samples) used external standards.

Clay mineral analyses were based on methods by Kübler (1987). Ground chips were mixed with deionized water (pH 7–8) and agitated. The carbonate fraction was removed with the addition of HCl 10% (1.25 N) at room temperature for 20 minutes or more until all the carbonate was dissolved. Ultrasonic disaggregation was accomplished during 3-minute intervals. The insoluble residue was washed and centrifuged (five to six times) until a neutral suspension was obtained (pH 7–8). Separation of different grain-size fractions (< 2 μm and 2–16 μm) was obtained by the timed settling method based on Stokes law. The selected fraction was then pipetted onto a glass plate and air-dried at room temperature. XRD analyses of oriented clay samples were made after air-drying at room temperature and with ethylene-glycol-solvated conditions. The intensities of selected XRD peaks characterizing each clay mineral present in the size fraction (e.g., chlorite, mica, and mixed layers) were measured for a semiquantitative estimate of the proportion of clay minerals present in the size fractions < 2 μm and 2–16 μm. Therefore, clay minerals are given in relative-percent abundance without correction factors. Content in swelling (% smectite) is estimated using the method of Moore and Reynolds (1989).

For stable isotope measurements, subsamples were taken using a dental drill. As with the preparation technique applied, only 60 to 80 μg of carbonate is needed; therefore, care was taken to prepare only lithologies showing no alteration or secondary calcite veinlets (checked by thin-section analyses). The measurements were carried out on a Finnigan MAT 251 spectrometer coupled on-line to an automatic carbonate preparation device (Finnigan MAT). The samples were reacted at 75°C with orthophosphoric acid onto individual samples.

Standard deviations of the measurements are < 0.04‰ and < 0.06‰ for carbon and oxygen, respectively (Hubberten and Meyer, 1989). Isotope ratios are given in δ notation versus Peedee Belemnite Standard (PDB); i.e., the δ scale was calibrated using NBS19 as a reference sample (Hut, 1987).

The content of total organic carbon (TOC) was determined by a Carlo-Erba EA 1108 Elemental Analyser CHN from the insoluble residue of the samples, in which significant positive excursions in $\delta^{13}C$ were detected. The content in organic matter does not show any significant correlation with $\delta^{13}C$.

RESULTS

Lithology

The San Lucas section spans approximately 700 m of rocks (Figure 4). In this paper, names of formations are those used by earlier authors (e.g., Imlay, 1936, 1937). The basal 25 m belong to the La Casita Formation and are characterized by alternating silts, marls, and thin-bedded limestones representing outer-shelf to pelagic environments. The contact between the La Casita and the Taraises is gradational. The Taraises consists of approximately 600 m of pelagic lime mudstones and intercalated shales, which correspond to a deep shelf to basinal setting. This unit is divided into a lower calcareous and upper marly member. The Taraises underlies the Cupido Formation, which consists of thick-bedded basinal limestones with chert layers. We sampled only the lowermost 50 m.

Compared with San Lucas, the thickness of the La Huasteca section is much reduced (Figure 5). The basal 33 m of the section belong to the La Casita Formation, which is here represented by shallow-marine sandstones and microconglomerates. Coarsening-upward cycles indicate progradation of delta lobes into a submarine shelf environment. The transition to the overlying Taraises Formation is gradual and consists of 10 m of thin-bedded sandy limestones. The basal Taraises consists of 3 m of coral-rich limestone, equivalent to the San Juan lentil of Frame and Ward (1987) and representing shallow outer-shelf environments overlain by 80 m of pelagic sandy and clayey marls. The carbonate content of the Taraises gradually increases upsection. These strata underlie the Cupido Formation, which consists of thick-bedded basinal limestones with chert nodules. Only the lowermost 15 m were sampled here.

Whole-rock analyses confirm the observed lithologies (Figures 4 and 5). The top of the La Casita Formation in the San Lucas section is predominantly of phyllosilicates, quartz, and feldspar, with only a little calcite. The Taraises and the basal Cupido Formations are monotonously dominated by calcite and, to a somewhat lesser extent, quartz and phyllosilicate.

In the La Huasteca section, the La Casita Formation is more detrital than in the San Lucas section, with high content of quartz, phyllosilicates, K-feldspar, and plagioclase (Figure 5). The transition to the Taraises Formation is characterized by increased calcite, reaching a maximum in the coral-rich limestone. Marly siltstone of the Taraises is characterized by a high content of phyllosilicates and calcite; carbonates dominate the upper part of this unit and the overlying Cupido Formation.

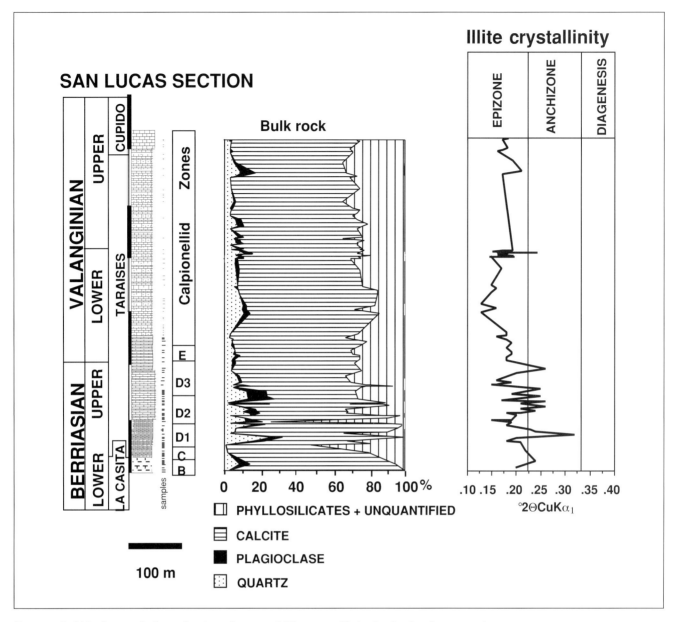

FIGURE 4. Lithology, whole-rock mineralogy, and illite crystallinity in the San Lucas section.

Clay Mineralogy

Clay mineral assemblages in the two sections consist of chlorite, mica, and rare mixed layers (illite-smectite and chlorite-smectite characterized by a low content in expandable layers). Values of illite crystallinity (Figures 4 and 5) are low (0.13–0.32°2ΘCuKα₁), thus indicating anchimetamorphic to epimetamorphic zones (Kübler, 1987). In general, illite crystallinity is higher in the San Lucas section (Figure 4, mean value = 0.19°2ΘCuKα₁) than in the La Huasteca profile (Figure 5, mean value = 0.23°2ΘCuKα₁).

Biostratigraphy

Calpionellids

In both the La Huasteca and the San Lucas sections, calpionellids are very rare and rather badly preserved, as a result of

incipient recrystallization of the limestones. More favorable profiles (Blauser and McNulty, 1980; Adatte et al., 1992, 1994, 1996 a, b) show, however, that the Mediterranean zonation of Remane (1985), which is agreed on by a great number of calpionellid workers (Remane et al., 1986), can also be applied to eastern Mexico.

Calpionellid zones B to E were identified in the San Lucas profile (Figure 6). As in the southern Mediterranean province, *Calpionellopsis oblonga* occurs with *Calpionellites darderi* in the lower part of zone E. With respect to ammonite zones, the base of zone E coincides for all practical purposes with the top of the basal Valanginian *Otopeta* zone (Allemann and Remane, 1979). In the La Huasteca section (Figure 6), the coral-rich limestone corresponds to late Berriasian calpionellid zone D, probably subzone D2, where the appearance of *C. oblonga* is

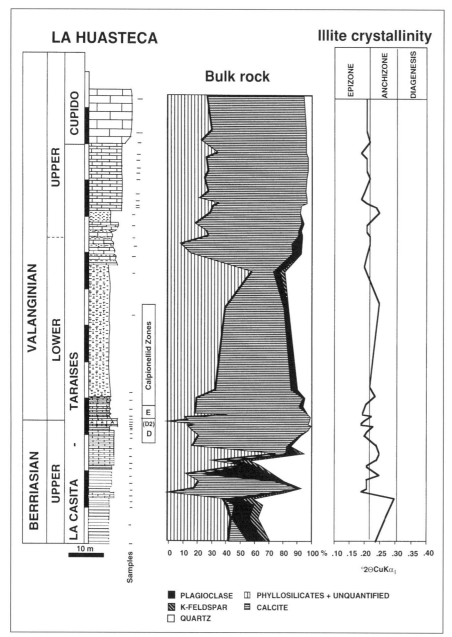

FIGURE 5. Lithology, whole-rock mineralogy, and illite crystallinity in the La Huasteca section.

(Figure 6). Early Valanginian ammonite genera *Thurmanniceras, Olcostephanus, Distoloceras,* and *Bochianites* (Imlay, 1938, 1940) occur in the middle Taraises Formation, whereas the presence of *Karakaschiceras* implies a late Valanginian age (Young, 1988). *Oosterella* identified in the upper Taraises and the basal Cupido Formations corresponds to the late Valanginian or early Hauterivian (Bulot et al., 1992).

Ammonites are rare in the La Huasteca section except for *Olcostephanus* and *Bochianites* (Figure 6). These ammonites indicate a Valanginian age for most of the Taraises. An interval with abundant *Bochianites* was recognized in both sections in overlying rocks of the calpionellid zone E.

Microfacies

Marine microfacies (MF) in the study area characterize deltaic to prodeltaic detrital environments (MF 10 and 11), and typical outer-platform and slope-to-basin environments (MF 1–9, Figure 8).

Microfacies 1: Micrite or silty clay with fine-grained quartz, very rare radiolarians, "filaments," and nannofossils. This microfacies suggests hemipelagic environments.

Microfacies 2: Biomicrite with well-preserved radiolarians, pellets, "filaments," and rare calpionellids in a micritic matrix locally enriched in organic matter. This microfacies indicates pelagic to hemipelagic environments under suboxic conditions.

Microfacies 3: Biomicrite rich in radiolarians and calcispheres, commonly recrystallized.

Microfacies 4: Micrite with rare calpionellids and radiolarians, rare small agglutinated foraminifera and echinoderm debris. This microfacies indicates a hemipelagic environment with sporadic and attenuated outer-platform (circalittoral) input.

Microfacies 5: Biomicrite with abundant calpionellids, "filaments," few calcispheres, and ammonite debris. This microfacies characterizes pelagic to hemipelagic environments

Microfacies 6: Biomicrite with numerous sponge spicules, calpionellids, abundant "filaments," ammonite debris, rare small agglutinated foraminifera, Nodosariidae, and some echinoderm debris. This microfacies indicates hemipelagic environments with reduced outer-platform (circalittoral) input.

considerably below that of *Ct. Darderi*. The absence of *L. hungarica* could, however, be explained by the scarceness of the fauna. The absence of subzone D3 thus remains uncertain. The co-occurence of *C. oblonga* and *Ct. darderi* directly above the coral-rich limestone indicates the lower part of zone E (*Pertransiens* zone) and thus the base of the Valanginian (Figure 7), according to Busnardo and Thieuloy (1979) (see also Bulot, 1996).

Ammonites

In the San Lucas section, *Pseudosubplanites, Substeueroceras, Berriasella,* and *Negreliceras* indicate an early Berriasian age (Adatte et al., 1996a) for the top of the La Casita Formation

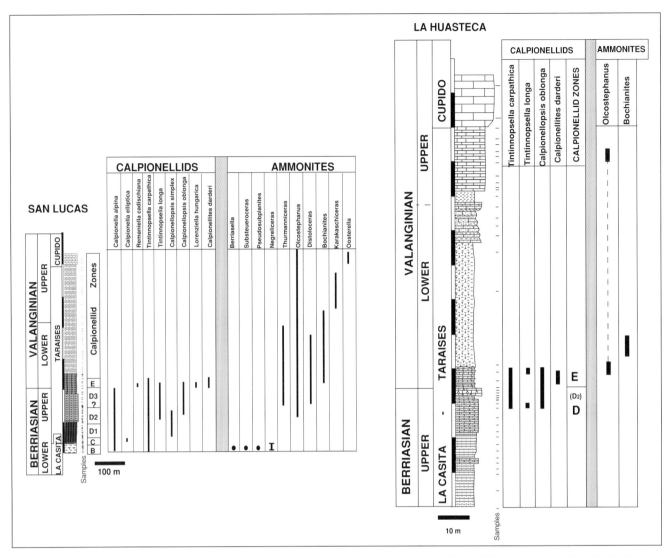

FIGURE 6. Biostratigraphy of the San Lucas and La Huasteca sections based on calpionellids and ammonite distribution (Calpionellid zonation from Remane, 1985; ammonite zonation after Imlay, 1938, 1940.) Note: 10 times difference in vertical scale.

Microfacies 7: Biomicrite with rare calpionellids, small agglutinated foraminifera, common echinoderm, and bivalve debris. Grain size is notably coarser than in microfacies 6. This microfacies indicates outer-platform environments, i.e., circalittoral conditions (slope).

Microfacies 8: Biopelmicrite with abundant echinoderm debris, bivalves, rare ammonites, agglutinated foraminifera and Nodosariidae, rare calpionellids, radiolaria, and quartz; grains are generally coarser and not well sorted. This microfacies reflects outer-platform environments, under circa- to infralittoral conditions.

Microfacies 9: Biomicrite with abundant corals (*Microsolena, Stylosmilia*; Michalzik, 1988) and stromatoporoids debris (cf. *Shugraia?*; Michalzik, 1988), echinoderms, gastropods, bivalves, and foraminifera (*Trocholina, Lenticulina,* Textulariidae); calpionellids and radiolarians are very rare. This microfacies represents forereef and reefal environments.

Microfacies 10: Well-sorted, coarse- to medium-grained

sandstone, composed mainly of quartz and feldspar, cemented by calcite or dolomite with rare Nodosariidae (*Lenticulina*), few bivalves and little echinoderm debris.

Microfacies 11: Coarse sandstone and microconglomerate composed mainly of polycrystalline quartz, feldspar, and rock fragments (well rounded to angular) of various origins (sedimentary, igneous, and metamorphic), cemented by calcite or dolomite. The occurrence of rare, small benthic foraminifera (Textulariidae and Nodosariidae) and echinoderm, bivalve, and ammonite debris indicates a shallow-marine environment that corresponds to marine delta-front–prodelta-front conditions.

At San Lucas, the top of the La Casita Formation corresponds to marine distal prodeltaic environments. The clastic sediments contain abundant radiolarians and few calpionellids (MF 1 and 2). Upsection, the basal Taraises Formation is characterized by more carbonate-rich microfacies with abundant calpionellids (MF 5 and 6). In calpionellid zone D, the presence

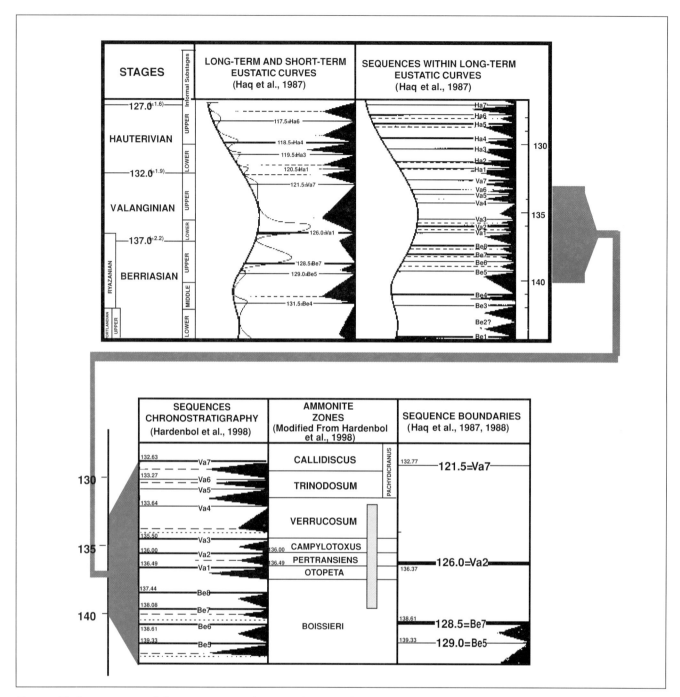

FIGURE 7. Long-term and short-term eustatic curves (Haq et al., 1987), sequences (Hardenbol et al., 1998) within long-term eustatic curves (Haq et al., 1988), and sequences chronostratigraphy and ammonite zones, modified from Hardenbol et al. (1998).

of small benthic foraminifera (*Lenticulina*, Textulariidae), echinoderm debris, and sponge spicules indicates the proximity of slope to outer-shelf environments (MF 6). Upward, most of the Valanginian stratal sequence is characterized by micritic mudstone with rare microfauna (MF 1), indicating basinal conditions.

The lower part of the La Huasteca section (lower upper Berriasian) is characterized by marine sandstones and micro-conglomerates indicating proximal delta-fan conditions (MF

10 and 11, Figure 8). In the upper Berriasian calpionellid zone D, the presence of coral-rich limestones indicates an outer platform environment (MF 9) overlain by fine detrital slope-to-basin strata (MF 6) of the lower Valanginian. Upper Valanginian rocks at La Huasteca are characterized by basinal mudstones similar to those of the San Lucas (MF 1). Environments in the two sections studied show a similar evolution. Microfacies at La Huasteca are, however, more proximal than at San Lucas, because it is in the vicinity of the Coahuila block (Figures 3a, b, and 8).

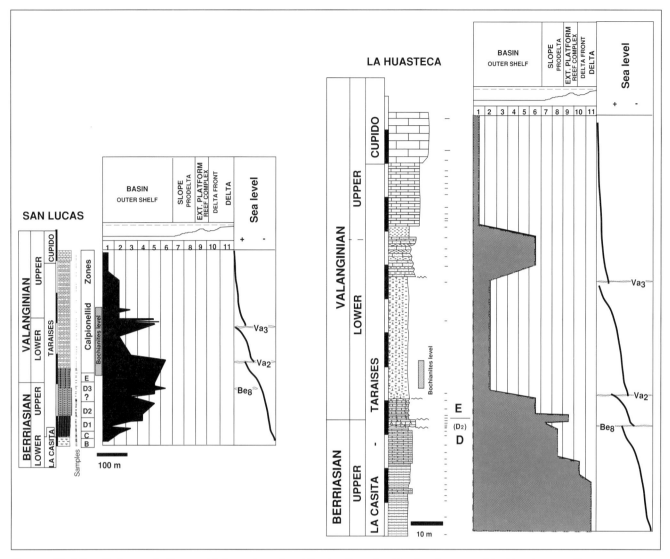

FIGURE 8. Microfacies evolution and inferred sea-level fluctuations for the San Lucas and La Huasteca section and tentative correlation with the sequence boundaries from Haq et al. (1988) and Hardenbol et al. (1998). Note: 10 times difference in vertical scale.

Stable Isotopes

The sediments investigated suffered anchimetamorphic to epimetamorphic alteration (see previous section titled *Clay Mineralogy*). This strong burial diagenesis may have altered the primary isotopic ratios of the carbonates. In addition, recrystallization of carbonate in the presence of meteoric water may alter the isotopic signal. Meteoric water is commonly depleted in $\delta^{18}O$ and carries $\delta^{12}C$ derived from organic matter. Strong diagenetic alteration may be indicated by a covariance of carbon and oxygen isotopes (Jenkyns and Clayton, 1986; Jenkyns, 1995). When plotted as a $\delta^{13}C$–$\delta^{18}O$ graph, the carbon and oxygen isotopic data show no evident covariance (no. of samples = 73, r^2 = 0.31 for the San Lucas section and, r^2 = 0.05; no. of samples = 33 in the La Huasteca section; Figure 9). But a lack in covariance under anchimetamorphic to epimetamorphic conditions does not necessarily reflect the absence of overprint, because large-scale $\delta^{18}O$ reequilibration during

recrystallization may not alter strongly the $\delta^{13}C$ signal (Früh-Green et al., 1990). Frequently, dissolution-recrystallization processes do not significantly influence the magnitude of the $\delta^{13}C$ shift, even though the primary oxygen isotope is overprinted by diagenesis and metamorphism (Veizer, 1983). This difference may be explained by much larger temperature-dependent fractionation during burial diagenetic cementation for oxygen than for carbon isotopes (Friedman and O'Neil, 1977). These observations suggest a degradation of the primary $\delta^{18}O$ signal, although the $\delta^{13}C$ signal remains relatively unaltered in our sections. Compared to coeval sections (e.g.; Bottacione section, Italy, Corfield et al., 1991; La Charce section, Hennig et al., 1999), $\delta^{18}O$ values in our sections are significantly more negative (–11 to –4‰) (Figures 10 and 11) and suggest strong diagenetic/metamorphic alteration. The more negative values in $\delta^{18}O$ have been detected in the lower part of the sections characterized by silt and sandstone lithogies in which carbonates are mainly of secondary origin. In these rocks, the range of $\delta^{18}O$ variability is quite high, and an analysis of detailed paleotemperature cannot be justified. $\delta^{13}C$ values, however, are mostly positive at San Lucas. Only a few samples of siltstones of the La Casita Formation and marly limestones of the Taraises Formation present negative values (-2 to $-1‰$ and thus reflect intense diagenetic alteration. Compared to San Lucas (Figure 10), values of $\delta^{13}C$ and $\delta^{18}O$ are systematically more negative in the La Huasteca section (Figure 11) and suggest a more intense diagenetic/metamorphic overprint. Note that the La Huasteca section is characterized by shallow conditions (Figure 8), diverse lithologies, and abrupt

lithologic breaks, implying a stronger diagenetic alteration. However, several geologic sections from distant locations and different lithologies show the same pattern of isotopic change. It is therefore difficult to concede that this general pattern is caused by local diagenetic alteration, although some samples may be contaminated. In consequence, the $\delta^{13}C$ at San Lucas and La Huasteca (values ranging from $0‰$ to $3‰$) probably preserve an original paleo-oceanograpic pattern (Jenkyns, 1995), especially at San Lucas. In contrast, $\delta^{18}O$ values cannot be used to infer detailed paleotemperatures, although general stratigraphic trends are clearly discernible.

In the San Lucas section, a significant change toward higher

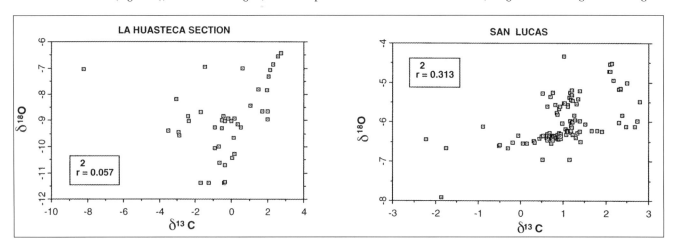

FIGURE 9. $\delta^{13}C$ and $\delta^{18}O$ crossplots for the two sections studied herein. Note that no significant correlation is present.

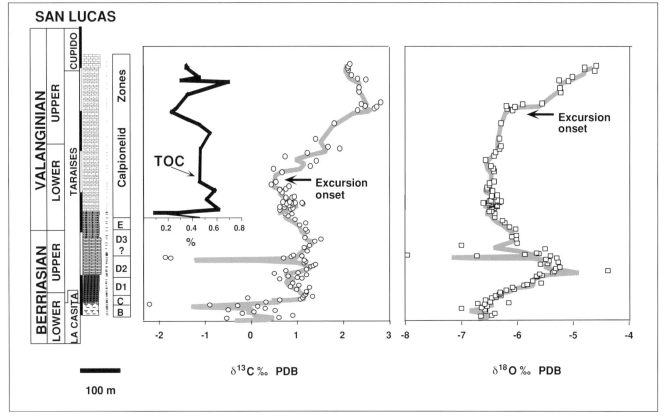

FIGURE 10. Carbon and oxygen isotope stratigraphy and TOC of the San Lucas section.

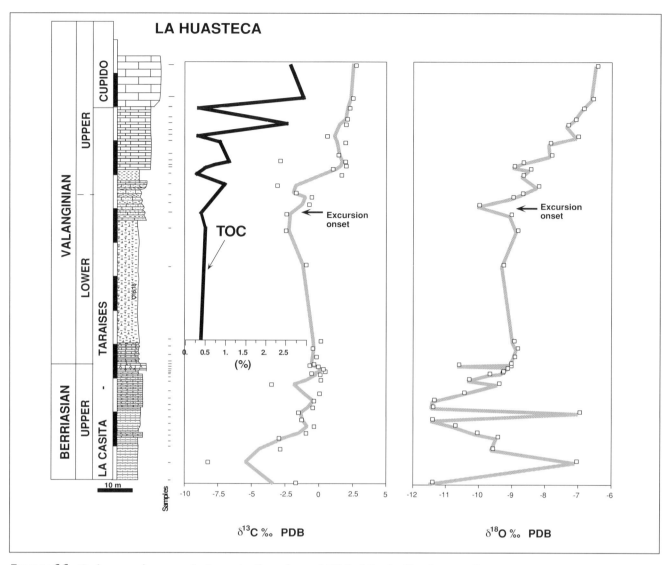

FIGURE 11. Carbon- and oxygen-isotope stratigraphy and TOC of the La Huasteca section.

values of $\delta^{13}C$ (0–1‰) and $\delta^{18}O$ (> –6‰) is observed in the upper part of calpionellid zones B and C (Figure 10). This corresponds to a major paleogeographic change (opening to European Tethys) in northeastern Mexico, as previously documented by Adatte et al. (1994, 1996a, b). A second shift toward higher values of $\delta^{13}C$ (> 2‰) and $\delta^{18}O$ (> –6‰) is observed in the upper part of the Taraises Formation. A shift toward more positive values of $\delta^{18}O$ (from –6.5‰ up to 4.5‰) is observed near the top of the Taraises, approximately 200 m above the onset of the $\delta^{13}C$ excursion. The TOC ranges from 0.3 to 0.6% and does not vary significantly before and during the $\delta^{13}C$ excursion (Figure 10). No changes in lithology were observed in this interval.

Similar excursions of both $\delta^{13}C$ and $\delta^{18}O$ are observed in the La Huasteca section (Figure 11). The shift of $\delta^{13}C$ and $\delta^{18}O$ toward more positive values is recognized in the upper part of the Taraises Formation and corresponds here to the interval of more proximal facies (Figure 8). Unlike the San Lucas section, there is no apparent spatial lag between the

onsets of the carbon and oxygen isotopic excursions. This difference may be a result of sedimentary gaps in the La Huasteca section (see "Discussion"). The increase in TOC amounts (from 0.5 to 2.5%) in the upper part of the Taraises and the base of the Cupido Formations coincides with the stable isotopes excursion (Figure 11).

DISCUSSION

The San Lucas and La Huasteca sections can be correlated by calpionellid and ammonite occurrences and by microfacies. Important correlative horizons based on microfacies (Figure 8) are (1) the change from more proximal to basinal environments above calpionellid zone E and (2) an interval of proximal facies near the lower/upper Valanginian boundary. There are, however, important differences in the thickness of the sediments.

The La Huasteca section was closer to the Coahuila block and was in a more proximal environment (Figure 3a, b); conse-

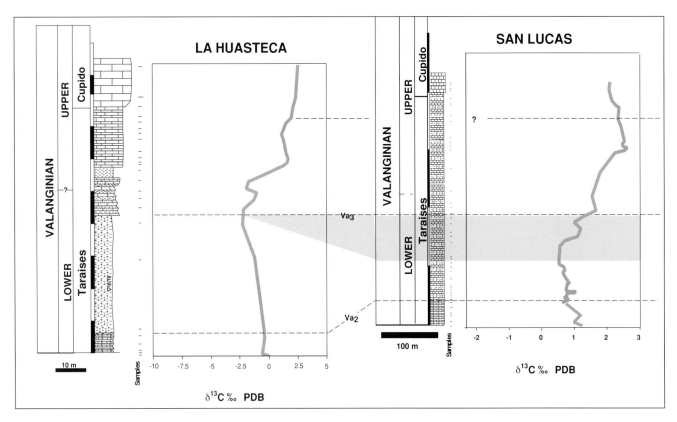

FIGURE 12. Distribution of δ¹³C in the upper part of the two sections studied. Dashed lines = correlative levels. Note: 10 times difference in vertical scale.

quently, the units are thinner. Interruptions of sedimentation and hiatuses are indicated by (1) erosional surfaces and intense bioturbation of the coral-rich limestones in upper Berriasian calpionellid zone D (Figure 8), (2) the reduced thickness of calpionellid zone E and the interval with abundant *Bochianites*, (3) the abrupt change in lithology and microfacies at the top of the *Bochianites*-rich level, and (4) the reduced thickness of the interval of proximal microfacies close to the lower/upper Valanginian boundary.

The upper Berriasian coral-algal buildup along the southern margin of the Coahuila block (Michalzik, 1987; Frame and Ward, 1987) is here interpreted as a sea-level highstand corresponding to Be5–Be7 sequence boundaries (Figures 7 and 8) of Hardenbol et al. (1998). These buildups represent the first episode of carbonate platform conditions in the Mexican Cretaceous and precede the extensive Cupido platform of Hauterivian to Aptian times (Figure 3b).

Correlations between San Lucas and La Huasteca can also be achieved by the stable isotopic curves (Figure 12). The combination of stable isotopic data and micro/macropaleontology allows a confident correlation between the two sections (Figures 6, 8, and 12). The Valanginian rocks at the La Huasteca section are less than 100 m thick, compared with more than 400 m at San Lucas, and they contain depositional hiatuses. These are also indicated by the microfacies, the biostratigraphy, and the δ¹³C curves. At La Huasteca, the onset of the δ¹³C excursion is absent; therefore, both δ¹³C and δ¹⁸O shift simul-

taneously, whereas at San Lucas, the positive shift in δ¹³C precedes the shift in δ¹⁸O (Figures 10–12).

It is also remarkable to note that the strong burial diagenesis observed in our Mexican sites has little influence on the record of stable isotopes. This is illustrated by the general isotopic trends, which agree well in both sections. A similar trend in δ¹³C has been detected worldwide in early/late Valanginian sections and appears to be independant of the lithology (Figure 13). Mexican data match well with data from the southern Italian Alps and Appennines (Weissert et al., 1985; Weissert and Channel; 1989; Weissert and Lini, 1991; Lini et al., 1992), the northern Tethys margin (Föllmi et al., 1994), the Gulf of Mexico (Patton et al., 1984), and the North Atlantic (Robertson and Bliefnick, 1983) and Pacific (Douglas and Savin, 1973) Oceans (Figure 13). Furthermore, isotopic trends in the Berriasian part of the San Lucas and La Huasteca sections are comparable with other Mexican sections that are diagenetically less affected and belong to different tectonic domains (Adatte et al., 1992, 1996a, 1996b). Although the δ¹⁸O values are less reliable because of strong diagenetic overprint, the general trend toward more positive values observed at San Lucas correlates well with data from other areas (Podlaha et al., 1998) (Figure 15).

Recently, Hennig et al. (1999) identified a similar excursion of the stable isotopes in the Vocontian Basin of France, where detailed biostratigraphy based on ammonite occurences places the onset in the *Campylotoxus* zone of the early/late Valangin-

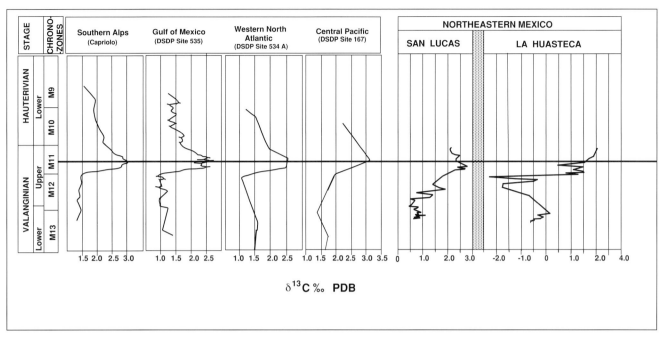

Figure 13. Compilation of carbon-isotope data by locality (after Lini et al., 1992), compared with the present study.

ian (Figures 7 and 14). Maximum values are reached in the *Verrucosum* zone of the late Valanginian. Our integrated data allow the correlation of our Mexican sections to these international chronostratigraphic zones. The upper/lower Valanginian boundary (*Campylotoxus/Verrucosum*) is placed in the upper part of the Taraises Formation, at the top of the interval of proximal microfacies (Figures 6 and 8). Integration of our biostratigraphic and isotopic data indicates the presence at San Lucas of a complete Valanginian succession in terms of European ammonite and calpionellid zones, whereas at La Huasteca, some of the zones may be absent. These data are contrary to the interpretation of Goldhammer et al. (1991) and Goldhammer (1999), who proposed that a major stratigraphic hiatus spanning most of the Valanginian is present in northeastern Mexico.

Our data on the lithology, microfacies, and mineralogy (i.e. increasing calcite) can be used to infer sea-level fluctuations (Figures 4–8). Sea-level changes are commonly recognized from field-based observations of lithologic characteristics and laboratory analysis of bulk-rock compositions. In open-marine settings, the carbonate/detritus (mainly phyllosilicates and quartz) ratio is a useful index for sea-level fluctuations. Increased carbonate content reflects a more distant detrital source and thus deeper water conditions, whereas an increase in detritus indicates a more proximal source and consequently a lower sea level or shallower-water environment. Hiatuses or disconformities are indicated by nondeposition and erosion surfaces, including burrowed or semilithified omission surfaces in deeper waters. Bored and encrusted hardgrounds with phosphate and/or glauconite indicate shallow-water environments, nondeposition and, less commonly, flooding surfaces (Donovan et al., 1988; Baum and Vail, 1988; Robaszynski et al., 1998;

Vincent at al., 1998). Our data set suggests a general transgressive trend in the late Berriasian, culminating in subzone D2, as indicated by the coral-rich limestone unit at La Huasteca (Figures 1 and 3b) and other sections between Monterrey and Saltillo (Michalzik, 1988; Fortunato, 1982; Michalzik and Schuman, 1994). This unit is tentatively correlated with the interval of sequence boundaries Be7–Be8 of Hardenbol et al. (1998). The base of the Valanginian at La Huasteca is marked by an erosional disconformity, as indicated by an abrupt change in lithofacies and microfacies (Figure 8). This contact is correlated with the Va1 sequence boundary of Hardenbol et al. (1998). Upward, a general transgressive trend is initiated above the abrupt change in microfacies and the corresponding hiatus at La Huasteca. This transgression corresponds to the onset of the $\delta^{13}C$ excursion located in the lower part of the *Campylotoxus* zone and may correlate to the sequence boundary Va2 of Hardenbol et al. (1998) (Figures 7 and 14). The upper part of both sections is mostly deepening; only the occurrence of more proximal microfacies observed in the upper part of the *Campylotoxus* zone may correlate to the minor sequence boundary Va3 of Hardenbol et al. (1998).

The $\delta^{13}C$ excursion is explained by numerous authors (e.g., Lini et al., 1992; Föllmi et al., 1994; Weissert et al., 1985) as major changes in the carbon budget as a result of drowning events. The positive $\delta^{13}C$ shift is driven by changes in the burial ratio of C organic to C carbonate. In their model, eutrophication results from increased weathering and runoff and is the causal mechanism for the platform drowning. Increased weathering and runoff are thought to be the products of greenhouse climate conditions that were initiated by flood basalt volcanism and increases in atmospheric CO_2 (Weissert et al., 1985; Föllmi et al., 1994). Our Mexican sections do not present

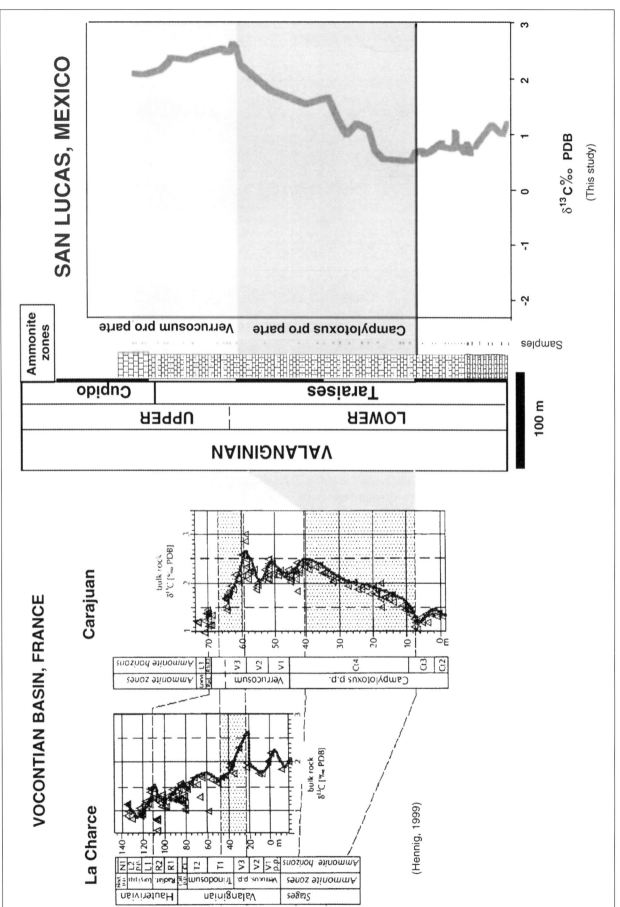

Figure 14. Carbon-isotope stratigraphy and ammonite biozonation in the La Charce and Pont de Carajuan sections of France (from Hennig et al., 1999) and correlation with the San Lucas section.

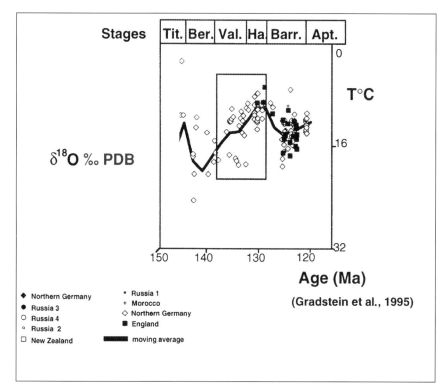

FIGURE 15. Geographic variations in $\delta^{18}O$ values for the Late Jurassic and Early Cretaceous belemnite rostra (Podlaha, 1998, modified).

decreased calcite content in the interval of the isotopic excursion, and TOC significantly increases only in the La Huasteca section. In the absence of widespread black shales during the Valanginian, it is likely that carbon was mainly of terrestrial origin (Weissert et al., 1991). A decrease in carbonate carbon production and increased burial of organic carbon are not observed in our sections and cannot be the only explanations for the $\delta^{13}C$ shift.

During the Early Cretaceous, major sea-level changes were driven by large-scale changes in oceanic ridge volumes. During the late Valanginian, pulses of volcanic activity in the Paraná igneous province (Stewart et al., 1996) coincide with a global trangression (Figure 7). These continental flood basalt events reflect an important tectonic phase that probably led to a considerable increase in ridge volume (Larson, 1991) and to increased CO_2 and subsequent greenhouse conditions. The sea-level fall observed in the lower Valanginian, below the $\delta^{13}C$ excursion, corresponds to one of the more regressive events recognized in the Early Cretaceous (SB Va2, Figures 7 and 8; Haq, 1988; Hardenbol et al., 1998). The area of exposed landmass therefore may have increased and enhanced the weathering flux.

Considering the original oxygen trend as globally preserved, the temporal lag observed between the onset of the late Valanginian $\delta^{13}C$ excursion and the $\delta^{18}O$ shift may be related to increased weathering reactions during greenhouse conditions, which consumed CO_2 over the long term (Holland, 1984) and may have contributed to cooling by effecting a

drawdown of atmospheric CO_2 (Compton et al., 1990; Frank and Arthur, 1999). The cooling trend indicated by $\delta^{18}O$ values also is contemporaneous with second-order sea-level rise (Figure 7; Hardenbol et al., 1998) and could be the result of mixing of the Tethyan and Boreal oceans.

CONCLUSIONS

In the Sierra Madre Oriental of northeastern Mexico, two sections (La Huasteca and San Lucas) spanning Berriasian to lower Hauterivian rocks were analysed and are correlated based on calpionellid and ammonite occurrences, microfacies, and stable isotopes (bulk rock). A major isotopic excursion (approximately 3‰) of both $\delta^{13}C$ and $\delta^{18}O$ was recognized in an interval of pelagic mudstone corresponding to the late Valanginian. It is also remarkable to note that strong burial diagenesis/metamorphism observed in our Mexican sites had little influence on the record of stable isotopes. This is illustrated by the general isotopic trends, which agree well in both our sections, and by the Berriasian isotopic curves, which agree well with other sections in the region that are less affected by burial diagenesis (Adatte et al., 1994, 1996a).

The Valanginian positive $\delta^{13}C$ excursion can be detected worldwide and was also identified in the Vocontian Basin, France (Hennig et al., 1999). In that region, it was calibrated in terms of ammonite zones, the *Campylotoxus-Verrucosum* zones of the early/late Valanginian. Integration of biostratigraphic and isotopic data indicates the presence at San Lucas of a complete Valanginian sequence in terms of European ammonite and calpionellid zones, whereas at La Huasteca, some of the zones may be absent. These data are contrary to the interpretation of Goldhammer et al. (1991) and Goldhammer (1999), who proposed a major stratigraphic hiatus in northeastern Mexico that spans most of the Valanginian. Combined with biostratigraphy, stable isotope stratigraphy is therefore a promising tool to solve correlation problems, especially between widely separated sections located in different basins and characterized by different biostratigraphic zonations.

A general transgressive trend began above the abrupt change in microfacies and the corresponding hiatus at La Huasteca. This change correlates with the SB Va2 and the onset of $\delta^{13}C$ excursion (lower part of the *Campylotoxus* zone, Figures 7 and 14). The upper part of both sections is mostly transgressive and coincides with the maximum in $\delta^{13}C$ (*Verrucosum* zone). This general sea-level rise is documented worldwide (Hardenbol, 1998; Haq et al., 1988).

This $\delta^{13}C$ excursion generally is interpreted as the first episode of Cretaceous greenhouse conditions. It reflects a major change in the global carbon budget and probably is linked to increased tectonic/volcanic activity (Paraná continental flood basalts) that led to increasing atmospheric CO_2 and, in consequence, a greenhouse effect. The sea-level fall observed in the lower Valanginian, below the $\delta^{13}C$ excursion, also may have enhanced weathering fluxes.

Oxygen-isotope data in the Mexican sections, however, trend toward higher values with a significant temporal lag, compared to the $\delta^{13}C$ curve. This indicates seawater and/or climatic cooling triggered by negative feedback reactions (e.g., increase in weathering fluxes) leading to a decrease in atmospheric CO_2. We suggest that the Late Valanginian $\delta^{13}C$ excursion is related to oceanic spreading, enhanced volcanism, and weathering flux, and the subsequent $\delta^{18}O$ excursion and cooling are related to global sea-level rise and decreased atmospheric CO_2.

ACKNOWLEDGMENTS

This research was supported by the Swiss National Fund (Grant No. 8220-028367, T. Adatte) and the Secretaría de Educación Publica of the Mexican State (Grant No. 90-01-0212-543-01, W. Stinnesbeck). Technical support was provided by the Facultad de Ciencias de la Tierra, Universidad Autónoma de Nuevo León, Linares, Mexico, and the Laboratoire de Géochimie et Pétrographie de l'Institut de Géologie de Neuchâtel, Switzerland. Sampling of the San Lucas section by R. Oviedo L. and V. H. Peña P. is gratefully acknowledged. The authors thank the reviewers J. Hardenbol, R. Scott, and J. L. Wilson for their comments and helpful suggestions.

REFERENCES CITED

Adatte, T., W. Stinnesbeck, H. Hubberten, and J. Remane, 1992, The Jurassic-Cretaceous boundary in northeastern and central Mexico: A multistratigraphical approach: 8.° Congreso Latinoamericano de Geología, Salamanca 1992, Proceedings, v. 4, p. 23–29.

Adatte, T., W. Stinnesbeck, and J. Remane, 1994, The Jurassic-Cretaceous boundary in northeastern Mexico: Confrontation and correlations by microfacies, clay mineral mineralogy, calpionellids and ammonites: Geobios, Mémoire Spécial 17, p. 37–56.

Adatte, T., W. Stinnesbeck, J. Remane, and H. Hubberten, 1996a, The western end of the Tethys at the Jurassic-Cretaceous boundary: Evidence for a paleogeographic change in NE Mexico: Cretaceous Research, v. 27, p. 671–689.

Adatte, T., W. Stinnesbeck, H. Hubberten, and J. Remane, 1996b, Correlation between northeastern and central Mexico at the Jurassic/Cretaceous boundary: A tentative paleogeographical reconstruction: Mitteilungen dem Geologische-Paläontologischen Institut der Universität, Hamburg, v. 77, p. 379–392.

Adatte, T., W. Stinnesbeck, and G. Keller, 1996c, Lithologic and mineralogic correlations of near K/T boundary clastic sediments in NE Mexico: Implication for origin and nature of deposition. Geological Society of America Special Publication 307, p. 211–226.

Alleman, F., and R. Remane, 1979, Les faunes de calpionelles du Berriasien Supérieur/Valanginian, in R. Busnardo, J. P. Thieuloy, and M.

Moullade, eds., Hypostratotype mésogéen de l'étage Valanginien (Sud-est de la France), les stratotypes français: Centre National de la Recherche Scientifique, v. 6, p. 99–109.

Baum, G. R., and P. R. Vail, 1988, Sequence stratigraphic concepts applied to Paleogene outcrops, Gulf and Atlantic basins, in C. K. Wilgus, B. S. Hastings, G. G. Kendall, H. W. Posamantier, C. Ross, and J. C. van Wagoner, eds., Sea level changes: An integrated approach: Society for Sedimentary Geology (SEPM) Special Publication 42, p. 309–328.

Blauser, W. H., and C. L. McNulty, 1980, Calpionellids and nannoconids of the Taraises Formation (Early Cretaceous) in Santa Rosa Canyon, Sierra de Santa Rosa, Nuevo León, Mexico: Gulf Coast Section, Society for Sedimentary Geology (SEPM) Foundation Transactions, 30th Annual Research Conference Proceedings, p. 263–271.

Bulot, L. G., J. P. Thieuloy, E. Blanc, and J. Klein, 1992, Le cadre stratigraphique du Valanginien Supériéur et de l'Hauterivien du Sud-Est de la France: Définition des biochronozones et caractérisation de nouveaux biohorizons: Geologie Alpine, v. 68, p. 13–56.

Bulot, L. G., 1996, The Valanginian stage, in P. Rawson et al., eds., Second International Symposium on Cretaceous Stage Boundaries, Brussels: Bulletin Institut Royal Sciences Naturelles, Belgique, v. 66, p. 11–18.

Busnardo, R., and J. P. Thieuloy, 1979, Les Zones d'ammonites du Valanginien du Sud-Est de la France, in R. Busnardo, J. P. Thieuloy, and M. Moullade, eds., Hypostratotype mésogéen de l'étage Valanginien (Sud-est de la France), les stratotypes français: Centre National de la Recherche Scientifique, v. 6, no. 58–68, p. 127–134.

Compton, J. S., S. Snyder, and D. Hodell, 1990, Phosphogenesis and weathering of shelf sediments from the southeastern United States: Implications for Miocene $\delta^{13}C$ excursions and global cooling: Geological Society of America, v. 18, no. 12, p. 1227–1230.

Corfield, R. M., J. E. Cartlidge, I. Premoli-Silva, and R. A. Housley, 1991, Oxygen and carbon isotope stratigraphy of the Palaeogene and Cretaceous limestones in the Bottacione gorge and the Contessa Highway sections, Umbria, Italy: Terra Nova, v. 3, no. 4, p. 414–422.

Cotillon, P., and M. Rio, 1974, Cyclic sedimentation in the Cretaceous of DSDP site 535 and 540 (Gulf of Mexico), 534 (Central Atlantic) and the Vocontian Basin (France): Initial Reports of the Deep-Sea Drilling Project 77: Washington, D. C., U. S. Government Printing Office, p. 339–376.

Donovan, A. D., G. R. Baum, G. L. Blechschmidt, T. S. Loutit, C. E. Pflumand, and P. R. Vail, 1988, Sequence stratigraphic setting of the Cretaceous-Tertiary boundary in central Alabama, in C. K. Wilgus, H. Posamentier, C. A. Ross, and C. G. Kendall, eds., Sea-level changes: An integrated approach: Society for Sedimentary Geology (SEPM) Special Publication 42, p. 299–307.

Douglas, R. G., and S. M. Savin, 1973, Oxygen and carbon isotope analyses of Cretaceous and Tertiary foraminifera from the central North Pacific: Initial Report of the Deep-Sea Drilling Project 17: Washington, D. C., U. S. Government Printing Office, p. 591–605.

Ferrero, J., 1966, Nouvelle méthode empirique pour le dosage des minéraux par diffraction R.X. Rapport, Centre Français du Pétrole, Bordeaux, inédit [unpublished].

Föllmi, K. B., H. Weissert, M. Bisping, and H. P. Funk, 1994, Phosphogenesis, carbon-isotope stratigraphy, and carbonate-platform evolution along the Lower Cretaceous northern Tethyan margin: Geological Society of America Bulletin 106, no. 6, p. 729–746.

Fortunato, K. S., 1982, Depositional framework of the La Casita Formation (Upper Jurassic–lowermost Cretaceous) near Saltillo, Coahuila, Mexico: Master's thesis, University of New Orleans, 198 p.

Frame, A., and W. C. Ward, 1987, Lowermost Cretaceous coral-rich limestone in Nuevo León and Coahuila, Mexico: Actas Facultad de Ciencias de la Tierra, Universidad Autónoma de Nuevo León, v. 2, p. 33–39.

Frank, T. D., and M. A. Arthur, 1999, Tectonic forcings of Maastrichtian ocean-climate evolution: Paleoceanography, v. 4, no. 2, p. 103–117.

Friedman, I., and J. R. O'Neil, 1977, Compilation of stable isotope fractionation factors of geochemical interest, in M. Fleischer, ed., Data of geochemistry: U. S. Geological Survey Professional Paper, 440 p.

Früh-Green, G., H. Weissert, and D. Bernoulli, 1990, A multiple fluid history recorded in alpine ophiolites: Journal of the Geological Society, London, v. 147, p. 959–970.

Goldhammer, R. K., P. J. Lehmann, R. G. Todd, J. L.Wilson, W. C. Ward, and C. R. Johnson, 1991, Sequence stratigraphy and cyclostratigraphy of the Mesozoic of the Sierra Madre Oriental, Northeast Mexico—a Field Guide Book: Gulf Coast Section, Society for Sedimentary Geology (SEPM), 85 p.

Goldhammer, R. K., 1999, Mesozoic sequence stratigraphy and paleogeographic evolution of northeast Mexico, in C. Bartolini, J. L. Wilson, and T. F. Lawton, eds., Sedimentary and tectonic history of north central Mexico: Geological Society of America Special Paper 340, p. 1–31.

Gradstein, F. M., F. P. Agterberg, J. Ogg, J. Hardenbol, P. van Veen, J. Thierry, and Z. Huang, 1995; A Triassic, Jurassic and Cretaceous time scale, in W. Berggren, D. V. Kent, M. P. Aubry, and J. Hardenbol, eds., Geochronology, time scales and global stratigraphic correlation: Society for Sedimentary Geology (SEPM), Special Publication 4, p. 95–126.

Haq, B. U., J. Hardenbol, and P. R. Vail, 1987, The new chronostratigraphic basis of Cenozoic and Mesozoic sea level cycles, in C. A. Ross and D. Haman, eds., Timing and depositional history of eustatic sequences constraints on seismic stratigraphy: Cushman Foundation for Foraminiferal Research, Special Publications 24, p. 7–13.

Haq, B. U., J. Hardenbol, and P. R. Vail, 1988, Mesozoic and Cenozoic chronostratigraphy and cycles of sea-level change, in C. Wilgus, B. S. Hastings, C. A. Ross, H. W. Posamentier, J. Van Wagoner, and C. Kendall, eds., Sea-level changes: Society for Sedimentary Geology (SEPM) Special Publication 42, p. 26–108.

Hardenbol, J., J. Thierry, M. B. Farley, P. C. De Graciansky, and P. R. Vail, 1998, Mesozoic and Cenozoic sequence chronostratigraphic framework of European basins, in P. C. De Graciansky, J. Hardenbol, T. Jacquin, and P. R. Vail, eds., Mesozoic and Cenozoic sequence stratigraphy of European basins: Society for Sedimentary Geology (SEPM) Special Publication 60, p. 3–13.

Hennig, S., H. Weissert, and L. Bulot, 1999, C-Isotope stratigraphy, a calibration tool between ammonite and magnetostratigraphy: The Valanginian-Hauterivian transition: Geologica Carpathica, v. 50, no. 1, p. 91–96.

Holland, H. D., 1984, The chemical evolution of the atmosphere and oceans, Princeton series in geochemistry: Princeton, New Jersey, Princeton University Press, 582 p.

Hubberten, H. W., and G. Meyer, 1989, Stable isotope measurements on foraminifera tests: Experiences with an automatic commercial carbonate preparation device: Terra Abstracts, v. 1, no. 2, p. 80–81.

Hut, G., 1987, Stable isotope reference samples for geochemical and hydrological investigations: Vienna, International Atomic Energy Agency Report, 42 p.

Imlay, R., 1936, Evolution of the Coahuila Peninsula, Mexico, part IV: Geology of the western part of the Sierra de Parras: Geological Society of America Bulletin, v. 47, p. 1091–1152.

Imlay, R., 1937, Geology of the middle part of the Sierra de Parras, Coahuila, Mexico: Geological Society of America Bulletin, v. 48, p. 587–630.

Imlay, R., 1938, Ammonites of the Taraises Formation of northern Mexico: Geological Society of America Bulletin, v. 49, p. 539–602.

Imlay, R., 1940, Neocomian faunas of northern Mexico: Geological Society of America Bulletin, v. 51, p. 117–190.

Jenkyns, H. C., 1995, Carbon-isotope stratigraphy and paleoceanography: Significance of the Lower Cretaceous shallow water carbonates of Resolution Guyot, Mid Pacific Mountains: Proceedings of Oceanic Drilling Progress Scientific Report 143, p. 99–104.

Jenkyns, H. C., and C. J. Clayton, 1986, Black shales and carbon isotopes in pelagic sediments from the Tethyan Lower Jurassic: Sedimentology, v. 33, p. 87–106.

Kübler, B., 1983, Dosage quantitatif des minéraux majeurs des roches sédimentaires par diffraction X: Cahier de l'Institut de Géologie de Neuchâtel, Série AX no. 1.1 and 1.2.

Kübler, B., 1987, Cristallinité de l'illite, methodes normalisées de preparations, méthodes normalisées de measures: Cahiers de l'Institut de Géologie de Neuchâtel, Série ADX no. 1.3.

Larson, R. L., 1991, Latest pulse of Earth; evidence for a Mid-Cretaceous super plume: Geology, v. 19, no. 6, p. 547–550.

Lini, A., H. Weissert, and E. Erba, 1992, The Valanginian carbon isotope event: A first episode of greenhouse climate conditions during the Cretaceous: Terra Nova, v. 4, no. 3, p. 374–385.

McFarlan, E., and L. S. Menes, 1991, Lower Cretaceous, in A. Salvador, ed., The Gulf of Mexico Basin: Geological Society of America, The Geology of North America, v. J, p. 181–204.

Michalzik, D., 1988, Trias bis tiefste Unter-Kreide der nordöstlichen Sierra Madre Oriental, Mexiko: Inaugural dissertation thesis, Darmstadt, Germany, 247 p.

Michalzik, D., and D. Schumann, 1994, Lithofacies relations and palaeoecology of a Late Jurassic to Early Cretaceous fan delta to shelf depositional system in the Sierra Madre Oriental of Northeast Mexico: Sedimentology, v. 41, no. 3, p. 463–477.

Moore, D., and R. Reynolds, 1989, X-ray diffraction and the identification and analysis of clay-minerals: Oxford, Oxford University Press, 332 p.

Patton, J. W., P. W. Choquette, G. K. Guennel, A. J. Kaltenback, and A. Moore, 1984, Organic geochemistry and sedimentology of lower to mid-Cretaceous deep-sea carbonates, sites 535 and 540, Leg 77.: Initial Reports of the Deep-Sea Drilling Projects: Washington, D. C., U. S. Government Printing Office, p. 417–443.

Podlaha, O. G, J. Mutterlose, and J. Veizer, 1998, Preservation of $\delta^{18}O$ and $\delta^{13}C$ in belemnite rostra from the Jurassic/Early Cretaceous successions: American Journal of Science, v. 298, no. 4, p. 324–347.

Remane, J., 1985, Calpionellids, in H. Bolli, J. Saunders, and K. Perch-Nielsen, eds., Plankton stratigraphy: New York, Cambridge University Press, p. 555–572.

Remane, J., D. Bakalova-Ivanova, K. Borza, J. Knauer, I. Nagy, G. Pop, and E. Tardi-Filacz, 1986, Agreement on the subdivision of the standard calpionellid zones defined at the 2nd planctonic conference, Rome 1970: Acta Geologica Hungarica, v. 29, no. 1–2, p. 5–14.

Robaszynski, F., A. Gale, P. Juignet, F. Amedro, and J. Hardenbohl, 1998, Sequence stratigraphy in the upper Cretaceous of the Anglo-Paris Basin: Exemplified by the Cenomanian stage, *in* P. C. De Graciansky, J. Hardenbol, T. Jacquin, and P. R. Vail, eds., Mesozoic and Cenozoic sequence stratigraphy of European basins: Society for Sedimentary Geology (SEPM) Special Publication 60, p. 364–386.

Robertson, A. H. F., and D. M. Bliefnick, 1983, Sedimentology and origin of Lower Cretaceous pelagic carbonates and redeposited clastics, Blake-Bahama Formation—Deep Sea Drilling Project, site 534, western equatorial Atlantic: Initial Reports of the Deep-Sea Drilling Project, 76, Washington, D. C., U. S. Government Printing Office, p. 795–820.

Salvador, A, 1991, Triassic-Jurassic, *in* A. Salvador, ed., The Gulf of Mexico Basin: Geological Society of America, The geology of North America, v. J, p. 131–180.

Smith, C. L., 1981, Lower Cretaceous stratigraphy and structure, northern Mexico: West Texas Geological Society Publication No 81–74, p. 1–28.

Stewart, K., S. Turner, S. Kelley, C. Hawkesworth, L. Kirstein, and M. Mantovani, 1996, $K^{40}Ar^{39}Ar$ geochronology in the Paraná continental flood basalt province: Earth and Planetary Science Letters, v. 143, no. 1–4, p. 95–109.

Veizer, J., 1983, Trace elements and isotopes in sedimentary carbonates, *in* R. J. Reeder, ed., Carbonates: Mineralogy and chemistry: Mineralogical Society of America, Reviews in Mineralogy, v. 11, p. 265–299.

Vincent, S. J., D. I. M. Macdonald, and P. Gutterridge, 1998, Sequence stratigraphy, *in* P. Doyle and M. R. Bennet, eds., Unlocking the stratigraphical record: Advances in modern stratigraphy: Chichester, U.K., John Wiley & Sons, p. 299–351.

Weissert, H., J. Mckenzie, A. Judith, J. E. T. Channell, 1985, Natural variations in the carbon cycle during the Early Cretaceous: Geophysical Monography, v. 32, p. 531–545.

Weissert, H., and J. E. T. Channell, 1989, Tethyan carbonate carbon isotope stratigraphy across the Jurassic-Cretaceous boundary: An indicator of decelerated global carbon cycling?: Paleoceanography, v. 4, no. 4, p. 483–494.

Weissert, H., and A. Lini, 1991, Ice Age interludes during the time of greenhouse climate?, *in* D. W. Müller, J. A. McKenzie, and H. Weissert, eds., Controversies in modern geology: Academic Press, London, p. 173–191.

Wilson, J. L., and G. Pialli, 1977, A Lower Cretaceous shelf margin in northern Mexico, *in* D. G. Bebout, and R. G. Loucks, eds., Cretaceous carbonates of Texas and Mexico: Application to subsurface exploration: University of Texas at Austin, Bureau of Economic Geology, Report of Investigations 89, p. 286–290.

Young, K., 1988, *Karakaschiceras* and the late Valanginian of northern Mexico and Texas, *in* J. Wiedmann and J. Kullmann, eds., Cephalopods—Present and past: Schweitzerbart'sche Verlagsbuchhandlung, Stuttgart, p. 621–632.

Cantú-Chapa, A., and R. Landeros-Flores, 2001, The Cretaceous-Paleocene boundary in the subsurface Campeche shelf, southern Gulf of Mexico, *in* C. Bartolini, R. T. Buffler, and A. Cantú-Chapa, eds., The western Gulf of Mexico Basin: Tectonics, sedimentary basins, and petroleum systems: AAPG Memoir 75, p. 389–395.

17

The Cretaceous-Paleocene Boundary in the Subsurface Campeche Shelf, Southern Gulf of Mexico

Abelardo Cantú-Chapa
Instituto Politécnico Nacional, Mexico City, Mexico

Román Landeros-Flores (deceased)
Petróleos Mexicanos, Ciudad del Carmen, Campeche, Mexico

Dedicated to Mr. Rudecindo Cantarell, a fisherman who discovered and reported the existence of oil seeps in waters of the Campeche Shelf.

ABSTRACT

The Cretaceous-Paleocene boundary on the Campeche Shelf is established with planktonic foraminiferal biostratigraphy from core and chip samples and is correlated with gamma-ray well logs. The Campanian-Maestrichtian Cantarell Formation is herein formally described as a breccia formed of calcareous or dolomitized fragments cemented by micrite. Its age is established using *Globotruncana* species. The Abkatun Formation is formally described as brown shale and shaly limestone that contains *Parvularugoglobigerina trinidadensis* and *Pa. pseudobulloides* of the lower Paleocene. The upper Paleocene is represented by greenish-gray shale that contains *Morozovella velascoensis*, *M. formosa formosa*, *M. aragonensis*, and *Globoanomalina pseudomenardii*.

The broad regional distribution and distinctive lithology of the Cantarell and Abkatun Formations allow them to be differentiated from overlying and underlying formations. The boundary between the lower and upper Paleocene is conformable.

INTRODUCTION

Exploration offshore in the Campeche shelf began in 1974 (Figure 1). The Chac-1 well penetrated a Cenozoic shale sequence overlying a rock unit that Petróleos Mexicanos (PEMEX) named the *Paleocene breccia*, based on lithology and preliminary foraminiferal biostratigraphy. The authors conducted a stratigraphic study of the Paleocene breccia in 1982 and integrated well logs with microfossils. These new data determined the precise stratigraphic relationships of the Upper Cretaceous

and Paleocene rocks in the southeastern part of the Gulf of Mexico. The planktonic foraminiferal biostratigraphy was accomplished using the zonation scheme of Berggren et al. (1995).

Four stratigraphic sections were constructed. The main stratigraphic cross section trends northwest-southeast, is 250 km long, and is built with 10 offshore boreholes located northwest of Ciudad del Carmen, southern Gulf of Mexico (Figure 1).

Locally, boreholes outside the main section line were projected into the section to aid in the interpretation. Three addi-

389

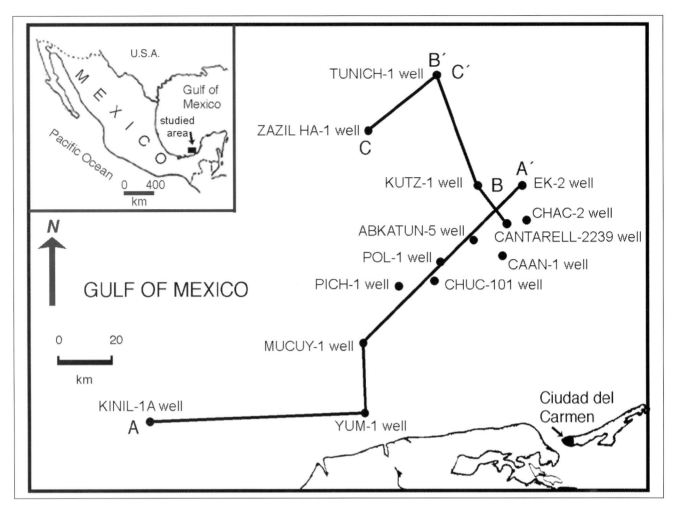

FIGURE 1. Location map of the Campeche Shelf, southern Gulf of Mexico, showing PEMEX boreholes and lines of cross section used in this study.

tional stratigraphic cross sections located northwest of the study area were also completed to support the interpretations and to depict the stratigraphic relationships between the Upper Cretaceous and Paleocene rock units. In addition, one structural cross section that comprises the Upper Jurassic, the Cretaceous, and the lower Paleocene strata was constructed to complement the stratigraphic study. The main goals of this investigation were: (1) to date the Upper Cretaceous breccia (called here the Cantarell Formation), (b) to establish the Cretaceous-Paleocene and the lower–upper Paleocene boundaries, (c) to determine Upper Cretaceous lateral facies in the southwestern part of the study area, and (d) to formally describe the lower Paleocene Abkatun Formation.

STRATIGRAPHY

The Upper Cretaceous and lower Cenozoic subsurface stratigraphy in the Campeche Shelf, southern Gulf of Mexico, is determined with well logs and core and ditch samples. The ages of the rock units were established based on planktonic foraminiferal biostratigraphy (Table 1).

Upper Eocene

South of the main cross-section line, brecciated reef structures have been dated late Eocene (Hernández-García, 1993, 1994). Also, a bentonitic matrix from core 2 (4039–4040 m) in the Kinil-1 well contains the benthic foraminifera *Amphistegina parbula*, *Helicostegina dimorpha*, *Lepidocyclina* sp., *Neolepidina postulosa*, and *Heterostegina* sp., indicating a late Eocene age, after the unpublished study of R. Landeros of PEMEX. These same rocks are known from other boreholes in the southeastern region of the Gulf of Mexico (e.g., Ixim-1 well, core 1, 960–968 m).

Upper Paleocene

This is a very distinctive lithostratigraphic and chronostratigraphic unit that consists of 50–450 m of bentonitic, soft to semihard, laminated, and locally calcareous greenish-gray shale (Figure 2). Foraminifera identified from the shale include *Morozovella velascoensis*, *M. formosa formosa*, *M. aragonensis*, and *Globanomalina pseudomenardii*, of late Paleocene–Eocene age (Berggren et al., 1995). In the southeastern region of the Gulf of Mexico, a core sample (3203–3212 m) from Kutz-1 well yielded the foraminifera *Globoanomalina* cf. *pseudomenardii* of the late Paleocene (Figure 3, B-B′).

Lower Paleocene (Abkatun Formation)

The Abkatun Formation, formally described herein for the first time, is a characteristic and regionally widespread unit in the Campeche Shelf, southern Gulf of Mexico. The name is taken from the Abkatun field. In Mayan dialect, the term *Abkatun* means "god who holds the world." The type section is the Abkatun-5 well, where the top of the unit lies at a depth of 3278 m (Figure 4). This formation consists of 30 to 60 m of interbedded reddish-brown shale and beige shaly limestone that contain the foraminifera *Parvularugoglobigerina trinidadensis* and *Pa. pseudobulloides* of early Paleocene age (Berggren et al., 1995). Because of its shaly nature, the gamma-ray log deflects to the right when this formation is penetrated and shows a slight deflection to the left near the base because of its calcareous shale content. The upper contact is with greenish-gray calcareous shale of the upper Paleocene. The lower contact is sharp at a depth of 3325 m with Upper Cretaceous dolomite, calcareous breccia, or marly sediments.

Both the top and the basal contacts of the Abkatun Formation are distinct gamma-ray log shifts to the right, indicating higher values. The upper–lower Paleocene boundary is characterized by an even more dramatic deflection of the gamma-ray log toward the right (Figure 2).

Historical background of this characteristic shaly unit is related to exploration in the offshore Campeche area. The lower Paleocene Abkatun Formation is present in the Campeche area (Figures 2–4) but is missing in the Chuc-101 well (Figure 2). Because of its shaly nature and the presence of planktonic foraminifera, the Abkatun Formation records deep-marine sedimentation. Other wells are listed in Table 1.

Campanian-Maestrichtian (Cantarell Formation)

In this study, the Cantarell Formation is formally described as approximately 260 m of breccia consisting of dolomitized, calcareous, angular fragments of diverse sizes cemented by a beige micrite. The type section is in the Cantarell-82 well (1972–2255 m), offshore Campeche, southern Gulf of Mexico

TABLE 1. Foraminiferal distribution in the Upper Cretaceous–Paleocene rocks based on PEMEX boreholes, southern Gulf of Mexico.

Age	Wells	Microfossils
Eocene	Kinil-1	**Reef breccia.** *Amphistegina parbula, Helicostegina dimorpha? Lepidocyclina* sp., *Neolepidina postulosa, Heterostegina* sp. Core 2 (4039–4040 m).
Lower Paleocene	Kinil-1	*Pr. trinidadensis, Pa. pseudobulloides.* 5640 m
Maestrichtian-Campanian	Kinil-1	**Marl.** *Globotruncana elevata, G. contusa, G. linneiana.* 5680 m
Maestrichtian Campanian	Mucuy-1	**Marl.** *Globotruncana leupoldi, G. stuartiformis, G. campanianarca, G. lapparenti.* 4890 m
Upper Paleocene	Pich-1	*Globanomalina pseudomenardii.* 4040 m
Lower Paleocene	Pich-1	*P. eugebina, P. fringa.* 4130 m
Maestrichtian-Campanian	Pich-1	**Dolomite breccia.** *G. elevata, G. havanensis, G. arca, G. stuartiformis.* 4155 m
Upper Paleocene	Pol-1	*M. velascoensis.* 3640 m
Maestrichtian-Campanian	Pol-1	**Dolomite breccia.** *G. conica, G. arca, G.* sp., *Globigerinoides* sp., *Rugoglobigerina* cf. *R. rugosa, G. stuartiformis, G. elevata, Abathomphalus mayaroensis, Planoglobulina, Heterohelix globulosa, Rugoglobigerina scotti.* Core 2 (3801-3804 m).
Upper Paleocene	Abkatun-5	*M. velascoensis, M. formosa-formosa, M. aragonensis.* 3160 m
Lower Paleocene	Abkatun-5	*Pr. trinidadensIs, Pa. pseudobulloides.* 3275 m
Maestrichtian-Campanian	Cantarell-2239	**Dolomite breccia.** *G. stuartiformis, G. arca, G. elevata, G. linneiana, G. lapparenti, G. calcarata, Calcisphaerula inominata, Sulcoperculina globosa, Heterohelix* sp., *Globigerinelloide* sp. Core 1 (2970–2979 m) and core 4 (3205–3212 m).
Lower Paleocene	Chac-2 well	*Pr. trinidadensis.* 3446 m.

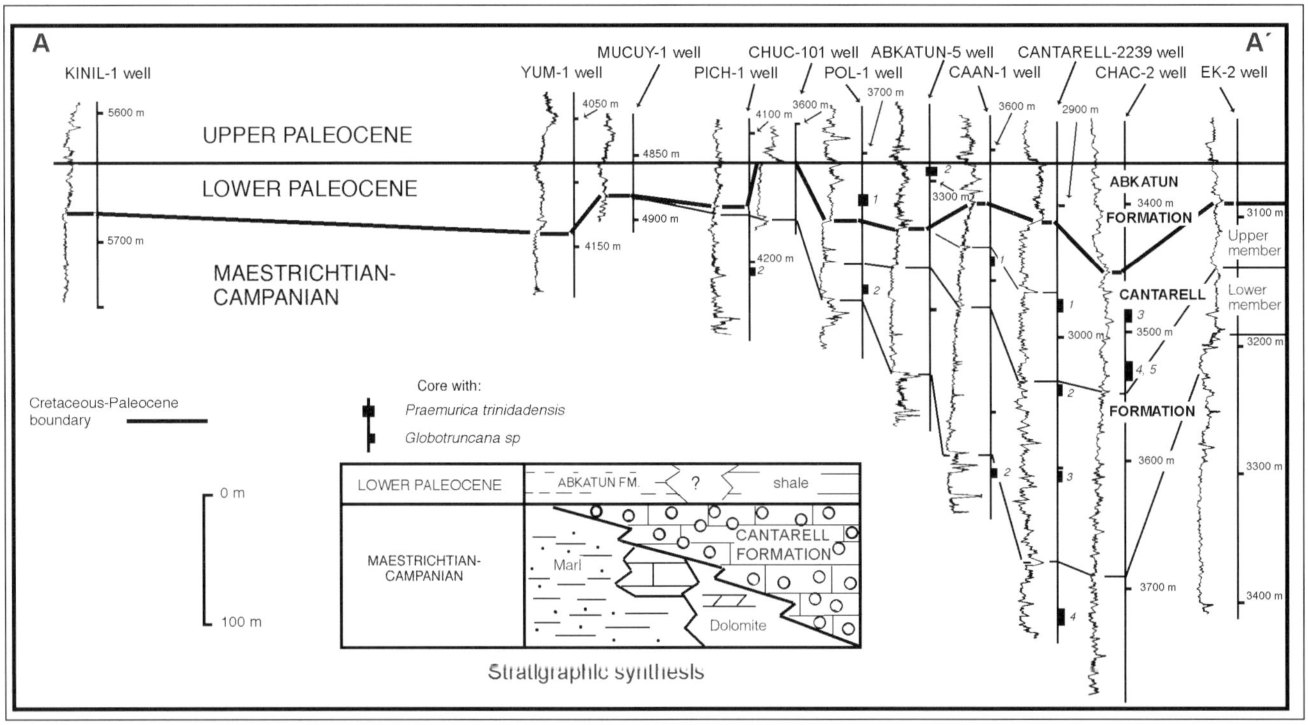

FIGURE 2. Stratigraphic cross section A-A'. Datum: lower–upper Paleocene boundary, showing the Cantarell Formation and Abkatun Formation. The Cretaceous-Paleocene boundary is based on gamma-ray logs and foraminifera-bearing ditch and core samples. Location is shown in Figure 1.

(Figure 4). During initial regional studies by PEMEX, strata of the Cantarell Formation were originally called the Paleocene breccia on the basis of incorrect micropaleontologic determinations. The erroneous Paleocene age originally proposed for strata herein assigned to the Cantarell Formation in the southern Gulf of Mexico region still is accepted by some authors (Stanford-Bestt, 1989).

The Cantarell Formation has a brown color caused by oil impregnation. Locally, bentonite layers occur. In the central part of the main cross section, this formation changes laterally to shaly limestone, and in the southwestern part of the cross section, the breccia changes to calcareous shale (Figure 2). In limited areas in the northeast, the Cantarell Formation displays four horizons. The upper three horizons are considered the upper member, although they are lacking in nearby boreholes (Figure 4). This formation is thickest in the north (approximately 270 m), and it thins gradually toward the southwest, where it is absent from the section.

Stratigraphic Relationships

In wells in the middle part of the main cross section, the lower contact of the Cantarell Formation is conformable on Campanian-Maestrichtian dolostone and interbedded limestone. No evidence for an unconformity between these two units was recognized, as previously suggested by Santiago et al. (1984), in the southern Gulf of Mexico.

The Cantarell Formation is overlain by either lower or upper Paleocene sediments. The contact with the upper Pale-

ocene is an unconformity observed in the central part of the area. The contact with the lower Paleocene is sharp, but the fossil succession does not suggest a hiatus. This contact is represented typically by a strong deflection to the left on the gamma-ray log (Figure 4).

Age of the Cantarell Formation

Planktonic foraminifera collected from the micritic matrix of breccia in the Cantarell Formation are primarily species of *Globotruncana* that indicate a Campanian-Maestrichtian age. In addition, the conformable stratigraphic position below the early Paleocene shaly horizon is consistent with a Cretaceous age for the breccia. The faunas from these Cretaceous and Paleocene rocks are shown in Table 1. The angularity of different-sized clasts in the breccia record the chaotic origin of the Cantarell Formation during the Late Cretaceous.

Structural Deformation

Jurassic to Cenozoic rocks in the Campeche Shelf were affected by extensional deformation. Normal faults with vertical displacements of 250 m in the central part of the cross section and 750 m in the northwest and southeast parts are depicted in Figure 5. The Cretaceous-Paleocene boundary is deeper toward the southwest, where it is 5000 m below sea level. In the Cantarell-2239 well, the Cretaceous-Paleocene boundary occurs at a depth of 2930 m on the upthrown side of southwest-northeast en-echelon structures (Figure 5).

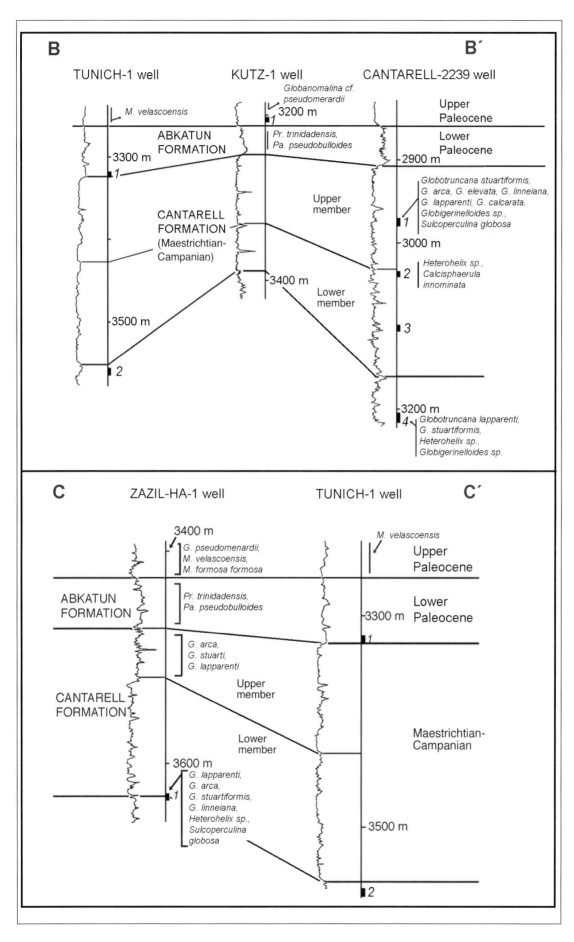

FIGURE 3. Stratigraphic cross sections B-B' and C-C' showing Upper Cretaceous and lower Paleocene rocks from the northeastern part of the study area, based on gamma-ray logs and foraminifera-bearing ditch and core samples. Location is shown in Figure 1.

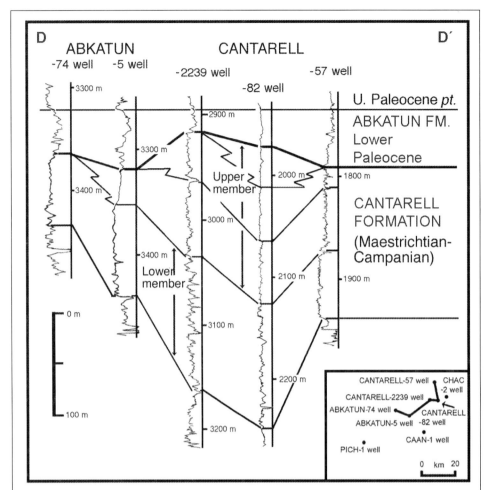

FIGURE 4. Stratigraphic cross section D-D' in the Cantarell and Abkatun fields. Cantarell-82 well and Abkatun-5 well represent the type sections for the Cantarell and Abkatun Formations, respectively. Location is shown in Figure 1.

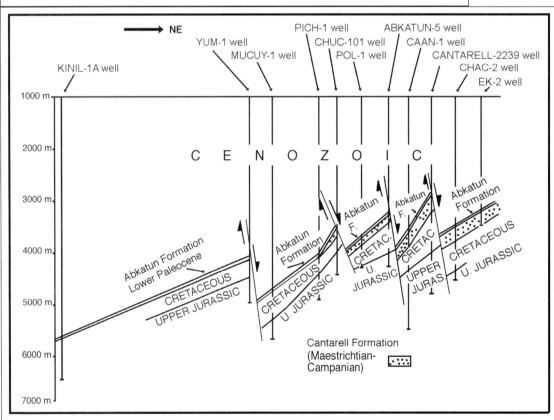

FIGURE 5. Structural cross section A-A' based on PEMEX borehole data, southern Gulf of Mexico. Location is shown in Figure 1.

CONCLUSIONS

The Cantarell Formation in the Campeche Shelf, southern Gulf of Mexico, is of Campanian-Maestrichtian age. The Cretaceous-Paleocene boundary is established on the basis of planktonic foraminiferal biostratigraphy from core and ditch samples and gamma-ray well logs. The wide regional distribution and distinctive lithology of two new lithostratigraphic units, the Cantarell and Abkatun Formations, allow them to be differentiated from overlying and underlying formations. Foraminiferal biostratigraphy also established the lower–upper Paleocene boundary.

ACKNOWLEDGMENTS

We thank Claudio Bartolini for the translation of the manuscript to English. We are grateful to Harold Lang and Robert Scott for providing constructive reviews of the manuscript.

REFERENCES CITED

Berggren, W. A., D. V. Kent, C. C. Swisher III, and M.-P. Aubry, 1995, A revised Cenozoic geochronology and chronostratigraphy, *in* W. A. Berggren, D. V. Kent, M-P. Aubry, and J. Hardenbol, eds., Geochronology time scales and global stratigraphic correlation: Society for Sedimentary Geology (SEPM) Special Publication 54, p. 129–212.

Hernández García, S., 1993, Subsistema generador en los horizontes Cretácico Superior y Eoceno Medio de la Región Marina, Primer Simposio de Geología de Subsuelo: El sistema petrolífero y su evaluación: Asociación Mexicana de Geólogos Petroleros, Memoria, Ciudad del Carmen, Campeche, p. 79–83.

Hernández García, S., 1994, Cretaceous carbonate breccia reservoirs of the Campeche Area, Mexico, *in* Geologic aspects of petroleum systems: First Joint AAPG/Asociación Mexicana de Geólogos Petroleros Hedberg Research Conference, p. 19–22.

Santiago A., J. Carrillo B., and B. Martell A., 1984. Geología Petrolera de México, *in* Evaluación de formaciones en México: Schlumberger, Paris, France, p. 1–36.

Stanford-Bestt, J., 1989, Sismología estratigráfica en la exploración y desarrollo de campos de la Sonda de Campeche: Ingeniería Petrolera, v. 24, no. 8, p. 7–18.

18

Salazar Medina, G., 2001, Tertiary zonation based on planktonic foraminifera from the marine region of Campeche, Mexico, *in* C. Bartolini, R. T. Buffler, and A. Cantú-Chapa, eds., The western Gulf of Mexico Basin: Tectonics, sedimentary basins, and petroleum systems: AAPG Memoir 75, p. 397–419.

Tertiary Zonation Based on Planktonic Foraminifera from the Marine Region of Campeche, Mexico

Gisela Salazar Medina
Consultant, Mexico City, Mexico

ABSTRACT

A study of foraminifera selected from the sediments of 17 wells located in the Campeche Sound, Gulf of Mexico, revealed that two wells have a column encompassing almost the complete Tertiary, whereas the remaining wells produced only samples corresponding to the Oligocene-Pleistocene.

Assessment of planktonic foraminifera allowed the establishment of a composited chronostratigraphical column for the Tertiary that encompasses the mid-Paleocene to Pleistocene. Twenty-six biozones were identified and described from oldest to youngest, 15 for the Paleogene and 11 for the Neogene.

The zonation is based on the outlines proposed by Toumarkine and Luterbacher (1985) and Bolli and Saunders (1985) for the Paleocene-Eocene and the Oligocene-Holocene, respectively. Some modifications are introduced, mainly because of the nature of the samples available for this work, as well as the local conditions of the study area.

This contribution compares the zonation described in the present study with those proposed by the authors mentioned above, as well as those of Blow (1969, 1979), Stainforth et al. (1975), and Berggren et al. (1995).

In two of the studied wells, the absence of zones at the Paleocene-Eocene boundary and at the Eocene-Oligocene boundary (*Morozovella velascoensis* zone at the late Paleocene and *Cassigerinella chipolensis/Pseudohastigerina micra* zone at the lower Oligocene) was noticed; therefore, both boundaries are unconformable. Furthermore, in all the studied wells, the *Globigerinoides ruber, Globorotalia mayeri,* and *Globorotalia menardii* zones in the upper part of the middle Miocene are missing; consequently, the middle Miocene–upper Miocene boundary is also unconformable.

Depositional environments were interpreted using the association of characteristic species of benthonic foraminifera and benthonic/planktonic ratios.

From the detailed study of one well and by applying the graphic-correlation method, a two-axis graph was generated. Four "terraces" were observed, corresponding to missing zones, which indicates that some relevant geologic events may be present in the study area, such as condensed sections, unconformity or hiatus, local structural problems (folds and faults), regional tectonic events, and diapiric intrusions.

The proposed zonation may be used as a base to carry out regional biostratigraphic correlations, as well as to provide support for future Tertiary geologic studies of the Gulf of Mexico and the Caribbean.

INTRODUCTION

The Campeche Sound constitutes an area of enormous economic interest to Petróleos Mexicanos (PEMEX) because of its hydrocarbon production (oil and gas); it represents a high percentage of total Mexican extraction.

Since the 1970s, intense geologic and geophysical exploration of this important area has been carried out; however, no detailed biostratigraphic study of the Cenozoic sediments has been published. Since 1990, the author has completed several projects (unpublished) for the Instituto Mexicano del Petróleo (IMP) which, through the study of foraminifera present in the cutting samples of several wells in the marine zone, have identified and described different biozones for this area (Salazar, 1991, 1994, 1995, 1998a, 1998b; Salazar and Alvarado, 1997, 1998; Salazar and Gómez, 1990).

The present study constitutes a summary of those projects, which aimed to establish a type zonation for the Campeche offshore, based on planktonic foraminifera. This may be used as a base to carry out regional biostratigraphic correlations as well as to determine depositional environments, depending on the paleoecologic value of the benthonic and planktonic foraminifera.

Location

The study area is located in the Mexican Gulf, at the western part of the Yucatán platform, north of Campeche and Tabasco states, and approximately 80 km from the coast. It is located on the continental platform in a strip encompassing isobaths 20 m to 200 m deep, with an approximate area of 15,000 km² and with structural features parallel to those of the Chiapas-Tabasco area (Santiago et al., 1984).

The tectonic framework is defined by the Yucatán platform, the Comalcalco Basin, the Macuspana Basin, and the Akal folded strip. The Chiapas massif to the south-southeast and the Yucatán platform are the elements that gave rise to the tectonic structure of the study area (Meneses de Gyves, 1980).

The study wells are located between longitudes 93°20' and 91°40' West and latitudes 18°35' and 20°00' North (Figure 1).

Regional Geologic Setting

The Tertiary sedimentary column in the Campeche's marine region is more than 6000 m thick but varies considerably, especially for the Neogene, because of synsedimentary fault activity, the influence of tectonic activity, and the presence of salt structures. Frequent hiatuses or unconformities are present in the Tertiary sequence, the causes of which have been attributed to the Laramide orogeny during the early Tertiary, Miocene tectonic events, and diapiric intrusions (Santiago et al., 1984).

The Cenozoic section is composed almost completely of siliciclastic sedimentary rocks, that is, sandy intervals interbedded with argillaceous sequences and sometimes with thin layers of bentonite. In the Paleocene base, however, breccias with fragments of Cretaceous and Paleocene rocks have been identified, although the presence of calcarenites facies deposited as grain flux from the Campeche platform have been reported in the middle Eocene, and carbonate breccias with shales also have been reported during the lower Miocene in certain wells of that area. Allochthonous salt emplacements also have been observed as sills between the upper Oligocene and lower Miocene and between different stratigraphic levels (Robles et al., 1998).

PLANKTONIC FORAMINIFERAL BIOSTRATIGRAPHY

For the present study, the schemes proposed by Bolli and Saunders (1985) for the Oligocene-Holocene and by Toumarkine and Luterbacher (1985) for the Paleocene-Eocene were used. Some modifications were introduced, because no core samples were available, only cutting samples; therefore, the first evolutionary appearance of index species—the principal criteria used by these investigators for defining their zonal boundaries—cannot be defined. For this reason, extinctions (or first downhole occurrences) of selected microfossils (espe-

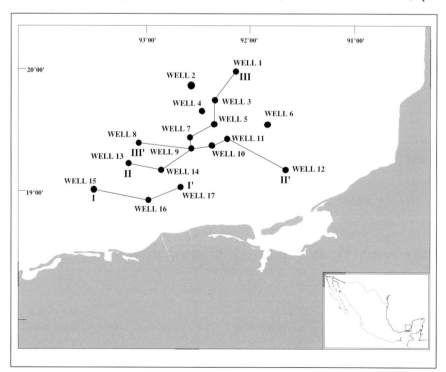

FIGURE 1. Study area showing locations of the wells and the cross sections.

cially those with tops coincident with or close to the boundaries specified by the mentioned authors) were used.

Identified in this study were 24 zones and two subzones: three zones for the Paleocene, nine for the Eocene, three for the Oligocene, five for the Miocene, three for the Pliocene, and one for the Pleistocene, with two subzones (Figure 2).

Figure 3 shows the distribution of the identified zones in the study wells in the Campeche area.

A comparison of the zones identified for the study area with the zonal schemes described by Bolli and Saunders (1985), Toumarkine and Luterbacher (1985), Blow (1969, 1979), Berggren et al. (1995) and Stainforth et al. (1975) is shown in Figure 4.

Figures 5, 6, and 7 show the correlation and thickness variations of the different Tertiary biozones in the study wells.

To determine depositional environments, paleoecological models proposed by different researchers were taken into account and a variety of parameters was used, such as benthonic foraminifera assemblages and planktonic/benthonic ratios.

ZONAL DESCRIPTIONS

Paleocene

Morozovella uncinata *Zone*

Age: Middle Paleocene.
Author: Bolli (1957a), emended Bolli (1966).
Definition: Interval with the common occurrence of *Morozovella uncinata.* The base is undefined in this study and corresponds to the base of the section. The top is marked with the last occurrence (LO) of *Morozovella trinidaden*sis.
Characterization: The joint presence of *Morozovella trinidadensis, Morozovella uncinata, Morozovella pseudobulloides,* and *Morozovella praecursoria* characterizes this zone.
Remarks: Abundant planktonic microfauna, generally well preserved.

The zone here described is similar in its planktonic foraminiferal content to the *Globorotalia uncinata* zone in Bolli (1957a, 1966) and Stainforth et al. (1975), the *Morozovella uncinata* zone in Toumarkine and Luterbacher (1985), zone P2 *Globorotalia (Acarinina) praecursoria* described by Blow (1979), and zone P2 *Praemurica uncinata–Morozovella angulata* in Berggren et al. (1995).

In this study, the top of the *Morozovella uncinata* zone is marked by the LO of *Morozovella trinidadensis* instead of the FO (first appearence) of *Morozovella angulata,* which is the criteria used by the mentioned authors.

Morozovella angulata *Zone*

Age: Middle Paleocene.
Author: Alimarina (1963) as *Acarinina angulata* zone (in Stainforth et al., 1975).
Definition: Interval from the LO of *Morozovella trinidadensis* (base) to the LO of *Morozovella uncinata* (top of the zone).

Characterization: This interval is characterized by abundant planktonic microfauna, represented mainly by *Morozovella uncinata, Morozovella praecursoria, Morozovella angulata, Morozovella conicotruncata,* and *Morozovella pseudobulloides.*

Remarks: The zone as described here would correlate with the *Globorotalia angulata* zone defined by Bolli (1966) and by Stainforth et al. (1975) and with the *Morozovella angulata* zone in Toumarkine and Luterbacher (1985), as well as with the lower part of zone P3 *Globorotalia (Morozovella) angulata angulata* proposed by Blow (1979) and subzone P3a in Berggren et al. (1995), based mainly on faunal content. However, the present study has used criteria different from those used by the authors mentioned—the FO of *Morozovella angulata* for the lower limit, and the FO of *Planorotalites pusilla pusilla* (or of *Igorina albeari* (= *Ig. pusilla laevigata*) in Berggren et al., 1995) for the upper limit. Because these events cannot be defined in our cuttings samples, the present study stresses the LO of *Morozovella uncinata, Morozovella praecursoria,* and *Morozovella pseudobulloides.* They are found at almost the same level in which *Planorotalites pusilla pusilla* appears, as Toumarkine and Luterbacher (1985) indicate, to mark the upper limit of the zone here described.

Planorotalites pseudomenardii *Zone*

Age: Late Paleocene.
Author: Bolli (1957a).
Definition: Interval with zonal marker from the LO of *Morozovella uncinata* and *Morozovella praecursoria* to the LO of *Planorotalites pseudomenardii.*
Characterization: This interval shows the consistent occurrence of *Planorotalites pseudomenardii* along with some heavily ornamented species, including *Morozovella velascoensis, Morozovella acuta,* and *Morozovella occlusa.*
Remarks: This zone is widely recognized in tropical and subtropical areas by the presence of the nominate taxon and by the *Morozovella* species mentioned above. In wells 1 and 8, these species occur consistently throughout the interval representing the zone.

In zonations described by Bolli (1957a, 1966), by Stainforth et al. (1975), and by Toumarkine and Luterbacher (1985), the youngest zone of the middle Paleocene is that of *Globorotalia pusilla pusilla* (= *Planorotalites pusilla pusilla* in Toumarkine and Luterbacher, 1985). Its lower limit is marked with the FO of *Planorotalites pusilla pusilla* and the upper limit with the FO of *Planorotalites pseudomenardii.*

The study wells have no occurrence of *Planorotalites pusilla pusilla,* a fossil that, as indicated by Toumarkine and Luterbacher (1985), may be very scarce or absent; therefore, recognition of the zone based on this species may be difficult. In samples studied herein, the *Planorotalites pusilla pusilla* zone could be included in the lower part of the *Planorotalites pseudomenardii* zone (this paper), because it would overlap the extinction level of *Morozovella uncinata* that, as indicated by Tou-

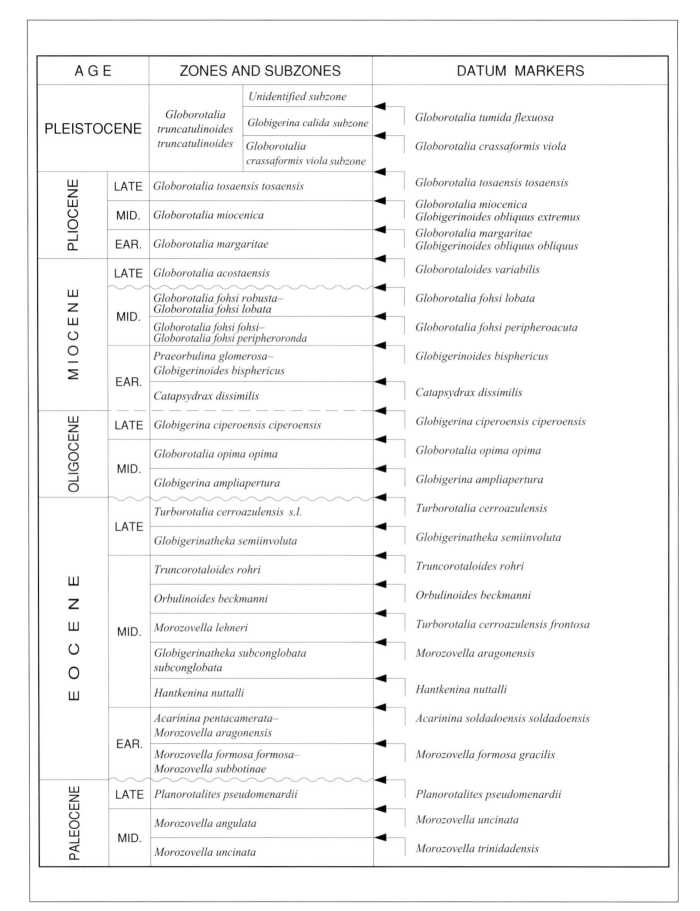

AGE		ZONES AND SUBZONES		DATUM MARKERS
PLEISTOCENE		*Globorotalia truncatulinoides truncatulinoides*	*Unidentified subzone*	*Globorotalia tumida flexuosa*
			Globigerina calida subzone	
			Globorotalia crassaformis viola subzone	*Globorotalia crassaformis viola*
PLIOCENE	LATE	*Globorotalia tosaensis tosaensis*		*Globorotalia tosaensis tosaensis*
	MID.	*Globorotalia miocenica*		*Globorotalia miocenica Globigerinoides obliquus extremus*
	EAR.	*Globorotalia margaritae*		*Globorotalia margaritae Globigerinoides obliquus obliquus*
MIOCENE	LATE	*Globorotalia acostaensis*		*Globorotaloides variabilis*
	MID.	*Globorotalia fohsi robusta– Globorotalia fohsi lobata*		*Globorotalia fohsi lobata*
		Globorotalia fohsi fohsi– Globorotalia fohsi peripheroronda		*Globorotalia fohsi peripheroacuta*
	EAR.	*Praeorbulina glomerosa– Globigerinoides bisphericus*		*Globigerinoides bisphericus*
		Catapsydrax dissimilis		*Catapsydrax dissimilis*
OLIGOCENE	LATE	*Globigerina ciperoensis ciperoensis*		*Globigerina ciperoensis ciperoensis*
	MID.	*Globorotalia opima opima*		*Globorotalia opima opima*
		Globigerina ampliapertura		*Globigerina ampliapertura*
EOCENE	LATE	*Turborotalia cerroazulensis s.l.*		*Turborotalia cerroazulensis*
		Globigerinatheka semiinvoluta		*Globigerinatheka semiinvoluta*
	MID.	*Truncorotaloides rohri*		*Truncorotaloides rohri*
		Orbulinoides beckmanni		*Orbulinoides beckmanni*
		Morozovella lehneri		*Turborotalia cerroazulensis frontosa*
		Globigerinatheka subconglobata subconglobata		*Morozovella aragonensis*
		Hantkenina nuttalli		*Hantkenina nuttalli*
	EAR.	*Acarinina pentacamerata– Morozovella aragonensis*		*Acarinina soldadoensis soldadoensis*
		Morozovella formosa formosa– Morozovella subbotinae		*Morozovella formosa gracilis*
PALEOCENE	LATE	*Planorotalites pseudomenardii*		*Planorotalites pseudomenardii*
	MID.	*Morozovella angulata*		*Morozovella uncinata*
		Morozovella uncinata		*Morozovella trinidadensis*

FIGURE 2. Identified zones and datum markers in the marine wells.

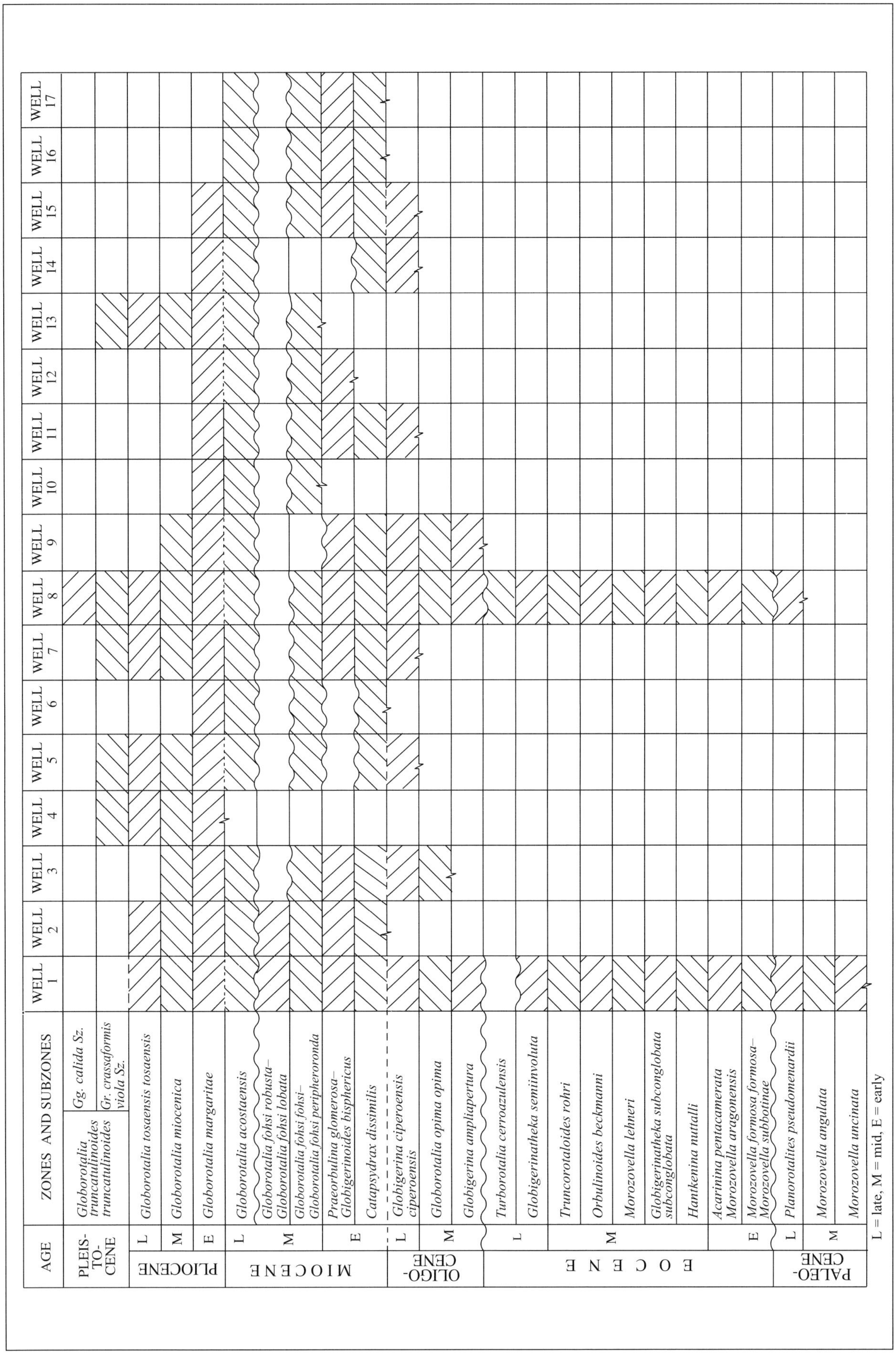

FIGURE 3. Identified Cenozoic zones in the study wells from the marine region.

Age			This work	Bolli and Saunders (1985) Toumarkine and Luterbacher (1985)	Blow (1969, 1979)	Berggren et al. (1995)	Stainforth et al. (1975)
HOL.			*Unidentified*	*Globorotalia fimbriata*	N 23	Pt1 b	*Globorotalia truncatulinoides*
NEOGENE	PLEIST.		*Globorotalia truncatulinoides truncatulinoides*	*Globigerina bermudezi*	N 23	Pt1 b	
			Globigerina calida	*Globorotalia truncatulinoides truncatulinoides*: *Globigerina calida calida*	N 22	Pt1 b	
			Globorotalia crassaformis viola	*Globorotalia crassaformis hessi*	N 22	Pt1 b	
				Globorotalia crassaformis viola	N 22	Pt1 a	
	PLIOCENE	L	*Globorotalia tosaensis tosaensis*	*Globorotalia tosaensis tosaensis*	N 21	Pl 6	
		M	*Globorotalia miocenica*	*Globorotalia miocenica*: *Globorotalia exilis*	N 21	Pl 5	
				Globigerinoides trilobus fistulosus	N 20	Pl 4 / Pl 3	*Pulleniatina obliquiloculata*
		E	*Globorotalia margaritae*	*Globorotalia margaritae*: *Globorotalia margaritae evoluta*	N 19	Pl 2 / Pl 1 b	*Globorotalia margaritae*
				Globorotalia margaritae margaritae	N 18	Pl 1 a / M 14	
	MIOCENE	L	*Globorotalia acostaensis*	*Globorotalia humerosa*	N 17	M 13 b	*Globorotalia acostaensis*
				Globorotalia acostaensis	N 16	M 13 a	
		M		*Globorotalia menardii*	N 15	M 12	*Globorotalia menardii*
				Globorotalia mayeri	N 14	M 11	*Globorotalia siakensis*
				Globigerinoides ruber	N 13	M 10	
			Globorotalia fohsi robusta–Globorotalia fohsi lobata	*Globorotalia fohsi robusta*	N 12	M 9 b	*Globorotalia fohsi lobata-robusta*
				Globorotalia fohsi lobata	N 11	M 9 a / M 8	
			Globorotalia fohsi fohsi–Globorotalia fohsi peripheroronda	*Globorotalia fohsi fohsi*	N 10	M 7	*Globorotalia fohsi fohsi*
				Globorotalia fohsi peripheroronda	N 9	M 6	*Globorotalia fohsi peripheroronda*
		E	*Praeorbulina glomerosa–Globigerinoides bisphericus*	*Praeorbulina glomerosa*	N 8	M 5 b / a	*Praeorbulina glomerosa*
				Globigerinatella insueta	N 7	M 4 b / a	*Globigerinatella insueta*
			Catapsydrax dissimilis	*Catapsydrax stainforthi*	N 6	M 3	*Catapsydrax stainforthi*
				Catapsydrax dissimilis	N 5	M 2	*Catapsydrax dissimilis*
				Globigerinoides primordius	N 4	M 1 b / a	*Globorotalia kugleri*
PALEOGENE	OLIGOCENE	L	*Globigerina ciperoensis ciperoensis*	*Globorotalia kugleri*	P 22 = N 3	P 22	*Globigerina ciperoensis*
				Globigerina ciperoensis ciperoensis			
		M	*Globorotalia opima opima*	*Globorotalia opima opima*	P 21 = N2	P 21 b / a	*Globorotalia opima opima*
			Globigerina ampliapertura	*Globigerina ampliapertura*	P 19 / 20	P 20 / P 19	*Globigerina ampliapertura*
		E		*Cassigerinella chipolensis / P. micra*	P 19 / 20 / P 18	P 18	*Cassigerinella chipolensis/P. micra*
	EOCENE	L	*Turborotalia cerroazulensis s.l.*	*Turborotalia cerroazulensis s.l.*	P 17 / P 16	P 17 / P 16	*Globorotalia cerroazulensis*
			Globigerinatheka semiinvoluta	*Globigerinatheka semiinvoluta*	P 15	P 15	*Globigerapsis semiinvoluta*
		M	*Truncorotaloides rohri*	*Truncorotaloides rohri*	P 14	P 14	*Truncorotaloides rohri*
			Orbulinoides beckmanni	*Orbulinoides beckmanni*	P 13	P 13	*Orbulinoides beckmanni*
			Morozovella lehneri	*Morozovella lehneri*	P 12	P 12	*Globorotalia lehneri*
			Globigerinatheka subconglobata subconglobata	*Globigerinatheka subconglobata subconglobata*	P 11	P 11	*Globigerinatheka subconglobata*
			Hantkenina nuttalli	*Hantkenina nuttalli*	P 10	P 10	*Hantkenina aragonensis*
		E	*Acarinina pentacamerata–Morozovella aragonensis*	*Acarinina pentacamerata*	P 10	P 9	*Globorotalia pentacamerata*
				Morozovella aragonensis	P 9	P 9 / P 8	*Globorotalia aragonensis*
			Morozovella formosa formosa–Morozovella subbotinae	*Morozovella formosa formosa*	P 8 b	P 7	*Globorotalia formosa formosa*
				Morozovella subbotinae	P 8 a / P 7	P 7 b / P 6 a	*Globorotalia subbotinae*
				Morozovella edgari	P 7	P 6 a	
	PALEOCENE	L		*Morozovella velascoensis*	P 6 / P 5	P 5	*Globorotalia velascoensis*
			Planorotalites pseudomenardii	*Planorotalites pseudomenardii*	P 4	P 4 c / b / a	*Globorotalia pseudomenardii*
				Planorotalites pusilla pusilla	P 3	P 3 b	*Globorotalia pusilla pusilla*
		M	*Morozovella angulata*	*Morozovella angulata*	P 3	P 3 a	*Globorotalia angulata*
			Morozovella uncinata	*Morozovella uncinata*	P 2	P 2	*Globorotalia uncinata*

L = late, M = mid, E = early

FIGURE 4. Comparison of the Cenozoic zonal scheme identified in this work with those of Bolli and Saunders (1985), Toumarkine and Luterbacher (1985), Blow (1969, 1979), Berggren et al. (1995), and Stainforth et al. (1975).

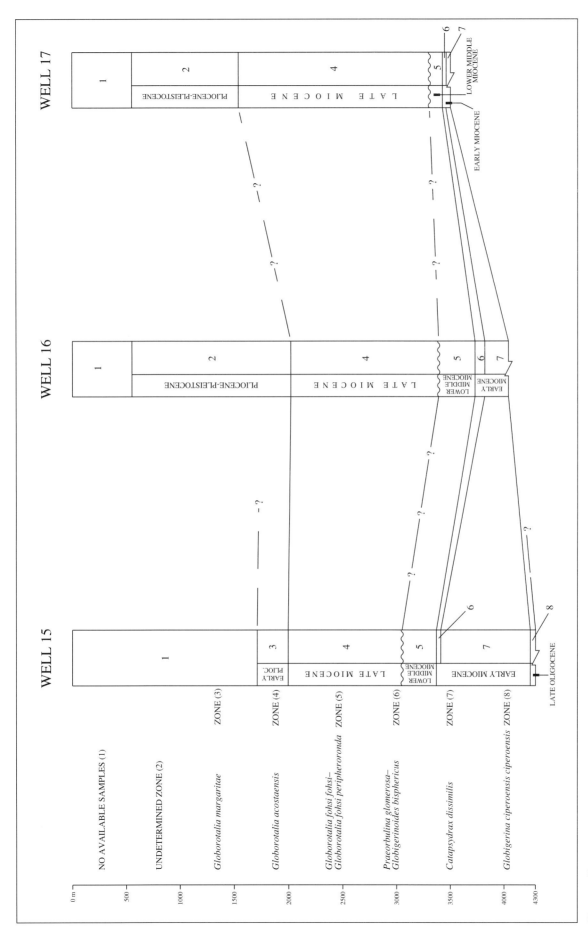

Figure 5. Biostratigraphic correlation—section I-I' from Figure 1.

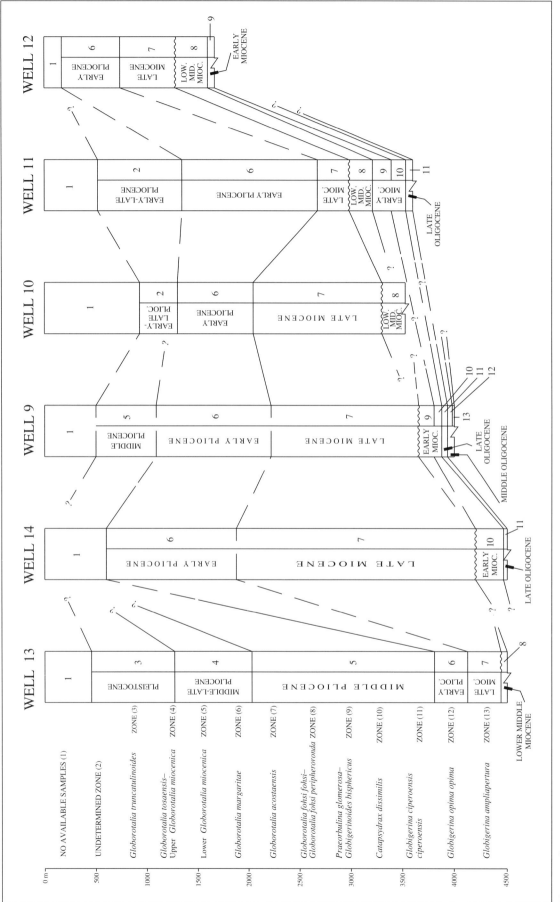

FIGURE 6. Biostratigraphic correlation—section II-II′ from Figure 1.

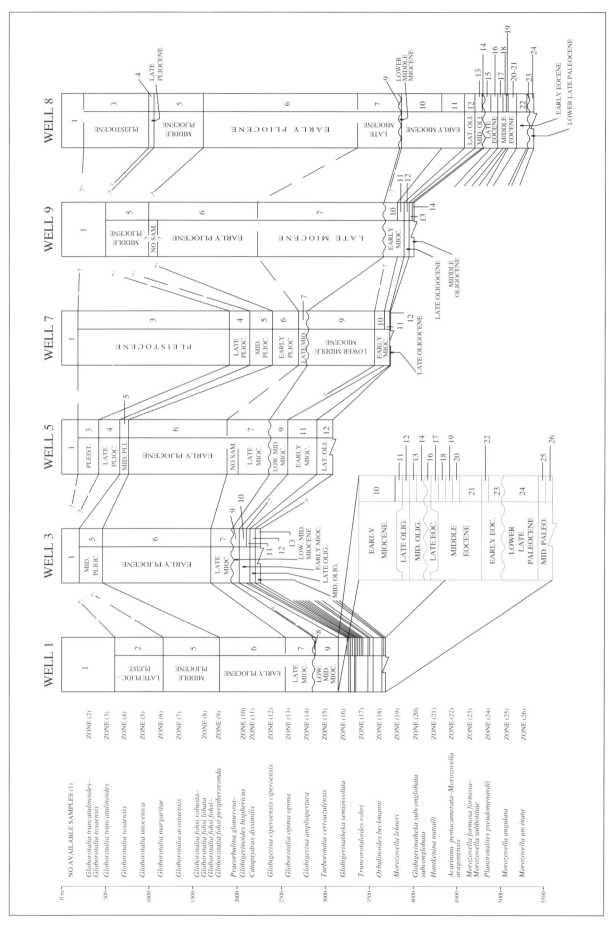

FIGURE 7. Biostratigraphic correlation—section III-III' from Figure 1.

markine and Luterbacher (1985, p. 100), occurs before the first appearance of *Planorotalites pusilla pusilla*.

From the current samples, it was impossible to define precisely the horizon of the FO of *Planorotalites pseudomenardii*, an event that would mark the limit between the middle and upper Paleocene. This limit would be included in the interval of the *Planorotalites pseudomenardii* zone. For practical purposes, the zone, as described here, is considered to be upper Paleocene.

The interval being described may be correlated with the *Globorotalia pusilla pusilla* and *Globorotalia pseudomenardii* zones proposed by Bolli (1957a, 1966) and by Stainforth et al. (1975), with the *Planorotalites pusilla pusilla* and *Planorotalites pseudomenardii* zones described by Toumarkine and Luterbacher (1985), and with part of the zone P3 *Globorotalia (Morozovella) angulata angulata* and zone P4 *Globorotalia (Globorotalia) pseudomenardii* in Blow (1979) and subzone P3b and zone P4 in Berggren et al. (1995).

Eocene

Morozovella subbotinae–Morozovella formosa formosa *Zone*

Age: Early Eocene.

Author: The interval corresponds to the zones that Bolli (1957a, 1966) called the *Globorotalia rex* zone and the *Globorotalia formosa formosa* zone, although criteria used in this study for defining limits are different from those specified by that author.

Definition: The lower limit of the zone is marked by the LO of *Planorotalites pseudomenardii*, and the upper limit is marked with the LO of *Morozovella formosa gracilis*.

Characterization: This interval is characterized by the regular occurrence of *Morozovella subbotinae, Morozovella formosa gracilis, Morozovella aequa, Morozovella quetra, Morozovella marginodentata, Acarinina soldadoensis soldadoensis,* and *Acarinina soldadoensis angulosa*.

Remarks: This interval includes the *Morozovella subbotinae* and *Morozovella formosa formosa* zones, and the species it contains makes it similar to zones with those nominate fossils described by several authors for the lower Eocene, such as Bolli (1957a, 1966), Beckmann et al. (1969), Premoli-Silva and Bolli (1973), Stainforth et al. (1975), Luterbacher and Premoli-Silva (in Caro et al., 1975), and Toumarkine and Luterbacher (1985).

Criteria for defining limits of the *Morozovella subbotinae–Morozovella formosa formosa* zone as described here are different from those specified by the above-mentioned authors, mainly because they use index species' first occurrences, which are impossible to determine from the samples available to us.

The *Morozovella edgari* zone, introduced by Premoli-Silva and Bolli in 1973 and adopted in the Toumarkine and Luterbacher work in 1985, could be included in the zone being described, because the microfaunal association is similar in the zone of those authors and in the Campeche samples, although *Morozovella edgari* was not found.

In this work, this zone is the first in the Eocene and is un-conformable with the *Planorotalites pseudomenardii* zone, because the *Morozovella velascoensis* zone mentioned by most researchers as the youngest in the upper Paleocene is missing. This latter zone is characterized by the presence of nominate species along with *Morozovella occlusa* and *Morozovella acuta,* as well as the absence of *Planorotalites pseudomenardii*, whose extinction marks the base of the *Morozovella velascoensis* zone, and the top is determined by the extinction of the nominate taxon.

In our samples, we find these species of *Morozovella* associated with *Planorotalites pseudomenardii*, indicating that the sediments represent the *Planorotalites pseudomenardii* zone as well.

The upper Paleocene–lower Eocene boundary has been placed by several authors, such as Bolli (1957a, 1966), Beckmann et al. (1969), Stainforth et al. (1975), and Berggren et al. (1995), between the *Morozovella velascoensis* and *Morozovella subbotinae* (= *Globorotalia rex, in* Bolli, 1957a, 1966) zones. Premoli-Silva and Bolli (1973), as well as Toumarkine and Luterbacher (1985), include the *Morozovella edgari* zone as basal Eocene.

In any case, the event used for delimiting the final zone of the Paleocene from the first of the Eocene is the extinction of *Morozovella velascoensis*.

In the samples used for the present study, it was impossible to identify the *Morozovella velascoensis* zone because the fossil was found in association with *Planorotalites pseudomenardii*; thus, the Paleocene-Eocene boundary is considered unconformable.

Our zone may be correlated with the *Globorotalia rex* and *Globorotalia formosa formosa* zones proposed by Bolli (1957a, 1966); the *Globorotalia subbotinae* and *Globorotalia formosa formosa* zones described by Stainforth et al. (1975); the *Morozovella edgari, Morozovella subbotinae,* and *Morozovella formosa formosa* zones in Toumarkine and Luterbacher (1985); the zone P7 *Globorotalia (Acarinina) wilcoxensis berggreni* and the subzone P8a *Globorotalia (Morozovella) formosa* in Blow (1979); and zones P6 *Morozovella subbotinae* and P7 *Morozovella aragonensis–Morozovella formosa formosa* in Berggren et al. (1995).

Acarinina pentacamerata–Morozovella aragonensis *Zone*

Age: Early Eocene.

Author: In the present study, it corresponds to the *Acarinina pentacamerata* zone introduced by Krasheninnikov, 1965 (*in* Toumarkine and Luterbacher, 1985) and to the *Globorotalia aragonensis* zone proposed by Bolli (1957a).

Definition: Interval between the LO of *Morozovella formosa gracilis* and the LO of *Acarinina soldadoensis soldadoensis*.

Characterization: This interval can be recognized by the presence of *Acarinina soldadoensis soldadoensis, Morozovella aragonensis,* and *Acarinina pentacamerata*.

Remarks: The evolutionary occurrence of *Turborotalia cerroazulensis frontosa* has been used by Toumarkine and Luterbacher (1985) to mark the boundary between the *Morozovella aragonensis* and *Acarinina pentacamerata* zones. These criteria cannot be used in our material, so the present study used a combination of both zones.

The *Acarinina pentacamerata–Morozovella aragonensis* zone, as defined here, may be compared with the *Globorotalia aragonensis* and *Globorotalia palmerae* zones proposed by Bolli (1957a, 1957c, 1966), the *Globorotalia aragonensis* and *Globorotalia pentacamerata* zones of Stainforth et al. (1975), the *Morozovella aragonensis* and *Acarinina pentacamerata* zones reported by Toumarkine and Luterbacher (1985), zone P9 *Globorotalia (Acarinina) aspensi/Globigerina lozanoi prolata* and the lower part of zone P10 *Subbotina frontosa frontosa/Globorotalia (T.) pseudomayeri* proposed by Blow (1979), and zones P8 *Morozovella aragonensis* and P9 *Planorotalites palmerae–Hantkenina nuttalli* of Berggren et al. (1995).

Hantkenina nuttalli *Zone*

Age: Middle Eocene.

Author: Bolli (1957c), emended by Stainforth et al. (1975), renamed by Toumarkine (1981).

Definition: Interval with *Hantkenina nuttalli* between the LO of *Acarinina soldadoensis soldadoensis* and the LO of the zonal marker.

Characterization: This zone is characterized by the presence of *Hantkenina nuttalli* and the absence of *Acarinina soldadoensis soldadoensis*. Other species found in this interval are those of the *Truncorotaloides* genus, along with *Turborotalia cerroazulensis frontosa, Morozovella spinulosa, Morozovella aragonensis, Acarinina spinuloinflata, Acarinina broedermanni, Acarinina bullbrooki, Guembelitrioides higginsi, Globigerina lozanoi, Globigerina cryptomphala,* and *Globigerina linaperta.*

Remarks: *Hantkenina nuttalli* is scarce in the wells studied. The last occurrence of *Globigerina lozanoi* served as additional criteria for identifying the zone, because, as indicated by Toumarkine and Luterbacher (1985, p. 100), this event occurs near the zonal top. Characteristic species of the middle Eocene, such as those listed immediately above, are also observed.

The *Hantkenina nuttalli* zone is the oldest zone of the middle Eocene, and most researchers mention the FO of the *Hantkenina* genus as a useful marker for the middle Eocene. This criterion could not be used in our material; the extinction of *Acarinina soldadoensis soldadoensis* was used instead for marking the lower Eocene–middle Eocene boundary, which is an alternative criterion specified by Stainforth et al. (1975) for the base of the middle Eocene.

The *Hantkenina nuttalli* zone identified for the study wells is equivalent to the *Hantkenina aragonensis* zone of Bolli (1957c, 1966) and Stainforth et al. (1975), as well as the *Hantkenina nuttalli* zone mentioned by Toumarkine and Luterbacher (1985). It may also be equivalent to the upper part of zone P10 *Subbotina frontosa/Globorotalia (Turborotalia) pseudomayeri* proposed by Blow (1979) and to zone P10 *Hantkenina nuttalli* in Berggren et al. (1995), because the microfaunal association is similar to the one identified by those researchers for this zone.

Globigerinatheka subconglobata subconglobata *Zone*

Age: Middle Eocene.

Author: Bolli (1957c), emended by Proto Decima and Bolli (1970), renamed by Bolli (1972), and redefined by Stainforth et al. (1975).

Definition: Interval with the zonal marker between the LO of *Hantkenina nuttalli* and the LO of *Morozovella aragonensis.*

Characterization: Overlapping occurrences of *Morozovella aragonensis, Morozovella lehneri,* and *Morozovella spinulosa* characterize this zone. Other typical species from the middle Eocene are also found, such as *Truncorotaloides rohri, Truncorotaloides topilensis, Acarinina spinuloinflata, Turborotalia cerroazulensis frontosa, Turborotalia cerroazulensis possagnoensis,* and some species of *Globigerinatheka,* including *Globigerinatheka subconglobata subconglobata* and *Globigerinatheka mexicana barri.*

Remarks: This zone is identified by the first downhole occurrence of *Morozovella aragonensis.* Most authors use this level to define the upper limit of this zone in the middle Eocene. This criterion is also used in the present study.

The zone's lower limit is marked by the LO of *Hantkenina nuttalli,* a species that although scarce and badly preserved in our samples, is easily identified. This criterion is not the same as that used by authors such as Stainforth et al. (1975) and Toumarkine and Luterbacher (1985). They use the FO of *Globigerinatheka mexicana mexicana.* However, that event is the criterion defined by Bolli (1957c, 1966) for describing the *Globigerapsis kugleri* zone.

This latter interval is correlative to the *Globigerapsis kugleri* zone described by Bolli (1957c, 1966), the *Globigerinatheka subconglobata* zone of Stainforth et al. (1975), and the *Globigerinatheka subconglobata subconglobata* zone of Toumarkine and Luterbacher (1985). It may also be compared with zone P11 *Globigerapsis kugleri/Subbotina frontosa boweri* proposed by Blow (1979), as well as zone P11 *Globigerapsis kugleri/Morozovella aragonensis* described by Berggren et al. (1995).

Morozovella lehneri *Zone*

Age: Middle Eocene.

Author: Bolli (1957b).

Definition: Interval containing the zonal marker between the LO of *Morozovella aragonensis* and the LO of *Turborotalia cerroazulensis frontosa.*

Characterization: The planktonic foraminiferal association is very similar to the one observed in the *Globigerinatheka subconglobata subconglobata* zone, without the presence of *Morozovella aragonensis.*

Other species found in this zone include *Turborotalia cerroazulensis possagnoensis, Globigerinatheka subconglobata euganea, Guembelitrioides higginsi,* and *Globigerinatheka mexicana mexicana,* among others.

Remarks: The first downhole occurrence of *Turborotalia cerroazulensis frontosa* and *Turborotalia cerroazulensis possagnoensis* indicates the presence of this zone. Extinction of these species has been indicated by most authors as events that take place at the zonal top.

Other data useful for recognizing this zone are the extinctions of *Guembelitrioides higginsi* and *Acarinina broedermanni*, as reported by Toumarkine and Luterbacher (1985, p. 100–101), at the top of the zone.

The *Morozovella lehneri* zone, as proposed for the wells studied, is equivalent to the *Globorotalia lehneri* zone proposed by Bolli (1957c, 1966) and by Stainforth et al. (1975), the *Morozovella lehneri* zone defined by Toumarkine and Luterbacher (1985), zone P12 *Globorotalia (Morozovella) lehneri* of Blow (1979), and zone P12 *Morozovella lehneri* of Berggren et al. (1995).

The criterion used for indicating the base of the zone here described, the LO of *Morozovella aragonensis*, is the same used by the authors mentioned, and the extinction of *Turborotalia cerroazulensis frontosa* was used for marking the zonal top, instead of the FO of *Orbulinoides beckmanni*, as specified by these same authors.

Orbulinoides beckmanni *Zone*

Age: Middle Eocene.

Author: Bolli (1957c) as *Porticulasphaera mexicana* zone, renamed by Cordey (1968) and by Blow and Saito (1968).

Definition: Interval between the LO of *Turborotalia cerroazulensis frontosa* and the LO of *Orbulinoides beckmanni*.

Characterization: In this zone, in addition to *Orbulinoides beckmanni*, we find the species *Globigerinatheka subconglobata subconglobata*, *Globigerinatheka index index*, *Globigerinatheka mexicana barri*, *Hantkenina alabamensis*, *Turborotalia cerroazulensis pomeroli*, *Truncorotaloides topilensis*, and *Truncorotaloides rohri*, among others.

Remarks: As mentioned in the description of the *Morozovella lehneri* zone, the present study used the LO of *Turborotalia cerroazulensis frontosa* as the event for marking the base of this zone, instead of the FO of *Orbulinoides beckmanni*, a criterion used by most authors, because it is a datum level difficult to determine in cutting samples. The upper limit of this zone was defined in the same manner as by the authors cited.

The *Orbulinoides beckmanni* zone here described may be correlated with the *Porticulasphaera mexicana* zone proposed by Bolli (1957c, 1966); with the *Orbulinoides beckmanni* zone of Stainforth et al. (1975) and Toumarkine and Luterbacher (1985); and with zone P13 *Globigerapsis beckmanni* described by Blow (1979) and Berggren et al. (1995).

Truncorotaloides rohri *Zone*

Age: Middle Eocene.

Author: Bolli (1957c).

Definition: Interval with zonal marker from the LO of *Orbulinoides beckmanni* to the LO of *Truncorotaloides rohri*.

Characterization: In addition to the nominate taxon, other common species include *Truncorotaloides topilensis*, *Morozovella spinulosa*, *Acarinina spinuloinflata*, *Acarinina bullbrooki*, *Moro-*

zovella lehneri, *Turborotalia cerroazulensis cerroazulensis*, *and Turborotalia cerroazulensis pomeroli*.

Remarks: Downhole, this zone is recognized by the occurrence of several species of planktonic foraminifera typical of the middle Eocene, including forms of spiny ornamentation such as *Truncorotaloides rohri*, *Truncorotaloides topilensis*, and *Acarinina spinuloinflata*, as stated in the zone's characterization.

The extinction of almost all spiny or hispid species of planktonic foraminifera has been recognized as defining a level that marks the boundary between the middle Eocene and the upper Eocene; this criterion is adopted in this research.

The *Truncorotaloides rohri* zone identified in the study wells may be correlated with the same zone proposed by Bolli (1957c, 1966) and described by Stainforth et al. (1975) and by Toumarkine and Luterbacher (1985), as well as with zone P14 *Globorotalia (Morozovella) spinulosa spinulosa* in Blow (1979) and with zone P14 *Truncorotaloides rohri–Morozovella spinulosa* in Berggren et al. (1995).

Globigerinatheka semiinvoluta *Zone*

Age: Late Eocene.

Author: Bolli (1957c), emended by Proto Decima and Bolli (1970).

Definition: Interval containing the zonal marker between the LO of *Truncorotaloides rohri* at its base and the LO of *Globigerinatheka semiinvoluta* at its top.

Characterization: *Globigerinatheka semiinvoluta*, *Turborotalia cerroazulensis cerroazulensis*, and *Turborotalia cerroazulensis pomeroli* are representative species of this zone.

Large globigerinas are observed, including *Globigerina yeguaensis*, *Globigerina venezuelana*, *Globigerina cryptomphala*, *Globigerina eocaena*, and *Globigerina tripartita*.

Remarks: Events used in this study for marking the lower and upper limits of the *Globigerinatheka semiinvoluta* zone are the same as those used by Stainforth et al. (1975) and by Toumarkine and Luterbacher (1985).

The *Globigerapsis semiinvoluta* zone proposed by Bolli (1957c, 1966) represents the same interval as for the zone determined in this work, except that the author used, in 1966, the FO of the index fossil of the zone to mark the base, although he mentions the extinction of *Truncorotaloides rohri* at the top of the preceding zone. This zone may also be correlated with part of zone P15 *Porticulasphaera semiinvoluta* described by Blow (1979) and by Berggren et al. (1995) and with the lower part of zone P16 *Cribrohantkenina inflata* in Blow (1979) and P16 *Turborotalia cunialensis/Cribrohantkenina inflata* in Berggren et al. (1995).

Turborotalia cerroazulensis s.l. *Zone*

Age: Late Eocene.

Author: Bolli (1957b), renamed by Bolli (1966, 1972).

Definition: Interval between the LO of *Globigerinatheka semiinvoluta* and the LO of *Turborotalia cerroazulensis* s.l.

Characterization: The presence of *Globorotalia cerroazulensis cerroazulensis*, *Globorotalia cerroazulensis cocoaensis*, *Globoro-*

talia cerroazulensis pomeroli, Cribrohantkenina inflata, and *Hantkenina alabamensis* is characteristic of this zone. Also present are *Globigerina ampliapertura, Globigerina gortanii praeturritilina,* and *Globorotalia opima nana,* species that persisted into the Oligocene.

Remarks: The upper limit of the *Turborotalia cerroazulensis* s.l. zone is accepted by most researchers as defining the Eocene-Oligocene boundary. This criterion is followed in this work.

This zone represents the same interval as the *Turborotalia cerroazulensis* s.l. zone of Toumarkine and Luterbacher (1985) and Stainforth et al. (1975) (therein called *Globorotalia cerroazulensis*). It may also be correlated with part of zone P16 *Cribrohantkenina inflata* and zone P17 *Globigerina gortanii gortanii–Globorotalia (Turborotalia) centralis* of Blow (1969, 1979), and with part of zone P16 *Turborotalia cunialensis/Cribrohantkenina inflata* and zone P17 *Turborotalia cerroazulensis* of Berggren et al. (1995).

Oligocene
Globigerina ampliapertura *Zone*

Age: Middle Oligocene.
Author: Bolli (1957b).
Definition: Interval between the LO of *Turborotalia cerroazulensis* s.l. at its base and the LO of *Globigerina ampliapertura* at its top.
Characterization: Along with abundant *Globigerina ampliapertura,* found are the frequent to abundant species *Globigerina venezuelana, Globigerina tripartita, Globigerina gortanii, Globigerina selli, Globigerina euapertura, Globigerina yeguaensis, Globigerina praebulloides* s.l., *Globigerina ouachitaensis, Globorotaloides suteri,* and *Catapsydrax dissimilis.*
Remarks: In the scheme proposed by Blow and Banner (1962), Bolli (1966), Stainforth et al. (1975), and Toumarkine and Luterbacher (1985), the first zone of the Oligocene is *Cassigerinella chipolensis/Pseudohastigerina micra,* based on the overlap in ranges of both species. This interval was proposed originally as the *Globigerina oligocaenica* zone by Blow and Banner (1962), also based on the overlap of these same species, which coincides, according to these authors, with the total range of *Globigerina oligocaenica.*

In our samples, the *Cassigerinella chipolensis/Pseudohastigerina micra* zone was not identified, because the first downhole appearance of *Pseudohastigerina micra* was found with *Turborotalia cerroazulensis,* whose extinction marks the top of the zone with that name. Thus, the first zone of the Oligocene is missing in the Campeche samples, suggesting an unconformity across the Eocene-Oligocene boundary.

The *Globigerina ampliapertura* zone described in this study would be equivalent to the zone of the same name reported by Bolli (1957b, 1966), Stainforth et al. (1975), Bolli and Saunders (1985), and Berggren et al. (1995). It may also be correlated with zone P19/20 *Globigerina selli–Globigerina ampliapertura,* and with the lowest part of zone P21 (=N2) *Globigerina angu-*

lisuturalis/Globorotalia (T.) opima opima, as proposed by Blow (1979).

Globorotalia opima opima *Zone*

Age: Middle Oligocene.
Author: Bolli (1957b).
Definition: Interval between the LO of *Globigerina ampliapertura* and the LO of *Globorotalia opima opima.*
Characterization: The presence of *Globorotalia opima opima* and the absence of *Globigerina ampliapertura,* a species that becomes extinct at the top of the *Globigerina ampliapertura* zone, are useful to identify this zone.

Other species such as *Globorotalia opima nana, Globigerina praebulloides* s.l., *Globigerina tripartita, Globigerina euapertura,* and *Globigerina ciperoensis* s.l. are also associated.
Remarks: In traditional zonal schemes that follow the Bolli criteria (1957b), the *Globorotalia opima opima* zone is based on the total range of the nominate taxon. In the samples studied from the marine zone wells, the top was marked by the extinction of *Globorotalia opima opima,* but because of the nature of our samples, the base was defined at the last occurrence of *Globigerina ampliapertura,* not with the first appearance of *Globorotalia opima opima,* which was the datum used by Bolli (1957b).

The zone we describe may be correlated with the *Globorotalia opima opima* zone of Bolli (1957b, 1966), Stainforth et al. (1975), and Bolli and Saunders (1985), with the middle and upper parts of zone P21 (= N2) *Globigerina angulisuturalis/Globorotalia (Turborotalia) opima opima* of Blow (1969, 1979), and with zones P20 (part) *Globigerina selli* and P21 *Globigerina angulisuturalis/Paragloborotalia opima opima* of Berggren et al. (1995).

Globigerina ciperoensis ciperoensis *Zone*

Age: Late Oligocene.
Author: Cushman and Stainforth (1945) as *Globigerina concinna* zone, emended by Bolli (1957b).
Definition: Interval with common and constant occurrence of *Globigerina ciperoensis* s.l. between the LO of *Globorotalia opima opima* at its base and the LO of the nominate taxon at its top.
Characterization: The frequent occurrence of *Globigerina ciperoensis ciperoensis, Globigerina ciperoensis angustiumbilicata,* and *Globigerina ciperoensis angulisuturalis,* and the absence of *Globorotalia opima opima* allow identification of this zone.
Remarks: The foraminiferal assemblage in this interval is dominated by subspecies of *Globigerina ciperoensis,* as well as *Globigerina praebulloides* s.l. and *Globigerina tripartita.* Some species characteristic of the Miocene also are found, such as *Orbulina* spp, *Globigerinoides primordius, Globigerinoides bisphericus,* and *Globigerinoides subquadratus.* These specimens probably represent downhole contamination.

In this study, this is the last zone of the Oligocene.

The Bolli and Saunders zonal scheme (1985) includes the *Globorotalia kugleri* zone in the late Oligocene; however, that species was not found in the study samples. The absence of this microfossil was probably caused by climatic factors.

Bandy (1964) noted the importance of the evolutionary occurrence of *Globigerinoides triloba* (*Globigerinoides* datum) to define the base of the Miocene. This criterion has been accepted to define the Oligocene-Miocene boundary by researchers such as Blow (1969), Berggren (1970, 1971), Postuma (1971), Bizon et al. (1974), Stainforth et al. (1975), and Bolli and Saunders (1985). The latter authors, however, suggest using the explosive occurrence of *Globigerinoides trilobus* instead of the scarce occurrence of primitive forms of *Globigerinoides primordius*.

The occurrence of this genus has been reported in the *Globigerina ciperoensis ciperoensis* zone and is therefore in the Oligocene (Lamb and Stainforth, 1976; Stainforth and Lamb, 1981; Berggren and Miller, 1988; Salazar, 1991; Salazar and Gómez, 1990).

Stainforth and Lamb (1981) suggest using the FO of *Globoquadrina altispira* aff. *altispira* and *Globorotalia fohsi peripheroronda* as a substitute for the *Globigerinoides* Datum.

The occurrence of *Globigerinoides* at different stratigraphic levels may be caused by rising temperatures after the Oligocene cold period. Although this was a worldwide event, it took place at different times locally (Bolli and Saunders, 1985).

In our samples, we observed the common presence of specimens of *Globigerinoides trilobus* s.l. in the *Globigerina ciperoensis ciperoensis* zone; as mentioned before, other *Globigerinoides* species were also found that were considered downhole contamination.

In the study wells, the *Globigerina ciperoensis ciperoensis* zone lies below the *Catapsydrax dissimilis* zone of the lower Miocene. The *Globorotalia kugleri* zone (Oligocene) and the *Globigerinoides primordius* zone (Miocene) described by Bolli and Saunders (1985) are missing, so the Oligocene-Miocene limit is drawn with a discontinuous line (Figure 2).

The *Globigerina ciperoensis ciperoensis* zone identified in the present study may be correlated with the zone of the same name proposed by Bolli (1957b, 1966), Stainforth et al. (1975), Stainforth and Lamb (1981), and Bolli and Saunders (1985), as well as with zone P22/N3 *Globigerina angulisuturalis* proposed by Blow (1969) and zone P22 *Globigerina ciperoensis* of Berggren et al., 1995.

Miocene

Catapsydrax dissimilis *Zone*

Age: Early Miocene.
Author: Cushman and Renz (1947) as *Globigerina dissimilis* zone, modified by Bolli (1957b).
Definition: Interval between the LO of *Globigerina ciperoensis ciperoensis* and the LO of *Catapsydrax dissimilis*.
Characterization: This zone features the common occurrence of *Catapsydrax dissimilis*, along with *Catapsydrax stainforthi* and *Globoquadrina dehiscens praedehiscens*. Other species present include *Globorotalia fohsi peripheroronda*, *Globorotalia mayeri*, and *Globigerinoides subquadratus*.
Remarks: In the zonal scheme proposed by Bolli (1957b,

1966) and described by Stainforth et al. (1975) and Bolli and Saunders (1985), the base of the *Catapsydrax dissimilis* zone is marked by the extinction of *Globorotalia kugleri*, a fossil generally not observed in the marine zone (as defined in other works: Salazar, 1991; Salazar and Gómez, 1990). Therefore, in the study samples, the base of the zone is marked by the continuous, common occurrence of *Globigerina ciperoensis ciperoensis*.

As mentioned previously, the *Globigerinoides primordius* zone could not be identified in our samples, because the first appearance of this fossil could not be determined with the available samples. The *Globigerinoides primordius* zone, therefore, is included in the concept of the *Catapsydrax dissimilis* zone, as defined here.

The *Catapsydrax dissimilis* zone encompasses the interval corresponding to the *Catapsydrax stainforthi* zone defined by Stainforth et al. (1975) and by Bolli and Saunders (1985). These authors use the FO of *Globigerinatella insueta* to mark the boundary between the zones. In the study samples, the latter taxon is not present, and these two zones were combined, because the top of the *Catapsydrax dissimilis* zone suggested in this work corresponds to the extinction of the nominate species, a datum used by most researchers to mark the upper limit of the *Catapsydrax stainforthi* zone.

This zone is correlative to the *Globigerinoides primordius, Catapsydrax dissimilis,* and *Catapsydrax stainforthi* zones described by Bolli and Saunders (1985); to the *Globorotalia kugleri, Catapsydrax dissimilis,* and *Catapsydrax stainforthi* zones in Stainforth et al. (1975); to zones N4 *Globigerinoides quadrilobatus primordius/Globorotalia (T.) kugleri,* N5 *Globoquadrina dehiscens praedehiscens–G. dehiscens dehiscens,* and N6 *Globigerinatella insueta/Globigerinita dissimilis* proposed by Blow (1969); and to zones M1 *Globorotalia kugleri,* M2 *Catapsydrax dissimilis,* and M3 *Globigerinatella insueta/Catapsydrax dissimilis* described by Berggren et al. (1995).

Praeorbulina glomerosa–Globigerinoides bisphericus *Zone*

Age: Early Miocene.
Author: Blow (1959), renamed by Bolli (1966).
Definition: Interval between the LO of *Catapsydrax dissimilis* and the LO of *Globigerinoides bisphericus*.
Characterization: The continuous occurrence of *Globigerinoides bisphericus, Praeorbulina glomerosa glomerosa, Praeorbulina glomerosa curva, Praeorbulina glomerosa circularis,* and *Praeorbulina glomerosa transitoria* allows the identification of this zone.
Remarks: As described here, this interval encompasses the zones of *Globigerinatella insueta* and *Praeorbulina glomerosa* of Bolli and Saunders (1985), because these authors defined the boundary between these two zones at the FO of *Praeorbulina glomerosa,* an event that could not be determined in this investigation.

Bolli and Saunders (1985) indicate that the top of the *Praeorbulina glomerosa* zone is identified at the extinction of *Globigerinatella insueta*.

This study uses the extinction of *Globigerinoides bisphericus* (= *Globigerinoides sicanus* of Blow 1959, 1969) instead of that of *Globigerinatella insueta*, because the latter was not present in the study samples. *Globigerinoides bisphericus* is placed at the same extinction level as *Globigerinatella insueta*, as indicated by the authors mentioned above.

The boundary between the lower and middle Miocene has been marked by the evolutionary appearance of the *Orbulina* genus (*Orbulina* datum). This is one of the most important events occurring in the Miocene, as recorded by several authors, such as Le Roy (1948), Bandy (1963, 1964), Blow (1969), Berggren (1969, 1971), Cita and Blow (1969), Bandy and Ingle (1970), Stainforth et al. (1975), and Bolli and Saunders (1985). It occurs almost at the same level as the extinction of *Globigerinoides bisphericus*.

The present study could not define the *Orbulina* datum because of the use of cutting samples.

The zone described may be correlated with the *Globigerinatella insueta* and *Praeorbulina glomerosa* zones in Bolli (1957b, 1966), Stainforth et al. (1975), and Bolli and Saunders (1985); with zones N7 *Globigerinatella insueta–Globigerinoides quadrilobatus trilobus* and N8 *Globigerinoides sicanus–Globigerinatella insueta* proposed by Blow (1969); and with zones M4 *Catapsydrax dissimilis–Praeorbulina sicana* and M5 *Praeorbulina sicana–Orbulina suturalis* in Berggren et al. (1995).

Globorotalia fohsi fohsi–Globorotalia fohsi peripheroronda *Zone*

Age: Middle Miocene.
Author: Bolli (1957b), renamed by Blow and Banner (1966) and by Bolli (1967).
Definition: Interval between the LO of *Globigerinoides bisphericus* at its base and the LO of *Globorotalia fohsi peripheroacuta* at its top.
Characterization: This zone is easily recognized by the presence of *Globorotalia fohsi peripheroacuta, Globorotalia fohsi peripheroronda, Globorotalia fohsi praefohsi,* and *Globorotalia fohsi fohsi. Globorotalia fohsi lobata* and *Globorotalia fohsi robusta* are also present in the Campeche samples as a result of downhole contamination.
Remarks: Zones based on the phylogenetic evolution of the "*fohsi*" group species are divided at the evolutionary appearance of each nominate species. In the marine zone wells, it was not possible to make this subdivision, so the extinction of *Globorotalia peripheroacuta* was used. This event took place at approximately the same level as the FO of *Globorotalia fohsi lobata*, as indicated by Bolli and Saunders (1985, p. 168). Therefore, in this work, the interval representing this zone is equivalent to the *Globorotalia fohsi peripheroronda* and *Globorotalia fohsi fohsi* zones described by Bolli and Saunders (1985) and by Stainforth et al. (1975), to zones N9 (*partim*) *Orbulina suturalis–Globorotalia (Turborotalia) peripheroronda,* N10 *Globorotalia (Turborotalia) peripheroacuta,* and to the lowest part of zone N11 *Globorotalia (Globorotalia) praefohsi* in Blow (1969). It also correlates with zones M6 *Orbulina suturalis–Globorotalia peripheroronda,* M7 *Globorotalia peripheroacuta,* and the lowest part of zone M8 *Globorotalia fohsi* s.s. in Berggren et al. (1995).

Globorotalia fohsi robusta–Globorotalia fohsi lobata *Zone*

Age: Middle Miocene.
Author: Bolli (1957b).
Definition: Interval between the LO of *Globorotalia fohsi peripheroacuta* and the LO of *Globorotalia fohsi lobata.*
Characterization: This interval is recognized by the presence of the nominate taxa, although without *Globorotalia fohsi peripheroacuta,* which is restricted to the underlying zone.
Remarks: Sediments of the study wells that correspond to this zone underlie samples containing a characteristic association of the late Miocene (*Globorotalia acostaensis* zone, described later). Thus, the zones corresponding to the upper part of the middle Miocene are missing (*Globigerinoides ruber, Globorotalia mayeri,* and *Globorotalia menardii* zones, described in the zonal scheme proposed by Bolli and Saunders, 1985). Therefore, the middle Miocene to late Miocene boundary is represented by an unconformity.

This zone is equivalent to the *Globorotalia fohsi lobata* and *Globorotalia fohsi robusta* zones described by Bolli and Saunders (1985) and by Stainforth et al. (1975), to zones N11 (*partim*) *Globorotalia (G.) praefohsi* and N12 *Globorotalia (G.) fohsi* proposed by Blow (1969), and to zone M8 (*partim*) *Globorotalia fohsi* s.s. and part of zone M9 *Globorotalia fohsi lobata–Globorotalia fohsi robusta* described in Berggren et al. (1995).

Globorotalia acostaensis *Zone*

Age: Late Miocene.
Author: Bolli and Bermúdez (1965).
Definition: Interval containing the zonal marker between the first downhole occurrence of *Globorotalia fohsi lobata* and the LO of *Globorotaloides variabilis.*
Characterization: The interval corresponding to the *Globorotalia acostaensis* zone in the study wells has a more diverse faunal association than the zones previously described.

Some species of planktonic foraminifera found in the upper Miocene are: *Globorotalia acostaensis, Globorotalia humerosa humerosa, Globorotalia humerosa praehumerosa, Globorotalia merotumida, Globorotalia plesiotumida, Globorotalia pseudopima, Sphaeroidinellopsis seminulina, Globorotalia menardii* "B," and *Globigerinoides obliquus extremus.*

Microfauna in this interval are generally scarce and badly preserved, with abundant pyritized fossils.
Remarks: In this study, I follow Lamb and Beard (1972) and Stainforth et al. (1975), who used a single zone for the upper Miocene, as opposed to two (*Globorotalia acostaensis* and *Globorotalia humerosa*), as defined by Bolli and Saunders (1985). Because these authors use the evolutionary appearance of *Globorotalia humerosa* to delimit both zones, this event could not be recognized in the Campeche samples.

The Miocene-Pliocene boundary is a matter of discussion

among many authors, who have also used many species for its definition. Some have stressed the evolutionary appearance of *Sphaeroidinella dehiscens* (Banner and Blow, 1965; Blow, 1969; Cita and Blow, 1969; Parker, 1973). Others have observed that this species appears at younger levels in the *Globorotalia margaritae evoluta* subzone or near the top of the lower Pliocene (Bolli and Saunders, 1985). Berggren and Amdurer (1973) and Berggren et al. (1995) place the base of the Pliocene at the evolutionary appearance of *Globorotalia tumida*. Cita (1973, 1975, 1976) marks the limit at the base of her *Sphaeroidinellopsis* Acme zone, for which she describes the first appearance of *Sphaeroidinella dehiscens immaturus*. This zone underlies the *Globorotalia margaritae margaritae* zone, defined by an abundant occurrence of this subspecies. Other authors, including Lamb and Beard (1972), Stainforth et al. (1975), Berggren (1977), and Bolli and Saunders (1985) define this boundary at the first appearance of *Globorotalia margaritae*.

In this study, the Miocene-Pliocene boundary was marked with the first downhole occurrence of *Globorotaloides variabilis* and/or *Sphaeroidinellopsis disjuncta* or *Sphaeroidinellopsis multiloba*. The last downhole occurrence of *Globorotalia margaritae* was used in well 1 to mark the Miocene-Pliocene boundary.

The *Globorotalia acostaensis* zone described in this work may be compared with the zone of the same name defined by Lamb and Beard (1972) and by Stainforth et al. (1975). It is equivalent also to the *Globorotalia acostaensis* and *Globorotalia humerosa* zones proposed by Bolli and Saunders (1985). It may be correlated to zones N16 *Globorotalia (Turborotalia) acostaensis–Globorotalia (G.) merotumida* and N17 *Globorotalia (G.) tumida plesiotumida* in Blow (1969), and with zones M13 *Neogloboquadrina acostaensis/Globorotalia lenguaensis* and M14 *Globorotalia lenguaensis/Globorotalia tumida*, as described by Berggren et al. (1995).

Pliocene

Globorotalia margaritae *Zone*

Age: Early Pliocene.
Author: Bolli and Bermúdez (1965).
Definition: Interval with consistent occurrence of *Globorotalia margaritae*. The zone's base is marked by the LO of *Globorotaloides variabilis* and/or the (FO) of the nominate species, and the top corresponds to the LO of *Globigerinoides obliquus* or the LO of *Globorotalia margaritae*.
Characterization: In addition to *Globorotalia margaritae*, the following species that are not present in the *Globorotalia acostaensis* zone are found: *Globorotalia crassaformis crassaformis*, *Globorotalia tumida*, *Globorotalia pseudomiocenica*, *Globigerinoides conglobatus*, *Neogloboquadrina dutertrei*, *Globigerinoides elongatus*, *Globorotalia miocenica*, *Sphaeroidinella dehiscens dehiscens*, *Sphaeroidinella dehiscens immaturus*, and *Sphaeroidinellopsis sphaeroides*. Most of the species present in the *Globorotalia acostaensis* zone also occur in the *Globorotalia margaritae* zone.
Remarks: The base of the *Globorotalia margaritae* zone is

marked by the extinction of *Globorotaloides variabilis*, which occurs, according to Bolli and Saunders (1985, p. 169), at the top of the Miocene.

The upper limit of the *Globorotalia margaritae* zone in some of the study wells is marked by the extinction of *Globigerinoides obliquus obliquus*, which occurs at the top of the *Globorotalia margaritae* zone (see Bolli and Saunders, 1985, p. 169), and in other wells by the extinction of *Globorotalia margaritae*.

The *Globorotalia margaritae* zone, as described in this study, may be correlated with the zone of the same name described by Stainforth et al. (1975) and by Bolli and Saunders (1985). It also may be correlated with zones N18 *Globorotalia (G.) tumida tumida–Sphaeroidinellopsis subdehiscens paenedehiscens* and N19 (*partim*) *Sphaeroidinella dehiscens dehiscens–Globoquadrina altispira altispira* proposed by Blow (1969), and Pl1 *Globorotalia tumida/Globoturborotalita nepenthes* and Pl2 (*partim*) *Globoturborotalita nepenthes–Globorotalia margaritae* described by Berggren et al. (1995).

Globorotalia miocenica *Zone*

Age: Middle Pliocene.
Author: Bolli (1970), renominated by Bolli and Premoli Silva (1973), redefined by Bolli and Saunders (1985).
Definition: Interval containing *Globorotalia miocenica*, between the LO of *Globigerinoides obliquus obliquus* or the LO of *Globorotalia margaritae* at the base and the LO of *Globigerinoides obliquus extremus* or the LO of *Globorotalia miocenica* at the upper limit.
Characterization: The presence of the nominate species associated with *Globigerinoides obliquus extremus*, in the absence of *Globigerinoides obliquus obliquus*, and *Globorotalia margaritae* characterizes this zone. Other species not found in either the *Globorotalia margaritae* zone or the *Globorotalia miocenica* zone are *Globigerina nepenthes* and *Globigerina venezuelana*. Except for the two species just mentioned, the faunal association of this zone is very similar to that of the *Globorotalia margaritae* zone described previously.
Remarks: In the present study, the criteria for defining the upper and lower boundaries of this zone are different from those indicated by previous researchers. Extinctions of *Globigerinoides obliquus obliquus* at the base and *Globigerinoides obliquus extremus* at the top are used in this study, in addition to or instead of the extinction of *Globorotalia margaritae* and *Globorotalia miocenica* (and/or *Globorotalia exilis*). In fact, the species chosen for this work are present in the Campeche samples more abundantly and consistently than those suggested by Bolli (1970), Bolli and Premoli-Silva (1973), and Bolli and Saunders (1985). Their evolutionary disappearances are almost synchronous with those of the microfossils used by those authors.

Bolli and Saunders (1985) divide the *Globorotalia miocenica* zone into two subzones—the *Globigerinoides trilobus fistulosus* and the *Globorotalia exilis*— placing the subzonal boundary at the extinction of *Globigerinoides trilobus fistulosus*.

This microfossil was not found in the wells studied, so no subdivision could be carried out.

The *Globorotalia miocenica* zone described in this research may be correlated with the *Pulleniatina obliquiloculata* zone and part of the *Globorotalia truncatulinoides* zone of Stainforth et al. (1975), and with the *Globorotalia miocenica* zone described by Bolli and Saunders (1985). It may also correlate with the uppermost part of zone N19 *Sphaeroidinella dehiscens dehiscens–Globoquadrina altispira altispira;* with zones N20 of *Globorotalia (G.) multicamerata–Pulleniatina obliquiloculata obliquiloculata* and N21 *Globorotalia (T.) tosaensis tenuitheca (partim)* in Blow (1969); and with the upper part of zone Pl2 *Globoturborotalita nepenthes–Globorotalia margaritae* and zones Pl3 *Globorotalia margaritae–Sphaeroidinellopsis seminulina,* Pl4 *Sphaeroidinellopsis seminulina–Dentoglobigerina altispira,* and Pl5 *Dentoglobigerina altispira–Globorotalia miocenica* in Berggren et al. (1995).

Globorotalia tosaensis tosaensis *Zone*

Age: Late Pliocene–Pleistocene.

Author: Bolli (1970), renamed in Bolli and Saunders (1985).

Definition: Interval between the LO of *Globigerinoides obliquus extremus* at the base and the LO of *Globorotalia tosaensis tosaensis* at the top.

Characterization: Many species found in the underlying zone are also present in the interval we describe, but *Globorotalia miocenica, Globigerinoides obliquus extremus,* and *Sphaeroidinellopsis seminulina* are absent. *Globorotalia truncatulinoides* was found only in the highest part of this interval.

Remarks: The Pliocene-Pleistocene boundary has been marked by several researchers at the FO of *Globorotalia truncatulinoides* (Berggren et al., 1967; Blow, 1969; Brönnimann and Resig, 1971; Bandy, 1972; Berggren and Amdurer, 1973; Bolli and Premoli-Silva, 1973; Huang, 1976; Natori, 1976; Bolli and Saunders, 1985). Other authors, such as Lamb and Beard (1972) and Stainforth et al. (1975), use the *Globoquadrina altispira* extinction horizon.

In this study, the Pliocene-Pleistocene boundary is included in this interval because the extinction of *Globorotalia tosaensis* occurs after the evolutionary appearance of *Globorotalia truncatulinoides,* as indicated by several authors, such as Bolli and Saunders (1985) and Berggren et al. (1995), among others.

This zone may be correlated with the *Globorotalia tosaensis tosaensis* zone and part of the *Globorotalia crassaformis viola* subzone described by Bolli and Saunders (1985); with part of the *Globorotalia truncatulinoides* zone described by Stainforth et al. (1975); with the upper part of zone N21 *Globorotalia (T.) tosaensis tenuitheca;* and the lowermost part of zone N22 *Globorotalia (G.) truncatulinoides truncatulinoides,* as defined by Blow (1969). It may also correlate with zone Pl6 *Globorotalia pseudomiocenica–Globigerinoides fistulosus* and subzone Pt1a *Globigerinoides fistulosus–Globorotalia tosaensis* in Berggren et al. (1995).

Pleistocene

Globorotalia truncatulinoides *Zone*

Age: Pleistocene.

Author: Bolli and Bermúdez (1965), renamed by Bolli (1966).

Definition: The base is marked with the LO of *Globorotalia tosaensis tosaensis.* The top is undefined in this study material.

Characterization: The presence of *Globorotalia truncatulinoides* and the absence of *Globorotalia tosaensis tosaensis* characterize this zone.

Remarks: In the present study, two subzones were identified in this interval: the *Globorotalia crassaformis viola* and the *Globigerina calida.*

Globorotalia crassaformis viola *Subzone*

Age: Pleistocene.

Author: Bolli and Premoli-Silva (1973).

Definition: Interval containing the subzonal marker between the LO of *Globorotalia tosaensis tosaensis* and the LO of *Globorotalia crassaformis viola.*

Characterization: The microfauna association is very similar to that of the previous zone; it is differentiated by the absence of *Globorotalia tosaensis tosaensis.*

Remarks: The presence of *Globorotalia crassaformis viola* indicates that samples of this subzone are not younger than the lower Pleistocene; in fact, this species disappears at the base of the *Globorotalia truncatulinoides truncatulinoides* zone (Bolli and Saunders, 1985).

This interval is equivalent to the *Globorotalia crassaformis viola* and *Globorotalia crassaformis hessi* subzones described by Bolli and Saunders (1985). These authors delineate the subzones at the evolutionary appearance of *Globorotalia crassaformis hessi.* This criterion could not be used in the present study.

This subzone also may be correlated with part of the *Globorotalia truncatulinoides* zone in Stainforth et al. (1975), with part of zone N22 *Globorotalia (G.) truncatulinoides truncatulinoides* in Blow (1979), and with part of subzone Pt1b *Globorotalia truncatulinoides* in Berggren et al. (1995).

Globigerina calida *Subzone*

Age: Pleistocene.

Author: Bolli and Premoli-Silva (1973).

Definition: Interval between the LO of *Globorotalia crassaformis viola* and the LO of *Globorotalia tumida flexuosa.*

Characterization: This subzone is identified by the presence of *Globorotalia tumida flexuosa* and the absence of *Globorotalia crassaformis viola.*

Remarks: The present study used the criterion specified by Bolli and Saunders (1985) to define the subzone's top, which is the extinction of *Globorotalia tumida flexuosa.*

This interval may be correlated with the subzone of the same name described by Bolli and Saunders (1985), with part of the *Globorotalia truncatulinoides* zone described by Stainforth

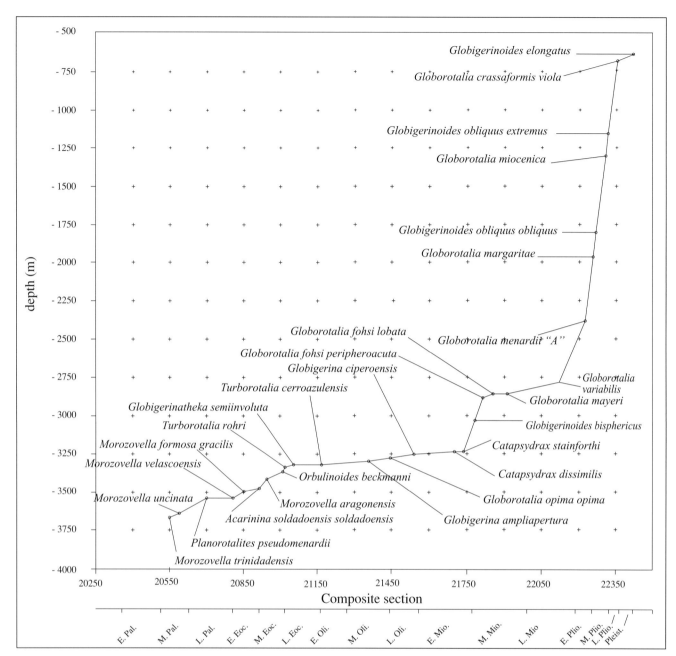

FIGURE 8. Graphic correlation of well 1 against the composite standard section. Data points are first downhole occurrences or tops of selected planktonic foraminifera.

et al. (1975), with part of zone N22 *Globorotalia* (*G.*) *truncatulinoides truncatulinoides* in Blow (1979), and with part of subzone Pt1b *Globorotalia truncatulinoides* in Berggren et al. (1995).

GRAPHIC CORRELATION

Graphic correlation was proposed by Shaw (1964) as a method of biostratigraphic correlation.

A two-axis graph was generated for the present research using foraminifera data obtained from the detailed study of well 1 and plotting these data against the composite standard section (Figure 8).

Only the tops of foraminiferal ranges were taken into account.

In the graph obtained for well 1, we can observe four "terraces" that represent the absence of one or several zones (Figure 9). A terrace, used in the sense of the graphic-correlation method, may refer to an unconformity, condensed section, hiatus, absence of strata of a certain age, etc. (Miller, 1977; Neal et al., 1995).

The first terrace corresponds to an interval between CSU (composite standard unit) 21910 and CSU 21855 and encompasses part of the middle Miocene (*Globigerinoides ruber, Globorotalia mayeri,* and *Globorotalia menardii* zones). The sec-

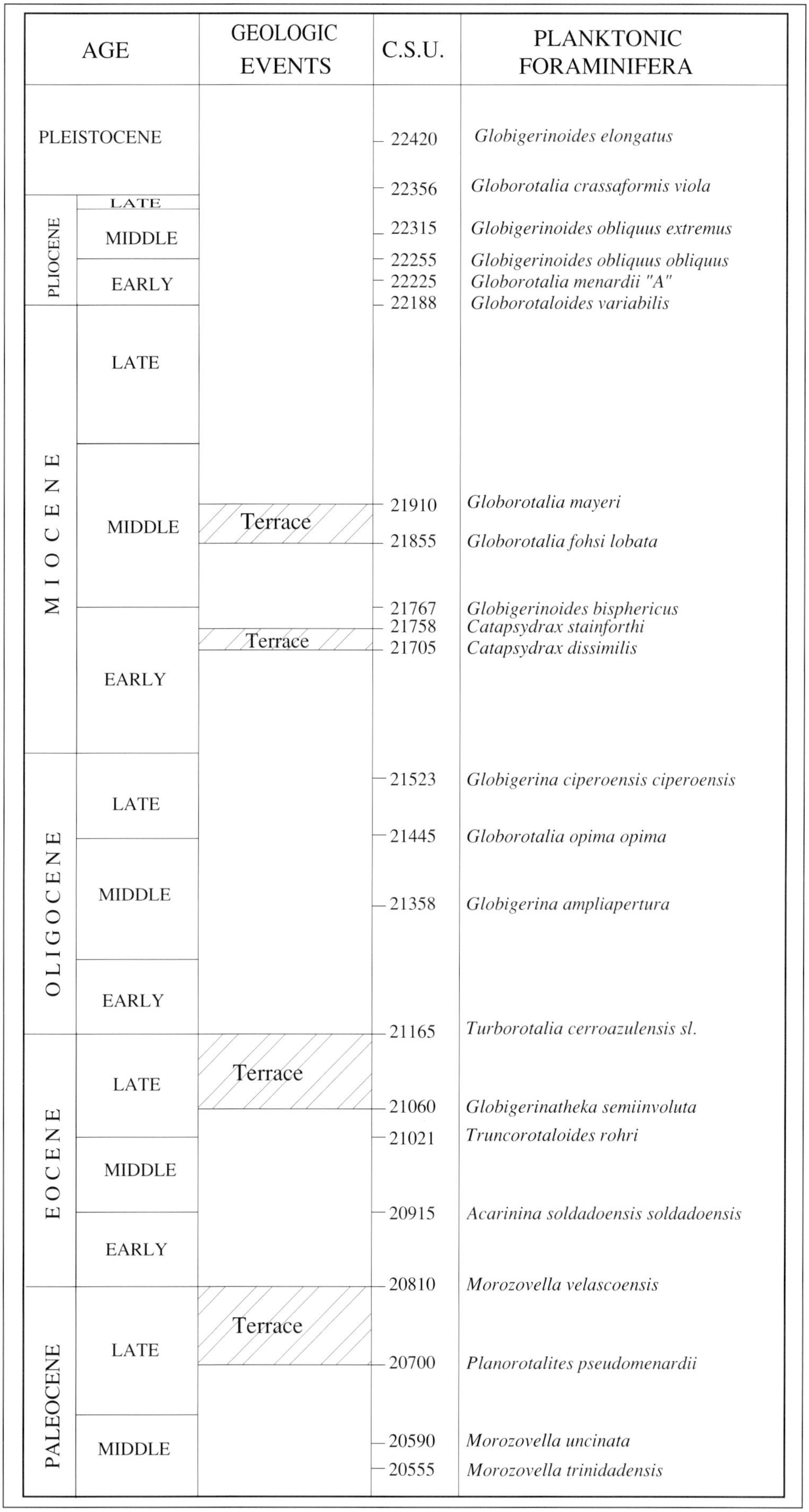

AGE		GEOLOGIC EVENTS	C.S.U.	PLANKTONIC FORAMINIFERA
PLEISTOCENE			— 22420	*Globigerinoides elongatus*
			— 22356	*Globorotalia crassaformis viola*
PLIOCENE	LATE		— 22315	*Globigerinoides obliquus extremus*
	MIDDLE		— 22255	*Globigerinoides obliquus obliquus*
	EARLY		— 22225	*Globorotalia menardii "A"*
			— 22188	*Globorotaloides variabilis*
M I O C E N E	LATE			
	MIDDLE	Terrace	— 21910	*Globorotalia mayeri*
			— 21855	*Globorotalia fohsi lobata*
	EARLY	Terrace	— 21767	*Globigerinoides bisphericus*
			— 21758	*Catapsydrax stainforthi*
			— 21705	*Catapsydrax dissimilis*
O L I G O C E N E	LATE		— 21523	*Globigerina ciperoensis ciperoensis*
			— 21445	*Globorotalia opima opima*
	MIDDLE		— 21358	*Globigerina ampliapertura*
	EARLY		— 21165	*Turborotalia cerroazulensis sl.*
E O C E N E	LATE	Terrace	— 21060	*Globigerinatheka semiinvoluta*
			— 21021	*Truncorotaloides rohri*
	MIDDLE		— 20915	*Acarinina soldadoensis soldadoensis*
	EARLY		— 20810	*Morozovella velascoensis*
PALEOCENE	LATE	Terrace	— 20700	*Planorotalites pseudomenardii*
	MIDDLE		— 20590	*Morozovella uncinata*
			— 20555	*Morozovella trinidadensis*

FIGURE 9. Chronostratigraphic chart of well 1, based on graphic-correlation method.

ond terrace is present from CSU 21758 to CSU 21705, corresponding to part of the early Miocene (*Catapsydrax stainforthi* zone). Another terrace is delimited from CSU 21165 to CSU 21060 in the late Eocene, corresponding to the *Turborotalia cerroazulensis* zone. The fourth terrace occurs from CSU 20810 to CSU 20700, indicating the absence of the *Morozovella velascoensis* zone in the late Paleocene.

The slope of the line of correlation (LOC) indicates the rock-accumulation rate.

In the study well, it can be deduced that the rock-accumulation rate during the Pliocene, upper part of the late Miocene, lower part of the middle Miocene, and upper part of the early Miocene had a rapid rate of sediment accumulation, causing an increase in the rate of sedimentation. For the early Miocene to the late Eocene, an almost horizontal slope for the LOC indicates a slow rate of sediment accumulation.

In the upper part of the middle Eocene, the LOC again becomes almost vertical, indicating an increase in the accumulation rate. Finally, during the lower part of the middle Eocene, the lower part of the early Eocene, and the lower part of the late Paleocene to middle Paleocene, the LOC inclination indicates an increased rock-accumulation rate, although not as fast as for the Pliocene.

Other changes in the thicknesses of Tertiary rocks may be related to local and regional tectonic problems, because the study area records structural complications (folds, faults, saline intrusions, etc.). Furthermore, many wells drilled in the upper parts of these geologic structures encounter variations in thickness from well to well.

CONCLUSIONS

Chronostratigraphy and Biostratigraphy

The most important conclusions of this work may be summarized as follows:

- A zonation of Tertiary sediments of the marine region of the Campeche subsurface has been established and can be used in future biostratigraphic studies in the area. Twenty-six biozones have been identified.
- Sediments from the middle Paleocene to late Eocene were identified only in wells 1 and 8. No samples in the remaining wells had corresponding ages.
- The Paleocene-Eocene boundary is unconformable in wells 1 and 8. The *Morozovella velascoensis* zone is missing in both.
- The Eocene-Oligocene boundary is also unconformable in wells 1 and 8.
- Strata of middle Oligocene age were found in wells 1, 3, 8, and 9.
- Late Oligocene sediments were found in almost all wells (material representative of the *Globigerina ciperoensis ciperoensis* zone).
- The early Miocene (*Catapsydrax dissimilis* and *Praeor-*

bulina glomerosa–Globigerinoides bisphericus zones) is present in the great majority of the study wells.
- With respect to middle Miocene, only the lower part is present (the *Globorotalia fohsi robusta–lobata* and *Globorotalia fohsi fohsi–peripheroronda* zones).
- Sediments of middle Miocene age (upper part) are absent in the study wells. It is inferred that the middle to late Miocene boundary is unconformable through the entire study area.
- The late Miocene is represented by the *Globorotalia acostaensis* zone, which was identified in almost all wells.
- The late Miocene–early Pliocene boundary is marked by the last appearance of *Globorotaloides variabilis*; however, this boundary could not be precisely defined in some wells, because this species is often rare or not present. Therefore, the presence of *Globorotalia margaritae* and other species characteristic of the Pliocene were used to differentiate the Miocene section from the Pliocene.
- Sediments of the early Pliocene were recognized in almost all wells, but the middle Pliocene interval was identified in only 10 wells, and the late Pliocene section could be recognized in only seven wells.
- Pleistocene samples were identified in five wells.

Paleoecology

Characteristic benthonic foraminiferal assemblages, as well as benthonic/planktonic ratios and other criteria, allow the following paleoenvironmental interpretations:

- In general, middle Paleocene to middle Miocene sediments were deposited in a lower bathyal environment.
- In the late Miocene, fluctuations from lower bathyal to middle neritic were observed.
- In the lower and middle part of the early Pliocene, the depositional environment fluctuated from lower to middle bathyal.
- Sediments of the upper part of the early Pliocene and those of the middle Pliocene were deposited in shallower depths, which ranged from middle bathyal to outer-middle neritic.
- Finally, the time interval from late Pliocene to Pleistocene still evidenced shallower conditions, from outer to inner neritic.

ACKNOWLEDGMENTS

I would like to express my gratitude to PEMEX Exploración-Producción (PEP) for its support in carrying out this research. I am very grateful to Thomas W. Dignes and Brian J. O'Neill for their constructive review of the manuscript and to Claudio Bartolini, Abelardo Cantú-Chapa, and Richard Buffler for inviting me to participate in this volume. Finally, I would like to thank Carlos Cantú-Chapa for his many insightful comments, which helped me greatly to improve the original manuscript.

REFERENCES CITED

Alimarina, V. P., 1963, Some peculiarities in the development of planktonic foraminifers in connection with the zonal subdivision of the Lower Paleogene in the northern Caucasus: Academy Nauk SSSR Voprosy Mikropaleontologii, n. 7, p. 158–185 (in Russian).

Bandy, O. L., 1963, Cenozoic planktonic foraminiferal zonation and basinal development in Philippines: AAPG Bulletin, v. 47, no. 9, p. 1733–1745.

Bandy, O. L., 1964, Cenozoic planktonic foraminiferal zonation: Micropaleontology, v. 10, no. 1, p. 1–17.

Bandy, O. L., 1972, Neogene planktonic foraminiferal zones, California, and some geologic implications: Palaeogeography, Palaeoclimatology, Palaeoecology, v. 12, no. 1–2, p. 131–150.

Bandy, O. L., and J. C. Ingle, 1970, Neogene planktonic events and radiometric scale, California, in O. L. Bandy, ed., Radiometric dating and paleontologic zonation: Geological Society of America, Special Paper 124, p. 131–172.

Banner, F. T., and W. H. Blow, 1965, Progress in the planktonic foraminiferal biostratigraphy of the Neogene: Nature, v. 208, p. 1164–1166.

Beckmann, J. P., I. El-Heiny, M. T. Kerdany, R. Said, and C. Viotti, 1969, Standard planktonic zones in Egypt, in P. Brönnimann and H. H. Renz, eds., Proceedings of the First International Conference on Planktonic Microfossils, Geneva, 1967, v. 1: Leiden, Netherlands, E. J. Brill, p. 92–103.

Berggren, W. A., 1969, Biostratigraphy and planktonic foraminiferal zonation of the Tertiary System of the Sirte Basin of Libya, North Africa, in P. Brönnimann and H. H. Renz, eds., Proceedings of the First International Conference on Planktonic Microfossils, Geneva, 1967, v. 1: Leiden, Netherlands, E. J. Brill, p. 104–120.

Berggren, W. A., 1970, Multiple phylogenetic zonations, in A. Farinacci, ed., Proceedings of the Second Planktonic Conference, Rome, 1970: Roma Tecnoscienza, v. 1, p. 41–56.

Berggren, W. A., 1971, Tertiary boundaries and correlations, in B. M. Funnell and W. R. Riedel, eds., Micropaleontology of the oceans: New York, Cambridge University Press, p. 693–809.

Berggren, W. A., 1977, Late Neogene planktonic foraminiferal biostratigraphy of the Rio Grande Rise (South Atlantic): Marine Micropaleontology, v. 2, p. 265–313.

Berggren, W. A., and M. Amdurer, 1973, Late Paleogene (Oligocene) and Neogene planktonic foraminiferal biostratigraphy of the Atlantic Ocean (lat. 30°N to lat. 30°S): Rivista Italiana di Paleontologia, v. 79, no. 3, p. 337–392.

Berggren, W. A., and K. G. Miller, 1988, Paleogene tropical planktonic foraminiferal biostratigraphy and magnetobiochronology: Micropaleontology, v. 34, no. 4, p. 362–380.

Berggren, W. A., J. D. Phillips, A. Bertels, and D. Wall, 1967, Late Pliocene–Pleistocene stratigraphy in deep-sea cores from the south central North Atlantic: Nature, v. 216, p. 253–254.

Berggren, W. A., D. V. Kent, C. C. Swisher III, and M.-P. Aubry, 1995, A revised Cenozoic geochronology and chronostratigraphy, in W. A., Berggren, D. V. Kent, M.-P. Aubry, and J. Hardenbol, Geochronology, time scales and global stratigraphic correlation: Society for Sedimentary Geology (SEPM) Special Publication 54, p. 129–212.

Bizon, G. J., J. Bizon, and A. Durand, 1974, Remaniements de l'Oligocène lors de la transgression aquitanienne sur le plateau continental basque a proximité d'Escorne béou. Présence de *Globigerina ampliapertura* (Eocène supérieur–Oligocène inférieur) et de *Globigerinoides primordius* (Aquitanien): Revue Institut Français du Pétrole, v. 29, p. 135–153.

Blow, W. H., 1959, Age, correlation, and biostratigraphy of the Upper Tocuyo (San Lorenzo) and Pozón Formations, eastern Falcón, Venezuela: Bulletin of American Paleontology, v. 39, no. 178, p. 67–251.

Blow, W. H., 1969, Late middle Eocene to recent planktonic foraminiferal biostratigraphy, in P. Brönnimann and H. H. Renz, eds., Proceedings of the First International Conference on Planktonic Microfossils, Geneva, 1967, v. 1: Leiden, Netherlands, E. J. Brill, p. 199–422.

Blow, W. H., 1979, The Cainozoic Globigerinida (3 vols.): Leiden, Netherlands, E. J. Brill, 1413 p.

Blow, W. H., and F. T. Banner, 1962, The Mid-Tertiary (Upper Eocene to Aquitanian) Globigerinaceae, in F. E. Eames, F. T. Banner, W. H. Blow, and W. J. Clarke, eds., Fundamentals of mid-Tertiary stratigraphical correlation: Cambridge University Press, p. 61–151.

Blow, W. H., and F. T. Banner, 1966, The morphology, taxonomy and biostratigraphy of *Globorotalia barisanensis* LeRoy, *Globorotalia fohsi* Cushman and Ellisor, and related taxa: Micropaleontology, v. 12, no. 3, p. 286–302.

Blow, W. H., and T. Saito, 1968, The morphology and taxonomy of *Globigerina mexicana* Cushman, 1925: Micropaleontology, v. 14, no. 3, p. 357–360.

Bolli, H. M., 1957a, The genera *Globigerina* and *Globorotalia* in the Paleocene–lower Eocene Lizard Springs Formation of Trinidad, B.W.I.: Washington, D. C., Smithsonian Institution, United States National Museum Bulletin 215, p. 61–81.

Bolli, H. M., 1957b, Planktonic foraminifera from the Oligocene-Miocene Cipero and Lengua Formations of Trinidad, B.W.I.: Washington, D. C., Smithsonian Institution, United States National Museum Bulletin 215, p. 97–123.

Bolli, H. M., 1957c, Planktonic foraminifera from the Eocene Navet and San Fernando Formations of Trinidad, B.W.I.: Washington, D. C., Smithsonian Institution, United States National Museum Bulletin 215, p. 155–172.

Bolli, H. M., 1966, Zonation of Cretaceous to Pliocene marine sediments based on planktonic foraminifera: Boletín Informativo de la Asociación Venezolana de Geología, Minería y Petróleo, v. 9, no. 1, p. 2–32.

Bolli, H. M., 1967, The subspecies of *Globorotalia fohsi* Cushman and Ellisor and the zones based on them: Micropaleontology, v. 13, p. 502–512.

Bolli, H. M., 1970, The foraminifera of sites 23–31, leg 4, in R. G. Bader, R. D. Gerard, W. E. Benson, H. M. Bolli, W. W. Hay, W. T. Rothwell Jr., M. H. Ruef, W. R. Riedel, and F. S. Sayles, eds., Initial reports of the Deep-Sea Drilling Project: Washington, D. C., U. S. Government Printing Office, v. 4, p. 577– 643.

Bolli, H. M., 1972, The genus *Globigerinatheka* Brönnimann: Journal of Foraminiferal Research, v. 2, no. 3, p. 109–136.

Bolli, H. M., and P. J. Bermúdez, 1965, Zonation based on planktonic foraminifera of middle Miocene to Pliocene warm-water sediments: Boletín Informativo de la Asociación Venezolana de Geología, Minería y Petróleo, v. 8, no. 5, p. 119–149.

Bolli, H. M., and I. Premoli-Silva, 1973, Oligocene to recent planktonic foraminifera and stratigraphy of the leg 15 sites in the Caribbean Sea, *in* N. T. Edgar, A. G. Kaneps, and J. R. Herring, eds., Initial reports of the Deep-Sea, Drilling Project: Washington, D. C., U. S. Government Printing Office, v. 15, p. 475–497.

Bolli, H. M., and J. B. Saunders, 1985, Oligocene to Holocene low latitude planktic foraminifera, *in* H. M. Bolli, J. B. Saunders, and K. Perch-Nielsen, eds., Plankton stratigraphy: New York, Cambridge University Press, p. 155–262.

Brönnimann, P., and J. Resig, 1971, A Neogene globigerinacean biochronologic time-scale of the southwestern Pacific, *in* E. L. Winterer, W. R. Riedel, P. Brönnimann, E. L. Gealy, G. R. Heath, L. Kroenke, E. Martini, R. Moberly Jr., J. Resig, and T. Worsley, eds., Initial Reports of the Deep-Sea Drilling Project: Washington D. C., U. S. Government Printing Office, v. 7, pt. 2, p. 1235–1469.

Caro, Y., H. Luterbacher, K. Perch-Nielsen, I. Premoli-Silva, W. R. Riedel, and A. Sanfilippo, 1975, Zonations à l'aide de microfossiles pélagiques du Paléocène Supérieur et de l'Eocène Inférieur: Bulletin de la Société Géologique de France, v. 17, no. 2, p. 125–147.

Cita, M. B., 1973, Pliocene biostratigraphy and chronostratigraphy, *in* W. B. F. Ryan, K. J. Hsü, M. B. Cita, P. Dumitrica, J. M. Lort, W. Maync, W. D. Nesteroff, G. Pautot, H. Stradner, and F. C. Wezel, eds., Initial reports of the Deep-Sea Drilling Project: Washington D. C., U. S. Government Printing Office, v. 13, p. 1343–1379.

Cita, M. B., 1975, The Miocene/Pliocene boundary: History and definition, *in* T. Saito and L. H. Burckle, eds., Late Neogene epoch boundaries: New York, American Museum of Natural History, Micropaleontology Press, Special Publication no. 1, p. 1–30.

Cita, M. B., 1976, Planktonic foraminiferal biostratigraphy of the Mediterranean Neogene, *in* Y. Takayanagi and T. Saito, eds., Progress in micropaleontology: New York, American Museum of Natural History, Micropaleontology Press, Special Publication, p. 47–68.

Cita, M. B., and W. H. Blow, 1969, The biostratigraphy of the Langhian, Serravallian and Tortonian stages in the type sections in Italy: Rivista Italiana di Paleontologia e Stratigrafia, v. 75, no. 3, p. 549–603.

Cordey, W. G., 1968, Morphology and phylogeny of *Orbulinoides beckmanni* (Saito, 1962): Paleontology, v. 11, no. 3, p. 371–375.

Cushman, J. A., and H. H. Renz, 1947, The foraminiferal fauna of the Oligocene Ste. Croix Formation of Trinidad, B.W.I.: Sharon, Massachusetts, Cushman Laboratory for Foraminiferal Research, Special Publication 22, 46 p.

Cushman, J. A., and R. M. Stainforth, 1945, The foraminifera of the Cipero Marl Formation of Trinidad, B.W.I.: Sharon, Massachusetts, Cushman Laboratory for Foraminiferal Research, Special Publication 14, 75 p.

Huang, T., 1976, Some significant biostratigraphic events in the Neogene formations of Taiwan, *in* Y. Takayanagi and T. Saito, eds., Progress in micropaleontology: New York, American Museum of Natural History, Micropaleontology Press, Special Publication, p. 103–109.

Lamb, J. L., and J. H. Beard, 1972, Late Neogene planktonic foraminifers in the Caribbean, Gulf of Mexico, and Italian stratotypes: University of Kansas Paleontological Contributions, Article 57 (Protozoa 8), 67 p.

Lamb, J. L., and R. M. Stainforth, 1976, Unreliability of *Globigerinoides* datum: AAPG Bulletin, v. 60, no. 9, p. 1564–1569.

LeRoy, L. W., 1948, The foraminifer *Orbulina universa* d'Orbigny, a suggested middle Tertiary time indicator: Journal of Paleontology, v. 22, no. 4, p. 500–508.

Meneses de Gyves, J., 1980, Geología de la Sonda de Campeche, Boletín de la Asociación Mexicana de Geólogos Petroleros, v. 32, no. 1, p. 1–26.

Miller, F. X., 1977, The graphic correlation method in biostratigraphy, *in* E. G. Kauffman and J. E. Hazel, eds., Concepts and methods in biostratigraphy: Stroudsburg, Pennsylvania, Dowden, Hutchinson and Ross, p. 165–186.

Natori, H., 1976, Planktonic foraminiferal biostratigraphy and datum planes in the Late Cenozoic sedimentary sequence in Okinawa-jima, Japan, *in* Y. Takayanagi and T. Saito, eds., Progress in micropaleontology: New York, American Museum of Natural History, Micropaleontology Press, Special Publication, p. 214–243.

Neal, J. E., J. A. Stein, and J. H. Gamber, 1995, Integration of the graphic correlation methodology in a sequence stratigraphic study: Examples from North Sea paleogene sections, *in* K. O. Mann and H. R. Lane, eds., Graphic correlation: Society for Sedimentary Geology (SEPM) Special Publication 53, p. 95–113.

Parker, F. L., 1973, Late Cenozoic biostratigraphy (planktonic foraminifera) of tropical Atlantic deep-sea sections: Revista Española de Micropaleontología, v. 5, no. 2, p. 253–289.

Postuma, J. A., 1971, Manual of planktonic foraminifera: Amsterdam, Elsevier Science Publishing Company, 420 p.

Premoli-Silva, I., and H. M. Bolli, 1973, Late Cretaceous to Eocene planktonic foraminifera and stratigraphy of leg 15 sites in the Caribbean Sea, *in* N. T. Edgar, A. G. Kaneps, and J. R. Herring, eds., Initial Reports of the Deep-Sea Drilling Project: Washington D. C., U. S. Government Printing Office, v. 15, p. 499–547.

Proto Decima, F., and H. M. Bolli, 1970, Evolution and variability of *Orbulinoides beckmanni* (Saito): Eclogae Geologicae Helvetiae, v. 63, n. 3, p. 883–905.

Robles, N. J., L. M. Pimienta, L. L. Villanueva González, and B. D. Jiménez, 1998, Structural-seismic and biostratigraphic analysis and its importance in exploration of hydrocarbons in the Sound of Campeche, Mexico, *in* J. F. Longoria and M. A. Gamper eds., International symposium on foraminifera, Forams '98, July 5, 1998 (abs): Sociedad Mexicana de Paleontología, A. C, Special Publication, p. 90–91.

Salazar M., G., 1991, Bioestratigrafía del Neógeno en la planicie costera y plataforma continental del sureste de México (Area Marina de Campeche) (Proyecto CAO-5509): Instituto Mexicano del Petróleo, internal report, p. 150–216.

Salazar M., G., 1994, Banco de Datos Paleontológicos del Terciario para la Sonda de Campeche: Foraminíferos (Proyecto CAC-0504): Instituto Mexicano del Petróleo, internal report, 54 p.

Salazar M., G., 1998a, Tertiary zonation based on planktonic foraminifera, from the Marine Region of Campeche, Mexico (poster presentation), *in* J. F. Longoria and M. A. Gamper, eds., International symposium on foraminifera, Forams '98, July 5, 1998 (abs.): Sociedad Mexicana de Paleontología, A. C, Special Publication, p. 94.

Salazar M., G., 1998b, Manual de foraminíferos planctónicos del Neógeno y su aplicación en la Industria del petróleo, *in* A. S. A. Alaniz, L. Ferrari, S. A. F. Nieto, eds., Primera reunion nacional de Ciencias de la Tierra, 21–25 de septiembre de 1998: Libro de Resúmenes, México, D. F., p. 90.

Salazar M., G., and J. Alvarado, 1997, Manual de Foraminíferos Planctónicos del Neógeno (Pleistoceno–Mioceno Superior) (Proyecto CAB-0506): Instituto Mexicano del Petróleo, internal report, 191 p.

Salazar M., G., and J. Alvarado, 1998, Manual de Foraminíferos Planctónicos del Paleógeno (Oligoceno–Eoceno Medio) (Proyecto CAB-0506): Instituto Mexicano del Petróleo, internal report, 160 p.

Salazar, M., G., and J. A. Gómez, 1990, Estudio bioestratigráfico del

Terciario en el área Kambul-Chakay, Campeche (Proyecto C-5023): Instituto Mexicano del Petróleo, internal report, 118 p.

Santiago, A. J., J. Carrillo, and B. Martell, 1984, Geología Petrolera de México, *in* D. Marmisosolle-Daguerre et al., eds., Evaluación de Formaciones en México: México and Central America, chapter 1: Paris, France, Schlumberger, p. 1–36.

Shaw, A. B., 1964, Time in stratigraphy: New York, McGraw-Hill, 365 p.

Stainforth, R. M., and J. L. Lamb, 1981, An evaluation of planktonic foraminiferal zonation of the Oligocene: University of Kansas Paleontological Contributions, Lawrence, Paper 104, p. 1–34.

Stainforth, R. M., J. L. Lamb, H. P. Luterbacher, J. H. Beard, and R. M.

Jeffords, 1975, Cenozoic planktonic foraminiferal zonation and characteristics of index forms: University of Kansas Paleontological Contributions, Lawrence, Article 62, p. 1–425.

Toumarkine, M., 1981, Discussion de la validité de l'espèce *Hantkenina aragonensis* Nuttall, 1930, Description de *Hantkenina nuttalli,* n. sp.: Cahiers de Micropaléontologie, Livre Jubilaire en l'honneur de Madame Le Calvez, fasc. 4, p. 109–119.

Toumarkine, M., and H. P. Luterbacher, 1985, Paleocene and Eocene planktic foraminifera, *in* H. M. Bolli, J. B. Saunders and K. Perch-Nielsen, eds., Plankton stratigraphy: New York, Cambridge University Press, p. 87–154.

Cantú-Chapa, A., 2001, Paleocanyons in the subsurface of eastern Mexico: Facts and uncertainties, *in* C. Bartolini, R. T. Buffler, and A. Cantú-Chapa, eds., The western Gulf of Mexico Basin: Tectonics, sedimentary basins, and petroleum systems: AAPG Memoir 75, p. 421–430.

19

Paleocanyons in the Subsurface of Eastern Mexico: Facts and Uncertainties

Abelardo Cantú-Chapa

Instituto Politécnico Nacional, Mexico City, Mexico

ABSTRACT

The misuse of stratigraphic information from oil wells in eastern Mexico generated the concept of the Chicontepec paleocanyon (Busch and Govela, 1975, 1978). Its original definition, a paleogeographic entity parallel to the Gulf coastline in the subsurface of the Poza Rica district, Veracruz, was based on the use of petrophysical logs for correlation of stratigraphic sections. The methods used to define the Chicontepec paleocanyon are questioned here, and its existence at the original proposed location is rejected on the basis of the following criteria: (1) the unconformity surface does not reveal the absence of lithologic formations; (2) the correlated stratigraphic sections lack accurate chronostratigraphic control; (3) Paleocene-Eocene lateral facies changes are erroneously considered unconformities; (4) the so-called reference horizon, Horizon C (top lower Eocene), for the stratigraphic correlation is not documented with lithostratigraphic and biostratigraphic data; and (5) the marks on check-shot points used to determine the contact between the Paleocene–lower Eocene formations are not accurate.

Two other areas, however, do contain identifiable paleocanyons: (1) The San Andrés paleocanyon is a paleogeographic province that lies perpendicular to the Gulf coastline, southeast Poza Rica district, Veracruz, eastern Mexico. This paleocanyon is characterized by the absence of upper Tithonian to Maestrichtian strata, and lower Tithonian rocks are unconformably overlain by Paleocene strata. (2) The Bejuco paleocanyon is another such feature that lies perpendicular to the Gulf coastline, south of Tampico, eastern Mexico. This paleocanyon is characterized by the absence of Cretaceous to Paleocene rocks, and Lower Cretaceous rocks are unconformably overlain by Eocene strata. The San Andrés and Bejuco paleocanyons are contemporaneous, with similar morphology and depth. In these two paleocanyons, hydrocarbon reservoirs are confined to the Upper Jurassic and Lower Cretaceous rocks.

The structural nature of the crystalline basement influenced the configuration of the paleocanyons. In fact, successive crustal movements of igneous and metamorphic blocks seemed to have impeded the consolidation of the Jurassic and Cretaceous sedimentary columns in the San Andrés paleocanyon, southeast Poza Rica, and in the Bejuco paleocanyon, south of Tampico, eastern Mexico.

INTRODUCTION

The marine unconformities and paleocanyons in eastern Mexico have been determined through stratigraphic studies of the Upper Jurassic–lower Eocene subsurface stratigraphy (Cantú-Chapa, 1985, 1987, 1992). The Chicontepec paleocanyon, in particular, has been described as a northwest-southeast-trending paleogeographic feature parallel to the present coastline and with a surface area of 3300 km^2 (Busch and Govela, 1975, 1978; Busch, 1992) (Figure 1b).

The original definition of the Chicontepec paleocanyon was based on the correlation of stratigraphic sections constructed with electric well logs. However, the paleogeographic and stratigraphic analysis that gave rise to this model (Busch and Govela, 1975, 1978; Busch, 1992) contains the following stratigraphic errors: (1) the so-called Horizon C of the top of the lower Eocene is not representative in the logs they studied, (2) the supposed unconformity surface does not reveal the absence of lithologic formations, and (3) Paleocene-Eocene lateral facies changes are erroneously considered unconformities. Unfortunately, these interpretations have been used as a reference to explain the petroleum potential of the Poza Rica area (Treviño-Rodríguez, 1995; Yañez-Mondragón, 1998).

USE AND MISUSE OF STRATIGRAPHIC METHODS TO CHARACTERIZE A PALEOCANYON

The misuse of stratigraphic information from oil wells in eastern Mexico generated the concept of the Chicontepec paleocanyon (Figure 1). The concept of a paleocanyon was questioned by Cantú-Chapa (1985), who noted: (1) the nonexistence of an unconformity between lower-Paleocene–middle Paleocene formations; (2) the marks on check-shot points used to determine the contact between the Paleocene–lower Eocene formations are not systematic and accurate; (3) the geographic position of the so-called Chicontepec paleocanyon has important petroleum implications because of lateral facies changes and structures that control the entrapment of hydrocarbons in the Paleocene–lower Eocene formations; and (4) there are no stratigraphic elements that support the existence of the proposed paleogeographic feature (Busch and Govela, 1975, 1978; Carrillo-Bravo, 1980a, 1980b; Cuevas, 1980b).

Despite the lack of evidence to support the existence of such a paleocanyon, recent literature still suggests its presence (Busch, 1992; Treviño-Rodríguez, 1995; Yañez-Mondragón, 1998). In the following section, I discuss and compare my own data with that of the aforementioned authors.

1) The Pardo-1 well in the Poza Rica oil district cut sediments of the Mendez Formation (Campanian-Maestrichtian) and lower Cenozoic (Figure 2). Busch and Govela (1975, Figure 2) indicated the existence of an unconformity between the Velasco (lower Paleocene) and Chicontepec (middle Paleocene) Formations. However, this interpretation is incorrect because the sedimentary sequence in this well is normal. The electric log, furthermore, does not show significant lithologic changes at the level of the supposed unconformity; no lithologic formations are missing. The Pardo-1 well was basically correlated using the so-called Horizon C, which could represent the top of the lower Eocene (Busch and Govela, 1975, 1978; Busch, 1992). This horizon was intended to configure and correlate wells along the Chicontepec paleocanyon, but the definition of Horizon C is not documented by lithostratigraphic and biostratigraphic data. It does not possess a particular electric signature, and it cannot be differentiated from the overlying horizon, which is not included in the sections studied by Busch and Govela (1975, 1978), Cuevas (1980a), and Treviño-Rodríguez (1995). Despite these inconsistencies, the supposed Horizon C frequently is used to correlate in the literature of the Poza Rica area (Cuevas, 1980a).

2) The same problem is observed in a structural section across the Chicontepec paleocanyon where three wells located in the southwestern edge of the section show wrong stratigraphic relations in the Paleocene-Eocene formations. The middle to upper Paleocene Chicontepec Formation is shown unconformably underlying Eocene strata (Treviño-Rodríguez, 1995). In Escobal-103 and Antares-1 wells, the unconformity cannot be docu-

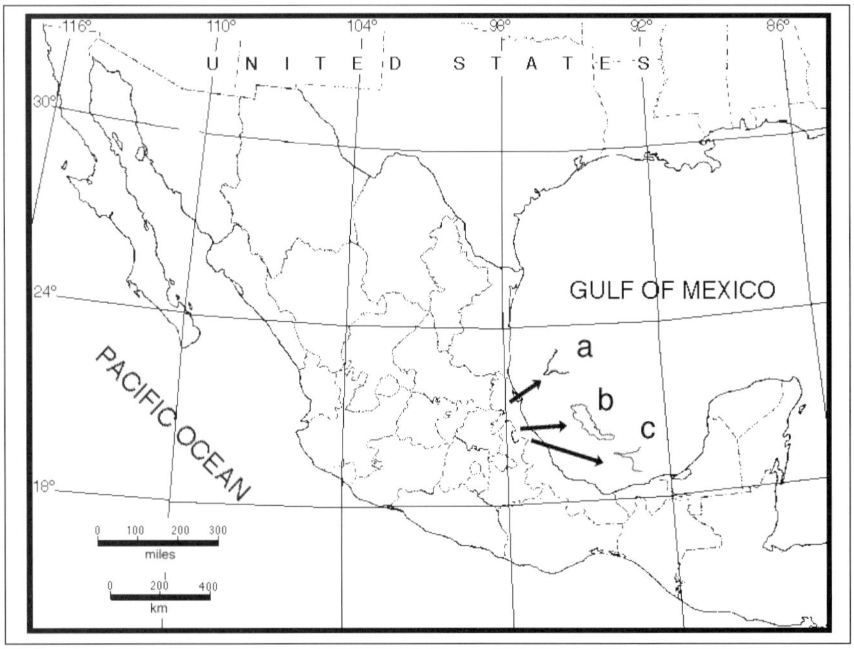

FIGURE 1. Location of the (a) Bejuco, (b) Chicontepec, and (c) San Andrés paleocanyons, east-central México (Busch, 1992; Cantú-Chapa, 1987; Carrillo-Bravo, 1980b; Cuevas, 1980b).

mented; here the entire sedimentary sequence is continuous and complete (Figure 3).

3) The configuration of stratigraphic sections based on electric logs to define the Chicontepec paleocanyon is not appropriate, as shown in Figure 4. Horizon C, which is the top of the lower Eocene after Busch and Govela (1975), is used by them as a marker horizon in the section I-I' (Busch, 1992, Figure 2; Busch and Govela, 1975). This horizon is shown in the Amixtlan-3A, Coyula-1, and Agua Fría-1 wells in the southwestern edge of the section (Figure 4). In this particular section, Horizon C is not characteristic in the wells studied by Busch and Govela (1975). The Agua Fría-1 well cut through the complete Upper Jurassic–Eocene sequences.

In the Amixtlán-1A well, the marker is 700 m above ground level, and it should be assumed that the deposition of the lower Eocene has not yet occurred in the area where this well was drilled (Busch and Govela, 1975, 1978). The Paleocene is characterized by the Velasco and Chicontepec Formations, but no unconformity exists between these two formations, as suggested by Busch and Govela (1975). On the contrary, the sequence is normal in that part of east-central Mexico, as was shown by these same authors (Figure 4).

4) The stratigraphic basis for delimiting the geographic position of the Chicontepec paleocanyon was the possible change of facies contact between the Velasco and Chicontepec Formations. However, sedimentary facies changes were never explained by Busch and Govela (1975) nor shown by the electric logs from the sections they provided. These authors did not explain whether the paleocanyon is an erosional unconformity or was caused simply by lateral facies changes that occurred in the Paleocene. To propose the Chicontepec paleocanyon as a paleogeographic feature, Busch and Govela (1975), Carrillo-Bravo (1980b), and Cuevas (1980b) used the isopach map of the Chicontepec Formation and interpreted it as the fill of an eroded area (Figure 5).

I, on the other hand, projected four traverses across the same isopach map that clearly show the abrupt change in thicknesses of the Chicontepec Formation in distances that do not exceed 5 km (Figure 5b, c). The thicknesses are constant in the western and southwestern edges of the paleocanyon. According to these sections, it may be inferred that the thickness of the Chicontepec Formation would range from 0 to 600 m in less than 5 km, which is indeed feasible (Figure 5). In fact, outside the supposed paleocanyon, the Chicontepec Formation is cut by the La Flor-2, Cañas-101, Brinco-1, Totomoxtle-1, and Sultepec-1 wells (Figure 5).

Thus, the delineation of the paleocanyon is dubious, because it is based solely on the isopach map of the Chicontepec Formation that was configured arbitrarily by Busch and Govela (1975). In addition, there are insufficient wells to define the north-central flank of the paleocanyon or to support the lobe-shaped body of sediment they suggested.

These four stratigraphic misinterpretations have deeply affected the concept of petroleum exploration in this area, par-

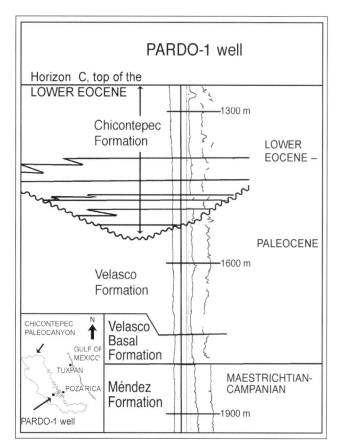

FIGURE 2. A typical misuse of the unconformity concept in a normal early Cenozoic sequence in Pardo-1 well, Poza Rica district (Busch and Govela, 1975a).

ticularly in the lower Cenozoic section of central Mexico. These erroneous stratigraphic concepts and interpretations cannot be used to explain the presence of hydrocarbons in this region. The following analysis does not support the existence of the Chicontepec paleocanyon in this area, as suggested by several authors (Busch, 1992; Busch and Govela, 1975, 1978; Carrillo-Bravo, 1980b; Cuevas, 1980b; Treviño-Rodríguez, 1995; Yañez-Mondragón, 1998).

ANALYSIS OF SEDIMENTARY SEQUENCES OF SEVERAL WELLS IN THE CHICONTEPEC PALEOCANYON: REDEFINITION AND RELOCATION

Wells in the northern part of the Chicontepec paleocanyon show a normal biostratigraphic sequence (based on planktonic foraminifera) that does not support an unconformity at the Maestrichtian–lower Eocene level (Cantú-Chapa, 1985, 1988) (Figure 6).

Figure 7 shows the Chicontepec paleocanyon after Busch and Govela (1975), Carrillo-Bravo (1980b), and Cuevas (1980b). No evidence of erosion is found in five wells drilled along the axis of the presumed paleocanyon. Canoas-101, Coyotes-3, Zontla-1, Agua Fría-1, and Tajín-101 wells cut a nor-

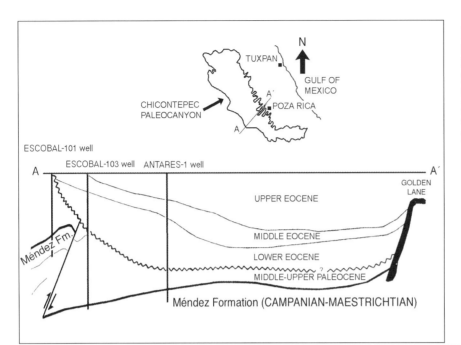

FIGURE 3. A supposed unconformity at the upper Paleocene–lower Eocene level in wells located perpendicular to the central section of the so-called Chicontepec paleocanyon, south Poza Rica district; without scale, after Treviño-Rodríguez (1995). In this case, the sequence is normal.

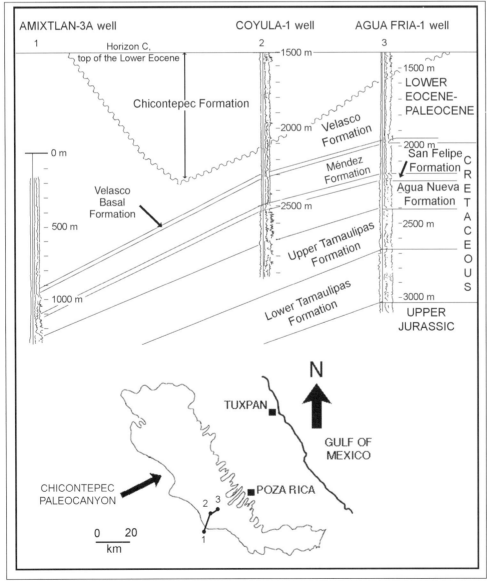

FIGURE 4. Stratigraphic cross section showing the so-called lower Eocene Horizon C above the surface in Amixtlán-1A well. According to Busch (1992) and Busch and Govela (1975), the deposition of the lower Eocene has not yet taken place in that well. A supposed unconformity at the Paleocene level between Velasco and Chicontepec Formations was proposed by Busch and Govela (1975) to justify the Chicontepec paleocanyon concept.

mal sequence that ranges in age from Late Jurassic to Cenozoic, as well as even older rocks that form the basement in this area. Mesa Chica-10 well, located in the southeastern edge and outside the Chicontepec paleocanyon, shows the absence of most

of the Cretaceous sequences. Thus, if a paleocanyon exists, it should be southeast of Poza Rica, where it must be redefined and its location constrained (Figure 7).

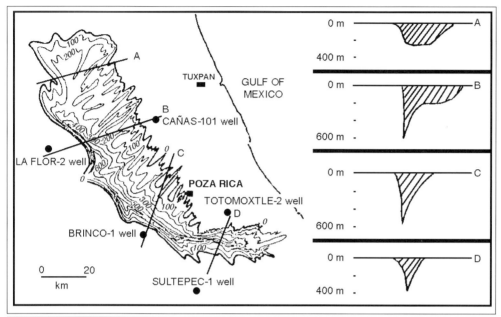

FIGURE 5. The Chicontepec paleocanyon was configured with isopachs of the Chicontepec Formation (Busch, 1992; Busch and Govela, 1975, 1978; Carrillo-Bravo, 1980a, b; Cuevas, 1980b). The sections projected here (B and C) depict the supposed abrupt change in thicknesses of the Chicontepec Formation, in short distances. La Flor-2, Cañas-101, Brinco-1, Sultpec-1, and Totomoxtle-2 wells, located outside the presumed paleocanyon, also cut through the Chicontepec Formation.

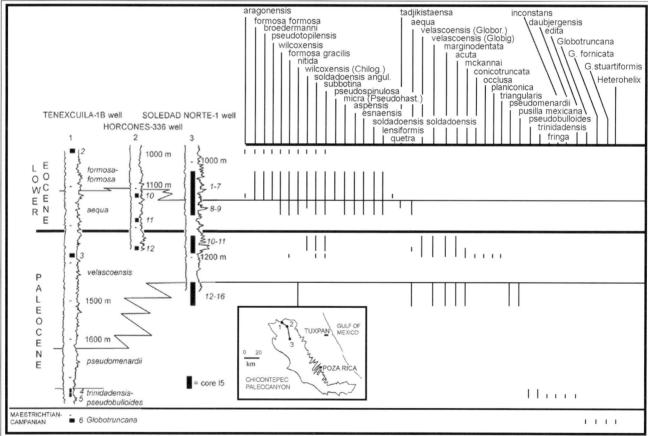

FIGURE 6. Stratigraphic cross section in the northern part of the so-called Chicontepec paleocanyon established with radioactivity log wells. Tenexcuila-1B well shows a normal biostratigraphic sequence, based on planktonic foraminifera at the Maestrichtian–lower Eocene (Cantú-Chapa, 1988). Paleontological determination by Lourdes Omaña; stratigraphic configuration, this work.

SAN ANDRÉS
PALEOCANYON

The Mesa Chica-10 well is in a region that underwent submarine erosion, as supported by the unconformable contact between the Lower Cretaceous Lower Tamaulipas Formation and the Coniacian San Felipe Formation (Figure 7). The stratigraphic analysis of wells drilled near Mesa Chica-10 well, southeast of Poza Rica, clearly show the hiatus known from another paleocanyon, the San Andrés paleocanyon (Cantú-Chapa, 1985, 1987).

This paleogeographic province formed after submarine erosion at different geologic times in the Upper Jurassic and Cretaceous (Figure 8). In a restricted area, there are no rocks of late Tithonian to Maestrichtian age, and the erosion becomes restricted in time, from the Early Cretaceous to the Late Cretaceous (Maestrichtian) (Figure 8a, b, d). The entire area is covered by Paleocene sedimentary rocks, indicating that submarine erosion is limited to the pre-Paleocene. Geographically, the area of erosion is relatively small, approximately 70 km in an east-west direction, but its maximum width is toward the east (Figure 8a). Successive crustal movements of igneous and metamorphic blocks seem to have impeded the consolidation of the Upper Jurassic and Cretaceous sedimentary columns in the San Andrés paleocanyon, southeastern Poza Rica (Figure 8c).

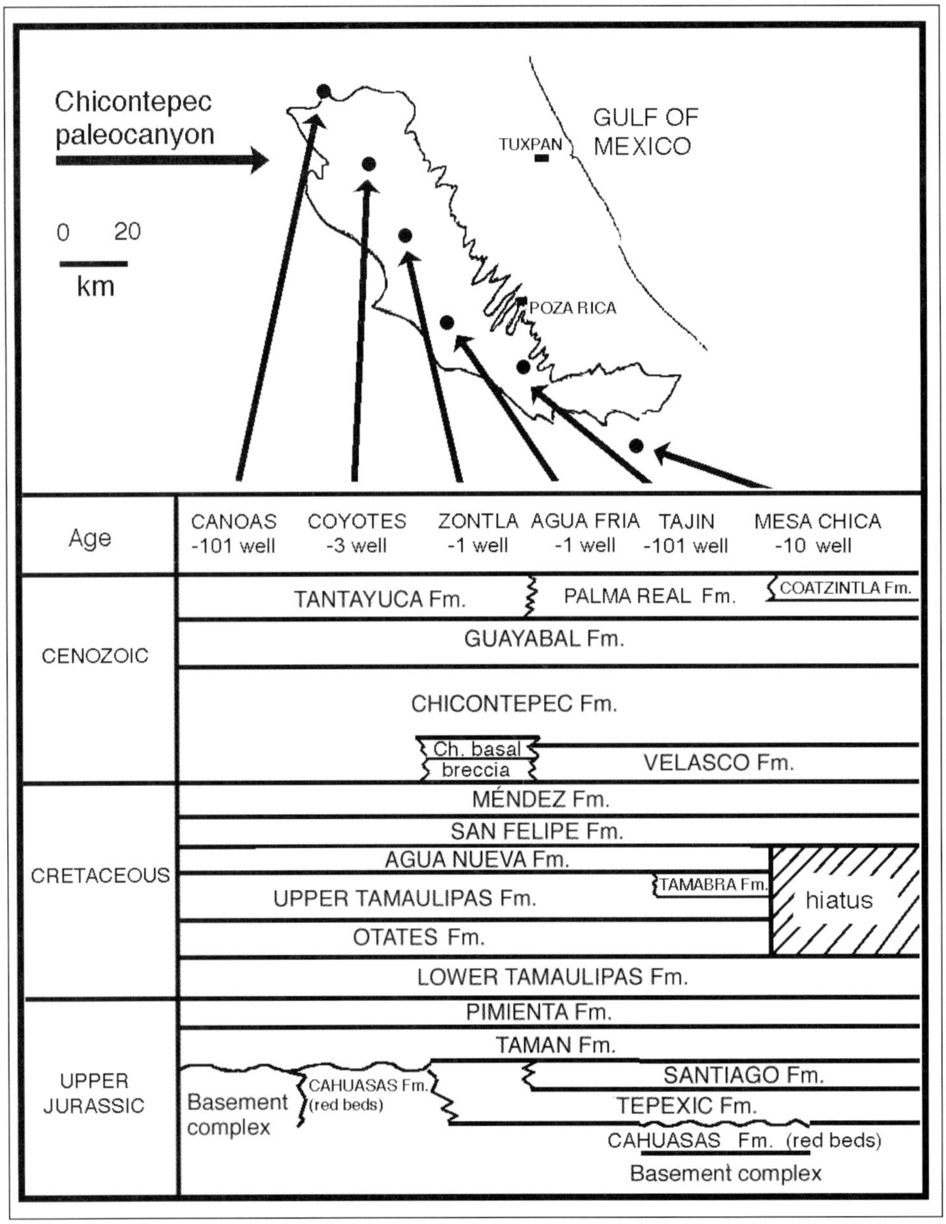

The unconformable contact between two formations is at depths that range from 1800 m to 3600 m. The greatest depth occurs northeast of the study area and corresponds to a syncline structure (Figure 8c). This paleogeographic entity, originally named the San Andrés paleocanyon after the oil field of the same name (Cantú-Chapa, 1987), formed as the result of submarine erosion; this is supported by the absence of the upper Tithonian–Maestrichtian strata (Figure 8b). The area of erosion is perpendicular to the Sierra Madre Oriental and to the present coastline, not parallel to them, as proposed for the Chicontepec paleocanyon (Busch and Govela, 1975; Carrillo-Bravo, 1980b; Cuevas, 1980b). The unconformity at different stratigraphic levels is observed in the radioactivity logs of the San Andrés-243 well and the Tlaloc-1 well (Figure 8b, d). The maximum stage of erosion in the San Andrés paleocanyon is present in its north-central part, in the San Andrés field (Figure 8a). The paleocanyon is evidenced by the contact between the lower Tithonian San Andrés Member and the overlying Paleocene Chicontepec Formation.

Figure 9 also depicts the structural controls in the San Andrés and Hallazgo oil fields, in the southeastern area of Poza Rica. The Chicontepec Formation is the

FIGURE 7. The Chicontepec paleocanyon according to Busch and Govela (1978), Carrillo-Bravo (1980a, b), and Cuevas (1980b). Five wells along the axis of the paleocanyon show a normal sequence, from Upper Jurassic to Eocene. Mesa Chica-10 well is the only one that shows a hiatus.

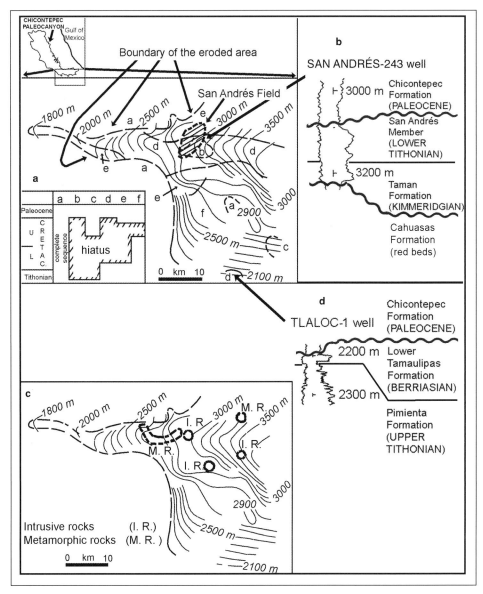

FIGURE 8. (a) Structural configuration of the surface of erosion in the San Andrés paleocanyon; insert at the bottom left depicts the hiatus by areas along the paleocanyon. (b) The San Andrés-243 well and (d) the Tlaloc-1 well show the unconformities at different stratigraphic levels, southeast of Poza Rica. (c) Distribution of metamorphic and intrusive rocks in the paleocanyon.

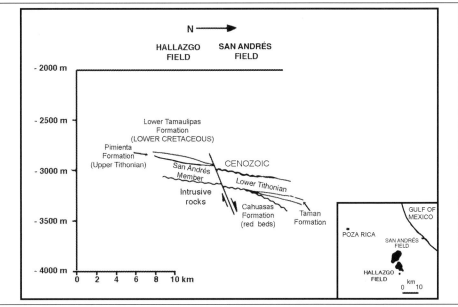

FIGURE 9. Structural cross section between Hallazgo field and San Andrés field, southeastern Poza Rica. It shows the unconformity between Cenozoic and lower Tithonian rocks in the San Andrés field.

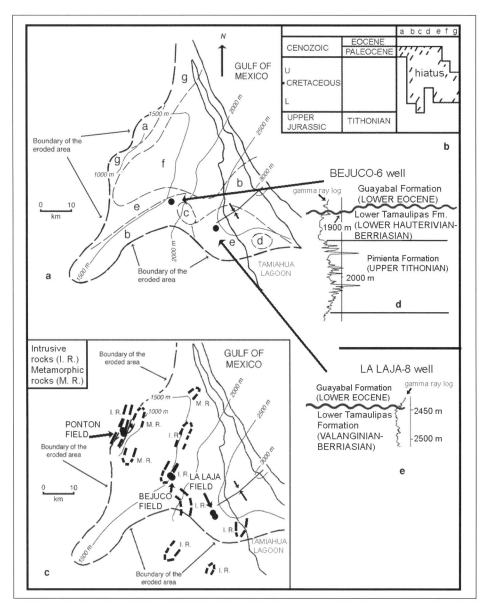

FIGURE 10. (a) Structural configuration of the surface of erosion in the Bejuco paleocanyon. **(b)** Stratigraphic hiatus synthesis by areas along the paleocanyon. **(c)** Distribution of metamorphic and intrusive rocks and Upper Jurassic–Lower Cretaceous oil fields in the paleocanyon. **(d)** The Bejuco-6 well and **(e)** the La Laja-8 well show the unconformities at different stratigraphic levels; southeast of Tampico, east-central Mexico (Cantú-Chapa, 2001).

seal rock for the oil fields in this area. It shows the unconformity between Lower Cenozoic (Chicontepec Formation) and lower Tithonian rocks (San Andrés Member) in the San Andrés field (Cantú-Chapa, 1999, 2001).

BEJUCO PALEOCANYON

Stratigraphic analysis of the Bejuco-6 well resulted in the recognition and location of an unconformity whose chronostratigraphic distribution spans the Neocomian to Eocene. Later on, the unconformities from 65 oil wells were used to configure the structural map of this paleocanyon, called the Bejuco paleocanyon, in eastern Mexico (Figure 10a). The central part of the unconformity has a wider geographic distribution, and it is in this area that the Cretaceous-Paleocene section is missing (Figure 10a, b). The Bejuco paleocanyon represents the eroded area, and the Bejuco field and La Laja field include the area of maximum erosion (Cantú-Chapa, 1987, 2001) (Figure 10c).

THE NEOCOMIAN-EOCENE UNCONFORMITY IN THE BEJUCO PALEOCANYON: ITS COMPARISON WITH THE SAN ANDRÉS PALEOCANYON

The San Andrés and Bejuco paleocanyons are stratigraphically and chronostratigraphically similar. Both are located along the Gulf of Mexico coastal region and are east-west-oriented and thus perpendicular to the Sierra Madre Oriental and the actual Gulf coastline (Figures 8 and 10). Both have a delta-like shape and are approximately 70 km long with a narrow half area (5–10 km) and a wide area (40–50 km) in the east. The depths of the erosion surfaces vary from 1300 m to 3000 m in the Bejuco paleocanyon, and from 1800 m to 3600 m in the San Andrés paleocanyon. The eroded section in the Bejuco paleocanyon comprises the upper Hauterivian–Maestrichtian; in the San Andrés paleocanyon, it includes the Cretaceous-Paleocene section (Figures 8, 10, and 11).

Reservoir rocks in both paleocanyons are early Tithonian in

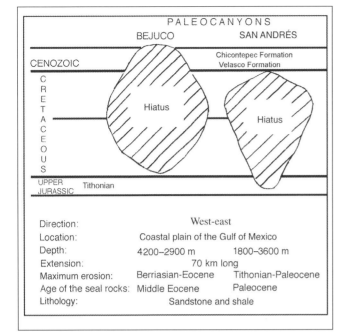

FIGURE 11. Comparative stratigraphic synthesis of the Bejuco and San Andrés paleocanyons in east-central Mexico.

age (San Andrés Member) and also Early Cretaceous in the Bejuco paleocanyon (Cantú-Chapa, 1987). Basin fills of middle Eocene age in the Bejuco paleocanyon and of Paleocene age in the San Andrés paleocanyon contain locally commercial hydrocarbons. Seal rocks in the Bejuco paleocanyon consist of middle Eocene shale (Guayabal Formation); in the San Andrés paleocanyon, they consist of an alternating sequence of lower Paleocene–Eocene shale and sandstone (Chicontepec Formation) (Figures 8, 10, and 11).

Metamorphic and igneous rocks are distributed in the central portion and in the margins of the Bejuco paleocanyon, whereas in the San Andrés paleocanyon, they are confined to the north-central part. Metamorphic and igneous rocks in both paleocanyons are located near the oil fields (Figures 8 and 10). No data have yet been obtained from the deepest central parts of the canyons, but most of the oil wells have bottomed in red beds of the Middle to Upper Jurassic Cahuasas Formation (Figures 8, 9, and 10).

CONCLUSIONS

The inappropriate and arbitrary application of stratigraphic analyses led several authors to propose the Chicontepec paleocanyon in eastern Mexico. However, a new analysis and interpretation of subsurface, stratigraphic, biostratigraphic, and lithostratigraphic data from wells along the axis of the presumed Chicontepec paleocanyon do not support the existence of such a paleogeographic feature. Rather, two paleocanyons can be recognized in different areas, here named the San Andrés and Bejuco paleocanyons.

The stratigraphic basis for the recognition of paleocanyons

along the Gulf coastal region of eastern Mexico is the absence, caused by submarine erosion, of thick Cretaceous sedimentary sections in the subsurface. The unconformable contact between marine sequences of different ages establishes a surface of erosion whose structural configuration characterizes a paleocanyon.

These paleogeographic structures are not parallel to the Gulf of Mexico coastline or to the Sierra Madre Oriental, as proposed for the supposed Chicontepec paleocanyon. The paleocanyons are perpendicular to both the Gulf of Mexico coastline and the Sierra Madre Oriental, and they display an open synclinal structure where maximum depths occur toward the east, in the Gulf of Mexico. Erosion cut down carbonate rocks with high porosities and permeabilities. These carbonates, which are important hydrocarbons reservoirs in this area, are part of the Upper Jurassic San Andrés Member and the Lower Cretaceous Lower Tamaulipas Formation. Cenozoic sedimentary rocks that overlie the surface of erosion constitute the seal rock; however, they act locally as reservoirs.

The Bejuco and San Andrés paleocanyons in east-central Mexico are similar in structure, size, age, and depth and also share the same relationship to Upper Jurassic and Lower Cretaceous hydrocarbon deposits.

ACKNOWLEDGMENTS

I thank Claudio Bartolini for the translation of the manuscript to English. I express my gratitude to Katherine A. Giles and James L. Wilson for providing constructive reviews of the manuscript.

REFERENCES CITED

Busch, D. A., 1992, Chicontepec field—Mexico, Tampico-Misantla Basin, *in* N. H. Foster and E. A. Beaumont, compilers, Stratigraphic traps III: AAPG Treatise of petroleum geology, Atlas of oil and gas fields, p. 113–128.

Busch, D. A., and S. A. Govela, 1975, Estudio estratigráfico-estructural de las turbiditas de la Formación Chicontepec en la porción sureste de la Cuenca Tampico–Misantla: Petróleos Mexicanos, Superintendencia General de Exploración, Distrito de Poza Rica, 28 p., internal report.

Busch, D. A., and S. A. Govela, 1978, Stratigraphy and structure of Chicontepec turbidites, Southeastern Tampico–Misantla Basin, Mexico: AAPG Bulletin, v. 62, no. 2, p. 235–246.

Cantú-Chapa, A., 1985, Is there a Chicontepec Paleocanyon in the Paleogene of Eastern Mexico?: Journal of Petroleum Geology, v. 8, no. 4, p. 423–434.

Cantú-Chapa, A., 1987, The Bejuco Paleocanyon (Cretaceous-Paleocene) in the Tampico District, Mexico: Journal of Petroleum Geology, v. 10, no. 2, p. 207–218.

Cantú-Chapa, A., 1988, ¿Existe el Paleocañón de Chicontepec en el Paleógeno del Este de México?: Publicaciones Especiales no. 1, Sociedad Mexicana de Paleontología, 19 p.

Cantú-Chapa, A., 1992, The Jurassic Huasteca Series in the subsurface of Poza Rica, Eastern Mexico: Journal of Petroleum Geology, v. 15, no. 3, p. 259–282.

Cantú-Chapa, A., 1999, Two unconformable stratigraphic relationships between Upper Jurassic redbeds and marine sequences in northeastern and eastern Mexico subsurface. In Bartolini, C., J. Wilson, and T. Lawton, eds., Mesozoic sedimentary and tectonic history of north-central Mexico: Boulder, Colorado, Geological Society of America Special Paper 340, p. 1–5.

Cantú-Chapa, A., 2001, The Taraises Formation (Berriasian-Valanginian) in northeastern Mexico: Subsurface and outcrop studies: This volume, chapter 15.

Carrillo-Bravo, J., 1980a, Paleocañones terciarios de la planicie costera del Golfo de México: Boletín de la Asociación Mexicana de Geólogos Petroleros, v. 32, no. 1, p. 27–55.

Carrillo-Bravo, J., 1980b, Aplicación de criterios obtenidos en el Paleocañon de Chicontepec a otras areas: 35.° Reunión a nivel de expertos de ARPEL, Nuevos Conceptos Geológicos en Exploración Petrolera, México, v. 1, 10 p.

Cuevas, S. F., 1980a, Geometría de las areniscas del Eoceno Superior del Distrito de Poza Rica y sus posibilidades petrolíferas: Boletín de la Asociación Mexicana de Geólogos Petroleros, v. 32, no. 2, p. 33–58.

Cuevas, S. F., 1980b, Exploración petrolera en sedimentos terrígenos: 35.° Reunión a nivel de expertos de ARPEL, Nuevos Conceptos Geológicos en Exploración Petrolera, México, v. 1, 14 p.

Treviño-Rodríguez, A. F., 1995, Estudio diagenético de las areniscas de Chicontepec: Ingeniería Petrolera. v. 35, no. 6, p. 16–24.

Yañez-Mondragón, M., 1998, La tecnología de vanguardia mejora sustancialmente la rentabilidad de los proyectos: Caso Chicontepec, Exitep 98, Memorias-Proceedings, Exposición Internacional de tecnología petrolera, PEMEX Exploración y Producción, 15–18 noviembre 1998, México, p. 403–411.

Oil and Gas
Fields

Martínez Castillo, F. J. , 2001, Geologic study of the Miocene Rodador field and its exploitation possibilities, Tabasco state, southeastern Mexico, *in* C. Bartolini, R. T. Buffler, and A. Cantú-Chapa, eds., The western Gulf of Mexico Basin: Tectonics, sedimentary basins, and petroleum systems: AAPG Memoir 75, p. 433–441.

20

Geologic Study of the Miocene Rodador Field and its Exploitation Possibilities, Tabasco State, Southeastern Mexico

Francisco Javier Martínez Castillo

PEMEX Exploración y Producción, Agua Dulce, Veracruz, Mexico

ABSTRACT

The interpretation and analysis of depositional systems played an important role in the definition of hydrocarbon reservoirs in the Rodador field of the Isthmian Saline Basin in southeastern Mexico. The two major sandstone reservoirs of the middle Miocene–lower Pliocene Encanto Formation, which are referred to as the 17-A and 18 sands, are located at depths ranging from 2873 m to 2995 m and 2877 m to 3031 m, respectively. The thickness of these two sands ranges from 8 m to 28 m (17-A) and from 7 m to 48 m (18).

The stratigraphic distribution of the hydrocarbon reservoir units and seal rocks has been established, allowing the identification of new, potential areas for exploitation.

The prograding complex of the Encanto Formation is the most attractive from a petroleum perspective, because it hosts the hydrocarbon-producing Rodador field.

A detailed analysis of sand-top maps in this study was used to estimate the original hydrocarbon volume in place, to rank the potential areas for new exploitation plans in the Rodador field. Calculated hydrocarbon reserves are not included in this paper.

INTRODUCTION

The Rodador field is in the state of Tabasco, southeastern Mexico. This field is in the onshore portion of the Isthmian Saline Basin, part of the province known as the Southeastern Tertiary Basins (Figure 1). The purpose of this study is to estimate the original hydrocarbon volume in place, which will be used to rank the potential of the sands for new exploitation plans.

The sequence-stratigraphic concepts and terminology used in this study are based on Van Wagoner et al. (1988, 1990) and Mitchum et al. (1993).

Seismic line L-64/39 is crucial to the study, because it crosses through wells Rodador 173, 175, 177, 179, and 278 (Figures 2

and 3). The study includes 25 well logs (Figure 2), as well as the stratigraphic data from the Ají-1 well, located to the southwest of the field (1.4 km S29°W of the Rodador-1 well). The Ají-1 well was chosen because cutting samples from 1900 m to 3835 m contain fossils that were studied by Rodríguez (1999), who identified the species outlined in Table 1.

Based on well logs, the top of the Encanto Formation is established at the contact of a shalier section with a sandier one.

According to correlation of well logs and supported by paleontological data furnished by the Rodador-195D well drilled in 1997, the top of the Encanto Formation was established with the following microfossils: *Uvigerina hispida, Reophax encantoensis,* and *Melonis pompilioides.*

PREVIOUS STUDIES

By 1953, several unpublished gravity, refraction, and reflection seismology surveys had resulted in identification of several geologic blocks considered to be attractive for oil exploration.

Jiménez (1976) interpreted the northern portion of the Rodador field to evaluate available commercial hydrocarbon reservoirs. His study defined the structure as a faulted anticline with a southeast-northwest-trending axis. The long axis of the structure measures 3 km, and the width measures about 2 km. It has a small closure dipping to the northwest.

This study encompasses the southern area of the field and concurs with Jiménez (1976) in that the cumulative sand horizons in the southeastern part of the field appear to be better developed. The maps of sand tops provided in this study indicate that these sand bodies are potential reservoirs.

Guerrero (1993) estimated the original hydrocarbon volume and reserves for the Rodador field. He proposed the existence of eight reservoirs divided into 18 blocks in the Encanto Formation. These reservoirs are at depths that range from 2400 m to 3450 m.

The Discovery of the Rodador Field

The Rodador field was discovered by wildcat Rodador-1 in 1971. This well was drilled to a depth of 3647 m and became a dual producer of oil and gas from intervals 3220–3228 m and 3458–3470 m. These intervals correspond to middle Miocene–lower Pliocene sand bodies of the Encanto Formation.

The Rodador field was developed mainly to the south and southeast of the Rodador-1 well. Forty-six wells had been drilled by 1980, of which 36 were producers and 10 were dry (Figure 2). The spacing between wells was 400 m.

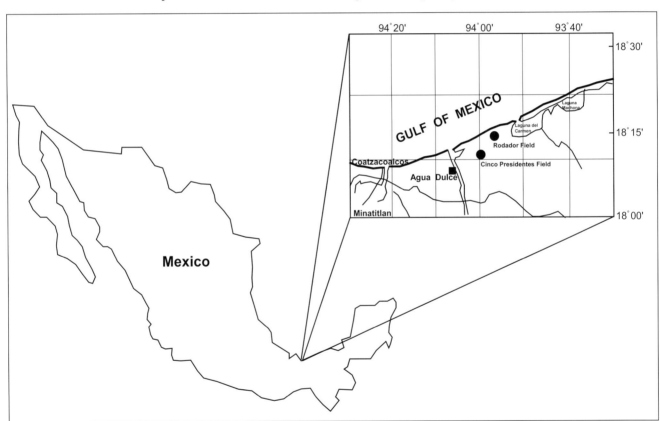

Figure 1. Location of the study area, Rodador field in the state of Tabasco, southeastern Mexico.

Table 1. Time-stratigraphic and biostratigraphic subdivision of the Ají-1 well. Modified after Rodríguez (1999), using data from Berggren et al. (1995).

Depth	Age	Species
2275 m	Upper Pliocene	*Globoquadrina altispira altispira*
2295 m	Upper Pliocene	*Globorotalia margaritae margaritae*
2370 m	Lower Pliocene	*Globorotalia scituala gigantea*
3105 m	Upper Miocene	*Globorotalia mayeri*
3805 m	Middle Miocene	*Globorotalia foshi lobata*

Seismic Analysis

Only seismic line L-64/39, which crosses the field in a southwest-northeast direction (Figure 2), was analyzed. This line ties to wells Rodador 173, 175, 177, 179, and 278 (Figure 3). The velocity survey used was that of the Magallanes-850 well, about 10 km southeast of Rodador field. This velocity survey was used to convert the seismic reflections from time to depth and to try to calibrate the seismic information with the wells, because the Rodador field does not have its own velocity survey.

Anhydrite and Salt in Rodador Field

Jurassic anhydrite and salt are common in the Southeastern Tertiary Basins of southern Mexico.

In Rodador field, 10 wells (Rodador 1, 2, 6, 20, 31, 32, 43, 52, 54, and 173) reached the anhydrite-salt top. These wells penetrated from 1 m below the top (Rodador-54) to 35 m (Rodador-6) and from 3453 m minimum depth (Rodador-173) to 3791 m maximum depth (Rodador-6). The Ají-1 well in the southwestern area reached the anhydrite top at 3853 m, penetrating 17 m.

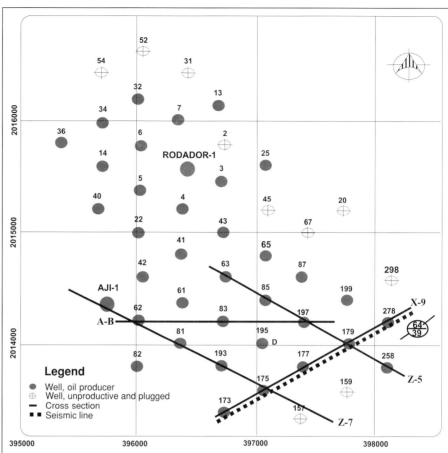

FIGURE 2. Rodador field layout, Tabasco, southeastern Mexico.

Seismic Line L-64/39

On seismic line L-64/39, at the position of wells 175 and 173 at approximately 2.1 to 2.3 s (equivalent to 2690 to 3010 m in depth), the following can be observed (Figure 3):

1) The pattern of imbricated turbiditic flows deposited on the slope forms stratigraphic traps in the Encanto Formation.

2) The line drawn across the lower part of wells Rodador 173, 175, 177, 179, and 278 corresponds to the base of the Encanto Formation at 2.48 s (3360 m) for Rodador-173 and at 2.55 s (3430 m) for Rodador-278. These lower sands are better developed than in the northern producing portion of the field. Below the base of the Encanto Formation, the section becomes more shaly.

3) Southwest of seismic line L-64/39 (Figure 3), from approximately 1.8 to 2.4 s, the seismic reflections are seen to rise upward, produced by the Cinco Presidentes salt dome to the southeast (Figure 1).

4) The main producing area with exploitation possibilities is located in the 2.0- to 2.5-s interval, from 2500 m to 3350 m deep. These producing intervals belong to the middle and lower part of the Encanto Formation, as illustrated on seismic line L-64/39 (Figure 3). Note that the 850-m thickness of the producing zone is a gross thickness, representing interbedded layers of sand and shale.

GEOLOGY

Depositional Model of the Producing Horizons

The Encanto Formation consists of interbedded sand and shale. Both lithologies have variable thickness through the formation. The thickness of the sand bodies ranges from 4 m to 65 m, and the thickness of shale horizons ranges from 3 m to 67 m. The sand is light to dark gray in color, fine to coarse grained, and the shale is soft to hard, green and dark gray.

The Concepción Inferior Formation overlies the Encanto Formation and is characterized by interbedded thick-shale and thin-sand bodies. This formation in the study area has an average thickness of 260 m.

Based on a correlation of electric logs, the top of the Encanto Formation is characterized by the presence of interbedded, well-defined sand and shale bodies located at an average depth of 2240 m (Figure 4).

The Encanto Formation in the study area has a minimum thickness of 1205 m (Rodador-67 well) and a maximum thickness of 1376 m (Rodador-177 well). The Ají-1 well, southwest

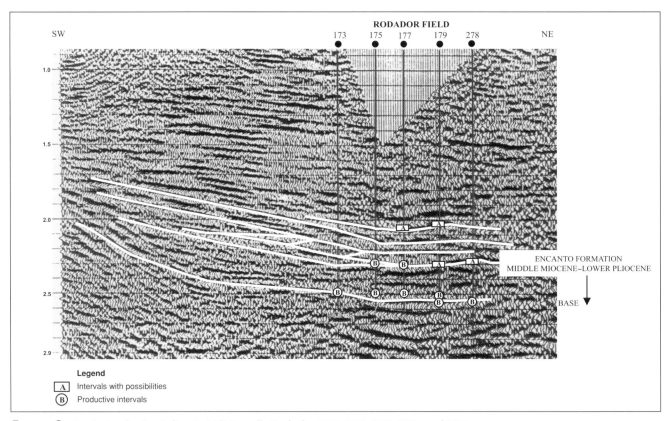

FIGURE 3. Analysis of seismic line L-64/39, wells Rodador 173, 175, 177, 179, and 278.

of the field and 1.4 km southwest of the Rodador-1 well (Figure 2), penetrated 1468 m of Encanto Formation before encountering the top of the anhydrite. On the basis of well logs, the base of the Encanto Formation is established at the contact of a sandier section with a shalier one.

Most of the oil produced from the Encanto Formation in southeastern Mexico is from shingled turbidites (PEP-BP Exploration Report, 1994). The producing and potential turbidite sands are found mainly in the middle and lower portions of the Encanto Formation.

In the southern area of Rodador field, two prograding complexes, referred to as A and B, have been recognized as part of the Encanto Formation (Figures 5 and 6).

Prograding complex A in the lower and middle portions of the Encanto Formation includes the 31 to 13 sands at depths ranging from 2700 m to 3450 m and whose thickness varies from 5 m to 69 m. Sands 22 to 17 (except 19) (Figure 5 and 6) show a good resistivity curve response because of the presence of hydrocarbons, and thus they have exploitation possibilities. Sand 17-A has been tested in wells 81, 82, and 175, and it also has good possibilities in nine more wells (83, 85, 87, 179, 193, 195D, 197, 199, and 278) (Figure 7).

The 18 sand in wells 195D and 278 is the main producer in the field. The sand has a good response on the log resistivity curve (well 278), and has an average thickness of 33 m and an average porosity of 22%. It is considered a potential future producer in wells 83, 85, 87, 179, 197, and 199 (Figure 8).

Prograding complex B comprises the sand bodies at depths of 2200 m to 2650 m. This is the shallowest complex, and it encompasses sands 12 to 8, whose thickness varies from 8 m to 65 m. Sand 12 in the Rodador-258 well has yielded good oil production and has exploitation possibilities in well 179 (Figure 9), as indicated by a high resistivity curve response. The 12 sand in well 179 reaches a gross thickness of 65 m. It is important to mention that in the Rodador field, there are intervals with exploitation possibilities, and they are shown with well logs in Figure 10.

Figure 11 depicts a conceptual sedimentary model of the Rodador field, showing the approximate location of the field in the prograding complex.

Reservoir Rocks

Hydrocarbon reservoirs are typically fine- to medium-grained gray sands, some of which are shaly. Coarse-grained sand rarely occurs in the formation. In the study area, sand reservoirs vary in thickness from 3 m to 33 m and have an average porosity of 20%.

Seal Rocks

Seal rocks in the Rodador field are shales of variable thickness interbedded with the reservoir sands of the Encanto Formation. In some cases, lateral seals are faults or changes in sedimentary facies (Figure 12; wells Rodador 63, 85, 197, 179, and 258).

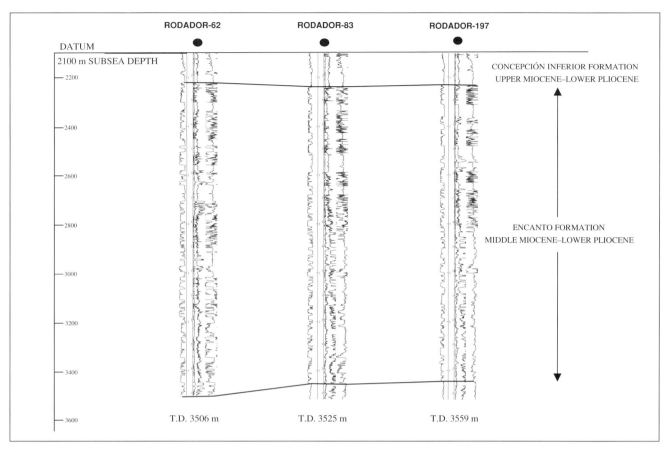

FIGURE 4. Cross section A-B among wells Rodador 62, 83, and 197, showing the Encanto Formation top and base.

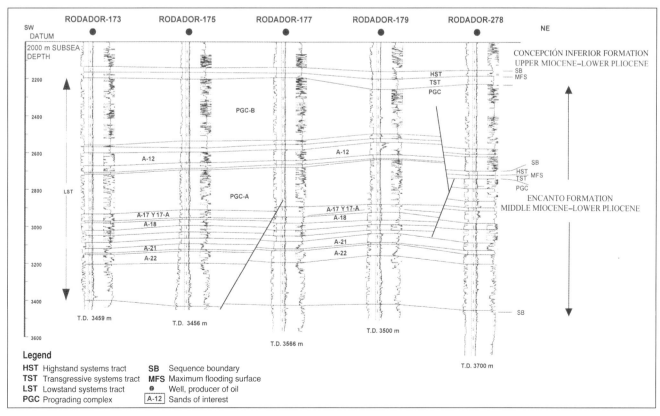

FIGURE 5. Cross section X-9 among wells Rodador 173, 175, 177, 179, and 278.

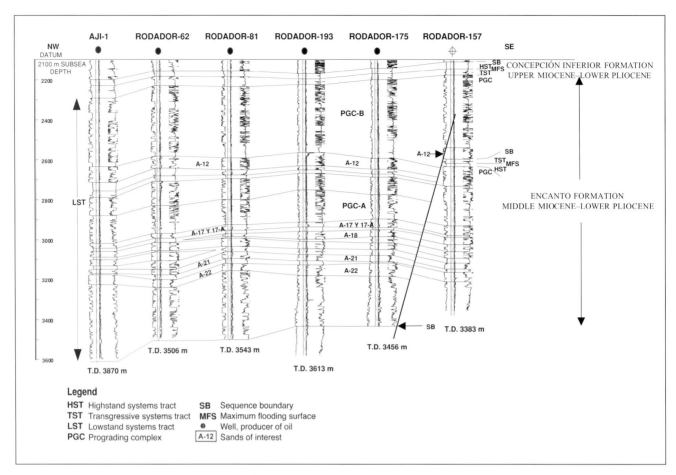

FIGURE 6. Cross section Z-7 among wells Ají 1, Rodador 62, 81, 193, 175, and 157.

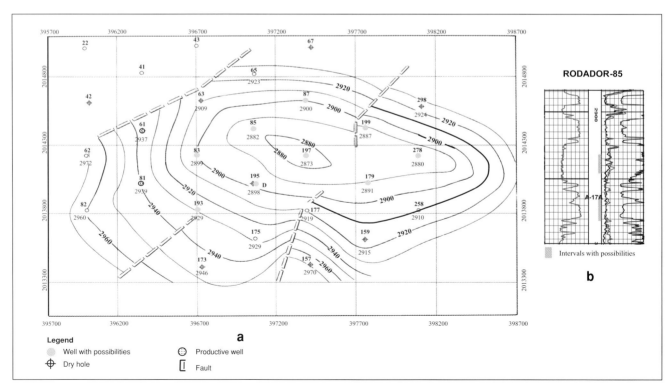

FIGURE 7. (a) Sand 17-A top configuration, Rodador field, Encanto Formation. (b) Rodador-85 well, showing detail of sand 17-A.

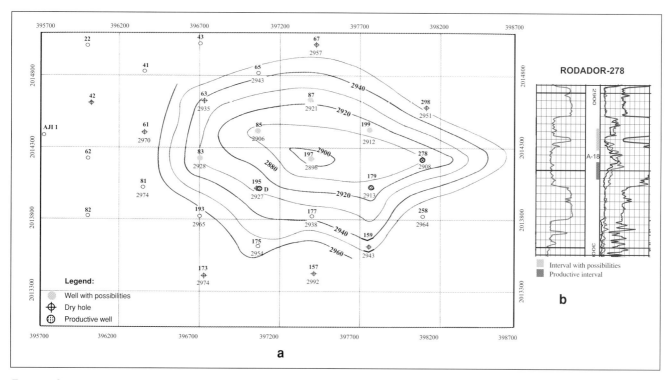

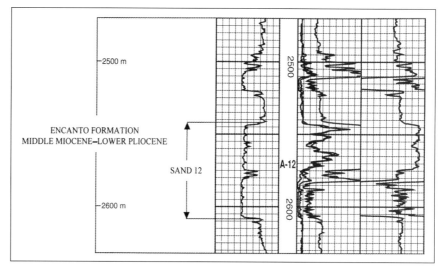

FIGURE 8. (a) Sand 18 top configuration, Rodador field, Encanto Formation. (b) Rodador-278 well showing detail of sand 18.

An example of a lateral fault seal is clearly seen between wells 179 and 258, where the impregnation of sand A-11 in well 179 no longer appears in well 258. An example of facies-change, lateral seal is shown where a sand body immediately below sand A-26 in well 197 becomes shalier toward well 179. Finally, a pinch-out lateral seal is illustrated with sands A-25 and A-26, where the pinching out of the sands acts as a seal.

Traps

Traps are mainly stratigraphic as a result of pinch-outs of sand bodies, as illustrated in Figure 12, where sands A-25 and A-26 pinch out to form traps. Structural and combination traps also occur between wells 179 and 278 for sand A-11.

FIGURE 9. Well in Rodador field, showing sand 12 (A-12).

CONCLUSIONS

1) Optimal hydrocarbon production in the Rodador field is from a prograding complex in the middle Miocene–lower Pliocene Encanto Formation.

2) The study area is not strongly affected by structural deformation; traps are mainly stratigraphic.

3) A conceptual sedimentary model for the Rodador field is proposed here on the basis of a multidisciplinary study.

4) The detailed analysis of sand-top maps was used to estimate the original hydrocarbon volume in place, and this helps to rank the areas or sand bodies for exploitation purposes. The calculated hydrocarbon reserves will remain confidential.

ACKNOWLEDGMENTS

I thank Dr. Abelardo Cantú-Chapa (Instituto Politécnico Nacional) for his suggestions for improving the manuscript. I also thank Petróleos Mexicanos, Activo de Producción Cinco Presidentes, Area Diseño de Explotación, Agua Dulce, for permitting me to publish this paper.

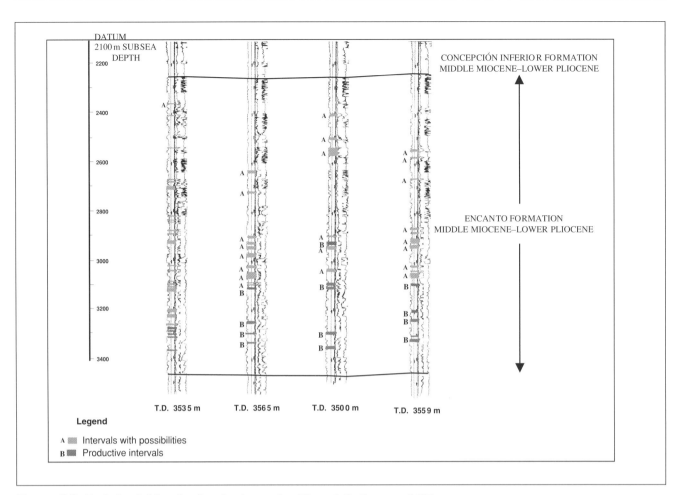

FIGURE 10. Rodador field wells, showing intervals with exploitation possibilities.

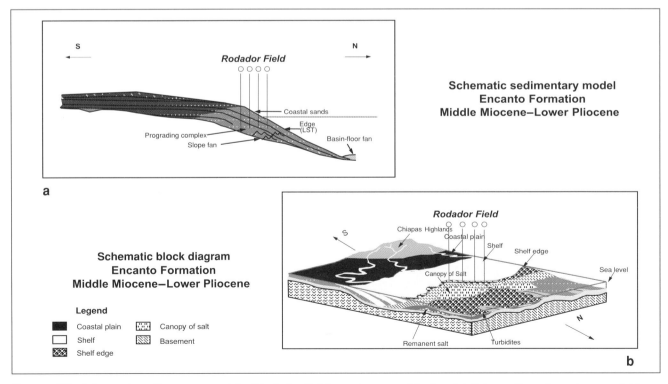

FIGURE 11. Schematic sedimentary model of Rodador field, southeastern Mexico. (a) Schematic sedimentary model for the Encanto Formation. (b) Schematic block diagram of the Encanto Formation.

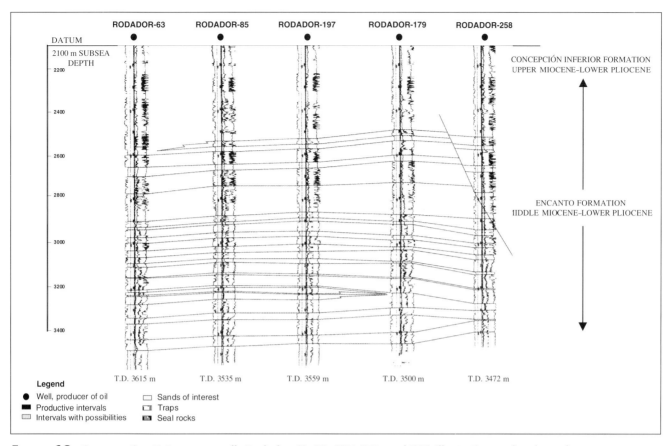

DATUM
2100 m SUBSEA DEPTH

RODADOR-63   RODADOR-85   RODADOR-197   RODADOR-179   RODADOR-258

CONCEPCIÓN INFERIOR FORMATION
UPPER MIOCENE-LOWER PLIOCENE

ENCANTO FORMATION
IIDDLE MIOCENE-LOWER PLIOCENE

T.D. 3615 m   T.D. 3535 m   T.D. 3559 m   T.D. 3500 m   T.D. 3472 m

Legend
● Well, producer of oil   ☐ Sands of interest
■ Productive intervals   ☐ Traps
☐ Intervals with possibilities   ▦ Seal rocks

FIGURE 12. Cross section Z-5 among wells Rodador 63, 85, 197, 179, and 258, illustrating seal rocks and traps.

REFERENCES CITED

Berggren, W. A., V. K. Dennis, C. Swisher, and M. P. Aubry, 1995, A revised Cenozoic geochronology and chronostratigraphy, *in* Geochronology time scales and global stratigraphic correlation: Society for Sedimentary Geology (SEPM) Special Publication 54, p. 129–212.

Guerrero, C. G., 1993, Cálculo del Volumen Original y Reservas de Hidrocarburos del Campo Rodador: PEMEX confidential report, p. 2.

Jiménez, T. A., 1976, Interpretación Geológica Campo Rodador: PEMEX confidential report, chapter 1, p. 1–11.

Mitchum, R. M., J. B. Sangree, and P. R. Vail, 1993, Recognizing sequences and systems tracts from well logs, seismic data, and biostratigraphy: Examples from the late Cenozoic of the Gulf of Mexico, *in* P. and H. W. Posamentier, eds., Siliciclastic sequence stratigraphy: Recent developments in applications: AAPG Memoir 58, p. 163–189.

Partida, M. P., and M. J. A. Zaldivar, 1997, Revisión Paleontológica del Pozo Rodador-1950: PEMEX confidential report.

PEP-BP Exploration, 1994, Proyecto Cuencas Terciarias del Sureste y Área Marbella, chapter 3, p. 37; chapter 4, p. 60–63, 139; chapter 5, p. 142–144, 148–149.

Rodríguez, C. R. L., 1999, Revisión Paleontológica del Pozo Ají-1: PEMEX confidential report.

Van Wagoner, J. C., R. M. Mitchum, K. M. Campion, and V. D. Rahmanian, 1990, Siliciclastic sequence stratigraphy in well logs, cores, and outcrops: Concepts for high-resolution correlation of time and facies: AAPG Methods in Exploration Series No. 7, 55 p.

Van Wagoner, J. C., H. W. Posamentier, R. M. Mitchum, P. R. Vail, J. F. Sarg, J. S. Loutit, and J. Hardenbol, 1988, An overview of the fundamentals of sequence stratigraphy and resolutions definitions, *in* C. K. Wilgus et al., eds., Sea-level changes: An integrated approach: Society for Sedimentary Geology (SEPM) Special Publication 42, p. 38–45.

Williams-Rojas, C. T., and N. F. Hurley, 2001, Geologic controls on reservoir performance in Muspac and Catedral gas fields, southeastern Mexico, *in* C. Bartolini, R. T. Buffler, and A. Cantú-Chapa, eds., The western Gulf of Mexico Basin: Tectonics, sedimentary basins, and petroleum systems: AAPG Memoir 75, p. 443–472.

21

Geologic Controls on Reservoir Performance in Muspac and Catedral Gas Fields, Southeastern Mexico

Carlos T. Williams-Rojas
PEMEX Exploración y Producción, Activo Salina del Istmo, Coatzacoalcos, Veracruz, Mexico

Neil F. Hurley
Colorado School of Mines, Golden, Colorado, U.S.A.

ABSTRACT

Muspac and Catedral are two of the most important gas- and condensate-producing fields in southern Mexico. They produce from Cretaceous fractured carbonates. The objective of this integrated study is to define the stratigraphic and structural controls that caused early water production in those fields.

Open-hole log correlation of 45 wells served to define eight reservoir zones, based on petrophysical characteristics. Petrophysical properties were mapped using a volumetric parameter to analyze the anisotropy of the gas-storage capacity in the fields. Dipmeter and borehole image logs were interpreted in 14 wells using cumulative dip and vector plot techniques to define unconformities, flooding surfaces, and faults. Borehole images from five wells were extremely useful in detecting evidence of sedimentologic and structural features. Fracture density depends on petrophysical properties of the reservoir rocks. In-situ stress directions were determined in 18 wells using borehole breakouts to define the predominant northwest-southeast orientation of the open fractures. Two dominant fracture sets, determined from seismic attributes and borehole images, are parallel to seismically determined faults.

According to this study, early water production is caused by coning through fractures, faults, and karstic zones. Some water-producing intervals depend on the location of perforations, especially when these are located in highly fractured rocks and close to the gas-water contact. To minimize early water production, the operator must avoid wells in fault zones, wells on the flanks close to the gas-water contact, and deviated wells drilled perpendicular to the direction of open fractures.

INTRODUCTION

Muspac and Catedral fields are two of the most important gas- and condensate-producing fields in the southern part of the petroleum province known as Area Chiapas-Tabasco, discovered in 1972 in southeastern Mexico. Like other fields in the area, they produce from shallow-water carbonates deposited during the middle and Late Cretaceous (Turonian-Maestrichtian). They are sealed by lower Tertiary shales, were deformed during the middle Miocene, and probably were charged during the Miocene-Pliocene.

The Area Chiapas-Tabasco is Mexico's second-largest petroleum province. At present, this petroleum province is considered a mature exploration area; therefore, reservoir characterization studies in this area have became an important tool, not only to understand reservoir performance but also to identify new exploration opportunities in the producing plays. This study, which is summarized from Williams-Rojas (2000), is focused on defining the stratigraphic and structural controls on reservoir performance in the Muspac and Catedral fields by following a methodology that could be applied to fields with similar characteristics in this province. A primary goal is to find the causes of early water production and to propose solutions and/or alternatives to avoid or minimize these problems. Four different methods were used in this research: (1) stratigraphic zonation of the reservoir, (2) analysis and interpretation of dipmeter and borehole image logs, (3) three-dimensional (3-D) seismic integration and interpretation, and (4) analysis and interpretation of production data. A combination

of structural and stratigraphic analysis suggests ways to avoid the coning of water into producing wells.

GEOLOGIC FRAMEWORK

The Area Chiapas-Tabasco is located in the Cuenca del Sureste (González-García and Holguín-Quiñones, 1972) where the Artesa Mundo–Nuevo isolated platform existed during the Cretaceous (Figure 1). In this basin, the oldest Mesozoic sedimentary environments evolved from continental (containing red beds deposited during the Bathonian-Bajocian) to marginal marine, where salt was deposited during the Callovian. During the Late Jurassic (Oxfordian and Kimmeridgian), a marine transgression produced shallow-water environments with high-energy carbonate deposition. Later, during the Tithonian, transgression combined with high subsidence produced pelagic carbonaceous shales, which constitute the most important source rocks in the basin. During the Early Cretaceous, subsidence and extension created horsts and grabens that led to the deposition of shallow-water carbonates on the horsts and shales and shaly carbonates in the grabens. It is believed by Rodríguez (personal communication, 1983) and Varela-Santamaría (1995) that the Artesa–Mundo Nuevo isolated platform was formed during this time. This isolated platform, which persisted throughout the entire Cretaceous, was drowned at the end of the Late Cretaceous and was covered by siliciclastic sediments during the Tertiary.

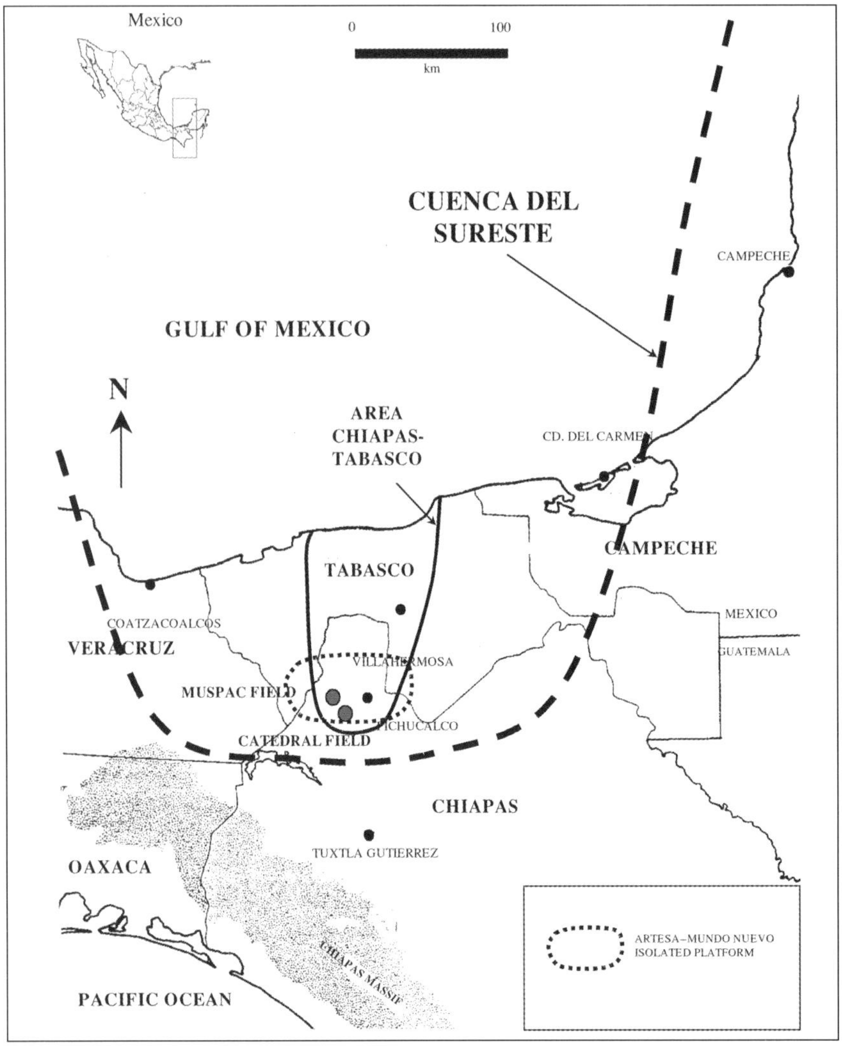

Sedimentary Model

The sedimentary model for the study area, which is located on the southern margin of the Artesa–Mundo Nuevo isolated platform, was proposed initially by Rodríguez (personal communication, 1983), detailed by Barceló-Duarte et al. (1994), Varela-Santamaría (1995), and Williams-Rojas (1995), and modified by PEMEX-NITEC personal communication, 1999). This model shows that the southern margin of the isolated platform first was a ramp during the Early Cretaceous and became a bypassed margin during the Albian-Cenomanian (known as the middle Cretaceous in Mexico). During the Turonian-Santonian, the platform was reduced because the sedimentary productivity was not able to keep pace with the gradual rise of sea level and subsidence (Figure 2). During the Campanian-Maestrichtian, the platform size was reduced even further, and the area of Muspac field was drowned because of a rapid

FIGURE 1. Map of southeastern Mexico showing the location of the basin Cuenca del Sureste, the petroleum province Area Chiapas-Tabasco, the isolated platform Artesa–Mundo Nuevo, and the Muspac and Catedral gas fields. Mexican states and important cities are labeled.

rise of sea level. The area of Catedral field kept up with the sea-level rise, maintaining high-energy shoal facies composed of oolitic and bioclastic grainstones. Most of the former platform gave up, except for some highs along the northern margin. Other zones, such as the area of Catedral field, provided some debris, grain flows, and turbidites to deeper areas of the drowned parts of the platform. Finally, the platform was eroded partially and drowned totally at the end of the Maestrichtian and was buried during the Tertiary under a thick siliciclastic section.

The original porosity was reduced by compaction and cementation in some cases and enhanced in others by other diagenetic processes such as dissolution, karstification, and dolomitization. The latter processes are related to subaerial exposure at several stratigraphic levels that are considered to be sequence boundaries.

Structural Model

Plate-tectonics evolutionary models of Scotese et al. (1988), Pindell et al. (1988), and Ross et al. (1988) agree that during the Campanian, the Chortis block, which is part of the Caribbean Plate, started to move eastward through the fault system known as Motagua-Polochic. This caused the subduction zone of the Pacific margin to move eastward during the Tertiary until it reached its current position on the southern coast of Mexico. Therefore, a combination of transform movement eastward and compressive stress northeastward caused compressive features and magmatic activity in Chiapas.

Although this event was important from the standpoint of trap formation (Figures 3 and 4), other tectonic events have been recognized in the area from the stratigraphic record and from outcrop studies. The tectonic evolution of the area can be summarized in four main events. The first was an extensional event during the Middle and Late Jurassic, involving the basement. This event produced a series of horsts and grabens that controlled later sedimentation. The second event was compressive and related to the Laramide orogeny that occurred during the Late Cretaceous–early Tertiary. Principal stress from the southwest produced gentle folds with low structural relief in a northwest-southeast orientation. The third event was the middle Miocene compressive deformation from the southwest, which reactivated the northwest-southeast structures and produced fault-bend folds and fault-propagation folds with salt detachment. This deformation is the most important event in the area and is responsible for the configuration of the traps in Catedral and Muspac fields. The fourth event was extensional and occurred during the middle and late Miocene, with normal faulting oriented perpendicular to the compressional structures.

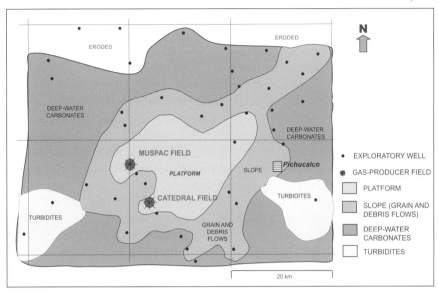

FIGURE 2. Sedimentary model for the Late Cretaceous (Turonian-Santonian) for the southern margin of the Artesa–Mundo Nuevo isolated platform where Catedral and Muspac fields are located. Interpretation is based on information from exploratory wells. Adapted from Alcántara and Solís (personal communication, 1999).

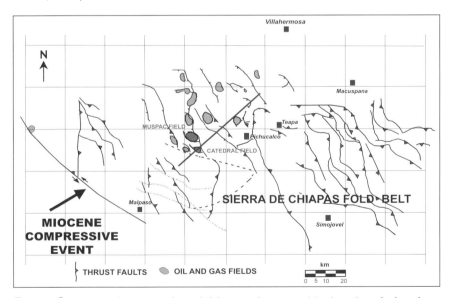

FIGURE 3. Regional structural model for southeastern Mexico. Catedral and Muspac fields are part of the Sierra de Chiapas folded belt, which formed during a middle Miocene compressive deformation event that created the structural traps for several oil and gas reservoirs. A black line shows the location of the cross section illustrated in Figure 4.

Field History

Muspac field was discovered in 1982 with the exploratory well Muspac-1. This well had initial production of 9 MMCFD (million cubic feet of gas per day) and 730 BOPD (barrels of oil per day) and initial pressure of 316 kg/cm². From 1982 to 1987, 11 more wells were drilled. From 1995 to 1999, the field was developed with 15 new wells (average spacing of 1000 m) after an integrated study was conducted at the end of 1994. The current production is about 350 MMCFD and 15,000 BOPD. Currently, 21 wells are producing, and there are six wells with high water-cut production in the field. Reservoir energy is supplied by an active water-drive system.

Catedral field was discovered in 1991 with the exploratory well Catedral-1. This well had initial production of 2.48 MMCFD and 377 BOPD and initial pressure of 297.8 kg/cm². From 1995 to 1998, the field was developed with 22 new wells (average spacing of 1000 m) after an integrated study. The current production is about 158 MMCFD and 7,000 BOPD. Currently, 18 wells are producing, and there are seven wells with high water-cut production. Reservoir energy is supplied by an active water-drive system.

STRATIGRAPHIC ZONATION OF THE RESERVOIR

The main purpose of this stage of the study was to define the stratigraphic zones and determine their quality, geometry, and petrophysical properties using open-hole log correlation, formation evaluation, and mapping techniques.

The available data set from which the stratigraphic zonation was defined in Catedral and Muspac fields includes open-hole logs, petrographic descriptions, digital photographs from core and thin-section samples, biostratigraphic information, and petrophysical data, such as average porosity and water saturation.

There are 23 wells in Catedral field and 28 wells in Muspac field. Only 2 wells (C-DL1 and M-63V) did not reach reservoir rocks. Of the 49 wells that penetrated the reservoir in both fields, only 45 have an open-hole log suite, in digital format, composed of gamma-ray, resistivity, neutron-porosity, and bulk-density logs.

A petrographic study done on core and thin-section samples by Webb (personal communication, 1999) includes descriptions and digital pictures from small pieces (10 to 20 cm) of 9 m total length for every core (34 cores in Catedral field and 12 cores in Muspac field). Despite the small portion of cores studied (which were not representative of all cores), this information was very useful in characterizing the lithology and sedimentary facies contained in every defined stratigraphic zone.

Biostratigraphic studies of the same samples used for petrographic studies were integrated in this study. Shallow-platform sediments do not contain many organisms that would allow age determination; however, a limited but very valuable biostratigraphic data set was available to support the stratigraphic zonation of Upper and middle Cretaceous reservoir rocks in Catedral and Muspac fields.

Two approaches have been used in the fields to obtain petrophysical properties of reservoir rocks: (1) effective porosity and water saturations were calculated by PEMEX using the ELANPlus module of GEOFRAME. The basic input information for ELANPlus is contained in gamma-ray, induction (ILD), bulk-density (RHOB), and neutron-porosity (PHI) logs; (2) Cluff and Cluff (personal communication, 1999) ran a deterministic log model that used the Archie water saturation model. This is a simple saturation model that is most effective in clean, nonshaly units such as the Cretaceous section at Catedral and Muspac fields. Effective porosity was calculated using a crossplot method with shale correction. Porosity measurements were corrected for shaliness. A similar calculation proce-

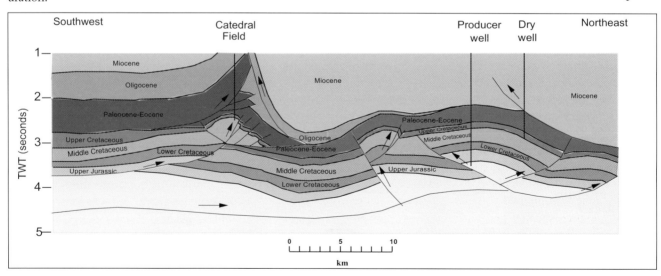

FIGURE 4. Diagrammatic structural cross section oriented southwest-northeast, obtained from a 2-D seismic section. The cross section shows the structural style and type of petroleum traps in the area. Catedral field is included in the cross section. TWT= two-way traveltime.

dure was used for density porosity. The volume of shale was calculated from the gamma-ray log.

The results of the deterministic approach produced a good match with the ELANPlus-determined porosity and water saturation done by PEMEX. That two totally different methods produced very similar results provides confidence in the obtained results.

After the correlation of open-hole logs was made and the reservoir zones were defined, the petrophysical properties were calculated using the PEMEX and Cluff and Cluff (personal communication, 1999) approaches. As a result, average porosity and water-saturation values for the total measured thickness in every defined reservoir zone were available to help analyze reservoir quality. No net pay or net/gross cutoffs were considered because these are gas reservoirs, and even the very low porosity carbonates have enough permeability to produce some gas.

Correlation of Conventional Open-hole Logs

All 45 wells with gamma-ray, resistivity, neutron-porosity, and bulk-density logs were correlated with neighboring wells to identify different stratigraphic zones of the reservoir with similar expression. The following considerations have been made:

1) The first analysis was based on the gamma-ray expression, looking for sequences containing stacked depositional cycles separated by widely persistent, thin, shaly carbonate successions that could be interpreted as flooding surfaces. These surfaces represent time lines or sharp contacts between clean carbonates and shaly carbonates that may be indicative of unconformities.

2) The neutron-porosity and bulk-density logs were analyzed, looking for zones with high and low porosity and observing how these units match the gamma-ray logs. Some zones have low gamma-ray expression and high porosity. Those zones were interpreted as clean carbonate intervals with high porosity. Other zones that showed high gamma-ray activity and high porosity were considered shaly carbonate sections with thin bedding. Zones with low gamma-ray and low porosity readings were interpreted as tight zones.

3) Because the well correlations were made using measured depth logs, the deviated wells gave a false impression of thickness changes. To solve that, the deviation surveys had to be considered, then true vertical depths and true stratigraphic thicknesses were calculated using trigonometric corrections.

4) The expected depositional sequences on an isolated platform are sediments composed of shallowing-upward stacked sequences and parasequences that may be interrupted by lowstand periods at different stratigraphic levels (Handford and Loucks, 1993). The most important control on deposition of sediments is relative sea-level changes, which are affected by eustatic sea-level variations and tectonic subsidence (Tucker and Wright, 1990). On the Artesa–Mundo Nuevo isolated platform, eustatic sea-level variations and tectonic subsidence cre-

ated the accommodation space needed for sedimentation. The middle Cretaceous reservoir rocks in Catedral and Muspac fields are thicker and contain higher sedimentation rates than those for the Upper Cretaceous. In general, the bases of the shallowing-upward parasequences identified in both fields are formed by shaly carbonates with high gamma-ray signature and high neutron-density porosity, and the tops are composed of clean carbonates with high porosity.

Stratigraphic Zones in the Reservoir

Every well was correlated with neighboring wells. Using biostratigraphic data and considering log markers as flooding surfaces or time lines allowed us to define a stratigraphic reservoir zonation with eight layers in every field. Some variations in gamma-ray, density, and neutron-porosity signatures are observed between the fields that reflect changes in gross petrophysical characteristics. Figures 5 and 6, which illustrate well-log correlations for both fields, show different log markers that served to interpret reservoir zones in the included wells in Catedral and Muspac fields.

Eight reservoir zones were defined, based on the gross petrophysical characteristics observed in gamma-ray and porosity logs. Zone 5 is the most homogeneous and easily recognized zone, because it contains clean carbonates (low gamma-ray) with high neutron-porosity and low density. Table 1 summarizes typical lithologies and porosity types, average thickness, average porosity, and other data for reservoir zones.

Petrophysical Properties and Quality of Reservoir Zones

Because one important objective of this study was to characterize the geometry and quality of the reservoir, it was necessary to divide the reservoir into stratigraphic zones and to construct maps for every defined zone to analyze how the petrophysical properties varied in time and space in the reservoir.

With data about true vertical thickness (H), average porosity (phi), and average gas saturation (1-Sw) computed for every zone, the product (H•phi•[1-Sw]) was calculated and called the *volumetric parameter*. This represents meters of porous space occupied by gas of the total reservoir thickness in every well. The volumetric parameter was contoured for every zone, and these maps tell us what reservoir zones have better storage capacities for gas and how this storage capacity varies through every field. The volumetric parameter is zero at the gas-water contact.

Volumetric-parameter contour maps were constructed for every zone. Figure 7 is an example of such a map for reservoir zone 5. According to statistics for reservoir zones, means for true vertical thickness (H), average porosity (phi), and volumetric parameter were obtained to determine which reservoir zones have the best storage capacities for gas. Table 1 shows the means for these parameters in every zone. In Catedral field, reservoir zones with the more important storage capacities are zones 7, 8, and 5, in that order. In Muspac field, the best stor-

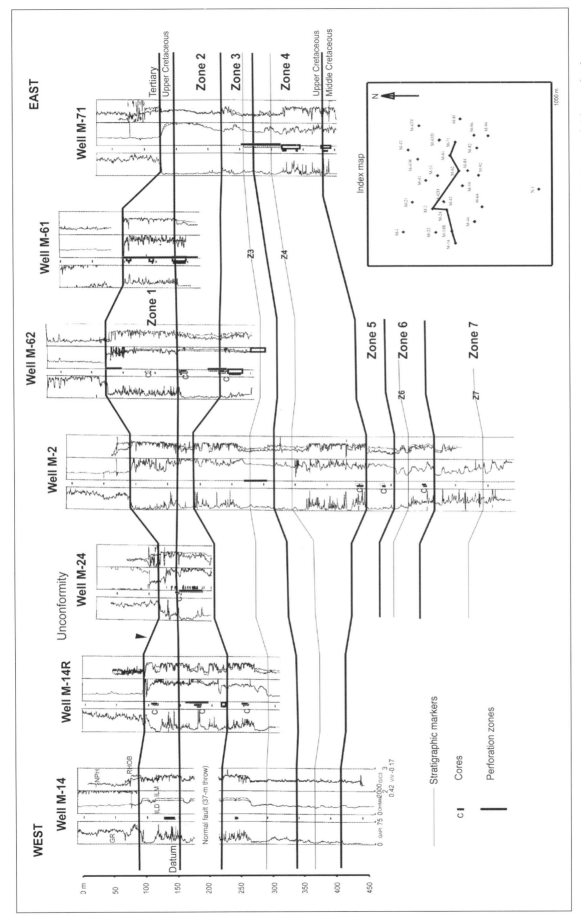

FIGURE 5. Stratigraphic cross section in Muspac field, using gamma-ray, resistivity, neutron-porosity, and bulk-density logs. Zones 2, 3, 4, and 5 have the best porosity values. Datum is the top of reservoir zone 2 in Upper Cretaceous rocks. Z3, Z4, Z6, and Z7 are stratigraphic markers.

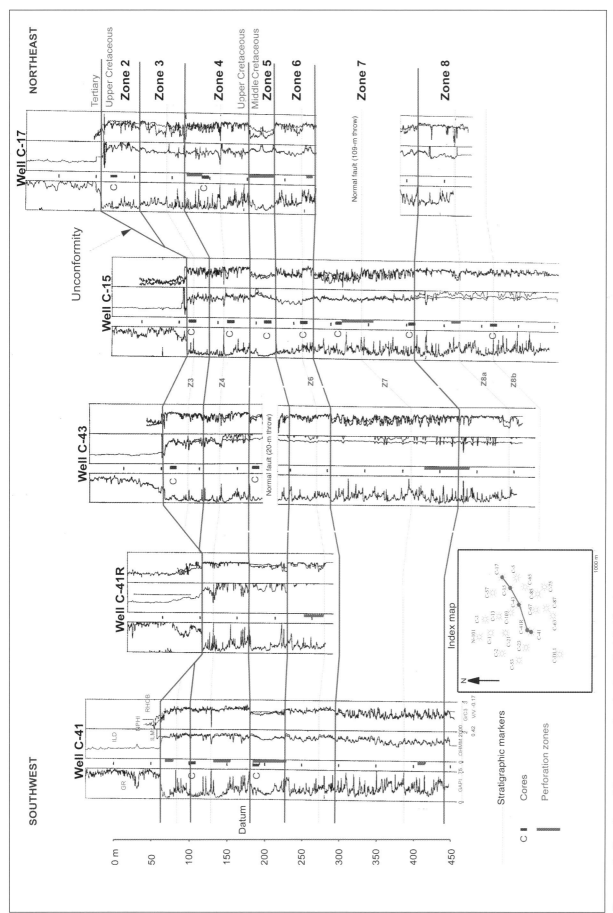

FIGURE 6. Stratigraphic cross section in Catedral field using gamma-ray, resistivity, neutron-porosity, and bulk-density logs. Zones 5, 7, and 2 have the best porosity values. Datum is the top of the middle Cretaceous. Z3, Z4, Z6, Z7, Z8a, and Z8b are stratigraphic markers.

age capacities are in zones 4, 3, 2, and 5, in that order (data below zone 5 are not representative because only three wells have information for zones 6, 7, and 8).

In addition, the percentage of the volumetric parameter with respect to the total true vertical thickness was calculated and taken as a *quality index* for every zone. This quality index is indicative of the homogeneity and cleanliness (shale-free nature) of the reservoir zones. Therefore, the best-quality reservoir zones for Catedral field are zones 5, 7, and 2, in that order. The best-quality reservoir zones for Muspac field are zones 2, 3, 5, and 4, in that order (Table 1).

ANALYSIS AND INTERPRETATION OF DIPMETER AND BOREHOLE IMAGE LOGS

The main objectives of this part of the study were to define unconformities, karst zones, faulting, fracturing, and fracture density using dipmeter and borehole image logs, and to obtain open-fracture orientations.

Dipmeters are HDT (high-resolution dipmeter tool) and SHDT (stratigraphic high-resolution dipmeter tool) logs. Borehole image logs are UBI (ultrasonic borehole image) logs, which are acoustic borehole televiewers.

The HDT dipmeter tool has four pads that record four resistivity (or microconductivity) curves spaced at 90° intervals around the borehole (Schlumberger, 1998). Data-sampling rate is one sample per 0.2 in. (0.5 cm), and vertical resolution is about 0.8 in. (2 cm).

The SHDT log records eight microresistivity curves with a high data-sampling rate (one sample per 0.1 in. [0.25 cm]) and good vertical resolution of 0.4 in. (1 cm). Microresistivity traces are recorded by two electrodes separated by 1.2 in. (3 cm) on each of the four orthogonal pads. This close spacing ensures high similarity between the two traces from individual pads and improves the confidence of correlations (Höcker et al., 1990).

The UBI log is an acoustic tool that scans the borehole wall with a rotating transducer that emits a pulsed ultrasonic beam (Schlumberger, 1998). As a result of ultrasonic reflections, two kinds of full coverage (360°) images are obtained: one from the reflected amplitude and another from the two-way traveltime (Luthi, 1993). A color convention indicates dark colors for lower reflected amplitudes and for higher two-way traveltime. Amplitude images are successfully used to characterize fractures in reservoirs (Luthi, 1997). The acoustic borehole televiewer is considered to be one of the best tools to derive maxi-

TABLE 1. Lithologic characteristics and porosity type of reservoir zones in Catedral and Muspac fields. Means values are shown for true vertical thickness (H), average porosity (phi), and volumetric parameter (H•phi•[1-water saturation]). Quality index indicates the percentage of the volumetric parameter with respect to the total thickness (H). The best-quality reservoir zones for Catedral field are zones 5, 7, and 2, and for Muspac field, zones 2, 3, 5, and 4, in that order.

Reservoir zones	Lithology and porosity	Catedral Field Thickness H (m)	Porosity phi (%)	Volumetric parameter VP (m)	Quality index (VP/H)•100	Muspac Field Thickness H (m)	Porosity phi (%)	Volumetric parameter VP (m)	Quality index (VP/H)•100
1	Mostly breccias formed by peloid and skeletal grainstone and packstone varying to floatstone, with well-rounded rudist fragments in grainstones and skeletal mudstones-wackestones. Matrix of mudstone and bentonitic shale containing planktonic forams and Globotruncanas. Moderate to good interparticle porosity.	71.2	4.3	2.46	**3.45**	51.2	7.8	2.98	**5.8**
2	In Catedral field, the reservoir is formed by skeletal wackestone-packstone that varies to rudist fragment grainstone and floatstone with moderate interparticle porosity and rare vuggy and moidic porosity. In Muspac field, the reservoir is formed by skeletal, peloid grainstone that varies to rudist fragment floatstone with some planktonic forams and Globotruncanas.	48.3	8.1	2.18	**4.5**	56.1	12.2	6.78	**12.1**
3	In Catedral field, the reservoir is formed by ostracode, peloid, and skeletal grainstones-packstones with very coarse, abraded, and rounded rudist fragments in wackestones and floatstones with moderate to good interparticle porosity. In Muspac, the reservoir is composed of peloid, ostracode, intraclastic, and skeletal grainstone, wackestone, packstone, rudist and mollusk fragments, and floatstone.	57.5	4.8	1.27	**2.2**	72.1	12.7	6.87	**9.5**
4	In Catedral field, the reservoir is formed by peloid, ostracode, and miliolid grainstone, packstone, and wackestone with laminated algal and rudist fragments. Poor to moderate interparticle, moldic, and open-fracture porosity. In Muspac field, the reservoir is formed by peloid and skeletal wackestone-packstone that changes to floatstone with rudist fragments, algal mats, and intraclasts.	65.2	3.8	1.23	**1.9**	109.9	12.3	9.46	**8.6**
5	Miliolid, peloid, and skeletal grainstone-packstone. Rocks are dolomitic in some wells with rudists, mollusks, and echinoderm fragments. Mudstones-wackestones are interbedded in some levels. Good to moderate interparticle, microintercrystalline, vuggy, moldic, and fenestral porosity and poor fracturing.	34.6	9	2.66	**7.7**	33.5	12.1	3.08	**9.2**
6	Dolomitic lime mudstones with scarce miliolids, planktonic forams, peloids, and ostracodes. These are traces of moldic pores and open fractures.	60.5	3.4	1.43	**2.4**	40	8.2	2.52	**6.3**
7	Miliolid, peloid, and skeletal grainstones, packstones, and floatstones with rudists and mollusks. There is moderate to good interparticle and moldic porosity, and moderate to poor fenestral and interparticle porosity.	131.7	7.8	8.32	**6.3**	108.8	7.7	6.1	**5.6**
8	Peloid, miliolid, intraclastic, and skeletal grainstone to packstone-wackestone and dolostones with poor moldic, intergranular, vuggy, and intercrystalline porosity.	133.4	3.5	4.24	**3.2**	168	4.4	5.89	**3.5**

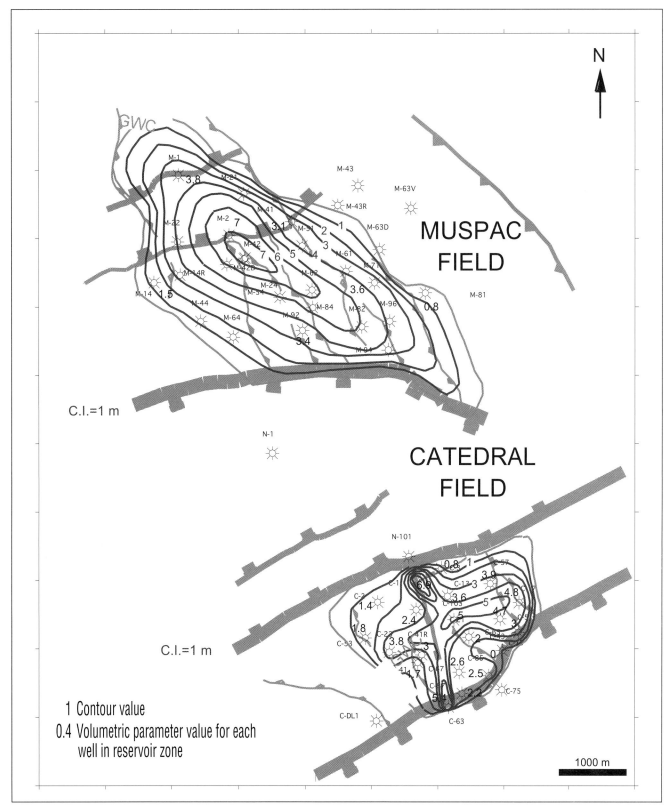

FIGURE 7. Volumetric parameter (H•phi•[1-Sw]) contour map for reservoir zone 5. Contour labeled with GWC indicates the gas-water contact for this zone. Contour interval is 1 m in both fields.

mum horizontal in-situ stress orientation from breakout analysis (Bell, 1993).

HDT, SHDT, and UBI logs were processed using Baker Atlas Recall/Review software, considering magnetic declination, accelerometer correction, and image generation. Two kinds of images were created during processing: static-normalized and dynamic-normalized. Static-normalized images are made with one contrast setting for the whole well; dynamic-normalized images are made with variable contrast. Dynamic images enhance definition of structural features such as faults and fractures and sedimentary features such as bedding, erosional structures (scours and lag deposits), unconformities, karst zones, breccias, and debris flows. For the UBI logs, static and dynamic images were processed for both amplitude and travel-time images.

Bed-dip Analysis Procedure

This kind of analysis was done for 14 wells. In three cases, the analysis was applied twice because of the availability of dipmeter and image logs in the same well. We took advantage of this situation to test the validity of the interpretation of dipmeter logs and to find out their limitations. The adopted procedure includes interpretation of bedding planes, quality control, construction and interpretation of cumulative dip plots, and dip azimuth vector plots.

Picking Bedding Planes

Manual picking, mostly on dynamic amplitude images, was made. Bedding dips, ranging from 0° to 36°, were seen as sinusoidal waves with varying amplitude, depending on the deviation of the borehole. Bed boundaries were fitted with a sine wave through amplitude contrasts on dynamic-amplitude images of UBI logs, or with similar resistivity-colored strips, in the case of HDT and SHDT logs.

Because they have lower resolution around the borehole than UBI logs, HDT and SHDT data were the most difficult to interpret, especially when the data quality was poor. The output ASCII file obtained from bed dips contains dip type, depth, dip azimuth, dip magnitude, and quality values.

Quality Control of the Bedding-plane Orientations

For quality-control purposes, we compared the obtained data with a structural contour map. The dipmeter bedding-plane orientations were plotted on a Schmidt stereonet lower-hemisphere projection to obtain the mean dip azimuth and dip magnitude for every studied well. The reference map for mean dip azimuth and dip magnitude is a structural contour map obtained from log tops and seismic interpretation. The consistency of the bedding-plane orientations is compared with the structural map to determine if they are reasonable. Of 14 studied wells, 12 showed a good match with present-day structure as mapped from seismic data. Wells C-17 and M-14 show different structural attitudes, where local drag folding and reverse fault-associated folding, respectively, have been interpreted.

Construction of Cumulative Dip and Dip Azimuth Vector Plots

The cumulative dip-plot technique developed by Hurley (1994) is a common method to identify stratigraphic and structural dip domains (Prosser et al., 1999). These plots are used to determine the depths of faults and unconformities.

Information contained in the ASCII file of bedding-plane orientations can be arranged in a spreadsheet with data sorted by depth. Starting at the shallowest dip, a sequential sample number is given to every data point, and cumulative dip magnitude is computed. The cumulative dip plot then can be constructed by plotting sample number (Y axis) versus cumulative dip magnitude (X axis). The result is a crossplot in which inflection points separate dip domains. Dip domains are intervals with constant, or relatively constant, dip magnitude and/or direction. Sample number is used rather than depth, because bed boundaries are normally sampled at irregular depths. Depth plotting could create false inflection points. According to the geologic setting, those inflection points could be interpreted as faults, sequence boundaries, flooding surfaces, or other sedimentological features. Figure 8 is the cumulative dip plot for well C-65, where nine inflection points (some of them representing very subtle changes) coincide with log markers and reservoir zone boundaries. Cumulative dip plots were made for 14 wells.

Dip azimuth vector plots (which represent a map view) were created with the dip-direction data by plotting every computed value as a unit vector, starting from the deepest bedding-plane orientation. Each succeeding dip orientation is plotted where the deeper one ends, forming a continuous plot. The result is a straight line for every dip domain, which is separated by inflection points. These could be interpreted as faults, sequence boundaries, flooding surfaces, or other sedimentologic features. Figure 9 illustrates a dip azimuth vector plot for well C-65, which contains seven dip domains. Some of them are interpreted as sequence boundaries and flooding surfaces, which match with reservoir zone boundaries and log markers defined previously using conventional well-log correlations. Others have been interpreted as faults, based on information from the cumulative dip plot. Dip domains defined by cumulative dip and vector plots do not always match because some inflection points have changes in dip magnitude but not in direction. The converse is also true. Dip azimuth vector plots have been constructed for 14 wells.

Cumulative dip and vector plots should be done before perforations to avoid perforating in fault zones, with the high risk of water production, as in Catedral and Muspac fields.

Interpretation of Sedimentologic and Structural Features from Borehole Images

UBI images are more useful than dipmeters because they have full borehole coverage. Borehole images of five wells (C-65, C-75, N-101, M-42D, and M-43) in Catedral and Muspac

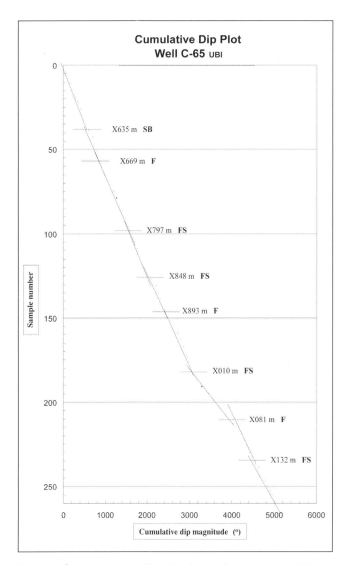

FIGURE 8. Cumulative dip plot for well C-65 for bedding planes obtained from UBI log. Axis Y is a sequential sample number, starting at the shallowest interpreted bed boundary. Axis X is the cumulative dip magnitude. Inflection points indicated at measured depths are interpreted as faults (F), flooding surfaces (FS), and sequence boundaries (SB); 259 bed boundaries were manually interpreted in 604 m of logged section.

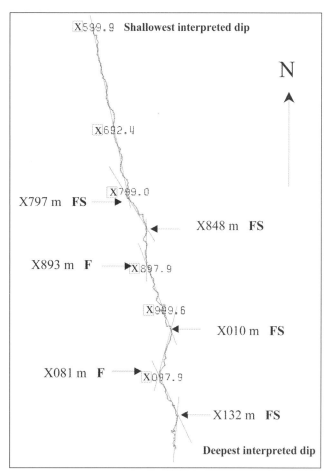

FIGURE 9. Dip azimuth vector plot for well C-65 for bedding planes derived from UBI log. Every interpreted dip, represented by a unit vector magnitude, is plotted, from the deepest to the shallowest reading. Dip azimuth changes (inflection points) are interpreted as faults (F), flooding surfaces (FS), and sequence boundaries (SB); 259 bed boundaries were manually interpreted in 604 m of logged section.

fields have been used to identify sedimentary and structural features.

Information about reservoir zone boundaries and their log markers and inflections from cumulative dip plots and vector plots at specific depths served as a framework for this analysis. The primary objective was to identify sedimentologic and structural features at those depths, looking for evidence to support or refute the presence of sequence boundaries, flooding surfaces, or faults. Examined evidence includes bed boundaries, karstic zones and collapse breccias, scour surfaces, debris flows, vuggy porosity structure, and open and healed fractures.

The analysis procedure consisted of displaying static- and dynamic-amplitude images and traveltime images on a computer workstation using Baker Atlas Recall/Review software. The interactive display and interpretation of these images made it possible to obtain results in a more reliable way. One technique consisted of selecting different color palette options, which permitted better contrast and enhanced some features. Another technique consisted of changing the vertical display interval to vary the resolution.

Bed boundaries refer to the acoustic contrast on static or dynamic images, which may indicate lithologic changes between clean carbonates and shaly carbonates (indicated by high gamma-ray values). Bed boundaries were interpreted using sine waves that were manually fitted with the above-mentioned contrasts. Figures 10, 11, and 12 show examples of bed boundaries.

Scour surfaces represent an erosional event on an underlying bedding surface. Such surfaces can be recognized on amplitude images or even on traveltime images by sharp lithologic

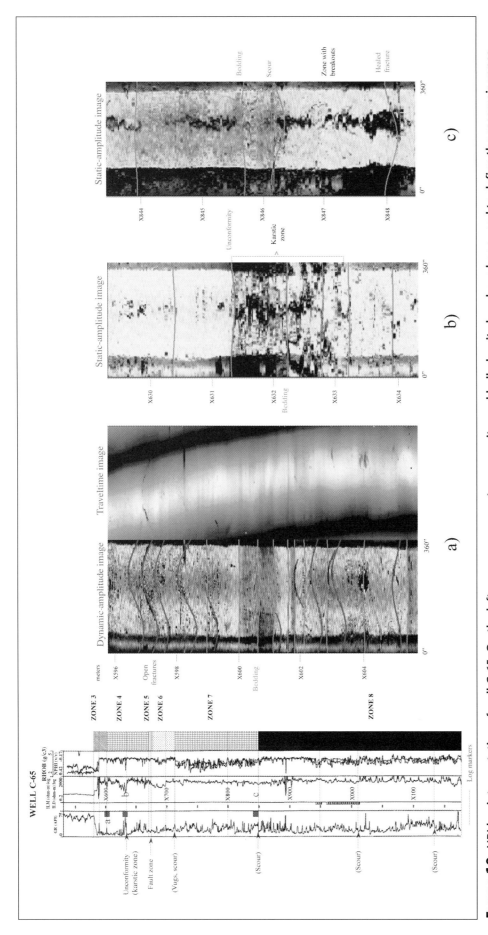

FIGURE 10. UBI log interpretation of well C-65. On the left, gamma-ray, neutron-porosity, and bulk-density logs have been used to define the reservoir zones. Some log markers are interpreted in zones 3, 4, 5, 6, 7, and 8. On the right, pictures of important features illustrate (a) highly fractured zone and bedding surfaces, with open fractures evident on both amplitude and traveltime images; (b) unconformity with karstic zone containing collapse breccias; and (c) scour and lithologic change at the top of zone 8 with breakout zone. Measured depths are in meters.

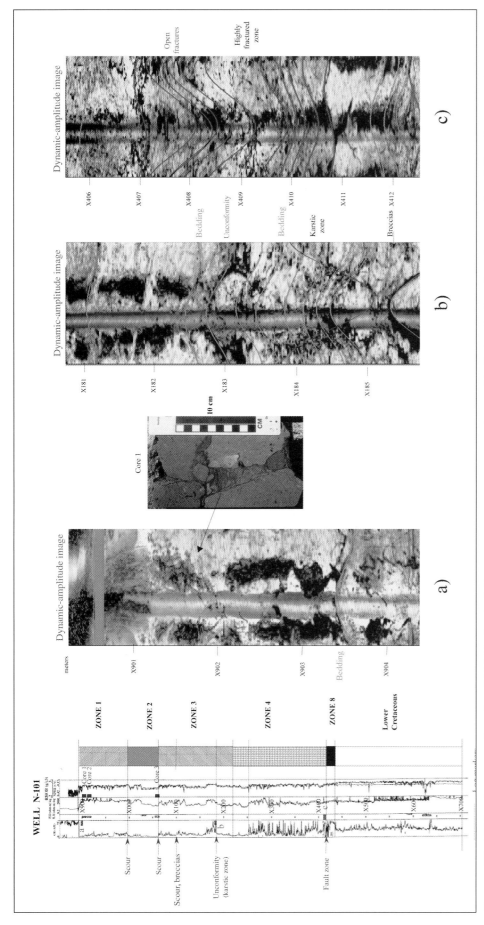

FIGURE 11. UBI log interpretation of well N-101. On the left, gamma-ray, neutron-porosity, and bulk-density logs have been used to define the reservoir zones. Some log markers are interpreted in zones 3 and 4. On the right, pictures of important sedimentologic and structural features illustrate (a) carbonates with debris flows with photograph of core 1 for comparison; (b) unconformity evidenced by changes in dip and karstic zone with collapse breccias; and (c) fault zone showing a highly fractured zone. Measured depths are in meters.

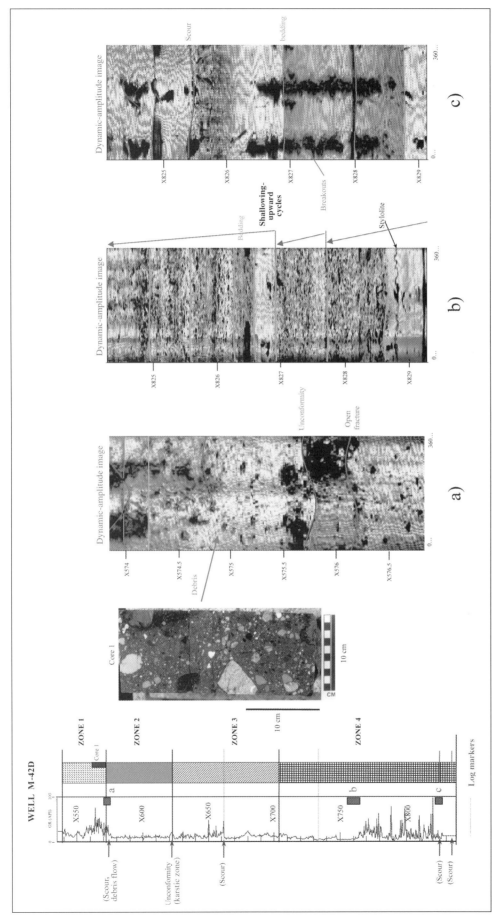

FIGURE 12. UBI log interpretation of well M-42D. On the left, the gamma-ray log has been used to define the reservoir zones. Some log markers are interpreted in zones 3 and 4. On the right, pictures of important features identified on the borehole image interpretation illustrate (a) unconformity (without karstic zone) covered by debris-flow deposits (breccias) with photograph of core 1 for comparison; (b) carbonate shallowing-upward cycles with high GR at the base and low GR at the top with vuggy porosity; and (c) scour indicating a sequence boundary and breakout zones. Measured depths are in meters.

contrasts. An additional criterion to recognize a scour surface is its commonly undulating shape that has a different dip orientation from nearby bedding. Figures 10c and 12c show examples of scours in wells C-65 and M-42D, respectively, at different stratigraphic levels. Some of these scours match with reservoir zone boundaries and log markers previously interpreted.

Flooding surfaces were assigned to lithologic changes identified as textural changes on the amplitude images, where high-energy carbonates with associated vuggy porosity were covered by a thin, shaly carbonate section (with high gamma-ray expression). These flooding surfaces are widely identified and correlated through Catedral and Muspac fields. Sometimes a small or subtle scour can be recognized at these surfaces. An example is illustrated in Figure 12b.

Vuggy porosity is defined as a textural feature on amplitude images with dotted, horizontally aligned patterns that have low gamma-ray values and are located usually at the top of shallowing-upward sequences, as illustrated in Figure 12b.

Karstic zones and collapse breccias were recognized on good-quality acoustic images. They are indicative of subaerial erosion in the presence of an unconformity. The typical brecciated texture is recognized on both amplitude and traveltime images. Good examples are depicted in Figures 10b (in well C-65) and 11b (in well N-101). Evidence of this kind of breccia has been reported from core descriptions from zones 2, 3, 4, and 5.

Debris flows (breccias) were identified as a disorganized, dotted texture with variable dot size. This kind of texture is pictured in Figure 11a (supported by a photograph of core 1 in well N-101) and in Figure 12a (supported by a photograph of core 1 in well M-42D).

Open fractures and healed fractures have a different dip attitude from bedding. Dynamic-amplitude images, in combination with traveltime images, differentiated between open and healed fractures. Open fractures appear as dark-colored traces caused by low amplitude and high two-way traveltime. Healed fractures may have amplitude contrast but no contrast on traveltime images. Examples of open fractures are shown in Figures 10a and 11c.

Faults have been identified where highly fractured zones occur in association with dip domain boundaries, as depicted in Figure 11c.

After borehole image interpretation of the results of bedding-plane analysis and reservoir zone framework, the information was integrated into two stratigraphic cross sections, illustrated in Figures 13 and 14. In these cross sections, seven wells in both Catedral and Muspac fields show reservoir zones and inflection points from cumulative dip and vector plots, along with sedimentologic and structural features identified from borehole image logs.

Sequence boundaries and associated unconformities have been observed. A sequence boundary is a widespread surface that represents a geometrically significant change in the pattern of sediment input and dispersal in a basin (Schlager, 1989). The limits of a depositional sequence are formed by two sequence boundaries, which represent unconformities and their correlative conformities. Therefore, the unconformities indicated in Figures 13 and 14 give the time and space distribution for interpreted sequence boundaries.

In addition to inflections on cumulative dip plots and vector plots, a decisive criterion to define unconformities was given by sedimentologic features identified on borehole images. When enough evidence—such as karstic zones and associated collapse breccias or scours and sharp lithologic changes on the same stratigraphic level—was found in different wells of Catedral and Muspac fields, that level was considered to be an unconformity. Five stratigraphic levels have been considered as unconformities: the top of zone 1, the top of zone 2, the top of zone 3, log marker Z3 (in zone 3), and log marker Z4 (in zone 4). Figures 13 and 14 show the indicators interpreted from dipmeter and borehole image logs. The karstic zones associated with these unconformities are interpreted to be highly permeable and can be paths for water coning or channeling, especially in wells on the flanks of the anticline structures where they are close to the gas-water contact. Examples include early water production in wells C-17, M-14, and M-43.

Marine flooding surfaces have also been identified. A marine flooding surface is defined as a surface that separates younger from older strata, across which there is evidence of an abrupt increase in water depth. This deepening is commonly accompanied by minor submarine erosion or nondeposition, with a minor hiatus indicated, but not by subaerial erosion (Van Wagoner et al., 1996).

In this study, 10 marine flooding surfaces were interpreted, based on picks identified from well-log correlations and inflection points from bedding analysis. They also contained evidence of scours and/or abrupt lithologic changes, suggesting a deepening in water depth during deposition of younger sediments on these flooding surfaces. Figures 13 and 14 show the distribution in time and space of those marine flooding surfaces.

Fracture Density and Fracture Orientation

Borehole image logs were a valuable source to identify open and healed fractures. Fractures wider than $1/32$ in. (0.08 cm) generally can be seen by the borehole televiewer. In some cases, damage may make the fracture appear to be wider than it is (Aguilera, 1997).

Some reservoir zones are more likely to contain fractures, depending on rock strength, shale content, bedding thickness, mineralogy, and other factors. The shaly carbonate sections, intervals with thin bedding, dolostone intervals, and fault zones are expected to contain more fracturing in the defined reservoir zones. Fracture density obtained from borehole-image log interpretation varies vertically through the eight reservoir zones. Highly fractured zones are clearly related to

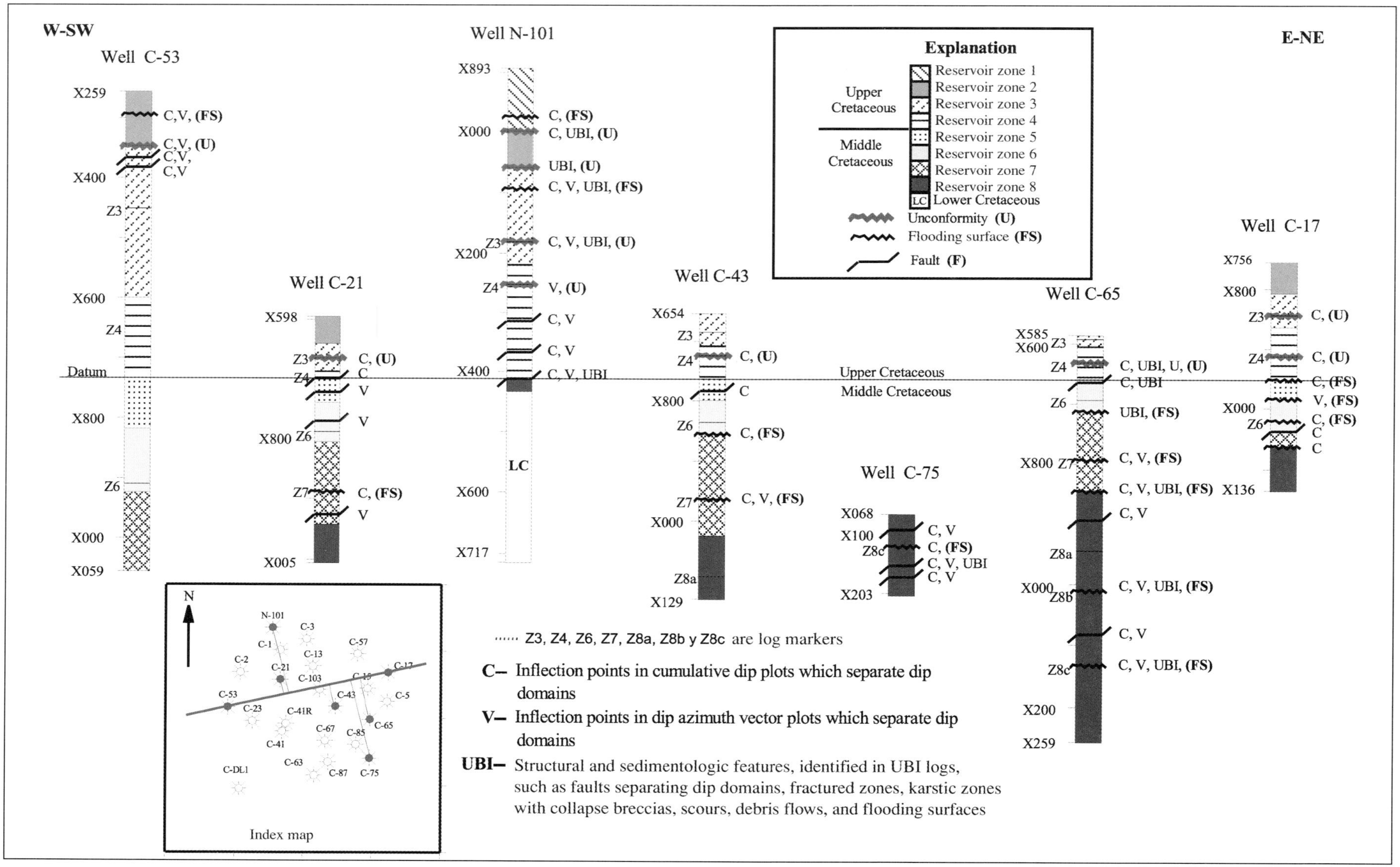

Figure 13. Stratigraphic correlation of reservoir zones showing the results of bedding-plane analysis in Catedral field. Unconformities, flooding surfaces, and faults were obtained from cumulative dip plots, dip azimuth vector plots, and identification of structural and sedimentologic features from UBI logs. Measured depths are in meters.

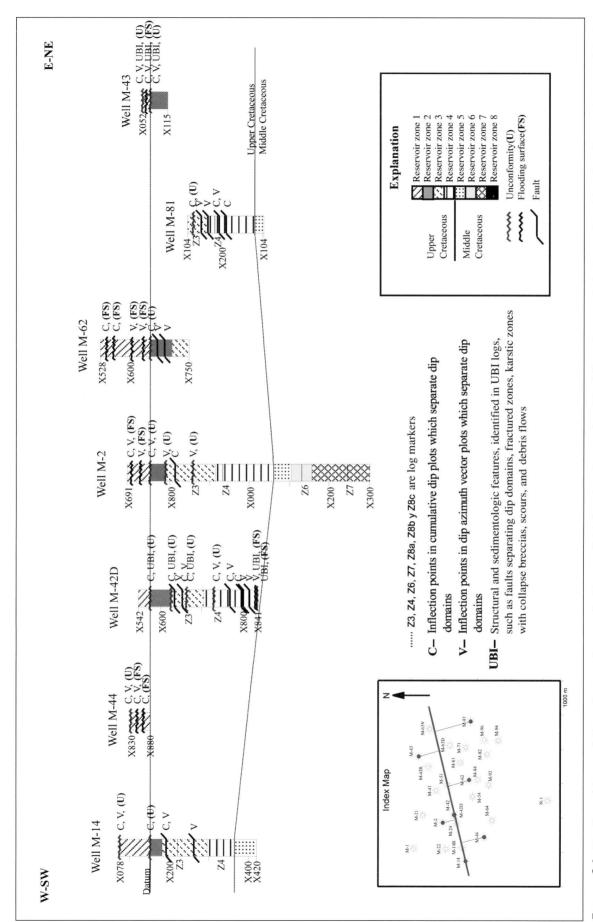

Figure 14. Stratigraphic correlation of reservoir zones in Muspac field showing the results of bedding-plane analysis. Unconformities, flooding surfaces, and faults were obtained from cumulative dip plots, dip azimuth vector plots, and identification of structural and sedimentologic features from UBI logs. Measured depths are in meters.

carbonates with low porosity, shaly carbonate sections, and thin bedding (of carbonates and shales).

Fracture populations interpreted in each well were plotted on a Schmidt stereonet lower-hemisphere projection to find out how many fracture sets were present and to determine the vector means for strike azimuth and dip magnitude.

Strike-azimuth rose diagrams were plotted on the structural map obtained from seismic interpretation by Kramer and Juárez (personal communication, 1999) to observe the relation of fracturing with seismically derived faults. Fracture sets derived from acoustic borehole televiewers are consistent with the orientation of these faults.

Fracture-density plots were built for studied wells. The plots included reservoir zonation obtained from open-hole logs. Figure 15 illustrates a fracture-density plot for well N-101; it plots the number of fractures contained in 10-m intervals versus depth. This example shows high fracture density associated with dolostones, fault zones, and low-porosity zones, especially in zones having high to moderate gamma-ray values. Thin-bedded carbonates and shales have more fracturing. Fracture-density plots for wells C-63, C-65, C-75, C-85, M-42D, M-43, and M-92 present similar results.

Maximum Horizontal In-situ Stress (SHmax) Orientations Using Borehole Breakouts

Any point located in the subsurface is being affected by three compressive stresses, which are usually of different magnitude and are commonly referred to as σ1 (maximum stress), σ2 (intermediate stress), and σ3 (minimum stress). One of the three stresses commonly is close to being vertical in flat-lying regions and nearly parallel to most boreholes (Bell, 1993). The borehole walls of any well are affected by these three stresses, and one of them (the maximum horizontal stress) may cause compressive failure in the rock and generate a borehole breakout. A breakout is defined as a wellbore elongation formed by nonuniform stress concentration in a homogeneous, isotropic medium (Springer, 1987). The elongation of the borehole is parallel to the minimum horizontal stress and perpendicular to the maximum horizontal stress. Breakout elongations are formed in the region of the borehole with smaller applied stress, where the tangential stress exceeds the shear strength, thereby causing failure of the rock.

All breakouts are borehole elongations, but not all borehole elongations are breakouts. Well-bore elongations can be washouts, key seats, or mud cake. A washout is an enlargement of the borehole in all directions; calipers C13 and C24 are at least 1 inch greater than the bit size (Bell, 1993). A key seat is formed when the drill string erodes an elongated well bore in a deviated well (Babcock, 1978, in Bell, 1993). Mud cake, which is the solid clay deposit formed in a borehole on a permeable layer when the liquid mud filtrate permeates the surrounding rocks, can be asymmetric, thereby creating a well-bore elongation (Bell, 1993).

Breakout Recognition and Maximum Horizontal In-situ Stress (SHmax) Orientations

Breakout recognition and measuring of in-situ stress orientations were done for 18 wells. The four-arm caliper logs included in dipmeter logs (HDT and SHDT) were used in 13 wells, and amplitude images from UBI logs were used in five more wells.

In addition to microresistivity curves, dipmeter logs record caliper size in C13 and C24 directions, as well as other measurements of borehole orientation such as hole azimuth (HAZI), vertical hole deviation (DEVI), and azimuth of pad 1 (P1AZ), which is the location of one of the four arms of the tool.

Four-arm caliper tools detect borehole elongations. When the tool is rising and logging, it rotates because of the cable torque. When the tool runs into an elongation, it stops rotating. Therefore, the important point is to identify intervals when the tool is rotating and to discard them. Akalin (1997) showed a technique to identify those intervals and look at the remaining intervals with potential breakout to obtain SHmax orientations. The first and most important step of this technique consists of constructing a graph that shows the well-bore geometry and tool and hole orientation. The plot includes, on the left, a gamma-ray curve, DEVI, and calipers (C13 and C24). On the right, differential caliper curves are plotted, azimuthally overlying P1AZ. HAZI is plotted at 90° offsets to visualize possible key seats. This graph is useful to identify at a glance potential breakout zones, zones of no elongation, key seats, washouts, and intervals where the tool is rotating.

Akalin (1997), Scuta (1997), and Knight (1999) used the following procedure to identify breakout zones and derive SHmax orientations. This procedure has been applied to 13 wells with dipmeter logs in Catedral and Muspac fields.

1) Digital data from C13, C24, DEVI, HAZI, and P1AZ curves were resampled to 0.5 m.

2) Using the graph that shows well-bore geometry, intervals were identified where the logging tool was rotating. These intervals were discarded because they do not contain breakouts.

3) Comparing caliper differences between C13 and C24, intervals with differences of 0.25 in. (.64 cm) or less were eliminated and labeled as intervals with no elongation.

4) Intervals in which the smallest caliper, either C13 or C24, is greater than or equal to 1 in. (2.54 cm) larger than the bit size were eliminated and labeled as washouts.

5) Comparisons between P1AZ and HAZI, HAZI+90°, HAZI+180°, and HAZI+270° were made to eliminate key seats. Intervals where P1AZ is within ±10° of those values were eliminated as possible key seats. Deviated wells with deviation greater than 30° were not analyzed because high deviation commonly causes key seats.

6) All remaining intervals were considered breakouts. The largest of the calipers between C13 and C24 was considered to be the elongation direction.

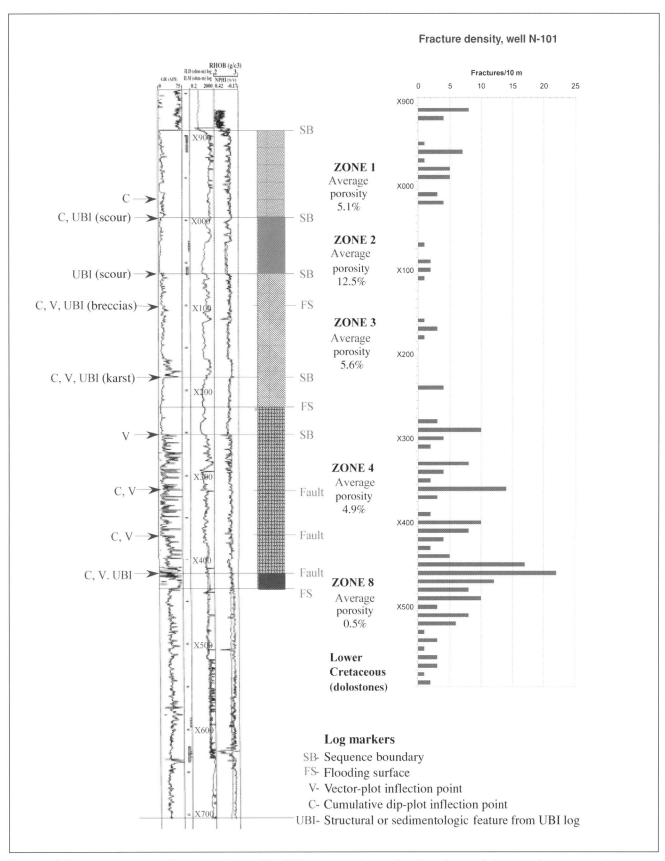

FIGURE 15. Fracture-density histogram for well N-101. Fracture density for 10-m intervals is plotted versus depth. Observe poor fracturing in zones 2 and 3, which are interpreted as high-porosity zones. The highest fracture density is associated with intervals containing thinly bedded carbonates and shales (high GR peaks) and dolostones (high density). The fault zones also have abundant fractures. Measured depths are in meters.

7) If the elongation direction was C13, the azimuth of SHmax is P1AZ+90°. If the elongation direction is C24, the azimuth of SHmax is P1AZ or P1AZ-180°. All SHmax orientations were converted to the range of 0° to 180° for plotting purposes.

This procedure has been applied to a spreadsheet of the output data composed of depth, calipers (C12 and C24), DEVI, HAZI, and P1AZ curves.

Identification of breakout zones in UBI logs was simpler than in dipmeter logs. The dynamic- or static-normalized amplitude images show the breakout zones as two dark strips separated by 180°, indicating continuous, discrete intervals with breakouts. Well-bore elongation causes a low-amplitude reflected signal, because the traveltime is greater than in the well-bore wall without elongation. To visualize breakout zones, displays with 10–15 m of vertical section on dynamic and traveltime images can be used. The azimuth of SHmax can be obtained as the direction perpendicular to elongations, which is computed by fitting a sinusoidal wave to the vertical dark strips that represent breakouts. SHmax orientations were interpreted from five wells with UBI logs.

SHmax orientations interpreted from UBI logs provide more continuous, consistent, and reliable data. Wells C-65, C-75, M-43, M-42D, and N-101 have 209 m, 85 m, 11 m, 76 m, and 364 m, respectively, of intervals with SHmax orientations. Vector means for every well, obtained from frequency rose diagrams, vary from 118° to 142°, which is a persistent orientation.

All SHmax orientation data were plotted on a structural map derived from a 3-D seismic interpretation of Kramer and Juárez (personal communication, 1999). The final map (Figure 16) shows a regionally persistent northwest-southeast SHmax orientation for most data of Catedral and Muspac fields.

Discussion

The observed orientation of SHmax is interpreted as a structural inversion that affects the area today. This is probably related to the active subduction zone located to the south of Mexico.

According to data obtained from UBI log interpretation, fractures and faults oriented in the range of ±30° of the SHmax orientations are interpreted to be open (Barton and Zoback, 1994) and affecting reservoir performance in both Catedral and Muspac gas fields. Evidence of this seems to be high-permeability zones with early water production associated with fault zones that have been detected in wells C-17, C-41, and C-57 in Catedral field and M-44, M-63D, and M-64 in Muspac field.

Open-fracture orientation is extremely important as a structural control on reservoir performance. Consequently, the best orientation for deviated wells drilled to avoid open fractures is northwest-southeast, thereby minimizing the water coning through them. Most of the deviated wells in Catedral and Muspac fields are perpendicular to SHmax orientation, which means that they may intersect several potential paths for water coning.

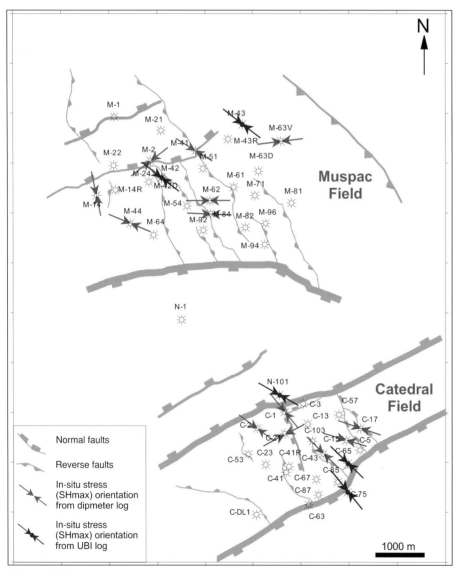

FIGURE 16. In-situ stress (SHmax) orientation map for Catedral and Muspac fields, interpreted from breakout analysis applied to dipmeter (HDT and SHDT) logs and acoustic borehole televiewers (UBI logs). Note the predominant northwest-southeast orientation, which is the direction of the open fractures in both fields. East-west orientation can be interpreted as a local rotation of SHmax associated with faulting.

INTEGRATION AND INTERPRETATION OF 3-D SEISMIC DATA

A 3-D seismic interpretation made by Kramer and Juárez (personal communication, 1999) constituted the input data of this study and consisted of time maps, depth maps, seismic sections (lines and traces), with well control and attribute maps such as edge enhancement and root-mean-square (rms) amplitude.

The 3-D seismic survey has a 20 m × 20 m bin size, the maximum stacking is 2250%, and the frequency content of seismic data is low (20–25 Hz). Structural resolution is good, but stratigraphic resolution is limited.

Two regional seismic sections are used to illustrate the structural model in Catedral and Muspac fields. In addition, five local seismic sections detailing the area of the fields, with well control in time, were used to show the structural framework in the reservoir and its relation to the reservoir zonation discussed in a previous section. Moreover, seismic attribute maps (edge enhancement and rms amplitude) were used to detect fracture zones and/or small-scale faults not imaged on conventional seismic sections.

Structural Model from 3-D Seismic Sections

Two regional seismic sections, one for Muspac field and another for Catedral field, were selected and interpreted to illustrate the structural configuration in both fields. Trace 700 in Muspac field shows a fault-bend fold structure with a well-defined ramp and two detachment levels (Figure 17). The lower detachment level probably corresponds to Tithonian shales, and the upper detachment level appears to be basal Paleocene shales. Salt (penetrated by well M-41) is present in the core structure. The resultant fold is a wide, well-defined anticline with three secondary thrust faults and three minor backthrust faults. The transport direction is from the southwest to northeast. Trace 400 in Catedral field depicts a fault-bend fold structure with two detachment levels similar to the Muspac field structure (Figure 17). The fold is a tight anticline with less displacement than the fold in Muspac field, and it is structurally more complex. The forelimb of this fold can be described as a trailing imbricate fan with two

younger thrust faults. The backlimb of this structure contains four back-thrust faults, which caused apparent normal faults on the crest of the anticline.

Seismic Sections with Control Wells

Five seismic sections were constructed from the 3-D survey through selected wells to illustrate the faults that constitute the structural framework in the reservoir. That structural framework was taken as a reference to create structural cross sections that include reservoir zones discussed in a previous section.

Open-hole log correlation of 45 wells served to identify 20

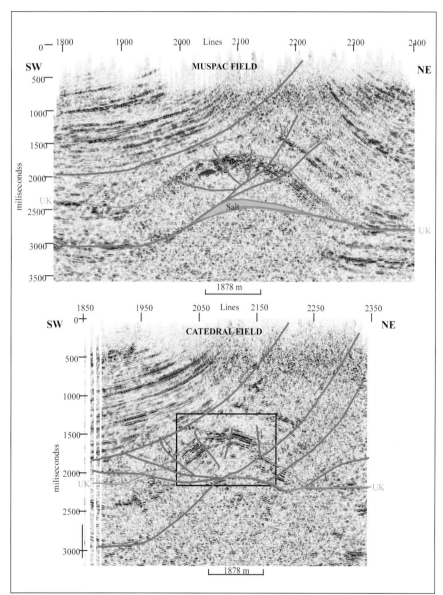

FIGURE 17. Seismic sections through Muspac and Catedral fields showing a fault-bend fold with a major ramp and secondary thrust and backthrust faults. The black outlined rectangle shows the location of Figure 18. Note that the anticlinal structure in Catedral field is tighter than the structure in Muspac field.

faults in both fields. The cumulative dip and azimuth vector plots identified 29 faults in both fields.

There is seismic evidence for eight of the 29 faults analyzed from logs; they are considered seismic-scale faults. In this study, seismic-scale faults are defined as those that are apparent using the available 3-D seismic data. Subseismic-scale faults cannot be seen on the existing 3-D seismic data. The other 21 faults detected by cumulative dip and vector plot techniques and borehole image interpretation are considered subseismic-scale faults.

Figure 18 shows a close-up of Catedral field, where some seismic-imaged faults are recognized. A diagrammatic structural cross section including reservoir zones depicts perforations located close to the gas-water contact. Wells C-53 and C-21 were studied with bedding analysis and borehole breakout analysis.

There appears to be a good match between seismic-imaged

faults and faults detected with well-log correlation and cumulative dip and vector plot techniques (Figure 18). Despite VSP control in two wells, the time-to-depth conversion is not exact.

Interpretation of Seismic-attribute Maps

Two kinds of seismic-attribute maps (edge enhancement and rms amplitude) obtained from a 3-D seismic interpretation made by Kramer and Juárez (personal communication, 1999) were analyzed and interpreted to look for fracture and fault zones.

A seismic attribute is necessarily a derivative of a basic seismic measurement (Brown, 1999). In the edge-enhancement attribute, the basic seismic information is two-way traveltime. In the rms-amplitude attribute, the basic seismic information is the amplitude of the seismic signal.

The edge-enhancement seismic attribute calculates abrupt changes in time for a picked horizon or a specific interval. This

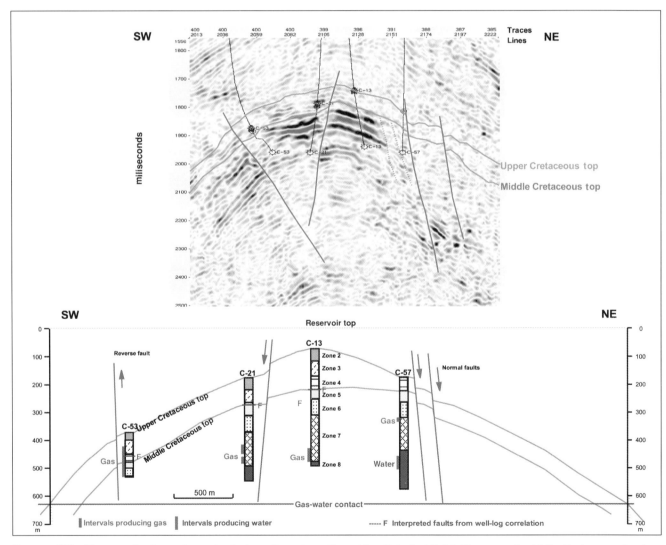

FIGURE 18. Above, seismic section oriented southwest-northeast for Catedral field. Upper and middle Cretaceous horizons and faults have been interpreted. Wells are displayed for control. Below, diagrammatic structural cross section showing reservoir zones and perforations with intervals producing water and gas. Observe erosion of reservoir zones 1, 2, and 3 eastward. Note that perforations are close to the gas-water contact. Some subtle faults have been interpreted to explain the early water production in well C-57.

approach compares every trace contained in the 3-D seismic survey. The reflection that corresponds with the top of the middle Cretaceous has the most consistent reflection character across the 3-D survey, and for that reason, it was chosen as the reference horizon. In Figure 19, the edge-enhancement attribute is displayed for a 20-ms window with zero offset from the top of the middle Cretaceous for Catedral field. When no abrupt changes existed, there was high coherence (measure-

ment of similarity between two or more traces), as indicated by a dark gray color. When abrupt changes occurred, there was low coherence, which is represented by a light gray color. Zones with low coherence are recognized as alignments that can be interpreted as subtle faults (small throw) or fracture zones. The orientation of those alignments fits with well-imaged seismic faults identified on seismic sections.

The rms-amplitude attribute tends to be more effective than

STRUCTURAL ALIGNMENTS FROM EDGE-ENHANCEMENT ATTRIBUTE

1 km

FIGURE 19. Interpretation of seismic-attribute map for Catedral field. Edge-enhancement seismic attribute (light gray for low coherence and dark gray for high coherence) for a 20-ms window with zero offset from the top of the middle Cretaceous. Structural alignments with low coherence are interpreted as faults or fracture zones. Seismic-attribute map was taken from Kramer and Juárez (personal communication, 1999).

absolute amplitude, because the high amplitudes are boosted by the squaring (Brown, 1999). The rms amplitude was calculated over a 20-ms window with 20-ms offset above the top of the middle Cretaceous in Muspac field and a 20-ms window with 20-ms offset below the top of the middle Cretaceous for Catedral field. The interpreted structural alignments obtained from the edge-enhancement attribute were overlaid on the rms-amplitude attribute map to analyze the correlation between them. Most showed a good match for both seismic attributes. Some evident structural alignments in rms-amplitude attribute maps, not resolved by edge enhancement, were interpreted also.

A map (Figure 20) including all structural alignments derived from both edge-enhancement and rms-amplitude attributes summarizes the results and presents an interpretation of the fracture and fault zones and density in each field.

Some acquisition artifacts exist in the seismic-attribute maps. These are nearly parallel to the interpreted lineations, and they should be identified carefully to avoid misinterpretations.

Discussion

Major faults at reservoir scale are well imaged on seismic section displays and give the orientation of minor faults and fracture zones that may affect the reservoir. They provide the structural framework to build structural sections with the reservoir zonation.

Seismic-attribute maps provide a method to detect the fracture and faulting network that is not resolved with seismic-section displays. This structural network gives an interpretation of the fracture-fault density in a spatial view. Edge-enhancement-derived alignments indicate that Catedral field is more densely fracture-faulted in the same interval than Muspac field (Figure 20). The structural fold in that field is tighter; thus, more fracturing is expected. Another probable reason is that a 20-ms window, which represents about 50–60 m of thickness, includes both reservoir zone 5 and upper reservoir zone 6. Zone 5 is similar in both fields, but zone 6 has higher shale content, less porosity, more intervals with thin bedding (with carbonates and shales), and less quality index than Muspac field. These properties could make the unit more brittle and consequently more fractured.

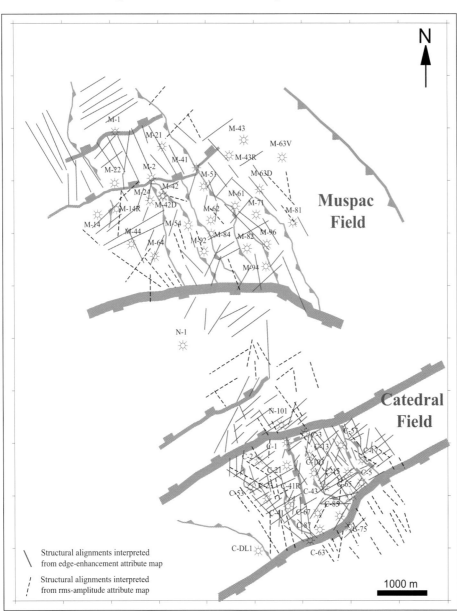

FIGURE 20. Structural alignments in Catedral and Muspac fields, interpreted from edge-enhancement and rms-amplitude attribute maps. Alignment orientations are consistent with major seismically derived faults.

ANALYSIS AND INTERPRETATION OF PRODUCTION DATA

A database containing perforated intervals and monthly production of gas, oil, and water was used in this analysis. In addition, well-log analysis, volumetric-parameter contour maps, and structural diagrams illustrating reservoir zonation and perforations for both fields were used to analyze and interpret the production data.

Structural Position of Perforated Intervals

The structural position with respect to the gas-water contact of every perforated interval is thought to be a critical control on early water production. Some of the wells close to the gas-water contact and some of the wells on the flanks of the structure had early water production in both fields, as is shown in Figures 21 and 22. In these figures, all wells are projected into a southwest-northeast line of cross section. Reservoir zone tops are in true vertical depth, and perforated intervals indicate gas-oil production or early water production. In general, wells without water problems are structurally high in the fields. Wells with early water production in Muspac field are M-1, M-2, M-14, M-43, M-44, and M-64. Wells with water problems in Catedral field are C-2, C-17, C-23, C-41, C-57, C-63, and C-75. Although Well N-101 does not belong to this reservoir, it was included in the diagram.

Production Data Analysis

A first step in this analysis was to organize production data and to compute an average monthly production rate for gas, oil, and water for each perforated interval in every reservoir zone. The source data contained information about every perforated interval, including start and end of production and monthly production (oil, gas, and water). Some wells have been perforated in several intervals at different times because they had high water cut. Average monthly productions were obtained for every interval. Intervals were grouped by reservoir zones when possible. Some wells have perforations that include two or more zones. They were not considered, because it was impossible to know the production for every zone.

Representative wells (wells producing over several months) in the same reservoir zone were chosen to construct histograms to analyze average monthly production of gas, oil, and water. (Wells recently drilled were not included, because they do not represent stable production.) The mean values were calculated to simplify the comparative analysis. Zone 1 seems to be the reservoir zone with the highest productivity for gas and oil and the minimum productivity for water in Muspac field. In Catedral field, zone 7 has the highest productivity for gas and oil and a medium value for water productivity.

The results of this simple analysis show only trends of productivity. They are not entirely valid because they depend on (1) the structural position of the wells in the field, (2) the position with respect to the gas-water contact, (3) the length of the perforated intervals, and (4) the anisotropy in the reservoir zones, as indicated on the contour maps by the volumetric parameter (Figure 7).

The Early Water Production Problem

Some wells that have been producing over long periods of time were selected to analyze water production. Graphs with barrels of water versus number of months producing were created. The water production does not show any pattern that could be associated with a specific reservoir zone. Several wells

in the same reservoir zone and in similar structural positions show different rates of water production at different times.

Discussion

Wells perforated at levels structurally low and close to the gas-water contact in zone 8, lower zone 7, and zone 6 have early water production. These zones contain a high percentage of low-porosity carbonates, shaly carbonate rocks with low porosity, and thin beds (of carbonates and shales) that cause high fracturing and have low storage capacity. This situation produces water coning through fractures with high permeability. Wells in this situation are C-2, C-17, C-57, C-63, C-75, and M-1 (Figures 21 and 22).

Early water production in wells perforated at structurally high levels, far from the gas-water contact, is attributed to the presence of faulting oriented parallel to SHmax (mostly northwest-southeast). Examples are wells C-17, C-41, C-57 (Figure 18), M-2, M-44, M-64, and M-63D (high water cut). Others wells, such as C-63, C-75, and N-101, which produced water production very quickly (in a few months) or never produced gas and oil, are associated with zones of fault intersection with high fracturing (reverse and major normal faults that form the structural limits in Catedral field) (Figure 16). Pressure data from both fields show similar pressures in wells perforated in different reservoir zones. Therefore, the reservoir is thought to be communicated through a fault and fracture network, which breaks depositional seals.

Early water production in wells on the flanks of both structures in Catedral and Muspac fields occurs in some wells that have been perforated in reservoir zones with high porosity and low fracturing. This is interpreted to have been caused by high-permeability karstic zones associated with the five unconformities interpreted in zones 1, 2, 3, and 4 (Figures 10, 11, 12, 13, and 14). These high-permeability surfaces are thought to cause water channelization in wells close to the gas-water contact, such as C-17, M-14, and M-43.

Production from Muspac field is obtained from clean carbonates with excellent porosity and low fracture density and from perforations located structurally high. In this field, zone 1, which has less porosity than zones 2 and 3, apparently is producing more oil and gas because it is more permeable as a result of fracturing, as illustrated on fracture-density plots (Figure 15). Zones 2 and 3, with high storage capacity, are the reservoirs of gas and oil for the high flow capacity of zone 1. According to fracture-density logs, zone 1 is moderately fractured. Its high permeability could be a problem in the future, when the gas-water contact moves close to the top of the reservoir.

In Catedral field, zone 5 (located structurally high in the reservoir) should be perforated. This zone is formed by homogeneous clean carbonates (shale free) with good to excellent porosity and low fracture density. This recommendation is proposed to avoid intersecting open fractures to increase the contact area with the reservoir, and to reduce coning rates by decreasing the vertical pressure gradients. Figure 23 depicts a

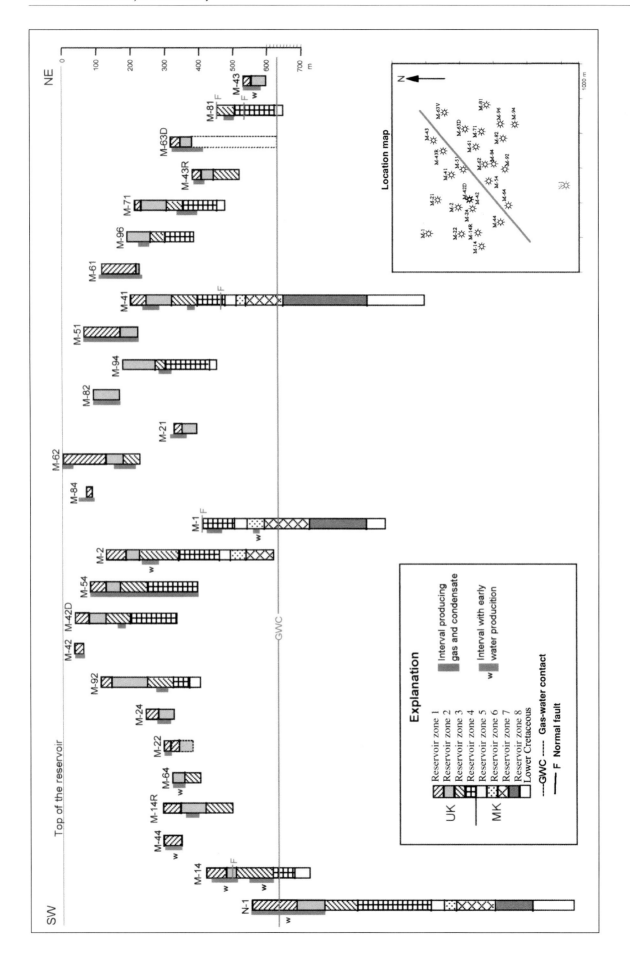

Figure 21. Diagram showing structural position with respect to the gas-water contact of wells in Muspac field. Wells are projected into a southwest-northeast line of cross section. Perforations are indicated for gas- and oil-producing wells and for early water production. Depths are true vertical depths below the structural top of the reservoir. Note that perforations are located mostly in zones 1, 2, and 3 in this field.

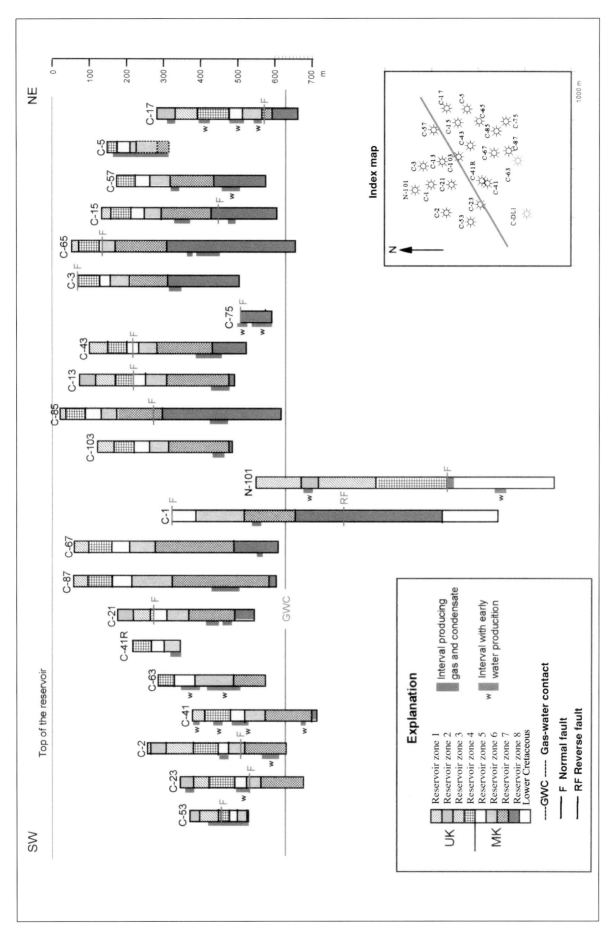

FIGURE 22. Diagram showing structural position with respect to the gas-water contact of wells in Catedral field. Wells are projected into a southwest-northeast line of cross section. Perforations are indicated for gas- and oil-producing wells and for early water production. Depths are true vertical depths below the structural top of the reservoir. Note that perforations are located mostly in zones 5, 6, 7, and 8 in this field.

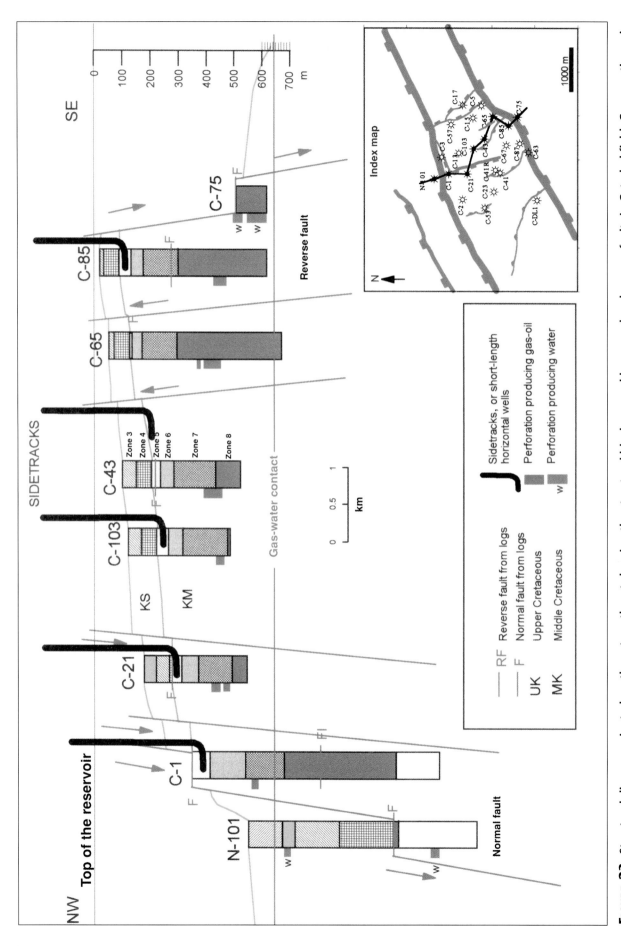

FIGURE 23. Structural diagram oriented northwest-southeast showing the structural blocks caused by normal and reverse faults in Catedral field. Cross-section orientation is parallel to maximum horizontal in situ stress orientation (SHmax). Locations for some sidetracks or short-length horizontal wells are sketched in reservoir zone 5.

structural diagrammatic cross section oriented parallel to SHmax with sidetracks, or short-length horizontal wells, in reservoir zone 5. A strategy is proposed in the following section to increase the gas-oil production, reduce the water production and water coning, and optimize the efficiency of the gas swept by the aquifer. The best example of the efficacy of this strategy is shown by reservoir performance in Muspac field, where the current production scenario is following a similar scheme.

CONCLUSIONS AND RECOMMENDATIONS

Structural and stratigraphic controls affect reservoir performance in Catedral and Muspac fields.

Varying rock properties in eight reservoir zones obtained from well-log correlation of 45 wells are considered to be a stratigraphic control affecting reservoir performance in these gas fields. The volumetric parameter (H•phi•[1-Sw]) has been mapped to help show this variability. Certain zones in Muspac and Catedral fields have the best reservoir properties.

Another example of stratigraphic controls is five unconformities with associated high-permeability karstic zones that have been interpreted from dipmeter and borehole-image logs.

Structural controls have an important effect on reservoir behavior in Catedral and Muspac fields. Fracture populations derived from interpretation of eight borehole-image logs indicate two preferential fracture sets oriented northwest-southeast and southwest-northeast. These are parallel to the orientation of seismically derived faults. Fracture networks obtained from interpretation of seismic-attribute maps (edge enhancement and rms amplitude) show a fracture-fault orientation similar to the seismically derived faults and to the fracture sets obtained from borehole-image log interpretation. Every fracture in this network is a potential path for water coning and gives a map view of the fracture density in these two fields.

Fracture density obtained from borehole-image log interpretation varies vertically through the eight reservoir zones. Highly fractured zones are clearly related to carbonates with low porosity, shaly carbonate sections, and thin bedding (of carbonates and shales). Fracture density is an important structural control affecting reservoir performance in Catedral and Muspac gas fields.

Results from maximum horizontal in-situ stress analysis (SHmax) for 18 wells indicate a predominant SHmax northwest-southeast orientation, which is the direction of open fractures. Therefore, every fracture in the range of the SHmax orientation ±30° degrees is interpreted to be open and affecting the reservoir performance in both Catedral and Muspac fields. Open-fracture orientation is extremely important as a structural control on reservoir performance.

Subseismic scale faults constitute an important structural control on reservoir performance; they can be evaluated with dipmeter or borehole-image logs applying the cumulative dip and vector plot techniques. These techniques should be done before perforations to avoid perforating in fault zones, which are high-risk zones to produce water.

This study has led to a set of recommendations for both fields. In Muspac field, the storage and flow capacities are associated with reservoir zones 1, 2, and 3. Perforations are located far from the gas-water contact. The production scenario seems to be working well except for some situations that could be modified by the following suggestions: (1) avoid wells in the fault zones, (2) avoid wells on the flanks close to the gas-water contact to minimize the risk of water channeling through karstic zones, and (3) avoid deviated wells drilled perpendicular to the direction of open fractures. The same suggestions mentioned for Muspac field apply to Catedral field. In addition, we propose that zone 5, located structurally high in the reservoir and formed by homogeneous clean carbonates (shale free) with good to excellent porosity and low fracture density, be perforated to minimize or avoid early water production in this field.

ACKNOWLEDGMENTS

Thanks go to PEMEX Exploración y Producción for permission to publish this work. We also thank Activo de Producción Muspac and Activo de Exploración Macuspana for open access to the database used in this research. Thanks go to Agustín Moreno, Julián Juárez, Ricardo Alcántara, and Hector Solís for their help in gathering the information. Special thanks go to Bob Cluff and Mark Kramer for sharing their ideas and for their support with the formation evaluation and seismic data. We also thank Javier Meneses for his review of this manuscript.

REFERENCES CITED

Aguilera, R., 1997, Formation evaluation of naturally fractured reservoirs, in D. Morton-Thompson and A. M. Woods, eds., Development geology reference manual: AAPG Methods in Exploration Series, no. 10, p. 92–93.

Akalin, A., 1997, Multi-scale characterization of fracture networks, Tokaris and Besikli oil fields, southeast Turkey: Master's thesis, Colorado School of Mines, Golden, 188 p.

Babcock, E. A., 1978, Measurement of subsurface fractures from dipmeter logs: AAPG Bulletin, v. 62, p. 1111–1126.

Barceló-Duarte, J., M. Varela-Santamaría, U. Hernández-Romano, and M. Martínez-Medrano, 1994, Facies sedimentarias de la porción oriental de la Plataforma Artesa–Mundo Nuevo: Área de exploración de recursos energéticos del subsuelo: DEPFI-Universidad Nacional Autónoma de México, Mexico City, 115 p.

Barton, C. A., and M. D. Zoback, 1994, Stress perturbations associated with active faults penetrated by boreholes: Evidence for near complete stress drop and a new technique for stress magnitude measurement, in B. Haimson, ed., Proceedings of 43th U. S. Symposium on Rock Mechanics, v. 1, p. 65–68.

Bell, J. S., 1993, Investigating stress regimes in sedimentary basins using information from oil industry wireline logs and drilling records, in A. Hurst, M. A. Lovell, and A. C. Morton, eds., Geological applications of wireline logs: Geological Society Special Publication Classics, Geological Society, London, p. 305–325.

Brown, R. A., 1999, Interpretation of three-dimensional seismic data: AAPG Memoir 42 and Society of Exploration Geophysicists Investigations in Geophysics, no. 9, 5th edition, Tulsa, Oklahoma, AAPG/SEG, 514 p.

González-García, R., and Holguín-Quiñones, N., 1992, Las rocas generadoras de México: Boletín de la Asociación Mexicana de Geólogos Petroleros (AMGP), v. 42, no. 1, p. 9–23.

Handford, C. R., and R. G. Loucks, 1993, Carbonate depositional sequences and systems tracts—responses of carbonate platforms to relative sea-level changes, *in* R. G. Loucks and J. F. Sarg, eds., Carbonate sequence stratigraphy—recent developments and applications: AAPG Memoir 57, p. 3–41.

Höcker, C., K. M. Eastwood, J. C. Herweijer, and J. T. Adams, 1990, Use of dipmeter data in clastic sedimentological studies: AAPG Bulletin, v. 74, p. 105–118.

Hurley, N. F., 1994, Recognition of faults, unconformities and sequence boundaries using cumulative dip plots: AAPG Bulletin, v. 78, p. 1173–1185.

Knight, C. N., 1999, Structural and stratigraphic controls on Mesaverde reservoir performance: North Labarge field, Sublette county, Wyoming: Ph.D. dissertation, Colorado School of Mines, Golden, 195 p.

Luthi, S. M., 1993, Borehole imaging devices, *in* D. Morton-Thompson and A. M. Woods, eds., Development geology reference manual: AAPG Methods in Exploration Series, no. 10, p. 163–166.

Pindell, J. L., S. C. Cande, W. C. Pitman III, D. B. Rowley, J. F. Dewey, J. Labrecque, and W. Haxby, 1988, A plate-kinematic framework for models of Caribbean evolution: Tectonophysics, v. 155, p. 27–48.

Prosser, J., S. Buck, S. Saddler, and V. Hilton, 1999, Methodologies for multi-well sequence analysis using borehole image and dipmeter data, *in* M. A. Lovell, G. Williamson, and P. K. Harvey, eds., Borehole imaging: Applications and case histories: Geological Society Special Publication No. 159, Geological Society, London, p. 91–121.

Ross, M. I., and C. R. Scotese, 1988, A hierarchical tectonic model of the Gulf of Mexico and Caribbean region: Tectonophysics, v. 155, p. 139–168.

Schlager, W., 1989, Drowning unconformities on carbonate platforms, *in* P. D. Crevello, J. L. Wilson, J. F. Sarg, and J. F. Read, eds., Controls on carbonate platforms and basin development: Society of Sedimentary Geology (SEPM) Special Publication 44, p. 15–25.

Schlumberger, 1998, Log interpretation—Principles and applications: Schlumberger Wireline and Testing, Sugar Land, Texas, 216 p.

Scotese, C. R., L. M. Cahagan, and R. L. Larson, 1988, Plate tectonic reconstructions of the Cretaceous and Cenozoic ocean basins: Tectonophysics, v. 155, p. 27–48.

Scuta, M. S., 1997, 3-D reservoir characterization of the Central Vacuum Unit, Lea County, New Mexico: Ph.D. dissertation, Colorado School of Mines, Golden, 274 p.

Springer, J., 1987, Stress orientation from well bore breakouts in the Coalinga region: Tectonics, v. 6, no. 5, p. 667–676.

Tucker, M. E., and V. P. Wright, 1990, Carbonate sedimentology: Oxford, Blackwell Science Ltd., 482 p.

Van Wagoner, J. C., R. M. Mitchum, K. M. Campion, and V. D. Rahmanian, 1996, Siliciclastic stratigraphy in well logs, cores, and outcrops: AAPG Methods in Exploration Series, no. 7, 55 p.

Varela-Santamaría, M., 1995, Una plataforma aislada en el sureste de México: Master's thesis, Universidad Nacional Autónoma de México, Mexico City, 194 p.

Williams-Rojas, C. T., 1995, Estudio estratigráfico-sedimentológico y diagénetico del área del campo Catedral, Edo. de Chiapas: Master's thesis, Universidad Nacional Autónoma de México, Mexico City, 171 p.

Williams-Rojas, C. T., 2000, Geological controls on reservoir performance in two gas fields, southeastern Mexico: Master's thesis, Colorado School of Mines, Golden, 251 p.

Index

A

Abkatun Formation, 391, 392, 394
Acarinina, 406–407
Adjuntas Formation, 234, 236
Akimpech Formation
    lithofacies distribution in, 347–349
    stratigraphy of, 345–349
Albian
    of Ixtapa graben, 200–202, 204
    paleogeographic map of northeast Mexico, 71
    sequence stratigraphy of Mexico, 70
Ammonites
    biostratigraphy of, 9
    France-to-Mexico correlation of, 384
    heteromorphs of, 358
    Jurassic of Mexico, 5–11, 23–25
    of Potrero de García, 361–362, 364, 368
    sexual dimorphism in, 360
    of Sierra Madre Oriental, 148, 376, 377
    systematic paleontology of, 360, 362–368
    of Taraises Formation, 355–368
    whorl measurements of, 363, 365
    of Zacatecas Formation, 298, 302
Ammonoids, Triassic of Mexico, 297, 298, 302, 305
Anatomites, 302
Ancient continent, of eastern Mexico, 16, 20
Angostura field, cross section of, 287
Anhydrite, in Rodador field, 435
Antimonio Formation, 304–305
Apatite geochronology, of Sierra Madre Oriental burial,
    167, 170–174
Aptian
    paleogeography of northeast Mexico, 67
    paleogeography of Sabinas Basin, 257, 258
    sequence stratigraphy of Mexico, 66–70
Area Chiapas-Tabasco, location of, 444
Arroyo La Tinaja, Viento Formation in, 237
Arteaga complex, 308
Artesa–Mundo Nuevo platform, sedimentary model for,
    444–446
Ataxioceras, 9, 20, 25
Austin Formation, 255

B

Baja California, Triassic of, 305
Bajocian, transgressions of, 5–7, 8
Balsas Portals, 3, 5
Barcodon wells, 12, 14
Barranco Group, 304
Barremian
    paleogeography of northeast Mexico, 67
    paleogeography of Sabinas Basin, 255, 257
    sequence stratigraphy of Mexico, 66–68
Basinal assemblages, of Mesozoic Mexico, 38
Bathonian, of Gulf of Mexico, 19

Bay of Campeche, 85
    burial-history chart of, 115
    charts of oil-field information for, 96–98, 102–103
    map and oil fields of, 88
    oil/source-rock correlation in, 131
    source rocks of, 112
Bejucoceras, 357, 358, 361
Bejuco paleocanyon
    location of, 422
    Neocomanian-Eocene unconformity in, 428–429
    stratigraphic analysis of, 428
Berriasian
    ammonites of Taraises Formation, 353–354
    paleogeography of Sabinas Basin, 253
    paleogeography of Sierra Madre Oriental, 373
Biomarkers, for Mexican Gulf Coast Basin oils, 132–133,
    135
Bivalves, Jurassic of Mexico, 24, 25
Bocacajeta wells, 13, 15
Boca La Carroza, lentil of, 226, 227, 229
Bochianites, 357, 358, 376
Bonanza fold nappe, 153, 156
Boquiapan-Balsam transgression, 10
Borehole-image analysis, for Muspac and Catedral fields,
    450–462
Bositra, 25
Bouguer gravity anomalies, of north-central Mexico,
    30–31
Burial history
    of Bay of Campeche, 115
    of Pimienta-Tamabra(!) petroleum system, 113–115
    of Sierra Madre Oriental, 175
Burro-Salado arch, tectonics of, 52

C

Caborca Block, 37
Cahuasas Formation, 14
Cahuayotes wells, 12, 13
Callovian
    of Gulf of Mexico, 19
    paleogeography of Mexico, 5, 62
    paleogeography of Sabinas Basin, 250
    rifting of Pangea, 54
    sequence stratigraphy of Mexico, 58
    transgressions of, 8, 17
Calpionellids, of Sierra Madre Oriental, 375–376, 377
Campanian
    of Campeche shelf, 391–392
    of Ixtapa graben, 202, 204
    microfossils of, 149
    paleogeography of Sabinas Basin, 260
Campeche shelf
    carbonate facies of, 349
    Cretaceous-Paleocene boundary of, 389–395
    cross sections of, 393, 394
    foraminifera of, 390–391